Nanotechnology and Microbiology for Sustainable Agriculture

Nanotechnology and Microbiology for Sustainable Agriculture

Edited by

Quang D. Nguyen
Institute of Bioengineering and Process Engineering, Faculty of Food Science, Szent István University, Budapest, Hungary

Ali Asghar Rastegari
Department of Molecular and Cell Biochemistry, Falavarjan Branch, Islamic Azad University, Isfahan, I. R. Iran

Ajar Nath Yadav
Department of Genetics, Plant Breeding and Biotechnology, Dr Khem Singh Gill Akal College of Agriculture, Eternal University, Baru Sahib, Sirmour, Himachal Pradesh, India

Mohamed Hijri
Institut de Recherche en Biologie Végétale, Universitéde Montréal & Jardin Botanique de Montréal, Québec, Canada

WILEY

This edition first published 2027

Registered Offices
John Wiley & Sons, Inc., 111 River Street, Hoboken, NJ 07030, USA
John Wiley & Sons Ltd, New Era House, 8 Oldlands Way, Bognor Regis, West Sussex, PO22 9NQ, UK

For details of our global editorial offices, customer services, and more information about Wiley products, visit us at www.wiley.com.

The manufacturer's authorized representative, according to the EU General Product Safety Regulation, is Wiley-VCH GmbH, Boschstr. 12, 69469 Weinheim, Germany, e-mail: Product_Safety@wiley.com.

Wiley also publishes its books in a variety of electronic formats and by print-on-demand. Some content that appears in standard print versions of this book may not be available in other formats.

Library of Congress Cataloging-in-Publication Data applied for:

Hardback ISBN: 9781119808497

Cover Design: Wiley
Cover Image: © oticki/stock.adobe.com

Set in 9.5/12.5pt STIXTwoText by Straive, Pondicherry, India

Contents

List of Contributors

Adegboyega Adeniji
State Key Laboratory for Biology of Plant Diseases and Insect Pests
Institute of Plant Protection
Chinese Academy of Agricultural Sciences
Beijing, China

Ali Khalaf Ahmed Albaggar
Department of Biology
Faculty of Science
Al-Baha University
Al Baha City, Al Baha Province
Saudi Arabia

Dalaq Aiysha
Department of Medical Laboratory Technology
Green International University
Lahore, Pakistan

Laith Khalil Tawfeeq Al-Ani
School of Biology Science
Universiti Sains Malaysia
Minden, Pulau Pinang, Malaysia

Abdulaziz M.S. Alqahtani
Department of Biology
Faculty of Science
University of Bisha
Bisha, Saudi Arabia

Subhikshaa Bagavathiappan
Department of Biology
Concordia University
Montreal, Quebec, Canada

Ritwik Banerjee
Department of Biotechnology
University of Engineering & Management
Kolkata, West Bengal, India

Alejandra G. Becerra
Instituto Multidisciplinario de Biología Vegetal-CONICET, FCEFyN-UNC.
Argentina

Joorie Bhattacharya
Genetic Gains
International Crops Research Institute for the Semi-Arid Tropics
Hyderabad, Telangana, India

and

Department of Genetics
Osmania University
Hyderabad, Telangana, India

Shivani Bhutani
Department of Biotechnology
Invertis Village
Bareilly- Lucknow National Highway, NH-24
Bareilly, Uttar Pradesh

Nitin Mahendra Chauhan
Department of Biotechnology
College of Natural and Computational Sciences
Dilla University
Dilla, Ethiopia

Taniya Chauhan
Department of Zoology
Akal College of Basic Sciences
Eternal University
Sirmour, Himachal Pradesh
India

Harshit Chavda
Faculty of Agriculture
Parul University
Vadodara, Gujarat, India

Sagar Chhabra
Founder & Independent Agronomy Advisor
AgCrest Bio Solutions, Saskatoon, Saskatchewan, Canada Professional Agrologist (PAg), Saskatchewan Institute of Agrologists (SIA) Director (Volunteer)
SIA Saskatoon Branch, Saskatchewan Institute of Agrologists

Hemant Dasila
Department of Microbiology
Akal College of Basic Sciences
Eternal University
Sirmaur, Himachal Pradesh, India

Ankita Dey
Department of Biotechnology
Adamas University
Kolkata, West Bengal, India

Saleha Diwan
Faculty of Agriculture
Parul University
Vadodara, Gujarat, India

Ayomide Emmanuel Fadiji
Plant Biosecurity Unit
Department of Agriculture and Fisheries
Darwin, Northern Territory, Australia

and

Food Security and Safety Focus Area
Faculty of Natural and Agricultural Sciences
North-West University, Private Bag
Mmabatho, South Africa

Guillermo Fernandez-Bunster
Laboratorio de Biotecnología (BioTecLab)
School of Medical Technology
Faculty of Medicine
Universidad de Valparaíso
Campus San Felipe
Camino la Troya S/N & El Convento
Circunvalación Oriente
San Felipe, Chile

and

Centro de Investigación Interdisciplinario en Salud Territorial del Valle de Aconcagua (CIISTe)
San Felipe, Chile

Sayantani Garai
Department of Biotechnology
University of Engineering & Management
Kolkata, West Bengal, India

Nitu Gautam
Department of Molecular Biology & Genetic Engineering
Lovely Professional University
Phagwara, Punjab, India

Sougata Ghosh
Department of Microbiology
School of Science
RK University
Rajkot, Gujarat, India

Arunava Goswami
Agricultural and Ecological Research Unit
Biological Sciences Division
Institute of National Importance
Indian Statistical Institute
West Bengal, India

Umesh Goutam
Department of Molecular Biology &
Genetic Engineering
Lovely Professional University
Phagwara, Punjab, India

Sunil Tulshiram Hajare
Department of Biotechnology
College of Natural and Computational Sciences
Dilla University
Dilla, Ethiopia

Lobna Hajji-Hedfi
Research Laboratory of Agricultural
Production Systems and Sustainable
Development (LR03AGR02)
Regional Centre of Agricultural Research
of Sidi Bouzid, CRRA
Carthage University
Sidi Bouzid, Tunisia

Karim Houali
Laboratoire de Biochimie appliquée et
Biotechnologie
Facuslté des Sciences Biologiques et des
Sciences Agrnomique
Université Mouloud Mammeri
Tizi-Ouzou, Algeria

Pratheek Jain
PathoGutOmics Laboratory
Alva's Traditional Medicinal Archive
(ATMA)
Department of Ayurveda Pharmacology
Alva's Ayurveda Medical College
Dakshina Kannada, Karnataka, India

Samiksha Jhamta
Department of Zoology
Akal College of Basic Sciences
Eternal University
Sirmour, Himachal Pradesh, India

Hanika Kapoor
Department of Zoology
Akal College of Basic Sciences
Eternal University
Sirmour, Himachal Pradesh
India

Jaspreet Kaur Brar
Department of Zoology
Akal College of Basic Sciences
Eternal University
Sirmour, Himachal Pradesh
India

Simranjeet Kaur
Department of Zoology
Akal College of Basic Sciences
Eternal University
Sirmour, Himachal Pradesh
India

Tanvir Kaur
Department of Biosciences
Graphic Era Deemed to Be University
Dehradun, Uttarakhand, India

Kounaina Khan
PathoGutOmics Laboratory
Alva's Traditional Medicinal Archive
(ATMA)
Department of Ayurveda Pharmacology
Alva's Ayurveda Medical College
Dakshina Kannada, Karnataka
India

and

TheOmic School Pvt. Ltd.
Mysore, Karnataka, India

Sofia Sharief Khan
Department of Biotechnology
Government Degree College for Boys
Sopore, Jammu and Kashmir, India

and

Department of Biotechnology
Shri Mata Vaishno Devi University
Katra, Jammu and Kashmir, India

Omar Khirennas
Laboratory of Geo-Environment and Spatial Development (LGEDE)
Faculty of Nature and Life Sciences
University Mustapha Stambouli of Mascara
Mascara, Algeria

and

Laboratory of Geomatics, Ecology and Environment (LGO2E)
Faculty of Nature and Life Sciences
University Mustapha Stambouli of Mascara
Mascara, Algeria

Divjot Kour
University Centre for Research and Development
Chandigarh University
Mohali, Punjab, India

Prameela Krishnan
Division of Agricultural Physics
Indian Agricultural Research Institute
Delhi, New Delhi, India

Sarvjeet Kukreja
Department of Agronomy
School of Agriculture
Lovely Professional University
Phagwara, Punjab, India

Monika Kundu
Division of Agricultural Physics
Indian Agricultural Research Institute
Delhi, New Delhi, India

Dibyajit Lahiri
Department of Biotechnology
University of Engineering & Management
Kolkata, West Bengal, India

Adedayo Ayodeji Lanrewaju
Biotechnology and Food Science
Faculty of Applied Science
Durban University of Technology
Durban, South Africa

Damini Maithani
Natural Resource Management, Plant Microbiology
College of Horticulture
Sardar Vallabhbhai Patel University of Agriculture and Technology
Meerut, India

Ivana Mitrović
Faculty of Technology Novi Sad
University of Novi Sad
Novi Sad, Serbia

Urvashi Mittal
Department of Molecular Biology & Genetic Engineering
Lovely Professional University
Phagwara, Punjab, India

Slimane Mokrani
Laboratoire de Maitrise Des Energies Renouvelables
Faculté des Sciences de la Nature et de la Vie
Université de Bejaia
Bejaia, Algeria

and

Laboratory of Research on Biological Systems and Geomatics
Faculty of Nature and Life Sciences
University Mustapha Stambouli of Mascara
Mascara, Algeria

Dipro Mukherjee
Department of Biotechnology
University of Engineering & Management
Kolkata, West Bengal, India

Salma Mukhtar
KAM School of Life Sciences
Forman Christian College (A Chartered University)
Lahore, Pakistan

and

Department of Plant Biology
Rutgers University
New Brunswick, NJ, USA

El-hafid Nabti
Laboratoire de Maitrise Des Energies Renouvelables
Faculté des Sciences de la Nature et de la Vie
Université de Bejaia
Bejaia, Algeria

Moupriya Nag
Department of Biotechnology
University of Engineering & Management
Kolkata, West Bengal, India

Nafiseh Sadat Naghavi
Department of Microbiology
Fal.C, Islamic Azad University
Isfahan, Iran

Rajeshwari Negi
Department of Biosciences
Graphic Era Deemed to be University
Dehradun, Uttarakhand, India

Yankita Negi
Department of Zoology
Akal College of Basic Sciences
Eternal University
Sirmour, Himachal Pradesh
India

Rahul Nitnavare
Division of Plant and Crop Sciences
School of Biosciences
University of Nottingham
Sutton Bonington, Leicestershire, United Kingdom

and

Department of Plant Sciences
Rothamsted Research
Harpenden, Hertfordshire, United Kingdom

Subrahmanya Padyana
PathoGutOmics Laboratory
Alva's Traditional Medicinal Archive (ATMA)
Department of Ayurveda Pharmacology
Alva's Ayurveda Medical College
Dakshina Kannada, Karnataka
India

Faezeh Parastesh
Department of Soil Science
University of Manitoba
Winnipeg, Manitoba
Canada

Shiv Prasad
Division of Environmental Sciences
ICAR-Indian Agricultural Research Institute
Delhi, New Delhi, India

Paridhi Puri
University Centre for Research and Development
Chandigarh University
Mohali, Punjab, India

Rina Rani Ray
Department of Biotechnology
Maulana Abul Kalam Azad University of Technology
Kolkata, West Bengal, India

Golnoosh Rezaeizadeh
Department of Microbiology
Fal.C, Islamic Azad University
Isfahan, Iran

Fahimeh Sarami
Department of Microbiology
Fal.C, Islamic Azad University
Isfahan, Iran

Nidhi Selwal
Department of Molecular Biology & Genetic Engineering
Lovely Professional University
Phagwara, Punjab, India

Babita Sharma
Department of Microbiology
Akal College of Basic Sciences
Eternal University
Sirmour, Himachal Pradesh
India

Chandra Kant Sharma
Faculty of Agriculture
Parul University
Vadodara, Gujarat, India

Parul Sharma
Department of Zoology
Akal College of Basic Sciences
Eternal University
Sirmour, Himachal Pradesh, India

Sakshi Sharma
Department of Zoology
Akal College of Basic Sciences
Eternal University
Sirmour, Himachal Pradesh, India

Vijay Bharti Sharma
Department of Zoology
Akal College of Basic Sciences
Eternal University
Sirmour, Himachal Pradesh, India

Avtar Kaur Sidhu
Zoological Survey of India
High Altitude Regional Center
Solan, Himachal Pradesh, India

Manali Singh
Department of Life Sciences
Parul Institute of Applied Sciences
Parul University
Post-Waghodia, Vadodara
Gujarat, India

Manoj Singh
Department of Biotechnology
Adamas University
Kolkata, West Bengal, India

Shipra Singh
Department of Chemistry
University of Delhi
Delhi, New Delhi, India

Maura Téllez-Téllez
Centro de Investigaciones Biológicas
Universidad Autónoma del Estado de Morelos
Cuernavaca, Morelos, Mexico

Neelam Thakur
Department of Zoology
Akal College of Basic Sciences
Eternal University
Sirmour, Himachal Pradesh, India

Samrendra Singh Thakur
Department of Biotechnology
School of Biological Sciences
Dr Harisingh Gour Vishwavidyalaya
(A Central University)
Sagar, Madhya Pradesh, India

Preety Tomar
Department of Zoology
Akal College of Basic Sciences
Eternal University
Sirmour, Himachal Pradesh, India

Vijay Upadhye
Faculty of Agriculture
Parul University
Vadodara, Gujarat, India

Ajar Nath Yadav
Department of Genetics, Plant Breeding and Biotechnology
Dr. Khem Singh Gill Akal College of Agriculture
Eternal University
Sirmour, Himachal Pradesh, India

Ashok Yadav
Department of Botany
Banaras Hindu University
Varanasi, Uttar Pradesh, India

Neelam Yadav
Centre of Research Impact and Outcome
Chitkara University Institute of Engineering and Technology
Chitkara University
Rajpura, Punjab, India

and

Chitkara Centre for Research and Development
Chitkara University
Baddi, Himachal Pradesh, India

Farhan Zameer
PathoGutOmics Laboratory
Alva's Traditional Medicinal Archive (ATMA)
Department of Ayurveda Pharmacology
Alva's Ayurveda Medical College
Dakshina Kannada, Karnataka
India

and

The Omic School Pvt. Ltd.
Mysore, Karnataka, India

Foreword

In the current era, where pressure on the world's agricultural systems to produce more with fewer resources is growing, the combination of cutting-edge technologies like nanotechnology and microbiology presents a promising way ahead. Once thought of as separate fields, these disciplines are now powerfully overlapping to address some of the most important issues facing modern agricultural systems, from increasing nutrient uptake and soil fertility. This book, "Nanotechnology and Microbiology for Sustainable Agriculture," is both relevant and necessary, as it provides a distinctive and thorough perspective on how technologies can transform the future of sustainable farming by capturing the dynamic interactions between microbial discoveries and nano-enabled technologies. The information in this volume is especially important as we move toward more resource-efficient and climate-resilient agriculture. In addition to highlighting recent developments, this book points out knowledge gaps and future directions, promoting innovation and interdisciplinary cooperation. I wholeheartedly endorse "Nanotechnology and Microbiology for Sustainable Agriculture" to the scientific and agricultural sectors. I applaud the editors for having the foresight to compile this diverse and thoughtful content. This book will definitely serve as an important resource for educators, researchers, students, policymakers, and industrial professionals seeking sustainable alternative solutions for the world's agricultural problems.

Dr. Davinder Singh
Eternal University
Baru Sahib, Himachal Pradesh, India

Foreword

The agriculture industry has to contend with a number of global obstacles involving climate change, as well as ecological concerns, including fertilizer and pesticide distributions, which will only get worse as the population grows and food supplies become restricted. Therefore, it is essential to replace traditional farming methods with new technologies, and using microbiology and nanotechnology, especially green technology, has a lot of promise to address these problems. In addition to bringing about advancements and changes in other technologies, microbiology and nanotechnology have the potential to revolutionize a number of agricultural industries, including food packaging, fertilizers, pesticides, biosensors, and other sectors. Thus, to support the increasing applications of nanoparticles in sustainable agriculture, recent research in nanotechnology has progressively shifted over the past decade towards eco-friendly and cost-effective approaches, integrating microbiological systems to enhance both environmental compatibility and agricultural productivity.

The present book, "Nanotechnology and Microbiology for Sustainable Agriculture," provides a comprehensive overview of the diverse roles of nanotechnology and microbiology for sustainable agriculture. The chapters in the book are contributed by experts in the field worldwide, which highlights the potential of synergistic applications of two rapidly advancing scientific domains to address pressing agricultural challenges. Together, these technologies offer novel solutions for soil health management, enhanced nutrient delivery, pest control, stress tolerance, and crop productivity.

The chapters in this book are organized into several thematic sections, each addressing a critical aspect of nanotechnology and microbiology in the context of sustainable agriculture. The book comprises of 21 chapters. The first chapter is an introduction to foundational concepts of nanotechnology and microbiology, highlighting their convergence and potential in promoting sustainable agriculture by Negi et al. Mokrani et al. describe the synthesis and application of nanomaterials to rejuvenate soil health, enhance nutrient use efficiency, and support crop resilience, while addressing environmental and regulatory challenges in Chapter 2. Chapter 3 by Kour et al. explores the synergistic effects of phosphorus-solubilizing microbes and nanoformulations to improve plant growth and soil fertility for agricultural sustainability. Chapter 4 by Al-Ani et al. highlights the role of fungi in the production of bioactive compounds and nanomaterial development. Chapter 5 by Negi et al. describes the role of cyanobacteria-derived nanoparticles in enhancing crop productivity, biocontrol, and stress mitigation in agricultural systems, whereas Sharma et al. highlights microbial interactions and nanotechnological applications in promoting plant health, reducing chemical inputs, and maintaining ecosystem stability in Chapter 6. In Chapter 7, Thakur et al. demonstrate how microbial nanotechnology can be applied to promote plant growth and protect crops from pathogens and abiotic stress factors. Tomar et al. outline microbial-derived nanoproducts used in the biocontrol of insect pests, highlighting mechanisms, formulations, and field applications in Chapter 8. Lahiri et al. explore the biosynthetic capabilities of actinomycetes in producing antimicrobial and growth-promoting nanomaterials for agricultural application in the next chapter. Chapter 10 by Nitnavare et al. highlights the role of nanotechnology in enhancing GMO functionality, safety assessment, and public perception in food systems. Chapter 11 by Zameer et al. covers nano-biotechnological solutions for remediating pesticides, hydrocarbons, and heavy metals from soils and water bodies. Chapter 12 by Aiysha and Mukhtar focuses on nanoparticle-based approaches to detoxify heavy metal contamination in agricultural lands to restore productivity. Chapter 13 by Mittal et al. describes green-synthesized nanomaterials from biological sources with potential for managing plant pathogens in a sustainable manner. Naghavi et al. discuss advanced nanostructures used for immobilizing beneficial microbes and enzymes to enhance their agricultural efficiency and stability in Chapter 14. Khan et al. evaluate the formulation, mode of action, and environmental benefits of nanopesticides compared to conventional chemical pesticides in the next chapter. Chapter 16 by Kundu et al. introduces nanobiosensor technologies for real-time monitoring of soil quality, pathogens, pollutants, and nutrient levels in precision agriculture. Chapter 17 by Fernandez-Bunster examines how nanotechnology can complement GMOs for traits' improvement, nutrient delivery, and safety evaluation. Chapter 18 by Khan et al. provides a global perspective on research advancements, trends, collaborations, and applications of nanotechnology in agriculture. Fadiji et al. explore the legal, environmental, and ethical considerations in the development and deployment of agricultural nanotechnologies in Chapter 19. The next chapter by Singh et al. highlights the role of genomics, transcriptomics, and bioinformatics in decoding beneficial microbial traits and their integration with nanotech platforms. Finally, the conclusion and future challenges in nanotechnology and microbiology for agricultural sustainability have been described by Negi et al. in the last chapter, summarizing current research trends, identifying knowledge gaps, and proposing future directions for research and application.

Overall, great efforts have been carried out by the editorial team and scientists from different countries to compile this volume as a highly unique and up-to-date source on "Nanotechnology and Microbiology for Sustainable Agriculture" for students, researchers, scientists, and academicians. I hope that the readers will find this book highly useful and interesting during their pursuit of nanomaterials and microbial treasure.

Dr. Jaswinder Singh
Eternal University
Baru Sahib, Himachal Pradesh, India

Preface

Nanotechnology is anticipated to become a major force in the future through its innovative applications and is presently offering a creative frontier in agricultural operations. The applications of nanotechnology in agriculture present a novel strategy with great potential for addressing global issues related to food supply, safety, security, and the environment, as well as the growing threat of climate change. In today's world, nanotechnology is widely used in the domains of agriculture, food processing, food safety, the dairy industry, packing, transportation, and agricultural product quality control. Integrating microbiology and green chemistry can help to establish more environmentally friendly industrial processes and facilitate the production of nanoparticles. In this context, the convergence of nanotechnology and microbiology presents transformative opportunities to enhance the sustainability, productivity, and resilience of agricultural systems. The chapters of this book have been contributed by leading researchers and experts from around the world, highlighting fundamental concepts, recent advances, and real-world applications. Key topics include the design of nano-enabled microbial biofertilizers, nanoformulations for plant disease management, microbial synthesis of nanoparticles, biosensors for soil and plant health monitoring, and regulatory perspectives.

Hungary	Prof. Quang D. Nguyen
I. R. Iran	Dr. Ali Asghar Rastegari
Baru Sahib, India	Dr. Ajar Nath Yadav
Canada	Dr. Mohamed Hijri

Acknowledgments

All the authors are sincerely acknowledged for contributing up-to-date information on nanotechnology and microbiology for sustainable agriculture. The editors are thankful to all the authors for their valuable contributions.

All editors would like to thank their families, who were very patient and supportive during this journey. Our sincere thanks to the whole Wiley team who were directly or indirectly involved in the compilation of this book. We are grateful to the many people who helped to bring this book to light. Editors wish to thank the Project Coordinator, Wiley, for generous assistance, constant support, and patience in initializing the volume.

The editor, Dr. Ajar Nath Yadav, is grateful to Dr. Samiksha Jhamta, Teaching Associate-IQAC; his Ph.D. research scholars Rajeshwari Negi, Babita Sharma, and colleagues for their support, love, and motivation in all my efforts during this project. We are very sure that this book will interest scientists, graduates, undergraduates, and postdocs who are investigating "Nanotechnology and Microbiology for Sustainable Agriculture."

1

Nanotechnology and Microbiology for Sustainable Agriculture: An Introduction

Rajeshwari Negi[1], *Babita Sharma*[2], *Neelam Yadav*[3], *Ashok Yadav*[4], *Paridhi Puri*[5], *Farhan Zameer*[6], *and Ajar Nath Yadav*[7]

[1] *Department of Biosciences, Graphic Era Deemed to be University, Dehradun, Uttarakhand, India*
[2] *Department of Microbiology, Akal College of Basic Sciences, Eternal University, Sirmaur, Himachal Pradesh, India*
[3] *Centre for Research Impact and Outcome, Chitkara University Institute of Engineering and Technology, Chitkara University, Rajpura, Punjab, India*
[4] *Department of Botany, Banaras Hindu University, Varanasi, Uttar Pradesh, India*
[5] *University Centre for Research and Development, Chandigarh University, Mohali, Punjab, India*
[6] *PathoGutOmics Laboratory, Alva's Traditional Medicinal Archive (ATMA), Department of Ayurveda Pharmacology, Alva's Ayurveda Medical College, Dakshina Kannada, Karnataka, India*
[7] *Department of Genetics, Plant Breeding and Biotechnology, Dr. Khem Singh Gill Akal College of Agriculture, Eternal University, Sirmour, Himachal Pradesh, India*

1.1 Introduction

One of the 17 Sustainable Development Goals of the United Nations is "Zero Hunger," which can only be attained through sustainable agriculture. The growing global population, environmental pollution, climate change, and increased energy and water demands are all placing significant stress on the world's food supply and delivery networks (Rasul 2016). The amount of resources used in modern agriculture is unanticipated (Negi et al. 2022a, 2022b). However, more than 3 billion tons of crops are produced globally each year, requiring the application of 187 million tons of fertilizers, 4 million tons of pesticides, 2.7 trillion cubic meters of water (roughly 70% of the world's freshwater consumption), along with more than 2 quadrillion British thermal units (BTU) of energy (Mateo-Sagasta et al. 2018). Specifically in rising economies, the Food and Agriculture Organization projects that by 2050, there will be 10 billion people on the planet, which would result in a 50% rise in food needs (Negi et al. 2024a). Furthermore, it is estimated that 815 million people are undernourished at the moment, and by 2050, an additional 2 billion people would fall into this group (Negi et al. 2024b). The present scenario calls for major changes to the global food production systems (Negi et al. 2024c). According to recent studies, nanotechnology and microbiology hold great promise for enhancing the agricultural industry by boosting the effectiveness of agricultural inputs and providing solutions for environmental and agricultural issues that will increase food security and production (Kim et al. 2018). In light of this, research on using nanotechnology and microbiology in agriculture has been receiving a lot of interest presently.

The fields of microbiology and nanotechnology individually have advanced science and technology by producing cutting-edge solutions for human welfare and preserving ecological and environmental balance (Prajapati et al. 2024). However, frequent drug use has been shown to cause antibiotic and multidrug resistance in microorganisms, and the food chain is being impacted by the delivery of metal nanoparticles (NPs) (Hasani et al. 2019). As a result, interdisciplinary approaches combining microbiology and nanotechnology are now required to combat secondary human health as well as ecological and environmental damage (Bano 2022). Combining these domains provides logical, creative, and sustainable solutions.

The emerging initiative to use nanotechnology in agriculture has sparked extensive research at multiple levels (Sekhon 2014). After being examined at the nanoscale, a number of well-characterized bulk materials were discovered to exhibit an array of intriguing properties. Environmentally friendly nanomaterial (NM) production technique is an alternate term for nano-biotechnology (Shahcheraghi et al. 2022). This innovative technology has plenty of potential for use in sustainable farming in the future (Malik et al. 2023b). Nano-biotechnology has been used in the field of engineering, food technology, and clinical research within the last few decades (Sivakami et al. 2021). By offering fresh perspectives and resources, nano-biotechnology is transforming the biosystems of several plants. While chemically synthesized NMs are currently thought to be hazardous, green nanoparticle (NP) synthesis is a profitable, sustainable process that has gained popularity due to its great stability and is safer to the environment (Malik et al. 2023a). Numerous investigations have revealed that NPs can be delivered by microbes and plants. Applications of microbial nanotechnology offer a safe way to advance agriculture in a sustainable manner (Mittal et al. 2020) (Figure 1.1).

1.2 Nanoparticles and Their Types

1.2.1 Organic Nanoparticles

Organic NPs are synthesized from organic molecules and have a size of 100 nm or less. Ferritin, liposomes, micelles, dendrimers, and other well-known organic NPs and polymers are representatives of this class (Aguilar et al. 2015). Micelles and liposomes, two types of NPs that have an internal hollow called a nanocapsule, are environmentally friendly, biodegradable, and sensitive to electromagnetic energies (Sapsford et al. 2013; Summer et al. 2024). They provide superior possibilities for medicine delivery because of their unique qualities. The stability, size, and surface form of the drug are important, but so are its delivery methods, either an adsorbed drug system or an entrapped drug system which determines its application range and efficacy (Adepu and Ramakrishna 2021). Organic NPs are widely used in the biomedical industry, such as in the medication delivery system, due to their effectiveness and the ability to be injected into specific bodily parts (a procedure known as targeted drug administration) (Zahin et al. 2020).

1.2.2 Metal-Based Metal Oxides

Both constructive and destructive techniques can be used to synthesize metal-based NPs at nanometric sizes (Shnoudeh et al. 2019). The synthesis of NPs commonly uses metals

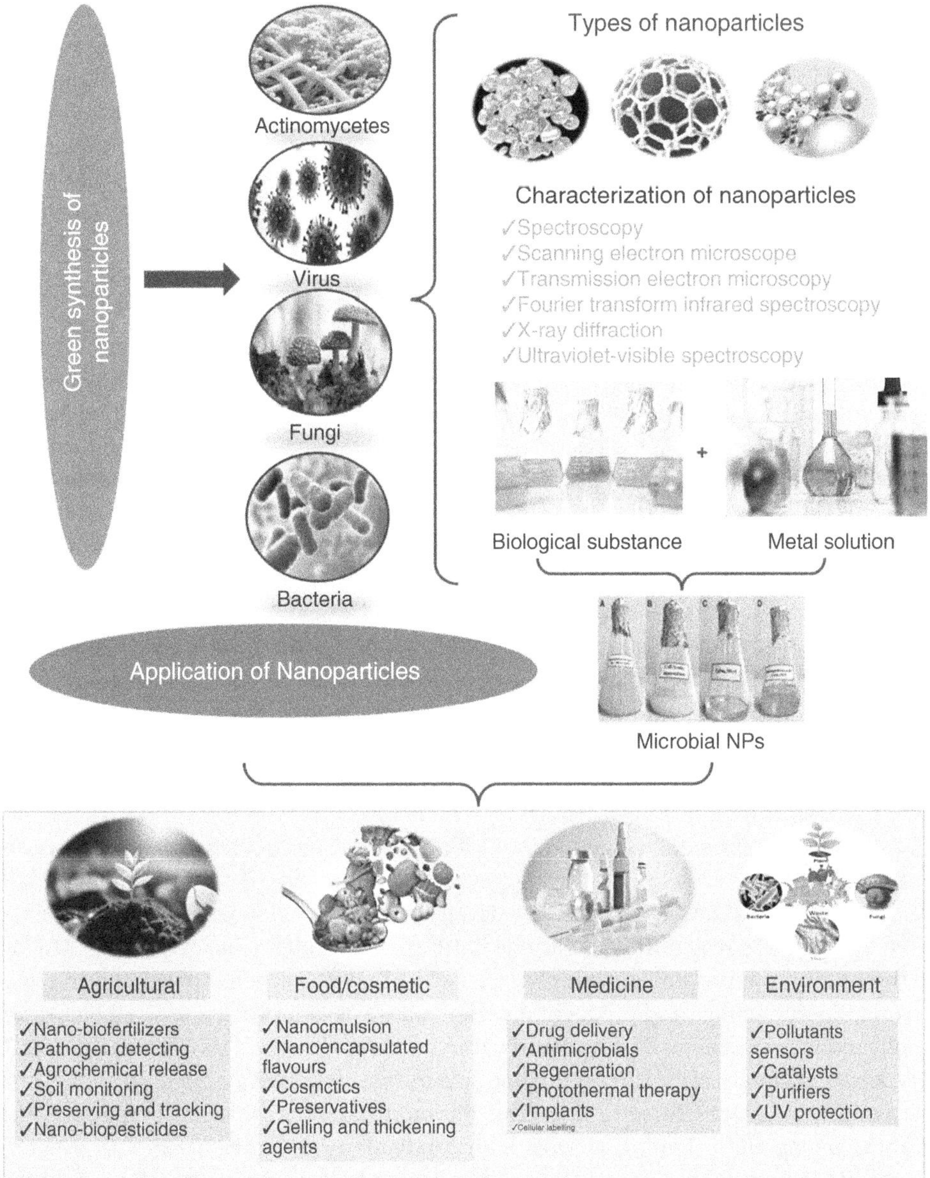

Figure 1.1 Microbial synthesis of nanoparticles and their applications. *Source:* Connect Images/ Adobe Stock Photos.

including aluminum, cadmium, cobalt, copper, gold, iron, lead, silver, and zinc (Kumar et al. 2021). Metal NPs' enormous surface-to-volume ratio and quantum effects render them remarkable ultraviolet (UV)–visible sensitivity as well as electrical, catalytic, thermal, and antibacterial properties (Das and Marsili 2011). The lower particle size means that there are significantly more atoms on the surface. The ratio of an NP's surface area to volume is determined by its size and structure, and this in turn affects properties like conductivity and UV light sensitivity (Khan and Hossain 2022). Changes in surface area affect a wide range of properties, including electronic energy levels, melting point, phase transition temperature, electron affinities and transitions, magnetic features, and affinities to polymeric, biological, and organic compounds (Zhang et al. 2016). When the quantum size and the Coulomb charge effect are linked together, a variety of fascinating features are obtained that are not observed in the same bulk material (Ediger et al. 2007). Quantum phenomena are especially prone to affect particles with sharp edges and spherical forms. The size-dependent effects of NPs make them useful in sensing, imaging, and catalysis (Dolai et al. 2021).

Metal oxides have garnered increased attention from scientists over the past few decades. Combining positive metallic ions with negative oxygen ions forms an ionized mixture of metal oxides (Pachaiappan et al. 2021). Ionic connections are strong and robust because of the electrostatic forces that exist between the positively charged metal ions and the negatively charged oxygen ions. In this manner, iron (Fe) NPs readily change into iron oxide (Fe_2O_3) NPs when exposed to oxygen at room temperature, which greatly boosts their reactivity in comparison to iron NMs (Leonel et al. 2021). The purpose of forming these oxide-based NPs is to alter the characteristics of their metal-based equivalents (Kumar et al. 2021). To benefit from their increased efficiency and reactivity, metal oxide NPs are frequently synthesized. Among the most commonly synthesized oxides are silicon dioxide (SiO_2), titanium dioxide (TiO_2), zinc oxide (ZnO), and aluminum oxide (Al_2O_3) (Shirsat et al. 2021). These NPs exhibited superior characteristics in contrast to their metal counterparts.

1.2.3 Carbon-Based Nanotubes

The evolution of human civilization on Earth has been significantly influenced by carbon. It forms unparalleled strength interactions when mixed with other materials (Caineng et al. 2022). Over the past few decades, a vast range of carbon-based NMs has been developed in recognition of various synthesis techniques (Rao et al. 2021). Owing to their peculiar forms and diverse characteristics, they have been utilized in numerous sectors. Energy storage and production, water and wastewater treatment, and biological applications are just a few of the potential uses for carbon-based NMs (Dutta et al. 2022). Carbon can assume several different allotropic forms, including diamond, graphite, and buckminsterfullerene (Singh 2023). The thermodynamic stability of graphite is the highest among all of them. Due to its high conductivity, it can be used in a variety of electronic devices, including electrodes, batteries, and solar panels. Graphite is made by layering graphene sheets. Graphene represents an innovative variation of carbon, consisting of a solitary layer of molecules arranged in a honeycomb configuration on a two-dimensional support (Sharif et al. 2021). In addition to its great strength, it can be used as a building block to create other types of carbon NPs (Lim et al. 2021). Another innovative kind of carbon is carbon nanotubes (CNTs). Although fullerenes, graphene, and CNTs are all synthesized differently, there is an association between their chemical and physical features (Georgakilas et al. 2015). Since fullerenes are the earliest known molecules, the production of derivatives and composites based on them seems the most improbable. Despite this, there is a great deal of untapped potential for CNTs and graphene research, and they appear to be promising substitutes in many different sectors (Dinadayalane and Leszczynski 2012). In numerous applications such as adsorption, separation, catalytic processes, and many more, activated carbon has proven to be an impressive adsorbent.

1.3 Methods of Nanomaterial Synthesis

1.3.1 Top-Down and Bottom-Up Approaches

A physical or chemical reduction in the size of the initial material is required for the top-down process (Arole and Munde 2014). There are various methods that can be utilized, such as the commonly used attrition method for the manufacture of metallic nanoparticles (MtNPs) (Hossain et al. 2023). An attrition process is used to reduce the size of macro- or micro-scale particles using a planetary or ordinary ball mill. The resulting particles are then air-classified to recover oxidized NPs. The material, milling time, ambient medium, and the combination of these parameters all have a substantial impact on the properties of the resulting NPs (Khan and Arasu 2019). The process of burning an organic precursor which could be a liquid or gas that has been squeezed through an aperture under extreme pressure is known as pyrolysis (Safdar et al. 2022). The oxidized NPs are recovered by air classification of the resultant ash. Another advantage is the average manufacturing rate. The imperfect surface structure is the downside (Srivastava et al. 2020). Such surface structural flaws have the major disadvantage of having a large aspect ratio, which can significantly change the mtNPs' surface chemistry and physical characteristics (Bahrulolum et al. 2021). The cost of keeping the high pressure and temperature necessary for the synthesis processes is also very significant, along with the energy usage.

In a bottom-up approach, a structured system is created atom by atom, cluster by cluster, or molecule by molecule (Šesták 2021). This method involves first synthesizing NP building blocks through a chemical or biological synthesis process in order to construct the final material. The traditional and most widely used techniques for the synthesis of mtNPs consist of wet technological procedures (Calvo et al. 2022).

1.3.2 Physical and Chemical Techniques

A variety of engineered nanomaterials (ENMs) can be produced via abrasion, condensation, melting, evaporation, and other physical processes that include high mechanical pressure, radiation, and thermal and/or electrical energy (Yadav et al. 2012). Solid-state precursors are used in top-down physical processes like milling, crushing, and etching, while liquid or gas precursors are used in bottom-up processes like chemical vapor deposition (CVD), flame pyrolysis, electrolysis, and many more. In these processes, molecules or atoms are carried to nucleation sites, where they combine to produce NPs (Diep and Schiffman 2023; Li et al. 2021). The formation of colloidal NPs in the liquid phase, which are easily sorted into powder form, is one advantage of employing chemical processes. ENMs can also be produced chemically in a variety of sizes and shapes (Haider et al. 2022). The disadvantages of physical and chemical techniques include the need for costly equipment, high energy consumption, significant heat generation, and limited production yield.

1.3.3 Green Synthesis of Nanomaterials

The synthesis of NMs from plants, microbes, and biomolecules is less expensive and safer for the environment when compared to alternative methods of production (Jegadeeswaran et al. 2012). Both internally and externally, plants and microbes synthesize metal NPs by consuming the metal molecules in their environment. Systemic metabolisms prevent all organisms from being able to synthesize NPs (Shedbalkar et al. 2014). These NPs have a wide range of uses in tissue engineering, pathogen detection, biology, and biopharmaceuticals (Koul et al. 2021). In a biological context, the following processes produce NPs. (i) Bioreduction: Stable metal ions can be obtained through bioreduction. The method of dissimilatory metal reduction is used to accomplish it (Vaidyanathan et al. 2010; Singh et al. 2015). This releases inactive forms of metal ions from the extracts or sample without causing any harm to the metal compounds. (ii) Biosorption: The metal ion is discovered adhered to the organism's cell (Khan et al. 2018). Peptide fragments from the cell wall or the metal ion that is directly attached to the walls are responsible for the NP arrangement. It is possible to create various metal NPs by employing bio extracts from living things (Andrade et al. 2024). The type of NPs either synthesized by or present in various microorganisms is the most crucial consideration to make during the production process (Negi and Singh 2018). There are many plant and microbial sources, but an organism's possibilities are limited if it is resistant to a small number of metals. The primary disadvantage of naturally occurring NPs is their unpredictability in terms of size and form throughout the synthesis phase (Bedlovičová 2022). A study reported that NPs of copper oxide were synthesized using the cell-free extract of *Stenotrophomonas* sp., and the synthesized NPs exhibited antibacterial, antioxidant, and cytotoxic effects (Talebian et al. 2023). Pallavi et al. (2022) reported that silver nanoparticles (AgNPs) synthesized using *Streptomyces hirsutus* exhibited antimicrobial and anticancer activities. A study concluded the biosynthesis of gold NPs from *Lysinibacillus odysseyi* (Cherian et al. 2022). A study documented that synthesis of AgNPs using *Fusarium oxysporum* showed an antifungal effect (Nasr Azadani et al. 2024). Another study documented that synthesis of magnesium oxide NPs using the metabolites of *Cystoseira crinita* showed promising antibacterial, larvicidal, and pupicidal properties (Fouda et al. 2022) (Table 1.1).

1.4 Why Use Nanotechnology in Agriculture

A wide range of issues affect agriculture worldwide, including a lack of nutrients, a halt in crop productivity, a reduction in soil organic matter, shifting climate patterns, a reduction in the amount of arable land and water available, opposition to genetically modified organisms, and a labor shortage (Kulkarni 2021). The green revolution practices are still causing pain and tiredness in Indian agriculture (John and Babu 2021). The amount of fertilizer used has increased dramatically over the past 50 years, from 0.5 tons in 1960 to 23 million tons in 2008. This increase is almost equal to a fourfold increase in the yield of food grains (Hasler 2017).

Table 1.1 Biological sources for the synthesis of various metal-based nanoparticles.

Source	Species	Nanoparticle	Size (nm)	References
Plant	*Acalypha indica*	Silver (Ag)	20–30	Krishnaraj et al. (2010)
	Anacardium occidental	Gold (Au)/Ag	6	(Sheny et al. 2011)
	Boswellia ovalifoliolata	Ag	30–40	Gebregiorgis and Nigussie (2018)
	Chenopodium ambrosioides	Au/Ag	10–30	Dwivedi and Gopal (2010)
	Cinnamomum camphora	Au/Ag	55–80	Huang et al. (2011)
	Garcinia mangostana	Ag	35	Veerasamy et al. (2011)
	Euphorbiaceae latex	Copper (Cu)/Ag	18	Patil et al. (2012)
	Jatropha curcas L. latex	Pb	10–12.5	Joglekar et al. (2011)
	Syzygium cumin	Ag	29–92	Sekhar et al. (2018)
	Trachyspermum copticum	Ag	6–50	Vijayaraghavan et al. (2012)
Bacteria	*Bacillus licheniformis*	Ag	50	Shanthi et al. (2016)
	Phaenerochaete chrysosporium	Ag	50–200	Vigneshwaran et al. (2006)
	Escherichia coli	Ag	50	Jiang et al. (2019)
	Escherichia coli	Au	15–35	Ranjan Sarker et al. (2019)
	Escherichia coli	Au/Pt	<10–50	Attard et al. (2012)
	Geobacillus sp.	Au	5–50	Correa-Llantén et al. (2013)
	Klebsiella aerogenes	Cd	28.2–122	Singaravelu et al. (2007)
	Lactobacillus strains	Ti	40–60	Prasad et al. (2007)
Fungi	*Trichoderma harzianum*	Ag	51.10	Gherbawy et al. (2013)
	Colletotrichum sp.	Ag	5–60	Azmath et al. (2016)
	Aspergillus fumigatus	Ag	322.8	Shahzad et al. (2019)
	Pycnoporus sanguineus	Ag	5–80	San Chan and Don (2013)
	Fusarium oxysporum	Ag	5–13	Khan and Jameel (2016)
	Penicillium purpurogenum	Ag	8.10	Nayak et al. (2011)
	Alternaria alternate	Au	7.17	Sarkar et al. (2012)
	Aspergillus clavatus	Au	12–20	Verma et al. (2010)
	Aspergillus niger	Au	5–15	Bhambure et al. (2009)
	Epicoccum nigrum	Au	5–50	Sheikhloo et al. (2011)
	Hormoconis resinae	Au	3–20	Mishra et al. (2010)
	Fusarium oxysporum	Ti/CdS	5–60	Gupta and Chundawat (2020)

Despite the explosive rise in grain production, stagnation in some crop yields has also been noted as a result of a decline in soil organic matter due to over-fertilization (Bharti et al. 2017). The ratio of nitrogen, phosphorous, potassium (NPK) fertilizers that is now maintained in India is 10:2.7:1, which is far greater than the 4:2:1 ratio that is recommended. (Pandey 2018). Experts are really concerned about the rapid deterioration of India's unevenly and over-fertilized soil (Mitra et al. 2021). The country will require 45 million tons of grains in addition to the current demand of around 23 million tons, in order to meet the target of producing 300 million tons of food for feeding the rapidly growing population of about 1.5 billion people until the year 2025 (Nadathur et al. 2024). There is a crop shortfall of roughly 25–30% as a result of the degree of multiple nutrient drainage that is always growing with each year (Majumdar et al. 2016). It is crucial to employ both organic and inorganic nutrients in a balanced manner to preserve and sustain the health of the soil, but this is becoming increasingly challenging due to the subsidized rates of inorganics and the fall in the supply of organics (Rao and Lenka 2020). Consideration should also be given to adapting some alternative ways in the agriculture sector due to drastic changes in climate, such as frequent occurrences of droughts, melting of the polar ice cap, and sudden rises in temperature (Grigorieva et al. 2023).

1.5 Microbial Nanotechnology and Sustainable Agricultural Development

Nanotechnology is essential for monitoring pesticides and water quality for sustainable agriculture, as well as for managing nutrients for higher agricultural output (Prasad et al. 2017). Nanotubes, nanocapsules, nanofertilizers, nanopesticides, biosensors, controlled delivery systems, and nanofiltration are widely used in the agricultural industry (Priyanka et al. 2020). Nanotechnology has been shown to be effective in managing plants, soil fertility and health, drug delivery systems, and the distribution of nutrients and water in agricultural areas (Vega-Vásquez et al. 2020). Since they are quick and strong, nanosensors are being utilized widely in agricultural areas nowadays to monitor soil and water contamination. The breakdown of waste and harmful compounds is greatly aided by NMs, which also help increase a microbe's efficiency in this process (Ahmed et al. 2022). In order to remove pollutants and hazardous compounds from agricultural soils and ensure sustainable development, nano–nano interactions are now thought to be fascinating phenomena (Bhardwaj et al. 2024).

1.5.1 Nanofertilizers

Fertilizers are vital for plant development and advancement because they provide nutrition for plants. Approximately 30–40% of the fertilizer applied is used by crop plants, while the remaining is lost through leaching, evaporation, and degradation (Ayub et al. 2020). Previous findings state that approximately 30–40% of the total fertilizers applied, 80–90% of the phosphorus, and 50–90% of the potassium are lost to the environment (Bindraban et al. 2020). This raises the expense of cultivation and reduces the effectiveness of fertilizers. Excess use of fertilizers results in pollution of the air, water, and soil (Tyagi et al. 2022). This raises the expense of agriculture and decreases the effectiveness of fertilizers.

Nanofertilizers could serve as a significant substitute for conventional fertilizers in order to increase fertilizer efficiency and address the issue of environmental contamination (Avila-Quezada et al. 2022). Depending on the needs of crops, regulated nutrient release is the basis for the synthesis of nanofertilizers (Rahman et al. 2021). It is thought that they are more effective than traditional fertilizers. While using nanofertilizers, plants are able to absorb the majority of nutrients without experiencing any loss, since they cause a delayed release of nutrients during crop growth (Thavaseelan and Priyadarshana 2021). Nanofertilizers are developed by encapsulating fertilizers inside NPs. The gradual release of nutrients is facilitated by fertilizer coating within NP membranes (Babu et al. 2022).

1.5.2 Nanoherbicides

Nanoherbicides offer an environmentally beneficial solution to get rid of weeds because of their microscopic size, which allows them to blend with the soil (Ghosh et al. 2022). Nanoherbicides can be used to remove weeds that are resistant to traditional herbicides. With the aid of tubers and deep roots, weeds proliferate and endure (Adarsh et al. 2024). When weeds are manually ploughed from diseased regions, they end up spreading to uninfected areas (Scavo and Mauromicale 2020). Although it is exclusively applied to the necessary agricultural sector, a combination of nano-sized active substances and an intelligent delivery system is very effective (Verma et al. 2017). Weed growth reduces or adversely affects crop productivity. Using nanotechnology to boost herbicide efficacy can increase crop yield without causing harmful impacts on the field (Shang et al. 2019).

1.5.3 Nanopesticides

Nanopesticides are microscopic particles with insecticidal properties. By increasing the dispersibility of agricultural formulations, nanopesticides can lessen the spread of dangerous pesticides (Asif et al. 2021). The qualities of NPs, including their solubility, permeability, stiffness, crystallinity, biodegradability, and thermal stability, are crucial for the synthesis of nanopesticides (Anandhi et al. 2020). Recently, the usage of nanopesticide subclasses such as nanocapsules, nano-emulsions, nanocontainers, and nanocages has also been observed in the agricultural sector (Jiang et al. 2024). In soil, NMs can break down more quickly. Due to its antibacterial activity against microorganisms and its nontoxicity toward human beings, silver has a wide range of applications in both its compound and metal forms (Bamal et al. 2021). AgNPs' efficiency can now be raised with the application of nanotechnology. Although AgNPs have a high surface-to-volume ratio and can penetrate, they enhance microbial interaction (Garg et al. 2020).

1.6 Regulatory Policies and Considerations

Leadership in an area where the majority of the information is still unknown or under research presents the highest challenge to policymakers (Oliver et al. 2014). In certain situations, excessive regulation could impede progress, but little regulation could have detrimental impacts on health. Despite the promising advancements in nanotechnology across various fields, its applications in the agriculture sector are still unable to meet the world's needs due to a lack of awareness and related safety issues (Prasad et al. 2017). The main reasons behind the lack of agri-nanotechnology commercialization are inadequate legislative frameworks, regulatory strategies, and negative public perceptions (Cummings et al. 2021). Therefore, before enforcing strict plans, it is imperative to carefully examine the risk assessment and risk management elements related to the applications of nanotechnology in the agricultural industry (Abady et al. 2024). Countries are adopting a variety of regulatory organizations, such as the Nanotechnology Initiative Division, DIT, Ministry of Law and Justice, Centers for Disease Control and Prevention, and Centre for Knowledge Management of Nano-science and Technology, to regulate the use and effects of nanotechnology in the food, feed, and agri-sectors (Pandey 2018). It is important to note that the growth and marketing of nano-based products intended for the agricultural sector are being hampered by the uncertainty in regulatory frameworks and the differences in viewpoints that exist globally (Jain et al. 2018). Encouraging worldwide discussion on risk assessment and management would facilitate the development of effective regulatory policies (Hopkin 2018).

According to a survey of Indian researchers working in the subject of nanotechnology, almost 95% of them have recognized the challenges pertaining to moral and ethical concerns in all areas of nanotechnology (Sandler 2009). A few of the concerns brought up deal with the negative impacts of nanotech research on human health and the environment, their usage as undetectable weapons in conflicts, and the implantation of innovative nano devices as enhancements of performance in human bodies (Boguszewicz et al. 2021). The illustrious status that science practitioners appreciate in India and the fact that policymakers spend a great deal of time celebrating scientific achievements rather than discussing the risks associated with new technologies and improving regulatory frameworks may be contributing factors to the dearth of discussions on ethical issues, safety concerns, health risks, and environmental concerns regarding new technologies (Sarwal et al. 2018). Due to gaps in the regulatory framework, chemicals such as nano-insecticides are traded in Indian marketplaces without the need for risk assessments. This is all being done despite the Indian government's acknowledgment that human cell-sized NPs have the potential to accumulate in the lungs and cause direct damage, as well as to spread to other body parts or enter the body through the bloodstream and absorb the particles (Jeevanandam et al. 2018).

A set of procedural and preventive guidelines for the safe and secure handling of NMs in businesses and research facilities was recently released by Department of Science and Technology (DST) in an effort to address these problems. This DST initiative could be a step toward safer nanotechnology applications across India's industrial sector (Kumar 2014). The Nano Mission has been taken the step to plan an authoritarian framework to handle concerns about the effects of environment, health, and safety and the risks connected to nanoproducts due to worries about safety issues with NPs (Hull and Bowman 2014). A team has recently been assembled to establish a framework for controlling the use of nanotechnology in the nation under the direction of the former head of the Indira Gandhi

Centre for Atomic Research (IGCAR) (Parthasarathi 2007). In keeping with the recommendations of this organization, the International Advanced Research Centre for Powder Metallurgy and New Materials and the Centre for Knowledge Management of Nanoscience and Technology are developing a set of regulatory guidelines to help technologists, researchers, and industries implement safe and responsible NM management practices (Charitidis et al. 2014).

1.7 Environmental Issues Associated with Microbial Nanoparticles

As the use of nanostructures grows globally, the pollution they cause, known as nanopollution, becomes an increasing environmental concern as these materials become more prevalent contaminants (Thiruvengadam et al. 2024). Nanopollution is frequently referred to as "invisible" pollution because of the minuscule size of nanostructures, making it very challenging to manage and regulate (Sharma and Sharma 2024). A multitude of issues, including environmental cleanup, are said to be resolved by NPs, which have a broad variety of uses. Chemicals are used in the development of synthetic nanostructures, which makes them a source of pollution from the moment of manufacture till disposal (Brar et al. 2022). The usage and disposal of NPs made from environmentally friendly sources, such as microorganisms, plant parts, and extracts, may be environmentally hazardous even though they do not employ harmful chemicals during their synthesis (Jeevanandam et al. 2022). After being released into the environment, the biogenic NPs experience a variety of processes that change their characteristics, including oxidation, adsorption, and aggregation (Ju-Nam and Lead 2016). Transformable green NPs undergo oxidation or reduction reaction to change into different forms. These new forms either release more harmful metal ions or show increased solubility and bioaccessibility, which allows them to invade biological cells (Wang et al. 2019).

Green NPs that are not biodegradable, however, exhibit environmental aggregation and accumulation (de Souza et al. 2019). By altering several parameters like pH and salinity, among others, catalytic NPs disrupt the ecological balance of the affected area by interacting with other environmental components like contaminants and macromolecules (Ahmed et al. 2021). It is critical to understand all ramifications, including possible downsides, while using nanostructures to optimize advantages in a sustainable manner (Ahmed et al. 2022). Through the destruction of cell membranes, disruption of the electron transport chain, creation of reactive oxygen species, and oxidative damage to DNA, they can result in significant harm to several cells and organs (Ozougwu 2016). Unlike NPs that are physicochemically synthesized, relatively little information is known on how green nanostructures affect the environment and biota. The toxicity of NPs is thought to be influenced by their size, shape, crystal structure, smoothness, solubility, surface energy, and chemical composition, as well as by whether or not they have surface coatings (Xu et al. 2018). The capacity of spherical NPs to undergo endocytosis surpasses that of star, triangle, or tubular forms. Additionally, the pore structure, charge, and reactivity of NPs coated with organic compounds are modified, and surface charge has been shown to regulate their aggregation ability (Shrestha et al. 2020). The majority of cells were shown to be toxic to NPs with a size less than 50 nm, indicating the significance of size in determining toxicity (Bahadar et al. 2016). In addition to causing genotoxicity in the form of DNA damage from strand breaks, point mutations, and altered gene expression, other hazardous effects caused by NPs include allergic responses, organ dysfunction, fibrosis, and splenic, hepatological, neurological, and hematological toxicities (Yamini et al. 2023).

The selection of agricultural materials is based on their nontoxicity, cost-effectiveness, biocompatibility, and biodegradability (Sah et al. 2022). Precision farming is expected to grow as a result of agri-nanotechnology advancements, which will also encourage and permit the responsible use of renewable resources for agricultural purposes (Ashraf et al. 2021). The environmentally friendly production of NPs with plant extracts and microbes is beneficial for advanced and secure agri-nanotechnology research (Pandey 2020). NPs are beneficial for the continuous release of fertilizers into the soil due to their huge surface area (Madzokere et al. 2021). This improves the release and surface protection of pesticides, herbicides, fungicides, and fertilizers by using nanocoated formulations (Kalia et al. 2020). Fertilizers with nanocoatings have a slower rate of breakdown, allowing for a more gradual and regulated release of nutrients that are taken by plant roots (Moradi et al. 2024). Phosphate and urea covered in nanotechnology aid in satisfying the soil's needs (Basavegowda and Baek 2021). It has been demonstrated that using nanosilver-encased nanofertilizers strengthens roots and promotes seed growth.

Fertilizers with regulated release of nitrogen, phosphorus, and potassium can be found in chitosan NPs that are made from polymethyl methacrylate (PMMA) (ABD GHAPAR 2018). Coating biofertilizers with NPs helps them overcome obstacles relating to storage, limited shelf life, and temperature sensitivity (Rai et al. 2023). Utilizing gold and AgNPs in conjunction with biofertilizers such as *Bacillus subtilis, Pseudomonas fluorescence*, and *Paenibacillus elgii* has demonstrated encouraging outcomes in terms of increased plant growth rate (Singh et al. 2023). Through the application of their nanoformulations as sprays on plants or in soil to increase soil vigor and health, nanotechnology helps make nutrients more accessible (Reddy and Chhabra 2022). Plant growth hormone, 1-naphthylacetic acid, has been proven to be positively impacted by chitosan NPs (Butova et al. 2023). Thus, in addition to boosting agricultural output through bio-remediation, sensing, and detection, the use of nanotechnology in agriculture offers an efficient way to (i) control and effectively release nutrients and water using nanoporous zeolites; (ii) distribute herbicides, pesticides, and insecticides continuously using nanocapsules; (iii) employ nanosensors to detect toxins and pests and monitor environmental conditions; and (iv) use NPs as photocatalysts in a variety of activities (Pandey 2018).

1.8 Conclusions

India is an agrarian country, but its rate of agricultural growth is struggling and needs to be significantly accelerated to increase productivity. This is an instance where nanotechnology comes in handy and aids in increasing production. The field of nanotechnology has applications in both on- and off-site agriculture, and Indian scientists and analysts are currently concentrating on research and development programs through education to establish and adapt nano applications that match India's socioeconomic environment. The impact of nanotechnology on the environment, its associated safety concerns, potential health risks, and societal implications will determine whether it is accepted or not. Future agri-nanotechnology research should focus on evaluating and resolving the risk issues associated with the usage of NPs, since the increased surface area of NMs may have harmful effects that are not apparent in bulk. The synthesis and application research are currently limited to the lab, which is why they are not helping the agribusiness industry recognize and embrace nanotechnology on a large scale. As a result, the scientific community must work to enhance research by implementing more useful and realistic methodologies. Research and clarification are needed about the impacts of NPs on future generations, as well as the acceptable dosages of these materials within safety bounds. Understanding the appropriate and safe dosage of NPs for farming may be achieved via concentration-dependent research in soil systems. In order to mitigate the harmful impacts of NPs and have a significant positive impact on the agro-ecosystem, further research is needed to improve the soil environment through proper soil management. The green synthesis of NPs may result in relatively low or no toxicity. Indian farmers want more information, assistance, counseling, and communication in order to comprehend and implement nano-solutions because the country's current nanotechnology situation is not particularly encouraging. Microbial nanotechnology is an intriguing and rapidly expanding field for the synthesis of revolutionary NPs. It holds great promise in agriculture, but further research is needed in this fascinating field to fully understand how NPs function and how to take benefit from all their advantages.

References

Abady, M.M., Imtiaz, S., Imtiaz, S., and Mohammed, D.M. (2024). Nanobiotechnology for the food industry: current scenario, risk assessment, and management. In: *Nanobiotechnology for Food Processing and Packaging* (ed. J. Singh, R.P. Singh, and A.K. Kaushik), 65–94. Elsevier.

ABD GHAPAR NFB (2018). *Preparation of Chitosan-g-pmma for Adsorption-Desorption Behavior of Urea*, 1–56. Universiti Technologi Malaysia.

Adarsh, S., Shilpa, S., Ameena, M. et al. (2024). A review on nano herbicides: the future of weed management. *Journal of Advances in Biology & Biotechnology* 27: 1244–1253.

Adepu, S. and Ramakrishna, S. (2021). Controlled drug delivery systems: current status and future directions. *Molecules* 26: 5905.

Aguilar, C.A.H., Jiménez, A.B.P., Silva, A.R. et al. (2015). Organic-inorganic hybrid nanoparticles for bacterial inhibition: synthesis and characterization of doped and undoped ONPs with Ag/Au NPs. *Molecules* 20: 6002–6021.

Ahmed, T., Noman, M., Manzoor, N. et al. (2021). Recent advances in nanoparticles associated ecological harms and their biodegradation: global environmental safety from nano-invaders. *Journal of Environmental Chemical Engineering* 9: 106093.

Ahmed, S.F., Mofijur, M., Rafa, N. et al. (2022). Green approaches in synthesising nanomaterials for environmental nanobioremediation: technological advancements, applications, benefits and challenges. *Environmental Research* 204: 111967.

Anandhi, S., Saminathan, V., Yasotha, P. et al. (2020). Nano-pesticides in pest management. *Journal of Entomology and Zoology Studies* 8: 685–690.

Andrade, F., Jenipher, C., Gurav, N. et al. (2024). Endophytic fungi-assisted biomass synthesis of eco-friendly formulated silver nanoparticles for enhanced antibacterial, antioxidant, and antidiabetic activities. *Journal of Drug Delivery Science and Technology* 97: 105749. https://doi.org/10.1016/j.jddst.2024.105749.

Arole, V. and Munde, S. (2014). Fabrication of nanomaterials by top-down and bottom-up approaches-an overview. *Journal of Materials Science* 1: 89–93.

Ashraf, S.A., Siddiqui, A.J., Abd Elmoneim, O.E. et al. (2021). Innovations in nanoscience for the sustainable development of food and agriculture with implications on health and environment. *Science of the Total Environment* 768: 144990.

Asif, M., Islam, S., Malik, M.A. et al. (2021). Nano pesticides application in agriculture and their impact on environment. In: *Agricultural Development in Asia-Potential Use of Nano-Materials and Nano-Technology* (ed. M. Asaduzzaman and M. Afroz), 43–74. Intech Open.

Attard, G., Casadesús, M., Macaskie, L.E., and Deplanche, K. (2012). Biosynthesis of platinum nanoparticles by *Escherichia coli* MC 4100: can such nanoparticles exhibit intrinsic surface enantioselectivity? *Langmuir* 28: 5267–5274.

Avila-Quezada, G.D., Ingle, A.P., Golińska, P., and Rai, M. (2022). Strategic applications of nano-fertilizers for sustainable agriculture: benefits and bottlenecks. *Nanotechnology Reviews* 11: 2123–2140.

Ayub, M.A., Usman, M., Faiz, T. et al. (2020). Restoration of degraded soil for sustainable agriculture. *Soil Health Restoration and Management* 31–81.

Azmath, P., Baker, S., Rakshith, D., and Satish, S. (2016). Mycosynthesis of silver nanoparticles bearing antibacterial activity. *Saudi Pharmaceutical Journal* 24: 140–146.

Babu, S., Singh, R., Yadav, D. et al. (2022). Nanofertilizers for agricultural and environmental sustainability. *Chemosphere* 292: 133451.

Bahadar, H., Maqbool, F., Niaz, K., and Abdollahi, M. (2016). Toxicity of nanoparticles and an overview of current experimental models. *Iranian Biomedical Journal* 20: 1.

Bahrulolum, H., Nooraei, S., Javanshir, N. et al. (2021). Green synthesis of metal nanoparticles using microorganisms and their application in the agrifood sector. *Journal of Nanobiotechnology* 19: 1–26.

Bamal, D., Singh, A., Chaudhary, G. et al. (2021). Silver nanoparticles biosynthesis, characterization, antimicrobial activities, applications, cytotoxicity and safety issues: an updated review. *Nanomaterials* 11: 2086.

Bano, I. (2022). Nanotechnology and its importance in the field of microbiology. In: *Methods in Microbiology*, vol. 51 (ed. V. Gurtler), 119–135. Elsevier.

Basavegowda, N. and Baek, K.-H. (2021). Current and future perspectives on the use of nanofertilizers for sustainable agriculture: the case of phosphorus nanofertilizer. *3 Biotech* 11: 357.

Bedlovičová, Z. (2022). Chapter 21: Green synthesis of silver nanoparticles using actinomycetes. In: *Green Synthesis of Silver Nanomaterials* (ed. K.A. Abd-Elsalam), 547–569. Elsevier https://doi.org/10.1016/B978.0-12-824508-8.00001-0.

Bhambure, R., Bule, M., Shaligram, N. et al. (2009). Extracellular biosynthesis of gold nanoparticles using *Aspergillus niger*–its characterization and stability. *Chemical Engineering & Technology: Industrial Chemistry-Plant Equipment-Process Engineering-Biotechnology* 32: 1036–1041.

Bhardwaj, P., Chauhan, A., Ranjan, A., and Tripathi, A. (2024). Nanomaterial-induced phytotoxicity and challenges to ongoing agricultural practices. In: *Emerging Contaminants* (ed. A. Kumari, V.D. Rajput, and E. van Hullebusch), 179–205. Elsevier.

Bharti, V., Dotaniya, M., Shukla, S., and Yadav, V. (2017). Managing soil fertility through microbes: prospects, challenges and future strategies. In: *Agro-Environmental Sustainability: Volume 1: Managing Crop Health* (ed. J.S. Singh and G. Seneviratne), 81–111. Springer.

Bindraban, P.S., Dimkpa, C.O., and Pandey, R. (2020). Exploring phosphorus fertilizers and fertilization strategies for improved human and environmental health. *Biology and Fertility of Soils* 56: 299–317.

Boguszewicz, C., Boguszewicz, M., Iqbal, Z. et al. (2021). The fourth industrial revolution-cyberspace mental wellbeing: Harnessing science & technology for humanity. *Global Foundation for Cyber Studies and Research*.

Brar, K.K., Magdouli, S., Othmani, A. et al. (2022). Green route for recycling of low-cost waste resources for the biosynthesis of nanoparticles (NPs) and nanomaterials (NMs) – a review. *Environmental Research* 207: 112202.

Butova, V.V., Bauer, T.V., Polyakov, V.A., and Minkina, T.M. (2023). Advances in nanoparticle and organic formulations for prolonged controlled release of auxins. *Plant Physiology and Biochemistry* 201: 107808.

Caineng, Z., Feng, M., Songqi, P. et al. (2022). Earth energy evolution, human development and carbon neutral strategy. *Petroleum Exploration and Development* 49: 468–488.

Calvo, V., González-Domínguez, J.M., Benito, A.M., and Maser, W.K. (2022). Synthesis and processing of nanomaterials mediated by living organisms. *Angewandte Chemie* 134: e202113286.

Charitidis, C.A., Georgiou, P., Koklioti, M.A. et al. (2014). Manufacturing nanomaterials: from research to industry. *Manufacturing Review* 1: 11.

Cherian, T., Maity, D., Rajendra Kumar, R.T. et al. (2022). Green chemistry based gold nanoparticles synthesis using the marine bacterium *Lysinibacillus odysseyi* PBCW2 and their multitudinous activities. *Nanomaterials* 12: 2940.

Correa-Llantén, D.N., Muñoz-Ibacache, S.A., Castro, M.E. et al. (2013). Gold nanoparticles synthesized by *Geobacillus* sp. strain ID17 a thermophilic bacterium isolated from Deception Island, Antarctica. *Microbial Cell Factories* 12: 1–6.

Cummings, C.L., Kuzma, J., Kokotovich, A. et al. (2021). Barriers to responsible innovation of nanotechnology applications in food and agriculture: a study of US experts and developers. *Nano Impact* 23: 100326.

Das, S.K. and Marsili, E. (2011). *Bioinspired Metal Nanoparticle: Synthesis, Properties and Application*. Croacia: InTech.

Diep, E. and Schiffman, J.D. (2023). Electrospinning living bacteria: a review of applications from agriculture to health care. *ACS Applied Bio Materials* 6: 951–964.

Dinadayalane, T.C. and Leszczynski, J. (2012). Fundamental structural, electronic, and chemical properties of carbon nanostructures: graphene, fullerenes, carbon nanotubes, and their derivatives. In: *Handbook of Computational Chemistry* (ed. T. Clark), 793–867. Wiley.

Dolai, J., Mandal, K., and Jana, N.R. (2021). Nanoparticle size effects in biomedical applications. *ACS Applied Nano Materials* 4: 6471–6496.

Dutta, V., Verma, R., Gopalkrishnan, C. et al. (2022). Bio-inspired synthesis of carbon-based nanomaterials and their potential environmental applications: a state-of-the-art review. *Inorganics* 10: 169.

Dwivedi, A.D. and Gopal, K. (2010). Biosynthesis of silver and gold nanoparticles using *Chenopodium album* leaf extract. *Colloids and Surfaces A: Physicochemical and Engineering Aspects* 369: 27–33.

Ediger, M., Bester, G., Badolato, A. et al. (2007). Peculiar many-body effects revealed in the spectroscopy of highly charged quantum dots. *Nature Physics* 3: 774–779.

Fouda, A., Eid, A.M., Abdel-Rahman, M.A. et al. (2022). Enhanced antimicrobial, cytotoxicity, larvicidal, and repellence activities of brown algae, *Cystoseira crinita*-mediated green synthesis of magnesium oxide nanoparticles. *Frontiers in Bioengineering and Biotechnology* 10: 849921.

Garg, D., Sarkar, A., Chand, P. et al. (2020). Synthesis of silver nanoparticles utilizing various biological systems: mechanisms and applications—a review. *Progress in Biomaterials* 9: 81–95.

Gebregiorgis, T. and Nigussie, B. (2018). Phytochemical, antimicrobial potential and antifungal activities stem bark extract of *Boswellia ovalifoliolata*. *Archives of Organic and Inorganic Chemical Sciences* 1: 1–7.

Georgakilas, V., Perman, J.A., Tucek, J., and Zboril, R. (2015). Broad family of carbon nanoallotropes: classification, chemistry, and applications of fullerenes, carbon dots, nanotubes, graphene, nanodiamonds, and combined superstructures. *Chemical Reviews* 115: 4744–4822.

Gherbawy, Y.A., Shalaby, I.M., Abd El-sadek, M.S. et al. (2013). The anti-fasciolasis properties of silver nanoparticles produced by *Trichoderma harzianum* and their improvement of the anti-fasciolasis drug triclabendazole. *International Journal of Molecular Sciences* 14: 21887–21898.

Ghosh, S., Sarkar, B., and Thongmee, S. (2022). Nanoherbicides for field applications. In: *Agricultural Nanobiotechnology* (ed. S. Ghosh, S. Thongmee, and A. Kumar), 439–463. Elsevier.

Grigorieva, E., Livenets, A., and Stelmakh, E. (2023). Adaptation of agriculture to climate change: a scoping review. *Climate* 11: 202.

Gupta, K. and Chundawat, T.S. (2020). Time and size-dependent biogenically synthesized nanoparticles using fungus *Fusarium oxysporum*: a review on their preparation, characterization and biological activities. *Nanoscience and Nanotechnology - Asia* 10: 95–108.

Haider, A.J., Alawsi, T., Haider, M.J. et al. (2022). A comprehensive review on pulsed laser deposition technique to effective nanostructure production: trends and challenges. *Optical and Quantum Electronics* 54: 488.

Hasani, A., Madhi, M., Gholizadeh, P. et al. (2019). Metal nanoparticles and consequences on multi-drug resistant bacteria: reviving their role. *SN Applied Sciences* 1: 1–13.

Hasler, K. (2017). Environmental impact of mineral fertilizers: possible improvements through the adoption of eco-innovations. Wageningen University and Research.

Hopkin, P. (2018). Fundamentals of risk management: understanding, evaluating and implementing effective risk management. Kogan Page Publishers.

Hossain, M.E., Saha, P., and Bezbaruah, A.N. (2023). Environmentally benign synthesis of metal nanoparticles for fertilizer applications in agriculture. In: *Nanofertilizers for Sustainable Agroecosystems: Recent Advances and Future Trends* (ed. K.A. Abd-Elsalam and M.A. Alghuthaymi), 125–150. Springer.

Huang, X., Wu, H., Pu, S. et al. (2011). One-step room-temperature synthesis of Au@ Pd core–shell nanoparticles with tunable structure using plant tannin as reductant and stabilizer. *Green Chemistry* 13: 950–957.

Hull, M., Bowman, D. (2014) Nanotechnology environmental health and safety: risks, regulation, and management. William Andrew.

Jain, A., Ranjan, S., Dasgupta, N., and Ramalingam, C. (2018). Nanomaterials in food and agriculture: an overview on their safety concerns and regulatory issues. *Critical Reviews in Food Science and Nutrition* 58: 297–317.

Jeevanandam, J., Barhoum, A., Chan, Y.S. et al. (2018). Review on nanoparticles and nanostructured materials: history, sources, toxicity and regulations. *Beilstein Journal of Nanotechnology* 9: 1050–1074.

Jeevanandam, J., Kiew, S.F., Boakye-Ansah, S. et al. (2022). Green approaches for the synthesis of metal and metal oxide nanoparticles using microbial and plant extracts. *Nanoscale* 14:2534–2571.

Jegadeeswaran, P., Shivaraj, R., and Venckatesh, R. (2012). Green synthesis of silver nanoparticles from extract of *Padina tetrastromatica* leaf. *Digest Journal of Nanomaterials and Biostructures* 7: 991–998.

Jiang, X., Fan, X., Xu, W. et al. (2019). Biosynthesis of bimetallic Au–Ag nanoparticles using *Escherichia coli* and its biomedical applications. *ACS Biomaterials Science & Engineering* 6: 680–689.

Jiang, X., Yang, F., Jia, W. et al. (2024). Nanomaterials and nanotechnology in agricultural pesticide delivery: a review. *Langmuir* 40 (36): 18806–18820.

Joglekar, S., Kodam, K., Dhaygude, M., and Hudlikar, M. (2011). Novel route for rapid biosynthesis of lead nanoparticles using aqueous extract of *Jatropha curcas* L. latex. *Materials Letters* 65: 3170–3172.

John, D.A. and Babu, G.R. (2021). Lessons from the aftermaths of green revolution on food system and health. *Frontiers in Sustainable Food Systems* 5: 644559.

Ju-Nam, Y. and Lead, J. (2016). Properties, sources, pathways, and fate of nanoparticles in the environment. *Engineered Nanoparticles and the Environment: Biophysicochemical Processes and Toxicity* 4: 95–117.

Kalia, A., Sharma, S.P., Kaur, H., and Kaur, H. (2020). Novel nanocomposite-based controlled-release fertilizer and pesticide formulations: prospects and challenges. In: *Multifunctional Hybrid Nanomaterials for Sustainable Agri-Food and Ecosystems* (ed. K.A. Abd-Elsalam), 99–134. Elsevier.

Khan, A.I. and Arasu, A.V. (2019). A review of influence of nanoparticle synthesis and geometrical parameters on thermophysical properties and stability of nanofluids. *Thermal Science and Engineering Progress* 11: 334–364.

Khan, S. and Hossain, M.K. (2022). Classification and properties of nanoparticles. In: *Nanoparticle-Based Polymer Composites* (ed. S.M. Rangappa, J. Parameswaranpillai, T.G. Yashas Gowda, et al.), 15–54. Elsevier.

Khan, N. and Jameel, J. (2016). Optimization of reaction parameters for silver nanoparticles synthesis from *Fusarium oxysporum* and determination of silver nanoparticles concentration. *Journal of Materials Science and Engineering* 5: 6–9.

Khan, A.U., Malik, N., Khan, M. et al. (2018). Fungi-assisted silver nanoparticle synthesis and their applications. *Bioprocess and Biosystems Engineering* 41: 1–20. https://doi.org/10.1007/s00449-017-1846-3.

Kim, D.Y., Kadam, A., Shinde, S. et al. (2018). Recent developments in nanotechnology transforming the agricultural sector: a transition replete with opportunities. *Journal of the Science of Food and Agriculture* 98: 849–864.

Koul, B., Poonia, A.K., Yadav, D., and Jin, J.-O. (2021). Microbe-mediated biosynthesis of nanoparticles: applications and future prospects. *Biomolecules* 11: 886.

Krishnaraj, C., Jagan, E., Rajasekar, S. et al. (2010). Synthesis of silver nanoparticles using *Acalypha indica* leaf extracts and its antibacterial activity against water borne pathogens. *Colloids and Surfaces B: Biointerfaces* 76: 50–56.

Kulkarni, S. (2021). Climate change, soil erosion risks, and nutritional security. In: *Climate Change and Resilient Food Systems: Issues, Challenges, and Way Forward* (ed. V.K.H. Mallappa and M. Shirur), 219–244. Springer Nature.

Kumar, A. (2014). Nanotechnology development in India: an overview. Research and Information System for Developing Countries New Delhi.

Kumar, J.A., Krithiga, T., Manigandan, S. et al. (2021). A focus to green synthesis of metal/metal based oxide nanoparticles: Various mechanisms and applications towards ecological approach. *Journal of Cleaner Production* 324: 129198.

Leonel, A.G., Mansur, A.A., and Mansur, H.S. (2021). Advanced functional nanostructures based on magnetic iron oxide nanomaterials for water remediation: a review. *Water Research* 190: 116693.

Li, Y., Zhu, J., Cheng, H. et al. (2021). Developments of advanced electrospinning techniques: a critical review. *Advanced Materials Technologies* 6: 2100410.

Lim, J.-V., Bee, S.-T., Tin Sin, L. et al. (2021). A review on the synthesis, properties, and utilities of functionalized carbon nanoparticles for polymer nanocomposites. *Polymers* 13: 3547.

Madzokere, T.C., Murombo, L., and Chiririwa, H. (2021). Nano-based slow releasing fertilizers for enhanced agricultural productivity. *Materials Today: Proceedings* 45: 3709–3715.

Majumdar, K., Sanyal, S.K., Dutta, S.K. et al. (2016). Nutrient mining: addressing the challenges to soil resources and food security. In: *Biofortification of Food Crops* (ed. U. Singh, C.S. Praharaj, S.S. Singh, and N.P. Singh), 177–198. Springer.

Malik, A.Q., Mir, T.G., Kumar, D. et al. (2023a). A review on the green synthesis of nanoparticles, their biological applications, and photocatalytic efficiency against environmental toxins. *Environmental Science and Pollution Research* 30: 69796–69823.

Malik, S., Muhammad, K., and Waheed, Y. (2023b). Nanotechnology: a revolution in modern industry. *Molecules* 28: 661.

Mateo-Sagasta, J., Zadeh, S.M., and Turral, H. (2018). *More People, More Food, Worse Water?: A Global Review of Water Pollution from Agriculture*. Food and Agriculture Organization.

Mishra, A.N., Bhadauria, S., Gaur, M.S., and Pasricha, R. (2010). Extracellular microbial synthesis of gold nanoparticles using fungus *Hormoconis resinae*. *JOM: The Journal of the Minerals, Metals & Materials Society* 62: 45–48.

Mitra, B., Chowdhury, A.R., Dey, P. et al. (2021). Use of agrochemicals in agriculture: alarming issues and solutions. In: *Input Use Efficiency for Food and Environmental Security* (ed. R. Bhatt, R.S. Meena, and A. Hossain), 85–122. Springer Nature.

Mittal, D., Kaur, G., Singh, P. et al. (2020). Nanoparticle-based sustainable agriculture and food science: recent advances and future outlook. *Frontiers in Nanotechnology* 2: 579954.

Moradi, S., Babapoor, A., Ghanbarlou, S. et al. (2024). Toward a new generation of fertilizers with the approach of controlled-release fertilizers: a review. *Journal of Coatings Technology and Research* 21: 31–54.

Nadathur, S., Wanasundara, J.P., and Scanlin, L. (2024). Feeding the globe nutritious food in 2050: obligations and ethical choices. In: *Sustainable Protein Sources* (ed. S.R. Nadathur, J.P.D. Wanasundara, and L. Scanlin), 649–668. Elsevier.

Nasr Azadani, F., Madani, M., Karimi, J., and Sepahvand, S. (2024). Green synthesis of silver nanoparticles by *Fusarium oxysporum* and its function against *Aspergillus* and *Fusarium* fungi. *Indian Journal of Microbiology* 64: 213–224. https://doi.org/10.1007/s12088-023-01162-w.

Nayak, R.R., Pradhan, N., Behera, D. et al. (2011). Green synthesis of silver nanoparticle by Penicillium purpurogenum NPMF: the process and optimization. *Journal of Nanoparticle Research* 13: 3129–3137.

Negi, S. and Singh, V. (2018). Algae: A potential source for nanoparticle synthesis. *Journal of Applied and Natural Science* 10: 1134–1140.

Negi, R., Kaur, T., Devi, R. et al. (2022a). First report on *Rahnella* sp. strain EU-A3SNfb, a plant growth promoting endophytic bacterium from wild wheat relative *Aegilops kotschyi*. *National Academy Science Letters* 45: 393–396.

Negi, R., Kaur, T., Devi, R. et al. (2022b). Assessment of nitrogen-fixing endophytic and mineral solubilizing rhizospheric bacteria as multifunctional microbial consortium for growth promotion of wheat and wild wheat relative *Aegilops kotschyi*. *Heliyon* 8 (12): e12579.

Negi, R., Sharma, B., Jan, T. et al. (2024a). Microbial consortia: promising tool as plant bioinoculants for agricultural sustainability. *Current Microbiology* 81: 222. https://doi.org/10.1007/s00284-024-03755-0.

Negi, R., Sharma, B., Kumar, S. et al. (2024b). Plant endophytes: unveiling hidden applications toward agro-environment sustainability. *Folia Microbiologica* 69: 181–206. https://doi.org/10.1007/s12223-023-01092-6.

Negi, R., Sharma, B., Parastesh, F. et al. (2024c). Microbial consortia mediated regulation of plant defense: a promising tool for sustaining crops protection. *Physiological and Molecular Plant Pathology* 134: 102393.

Oliver, K., Innvar, S., Lorenc, T. et al. (2014). A systematic review of barriers to and facilitators of the use of evidence by policymakers. *BMC Health Services Research* 14: 1–12.

Ozougwu, J.C. (2016). The role of reactive oxygen species and antioxidants in oxidative stress. *International Journal of Research* 1: 1–8.

Pachaiappan, R., Rajendran, S., Show, P.L. et al. (2021). Metal/metal oxide nanocomposites for bactericidal effect: a review. *Chemosphere* 272: 128607.

Pallavi, S.S., Rudayni, H.A., Bepari, A. et al. (2022). Green synthesis of silver nanoparticles using *Streptomyces hirsutus* strain SNPGA-8 and their characterization, antimicrobial activity, and anticancer activity against human lung carcinoma cell line A549. *Saudi Journal of Biological Sciences* 29: 228–238. https://doi.org/10.1016/j.sj.s.2021.08.084.

Pandey, G. (2018). Challenges and future prospects of agri-nanotechnology for sustainable agriculture in India. *Environmental Technology & Innovation* 11: 299–307.

Pandey, G. (2020). Agri-nanotechnology for sustainable agriculture. In: *Ecological and Practical Applications for Sustainable Agriculture* (ed. K. Bauddh, S. Kumar, R.P. Singh, and J. Korstad), 229–249. Springer Nature.

Parthasarathi, A. (2007). *Technology at the Core: Science & Technology with Indira Gandhi.* Pearson Education India.

Patil, S.V., Borase, H.P., Patil, C.D., and Salunke, B.K. (2012). Biosynthesis of silver nanoparticles using latex from few euphorbian plants and their antimicrobial potential. *Applied Biochemistry and Biotechnology* 167: 776–790.

Prajapati, D., Jabborova, D., Saharan, B.S. et al. (2024). Bionanotechnology: a paradigm for advancing environmental sustainability. *Indian Journal of Microbiology* 65 (1): 306–332.

Prasad, K., Jha, A.K., and Kulkarni, A. (2007). Lactobacillus assisted synthesis of titanium nanoparticles. *Nanoscale Research Letters* 2: 248–250.

Prasad, R., Bhattacharyya, A., and Nguyen, Q.D. (2017). Nanotechnology in sustainable agriculture: recent developments, challenges, and perspectives. *Frontiers in Microbiology* 8: 1014.

Priyanka, P., Kumar, D., Yadav, A., and Yadav, K. (2020). Nanobiotechnology and its application in agriculture and food production. In: *Nanotechnology for Food, Agriculture, and Environment* (ed. A. Kumari, R. Garg, and R. Garg), 105–134. CRC Press.

Rahman, M.H., Haque, K.S., and Khan, M.Z.H. (2021). A review on application of controlled released fertilizers influencing the sustainable agricultural production: a cleaner production process. *Environmental Technology & Innovation* 23: 101697.

Rai, P.K., Rai, A., Sharma, N.K. et al. (2023). Limitations of biofertilizers and their revitalization through nanotechnology. *Journal of Cleaner Production* 418: 138194.

Ranjan Sarker, S., Polash, S.A., Boath, J. et al. (2019). Functionalization of elongated tetrahexahedral Au nanoparticles and their antimicrobial activity assay. *ACS Applied Materials & Interfaces* 11: 13450–13459.

Rao, A.S. and Lenka, N.K. (2020). Developments on soil health management in India as mirrored through sustained researches and policy interventions. *Indian Journal of Fertilisers* 16: 1230–1242.

Rao, N., Singh, R., and Bashambu, L. (2021). Carbon-based nanomaterials: synthesis and prospective applications. *Materials Today: Proceedings* 44: 608–614.

Rasul, G. (2016). Managing the food, water, and energy nexus for achieving the sustainable development goals in South Asia. *Environmental Development* 18: 14–25.

Reddy, S. and Chhabra, V. (2022). Nanotechnology: its scope in agriculture. *Journal of Physics: Conference Series* 1: 012112.

Safdar, M., Kim, W., Park, S. et al. (2022). Engineering plants with carbon nanotubes: a sustainable agriculture approach. *Journal of Nanobiotechnology* 20: 275.

Sah, M.K., Mukherjee, S., Flora, B. et al. (2022). Advancement in “Garbage In Biomaterials Out (GIBO)” concept to develop biomaterials from agricultural waste for tissue engineering and biomedical applications. *Journal of Environmental Health Science and Engineering* 20: 1015–1033.

San Chan, Y. and Don, M.M. (2013). Biosynthesis and structural characterization of Ag nanoparticles from white rot fungi. *Materials Science and Engineering: C* 33: 282–288.

Sandler, R. (2009). *Nanotechnology: The Social and Ethical Issues*. The Pew Charitable Trusts.

Sapsford, K.E., Algar, W.R., Berti, L. et al. (2013). Functionalizing nanoparticles with biological molecules: developing chemistries that facilitate nanotechnology. *Chemical Reviews* 113: 1904–2074.

Sarkar, J., Ray, S., Chattopadhyay, D. et al. (2012). Mycogenesis of gold nanoparticles using a phytopathogen *Alternaria alternata*. *Bioprocess and Biosystems Engineering* 35: 637–643.

Sarwal, A., Singh, G., Singh, K., and Garg, S. (2018). Recent interventions for nanotechnology based drug products: insights into the regulatory aspects. *Current Pharmaceutical Design* 24: 5219–5228.

Scavo, A. and Mauromicale, G. (2020). Integrated weed management in herbaceous field crops. *Agronomy* 10: 466.

Sekhar, E.C., Rao, K.K., Rao, K.M.S., and Alisha, S.B. (2018). A simple biosynthesis of silver nanoparticles from *Syzygium cumini* stem bark aqueous extract and their spectrochemical and antimicrobial studies. *Journal of Applied Pharmaceutical Science* 8: 073–079.

Sekhon, B.S. (2014). Nanotechnology in agri-food production: an overview. *Nanotechnology, Science and Applications* 7: 31–53.

Šesták, J. (2021). Thermal physics of nanostructured materials: thermodynamic (top-down) and quantum (bottom-up) issues. *Materials Today: Proceedings* 37: 28–34.

Shahcheraghi, N., Golchin, H., Sadri, Z. et al. (2022). Nano-biotechnology, an applicable approach for sustainable future. *3 Biotech* 12: 65.

Shahzad, A., Saeed, H., Iqtedar, M. et al. (2019). Size-controlled production of silver nanoparticles by *Aspergillus fumigatus* BTCB10: likely antibacterial and cytotoxic effects. *Journal of Nanomaterials* 2019: 5168698.

Shang, Y., Hasan, M.K., Ahammed, G.J. et al. (2019). Applications of nanotechnology in plant growth and crop protection: a review. *Molecules* 24: 2558.

Shanthi, S., Jayaseelan, B.D., Velusamy, P. et al. (2016). Biosynthesis of silver nanoparticles using a probiotic *Bacillus licheniformis* Dahb 1 and their antibiofilm activity and toxicity effects in *Ceriodaphnia cornuta*. *Microbial Pathogenesis* 93: 70–77.

Sharif, S., Arbab, S., Saeed, A. et al. (2021). Introductory chapter: Introduction to advanced carbon materials and innovative engineering applications. In: *21st Century Advanced Carbon Materials for Engineering Applications – A Comprehensive Handbook* (ed. M. Ikram and A. Maqsood), 3–14. Intech Open.

Sharma, P. and Sharma, P. (2024). Micro (nano) plastics: invisible compounds with a visible impact. *F1000Research* 13: 69.

Shedbalkar, U., Singh, R., Wadhwani, S. et al. (2014). Microbial synthesis of gold nanoparticles: current status and future prospects. *Advances in Colloid and Interface Science* 209: 40–48. https://doi.org/10.1016/j.cis.2013.12.011.

Sheikhloo, Z., Salouti, M., and Katiraee, F. (2011). Biological synthesis of gold nanoparticles by fungus *Epicoccum nigrum*. *Journal of Cluster Science* 22: 661–665.

Sheny, D., Mathew, J., and Philip, D. (2011). Phytosynthesis of Au, Ag and Au–Ag bimetallic nanoparticles using aqueous extract and dried leaf of *Anacardium occidentale*. *Spectrochimica Acta Part A: Molecular and Biomolecular Spectroscopy* 79: 254–262.

Shirsat, S.D., Mane, R.S., and Dhuldhaj, U.P. (2021). Biogenic synthesis of silicon oxide, titanium oxide and cerium oxide nanomaterials. *Microbial Synthesis of Nanomaterials* 137.

Shnoudeh, A.J., Hamad, I., Abdo, R.W. et al. (2019). Synthesis, characterization, and applications of metal nanoparticles. In: *Biomaterials and Bionanotechnology* (ed. R.K. Tekade), 527–612. Elsevier.

Shrestha, S., Wang, B., and Dutta, P. (2020). Nanoparticle processing: understanding and controlling aggregation. *Advances in Colloid and Interface Science* 279: 102162.

Singaravelu, G., Arockiamary, J., Kumar, V.G., and Govindaraju, K. (2007). A novel extracellular synthesis of monodisperse gold nanoparticles using marine alga, *Sargassum wightii* Greville. *Colloids and Surfaces B: Biointerfaces* 57: 97–101.

Singh, A.K. (2023). Importance and contribution of carbon allotropes in a green and sustainable environment. In: *Carbon Allotropes and Composites: Materials for Environment Protection and Remediation* (ed. C. Verma and C.M. Hussain), 337–382. John Wiley & Sons.

Singh, R., Shedbalkar, U.U., Wadhwani, S.A., and Chopade, B.A. (2015). Bacteriagenic silver nanoparticles: synthesis, mechanism, and applications. *Applied Microbiology and Biotechnology* 99: 4579–4593. https://doi.org/10.1007/s00253-015-6622-1.

Singh, N.A., Narang, J., Garg, D. et al. (2023). Nanoparticles synthesis via microorganisms and their prospective applications in agriculture. *Plant Nano Biology* 5: 100047.

Sivakami, A., Sarankumar, R., and Vinodha, S. (2021). Introduction to nanobiotechnology: novel and smart applications. In: *Bio-manufactured Nanomaterials: Perspectives and Promotion* (ed. K. Pal), 1–22. Springer Nature.

de Souza, T.A.J., Souza, L.R.R., and Franchi, L.P. (2019). Silver nanoparticles: an integrated view of green synthesis methods, transformation in the environment, and toxicity. *Ecotoxicology and Environmental Safety* 171: 691–700.

Srivastava, N., Srivastava, M., Mishra, P., and Gupta, V.K. (2020). *Green Synthesis of Nanomaterials for Bioenergy Applications*. John Wiley & Sons.

Summer, M., Hussain, T., Ali, S. et al. (2024). Exploring the underlying modes of organic nanoparticles in diagnosis, prevention, and treatment of cancer: a review from drug delivery to toxicity. *International Journal of Polymeric Materials and Polymeric Biomaterials* 74: 829–845.

Talebian, S., Shahnavaz, B., Nejabat, M. et al. (2023). Bacterial-mediated synthesis and characterization of copper oxide nanoparticles with antibacterial, antioxidant, and anticancer potentials. *Frontiers in Bioengineering and Biotechnology* 11: 1140010.

Thavaseelan, D. and Priyadarshana, G. (2021). Nanofertilizer use for sustainable agriculture. *Journal of Research Technology and Engineering* 2: 41–59.

Thiruvengadam, M., Chi, H.Y., and Kim, S.-H. (2024). Impact of nanopollution on plant growth, photosynthesis, toxicity, and metabolism in the agricultural sector: an updated review. *Plant Physiology and Biochemistry* 207: 108370.

Tyagi, J., Ahmad, S., and Malik, M. (2022). Nitrogenous fertilizers: impact on environment sustainability, mitigation strategies, and challenges. *International Journal of Environmental Science and Technology* 19: 11649–11672.

Vaidyanathan, R., Gopalram, S., Kalishwaralal, K. et al. (2010). Enhanced silver nanoparticle synthesis by optimization of nitrate reductase activity. *Colloids and Surfaces B: Biointerfaces* 75: 335–341. https://doi.org/10.1016/j.colsurfb.2009.09.006.

Veerasamy, R., Xin, T.Z., Gunasagaran, S. et al. (2011). Biosynthesis of silver nanoparticles using mangosteen leaf extract and evaluation of their antimicrobial activities. *Journal of Saudi Chemical Society* 15: 113–120.

Vega-Vásquez, P., Mosier, N.S., and Irudayaraj, J. (2020). Nanoscale drug delivery systems: from medicine to agriculture. *Frontiers in Bioengineering and Biotechnology* 8: 79.

Verma, V.C., Kharwar, R.N., and Gange, A.C. (2010). Biosynthesis of antimicrobial silver nanoparticles by the endophytic fungus *Aspergillus clavatus*. *Nanomedicine* 5: 33–40.

Verma, D.K., Srivastava, S., Kumar, V. et al. (2017). Nano-particle based delivery systems: applications in agriculture. In: *Engineering Interventions in Agricultural Processing* (ed. M.R. Goyal and D.K. Verma), 107–130. Apple Academic Press.

Vigneshwaran, N., Kathe, A.A., Varadarajan, P. et al. (2006). *Biomimetics of silver nanoparticles by white rot fungus, Phaenerochaete chrysosporium*. *Colloids and Surfaces B: Biointerfaces* 53: 55–59.

Vijayaraghavan, K., Nalini, S.K., Prakash, N.U., and Madhankumar, D. (2012). One step green synthesis of silver nano/microparticles using extracts of *Trachyspermum ammi* and *Papaver somniferum*. *Colloids and Surfaces B: Biointerfaces* 94: 114–117.

Wang, Y., O'connor, D., Shen, Z. et al. (2019). Green synthesis of nanoparticles for the remediation of contaminated waters and soils: constituents, synthesizing methods, and influencing factors. *Journal of Cleaner Production* 226: 540–549.

Xu, L., Liang, H.-W., Yang, Y., and Yu, S.-H. (2018). Stability and reactivity: positive and negative aspects for nanoparticle processing. *Chemical Reviews* 118: 3209–3250.

Yadav, T.P., Yadav, R.M., and Singh, D.P. (2012). Mechanical milling: a top down approach for the synthesis of nanomaterials and nanocomposites. *Nanoscience and Nanotechnology* 2: 22–48.

Yamini, V., Shanmugam, V., Rameshpathy, M. et al. (2023). Environmental effects and interaction of nanoparticles on beneficial soil and aquatic microorganisms. *Environmental Research* 236: 116776.

Zahin, N., Anwar, R., Tewari, D. et al. (2020). Nanoparticles and its biomedical applications in health and diseases: special focus on drug delivery. *Environmental Science and Pollution Research* 27: 19151–19168.

Zhang, X.-F., Liu, Z.-G., Shen, W., and Gurunathan, S. (2016). Silver nanoparticles: synthesis, characterization, properties, applications, and therapeutic approaches. *International Journal of Molecular Sciences* 17: 1534.

2

Nanomaterials in Agriculture and Soil Revitalization: Applications and Challenges

Slimane Mokrani[1,2], El-hafid Nabti[1], Omar Khirennas[3,4], and Karim Houali[5]

[1] *Laboratoire de Maitrise Des Energies Renouvelables, Faculté des Sciences de la Nature et de la Vie, Université de Bejaia, Bejaia, Algeria*
[2] *Laboratory of Research on Biological Systems and Geomatics, Faculty of Nature and Life Sciences, University Mustapha Stambouli of Mascara, Mascara, Algeria*
[3] *Laboratory of Geo-Environment and Spatial Development (LGEDE), Faculty of Nature and Life Sciences, University Mustapha Stambouli of Mascara, Mascara, Algeria*
[4] *Laboratory of Geomatics, Ecology and Environment (LGO2E), Faculty of Nature and Life Sciences, University Mustapha Stambouli of Mascara, Mascara, Algeria*
[5] *Laboratoire de Biochimie appliquée et Biotechnologie, Facuslté des Sciences Biologiques et des Sciences Agrnomique, Université Mouloud Mammeri, Tizi-Ouzou, Algeria*

2.1 Introduction

Farmers worldwide will prioritize applying contemporary innovations and technology to enhance agricultural productivity via intensive and widespread farming practices. In order to maintain livable living standards and ensure food security, the nation must adopt newer technology that will increase production and reduce food waste (Neme et al. 2021). Conventional agricultural technologies are labor- and time-intensive; yet, they are not as productive and effective (Gupta et al. 2023). In order to harness both natural and artificial resources to support the growing global population, agricultural advances are essential. In particular, nanotechnology may be used to tackle a variety of issues pertaining to agriculture (Nandini et al. 2023). Nowadays, the majority of experts agree that one of the main tenets of developing science and technology in the 21st century is nanomaterials (NMs). More NMs could make their way into the neighboring area (Adam and Nowack 2017; Scaria et al. 2020) through a variety of channels, and as a result, the relationship between NMs and ecosystems has grown in sensitivity on a worldwide scale. NMs possess at least one structural dimension at the nanoscale, and because of their potential for use in several technological fields, these materials are the subject of increased scientific interest (Saleem and Zaidi 2020).

Structures having sizes between 1.0 and 100.0 nm are known as NMs (Saleh 2016). Their high surface-area-to-volume ratio and potential existence of quantum phenomena set them distinct from their bulk counterparts (Ding et al. 2017). The NMs are now categorized as follows based on their construction: (i) composites; (ii) carbon-based; (iii) metal-based; and (iv) dendrimers (Saleh and Gupta 2016). Norio Taniguchi coined the word "nanotechnology" for the first time in 1974 (Bulovic et al. 2004). Although there are many different definitions of nanotechnology in the literature, one definition is that it is a new, inventive, and multidisciplinary discipline of research that deals with the development, production, and use of materials and technologies at the molecular level in nanometric size (Parisi et al. 2015). Another definition of nanotechnology is the study and manipulation of materials at sizes between 1 and 100 nm, where phenomena allow for the invention of new applications (Park et al. 2006; Helal 2013). It involves the investigation and understanding of atoms or formations in which each atom has unique properties. Being able to manipulate an atom or other unit of matter at this nanoscale can aid in our understanding of how chemical, biological, and physical processes function (Gupta et al. 2023). There is no doubt that the development of better materials or constructions has positive effects. The atoms' modifications and increased magnetic strength cause the characteristic to alter at the nanoscale. It seems to reason that NMs in smaller sizes would be more active and have larger surface areas (Pokropivny et al. 2007). In order to manufacture new products that have been previously modified at the nanoscale and may have better conditions than the conventional materials, nanobiotechnology integrates nanotechnology, biotechnology, chemical and physical processing, material science, and system engineering (Rai and Ingle 2012).

Owing to the novel chemical, physical, and biological characteristics of NMs and nanoparticles, they have found extensive use in a wide range of fields, including energy, material science, food agriculture, and pharmacy medicine (Parisi et al. 2015; Rai and Ingle 2012). Particularly, nanotechnology has become an advanced tool for improving agricultural products and has a big influence on businesses and the global economy (Haris et al. 2023). Nanotechnology and NMs have been used more often in agriculture during the past 10 years to protect plants from pests and diseases, grow crops more productively, and enhance the way agricultural products are stored (Xu 2022). Modern agricultural systems with their designed smart nanotools have the potential to revolutionize agricultural operations and increase yields in terms of both quality and quantity (Singh Sekhon 2014; Liu and Lal 2015). It is one of the six "Key Enabling Technologies" identified by the European Commission as promoting sustainable competitiveness and growth in several industrial application sectors. It is found in nearly every aspect of our everyday lives (Ali et al. 2014). The aim of this chapter is to deal with use NMs in agriculture and soil revitalization, their applications, and challenges.

2.2 Effect of Nanomaterials on Soil and Plant Systems

2.2.1 Impacts of Nanomaterials on Soil

Nanomaterials may have physical, chemical, or biological changes once they reach soil, depending on their makeup and how they interact with different soil constituents (both organic and inorganic). The primary physical process that happens naturally when NMs are added to the soil environment is aggregation (Usman et al. 2020). Natural colloids and organic and mineral fractions may interact with NMs, causing them to partition in the soil system's solid and aqueous phases (Darlington et al. 2009; Ben-Moshe et al. 2010). Through the adherence of nanoparticles to clay lattice, which prevents the retention of nutrient ions, the use of NMs enhances soil fertility and promotes the balanced use of nutrients, allowing the minerals to enter the solution and become available to plants (Dubey and Mailapalli 2016). Their application in contemporary agriculture could have a big effect on the dynamics of SOM (soil organic matter). However, the type of SOM may vary; it can be hydrophilic like humic substances (HS) or hydrophobic (non-humic), and variations in its biochemical content can also affect how easily they decompose in soil (Grillo et al. 2015). These NMs can help reduce global warming by lowering soil CO_2 emissions, which in turn allows the soil to absorb organic carbon. While, Shi et al. (2018) established that lower SOM levels in paddy soils can also be the consequence of increased CuO nanoparticle concentrations. Moreover, the precipitation of the freely mobile nutritional ions is prevented by nanoparticles. This process lowers nitrogen loss while improving crops' capacity to use fertilizer more efficiently. The pH and soil structure are improved by the use of nanoparticles. It also reduces soil erosion and the cytotoxicity, bioavailability, and dispersion of heavy metals (Gupta et al. 2023) (Figure 2.1).

2.2.2 Effect of Nanomaterials on Soil Microorganisms

Understanding how NMs applied to soil might impact microbial diversity, structure, and activity is crucial. According to reports, the toxicity of newly introduced NMs might directly damage soil microorganisms, or it can indirectly impact them by increasing the

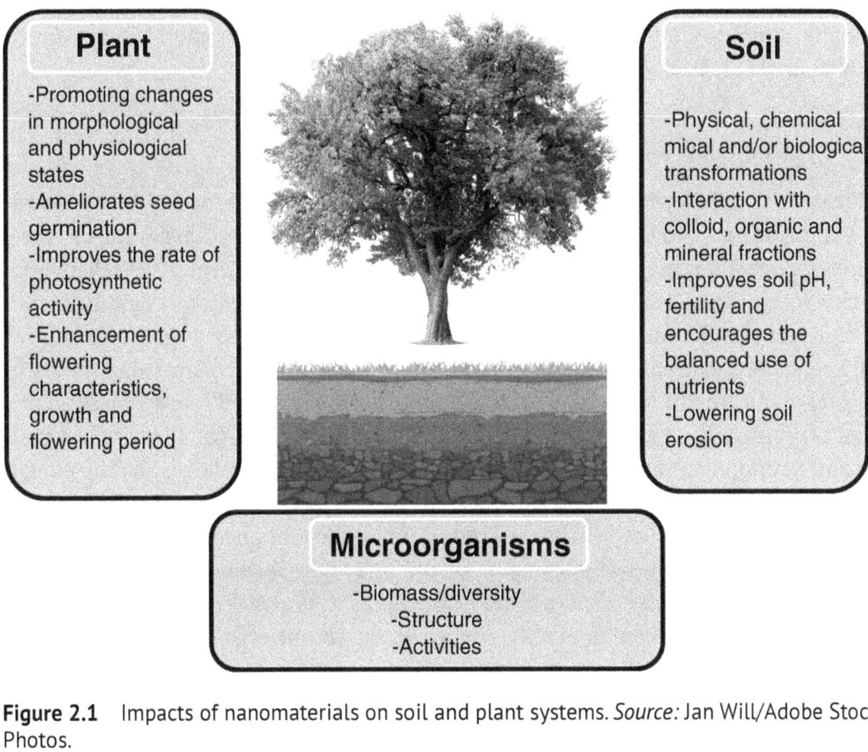

Figure 2.1 Impacts of nanomaterials on soil and plant systems. *Source:* Jan Will/Adobe Stock Photos.

bioavailability of other hazardous chemicals that are already present in the soil (Simonet and Valcárcel 2009). Due to the differences in their cell wall structures, Gram-negative bacteria are often less susceptible to ENMs than Gram-positive bacteria (McKee and Filser 2016). In addition, Ben-Moshe et al. (2013) reported that several microscopic soil characteristics have been shown to be impacted by nanoparticles. One of the main concerns in the field of sustainable soil usage is the preservation of soil microbial biomass and diversity (Torsvik and Øvreås 2002). The concentration, type, and enzymatic activity of the soil all affect how nanoparticles interact with the soil. Dehydrogenase enzyme activity was found to be negatively impacted by high nanoparticle concentrations (Jośko et al. 2014).

Nanomaterials may have an impact on the microbial ecology of the soil or the plant, which may subsequently directly or indirectly affect the development of the plant. For example, it has been noted that nanoscale ZnO and TiO_2 have an effect on soil microbial communities (Moll et al. 2017). After being treated with NMs, the population of bacteria that fix nitrogen and oxidize methane significantly diminished, but the population of bacteria that break down refractory organic pollutants—notable members of the Sphingomonadaceae family—greatly expanded. The influence of TiO_2, one of the prevalent soil NMs, on the wheat microbiome was shown to mostly not alter arbuscular mycorrhizal root colonization, even if the populations of a specific group of microorganisms changed (Grün et al. 2019). Nonetheless, comprehending the interaction between NMs and plant growth-promoting rhizobacteria (PGPR) in the rhizosphere may offer a promising opportunity to investigate low-cost, ecologically safe nanoformulations for uses in agriculture. Over the past 10 years, PGPR has become a viable option for boosting crop productivity and promoting soil health in challenging environmental circumstances (Ahkami et al. 2017; Backer et al. 2018).

2.2.3 Interactions of Nanomaterials and Plants

There are several ways for NPs to enter the plant system, but mostly through the roots and leaves. Following entryway, NPs interact with plants at the cellular and subcellular levels, causing changes to their morphology and physiological states (Khan et al. 2019). Depending on the characteristics of the plant species and the NPs, these interactions might be beneficial or detrimental. The impacts of NPs on plant systems may depend on their chemical make-up, reactivity, size, and, in particular, concentration in or on the plant (Ma et al. 2010).

In several plants, the function of NPs in photosynthesis, shoot and root biomass, germination of plant seeds, and growth has been partially demonstrated. The first signs of plant growth and development are the germination of seeds, which are followed by the elongation of roots and shoots (Nandini et al. 2023). In tomato, Siddiqui and Al-Whaibi (2014) demonstrated that Under NaCl stress conditions, nano-SiO_2 enhances squash seed germination and activates the antioxidant system. According to Bao-Shan et al. (2004), the effects of nano-SiO_2 on *Larix olgensis* seedlings are favorable. In addition to eliciting the synthesis of chlorophyll in the host plant, nano-SiO_2 increases the characteristics of seedling growth quality, including an enhanced mean plant height, increased root collar width, increased main root length, and an increased number of lateral roots. SiO_2NPs have been shown to increase the rate of photosynthetic activity by promoting the synthesis of photosynthetic pigments and carbonic anhydrase activity (Siddiqui et al. 2014). In marigold, *Tagetes erecta* L., simultaneous application of silica nanoparticles (SiNPs) at 250 and 600 mg/L through soil and foliage, respectively, improved flowering features, growth, and flowering duration (Attia and Elhawat 2021). The final effects of nanoparticles on plants and the related ecosystem/biodiversity are determined not only by the precursors utilized in their synthesis and by their size and/or shape, but also by the application route in agricultural land, the chemical makeup of the particles, their surface functionalization, and the payload they carry (Balusamy et al. 2023) (Table 2.1).

Table 2.1 Beneficial effects of nanoparticles on plant.

Plant	Nanoparticles	Effect of nanoparticles	References
Wheat (*Triticum aestivum* L.)	Silica nanoparticles (SiNPs)	Promoted wheat growth by mediating hormones and sugar metabolism Increased the expression levels of key pathway genes related to soluble sugars, indole-3-acetic acid (IAA), and abscisic acid (ABA)	Li et al. (2023)
Rice (*Oryza sativa* L.)	Selenium nanoparticles (SeNPs)	Stimulated antioxidant metabolism and mitigated reactive oxygen species	Freire et al. (2024)
Wheat (*Triticum aestivum* L.)	Cerium oxide nanoparticles (CeO_2 NPs)	Exhibited significant antifungal potential against *U. tritici* under field conditions	Alotaibi et al. (2023)
Sunflower (*Helianthus annuus* L.)	Silicon nanoparticles (SiNPs)	Improved root length, mean daily germination, seedling vigor index and final germination percentage	Janmohammadi and Sabaghnia (2015)
Tomato (*Solanum lycopersicum*)	Silver nanoparticles (AgNPs)	Increased nitrogen, phosphorus, and potassium concentrations in leaves and stimulate root length and number of roots	Guzmán-Báez et al. (2021)
Barley (*Hordeum vulgare* L.)	Magnetic nanoparticles (MNPs)	Enhanced chlorophyll content and growth of barley through the induction of photosystem genes	Tombuloglu et al. (2020)
Maize (*Zea mays* L.)	Biogenic iron oxide nanoparticles (Fe_3O_4NPs)	Enhanced germination and improved yield Increased cob weight, average tassel length, and average plant height	Jayarambabu et al. (2018)
Bean (*Phaseolus vulgaris*)	Silicon dioxide nanoparticles (SiO_2NPs)	Enhanced seed germination and seedlings development	Alsaeedi et al. (2017)

2.3 Applications in Agriculture

2.3.1 Nanofertilizers in Agriculture

The three primary categories of NFs are nanoscale fertilizers, nanoscale additives, and nanoscale coatings (Mikkelsen 2018). Among them, nutrient-containing nanoparticles make up nanoscale fertilizers. Nanoscale coating fertilizers, on the other hand, are conventional fertilizers coated or loaded with nanoparticles, whereas nanoscale additive fertilizers are known as conventional fertilizers incorporating nanoscale additives (Mejias et al. 2021). It's a new approach to maximize the availability of nutrients, either separately or in combination. NFs are a cheap substitute for typical chemical fertilizers that can boost food output worldwide in an environmentally friendly manner. NFs can function as nutrition transporters and are composed of both macro- and micronutrients (Avila-Quezada et al. 2022). According to several researches, NFs are more effective than conventional fertilizers. For example, when NFs of macronutrients are used instead of traditional fertilizers, plant growth is increased by 19%. Compared to standard fertilizers, NFs of micronutrients are superior by 18% while NFs of macronutrient carriers boost growth by 29% (Kah et al. 2018). Through supplying accessible elements with low bioavailability, such as phosphorus and zinc, and minimizing the loss of mobile nutrients to the soil, like nitrate, nanofertilizers are designed to be more effective than traditional fertilizers (Kah et al. 2018; Tarafder et al. 2020) (Figure 2.2).

2.3.2 Nanotechnology in Crop Protection

2.3.2.1 Nanopesticides

Nanoparticles are the active components of nanopesticides, which are used to protect crops from diseases (Javed et al. 2023). According to Athanassiou et al. (2018), nanoformulations are comparable to other usual pesticide formulations in that they aid in improving the apparent solubility of a weakly soluble active element or releasing the active element in a relaxed or guided way, thereby safeguarding the active component against early degradation. "Nanopesticides" refer to insecticides developed by the nanoformulation technique (Nandini et al. 2023). As an alternative to conventional pesticides, next-generation intelligent nanopesticides (such as metal-based NMs and nanocarrier-encapsulated nanopesticides) have a controlled, targeted, and steady discharge of active ingredients in IPM (IPM; pests, weeds, and pathogens control) based on crop demands and other biotic/abiotic stimuli (Beegum and Das 2022). For instance, against four stored product insects—*Rhyzopertha dominica*, *Tribolium castaneum*, *Sitophilus oryzae*, and *Oryzaephilus surinamensis*—SiNPs (SiO_2-NPs, 2 g/kg of stored grain) demonstrated 100% mortality (El-Naggar et al. 2020). Furthermore, imidacloprid-sodium alginate nanoformulations have been used to suppress sucking pests such as leafhoppers and jassids (Kumar et al. 2014) and Si nanoparticles to prevent tomato plant leaf worms (*S. littoralis*) (El-Bendary and El-Helaly 2013).

2.3.2.2 Nanofungicides

Developing NPs as a potent defense against fungal diseases is one of the best strategies. AgNPs are frequently utilized for disinfection because of their antibacterial qualities (Baker et al. 2005). It is recognized that a variety of mental oxide-based nanoparticles,

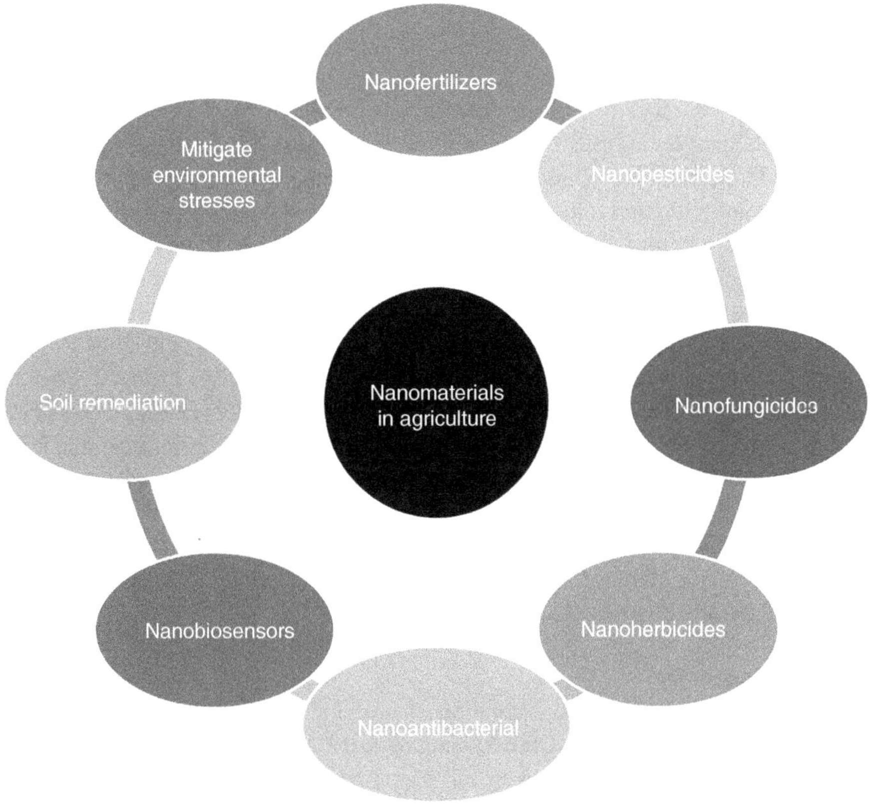

Figure 2.2 Applications of nanomaterials in agriculture.

such as ZnO against Alternaria, Fusarium, Mucor, and Rhizopus species, provide disease resistance against pathogens (Lyons 2010). An investigation of the effects of AgNPs and Ag ions on the plant-pathogenic fungus *Magnaporthe grisea* and *Bipolaris sorokiniana* was carried out. When used for three hours, the ionic and nanoparticulate forms greatly decreased the severity of the condition and were effective against fungi (Jo et al. 2009). Compared to commercial compounds, the development of a nanoformulated commercial fungicide (Trifloxystrobin 25% + Tebuconazole 50%) against the soil-borne fungal pathogen *Macrophomina phaseolina* shown higher effectiveness at different doses (5, 10, 15, and 25 ppm) (Kumar et al. 2016).

2.3.2.3 Nanoherbicides

The primary component of nanoherbicides is biodegradable polymeric material, which may increase the herbicides' efficacy. Atrazine has been encapsulated in poly(epsilon caprolactone) due to its favorable physico-chemical characteristics, improved bioavailability, and biocompatibility (Abigail and Chidambaram 2017). The active component enters the nanomatrix via being adsorbed, attached, encapsulated, or trapped. The slow-release characteristics of NMs, the bonding between the components and the material, and the surrounding circumstances all contribute to the regulated release of the active ingredient

(Mujtaba et al. 2020). A high specific surface area is another feature of nanoherbicides that increases their affinity for the target (Adarsh et al. 2024). Their gradual deterioration and regulated release in natural environments may mitigate the toxicity of herbicides and enhance the efficacy of bioactive compounds (Bhaskar et al. 2023). Nanoherbicides avoid leaching and volatilization of active components and premature degradation by photolysis, hydrolysis, and biodegradation. Studies show that herbicide nanoencapsulation results in more specialized and safer agricultural formulations (Khan et al. 2024). It is advantageous to use nanoherbicides at lower concentrations. There is a decrease in the toxicity of these herbicides to the environment and the long-term effects of herbicide residues in wheat fields. It also helps to get eliminate of weeds without ever coming into contact with the crop plants, which eventually leads to a higher harvest of wheat (Khan et al. 2024).

A research on nanoformulation of atrazine foliar contact with Indian mustard (*Brassica juncea* (L.) Czern.) plants indicated herbicidal activity directly via the vascular tissue of the leaves (Bombo et al. 2019). This demonstrated a significant increase in herbicidal efficacy and a high potential for maintaining herbicidal activity at low doses. Likewise, atrazine-loaded nanocapsules outperformed commercial atrazine products in their ability to suppress hairy beggarticks (*Bidens pilosa* L.) and slender amaranth (*Amaranthus viridis* L.) (Sousa et al. 2018). Thus, nanoformulations have emerged as excellent nanoherbicides for managing weeds as well (Usman et al. 2020).

2.3.2.4 Nanoantibacterial

Several studies have reported the proliferation of antibiotic-resistant microorganisms in agricultural soil (Udikovic-Kolic et al. 2014; Wang et al. 2022). This has clearly emphasized the urgent need to find alternate ways to kill bacteria without needing antibiotics, and nanoparticles may provide a solution (El Hajj et al. 2015). Some metal oxides, notably silver nanoparticles, have demonstrated significant antibacterial and wound healing potential (Kim et al. 2007; Azam et al. 2012). For instance, Xing et al. (2010) assessed the antibacterial activity of poly(3-hydroxybutyrate-co-3-hydroxyvalerate) (PHBV) nanofibrous scaffolds containing silver. It was discovered that silver-containing PHBV nanofibrous scaffolds have good antibacterial activities and great in vitro cell compatibility. Stoimenov et al. (2002) tested the efficiency of nanoparticles on three types of bacteria: *Escherichia coli*, *Bacillus megaterium*, and *Bacillus subtilis*. The nanoparticulate formulations efficiently eliminated *E. coli* and *B. megaterium* bacteria in around 20 minutes. Likewise, ZnO nanoparticles have been shown to have varying degrees of antibacterial activity against a variety of plant species (Daniel et al. 2024). According to El-Sawy et al. (2017), when applied to eggplant in a greenhouse setting, ZnO nanoparticles at varied doses (50, 100, and 200 mg/L) significantly inhibited the growth of the cucumber mosaic virus in comparison to other compounds like 2-nitromethyl phenol and seaweed extract. Also, Cui et al. (2009) reported that TiO_2 nanoparticles induce a 30% increase in photosynthetic activity in the cucumber plant by substantially inhibiting the development of *Psilocybe cubensis* and *Pseudomonas syringae*.

2.3.3 Utilization of Nanobiosensors

Biosensors are hybrid receptor-transducer systems that sense the physical and chemical characteristics of a medium in the presence of an organic or biological recognition element in order to identify the presence of a particular biological analyte (Sun et al. 2006). "Nanobiosensor" is a modified form of the term "biosensor," which is commonly used to describe a tiny analytical device or unit that has a physiologically sensitive component coupled to a physio-chemical transducer. In order to gather, interpret, and analyze data at the microscopic level, nanobiosensors are usually built at the nanoscale and comprise immobilized bioreceptor probes unique to certain analyte molecules (Gupta et al. 2023).

The next generation of biosensors, known as nanobiosensors, is smaller and connected to a sensitive element via a physico-chemical transducer to detect specific analytes at extremely low concentrations. Early detection and quick decision-making to increase agricultural yields through wise water, land, fertilizer, and pesticide management is made possible by nanobiosensor technology (Usman et al. 2020). They are useful for sensing a wide range of substances, including moisture, soil pH, pathogens, herbicides, pesticides, and fertilizers. When used responsibly, they may promote sustainable agriculture and increase crop output (Rai et al. 2012).

For example, organo-phosphates, neonicotinoids, carbamates, and atrazines are among of the most common pesticide groups. Because of their poor homogeneity, their residues, even at low concentrations, last longer in the soil. Nanobiosensors that use piezoelectric transducers with antigen-antibody interaction are used to detect these pesticides (Ivask et al. 2002; Přibyl et al. 2006). On the other hand, with the use of smart sensors, precision farming might boost agricultural output by enabling farmers to reduce inputs, better manage time and the environment, and better regulate fertilization. Precision farming might benefit from the effective use of natural resources in agriculture, such as water, fertilizers, and pesticides, due to nanosensors and nano-based smart delivery systems (Sekhon 2014).

2.3.4 Nanomaterials for Soil Remediation

Degradation and pollution of soil remain major environmental issues, and remediation is becoming a global priority. Degrading soil reservoirs need to be addressed immediately since they have a significant impact on food safety and agricultural productivity worldwide (Bakshi and Abhilash 2020). Major modifications to the primary soil characteristics result from the application of amendments, which also affect the pollutants' availability and movement (Aragón et al. 2019). The literature has examined a number of NM-based amendments to find an appropriate material while taking into account a number of different factors, including cost, efficiency, stability, environmental effects, and material reusability (Usman et al. 2020). Recently, the use of NMs in advanced biotechnology to remediate polluted soils has gained attention as a research area. While some researchers showed that the toxicity of NMs to organisms had a detrimental impact on the ability of polluted soils to heal, others discovered that the addition of NMs increased the bioremediation effectiveness of contaminated soils (Gong et al. 2018). Nevertheless, traditional cleanup methods can cause their own pollution, are very expensive, time-consuming, and ineffective (Mueller and Nowack 2010; Adeleye et al. 2013; Gil-Díaz et al. 2016).

The use of certain NMs, such as nanoparticles, which have the ability to react with pollutants and enable their transformation into less toxic compounds or immobilize them through adsorption, has made nanoremediation a promising alternative to conventional remediation that may be less expensive and more efficient (Adeleye et al. 2013). NMs are often more reactive chemically and biogeologically than other macroscale materials because of their high surface-area-to-volume ratio. They are especially well suited to perform these operations due to their enhanced qualities and effectiveness. Growing interest has been shown in using NMs to remediate polluted soils because of their larger surface area, smaller particle size, and well-established reactivity (Guerra et al. 2018).

2.3.5 Nanomaterials to Mitigate Environmental Stresses in Plants

In addition to promoting plant growth, NPs protect them against abiotic stress. Toxic metal attaches to the surface of the NP because of its huge surface area and tiny size, restricting its availability (Mittal et al. 2020). Abiotic stressors include drought, salinity, alkalinity, temperature changes, and mineral and metal toxicity. Nanoparticles can imitate the function of antioxidant enzymes in the form of nanoenzymes that can scavenge oxidative stress (Sharifi et al. 2020). Through a similar mechanism of oxidative stress, the salt stress likewise poses a threat to plant physiology. It was discovered that *Brassica napus*'s oxidative stress might be reduced by CeO NPs. NaCl (100 mM) altered physiobiochemical characteristics to have the desired impact, including higher plant biomass, more chlorophyll (which raises Mg^{2+} uptake), and an effective photosystem at concentrations of 200 and 1,000 mg/kg (Rossi et al. 2016). Insufficient water is a serious problem for plant cultivation since it makes plants less vigorous and damages crops severely. Through stimulating the nitrate reductase (NR) enzyme, which in turn promotes the accumulation of osmolytes, NanoTiO_2 can improve the hydration status of the plant. Nitric oxide (NO), which is produced as a result of enhanced NR enzyme activity, subsequently triggers the synthesis of proline and glycine betaine (Khan et al. 2020).

Ultimately, by reducing abiotic stress-related plant damage, carbon quantum dots combined with chemical priming agents demonstrated their plant priming effect (Gohari et al. 2021). Once yttrium doping-stabilized γ-Fe_2O_3 NPs were applied to *Brassica napus* plants in nutrient solutions, it was determined that the NPs enhanced plant growth during drought stress, decreased H_2O_2 and lipid peroxidation, and increased chlorophyll content. Therefore, when compared to chelated iron, maghemite NPs could be employed as an effective plant fertilizer and contribute in the control of drought stress (Palmqvist et al. 2017) (Table 2.2).

2.4 Challenges and Opportunities

Farmers in traditional agriculture confront a number of challenges, such as chemical toxicity from overuse of fungicides and pesticides, resistance to these products developing in farmers, and occasionally high costs that are unaffordable for marginal farmers, especially in developing nations (Avila-Quezada et al. 2022). Agricultural strategy also needs to take into account the particular difficulties faced by farming systems, which are frequently marked by inefficient use of inputs, high resource requirements, and intricate relationships between the flora, fauna, and corresponding microenvironment. These interactions could require for the use of nanoparticles, but they nevertheless effectively carry out tasks in farming production systems (Mukhopadhyay 2014). The first approach is to comprehend

Table 2.2 Beneficial impacts of nanoparticles on plant under various abiotic stresses.

Plant	Nanoparticles	Mode of action	Environnemental stress	References
Mungbean (*Vigna radiata* L.)	Zinc oxide nanoparticles (ZnO-NPs)	Inflates the chlorophyll contents and gas exchange attributes, seeds per pod (SPP) and pods per plant (PPP)	Heat	Kareem et al. (2022)
Radish (*Raphanus sativus* L.)	Magnetic nanoparticles (Fe_3O_4)	Increased growth parameters like root and shoot fresh weights	Cadmium	Aslam et al. (2024)
Tomato (*Solanum lycopersicum* L.)	Copper nanoparticles (CuNPs)	Increased performance and improved the Na^+/K^+ ratio Increased the phenols in the leaves and the content of vitamin C, glutathione, and phenols in the fruits The enzyme activity of phenylalanine ammonia lyase (PAL), ascorbate peroxidase (APX), glutathione peroxidase (GPX), superoxide dismutase (SOD), and catalase (CAT) increased in leaf tissues	Salinity	Pérez-Labrada et al. (2019)
Rice (*Oryza sativa* L. cv. EHR1)	Silica oxide nanoparticles (SiO_2NPs)	Improving growth, yield, and enzymatic antioxidant	Dought	Elshayb et al. (2021)
Wheat (*Triticum aestivum* L.) genotype Zlata	Gold nanoparticles (Au-NPs)	Boosted the intensity of growth processes, the quantity of photosynthetic pigments, sucrose in leaves, and the expressions of encoded RuBisCo and Wcor15 genes	Cold	Venzhik et al. (2024)
Lettuce (*Lactuca sativa*)	Water soluble carbon nanoparticles (CNPs)	Improved seed germination and post-germination growth	Salinity	Baz et al. (2020)

the toxicity to human health, which is mostly directed at organisms, tissues, and cells (such as immune cells and mice skin). The second is the harm to environmental health; societies are more worried about other parts of the life tree (fish, invertebrates, plant tissues, single-cell organisms, etc.). Plant-based nanosafety research has therefore concentrated on the intentional or direct exposure of food crops to certain NMs (e.g., nanopesticides, nanofertilizers, and nanoherbicides), as well as the indirect or inadvertent exposure (e.g., spray drifting, leaching, and runoff of NPs) (Mittal et al. 2020).

The production of NMs in large enough quantities, with standard nanoscale sizes and accepted standards, is highly technical and expensive, requiring a sophisticated infrastructure manned by technical personnel. The rapid advancement of nanotechnology raises concerns about the buildup of artificial nanoparticles and their possible penetration into the food supply chain (Pandey et al. 2023). Additionally, the destiny of NMs in the environment is unknown, but it is the same end as the thousands of tons of conventional fertilizers that are utilized today. In fact, industrialized nations (such as the United States, Europe, and others) formulate the laws pertaining to large natural forests (NFs); emerging nations—which supply a greater amount of food globally—develop regulations and execute them much later (Avila-Quezada et al. 2022). To date, many researchers in this field have not considered the ethical and social implications of nanobiotechnology to be a major concern. Consequently, to address potential public, media, and political issues in a proactive and responsible manner, it is suggested that efforts be developed to foresee the ethical and social issues that might arise as nanobiotechnology advances (Bennett and Schuurbiers 2005).

2.5 Recommendations and Expected Future Aspects

It is necessary to demonstrate the advancement of nanosafety research as well as risk assessment, risk management, policymaking, and risk communication support among the various stakeholders (such as society, industries, researchers, policymakers, insurance companies, and the general public) in order to exchange information amongst various nanotechnology fields. Mostly, researchers studying nanosafety employ a two-pronged strategy (Bos et al. 2015). For agricultural purposes, soil quality may be monitored using nanosensors. This technique might serve as a milestone in addressing the current challenges encountered in the agriculture sector. Nevertheless, the safety and health risks that the technology presents require more investigation (Gupta et al. 2023). For example, CuO and ZnO nanoparticles have the ability to pass through a variety of chemical and biological processes, which may harm plant cells, have an impact on soil biota and nitrogen fixation, or even cause serious health issues (Rajput et al. 2020). Nanofertilizers have the potential to transform agriculture, but before they are commercialized, a thorough investigation of the toxicity of nanoparticles at high concentrations, their cumulative effect, and biosafety issues must be completed (Tripathi et al. 2017; Khalifa and Hasaneen 2018; He et al. 2019).

Consequently, to increase their effectiveness, more work has to be done on inventing novel and enhanced synthesis techniques that provide exact control over the composition of the final product. Besides, there is little information known about the field use of nano-assisted agriculture, and the majority of studies on the subject rely on controlled experimentation (Usman et al. 2020). Through the integration of newly developed technologies and fields like chemical biology with nanotechnology to address present biological bottlenecks that impede further advancements, multidisciplinary methods have the potential to enhance food production (Sekhon 2014).

2.6 Conclusions

Nanomaterials are an important resource for agricultural products. They are a promising means of solving the various problems facing modern agriculture. However, major challenges remain to be solved, and a number of investigations must continue in order to resolve the problems linked to the use of these NMs in modern agriculture, including the development of innovative and improved synthesis methods with precise control of the composition of these products and their toxicity for humans, animals, and the environment.

References

Abigail, E.A. and Chidambaram, R. (2017). Nanotechnology in herbicide resistance. *Nanostructured Materials* 207: https://doi.org/10.5772/intechopen.68355.

Adam, V. and Nowack, B. (2017). European country-specific probabilistic assessment of nanomaterial flows towards landfilling, incineration, and recycling. *Environmental Science: Nano* 4: 1961–1973. https://doi.org/10.1039/c7en00487g.

Adarsh, S., Shilpa, S., Ameena, M. et al. (2024). A review on nano herbicides: the future of weed management. *Journal of Advances in Biology & Biotechnology* 27 (7): 1244–1253. https://doi.org/10.9734/jabb/2024/v27i71085.

Adeleye, A.S., Keller, A.A., Miller, R.J., and Lenihan, H.S. (2013). Persistence of commercial nanoscaled zero-valent iron (nZVI) and by-products. *Journal of Nanoparticle Research* 15: 1418. https://doi.org/10.1007/s110.1-013-1418-7.

Ahkami, A.H., White, R.A., Handakumbura, P.P., and Jansson, C. (2017). Rhizosphere engineering: Enhancing sustainable plant ecosystem productivity. *Rhizosphere* 3: 233–243. https://doi.org/10.1016/j.rhisph.2017.04.012.

Ali, M.A., Rehman, I., Iqbal, A. et al. (2014). Nanotechnology, a new frontier in agriculture. *Advancements in Life Sciences* 1: 129–138. http://dx.doi.org/10.62940/als.v1i3.58.

Alotaibi, M.O., Alotaibi, N.M., Ghoneim, A.M., and ul Ain, N., Irshad, M.A., Nawaz, R., and Ali, S. (2023). Effect of green synthesized cerium oxide nanoparticles on fungal disease of wheat plants: a field study. *Chemosphere* 339: 139731. https://doi.org/10.1016/j.chemosphere.2023.139731.

Alsaeedi, A., El-Ramady, H., Alshaal, T.A., and Almohsen, M. (2017). Enhancing seed germination and seedlings development of common bean (*Phaseolus vulgaris*) by SiO_2 nanoparticles. *Egyptian Journal of Soil Science* 57 (4): 407–415. https://doi.org/10.21608/ejss.2017.891.1098.

Aragón, M.S., Nakamaru, Y.M., García-Carmona, M. et al. (2019). The role of organic amendment in soils affected by residual pollution of potentially harmful elements. *Chemosphere* 237: 124549. https://doi.org/10.1016/j.chemosphere.2019.124549.

Aslam, A., Noreen, Z., Rashid, M. et al. (2024). Understanding the role of magnetic (Fe_3O_4) nanoparticle to mitigate cadmium stress in radish (*Raphanussativus* L.). *Botanical Studies* 65 (1): 20. https://doi.org/10.1186/s40529-024-00420-4.

Athanassiou, C.G., Kavallieratos, N.G., Benelli, G. et al. (2018). Nanoparticles for pest control: current status and future perspectives. *Journal of Pest Science* 91: 1–15. https://doi.org/10.1007/s10340-017-0898-0.

Attia, E.A. and Elhawat, N. (2021). Combined foliar and soil application of silica nanoparticles enhances the growth, flowering period and flower characteristics of marigold (*Tageteserecta* L). *Scientia Horticulturae* 282: 110015. https://doi.org/10.1016/j.scienta.2021.110.15.

Avila-Quezada, G.D., Ingle, A.P., Golińska, P., and Rai, M. (2022). Strategic applications of nano-fertilizers for sustainable agriculture: benefits and bottlenecks. *Nanotechnology Reviews* 11 (1): 2123–2140. https://doi.org/10.1515/ntrev-2022-0126.

Azam, A., Ahmed, A.S., Oves, M. et al. (2012). Antimicrobial activity of metal oxide nanoparticles against Gram-positive and Gram-negative bacteria: a comparative study. *International Journal of Nanomedicine* 7: 6003. https://doi.org/10.2147/IJN.S35347.

Backer, R., Rokem, J.S., Ilangumaran, G. et al. (2018). Plant growth-promoting Rhizobacteria: context, mechanisms of action, and roadmap to commercialization of biostimulants for sustainable agriculture. *Frontiers in Plant Science* 9: 1473. https://doi.org/10.3389/fpls.2018.01473.

Baker, C., Pradhan, A., Pakstis, L. et al. (2005). Synthesis and antibacterial properties of silver nanoparticles. *Journal of Nanoscience and Nanotechnology* 5: 244–249. https://doi.org/10.1166/jnn.2005.034.

Bakshi, M. and Abhilash, P.C. (2020). Nanotechnology for soil remediation: revitalizing the tarnished resource. In: *Nano-Materials as Photocatalysts for Degradation of Environmental Pollutants* (ed. P. Singh, A. Borthakur, P.K. Mishra, and D. Tiwary), 345–370. Amsterdam, The Netherlands: Elsevier https://doi.org/10.1016/B978.0-12-818598-8.00017-1.

Balusamy, S.R., Joshi, A.S., Perumalsamy, H. et al. (2023). Advancing sustainable agriculture: a critical review of smart and eco-friendly nanomaterial applications. *Journal of Nanobiotechnology* 21 (1): 372. https://doi.org/10.1186/s12951-023-02135-3.

Bao-shan, L., Chun-hui, L., Li-jun, F. et al. (2004). Effect of TMS (nanostructured silicon dioxide) on growth of Changbai larch seedlings. *Journal of Forest Research* 15: 138–140. https://doi.org/10.1007/bf02856749.

Baz, H., Creech, M., Chen, J. et al. (2020). Water-soluble carbon nanoparticles improve seed germination and post-germination growth of lettuce under salinity stress. *Agronomy* 10 (8): 1192. https://doi.org/10.3390/agronomy10081192.

Beegum, S. and Das, S. (2022). Nanosensors in agriculture. In: *Agricultural Nanobiotechnology* (ed. S. Ghosh, S. Thongmee, and A. Kumar), 465–478. Woodhead Publishing https://doi.org/10.1016/B978-0-323-91.08-1.00012-2.

Ben-Moshe, T., Dror, I., and Berkowitz, B. (2010). Transport of metal oxide nanoparticles in saturated porous media. *Chemosphere* 81: 387–393. https://doi.org/10.1016/j.chemosphere.2010.07.007.

Ben-Moshe, T., Frenk, S., Dror, I. et al. (2013). Effects of metal oxide nanoparticles on soil properties. *Chemosphere* 90: 640–646. https://doi.org/10.1016/j.chemosphere.2012.09.018.

Bennett, D.J. and Schuurbiers, D. (2005). Nanobiotechnology: responsible actions on issues in society and ethics. *NSTI Nanotechnology* 2: 765–768.

Bhaskar, M., Kumar, A., and Rani, R. (2023). Application of nano formulations in agriculture. *Biocatalysis and Agricultural Biotechnology* 54: 102934. https://doi.org/10.1016/j.bcab.2023.102934.

Bombo, A.B., Pereira, A.E.S., Lusa, M.G. et al. (2019). A mechanistic view of interactions of a nanoherbicide with target organisms. *Journal of Agricultural and Food Chemistry* 67: 4453–4462. https://doi.org/10.1021/acs.jafc.9b00806.

Bos, P.M., Gottardo, S., Scott-Fordsmand, J.J. et al. (2015). The MARINA risk assessment strategy: a flexible strategy for efficient information collection and risk assessment of nanomaterials. *International Journal of Environmental Research and Public Health* 12: 15007–15021. https://doi.org/10.3390/ijerph121214961.

Bulovic, V., Mandell, A., and Perlman, A. (2004). Molecular memory device. US 20050116256 A, 1, 388.

Cui, X., Kim, H.M., Kawashita, M. et al. (2009). Preparation of bioactive titania films on titanium metal via anodic oxidation. *Dental Materials* 25 (1): 80–86. https://doi.org/10.1016/j.dental.2008.04.012.

Daniel, A.I., Hüsselmann, L., Shittu, O.K. et al. (2024). Application of nanotechnology and proteomic tools in crop development towards sustainable agriculture. *Journal of Crop Science and Biotechnology* 1-21: https://doi.org/10.1007/s12892-024-00235-6.

Darlington, T.K., Neigh, A.M., Spencer, M.T. et al. (2009). Nanoparticle characteristics affecting environmental fate and transport through soil. *Environmental Toxicology and Chemistry* 28: 1191–1199. https://doi.org/10.1897/08-341.1.

Ding, Y., Kuhlbusch, T.A., Van Tongeren, M. et al. (2017). Airborne engineered nanomaterials in the workplace—a review of release and worker exposure during nanomaterial production and handling processes. *Journal of Hazardous Materials* 322: 17–28. https://doi.org/10.1016/j.jhazmat.2016.04.075.

Dubey, A. and Mailapalli, D.R. (2016). Nanofertilisers, nanopesticides, nanosensors of pest and nanotoxicity in agriculture. *Sustainable Agriculture Reviews* 19: 307–330. https://doi.org/10.1007/978-3-319-26777-7_7.

El Hajj, F., Hasan, A., Nakhleh, J. et al. (2015). Nanosilver loaded Gel MA hydrogel for antimicrobial coating of biomedical implants. *2015 International Conference on Advances in Biomedical Engineering (ICABME)*, Beirut, Lebanon (16–18 September 2015). IEEE. https://doi.org/10.1109/ICABME.2015.7323284.

El-Bendary, H. and El-Helaly, A. (2013). First record nanotechnology in agricultural: silica nanoparticles a potential new insecticide for pest control. *Applied Science Reports* 4 (3): 241–246.

El-Naggar, M.E., Abdelsalam, N.R., Fouda, M.M.G. et al. (2020). Soil application of nano silica on maize yield and its insecticidal activity against some stored insects after the post-harvest. *Nanomaterials* 10: 739. https://doi.org/10.3390/nano10040739.

El-Sawy, M.M., Elsharkawy, M.M., Abass, J.M., and Kasem, M.H. (2017). Antiviral activity of 2–nitromethyl phenol, zinc nanoparticles and seaweed extract against cucumber mosaic virus (CMV) in Eggplant. *Journal of Virology & Antiviral Research* 6: 2. https://doi.org/10.4172/2324-8955.1000173.

Elshayb, O.M., Nada, A.M., Ibrahim, H.M. et al. (2021). Application of silica nanoparticles for improving growth, yield, and enzymatic antioxidant for the hybrid rice ehr1 growing under water regime conditions. *Materials* 14 (5): 1150. https://doi.org/10.3390/ma14051150.

Freire, B.M., Lange, C.N., Cavalcanti, Y.T. et al. (2024). The dual effect of selenium nanoparticles in rice seedlings: from increasing antioxidant activity to inducing oxidative stress. *Plant Stress* 11: 100372. https://doi.org/10.1016/j.stress.2024.100372.

Gil-Díaz, M., Diez-Pascual, S., González, A. et al. (2016). A nanoremediation strategy for the recovery of an As-polluted soil. *Chemosphere* 149: 137–145. https://doi.org/10.1016/j.chemosphere.2016.02.048.

Gohari, G., Panahirad, S., Sepehri, N. et al. (2021). Enhanced tolerance to salinity stress in grapevine plants through application of carbon quantum dots functionalized by proline. *Environmental Science and Pollution Research* 28: 42877–42890. https://doi.org/10.1007/s11356-021-13794-w.

Gong, X., Huang, D., Liu, Y. et al. (2018). Remediation of contaminated soils by biotechnology with nanomaterials: bio-behavior, applications, and perspectives. *Critical Reviews in Biotechnology* 38 (3): 455–468. https://doi.org/10.1080/07388551.2017.1368446.

Grillo, R., Rosa, A.H., and Fraceto, L.F. (2015). Engineered nanoparticles and organic matter: a review of the state-of-the-art. *Chemosphere* 119: 608–619. https://doi.org/10.1016/j.chemosphere.2014.07.049.

Grün, A.-L., Manz, W., Kohl, Y.L. et al. (2019). Impact of silver nanoparticles (AgNP) on soil microbial community depending on functionalization, concentration, exposure time, and soil texture. *Environmental Sciences Europe* 31: 15. https://doi.org/10.1186/s12302-019-0196-y.

Guerra, F.D., Attia, M.F., Whitehead, D.C., and Alexis, F. (2018). Nanotechnology for environmental remediation: materials and applications. *Molecules* 23: 1760. https://doi.org/10.3390/molecules23071760.

Gupta, P., Sharma, M., and Sharma, S. (2023). Nanotechnology in agriculture: prospect, opportunities and challenges. *Emergent Life Science Research* 9 (1): 164–176. https://doi.org/10.31783/elsr.2023.91164176.

Guzmán-Báez, G.A., Trejo-Téllez, L.I., Ramírez-Olvera, S.M. et al. (2021). Silver nanoparticles increase nitrogen, phosphorus, and potassium concentrations in leaves and stimulate root length and number of roots in tomato seedlings in a hormetic manner. *Dose-Response* 19 (4): 15593258211044576. https://doi.org/10.1177/15593258211044576.

Haris, M., Hussain, T., Mohamed, H.I. et al. (2023). Nanotechnology—a new frontier of nano-farming in agricultural and food production and its development. *Science of the Total Environment* 857: 159639. https://doi.org/10.1016/j.scitotenv.2022.159639.

He, X., Deng, H., and Hwang, H.-M. (2019). The current application of nanotechnology in food and agriculture. *Journal of Food and Drug Analysis* 27: 1–21. https://doi.org/10.1016/j.jfda.2018.12.002.

Helal, N.A.S. (2013). Nanotechnology in agriculture: a review. *Poljoprivredai Sumarstvo* 59: 117–142.

Ivask, A., Virta, M., and Kahru, A. (2002). Construction and use of specific luminescent recombinant bacterial sensors for the assessment of bioavailable fraction of cadmium, zinc, mercury, and chromium in the soil. *Soil Biology and Biochemistry* 34: 1439–1447. https://doi.org/10.1016/S0038-0717(02)00088-3.

Janmohammadi, M. and Sabaghnia, N. (2015). Effect of pre-sowing seed treatments with silicon nanoparticles on germinability of sunflower (*Helianthus annuus*). *BotLith* 21 (1): 13–21.

Javed, R., Bilal, M., Ali, J.S. et al. (2023). Nanotechnology: a tool for the development of sustainable agroindustry. In: *Agricultural and Environmental Nanotechnology: Novel Technologies and Their Ecological Impact* (ed. F. Fernandez-Luqueno and J.K. Patra), 317–339. Springer https://doi.org/10.1007/978-981-19-5454-2_11.

Jayarambabu, N., Rao, K., Park, S.H., and Rajendar, V. (2018). Biogenic synthesized Fe_3O_4 nanoparticles affect the growth parameters of maize (*Zea mays* L.). *Digest Journal of Nanomaterials and Biostructures* 13: 903–913.

Jo, Y.K., Kim, B.H., and Jung, G. (2009). Antifungal activity of silver ions and nanoparticles on phytopathogenic fungi. *Plant Disease* 93: 1037–1043. https://doi.org/10.1094/PDIS-93-10.1037.

Jośko, I., Oleszczuk, P., and Futa, B. (2014). The effect of inorganic nanoparticles (ZnO, Cr_2O_3, CuO, and Ni) and their bulk counterparts on enzyme activities in different soils. *Geoderma* 232: 528–537. https://doi.org/10.1016/j.geoderma.2014.06.012.

Kah, M., Kookana, R.S., Gogos, A., and Bucheli, T.D. (2018). A critical evaluation of nanopesticides and nanofertilizers against their conventional analogues. *Nature Nanotechnology* 13: 677–684. https://doi.org/10.1038/s41565-018-0131-1.

Kareem, H.A., Saleem, M.F., Saleem, S. et al. (2022). Zinc oxide nanoparticles interplay with physiological and biochemical attributes in terminal heat stress alleviation in mungbean (*Vigna radiata* L.). *Frontiers in Plant Science* 13: 842349. https://doi.org/10.3389/fpls.2022.842349.

Khalifa, N.S. and Hasaneen, M.N. (2018). The effect of chitosan–PMAA–NPK nanofertilizer on *Pisumsativum* plants. *3 Biotech* 8: 193. https://doi.org/10.1007/s13205-018-1221-3.

Khan, M.R., Adam, V., Rizvi, T.F. et al. (2019). Nanoparticle–plant interactions: two-way traffic. *Small* 15 (37): 1901794. https://doi.org/10.1002/smll.201901794.

Khan, M.N., AlSolami, M.A., Basahi, R.A. et al. (2020). Nitric oxide is involved in nano-titanium dioxide-induced activation of antioxidant defense system and accumulation of osmolytes under water-deficit stress in *Viciafaba* L. *Ecotoxicology and Environmental Safety* 190: 110152. https://doi.org/10.1016/j.ecoenv.2019.110152.

Khan, B.A., Nadeem, M.A., NajeebAlawadi, H.F. et al. (2024). An overview of the role of nanoherbicides in tackling challenges of weed management in wheat: a novel approach. *Green Processing and Synthesis* 13 (1): 20240021. https://doi.org/10.1515/gps-2024-0021.

Kim, J. S., Kuk, E., Yu, K. N., Kim J-H, Park S.J, Lee H.J., Kim S.H., Park Y.A., Park, Y.H., Hwang, C.Y., Kim, Y.K., Lee, Y-S., Jeong, D.H., and Cho, M.H. (2007). Antimicrobial effects of silver nanoparticles. *Nanomedicine*, 3(1), 95–101. https://doi.org/10.1016/j.nano.2006.12.001

Kumar, S., Bhanjana, G., Sharma, A. et al. (2014). Synthesis, characterization, and on-field evaluation of pesticide-loaded sodium alginate nanoparticles. *Carbohydrate Polymers* 101: 1061–1067. https://doi.org/10.1016/j.carbpol.2013.10.025.

Kumar, G.D., Natarajan, N., and Nakkeeran, S. (2016). Antifungal activity of nanofungicide trifloxystrobin 25% + tebuconazole 50% against *Macrophominaphaseolina*. *African Journal of Microbiology Research* 10: 100–105. https://doi.org/10.5897/AJMR2015.7692.

Li, Y., Xi, K., Liu, X. et al. (2023). Silica nanoparticles promote wheat growth by mediating hormones and sugar metabolism. *Journal of Nanobiotechnology* 21 (1): 2. https://doi.org/10.1186/s12951-022-01753-7.

Liu, R. and Lal, R. (2015). Potentials of engineered nanoparticles as fertilizers for increasing agronomic productions. *Science of the Total Environment* 514: 131–139. https://doi.org/10.1016/j.scitotenv.2015.01.104.

Lyons, K. (2010). Nanotechnology: transforming food and the environment. *Food First Backgrounder* 16: 1–4.

Ma, Y., Kuang, L., He, X. et al. (2010). Effects of rare earth oxide nanoparticles on root elongation of plants. *Chemosphere* 78 (3): 273–279. https://doi.org/10.1016/j.chemosphere.2009.10.050.

McKee, M.S. and Filser, J. (2016). Impacts of metal-based engineered nanomaterials on soil communities. *Environmental Science. Nano* 3: 506–533. https://doi.org/10.1039/C6EN00007J.

Mejias, J.H., Salazar, F., Pérez Amaro, L. et al. (2021). Nanofertilizers: a cutting-edge approach to increase nitrogen use efficiency in grasslands. *Frontiers in Environmental Science* 9: 635114. https://doi.org/10.3389/fenvs.2021.635114.

Mikkelsen, R. (2018). Nanofertilizer and nanotechnology: a quick look. *Better Crops* 102: 18–19. https://doi.org/10.24047/BC102318.

Mittal, D., Kaur, G., Singh, P. et al. (2020). Nanoparticle-based sustainable agriculture and food science: recent advances and future outlook. *Frontiers in Nanotechnology* 2: 579954. https://doi.org/10.3389/fnano.2020.579954.

Moll, J., Klingenfuss, F., Widmer, F. et al. (2017). Effects of titanium dioxide nanoparticles on soil microbial communities and wheat biomass. *Soil Biology and Biochemistry* 111: 85–93. https://doi.org/10.1016/j.soilbio.2017.03.019.

Mueller, N.C. and Nowack, B. (2010). Nanoparticles for remediation: solving big problems with little particles. *Elements* 6: 395–400. https://doi.org/10.2113/gselements.6.6.395.

Mujtaba, M., Khawar, K.M., Camara, M.C. et al. (2020). Chitosan-based delivery systems for plants: a brief overview of recent advances and future directions. *International Journal of Biological Macromolecules* 154: 683–697. https://doi.org/10.1016/j.ij.iomac.2020.03.128.

Mukhopadhyay, S.S. (2014). Nanotechnology in agriculture: prospects and constraints. *Nanotechnology, Science and Applications* 7: 63–71. https://doi.org/10.2147/NSA.S39409.

Nandini, B., Mawale, K.S., and Giridhar, P. (2023). Nanomaterials in agriculture for plant health and food safety: a comprehensive review on the current state of agro-nanoscience. *3 Biotech* 13 (3): 73. https://doi.org/10.1007/s13205-023-03470-w.

Neme, K., Nafady, A., Uddin, S., and Tola, Y.B. (2021). Application of nanotechnology in agriculture, postharvest loss reduction and food processing: food security implication and challenges. *Heliyon* 7 (12): e08539. https://doi.org/10.1016/j.heliyon.2021.e08539.

Palmqvist, N.M., Seisenbaeva, G.A., Svedlindh, P., and Kessler, V.G. (2017). Maghemite nanoparticles act as nanozymes, improving growth and abiotic stress tolerance in *Brassica napus*. *Nanoscale Research Letters* 12: 631. https://doi.org/10.1186/s11671-017-2404-2.

Pandey, G., Tripathi, S., Bajpai, S., and Kamboj, M. (2023). Approaches, challenges, and prospects of nanotechnology for sustainable agriculture. In: *Agricultural and Environmental Nanotechnology: Interdisciplinary Biotechnological Advances* (ed. F. Fernandez-Luqueno and J.K. Patra), 83–103. Singapore: Springer https://doi.org/10.1007/978-981-19-5454-2_3.

Parisi, C., Vigani, M., and Rodríguez-Cerezo, E. (2015). Agricultural nanotechnologies: what are the current possibilities? *Nano Today* 10: 124–127. https://doi.org/10.1016/j.nantod.2014.09.009.

Park, H.J., Kim, S.H., Kim, H.J., and Choi, S.H. (2006). A new composition of nanosized silica-silver for control of various plant diseases. *Plant Pathology Journal* 22: 295–302. https://doi.org/10.5423/PPJ.2006.22.3.295.

Pérez-Labrada, F., López-Vargas, E.R., Ortega-Ortiz, H. et al. (2019). Responses of tomato plants under saline stress to foliar application of copper nanoparticles. *Plants* 8 (6): 151. https://doi.org/10.3390/plants8060151.

Pokropivny, V., Lohmus, R., Hussainova, I. et al. (2007). *Introduction to Nanomaterials and Nanotechnology*, 45–100. Tartu: Tartu University Press.

Přibyl, J., Hepel, M., and Skládal, P. (2006). Piezoelectric immunosensors for polychlorinated biphenyls operating in aqueous and organic phases. *Sensors and Actuators B: Chemical* 113: 900–910. https://doi.org/10.1016/j.snb.2005.03.077.

Rai, M. and Ingle, A. (2012). Role of nanotechnology in agriculture with special reference to management of insect pests. *Applied Microbiology and Biotechnology* 94: 287–293. https://doi.org/10.1007/s00253-012-3969-4.

Rai, V., Acharya, S., and Dey, N. (2012). Implications of nanobiosensors in agriculture. *Journal of Biomaterials and Nanobiotechnology* 3: 315–324. https://doi.org/10.4236/jbnb.2012.322039.

Rajput, V., Minkina, T., Sushkova, S. et al. (2020). ZnO and CuO nanoparticles: a threat to soil organisms, plants, and human health. *Environmental Geochemistry and Health* 42 (1): 147–158. https://doi.org/10.1007/s10653-019-00317-3.

Rossi, L., Zhang, W., Lombardini, L., and Ma, X. (2016). The impact of cerium oxide nanoparticles on the salt stress responses of *Brassica napus* L. *Environmental Pollution* 219: 28–36. https://doi.org/10.1016/j.envpol.2016.09.060.

Saleem, H. and Zaidi, S.J. (2020). Recent developments in the application of nanomaterials in agroecosystems. *Nanomaterials* 10 (12): 2411. https://doi.org/10.3390/nano10122411.

Saleh, T.A. (2016). Nanomaterials for pharmaceuticals determination. *Bioenergetics* 5: 2.

Saleh, T.A. and Gupta, V.K. (2016). *Nanomaterial and Polymer Membranes: Synthesis, Characterization, and Applications*. Amsterdam, The Netherlands: Elsevier.

Scaria, J., Nidheesh, P.V., and Kumar, M.S. (2020). Synthesis and applications of various bimetallic nanomaterials in water and wastewater treatment. *Journal of Environmental Management* 259: 110011. https://doi.org/10.1016/j.jenvman.2019.110.11.

Sekhon, B.S. (2014). Nanotechnology in agri-food production: an overview. *Nanotechnology, Science and Applications* 31-53: https://doi.org/10.2147/NSA.S39406.

Sharifi, M., Faryabi, K., Talaei, A.J. et al. (2020). Antioxidant properties of gold nanozyme: a review. *Journal of Molecular Liquids* 297: 112004. https://doi.org/10.1016/j.molliq.2019.112004.

Shi, J., Ye, J., Fang, H. et al. (2018). Effects of copper oxide nanoparticles on paddy soil properties and components. *Nanomaterials* 8: 839. https://doi.org/10.3390/nano8100839.

Siddiqui, M.H. and Al-Whaibi, M.H. (2014). Role of nano-SiO_2 in germination of tomato (*Lycopersicumesculentum* seeds Mill.). *Saudi Journal of Biological Sciences* 21: 13–17. https://doi.org/10.1016/j.sj.s.2013.04.005.

Siddiqui, M.H., Al-Whaibi, M.H., Faisal, M., and Al Sahli, A.A. (2014). Nanosilicon dioxide mitigates the adverse effects of salt stress on *Cucurbita pepo* L. *Environmental Toxicology and Chemistry* 33: 2429–2437. https://doi.org/10.1002/etc.2697.

Simonet, B.M. and Valcárcel, M. (2009). Monitoring nanoparticles in the environment. *Analytical and Bioanalytical Chemistry* 393: 17. https://doi.org/10.1007/s00216-008-2484-z.

Singh Sekhon, B. (2014). Nanotechnology in agri-food production: an overview. *Nanotechnology, Science and Applications* 7: 31–53. https://doi.org/10.2147/NSA.S39406.

Sousa, G.F., Gomes, D.G., Campos, E.V. et al. (2018). Post-emergence herbicidal activity of nanoatrazine against susceptible weeds. *Frontiers in Environmental Science* 6: 12. https://doi.org/10.3389/fenvs.2018.00012.

Stoimenov, P.K., Klinger, R.L., and Marchin, G.L. (2002). Metal oxide nanoparticles as bactericidal agents. *Langmuir* 18 (17): 6679–6686. https://doi.org/10.1021/la0202374.

Sun, Y.-P., Li, X.-Q., Cao, J. et al. (2006). Characterization of zero-valent iron nanoparticles. *Advances in Colloid and Interface Science* 120: 47–56. https://doi.org/10.1016/j.cis.2006.03.001.

Tarafder, C., Daizy, M., Alam, M.M. et al. (2020). Formulation of a hybrid nanofertilizer for slow and sustainable release of micronutrients. *ACS Omega* 5: 23960–23966. https://doi.org/10.1021/acsomega.0c03233.

Tombuloglu, H., Slimani, Y., Tombuloglu, G. et al. (2020). Engineered magnetic nanoparticles enhance chlorophyll content and growth of barley through the induction of photosystem genes. *Environmental Science and Pollution Research* 27: 34311–34321. https://doi.org/10.1007/s11356-020-09693-1.

Torsvik, V. and Øvreås, L. (2002). Microbial diversity and function in soil: from genes to ecosystems. *Current Opinion in Microbiology* 5: 240–245. https://doi.org/10.1016/S1369-5274(02)00324-7.

Tripathi, K.M., Bhati, A., Singh, A. et al. (2017). Sustainable changes in the contents of metallic micronutrients in first-generation gram seeds imposed by carbon nano-onions: life cycle seed to seed study. *ACS Sustainable Chemistry & Engineering* 5: 2906–2916. https://doi.org/10.1021/acssuschemeng.6b01937.

Udikovic-Kolic, N., Wichmann, F., Broderick, N.A., and Handelsman, J. (2014). Bloom of resident antibiotic-resistant bacteria in soil following manure fertilization. *Proceedings of the National Academy of Sciences of the United States of America* 111 (42): 15202–15207. https://doi.org/10.1073/pnas.1409836111.

Usman, M., Farooq, M., Wakeel, A. et al. (2020). Nanotechnology in agriculture: current status, challenges and future opportunities. *Science of the Total Environment* 721: 137778. https://doi.org/10.1016/j.scitotenv.2020.137778.

Venzhik, Y., Deryabin, A., and Zhukova, K. (2024). Au-based nanoparticles enhance low temperature tolerance in wheat by regulating some physiological parameters and gene expression. *Plants* 13 (9): 1261. https://doi.org/10.3390/plants13091261.

Wang, J., Wang, L., Zhu, L. et al. (2022). Antibiotic resistance in agricultural soils: source, fate, mechanism and attenuation strategy. *Critical Reviews in Environmental Science and Technology* 52 (6): 847–889. https://doi.org/10.1080/10643389.2020.1835438.

Xing, Z.C., Chae, W.P., Baek, J.Y. et al. (2010). In vitro assessment of antibacterial activity and cytocompatibility of silver-containing PHBV nanofibrous scaffolds for tissue engineering. *Biomacromolecules* 11 (5): 1248–1253. https://doi.org/10.1021/bm1000372.

Xu, Z.P. (2022). Material nanotechnology is sustaining modern agriculture. *ACS Agricultural Science & Technology* 2 (2): 232–239. https://doi.org/10.1021/acsagscitech.1c00204.

3

Phosphorus-Solubilizing Microbes and Nanotechnologies for Plant Growth Promotion and Soil Health

Divjot Kour[1], Babita Sharma[2], Damini Maithani[3], Ali Khalaf Ahmed Albaggar[4], Tanvir Kaur[5], Abdulaziz M.S. Alqahtani[6], Sofia Sharief Khan[7], Hemant Dasila[2], Neelam Yadav[8,9], and Ajar Nath Yadav[10]

[1] *University Centre for Research and Development, Chandigarh University, Mohali, Punjab, India*
[2] *Department of Microbiology, Akal College of Basic Sciences, Eternal University, Sirmour, Himachal Pradesh, India*
[3] *Natural Resource Management, Plant Microbiology, College of Horticulture, Sardar Vallabhbhai Patel University of Agriculture and Technology, Meerut, India*
[4] *Department of Biology, Faculty of Science, Al-Baha University, Al Baha City, Al Baha Province, Saudi Arabia*
[5] *Department of Biosciences, Graphic Era University (Deemed to be), Dehradun, Uttarakhand, India*
[6] *Department of Biology, Faculty of Science, University of Bisha, Bisha, Saudi Arabia*
[7] *Department of Biotechnology, Government Degree College for Boys, Sopore, Jammu and Kashmir, India*
[8] *Centre for Research Impact and Outcome, Chitkara University Institute of Engineering and Technology, Chitkara University, Rajpura, Punjab, India*
[9] *Chitkara Centre for Research and Development, Chitkara University, Baddi, Himachal Pradesh, India*
[10] *Department of Genetics, Plant Breeding and Biotechnology, Dr. Khem Singh Gill Akal College of Agriculture, Eternal University, Sirmour, Himachal Pradesh, India*

3.1 Introduction

The production of healthy food is one of the major issues to be overcome to feed the increasing population. The use of chemical fertilizers for increasing the productivity is degrading the ecosystems (Alavaisha et al. 2019). It is obvious that the chemicals lead to toxic effects on human health and the environment (Pelosi et al. 2014). Another day-by-day emerging challenge is environmental sustainability. Anthropogenic activities are deteriorating the environment and negatively influencing soil health and affecting plant growth, development, and yield (Harte 2007). The concept of healthy soil refers to its capability to sustain and support growth and maintain the long-standing quality of the environment.

The fundamental components that maintain the health of the soil consist of the high population of the organisms that possess plant growth-promoting traits, such as the soil microbes and the plant-associated microbes, such as the rhizospheric microbes. The microbes inhabiting the rhizosphere of plants have been named plant growth-promoting rhizospheric microbes and these play many beneficial and ecological roles along with maintaining healthy soil and crop productivity (Kumar and Verma 2019; Kaur et al. 2021). The plant and microbe interactions are an important component of healthy soil as well as for enabling more sustainable agriculture. These microbial communities have a role in nutrient cycling, combating biotic and abiotic stress, and enhancing plant growth by producing plant growth hormones, and various secondary metabolites (Khatoon et al. 2020; Kour et al. 2020a).

Phosphorus (P) is a vital nutrient for plants and is available in both organic and inorganic forms in soil. The P-deficiency severely impacts the root growth and declines yield. Due to its unavailability for plant uptake, phosphorus limits crop productivity, and is supplied in the form of chemicals (Kour et al. 2021). An alternate and eco-friendly approach to phosphorus fertilizers is the use of the phosphorus-solubilizing microbes (PSMs), which solubilize the insoluble forms of phosphorus and enhance the availability of phosphorus for plants (Khan et al. 2010b). P-solubilizing bacteria and fungi produce organic acids, which lower the pH and solubilize the phosphorus in soil and make it available for assimilation by plants (Zaidi and Khan 2007). Other mechanisms for microbial P-solubilization include the production of siderophores for chelation of metal ions and the production of exopolysaccharides (Yadav et al. 2020; Rawat et al. 2021). Numerous bacteria and fungi from plant internal tissues, soil, and rhizospheric soil with P-solubilizing activity have been reported (Kour et al. 2020b, 2020c).

Nanotechnology is an emerging area and innovative approach in the field of agriculture that is in use to increase crop productivity (Fatima et al. 2022). It has been foreseen that nanotechnology is a supportive and sustainable solution for replacing chemical fertilizers with nanoparticulate counterparts with superior qualities for overcoming the existing challenges of uptake and bioavailability of minerals and decreasing the waste of chemical fertilizers (Achari and Kowshik 2018). Most research on nanofertilizers has been done on metallic nutrients such as Mn, Zn, Cu, and so on, but the reports on macronutrients are limited and need further research (Chen 2018; Vázquez-Núñez et al. 2018). The potential of using nano-hydroxyapatite as a source of phosphorus nano-CaP is reported (Tiwari et al. 2022). This chapter highlights the role of phosphorus for plants, the biodiversity of PSMs and their mechanisms of P-solubilization, and phosphorus nanofertilizers in agriculture.

3.2 Phosphorus as an Important Macronutrient for Plants

Phosphorus is the most important element for plant growth. It exists in an insoluble state, and the biological growth and development of plants requires phosphorus in soluble form (Kouas et al. 2005). The amount of P that is soluble and available to plants in soil is about 0.1% (Goldstein 1994). Phosphorus is rarely present in adequate amounts in the soil for plants to grow and develop to their full potential (Malhotra et al. 2018). Plants usually absorb P from the soil as inorganic phosphate (Pi). After being transported into the plant, Pi is transformed into organic forms that are utilized as energy carriers, such as adenosine triphosphate (ATP) and adenosine diphosphate (ADP). Pi also contributes to the synthesis of other significant organic molecules, such as phospholipids, nucleic acids, and inorganic polyphosphates (poly P), which enable plant development and physiological and molecular adaptations (Plaxton and Lambers 2015). This nutritious element plays an important role in various physiological processes such as fruit and seed formation, energy storage, respiration, photosynthesis, cell division, cell expansion, and several other physiological activities (Fathi and Afra 2023).

A significant percentage of soils globally are deficient in P, which severely limits agricultural production and poses a serious threat to global food security (Heuer et al. 2017). Crop productivity is severely hampered by the restricted availability of P in the soil, predominantly due to the exposure of the plants to abiotic stressors, including water deficit, salinity, and extreme temperatures. For agricultural production systems to remain sustainable, it is imperative that the complex mechanisms by which P contributes to plant physiological responses to abiotic stresses be fully understood (Khan and Siddique 2023). The absorption of other nutrients such as Ca, K, and N is also impacted by insufficient P. This changes the plant architecture and results in shorter and thicker stems and reduced stem elongation (Tariq et al. 2023). P deficit has an adverse effect on the plant's carbon budget, which in turn impairs root development, biomass accumulation, and leaf expansion. The reduction in leaf area that follows the initial rise in carbohydrate availability for root growth limits light interception, which in turn slows down root growth (Mollier and Pellerin 1999).

3.3 Phosphate-Solubilizing Microbes

Soil P is mainly derived from biomolecules that include phosphides, co-enzymes, nucleotides, sugar phosphates, phosphoproteins, and phosphonates which play an essential role in soil P recycling (Tamburini et al. 2012). Elemental P derived from other compounds that include phosphonates, orthophosphate esters, and polyphosphates are short-lived compounds and are reported to comprise 65% of total P in soil (Dodd and Sharpley 2015; Fabiańska et al. 2019). Phosphate-solubilizing microbes (PSMs) are known to derive available P from their derivative form via solubilization of P by different means. PSM-mediated P mineralization from organic compounds incorporates P into the bacterial cells as cellular P (Dasila et al. 2023). PSMs are distributed heterogeneously, and different groups of microorganisms have the potential to solubilize P and thus participate in the soil P cycling pool. Among heterogeneously distributed groups in soil, a large number of PSM microflora are present, including phosphate-solubilizing bacteria (PSB), phosphate-solubilizing actinomycetes (PSA), phosphate-solubilizing fungi (PSF), and cyanobacteria (Khan et al. 2010a). Amongst PSB, common genera that are distributed in soil and reported to solubilize P include, *Bacillus, Rhizobium, Pseudomonas,* and *Escherichia*. PSF are reported to possess greater potential to solubilize P from organic compounds as compared to bacteria, and common genera of PSF present in soil include *Aspergillus, Mucor, Penicillium,* and *Rhizopus.* Several reports also suggest that arbuscular mycorrhizal fungi (AMF), for example, *Rhizophagus irregularis*, *Glomus mosseae*, and *Glomus aggregatum*, have the potential to solubilize P from organic sources (Mackay et al. 2017; Jiang et al. 2020). In addition to the production of important secondary metabolites that include antibiotics, actinomycetes are also known to solubilize P from the soil P pool. PSA includes genera of *Micromonospora* and *Streptomyces*, which are reported to solubilize insoluble P minerals during the weathering process and soil P cycle (Farhat et al. 2015; Chaiharn et al. 2018). Two diazotrophic cyanobacteria, for example, *Anabaena variabilis* and *Westiellopsis prolifica* have been reported to solubilize inorganic P from rock phosphate and tricalcium phosphate (Yandigeri et al. 2011). On periodic monitoring, it has been observed that both these cyanobacterial strains could significantly increase total P and available P in the medium, which is P-starved in nature. All the PSMs employ different means of solubilizing soil P which include lowering of soil pH via releasing protons in the environment, chelation, and production of organic acids. At the molecular level, the expression of certain genes is upregulated and downregulated during P solubilization (Table 3.1).

Table 3.1 Biodiversity of phosphorus-solubilizing microbes.

S.No.	Phosphorus solubilizers	References
1.	*Achromobacter xylosoxidans*	Bumunang and Babalola (2014)
2.	*Acinetobacter baylyi*	Yu et al. (2011)
3.	*Acinetobacter calcoaceticus*	Yadav et al. (2014)
4.	*Acinetobacter calcoaceticus*	Singh et al. (2019)
5.	*Acinetobacter radioresistens*	Yu et al. (2011)
6.	*Acinetobacter tandoii*	Gopalakrishnan et al. (2011)
7.	*Aeromonas hydrophila*	Martínez-Gallegos et al. (2018)
8.	*Agrobacterium tumefaciens*	Yu et al. (2011)
9.	*Alcaligenes faecalis*	Yadav et al. (2014)
10.	*Arthrobacter siderocapsulatus*	Yu et al. (2011)
11.	*Arthroderma cuniculi*	Karmakar et al. (2018)
12.	*Arthroderma cuniculi*	Karmakar et al. (2018)
13.	*Aspergillus aculeatus*	Narsian and Patel (2000)
14.	*Aspergillus allahabadii*	Ruangsanka (2014)
15.	*Aspergillus neoniger*	Zhang et al. (2018a)
16.	*Aspergillus neoniger*	Zhang et al. (2018a)
17.	*Aspergillus niger*	Ruangsanka (2014)
18.	*Bacillus aerius*	Singh et al. (2019)
19.	*Bacillus aerophilus*	Verma et al. (2016)
20.	*Bacillus alcalophilus*	Verma et al. (2016)
21.	*Bacillus altitudinis*	Gopalakrishnan et al. (2011)
22.	*Bacillus altitudinis*	Verma et al. (2016)
23.	*Bacillus amyloliquefaciens*	Verma et al. (2016)
24.	*Bacillus amyloliquefaciens*	Hariprasad and Niranjana (2009)
25.	*Bacillus amyloliquefaciens*	Singh et al. (2019)
26.	*Bacillus anthracis*	Agbodjato et al. (2018)
27.	*Bacillus aquimaris*	Verma et al. (2016)
28.	*Bacillus aryabhattai*	Verma et al. (2016)
29.	*Bacillus atrophaeus*	Verma et al. (2016)
30.	*Bacillus cereus*	Yu et al. (2011)
31.	*Bacillus circulans*	Agbodjato et al. (2018)
32.	*Bacillus circulans*	Verma et al. (2016)
33.	*Bacillus endophyticus*	Verma et al. (2016)
34.	*Bacillus flexus*	Verma et al. (2016)
35.	*Bacillus fusiformis*	Verma et al. (2016)
36.	*Bacillus licheniformis*	Verma et al. (2016)
37.	*Bacillus megaterium*	Verma et al. (2016)

Table 3.1 (Continued)

S.No.	Phosphorus solubilizers	References
38.	*Bacillus methylotrophicus*	Verma et al. (2016)
39.	*Bacillus mojavensis*	Verma et al. (2016)
40.	*Bacillus mucilaginosus*	Singh et al. (2019)
41.	*Bacillus nanhaiensis*	Verma et al. (2016)
42.	*Bacillus nealsonii*	Verma et al. (2016)
43.	*Bacillus rigui*	Verma et al. (2016)
44.	*Bacillus subtilis*	Singh et al. (2019)
45.	*Bacillus thuringiensis*	Yu et al. (2011)
46.	*Bacillus toyonensis*	Bahena et al. (2015)
47.	*Brevibacterium antiquum*	Gopalakrishnan et al. (2011)
48.	*Burkholderia cepacia*	Bumunang and Babalola (2014)
49.	*Burkholderia ambifaria*	Acevedo et al. (2014)
50.	*Burkholderia caledonica*	Bahena et al. (2015)
51.	*Candida railenensis*	Sarabia et al. (2018)
52.	*Chryseobacterium lathyri*	Scagliola et al. (2016)
53.	*Clavispora lusitaniae*	Sarabia et al. (2018)
54.	*Enterobacter hormaechei*	Inui-Kishi et al. (2012)
55.	*Erwinia rhapontici*	Muleta et al. (2013)
56.	*Ewingella americana*	Bumunang and Babalola (2014)
57.	*Exiguobacterium acetylicum*	Verma et al. (2016)
58.	*Meyerozyma guilliermondii*	Sarabia et al. (2018)
59.	*Paenibacillus amylolyticus*	Verma et al. (2016)
60.	*Paenibacillus dendritiformis*	Verma et al. (2016)
61.	*Paenibacillus durus*	Verma et al. (2016)
62.	*Paenibacillus mucilaginosus*	Goswami et al. (2015)
63.	*Paenibacillus polymyxa*	Verma et al. (2016)
64.	*Paenibacillus xylanexedens*	Verma et al. (2016)
65.	*Pantoea agglomerans*	Yu et al. (2011)
66.	*Pantoea cypripedii*	Kaur and Reddy (2013)
67.	*Penicillium calidicanium*	Ruangsanka (2014)
68.	*Penicillium oxalicum*	Ruangsanka (2014)
69.	*Penicillium purpurogenum*	Gizaw et al. (2017)
70.	*Pseudomonas aeruginosa*	Yadav et al. (2014)
71.	*Pseudomonas aspleni*	Yadav et al. (2014)
72.	*Pseudomonas aurantiaca*	Yu et al. (2011)
73.	*Pseudomonas azotoformans*	Bahena et al. (2015)
74.	*Pseudomonas baetica*	Bahena et al. (2015)

(*Continued*)

Table 3.1 (Continued)

S.No.	Phosphorus solubilizers	References
75.	*Pseudomonas chlororaphis*	Muleta et al. (2013)
76.	*Pseudomonas syncyanea*	Yadav et al. (2014)
77.	*Pseudomonas syringae*	Agbodjato et al. (2018)
78.	*Pseudomonas vancouverensis*	Yadav et al. (2014)
79.	*Stenotrophomonas maltophilia*	Singh et al. (2019)
80.	*Stenotrophomonas rhizophila*	Majeed et al. (2015)
81.	*Streptomyces cellulosae*	Jog et al. (2014)
82.	*Streptomyces cheonanensis*	Jog et al. (2014)
83.	*Streptomyces lomondensis*	Jog et al. (2014)
84.	*Streptomyces werraensis*	Jog et al. (2014)
85.	*Symmetrospora coprosmae*	Sarabia et al. (2018)
86.	*Talaromyces aurantiacus*	Zhang et al. (2018a)
87.	*Talaromyces aurantiacus*	Zhang et al. (2018a)
88.	*Talaromyces calidicanius*	Acevedo et al. (2014)

3.4 Mechanism of Microbial Phosphorus Solubilization

Phosphorus constitutes around 0.2–0.8% of the dry weight of plants and is important for physiological and developmental processes. It is one of the major limiting macronutrients after nitrogen (Sharma et al. 2013). Phosphate is present in soil in inorganic and organic forms abundantly, but most of it is complexed with metals, fixed, and precipitated, and thus not available to plants. P availability is affected by mineralization, solubilization, and immobilization processes carried out by various microorganisms (Rawat et al. 2021). Phosphate deficiencies are generally mitigated through the application of phosphatic fertilizers that negatively impact the environment and soil health. Thus, finding alternative eco-friendly solutions becomes imperative in agriculture. PSMs are able to increase the bioavailability of plant-usable phosphorus in a sustainable manner. PSMs with multifarious plant growth promoting (PGP) properties such as nutrient acquisition, phytohormone production, and combating abiotic and biotic stresses in plants play diverse roles in growth and development (Khan et al. 2009).

PSMs, through a variety of solubilization and mineralization mechanisms, are able to increase the available phosphorus pool for plants. Numerous bacterial and fungal genera are reported to solubilize insoluble phosphate, including *Bacillus*, *Pseudomonas*, *Arthrobacter*, *Serratia*, *Phyllobacterium*, *Azotobacter*, *Pantoea*, *Xanthomonas*, *Aspergillus*, and *Penicillium* (Khan et al. 2009). Organic acid (OA) production is one of the important

Table 3.2 P-solubilizing microbes and their mechanism of action.

P-solubilizing microbes	Mechanism of action	References
Arthrobacter sp.	EPS production	Yi et al. (2008)
Aspergillus niger	Gluconic and oxalic acid production	Chuang et al. (2007)
Azotobacter sp.	EPS production	Yi et al. (2008)
Burkholderia cepacia	Gluconic acid production	Song et al. (2008)
Dyella ginsengisoli	Sulfuric acid production	Anandham et al. (2008)
Enterobacter intermedium	2-ketogluconic acid production	Hwangbo et al. (2003)
Enterobacter sp.	EPS production	Yi et al. (2008)
Enterobacter sp.	EPS production	Yi et al. (2008)
Microbacterium phyllosphaerae	Sulfuric acid production	Anandham et al. (2008)
Micromonospora sp.	Production of siderophores	Hamdali et al. (2008)
Penicillium aurantiogriseum	Phytase production	Illmer and Schinner (1995)
Pseudomonas fluorescens	Phytase production	Li et al. (2013)
Pseudomonas sp.	Phytase production	Illmer and Schinner (1995)
Rahnella aquatilis	Phytase production	Li et al. (2013)
Streptomyces sp.	Production of siderophores	Hamdali et al. (2008)

attributes of PSMs, which helps in chelating ions. OA accumulation causes a drop in pH and thus acidification of the surrounding environment, leading to the release of phosphate ions from its minerals (Rashid et al. 2004) (Table 3.2).

The strength and type of organic acid are important factors that affect the amount of P solubilized. Microorganisms are reported to produce gluconic acid, acetic acid, fumaric acid, malonic acid, succinic acid, and various other organic acids. Inorganic acids such as nitric and sulfuric acids also help to solubilize P, but solubilization efficiency is less than that of organic acids. All these acids, gluconic acid is the major OA produced by PSMs (Deubel et al. 2000). OAs are able to chelate cations and thus are able to solubilize phosphorus complexed with calcium, iron, and aluminum. OAs chelate metal ions with the help of negatively charged hydroxyl and carboxyl groups. However, researchers have shown that OA production is not the only mechanism for P solubilization, as many microorganisms were reported to solubilize P but did not produce any OA (Chen et al. 2006). Acidification of the medium due to proton extrusion is also one of the important mechanisms leading to P solubilization (Prabhu et al. 2019). Protons in the medium are able to substitute cations such as Fe^{3+}, Al^{3+}, and Ca^{2+}.

Besides OA production and proton extrusion, studies have also revealed that bacterial isolates producing exopolysaccharides (EPS) were able to increase the availability of P in the medium (Yi et al. 2008). There exists a positive correlation between P solubilization and EPS production. Although EPS-producing bacteria are not able to release P from insoluble TCP, EPS causes variability in the homeostasis of P solubilization. It also holds metal ions in soil, which further pushes P dissolution from its insoluble forms. Other microorganisms lead to the enzymatic dissolution of insoluble organic and inorganic phosphates. Microbial phosphatases are widely produced exoenzymes that help in the release of P from its insoluble and complexed reserves. These are classified as acid, neutral, and alkaline phosphatases (Rodríguez et al. 2006). Phosphatases break down phosphoester and phosphoanhydride bonds and release P from organic P compounds. Phosphonates are broken down into hydrocarbon and P using phosphonatases and C—P lyase enzymes, which break down the C—P bond present in organophosphates (Richardson and Simpson 2011). Phytates are the most common organic form of phosphorus in soil. PSMs with phytase activity lead to significant release of P from phytate and myo-inositol phosphate compounds and make P available to plants (Vassilev et al. 2007). Although there is no clearly reported correlation between siderophore production and P solubilization, metal sequestration by siderophores may help in liberating P from its metal salts.

3.5 Phosphorus-Solubilizing Microbes as Bioinoculants for Sustainable Agriculture

The excessive application of conventional phosphorus (P) fertilizers to enhance agricultural productivity, in response to the ever-growing global food demands, may lead to surface and groundwater contamination, eutrophication of waterways, depletion of soil fertility, and the accumulation of toxic elements, including elevated concentrations of selenium (Se) and arsenic (As) in the soil (Alori et al. 2017). A significant number of soil microorganisms can solubilize or mineralize insoluble soil phosphate. Consequently, inoculating seeds, crops, or soil with PSMs represents a viable approach to enhance global food production while mitigating environmental risks (Alori et al. 2012). Although PSMs are fundamental for enhancing soil fertility, but they have not substituted traditional chemical fertilizers in commercial agriculture.

PSMs by increasing the utilization of P stored in soils may serve as biofertilizers as they convert insoluble form of P into soluble form. This enhances the soluble phosphorus concentration (Zhu et al. 2012). The application of phosphorus biofertilizers represents a promising strategy for augmenting food production by improving agricultural yield, as it advocates for an environmentally sustainable method that prioritizes biological additives to the soil over chemical alternatives to address soil infertility (Babalola and Glick 2012). PSMs provide microenvironments with abundant phosphorus availability in the plant rhizosphere or soil, hence improving the efficacy of phosphorus resource consumption (Sahu et al. 2018). *Pseudomonas* and *Bacillus* have been named as the main genera of PSB, which comprise 50% of PSMs (Kishore et al. 2015; Ameen et al. 2019; Bandeppa et al. 2022).

The non-mycorrhizal P-solubilizing fungi include the genera *Alternaria, Arthrobotrys, Aspergillus, Fusarium, Helminthosporium, Penicillium,* and *Trichoderma;* examples of mycorrhizal fungi include *Rhizophagus irregularis, Glomus mossea, G. fasciculatum,* and *Entrophospora colombiana* (Zhang et al. 2018b; Kalayu 2019; Silva et al. 2023; Fu et al. 2024). Examples of PSMs among actinobacteria include diverse species of *Streptomyces, Thermobifida,* and *Micrococcus,* whereas cyanobacteria examples are *Calothrix braunii, Westiellopsis prolifica, Anabaena variabilis,* and *Scytonema* sp. (Rasul et al. 2019; Silva et al. 2023). A study reported the immobilization of *Pseudomonas* sp. NK2 using cotton straw enhanced the availability of phosphorus in the soil, hence facilitating tomato development. These discoveries can enhance the progression of sustainable agriculture techniques (Bai et al. 2024).

Apart from releasing P into the environment, PSMs can also stimulate plant development through complementary mechanisms. They can help plants grow in both direct and indirect ways, such as by biologically fixing nitrogen and making phytohormones (Kudoyarova et al. 2019; Singh et al. 2020; Nadal et al. 2022); providing tolerance to environmental stressors, including drought, salinity, and low soil fertility (Marulanda et al. 2009; Zaheer et al. 2016). They can also stimulate host plant defense through the development of antibiotics, biosurfactants, and secondary metabolites (Moura et al. 2021; Buttrós et al. 2022). Field studies indicate that inoculation with PSM, such as *Pseudomonas*, *Aspergillus*, or *Penicillium* species enhance growth and grain yield of diverse crops, regardless of soil phosphorus availability and phosphorus fertilizer use (Yadav et al. 2023). Several PSMs inoculation investigations have demonstrated enhanced plant output and augmented phosphorus absorption in both pot tests and field trials. In an *in vitro* experiment utilizing *Aspergillus niger* as a biofertilizer, with wheat husks and 20% perlite as the carrier medium, the soil colonization rate was 5.6×10^6 spores per gram of soil (Wang et al. 2015).

The utilization of PSMs as biofertilizers enhance plant growth, increase plant-available phosphorus, boost yield, and promote the release of indole acetic acid and gibberellic acid, which facilitate cell growth and elongation (Mahmud et al. 2021). However, the establishment and efficacy of these laboratory-developed PSM inoculants in *in vitro* field trials are significantly hindered by environmental variables such as moisture, pH, temperature, salinity, and soil climatic conditions. Additionally, it is well known that inocula derived from a specific soil perform less effectively in soils with different characteristics (Sharma et al. 2013). Hence, as we gain a deeper understanding of PSMs and its processes, we can confidently predict that using PSM as bioinoculants will boost their effectiveness as essential components in the advancement of sustainable agriculture. People who buy agricultural products are mostly interested in how healthy, high-quality, and nutritious those products are. Thus, the application of PSMs as microbial inoculants represents a novel opportunity for enhanced plant productivity. The utilization of PSMs as bioinoculants is a viable approach to enhance food production without creating health risks while also preserving the environment. Researchers must continue to learn more about PSMs and promptly convert that knowledge into an approach that farmers can utilize. Therefore, it is essential to develop region-specific P-solubilizing microbial technologies and spread awareness among the farmers farmers (Table 3.3).

Table 3.3 Effect of various P-solubilizing microbes on different crops.

Crop treated	P-solubilizer used	Results	References
Beans	*Rhizobium tropici*	Growth of the root, dry weight, shoot, nodule mass, and number	Tajini et al. (2012)
Chickpea	*Aspergillus awamori*	Improved growth, chlorophyll content, grain yield, and straw yield	Waghmare et al. (2024)
Maize	*Azospirillum lipoferum*	Enhanced sprouting	Rozier et al. (2019)
Onion	*Serendipita indica*	Improved plant growth and higher nutrient concentrations in both leaves and bulbs	Arunachalam et al. (2024)
Rice	*Bacillus thuringiensis*	Increased shoot length	Raj et al. (2014)
Rice	*Acinetobacter* sp.	Increased P uptake, growth, and yield	Rasul et al. (2019)
Rice	*Pseudomonas* sp.	Increased P uptake, growth, and yield	Rasul et al. (2019)
Soybean	*Trichoderma longibrachiatum*	Increased biological nitrogen fixation and plant growth	Castro-Barquero (2024)
Sugarcane	*Enterobacter cloacae*	Increased root length and biomass	Safirzadeh et al. (2019)
Tomato	*Pseudomonas* sp.	Improved growth and managed bacterial canker disease	Bakki et al. (2024)
Tomato	*Bacillus* sp.	Improved growth and managed bacterial canker disease	Bakki et al. (2024)
Wheat	*Serratia* sp.	Increased growth	Swarnalakshmi et al. (2013)

3.6 Phosphorus Nanofertilizers in Agriculture

Nanotechnology and nanoscience are the revolutionary technologies of the century after biotechnology. They have been applied in different sectors such as agriculture, physics, chemistry, materials science, biology, and the environment. In agriculture, nanotechnology helps in converting conventional agriculture to precision farming in which balance is maintained between crop yield and environmental sustainability (Basavegowda and Baek 2021). Nanotechnology techniques help in the production of nanoformulations, functionalized nanomaterials, and nanoencapsulations, which can be the next generation of fertilizers, and they can deliver the active ingredients to the specific site at a controlled rate. Nutrients delivered by the nanofertilizer help in the improvement of production and growth of the plants. Nanofertilizers are categorized into two types based on the nutrients, that is, macronutrient nanofertilizers (N, P, K, Mg, S, and Ca) and micronutrient nanofertilizers (Zn, Fe, and Mn) (Chhipa 2017).

Phosphorus nanofertilizer, a category of macronutrient nanofertilizer, is one of the most significant types, as P is the second vital nutrient to plants. Phosphorus has been a limiting nutrient to the plants and it has been supplied through chemical fertilizers discovered 174 years ago. Its use was discovered to be harmful for the environment and disturbance soil sustainability (Guelfi et al. 2022). Therefore, a search for alternatives began, and the use of nano-phosphorus was found to enhance precision agriculture. Different studies have shown the positive effect of the phosphorus nanofertilizer on the growth promotion of the crops. In an investigation, the combined application of calcium phosphate nanoparticles and fungal strains (*Piriformospora indica* and *Glomus mosseae*) in maize plants resulted in the increased performance index and chlorophyll content of the plants (Rane et al. 2015). In a report, the inoculation of nano-KH_2PO_4 in rice was reported for the increment of P accumulation and photosynthesis parameters in the treated plants (Miranda-Villagómez et al. 2019). In a report, the inoculation of *Pseudomonas mallei*, *P. cepaceae*, and nano-phosphorus (mono-ammonium phosphate) in *Phaseolus vulgaris* crops resulted in the increase of yield, efficiency of photosynthesis, relative water content, gas exchange, membrane stability, glutathione contents, glycine betaine content, and content of P in the treated plants (Rady et al. 2020).

In a study, nanophos inoculation in maize seeds resulted in the enhancement of plant height, germination rate, leaf number, total sugar content, protein level, and photosynthetic pigments (Chaudhary et al. 2021). The application of nano calcium phosphate fertilizers in *Phaseolus vulgaris* L. resulted in the increased shoot and root dry weight, nutrient concentration, yield, and crude protein in pods of the treated plants (El-Ghany et al. 2021). Nano-hydroxyapatite application to soil was found to enhance the soil quality by increasing the low molecular weight metabolites such as phenolic acids, amino acids, sugar, and sugar alcohols, along with increasing the bacterial species belonging to *Massilia*, *Streptomyces*, and *Sinomonas* (Jia et al. 2022). The foliar application of phosphate nano-fertilizers (nano-hydroxylapatite, macro-sized parascholzite, and calcium zinc phosphate) resulted in the plant growth promotion by increasing the vegetation index, water stress index, and stomatal conductance index of sunflower, an oilseed plant (Ernst et al. 2023).

3.7 Conclusion

The demands for alternative, cost-efficient, and sustainable strategies have greatly increased for meeting the global food demands. In this regard, the plant-microbe interactions have played a major role in maintaining soil health, fertility, nutrient availability, and enhancing productivity in an environmentally conscious manner. The insolubility of phosphorus makes it a limiting factor for plants, and thus, chemicals are used for plants to fulfill their phosphorus requirements, but chemical fertilizers are harmful to the environment. PSMs have an important role in agricultural systems, as they solubilize the insoluble forms of phosphorus and convert them into soluble forms for plant uptake. The use of PSMs in agriculture is an eco-friendly approach for phosphorus biofortification. There is a need for more field trials for the use of PSMs as bioinoculants due to variability in results under diverse conditions. Developing nano-based phosphorus fertilizers using PSMs is a new horizon for research to boost crop production and bring a revolution in agriculture, and a step forward for a greener environment. Further research on elucidating the interactions of nano-P fertilizers with plants and soil could be taken into consideration.

Acknowledgments

The authors are grateful to Chandigarh University for providing the library support.

References

Acevedo, E., Galindo-Castaneda, T., Prada, F. et al. (2014). Phosphate-solubilizing microorganisms associated with the rhizosphere of oil palm (*Elaeis guineensis* Jacq.) in Colombia. *Applied Soil Ecology* 80: 26–33.

Achari, G.A. and Kowshik, M. (2018). Recent developments on nanotechnology in agriculture: plant mineral nutrition, health, and interactions with soil microflora. *Journal of Agricultural and Food Chemistry* 66 (33): 8647–8661.

Agbodjato, N.A., Amogou, O., Noumavo, A.D.P. et al. (2018). Biofertilising, plant-stimulating and biocontrol potentials of maize plant growth promoting rhizobacteria isolated in central and northern Benin. *African Journal of Microbiology Research* 12 (28): 664–672.

Alavaisha, E., Manzoni, S., and Lindborg, R. (2019). Different agricultural practices affect soil carbon, nitrogen and phosphorous in Kilombero-Tanzania. *Journal of Environmental Management* 234: 159–166.

Alori, E.T., Fawole, O., and Afolayan, A. (2012). Characterization of arbuscular mycorrhizal spores isolated from Southern Guinea Savanna of Nigeria. *Journal of Agricultural Science* 4 (7): 13–19.

Alori, E.T., Glick, B.R., and Babalola, O.O. (2017). Microbial phosphorus solubilization and its potential for use in sustainable agriculture. *Frontiers in Microbiology* 8: 971.

Ameen, F., AlYahya, S.A., AlNadhari, S. et al. (2019). Phosphate solubilizing bacteria and fungi in desert soils: species, limitations and mechanisms. *Archives of Agronomy and Soil Science* 65 (10): 1446–1459.

Anandham, R., Gandhi, P.I., Madhaiyan, M., and Sa, T. (2008). Potential plant growth promoting traits and bioacidulation of rock phosphate by thiosulfate oxidizing bacteria isolated from crop plants. *Journal of Basic Microbiology* 48 (6): 439–447.

Arunachalam, T., Gade, K., Mahadule, P.A. et al. (2024). Optimizing plant growth, nutrient uptake, and yield of onion through the application of phosphorus solubilizing bacteria and endophytic fungi. *Frontiers in Microbiology* 15: 1442912.

Babalola, O.O. and Glick, B.R. (2012). Indigenous African agriculture and plant associated microbes: Current practice and future transgenic prospects. *Scientific Research and Essays* 7 (28): 2431–2439.

Bahena, M.H.R., Salazar, S., Velázquez, E. et al. (2015). Characterization of phosphate solubilizing rhizobacteria associated with pea (*Pisum sativum* L.) isolated from two agricultural soils. *Symbiosis* 67 (1–3): 33–41.

Bai, K., Wang, W., Zhang, J. et al. (2024). Effects of phosphorus-solubilizing bacteria and biochar application on phosphorus availability and tomato growth under phosphorus stress. *BMC Biology* 22 (1): 211.

Bakki, M., Banane, B., Marhane, O. et al. (2024). Phosphate solubilizing *Pseudomonas* and *Bacillus* combined with rock phosphates promoting tomato growth and reducing bacterial canker disease. *Frontiers in Microbiology* 15: 1289466.

Bandeppa, S., Kumar, K., Latha, P.C. et al. (2022). 19—Phosphate-solubilizing microbial inoculants for sustainable agriculture. In: *Trends of Applied Microbiology for Sustainable Economy* (ed. R. Soni, D.C. Suyal, A.N. Yadav, and R. Goel), 501–524. Academic Press.

Basavegowda, N. and Baek, K.-H. (2021). Current and future perspectives on the use of nanofertilizers for sustainable agriculture: The case of phosphorus nanofertilizer. *3 Biotech* 11 (7): 357.

Bumunang, E.W. and Babalola, O.O. (2014). Characterization of rhizobacteria from field grown genetically modified (GM) and non-GM maizes. *Brazilian Archives of Biology and Technology* 57 (1): 1–8.

Buttrós, V.H., Araújo, N.A.F., D'Ávila, V.d.A. et al. (2022). A little helper: Beneficial bacteria with growth-promoting mechanisms can reduce asian soybean rust severity in a cell-free formulation. *Agronomy* 12 (11): 2635.

Castro-Barquero, L. (2024). Increase the biological nitrogen fixation and the growth of soybean (*Glycine max*) using phosphorus-solubilizing fungi. *Agronomia Costarricense* 48 (1): 93–109.

Chaiharn, M., Pathom-Aree, W., Sujada, N., and Lumyong, S. (2018). Characterization of phosphate solubilizing *Streptomyces* as a biofertilizer. *Chiang Mai Journal of Science* 45 (2): 701–716.

Chaudhary, P., Chaudhary, A., Parveen, H. et al. (2021). Impact of nanophos in agriculture to improve functional bacterial community and crop productivity. *BMC Plant Biology* 21 (1): 519.

Chen, H. (2018). Metal based nanoparticles in agricultural system: Behavior, transport, and interaction with plants. *Chemical Speciation and Bioavailability* 30 (1): 123–134.

Chen, Y.P., Rekha, P.D., Arun, A.B. et al. (2006). Phosphate solubilizing bacteria from subtropical soil and their tricalcium phosphate solubilizing abilities. *Applied Soil Ecology* 34 (1): 33–41.

Chhipa, H. (2017). Nanofertilizers and nanopesticides for agriculture. *Environmental Chemistry Letters* 15 (1): 15–22.

Chuang, C.-C., Kuo, Y.-L., Chao, C.-C., and Chao, W.-L. (2007). Solubilization of inorganic phosphates and plant growth promotion by *Aspergillus niger*. *Biology and Fertility of Soils* 43 (5): 575–584.

Dasila, H., Sah, V., Jaggi, V. et al. (2023). Cold-tolerant phosphate-solubilizing *Pseudomonas* strains promote wheat growth and yield by improving soil phosphorous (P) nutrition status. *Frontiers in Microbiology* 14: 1135693.

Deubel, A., Gransee, A., and Merbach, W. (2000). Transformation of organic rhizodepositions by rhizosphere bacteria and its influence on the availability of tertiary calcium phosphate. *Journal of Plant Nutrition and Soil Science* 163 (4): 387–392.

Dodd, R.J. and Sharpley, A.N. (2015). Recognizing the role of soil organic phosphorus in soil fertility and water quality. *Resources, Conservation and Recycling* 105: 282–293.

El-Ghany, M.F.A., El-Kherbawy, M.I., Abdel-Aal, Y.A. et al. (2021). Comparative study between traditional and nano calcium phosphate fertilizers on growth and production of snap bean (*Phaseolus vulgaris* L.) plants. *Nanomaterials* 11: https://doi.org/10.3390/nano11112913.

Ernst, D., Kolenčík, M., Šebesta, M. et al. (2023). Significance of phosphate nano-fertilizers foliar application: A brief real-field study of quantitative, physiological parameters, and agro-ecological diversity in sunflower. *Agronomy* 13: https://doi.org/10.3390/agronomy13102606.

Fabiańska, M.J., Kozielska, B., Konieczyński, J., and Bielaczyc, P. (2019). Occurrence of organic phosphates in particulate matter of the vehicle exhausts and outdoor environment—a case study. *Environmental Pollution* 244: 351–360.

Farhat, M.B., Boukhris, I., and Chouayekh, H. (2015). Mineral phosphate solubilization by *Streptomyces* sp. CTM396 involves the excretion of gluconic acid and is stimulated by humic acids. *FEMS Microbiology Letters* 362 (5): fnv008.

Fathi, A. and Afra, J.M. (2023). Plant growth and development in relation to phosphorus: A review. *Bulletin of University of Agricultural Sciences and Veterinary Medicine Cluj-Napoca Agriculture* 80 (1): 1–7.

Fatima, F., Ahmad, M.M., Verma, S.R., and Pathak, N. (2022). Relevance of phosphate solubilizing microbes in sustainable crop production: A review. *International Journal of Environmental Science and Technology* 19 (9): 9283–9296.

Fu, S.-F., Balasubramanian, V.K., Chen, C.-L. et al. (2024). The phosphate-solubilising fungi in sustainable agriculture: unleashing the potential of fungal biofertilisers for plant growth. *Folia Microbiologica* 69 (4): 697–712.

Gizaw, B., Tsegay, Z., Tefera, G. et al. (2017). Phosphate solubilizing fungi isolated and characterized from Teff rhizosphere soil collected from North Showa zone, Ethiopia. *African Journal of Microbiology Research* 11 (17): 687–696.

Goldstein, A. (1994). Involvement of the quinoprotein glucose dehydrogenase in the solubilization of exogenous phosphates by Gram-negative bacteria. In: *Phosphate in Microorganisms: Cellular and Molecular Biology* (ed. A. Torriani-Gorini, E. Yagil, and S. Silver), 197–203. Washington, DC: ASM Press.

Gopalakrishnan, S., Humayun, P., Kiran, B.K. et al. (2011). Evaluation of bacteria isolated from rice rhizosphere for biological control of charcoal rot of sorghum caused by *Macrophomina phaseolina* (Tassi) Goid. *World Journal of Microbiology and Biotechnology* 27 (6): 1313–1321.

Goswami, D., Parmar, S., Vaghela, H. et al. (2015). Describing *Paenibacillus mucilaginosus* strain N3 as an efficient plant growth promoting rhizobacteria (PGPR). *Cogent Food & Agriculture* 1 (1): 1000714.

Guelfi, D., Nunes, A.P., Sarkis, L.F., and Oliveira, D.P. (2022). Innovative phosphate fertilizer technologies to improve phosphorus use efficiency in agriculture. *Sustainability* 14: https://doi.org/10.3390/su142114266.

Hamdali, H., Bouizgarne, B., Hafidi, M. et al. (2008). Screening for rock phosphate solubilizing *Actinomycetes* from Moroccan phosphate mines. *Applied Soil Ecology* 38 (1): 12–19.

Hariprasad, P. and Niranjana, S. (2009). Isolation and characterization of phosphate solubilizing rhizobacteria to improve plant health of tomato. *Plant and Soil* 316 (1–2): 13–24.

Harte, J. (2007). Human population as a dynamic factor in environmental degradation. *Population and Environment* 28 (4): 223–236.

Heuer, S., Gaxiola, R., Schilling, R. et al. (2017). Improving phosphorus use efficiency: A complex trait with emerging opportunities. *The Plant Journal* 90 (5): 868–885.

Hwangbo, H., Park, R.D., Kim, Y.W. et al. (2003). 2-Ketogluconic acid production and phosphate solubilization by *Enterobacter intermedium*. *Current Microbiology* 47 (2): 0087–0092.

Illmer, P. and Schinner, F. (1995). Solubilization of inorganic calcium phosphates—solubilization mechanisms. *Soil Biology and Biochemistry* 27 (3): 257–263.

Inui-Kishi, R.N., Kishi, L.T., Picchi, S.C. et al. (2012). Phosphorus solubilizing and IAA production activities in plant growth promoting rhizobacteria from brazilian soils under sugarcane cultivation. *Journal of Engineering and Applied Science* 7 (11): 1446–1454.

Jia, X., Shi, N., Tang, W. et al. (2022). Nano-hydroxyapatite increased soil quality and boosted beneficial soil microbes. *Plant Nano Biology* 1: 100002.

Jiang, Y., Tian, J., and Ge, F. (2020). New insight into carboxylic acid metabolisms and ph regulations during insoluble phosphate solubilisation process by *Penicillium oxalicum* PSF-4. *Current Microbiology* 77 (12): 4095–4103.

Jog, R., Pandya, M., Nareshkumar, G., and Rajkumar, S. (2014). Mechanism of phosphate solubilization and antifungal activity of *Streptomyces* spp. isolated from wheat roots and rhizosphere and their application in improving plant growth. *Microbiology* 160 (4): 778–788.

Kalayu, G. (2019). Phosphate solubilizing microorganisms: promising approach as biofertilizers. *International Journal of Agronomy* 2019 (1): 4917256.

Karmakar, P., Sharma, D., Das, P., and Saha, A.K. (2018). Phosphate solubilizing capacity and siderophore production by *Arthroderma cuniculi* Dawson 1963 isolated from rhizospheric soil. *Research Journal of Life Sciences Bioinformatics Pharmaceutical and Chemical Sciences* 4 (3): 330–336.

Kaur, G. and Reddy, M.S. (2013). Phosphate solubilizing rhizobacteria from an organic farm and their influence on the growth and yield of maize (*Zea mays* L.). *The Journal of General and Applied Microbiology* 59 (4): 295–303.

Kaur, T., Devi, R., Kour, D. et al. (2021). Plant growth promoting soil microbiomes and their potential implications for agricultural and environmental sustainability. *Biologia* 76 (9): https://doi.org/10.1007/s11756-021-00806-w.

Khan, F. and Siddique, A.B. (2023). Phosphorus plays key roles in regulating plants. *Physiological Responses to Abiotic Stresses* 12 (15).

Khan, M.S., Zaidi, A., and Wani, P.A. (2009). Role of phosphate solubilizing microorganisms in sustainable agriculture—a review. In: *Sustainable Agriculture* (ed. E. Lichtfouse, M. Navarrete, P. Debaeke, et al.), 551–570. Dordrecht: Springer Netherlands.

Khan, M.S., Zaidi, A., Ahemad, M. et al. (2010a). Plant growth promotion by phosphate solubilizing fungi–current perspective. *Archives of Agronomy and Soil Science* 56 (1): 73–98.

Khan, M.S., Zaidi, A., and Musarrat, J. (2010b). *Microbes for Legume Improvement*. Springer, Vienna.

Khatoon, Z., Huang, S., Rafique, M. et al. (2020). Unlocking the potential of plant growth-promoting rhizobacteria on soil health and the sustainability of agricultural systems. *Journal of Environmental Management* 273: 111118.

Kishore, N., Pindi, P.K., and Reddy, S.R. (2015). Phosphate-solubilizing microorganisms: A critical review. In: *Plant Biology and Biotechnology: Volume I: Plant Diversity, Organization, Function and Improvement* (ed. B. Bahadur, M.V. Rajam, L. Sahijram, and K.V. Krishnamurthy), 307–333. New Delhi: Springer India.

Kouas, S., Labidi, N., Debez, A., and Abdelly, C. (2005). Effect of P on nodule formation and N fixation in bean. *Agronomy for Sustainable Development* 25 (3): 389–393.

Kour, D., Kaur, T., Devi, R. et al. (2020a). Biotechnological applications of beneficial microbiomes for evergreen agriculture and human health. In: *Trends of Microbial Biotechnology for Sustainable Agriculture and Biomedicine Systems: Perspectives for Human Health* (ed. A.A. Rastegari, A.N. Yadav, and N. Yadav), 255–279. Amsterdam: Elsevier.

Kour, D., Rana, K.L., Kaur, T. et al. (2020b). Microbe-mediated alleviation of drought stress and acquisition of phosphorus in great millet (*Sorghum bicolour* L.) by drought-adaptive and phosphorus-solubilizing microbes. *Biocatalysis and Agricultural Biotechnology* 23: 101501.

Kour, D., Rana, K.L., Sheikh, I. et al. (2020c). Alleviation of drought stress and plant growth promotion by *Pseudomonas libanensis* EU-LWNA-33, a drought-adaptive phosphorus-solubilizing bacterium. *Proceedings of the National Academy of Sciences, India Section B* 90: 785–795.

Kour, D., Rana, K.L., Kaur, T. et al. (2021). Biodiversity, current developments and potential biotechnological applications of phosphorus-solubilizing and -mobilizing microbes: A review. *Pedosphere* 31 (1): 43–75.

Kudoyarova, G., Arkhipova, T., Korshunova, T. et al. (2019). Phytohormone mediation of interactions between plants and non-symbiotic growth promoting bacteria under edaphic stresses. *Frontiers in Plant Science* 10: 1368.

Kumar, A. and Verma, J.P. (2019). The role of microbes to improve crop productivity and soil health. In: *Ecological Wisdom Inspired Restoration Engineering* (ed. V. Achal and A. Mukherjee), 249–265. Singapore: Springer Singapore.

Li, G.-E., Wu, X.-Q., Ye, J.-R. et al. (2013). Isolation and identification of phytate-degrading rhizobacteria with activity of improving growth of poplar and Masson pine. *World Journal of Microbiology and Biotechnology* 29: 2181–2193.

Mackay, J.E., Cavagnaro, T.R., Stöver, D.S.M. et al. (2017). A key role for arbuscular mycorrhiza in plant acquisition of P from sewage sludge recycled to soil. *Soil Biology and Biochemistry* 115: 11–20.

Mahmud, A.A., Upadhyay, S.K., Srivastava, A.K., and Bhojiya, A.A. (2021). Biofertilizers: A Nexus between soil fertility and crop productivity under abiotic stress. *Current Research in Environmental Sustainability* 3: 100063.

Majeed, A., Abbasi, M.K., Hameed, S. et al. (2015). Isolation and characterization of plant growth-promoting rhizobacteria from wheat rhizosphere and their effect on plant growth promotion. *Frontiers in Microbiology* 6: 198.

Malhotra, H., Vandana, S.S., and Pandey, R. (2018). *Phosphorus Nutrition: Plant Growth in Response to Deficiency and Excess. Plant Nutrients and Abiotic Stress Tolerance* (ed. M. Hasanuzzaman, M. Fujita, H. Oku, et al.), 171–190. Singapore: Springer Singapore.

Martínez-Gallegos, V., Bautista-Cruz, A., Martínez-Martínez, L., and Soledad Sanchez-Medina, P. (2018). First report of phosphate-solubilizing bacteria associated with Agave angustifolia. *International Journal of Agriculture and Biology* 20 (6): 1298–1302.

Marulanda, A., Barea, J.-M., and Azcón, R. (2009). Stimulation of plant growth and drought tolerance by native microorganisms (AM fungi and bacteria) from dry environments: Mechanisms related to bacterial effectiveness. *Journal of Plant Growth Regulation* 28 (2): 115–124.

Miranda-Villagómez, E., Trejo-Téllez, L.I., Gómez-Merino, F.C. et al. (2019). Nanophosphorus fertilizer stimulates growth and photosynthetic activity and improves P status in rice. *Journal of Nanomaterials* 2019 (1): 5368027.

Mollier, A. and Pellerin, S. (1999). Maize root system growth and development as influenced by phosphorus deficiency. *Journal of Experimental Botany* 50 (333): 487–497.

Moura, G.G.D.d., Barros, A.V.d., Machado, F. et al. (2021). Endophytic bacteria from strawberry plants control gray mold in fruits via production of antifungal compounds against *Botrytis cinerea* L. *Microbiological Research* 251: 126793.

Muleta, D., Assefa, F., Börjesson, E., and Granhall, U. (2013). Phosphate-solubilising rhizobacteria associated with *Coffea arabica* L. in natural coffee forests of southwestern Ethiopia. *Journal of the Saudi Society of Agricultural Sciences* 12 (1): 73–84.

Nadal, M.C., Ferreira, G.M.d.R., Andrade, G.V.S. et al. (2022). Endophytic bacteria can replace the need for synthetic auxin during in vitro rooting of *Pyrus communis*. *Agronomy* 12 (5): 1226.

Narsian, V. and Patel, H. (2000). *Aspergillus aculeatus* as a rock phosphate solubilizer. *Soil Biology and Biochemistry* 32 (4): 559–565.

Pelosi, C., Barot, S., Capowiez, Y. et al. (2014). Pesticides and earthworms. A review. *Agronomy for Sustainable Development* 34 (1): 199–228.

Plaxton, W. and Lambers, H. (2015). *Annual Plant Reviews, Phosphorus Metabolism in Plants.* John Wiley & Sons.

Prabhu, N., Borkar, S., and Garg, S. (2019). Chapter 11—Phosphate solubilization by microorganisms: overview, mechanisms, applications and advances. In: *Advances in Biological Science Research* (ed. S.N. Meena and M.M. Naik), 161–176. Academic Press.

Rady, M.M., El-Shewy, A.A., Seif El-Yazal, M.A., and Abd El-Gawwad, I.F.M. (2020). Integrative application of soil P-solubilizing bacteria and foliar nano P improves *Phaseolus vulgaris* plant performance and antioxidative defense system components under calcareous soil conditions. *Journal of Soil Science and Plant Nutrition* 20 (3): 820–839.

Raj, D., Linda, R., and Babyson, R.S. (2014). Molecular characterization of phosphate solubilizing bacteria (PSB) and plant growth promoting rhizobacteria (PGPR) from pristine soils. *International Journal of Innovative Science Engineering and Technology* 1: 317–324.

Rane, M., Bawskar, M., Rathod, D. et al. (2015). Influence of calcium phosphate nanoparticles, *Piriformospora indica* and *Glomus mosseae* on growth of *Zea mays*. *Advances in Natural Sciences: Nanoscience and Nanotechnology* 6 (4): 045014.

Rashid, M., Khalil, S., Ayub, N. et al. (2004). Organic acids production and phosphate solubilization by phosphate solubilizing microorganisms (PSM) under *in vitro* conditions. *Pakistan Journal of Biological Sciences* 7 (2): 187–196.

Rasul, M., Yasmin, S., Suleman, M. et al. (2019). Glucose dehydrogenase gene containing phosphobacteria for biofortification of phosphorus with growth promotion of rice. *Microbiological Research* 223–225: 1–12.

Rawat, P., Das, S., Shankhdhar, D., and Shankhdhar, S.C. (2021). Phosphate-solubilizing microorganisms: mechanism and their role in phosphate solubilization and uptake. *Journal of Soil Science and Plant Nutrition* 21 (1): 49–68.

Richardson, A.E. and Simpson, R.J. (2011). Soil microorganisms mediating phosphorus availability update on microbial phosphorus. *Plant Physiology* 156 (3): 989–996.

Rodríguez, H., Fraga, R., Gonzalez, T., and Bashan, Y. (2006). Genetics of phosphate solubilization and its potential applications for improving plant growth-promoting bacteria. *Plant and Soil* 287 (1): 15–21.

Rozier, C., Gerin, F., Czarnes, S., and Legendre, L. (2019). Biopriming of maize germination by the plant growth-promoting rhizobacterium *Azospirillum lipoferum* CRT1. *Journal of Plant Physiology* 237: 111–119.

Ruangsanka, S. (2014). Identification of phosphate-solubilizing fungi from the asparagus rhizosphere as antagonists of the root and crown rot pathogen *Fusarium oxysporum*. *ScienceAsia* 40: 16–20.

Safirzadeh, S., Chorom, M., and Enayatizamir, N. (2019). Effect of phosphate solubilising bacteria (*Enterobacter cloacae*) on phosphorus uptake efficiency in sugarcane (*Saccharum officinarum* L.). *Soil Research* 57 (4): 333–341.

Sahu, A., Bhattacharjya, S., Mandal, A. et al. (2018). Microbes: a sustainable approach for enhancing nutrient availability in agricultural soils. In: *Role of Rhizospheric Microbes in*

Soil: Volume 2: Nutrient Management and Crop Improvement (ed. V.S. Meena), 47–75. Singapore: Springer Singapore.

Sarabia, M., Cazares, S., González-Rodríguez, A. et al. (2018). Plant growth promotion traits of rhizosphere yeasts and their response to soil characteristics and crop cycle in maize agroecosystems. *Rhizosphere* 6: 67–73.

Scagliola, M., Pii, Y., Mimmo, T. et al. (2016). Characterization of plant growth promoting traits of bacterial isolates from the rhizosphere of barley (*Hordeum vulgare* L.) and tomato (*Solanum lycopersicon* L.) grown under Fe sufficiency and deficiency. *Plant Physiology and Biochemistry* 107: 187–196.

Sharma, S.B., Sayyed, R.Z., Trivedi, M.H., and Gobi, T.A. (2013). Phosphate solubilizing microbes: sustainable approach for managing phosphorus deficiency in agricultural soils. *Springerplus* 2 (1): 587.

Silva, L.I.d., Pereira, M.C., Carvalho, A.M.X.d. et al. (2023). Phosphorus-solubilizing microorganisms: a key to sustainable agriculture. *Agriculture* 13 (2): 462.

Singh, R., Kumar, A., Singh, M., and Pandey, K.D. (2019). Isolation and characterization of plant growth promoting rhizobacteria from *Momordica charantia* L. In: *PGPR Amelioration in Sustainable Agriculture* (ed. A.K. Singh, P.K. Singh, and A. Kumar), 217–238. Elsevier.

Singh, R.K., Singh, P., Li, H.-B. et al. (2020). Diversity of nitrogen-fixing rhizobacteria associated with sugarcane: A comprehensive study of plant-microbe interactions for growth enhancement in *Saccharum* spp. *BMC Plant Biology* 20 (1): 220.

Song, O.-R., Lee, S.-J., Lee, Y.-S. et al. (2008). Solubilization of insoluble inorganic phosphate by *Burkholderia cepacia* DA23 isolated from cultivated soil. *Brazilian Journal of Microbiology* 39: 151–156.

Swarnalakshmi, K., Prasanna, R., Kumar, A. et al. (2013). Evaluating the influence of novel cyanobacterial biofilmed biofertilizers on soil fertility and plant nutrition in wheat. *European Journal of Soil Biology* 55: 107–116.

Tajini, F., Trabelsi, M., and Drevon, J.-J. (2012). Combined inoculation with *Glomus intraradices* and *Rhizobium tropici* CIAT899 increases phosphorus use efficiency for symbiotic nitrogen fixation in common bean (*Phaseolus vulgaris* L.). *Saudi Journal of Biological Sciences* 19 (2): 157–163.

Tamburini, F., Pfahler, V., Bünemann, E.K. et al. (2012). Oxygen isotopes unravel the role of microorganisms in phosphate cycling in soils. *Environmental Science & Technology* 46 (11): 5956–5962.

Tariq, A., Zeng, F., Graciano, C. et al. (2023). Regulation of metabolites by nutrients in plants. *Plant Ionomics* 1–18.

Tiwari, K., Kumar, Y., Singh, T., and Nayak, R. (2022). Nano technology based P fertilizers for higher efficiency and agriculture sustainability. *Annals of Plant and Soil Research* 24 (2): 198–207.

Vassilev, N., Vassileva, M., Bravo, V. et al. (2007). Simultaneous phytase production and rock phosphate solubilization by *Aspergillus niger* grown on dry olive wastes. *Industrial Crops and Products* 26 (3): 332–336.

Vázquez-Núñez, E., López-Moreno, M.L., de la Rosa Álvarez, G., and Fernández-Luqueño, F. (2018). Incorporation of nanoparticles into plant nutrients: the real benefits. In: *Agricultural Nanobiotechnology: Modern Agriculture for a Sustainable Future* (ed. F. López-Valdez and F. Fernández-Luqueño), 49–76. Cham: Springer International Publishing.

Verma, P., Yadav, A.N., Khannam, K.S. et al. (2016). Molecular diversity and multifarious plant growth promoting attributes of *Bacilli* associated with wheat (*Triticum aestivum* L.) rhizosphere from six diverse agro-ecological zones of India. *Journal of Basic Microbiology* 56 (1): 44–58.

Waghmare, M.S., Ugile, S.K., Chavan, P.G., and Waghmode, B.G. (2024). Effect of phosphate solubilizing microorganisms and phosphorus levels on growth yield and quality of chickpea (*Cicer arietinum* L.) in inceptisol. *International Journal of Plant & Soil Science* 36 (5): 432–440.

Wang, H.-y., Liu, S., Zhai, L.-m. et al. (2015). Preparation and utilization of phosphate biofertilizers using agricultural waste. *Journal of Integrative Agriculture* 14 (1): 158–167.

Yadav, S., Yadav, S., Kaushik, R. et al. (2014). Genetic and functional diversity of fluorescent *Pseudomonas* from rhizospheric soils of wheat crop. *Journal of Basic Microbiology* 54 (5): 425–437.

Yadav, A.N., Rastegari, A.A., Yadav, N., and Kour, D. (2020). *Advances in Plant Microbiome and Sustainable Agriculture: Diversity and Biotechnological Applications*. Singapore: Springer.

Yadav, R.S., Kumar, M., Santra, P. et al. (2023). Plant growth-promoting microbes: the potential phosphorus solubilizers in soils of arid agro-ecosystem. In: *Plant Growth Promoting Microorganisms of Arid Region* (ed. R. Mawar, R.Z. Sayyed, S.K. Sharma, and K.S. Sattiraju), 71–92. Singapore: Springer Nature Singapore.

Yandigeri, M.S., Yadav, A.K., Srinivasan, R. et al. (2011). Studies on mineral phosphate solubilization by cyanobacteria *Westiellopsis* and *Anabaena*. *Microbiology* 80 (4): 558–565.

Yi, Y., Huang, W., and Ge, Y. (2008). Exopolysaccharide: a novel important factor in the microbial dissolution of tricalcium phosphate. *World Journal of Microbiology and Biotechnology* 24 (7): 1059–1065.

Yu, X., Liu, X., Zhu, T.H. et al. (2011). Isolation and characterization of phosphate-solubilizing bacteria from walnut and their effect on growth and phosphorus mobilization. *Biology and Fertility of Soils* 47 (4): 437–446.

Zaheer, A., Mirza, B.S., McLean, J.E. et al. (2016). Association of plant growth-promoting *Serratia* spp. with the root nodules of chickpea. *Research in Microbiology* 167 (6): 510–520.

Zaidi, A. and Khan, M.S. (2007). Stimulatory effects of dual inoculation with phosphate solubilising microorganisms and arbuscular mycorrhizal fungus on chickpea. *Australian Journal of Experimental Agriculture* 47 (8): 1016–1022.

Zhang, Y., Chen, F.-S., Wu, X.-Q. et al. (2018a). Isolation and characterization of two phosphate-solubilizing fungi from rhizosphere soil of moso bamboo and their functional capacities when exposed to different phosphorus sources and pH environments. *PloS One* 13 (7): e0199625.

Zhang, L., Feng, G., and Declerck, S. (2018b). Signal beyond nutrient, fructose, exuded by an arbuscular mycorrhizal fungus triggers phytate mineralization by a phosphate solubilizing bacterium. *The ISME Journal* 12 (10): 2339–2351.

Zhu, H.-J., Sun, L.-F., Zhang, Y.-F. et al. (2012). Conversion of spent mushroom substrate to biofertilizer using a stress-tolerant phosphate-solubilizing *Pichia farinose* FL7. *Bioresource Technology* 111: 410–416.

4

Role of Fungi in the Production of Bioactive Compounds and Nanotechnology

Laith Khalil Tawfeeq Al-Ani[1], Ivana Mitrović[2], Alejandra G. Becerra[3], Maura Téllez-Téllez[4], and Lobna Hajji-Hedfi[5]

[1] *School of Biology Science, Universiti Sains Malaysia, Minden, Pulau Pinang, Malaysia*
[2] *Faculty of Technology Novi Sad, University of Novi Sad, Novi Sad, Serbia*
[3] *Instituto Multidisciplinario de Biología Vegetal-CONICET, FCEFyN-UNC., Argentina*
[4] *Centro de Investigaciones Biológicas, Universidad Autónoma del Estado de Morelos, Cuernavaca, Morelos, Mexico*
[5] *Research Laboratory of Agricultural Production Systems and Sustainable Development (LR03AGR02), Regional Centre of Agricultural Research of Sidi Bouzid, CRRA, Carthage University, Sidi Bouzid, Tunisia*

4.1 Introduction

Fungi produce a lot of bioactive compounds from beneficial to harmful. Some of them are harmful fungi to animals, plants, and human (De Lucca 2007). Fungal compounds caused damage to all parts of plants and caused different diseases to plants. *Fusarium oxysporum* f. sp. *cubensis* tropical race 4 (FocTR4) caused high damage to the vascular of the banana tree and led to wilt symptoms in the banana (Al-Ani 2017). Plant fungal pathogens affected the increase in permeability of plant cells by producing phytotoxins as secondary compounds (Möbius and Hertweck 2009; Mohammed and Al-Ani 2021). Some fungi contaminated feed, foodstuffs, and grains, producing high levels of mycotoxins, such as *Fusarium, Alternaria, Aspergillus,* and *Penicillium* (De Lucca 2007; Attitalla et al. 2010a, 2010b; Greeff-Laubscher et al. 2019).

On the other hand, some compounds produced by fungi are beneficial secondary compounds. Beneficial fungi are available around and inside of plants (Al-Ani 2019a, 2019b; Hajji-Hedfi et al. 2025a). These include different groups of fungi such as *Trichoderma* (Al-Ani 2018a, 2019c; Al-Ani and Mohammed 2020; Al-Ani et al. 2025; Hajji-Hedfi et al. 2026 a,b), entomopathogenic fungi *Metarhizium* and *Beauveria* (Al-Ani et al. 2018; Al-Ani 2019d; Nyahe et al., 2025; Ameyaw et al. 2026), nematophagous fungi (Al-Ani et al. 2022; Villar-Luna et al. 2022; de Freitas Soares et al. 2023), dark septate endophytic fungi (Al-Ani et al. 2021; Dalimunthe et al. 2023), endophyte microbes (Aguilar-Marcelino et al. 2020 a,b; Al-Ani and Furtado 2020; Hajji-Hedfi et al. 2024; Jiang et al. 2026) and mycorrhizae (Mitra et al. 2023; Yelikbayev et al. 2024). This indicates the importance of fungi for utilization in the line of nanotechnology. Many bioactive compounds of fungi are utilized in the control of viruses, bacteria, fungi, insects, mites, and weeds (Al-Ani 2019e, 2019f; Al-Ani et al. 2020a), and enhancing the plant growth (Al-Ani et al. 2020b; Soumare et l., 2021; Hajji-Hedfi et al. 2025b). This study aimed to potentially utilize the synthesized NPs (Nanoparticles) with bioactive fungal compounds in the confrontation of plant diseases, pests, and weeds.

4.2 Fungal Diversity

Nanotechnology is defined as the engineering discipline that focuses on the design, synthesis, and application of technologically significant particles at the nanoscale (Sousa et al. 2020). These particles range in size from 0.1 to 100 nm and are termed "nanoparticles" (NPs), which can be synthesized in a variety of shapes, exhibiting different organizational levels (Castro-Longoria 2016). In recent years, there has been a significant increase in the focus on microorganisms, including bacteria, fungi, algae, yeast, and actinomycetes, as a result of their status as an environmentally friendly, cost-effective, and straightforward synthesis process for NPs (Rai et al., 2021a, 2021b). Among these microorganisms, the potential and applications of fungi in various scientific and technological fields have been particularly promising. Fungi are eukaryotic organisms found in a wide variety of natural environments (Mehra and Winged 1991). Based on sequencing methods, up to 5.1 million fungal species have been identified (Blackwell 2011). Furthermore, fungi have the ability to be cultivated on a large scale, allowing the production of NPs of defined length and morphology (Gade et al. 2008; Khan et al. 2017).

In addition, the synthesis of NPs by fungi is recognized as an environmentally friendly and sustainable approach (Gade and Rai 2021). The process of synthesizing NPs using fungi is referred to as *myconanotechnology* (Rai et al. 2009; Sousa et al. 2020; Jagtap et al. 2021). This process can occur both intracellularly and extracellularly, facilitated by enzymes and other biologically active molecules. Extracellular synthesis of NPs is a process that involves the capture of metal ions on the surface of cells, followed by their reduction in the presence of enzymes (Zhang et al. 2011). The extracellular synthesis of NPs is influenced by several parameters, including pH, temperature, and the composition of the growth medium. These parameters have been shown to influence the size and morphology of the resulting NPs (Alghuthaymi et al. 2015). In contrast, intracellular synthesis occurs within the fungal cell in the presence of enzymes (Siddiqi and Husen 2016). It has been observed that extracellular NP synthesis is faster than intracellular NP synthesis (Singh et al. 2018). In addition, NPs synthesized by extracellular methods are typically larger than those synthesized intracellularly (Zhang et al. 2020). Furthermore, extracellular secretion of large amounts of proteins by fungi has been shown to enhance NP synthesis (Diko et al. 2020). In comparison with bacterial systems, the utilization of fungi confers a number of distinct advantages, including the capacity to precipitate NPs outside the cell, which results in the elimination of cellular contaminants, and the direct employment in a range of applications.

It has been demonstrated that fungal diversity contributes to the synthesis of NPs, as illustrated by examples of fungal species and their synthesized NPs shown in Table 4.1. A range of metals, including silver (Ag), gold (Au), titanium (Ti), and zinc (Zn), have been used to produce NPs by fungi (Khandel et al., 2018). The ability to produce silver nanoparticles (AgNPs) has been detected in more than 120 species of fungi from different taxa produced extracellularly or intracellularly (Ahmad et al. 2003; Bhainsa and D'Souza 2006). The main phyla of fungi with the ability to biosynthesize Au and AgNPs are Ascomycota and Basidiomycota. Examples within the Ascomycota include *Fusarium*, *Aspergillus*, *Penicillium*, and *Trichoderma* (Bhainsa and D'Souza 2006; Honary et al. 2012; Shelar and

Table 4.1 Nanoparticles synthesized by fungi with antimicrobial activity.

Fungi	Nanomaterial/ size	Inhibition	Reference
Aspergillus clavitus	Silver (550–650 nm)	*Staphylococcus aureus, Staphylococcus epidermidis*	Saravanan and Nanda (2010)
Aspergillus flavus	TiO_2 (62–74 nm)	*Escherichia coli, Staphylococcus aureus, Klebsiella pneumoniae, Bacillus subtilis*	Rajakumar et al. (2012)
	Silver (50 nm)	*Bacillus cereus, Bacillus subtilis, Enterobacter aerogenes, Escherichia coli, Staphylococcus aureus*	Fatima et al. (2016)
Aspergillus niger	Silver (3–30 nm)	*Staphylococcus* sp., *Bacillus* sp., *Escherichia coli*	Jaidev et al. (2010)
	Copper (500 nm)	*Escherichia coli, Staphylococcus aureus, Klebsiella pneumoniae, Micrococcus luteus, Bacillus subtilis*	Noor et al. (2020)
Aspergillus terreus	Silver (16.54 nm)	*Salmonella typhi, Staphylococcus aureus, Escherichia coli*	Rani et al. (2017)
Amylomyces rouxii	Silver (5–27 nm)	*Shigella dysenteriae* type I, *Staphylococcus aureus, Citrobacter* sp., *Escherichia coli, Pseudomonas aeruginosa, Bacillus subtilis*	Musarrat et al. (2010)
Alternaria alternata	Silver (treated cotton)	*Escherichia coli, Staphylococcus aureus*	Ibrahim and Hassan (2016)
Fusarium oxysporum	Silver (1–50 nm)	*Escherichia coli, Pseudomonas aeruginosa*	Srivastava et al. (2019)
Penicillium italicum	Silver (33 nm)	*Staphylococcus aureus, Shewanella putrefaciens*	Nayak et al. (2018)
Penicillium corylophilum	Selenium (29.1–48.9 nm)	Gram +ve and Gram −ve bacteria	Salem et al. (2021)
Penicillium oxalicum	Silver	*Staphylococcus aureus, S. dysenteriae, Salmonella typhi*	Feroze et al. (2020)
Trichoderma asperellum	Zinc oxide	*Escherichia coli, Staphylococcus aureus*	Shobha et al. (2023)
Trichoderma harzianum	Silver (51.10 nm)	*Staphylococcus aureus, Klebsiella pneumoniae*	Ahluwalia et al. (2014)
Trichoderma viride	Silver (5–40 nm)	*Salmonella typhi, Escherichia coli, Staphylococcus aureus, Micrococcus luteus*	Fayaz et al. (2010)
Agaricus bisporus	Silver	*Salmonella typhi, Proteus* sp., *Enterobacter* sp., *Klebsiella* sp.	Dhanasekaran et al. (2013)
	Silver (67.45–102.7 nm)	*Bacillus cereus, Escherichia coli, Enterococcus faecalis, Klebsiella quasipneumoniae, Pseudomonas aeruginosa, Salmonella typhi, Shigella sonnei, Staphylococcus aureus*	Amr et al. (2023)

(*Continued*)

Table 4.1 (Continued)

Fungi	Nanomaterial/ size	Inhibition	Reference
Fomes fomentarius	Silver (24 nm)	*Escherichia coli, Pseudomonas aeruginosa, Bacillus subtilis, Staphylococcus aureus*	Pavić et al. (2024)
Ganoderma lucidum	Silver (2 nm)	*Staphylococcus aureus, Escherichia coli*	Karwa et al. (2011)
	Silver (40 nm)	*Staphylococcus aureus, Enterococcus hirae, Bacillus cereus, Escherichia coli, Pseudomonas aeruginosa, Legionella pneumophila* subsp. *Pneumophila*	Aygün et al. (2020)
	Silver	*Staphylococcus aureus, Escherichia coli, Pseudomonas aeruginosa*	Constantin et al. (2023)
	Silver	*Staphylococcus aureus, Streptococcus mutans*	Parameswari et al. (2024)
Ganoderma sessiliforme	Silver	*Escherichia coli, Bacillus subtilis, Streptococcus faecalis, Listeria innocua, Micrococcus luteus*	Mohanta et al. (2018)
Schizophyllum commune	Silver	*Escherichia coli, Bacillus subtilis, Klebsiella pneumoniae, Pseudomonas fluorescens*	Arun et al. (2014)
Phellinus adamantinus	Silver (50 nm)	*Bacillus subtilis, Staphylococcus aureus, Pseudomonas aeruginosa, Escherichia coli*	Vijayakumar et al. (2024)
Pleurotus giganteus	Silver (2–20 nm)	*Escherichia coli, Pseudomonas aeruginosa, Bacillus subtilis, Staphylococcus aureus*	Debnath et. al. (2019)
Pleurotus ostreatus	Silver (40 nm)	*Bacillus subtilis, Bacillus cereus, Staphylococcus aureus, Escherichia coli, Pseudomonas aeruginosa*	Al-Bahrani et al. (2017)
	Gold (10.3–38.7 nm)	*Enterococcus faecalis, E. coli, Klebsiella pneumoniae, P. aeruginosa, Staphylococcus aureus*	Bawadekji et al. (2018)

Chavan 2015; Al-Ani 2018b; Rai et al. 2021b). With respect to Basidiomycota, edible and medicinal mushrooms were studied for mycosynthesis, many of which were cultivated in an artificial culture. The primary research utilized the synthesis of NPs (producing nanofibers, also termed Schizophyllan), derived from the mushroom *Schizophyllum commune* (Numata et al. 2004). The initial study employed proteins isolated from the residual mushroom substrate of *Pleurotus ostreatus* (Vigneshwaran et al. 2007) employed as nano-drugs and medically tested against microbes in vitro.

The potential applications of fungal-produced NP are diverse, with a range of biological functions. The broad-spectrum antifungal properties of these NPs (Li et al. 2012), their capacity to act as antimicrobials against bacterial (Devi and Joshi 2012) and fungal strains

(Gajbhiye et al. 2009), and the possibility of their involvement in future pharmaceutical development (Ingle et al. 2008) are particularly significant. This demonstrates that fungal-mediated AgNPs could be considered as a potent antimicrobial agent, implicating their possible biomedical application. Fungi are the most suitable biological systems for synthesizing metal NPs. They are easy to cultivate, reliable, eco-friendly, and the extracellular synthesis methodology is extremely cost-effective and less time-consuming. NPs have a number of applications in medicine, agriculture (pesticides and herbicides), and textiles (Gade et al. 2010). Concurrently, the implementation of myconanotechnology in agriculture could prove both cost-effective and beneficial to the enhancement of crop yield (Table 4.1).

4.3 Types of Fungal Secondary Metabolites

Secondary metabolites, a term introduced by Albrecht Kossel (Hartmann 2007), are defined as organic compounds produced by bacteria, fungi, and plants. Plant and fungal sources are abundant in thousands of secondary metabolites (SMs). These compounds typically play a significant role in the inhibition of fungal growth, as well as in signaling molecules in systemic resistance and acquired systemic resistance processes (Pusztaheli et al. 2015; Al-Ani et al. 2024). A categorization of fungal SMs is performed into four primary chemical classes: terpenoids, polyketides, shikimic acid-derived compounds, and nonribosomal peptides. According to the most recent research, published up to 2023, 553 new compounds derived from fungal secondary polyketides play a dominant role, followed by terpenoids (Shi et al. 2024). Bioactivities of SMs include cytotoxic, anti-inflammatory, antibacterial, antifungal, antiviral, antiparasitic, and antioxidant effects. Among these bioactivities, antibacterial activities are particularly notable, followed by cytotoxic activities. Furthermore, terpenoids exhibit cytotoxic activities, while the number of polyketides with antibacterial activities is the highest (Shi et al., 2024).

In contrast to primary metabolites, SMs are individually produced compounds, often specific to one species or a limited group of species (Narayanan and Glick 2022). The functions of SMs are diverse and are typically associated with an enhancement of the fitness of the producing fungi in their environment, which often compete with other microorganisms or interact with host plants. A substantial number of SMs derived from fungi have been shown to possess advantageous biological properties or to have biotechnological applications (Demain 2014), while others have been demonstrated to have detrimental or disadvantageous effects (Marin et al. 2013; Wu et al. 2014). A significant proportion of SMs have beneficial applications, including as antibiotics or medical drugs; however, others, classified as mycotoxins, are detrimental to health (Avalos and Limón 2022). Exploration of bioactive compounds from fungi for use in biomedicine and agrochemistry has also been undertaken, with the SMs of fungi serving as a potential source of novel bioactive metabolites.

4.3.1 Antifungal Activity

In view of the continuing increase in drug-resistant microorganisms, it is important to explore the development of new molecules with antibacterial and/or antifungal properties (Hajji-Hedfi et al. 2025c). The term "antifungal activity" or "antifungal agent" is a drug that selectively eliminates fungal pathogens from a host with minimal toxicity to the host (Dixon and Walsh 1996; Devi et al. 2020; Sukmawati et al. 2024). The development of antifungal agents has been comparatively slow in relation to antibacterial agents due to the fact that fungi generally grow more slowly and are more difficult to quantify than bacteria (Aziz and Abdel-Azeem 2024). Despite these limitations, numerous advances have been made in developing new antifungal agents.

Antifungal agents can be derived from various sources, including soil fungi, marine fungi, and endophytic fungi (Al-Ani et al. 2027). The latter can benefit from the nutrients and protection of the plant and, in response, produce antifungal compounds that protect the plant from pathogens (Aziz and Abdel-Azeem 2024). Endophytic fungi have been studied as a sustainable means of converting the natural compounds in host plants to antifungal compounds, which are effective against both human and plant pathogens (Deshmukh et al. 2018). Fungi make certain antifungals as SMs, to help them survive. SMs produced by endophytic fungi, especially those isolated from native plants, exhibit a wide range of biological activities, including antimicrobial, antiviral, anticancer, and, in particular, antifungal effects (Strobel and Daisy 2003). These SMs from endophytic fungi inhibit cell wall synthesis, damage cell membranes, and suppress chemical reactions within fungal cells.

To treat systemic animal infections, the treatments are limited to four classes of antifungal agents (Fisher et al. 2018): *polyenes* e.g., Amphotericin B, nystatin (Hamilton-Miller 1973), which bind to ergosterol, a vital component of the fungal cell membrane, creating pores causing cell death. The earliest inhibitory agent specific to fungal species was griseofulvin, which was isolated from *Penicillium griseofulvum* in 1939. It has been shown to impede the growth of fungal cells by binding with tubulin and preventing microtubule function (Gupta et al. 2004) and to cause fungal hyphae to curl (Petersen et al. 2014). This drug is used to treat fungal infections of the hair, skin, and nails. The most widely used class of fungicides, the *azoles*, which include ketoconazole, fluconazole, and voriconazole (Singh et al. 2023). These agents inhibit ergosterol synthesis, thereby compromising the fungal cell membrane and leading to cell death. *Allylamines* (e.g., terbinafine and naftifine) inhibit ergosterol synthesis, reducing ergosterol levels in the fungal cell membrane, thus leading to cell death. *Echinocandins* (e.g., caspofungin, micafungin, and anidulafungin) inhibit the enzyme responsible for synthesizing beta-glucan, a crucial component for the formation and maintenance of the fungal cell wall. These agents have a narrower spectrum, primarily effective against *Candida* and *Aspergillus*, fungal species known to cause invasive infections.

To control crop diseases, approximately there are six main classes of fungicides: the *morpholines*, which inhibit the ergosterol biosynthetic pathway; the *azoles* (also used in animal infections), which target the ergosterol biosynthetic pathway; the *benzimidazoles*, which interfere with the cytoskeleton, preventing the assembly of microtubules; the *strobilurins* and *succinate dehydrogenase inhibitors*, both inhibit the electron transfer chain of mitochondrial respiration; and the *anilinopyrimidines*, which may target mitochondrial signaling pathways (Fisher et al. 2018). The emergence of fungicide-resistant pathogenic fungi has had a significant impact on the field of antifungal agents, creating a clear need for the development of new antifungal agents. In this context, nanotechnology is a promising strategy for the identification and use of new antifungal compounds (Brauer et al. 2019).

4.3.2 Nematicidal Activity

Plant parasitic nematodes (PPN) and animal parasitic nematodes (APN) are highly effective on animals and plant growth and cause economic losses in worldwide. The new technology is being utilized in identifying species and management (Al-Ani et al. 2026a). This is necessary for development in the field of plant nematology (Al-Ani et al. 2026b). The management of PPN by using synthetic chemical nematicides can have a high effect on ecosystem (Aguilar-Marcelino et al. 2023). Many alternative methods are available with the capability to control PPN, such as biological control and bioactive compounds, including volatile and nonvolatile compounds (Al-Ani et al. 2022; Mitra et al. 2023). Edible mushrooms are considered a repository of bioactive compounds as anthelmintic activity (Castañeda-Ramírez et al. 2020; Comans-Pérez et al. 2021; Aguilar-Marcelino et al. 2022). *Phyllosticta* sp. as an endophytic fungi, is a biocontrol agent against nematode by secreting some bioactive compounds as nematicidal, including metguignardic acid (Compound 4), guignardianones d (Compound 7), and phenguignardic acid methyl ester (Compound 8) (Díaz et al. 2024). Barbosa et al. (2019) mentioned anthelmintic activity for fungal extract treated with synthesized AgNPs. This bioactive compounds were extracted from nematophagous fungi *Duddingtonia flagrans*.

4.3.3 Insecticidal Activity

The usage of natural compounds in control of pests is required more to keep the environment safe (Adetunji et al. 2019). Many biological control factors are useful in damage and reduction of insect community (Sharma et al. 2020). Entomopathogenic fungi start with attaching the spores to the arthropod exoskeleton, then, activate SMss, and lytic enzymes to complete the infection (Skinner et al. 2014). Two genera, such as *Metarhizium* and *Beauveria*, are more entomopathogenic fungi for insects (Rhouma et al. 2024), as well as, some other fungi, like *Fusarium* (Al-Ani et al. 2018, 2021; Aguilar-Marcelino et al. 2021; Bhattacharyya et al. 2024). *Beauveria* produces many compounds affecting insects, such as cyclopeptides, pigments, alkaloids, and other volatile compounds (Patocka 2016). The metabolites of *Nomuraea rileyi* uniform with chitosan nanocomposite, showed the lethal effect on larvae of *Spodoptera litura* (Namasivayam et al. 2018).

In general, several metal NPs are used in creating nanoinsecticides, such as silver, titanium dioxide, selenium, iron, and zinc. Two metal NPs, such as silver and zinc, with *Beauveria bassiana* showed a highly significant effect in controlling *Aedes larvae* and *Trialeurodes vaporariorum* (Banu and Balasubramanian 2014; Khooshe-Bast et al. 2016). Wang et al. (2019) mentioned the potential use of nanoinsecticide against *Bemisia tabaci* that was synthesized from *Isaria fumosorosea* with iron NPs. *Trichoderma viride* is used to synthesize titanium dioxide NPs with high efficacy against *Helicoverpa armigera* (Chinnaperumal et al. 2018). Arunthirumeni et al. (2022) found *Trichoderma* with the synthesized selenium NPs useful in producing mycoinsecticide to control *S. litura*. The synthesis of AgNPs of *Fusarium pallidoroseum* could show high efficiency against *Holotrichia* sp. (Shukla et al. 2022).

4.3.4 Antiviral Activity

Fungi play a significant role in the production of bioactive compounds and have emerged as key players in the field of nanotechnology. Their unique metabolic pathways and diverse biochemical capabilities allow them to synthesize a wide range of bioactive compounds, including antibiotics, enzymes, and antioxidants, which have important applications in medicine, agriculture, and food industries. Viruses generate significant outbreaks on all continents, resulting in difficult symptoms, mortality, and a massive economic cost to society. There are two types of viruses: those with a lipid envelope and those with a protein shell covering their genome (Linnakoski et al. 2018). Viruses are an enhanced type of parasitism that reproduces by using the host cell's cellular apparatus. Its simple basic structure consists of a protein coat enclosing the nucleic acid, while certain viruses also include an envelope or lipoprotein membrane (dos Santos Lacerda et al. 2021). In order to combat viral infections, vaccinations as well as antivirals are therefore necessary (Linnakoski et al. 2018). Nevertheless, it is difficult to create vaccinations against a broad variety of recently discovered virus serotypes. Furthermore, if the infection is already established in the system, immunization will not assist (dos Santos Lacerda et al. 2021).

Bioactive chemicals obtained from natural sources offer an unknown and wide diversity of chemical structures. Natural products or their derivatives account for approximately 40% of contemporary pharmaceuticals and 49% of novel chemical products registered by the US Food and Drug Administration (FDA) (Linnakoski et al. 2018). Endophytic fungi are an abundant source of valuable compounds and biochemicals that have a wide range of biological applications. These organisms create a variety of metabolites, including alkaloids, steroids, phenols, xanthones, isocoumarins, terpenoids, quinones, peptides, cytochalasins, glycosides, lipids, and more (dos Santos Lacerda et al. 2021).

Fungi produce around 20,000 SMs, with bioactive compounds accounting for approximately 80%. Xanthoviridatins E and F were produced using the fungus *Penicillium chrysogenum*. The identification of active compounds from the cultivation broth produced by the cultivation of this fungus revealed the presence of xanthoviridatins E and F, which showed antiviral activity against the HIV virus (Singh et al. 2003). Also, the examination of the ethyl acetate extract of the fermentation liquid obtained after the cultivation of *Alternaria tenuissima* fungi revealed the presence of two compounds whose antiviral activity was tested against the HIV virus. The compound altertoxin V inhibited the virus at a concentration of 0.5 μM, while altertoxin VI did not show a significant effect (Bashyal et al. 2014). Brefeldin A is another important component that is a product of fungal metabolism. In addition to antiviral activity, this compound also exhibits antitumor and antibacterial activity. The most important fungi that produce it are *Penicillium cyaneum*, *P. brefeldianum*, and *Cladosporium* sp., and its effectiveness against Herpes simplex-1 and HIV viruses has been confirmed (dos Santos Lacerda et al. 2021). 28 polyketides were isolated by Luo et al. (2018) from the fungus *Diaporthe* sp. SCSIO 41011, and examined the effect of each of them on the influenza A virus. It was found that three components: pestalothiopsone B, F, and 5-chloroisorothiori show antiviral activity against three subtypes of this virus. On the other hand, a compound with a steroid structure, Dankasterone A, with pronounced antiviral activity against pseudovirus influenza A (H5N1), was produced using the fungus *Phomopsis* sp. YM355364 (Ma et al. 2014).

Li et al. (2008) conducted research using *Pestalotiopsis theae*, an endophytic fungus. They found that this fungus has the ability to produce four metabolites with a similar structure, Pestalotheols A–D. One of these compounds, Pestalotheol C, has already been found as a metabolic product of the fungus *Pestalotiopsis fici*. This compound inhibits the replication of the HIV-1LAI virus and shows anti-HIV activity at a concentration of 16.1 μM/mL. Naphthoquinones have been studied by many scientists because of their antibiotic, anti-inflammatory, analgesic, and antiviral effects. It was established that *Verticillium* sp. produces α-lapachone, while β-lapachone inhibits the LTR117 enzyme, which also inhibits HIV-1 virus replication. This compound also inhibits DNA polymerase and the reverse transcriptase enzyme, thus simultaneously acting against Rauscher murine leukemia and avian myeloblastosis virus. In general, naphthoquinones in concentrations of 5–10 μg/mL have been found to have an inhibitory effect on RNA viruses with and without envelopes (dos Santos Lacerda et al. 2021).

Pullularin A, B, and C are compounds produced by fungi of the genus *Pullularia*, and their antiviral activity has been confirmed against herpes simplex virus type 1 (Isaka et al. 2007). The fungus *Thielavia subthermophila* has the ability to produce emodin, an anthraquinone derivative with significant pharmacological activities, including antiviral activity. This compound affects the replication of the hepatitis B virus, and some studies show the great potential of using this compound against SARS-CoV (Ho et al., 2007). Zhang et al. (2011) investigated the metabolites of the fungus *Emericella* sp. and their action. They found that two metabolites of emerimidine A and B showed activity against the H1N1 virus, influenza virus, at concentrations of 42.07 and 62.05 μg/mL, respectively. These compounds are isoindolone derivatives.

Fungi from the genera *Aspergillus, Syncephalastrum,* and *Phomopsis* have the ability to produce another antiviral compound, betulinic acid. This pentacyclic triterpenoid may play a role in inhibiting the HIV virus (Peyrat et al. 2017). Yang et al. (2020) from the fungus *Phomopsis* sp. strain CGMCC 5416 isolated the compounds phomopsone B and C, which showed pronounced inhibitory activity against HIV virus. Another compound with antiviral activity, quinone, hinnuliquinone, was isolated from the fungus *Nodulisporium hinnuleum*. This compound inhibits the HIV protease of a clinically resistant strain and shows anti-HIV activity (Singh et al. 2004). Various sesquiterpenoids can be isolated from marine organisms, animals, plants, and microorganisms and exhibit exceptional bioactivity. 14-nordrimane is a rare sesquiterpenoid obtained by using the fungus *Phoma* sp., and its antiviral effect on the influenza A (H1N1) virus was also confirmed in the paper Liu et al. (2019).

Also, antiviral NPs produced by fungi represent a promising area of research in nanotechnology and medicine. Fungi can synthesize a variety of NPs, such as silver, gold, and zinc oxide, which have demonstrated antimicrobial properties. These NPs can interact with viral particles, potentially inhibiting their ability to infect host cells. The production of these NPs through fungal biosynthesis is advantageous due to its eco-friendly nature, cost-effectiveness, and the ability to produce NPs at room temperature, which preserves their functional properties. Additionally, the unique surface characteristics of fungal NPs can enhance their interaction with viruses, making them effective in targeting and neutralizing viral infections (Hussain et al. 2022).

The mechanism of metal NP or metal ion action involves their adhesion to the viral surface, thereby preventing the virus from binding to host cells; the generation of highly reactive oxygen species that interact with the viral spike or membrane, disrupting nucleic acid and protein functions; the stimulation of the host's immune system; and the destruction of host cells, which impedes viral propagation (Hussain et al. 2022). Over the past 10 years, nanotechnology has demonstrated potential in combating viruses, particularly AgNPs, which have garnered significant interest from the scientific community due to their broad-spectrum antimicrobial properties and prospective applications in other medicinal domains. AgNPs typically range in size from 1 to 100 nm. Recent studies have documented the antiviral efficacy of AgNPs against many virus families, highlighting their notable preventative effects, primarily by direct virucidal activity or modification of viral binding to host cells (Luceri et al. 2023).

The effect of AgNPs on the HIV virus and their anti-HIV action was explained by Elechiguerra et al. (2005) in their paper. On the other hand, human parainfluenza virus type 3 (HPIV-3) and herpes simplex virus types 1 (HSV-1) and 2 (HSV-2) are significant viruses because they can infect humans with devastating diseases. Gaikwad et al. (2013) isolated several fungi: *F. oxysporum, Chrysanthemum indicum, Curvularia* species, *Phoma* species and *Alternaria* species, from leaves of plants *Musa paradisiaca, Dryopteris* species, *Selaginella bryopteris* and *Catharanthus roseus*, and examined their ability to synthesize NPs. The results of these investigations showed that these types of fungi have the ability to synthesize AgNPs with different properties. It was determined that AgNPs synthesized using *Curvularia* species and *F. oxysporum* have sizes of 5–23 and 4–13 nm, respectively, and that these particles exhibit pronounced antiviral activity with an inhibition percentage between 80% and 90% against HPIV-3 and HSV-1. On the other hand, *Phoma* and *Alternaria* species produce AgNPs of smaller dimensions, 7–20 nm, and these particles showed less antiviral activity, but had the ability to inhibit the replication of the analyzed viruses. Certainly, it was established that with the increase in the concentration of applied NPs, the antiviral effect on the tested viruses is also enhanced (Gaikwad et al. 2013; Luceri et al. 2023).

It is believed that fungi produce proteins in large quantities that are important for the production of NPs because they participate in the hydrolysis of metal ions. Fungi are widely distributed in nature, which is their great advantage. The synthesis of NPs using fungi is a new nanotechnology; however, it is considered that all microorganisms used in nanobiotechnology should generally have GRAS (generally recognized as safe) status. Also, fungi can be cultivated on various types of media, which is an important segment in their use in the synthesis of NPs (Adebayo et al. 2021). The role of fungi in the production of bioactive compounds and their applications in nanotechnology is a rapidly evolving field with immense potential. As research continues to uncover the capabilities of fungi, we can expect innovative solutions that leverage their unique properties for health, industry, and environmental sustainability.

4.3.5 Antibacterial Activity

The indiscriminate use of antibiotics promotes resistance in the microbial population, becoming a global threat since bacterial infections are a serious problem in numerous areas of daily life, causing death, pain, and enormous added costs to healthcare worldwide

(Díez-Pascual et al. 2020). Nowadays, multi-resistant microorganisms have become a major problem; for example, *Staphylococcus aureus* is resistant to methicillin, and *Candida albicans* is resistant to fluconazole (Schaller et al. 2004). Bacteria have developed strategies for resistance, among which the expression of proteins that can favor structural modification or genes that modify the drug-binding surface and bacterial biofilms, which are a defense mechanism, stand out (Menichetti et al. 2024). Alternatives are being sought to counteract or control bacterial adhesion and proliferation. It has been suggested that approximately 70% of human microbial infections are due to biofilm-forming microorganisms and lead to various diseases, including chronic nonhealing wounds, endocarditis, periodontitis, fibrosis, osteomyelitis, renal infections, and prosthetic and device implant-related infections (Paharik and Horswill 2016).

The difficulty in inhibiting bacteria that inhabit biofilms is because they increase resistance to antibiotics and/or drugs 10–1,000 times compared to their planktonic state (Mah 2012). Nanotechnology provides an alternative to inhibit resistant bacteria (Rabiee et al. 2022), since nanomaterials can be designed according to requirements, such as size, surface modification, crystalloid change, and the response to the interaction of different bacterial cells (Makabenta et al. 2021). Most nanomaterials cannot enter the cell's cytoplasm but interact with the components of the cell wall or membrane through destabilization, oxidative damage, mechanical disruption, and thermal effects (Parra-Ortiz et al. 2022). Among the mechanisms of antibacterial activity of NPs, the following have been mentioned: (i) generation of reactive oxygen species, in which oxidation in bacterial cells induces peroxidation of the lipid membrane, thus destroying proteins and DNA, (ii) the release of metal ions from the NPs, which can enter the cell wall of bacteria and interact directly with the amino and carboxylic acid groups of proteins and nucleic acids, resulting in cell death, and (iii) membrane disruption due to accumulation of the NPs and internalization of the NPs (Díez-Pascual 2020).

Among the main nanomaterials that have been used for antimicrobials are: (i) metallic NPs (e.g., Ag, Cu, Au, ZnO, La_2O_3, CeO_2, V_2O_5, etc.), (ii) carbonaceous nanomaterials (e.g., graphene, carbon nanotubes), (iii) borides (e.g., BN), (iv) nanosized polymers (e.g., polycarbonate), and (v) nanocomposites (e.g., La_2O_3/Ag-GO) (Xie et al. 2023). Gold NPs are widely used in the medical area since they have high surface compatibility and can adsorb a wide variety of biomolecules, which is of great help for clinical medicine (Lee et al. 2020). On the other hand, Nayem et al. (2020) indicate that AgNPs have an inhibitory effect on Gram-negative and Gram-positive bacteria. They can also be used as photocatalysts in medical and electronic devices since the ionized form is released from the NPs and penetrates the cell membrane. The positive ions released by the NPs react with negatively charged molecules, which causes an increase in ion absorption, causing cellular damage. It has also been reported that silver intervenes in the process of cellular respiration (Kareem et al. 2020).

Because certain compounds used may pose health and environmental risks (Zhang et al. 2016), biogenic synthesis, which is the synthesis of NPs using organisms such as plants, bacteria, and fungi or the byproducts of their metabolism, which act as reducing and stabilizing agents, has been intensified. It is currently called myconanotechnology, which is a dynamic interface between mycology and nanotechnology, which is the creation and exploitation of NPs using fungi. Due to the biological characteristics of fungi, they are being used as biofactories to produce environmentally friendly NPs. Fungi are resistant to many adverse conditions, have a great capacity to produce extracellular reducing proteins that facilitate the subsequent process, have availability of biomass, are easy to handle, have a wide range of metabolites, and have the ability of fungi to produce antibiotics that could be contained in the layer and act in synergy with the core of the NP (Guilger-Casagrande and Lima 2019; Prabhu et al. 2024).

It is important to remember that the synthesis of NPs depends on the production conditions, so it is still necessary to characterize the different parameters depending on the organism used and the desired physicochemical characteristics so that the production of NPs on a large scale can be improved. Bhat et al. (2011) indicate that within the synthesis of NPs, light and water solubility are two important criteria; riboflavin from fungi acts as an enzyme in redox reactions, metal ions are reduced to NPs by the presence of flavoproteins (flavin) in the extracts, and photons of energy are absorbed in the presence of sunlight, which consequently excites the flavin to donate electrons or act as an important oxidizing agent in the production of NPs. Other works have indicated the importance of NADPH, anthraquinones, nitrate reductase, quinones, and extracellular NADPH-dependent nitrate reductase in the synthesis of NPs (Durán et al. 2005; Anil-Kumar et al. 2007; Hietzschold et al. 2019), so it is confirmed that fungi have great advantages to be an invaluable source of NPs.

However, Hulikere et al. (2019) reported the biosynthesis of AgNPs using an aqueous extract of *Cladosporium cladosporioides*, a marine endophytic fungus. It showed antimicrobial activity against *Staphylococcus aureus*, *S. epidermis*, *Bacillus subtilis*, and *Escherichia coli*, indicating that the activity was due to the surface-volume relationship and crystallographic structure. AgNPs have been shown to disrupt the cell membrane and release reactive oxygen species that damage DNA and proteins, followed by the death of the microorganism. Mycogenic synthesis of zinc oxide (ZnO) and copper oxide (CuO) NPs was performed using enzymes and extracellular proteins of *P. chrysogenum*. Hexagonal and spherical NPs were obtained with a size between 9.0 and 59.7 nm. ZnO and CuO NPs showed high antimicrobial activity against Gram-positive and Gram-negative bacteria and various antibiofilm effects, making them an alternative for use in various medical devices (Mohamed et al. 2021).

One of the medicinal and edible mushrooms that are among the most cultivated worldwide is the genus *Pleurotus*, and NPs have been made of which 66% are silver (AgNP); 25% are gold (AuNP) and zinc oxide (ZnONP); 9% are iron (FeNP), cadmium sulfide (CdSNP), and selenium (SeNP); the remaining 9% is approximately composed of nonmetallic NPs (Paul et al. 2015). Photoirradiated extracellular synthesis of AgNPs was performed using the aqueous extract of *Pleurotus florida* as a reducing agent; it presented an inhibitory effect against *Providencia alcalifaciens, Proteus mirabilis, Salmonella aureus*, and *S. typhimurium*. The AgNPs showed excellent antimicrobial biofunctionality, either mixing them with an antibiotic drug or using them directly (Bhat et al. 2011). AgNPs were synthesized using extracts of *Agaricus bisporus*, *Calocybe indica*, *Pleurotus florida*, and *P. platypus*. AgNPs from *P. platypus* showed higher antibacterial activity in all cultures. The zone of maximum inhibition was in *Aerobacter aerogenes* (18 mm), and the zone of moderate activity was against *Enterobacter aerogenes* (14 mm) (Sujatha et al. 2013).

Thus, Hassan (2016), who developed a cotton fabric with AgNPs, said the fabric presented excellent quantitative and qualitative antimicrobial activity against *E. coli* and

S. aureus, the activity was maintained after 20 wash cycles, and the presence of NPs did not modify the appearance (Table 4.1). In another work, a cotton fabric was made to which AgNPs were incorporated that were synthesized using the fruiting body of the medicinal mushroom *Ganoderma lucidum*. They indicate that it provided sterile properties and presented antibacterial activity against *Proteus* sp. (46% inhibition), *S. aureus* (33% inhibition), *B. subtilis* (53% inhibition), *Pseudomonas* sp. (36.3% inhibition), and *Klebsiella pneumoniae* (43.8% inhibition) (Paul et al. 2015). A novel tertiary compound (TC), made of TEMPO cellulose (TOC), chitosan, starch, and selenium NPs, was synthesized. The proteins secreted by *Rhizopus oryzae* played an effective role in reducing selenium ions, exhibited antimicrobial activity against *S. aureus, B. subtilis, P. aeruginosa, S. typhimurium,* and *C. albicans*, and inhibited the biofilm formed by *P. aeruginosa, S. aureus,* and *C. albicans* (Abu-Elghait et al. 2021).

Indeed, Shobha et al. (2020) performed mycosynthesis of zinc oxide nanoparticles (ZnO NPs) by co-cultivation of *Trichoderma* spp. (*T. harzianum* and *T. reesei*) to inhibit *Xanthomonas oryzae* pv. *oryzae*, the causative organism of bacterial leaf blight in rice. The maximum zone of inhibition was recorded with co-culture of *Trichoderma* spp. instead of monocultures, indicating that co-culture can stimulate the synthesis of new SMs. Fungi can be used for an ecologically and economically viable production of NPs that are important in the health field without causing side effects, and several works indicate that they help to reinforce the effect of several antibiotics.

4.3.6 Herbicidal Activity

Excessive and improper use of synthetic chemicals in agriculture has harmed natural resources and decreased food quality and food safety (Mitrović et al. 2021). Fungal metabolites with herbicidal activity are an exciting area of research, as they offer potential alternatives to synthetic herbicides in agriculture. These natural compounds can inhibit the growth of unwanted plants (weeds) while being less harmful to the environment and non-target organisms. Ongoing research is focused on isolating and characterizing these metabolites, understanding their mechanisms of action, and developing formulations for agricultural use. Some studies have shown promising results in greenhouse and field trials, indicating the potential for commercial applications (Namasivayam et al. 2015).

Weed management in agricultural regions is a primary challenge faced by farmers globally. Productivity losses may exceed 40%, contingent upon the invasive species (Daniel et al. 2018). Bioherbicides have been explored since the 1970s, with DeVine® becoming the first to be registered for commercial usage in the 1980s in the United States (Ulrich et al. 2021). Bioherbicides can replace chemical herbicides by utilizing microorganisms and their SMs. Phytotoxins are commonly produced by microorganisms, such as bacteria and fungi. Fungi have significant metabolic capabilities to generate a diverse array of enzymes and bioactive compounds for agricultural use. *Fusarium fujikuroi* is a fungus that synthesizes metabolites such as indole acetic acid, 2-(4-butylpicolinamide) acetic acid, fusaric acid, and the sesterterpene terpestacin, among others, classified as nonribosomal peptides and polyketides. Daniel et al. (2018), in their research, examined 132 fungi and their ability to produce bioherbicides. After separate detached leaf-punctured assays, 11 isolates were selected, one of which stood out with the most pronounced phytotoxic effect—*F. fujikuroi*. When this fungus was used for submerged fermentation, the metabolites showed 25% and 20% phytotoxicity against *Sorghum bicolor* and *Crocus sativus*, respectively. Another *Fusarium* species, *F. oxysporum*, was tested for its herbicidal effect against the *Ninidam theenjan* plant. By using the leaf necrosis assay, which demonstrates the development of necrotic lesions that were noticed 12–18 hours after the application, the herbicidal activity of both free and chitosan NP-coated *F. oxysporum* metabolites was determined. The same scientists confirmed the herbicidal activity of this *Fusarium* species against the rice weed *Cyperus iria* (Namasivayam et al. 2015). *Fusarium graminearum* is a fungus known as the causative agent of corn disease (Mitrović et al. 2023); however, studies have shown that this fungus produces metabolites that also have herbicidal activity (Ulrich et al. 2021).

Colletotrichum gloeosporiodes is a fungus well known as a causative agent of apple storage disease (Grahovac et al. 2014); however, metabolites of this fungus have been found to cause necrotic lesions on 14 different weeds from Western Nigeria. Findings of Ulrich et al. (2021) indicate that the fungus *Trichoderma koningiopsis* is a viable biocontrol agent due to its ability to create enzymes that boost its phytotoxicity. The application of a biocompound derived from *T. koningiopsis* resulted in up to 80% mortality in cucumber plants. On the other hand, the plants *Brachiaria plantaginea* and *Euphorbia heterophylla* compete for important nutrients, limiting crop productivity. The results of Ulrich et al. (2023) showed that the fungus *T. koningiopsis* as a potential bioherbicide can replace synthetic herbicides in the control of these two weeds.

One of the most invasive aquatic plants in the world, *E. crassipes,* is indigenous to the Amazon basin and is recognized to have serious ecological and socioeconomic impacts. Namasivayam et al. (2023) investigated the herbicidal activity of a fungal metabolite-loaded chitosan nanocomposite made using in situ green principles against the mentioned plant. The active herbicidal fraction was isolated from *Allophoma oligotrophica* ethyl acetate extract and combined with chitosan nanocomposite, resulting in a very stable nanodimensional structure. Nanocomposite materials enable regulated release of herbicidal metabolites, resulting in phytotoxic effects.

Fungal phytotoxins are crucial in causing disease symptoms in weeds and plants found in forests and agriculture. Cimmino et al. (2015) summarized in their paper the important phytotoxins that are produced by different fungi. Ascaulitoxin is produced by the fungus *Ascochyta caulina*, and it was found that this compound has a herbicidal effect against *Chenopodium album* L. a widespread weed that cause problems in maize, potato, sugar beet, vegetable crops, and cereal worldwide. Ascosonchine is a component isolated from *Ascochyta sonchi* culture filtrate and is a potential biocontrol agent for the dangerous weeds *Cirsium arvense* L. and *Sonchus arvensis* L. Stagonolide is a phytotoxin isolated from the fungus *Stagonospora cirsii* with confirmed biocontrol activity against *C. arvense* L. An autochthonous weed originating from North America that is increasingly present in the pastures and fields of Europe, *Erigeon annuus* L., can be controlled with putaminoxin, a phytotoxin isolated from *Phoma putaminum*. In addition to the three already known components, epoxycytochalasin H, cytochalasin H, and cytochalasin N, the solid culture of *Phomopsis* sp. HCCB03520 was isolated and trisubstituted nonenolide. This compound showed activity on the growth and germination of *Buchloe dactyloides*, *Medicago sativa* L. and *Trifolium hybridum* L. plants. Drazepinone is a new naphthofuroazepinone isolated from the liquid culture of the fungus *Drechslera siccans*. It was found that this compound is suitable for controlling the worldwide weed, *Lolium perenne* L. Also, two compounds, preussomerin and palmarumycin, were isolated from the mycelium of *Edenia gomezpompae*. These two compounds inhibit root growth and seed germination of *Echinochloa crusgalli* L., *Amaranthus hypochondriacus* L., and *Solanum lycopersicum* L. at a dose of 100 μg/L.

Nanoformulations of bioherbicides might enhance biocontrol efficacy due to the extensive NPs surface areas, which necessitate a reduced volume of bioherbicide, thereby concentrating it in a smaller package and lowering costs. One benefit of employing different types of NPs and other nanoformulation technologies is the ability to accomplish controlled release of phytotoxic metabolite(s). Stability under extreme environmental conditions is possible with lignin-derived NPs that are manufactured in diameters ranging from 45 to 250 nm and have surface characteristics altered by a polyelectrolyte coating. An additional option for the efficient delivery of bioherbicidal metabolites might be a nanoemulsion technology, which is a nonequilibrium colloidal system with particles that are usually between 20 and 200 nm in diameter. So, in order to create a variety of bioactive substances that are combined into NPs based on distinct chemical properties (such as lipids, alginate, chitosan, and nanoclay) for field delivery and controlled release, biopesticides have been created using nanotechnology (Kremer 2019). In summary, fungal metabolites with herbicidal activity represent a promising avenue for developing sustainable weed management strategies in agriculture, contributing to more environmentally friendly farming practices.

4.4 Fungi in Nanoparticle Biosynthesis

Fungi play a significant role in the biosynthesis of NPs, leveraging their unique metabolic pathways to produce various nanomaterials. The process typically involves the use of fungal biomass, which can reduce metal ions to form NPs through enzymatic and nonenzymatic mechanisms. The biosynthesis process is often environmentally friendly, as it can occur under mild conditions and does not require toxic chemicals, making it a sustainable alternative to traditional chemical methods. The biogenic synthesis of NPs can be conducted using organisms like bacteria, fungi, and plants, or their metabolic byproducts, which serve as reducing and stabilizing agents (Guilger-Casagrande and Lima 2019). The list of nanominerals and NPs generated by fungi has significantly increased in recent years, encompassing over 30 elements from the periodic table. This encompasses transition metals (Fe, Mn, Ti, Cu, Ni, Co, Zr, Au, Ag, Pt), alkaline earth metals (Ba, Mg, Sr), post-transition metals (Zn, Al, Bi, Cd), nonmetals (H, O, N, C, S, P, Se), metalloids (Te, Si), and lanthanides (Ce, La) (Li et al. 2022).

Many traditional techniques for NP production include drawbacks, including the utilization of hazardous chemicals and the creation of trash, which may lead to environmental contamination. As a result, there has been a growing interest in environmentally sustainable synthesis processes in recent years. So, new techniques utilize organisms such as fungi, bacteria, and plants to decrease metal salts, facilitating the synthesis of NPs with the required size and morphology. Due to their high metal tolerance and ease of handling, fungi are desirable agents for the biogenic production of AgNPs. Additionally, they release a lot of extracellular proteins that help keep the NPs stable (Guilger-Casagrande and Lima 2019).

Although several studies have been published on the biogenic production of AgNPs utilizing fungi, the particular mechanisms involved have yet to be fully understood. The synthesis of NPs at the extracellular level takes place as the enzymes present in the fungal filtrate reduce silver ions, thereby producing nanoscale elemental silver. The size of produced AgNPs varies based on synthesis circumstances, including pH, temperature, fungal species, dispersion media, and cappings. In the biogenic creation of metallic NPs, NADH and NADH-dependent nitrate reductase enzymes are thought to be the most crucial components (Guilger-Casagrande and Lima 2019). While the precise mechanisms underlying fungal synthesis of NPs remain ambiguous, extensive study has shown that NP formation may be associated with reducing enzymes. An NADH-dependent reductase was identified as the principal protein involved in the reduction of Ag^+ and the subsequent creation of AgNPs. For example, this reductase is specific for the *F. oxysporum* strain, as AgNPs were not synthesized by *Fusarium moniliforme* under identical experimental circumstances. Using gelatine as a capping agent to prevent aggregation, nitrate reductase was isolated from the supernatant of *F. oxysporum* cultures in order to synthesize AgNPs (~50 nm). In addition to reductases, quinine derivatives can serve as redox centers for the creation of NPs and metal reduction (Li et al. 2022).

AgNPs stand out among the various kinds of metallic NPs due to their broad-spectrum antibacterial properties. Three *Trichoderma* species—*Trichoderma harzianum*, *T. viride*, and *T. longibrachiatum*—were employed in the Elamawi et al. (2018) study to biosynthesize stable AgNPs—AgNPs. Each fungal isolate's filtrates and biomass were cultured with $AgNO_3$ with or without shaking in a dark environment. After 72 hours of incubation, the color was visually monitored. AgNPs produced by *T. longibrachiatum* were tested for their antifungal properties against nine fungal isolates. Very high inhibition percentages (>93%) showed that the generated AgNPs were most effective against *Fusarium verticillioides*, *F. moniliforme*, *Hirschmanniella oryzae*, *Alternaria alternata*, and *Pyricularia grisea* in this investigation. Also, *T. longibrachiatum* AgNPs increased the vigor index and seed germination percentage while lowering the disease incidence of *F. oxysporium*, a seed-borne pathogen that affects barley, tomato, and faba beans.

In the study of Wang et al. (2021), *Aspergillus sydowii* culture supernatants were used to extracellularly biosynthesize AgNPs. A color shift during synthesis served as a preliminary indicator of AgNP formation, and the UV peak at 420 nm provides additional proof of AgNP creation. Erlenmeyer flasks of 250 mL were used for the cultivation of *A. sydowii* in Sabouraud media inoculated with 1×10^6 spores. Cultivation was carried out on a shaker at 140 rpm for five days at 28 °C. Fungal biomass was collected by filtering it using Whatman filter paper and rinsing it with double-distilled water. After incubating 20 g of biomass for 24 hours at 28 °C in a flask with 100 mL of double-distilled water, the suspension was filtered. In a 200 mL flask containing 50 mL of fungal filtrate, $AgNO_3$ solution was added (9:1), and the flask was incubated for 48 hours. The generated AgNPs have strong antifungal activity against the tested fungi and exhibit good stability. Similar research was done by Soleimani et al. (2022) using entomopathogenic fungi *Beauveria bassiana*, and one strain of *Metarhizium anisopliae*.

To achieve desired NP qualities, fungus-specific parameters such as production time, cultivation medium, agitation, light, and temperature can be optimized. AbdelRahim et al. (2017) found NPs with diameters of 25.89, 2.86, and 48.43 nm at temperatures of 20, 40, and

60 °C. Husseiny et al. (2015) discovered that increasing the temperature to 50 °C reduced the size of NPs in the fungus *F. oxysporum*. According to Banu and Balasubramanian (2014), *I. fumosorosea* produces the most stable AgNPs at 30 °C. *Colleotrichum* sp. ALF2-6 synthesis was accomplished in about 20 minutes at a higher temperature of 50 °C and an alkaline pH, which was faster than at a lower pH (Azmath et al., 2016). According to Qian et al. (2013), adding $AgNO_3$ to the fungus *Epicoccum nigrum* filtrate promoted the creation of AgNPs due to its alkaline pH. The production of NPs peaked between pH 9 and 11 in the synthesis carried out by Birla et al. (2013) employing *F. oxysporum*, while production was decreased at pH 7, and aggregates developed between pH 3 and 5.

Phanjom and Ahmed (2017) investigated the production of NPs with *A. oryzae* with varying doses of $AgNO_3$ ranging from 1 to 10 mM. The NPs were found to have diameters ranging from 7 to 17 nm at $AgNO_3$ concentrations up to 8 mM, and at $AgNO_3$ concentrations of 9 and 10 mM, they have grown to almost 46 and 62 nm, respectively. *Rhizopus stolonifer* used in the research of AbdelRahim et al. (2017), produced very small NPs (2.86 nm) with 10 mM $AgNO_3$, whereas sizes of 14.23 and 54.67 nm were produced at 1 and 100 mM, respectively. Using *F. oxysporum*, Husseiny et al. (2015) observed similar results. The production and properties of silver NPs can also be influenced by the quantity of biomass employed. In their evaluation of the usage of 10, 7, 4, and 1 g quantities of *Aspergillus fumigatus* BTCB10 biomass, Shahzad et al. (2019) found that using 7 g of biomass resulted in higher production, smaller size, and better dispersion of the NPs. Studies where various media were evaluated for the growth of fungi showed varying behaviors. Using *Sclerotinia sclerotiorum* grown in different broths, Saxena et al. (2016) produced AgNPs, and the maximum NP production was accomplished using potato dextrose medium.

In addition to the most important AgNPs, fungi can also be producers of other types of NPs. Using an extract from *Alternaria* species, golden nanoparticles (AuNPs) with a variety of morphologies and significant antibacterial, antioxidant, and antimitotic properties were produced (Hemashekhar et al. 2019). Mycelial extract from *T. hamatum* was used to create hexagonal, pentagonal, and spherical NPs. The authors used *T. hamatum* to optimize the production of AuNPs with the lowest size. Mycelial biomass from *T. harzianum* was used to biosynthesize NPs, which were 30 nm and had a spherical shape (Loshchinina et al. 2023). *F. oxysporum* has been the subject of the most research on platinum nanoparticle (PtNP) biosynthesis. Riddin et al. (2006) identified the ideal circumstances for the highest yield of NPs and demonstrated that the *F. oxysporum* biomass is capable of generating NPs of different sizes and forms. In addition to the fact that *A. alternata* is a significant phytopathogen (Mitrović et al. 2021), this microorganism can also be used for the production of nanoplatinum. The generated particles were found to be variable in shape, with an overall morphology of polygonal, hexagonal, tetrahedral, rectangular, and spherical. Their size ranged from 50 to 315 nm (Sarkar and Acharya 2017). There have also been reports of palladium nanoparticle (PdNP) synthesis using the biomass of the baker's yeast *Saccharomyces cerevisiae*. PdNPs with diameters ranging from 10 to 20 nm were formed and deposited on the surfaces of *S. cerevisiae* cells, as reported by Saitoh et al. (2020). Besides the fact that *A. flavus* producer of a very dangerous mycotoxin, aflatoxin, this species has the ability to produce copper NPs with a diameter of 2–60 nm (Saitawadekar and Kakde 2020).

The utilization of fungi in the biosynthesis of NPs represents a promising and innovative approach in nanotechnology. Because of their capacity for metal bioaccumulation and tolerance, fungi have garnered increased interest in the study of biological synthesis of metallic NPs. Fungi possess unique metabolic pathways and enzymatic capabilities that facilitate the reduction and stabilization of metal ions, leading to the formation of NPs with desirable properties. As research in this field continues to advance, the integration of fungal biosynthesis in the production of NPs could open the way for novel applications in medicine, environmental remediation, and materials science, ultimately contributing to sustainable technological development.

4.5 Conclusions

Fungi produce various biologically active compounds that serve as novel scaffolds and can be utilized as bio-resources with distinct biological properties that have applications in pharmaceuticals, agriculture, and food processing. Bioactive compounds derived from fungi encompass alkaloids, coumarins, flavonoids, lignans, saponins, and terpenes, among others. Fungi secrete extracellular enzymes that aid in the biosynthesis of diverse bioactive substances. Nanotechnology enhances the stability of bioactive compounds with low bioavailability, preventing degradation and thereby increasing their half-life and potential biological efficacy against nematodes, insects, viruses, and fungi. Additionally, nano-biotechnology utilizes biological methods to generate nanomaterials from renewable biological resources, primarily fungi. Nanoparticles have garnered significant attention due to their extensive range of applications among nanomaterials. Fungi play a significant role in the green synthesis of NPs. Fungi act as biological agents in NP synthesis, utilizing their rapid growth and straightforward culturing for applications in nanotechnology.

References

AbdelRahim, K., Mahmoud, S.Y., Ali, A.M. et al. (2017). Extracellular biosynthesis of silver nanoparticles using *Rhizopus stolonifer*. *Saudi Journal of Biological Sciences* 24: 208–216. https://doi.org/10.1016/j.sj.s.2016.02.025.

Adebayo, E.A., Azeez, M.A., Alao, M.B. et al. (2021). Fungi as veritable tool in current advances in nanobiotechnology. *Heliyon* 7: e08480. https://doi.org/10.1016/j.heliyon.2021.e08480.

Adetunji, C.O., Egbuna, C., Tijjani, H. et al. (2019). Homemade preparations of natural biopesticides and applications. In: *Natural Remedies for Pest, Disease and Weed Control* (ed. C. Egbuna and B. Sawicka), 179–186. Elsevier Science.

Aguilar-Marcelino, L., Al-Ani, L.K.T., Soares, F.E.d.F. et al. (2021). Formation, resistance, and pathogenicity of fungal biofilms: current trends and future challenges. In: *Recent Trends in Mycological Research, Volume 1: Agricultural and Medical Perspective, Fungal Biology* (ed. A.N. Yadav), 411–438. Springer Nature Switzerland.

Aguilar-Marcelino, L., Al-Ani, L.K.T., Soares, F.E.d.F. et al. (2022). Wild mushrooms: characteristics, nutrition, and processing. In: *Wild Mushrooms: Characteristics, Nutrition, and Processing* (ed. S.B. Dhull, A. Bains, P. Chawla, and P.K. Sadh), 471–492. Taylor & Francis, CRC Press.

Aguilar-Marcelino, L., Al-Ani, L.K.T., Wong-Villarreal, A., and César Sotelo-Leyva, C. (2023). Persistence of pesticides residues with chemical food preservatives in fruits and vegetables. In: *Current Developments in Biotechnology and Bioengineering Pesticides: Human Health, Environmental Impacts and Management*, 1ste (ed. J. Singh, A. Pandey, S. Singh, et al.), 99–118. Elsevier India.

Ahmad, A., Mukherjee, P., Senapat, S. et al. (2003). Extracellular biosynthesis of silver nanoparticles using the fungus *Fusarium oxysporum*. *Colloids and Surfaces. B, Biointerfaces* 28: 313–318.

Al-Ani, L.K.T. (2017). Potential of utilizing biological and chemical agents in the control of *Fusarium wilt* of banana. PhD School of Biology Science, Universiti Sains Malaysia, Pinang, Malaysia, p. 259.

Al-Ani, L.K.T., Yonus, M.I., Mahdii, B.A. et al. (2018). First record of use *Fusarium proliferatum* fungi in direct treatment to control the adult of wheat flour *Tribolium confusum*, as well as, use the entomopathogenic fungi *Beauveria bassiana*. *Ecology, Environment and Conservation* 24 (3): 29–34.

Al-Ani, L.K.T. (2018a). *Trichoderma*: beneficial role in sustainable agriculture by plant disease management. In: *Plant Microbiome: Stress Response. Microorganisms for Sustainability*, vol. 5 (ed. D. Egamberdieva and P. Ahmad), 105–126. Singapore: Springer.

Al-Ani, L.K.T. (2018b). *Trichoderma* from extreme environments: physiology, diversity, and antagonistic activity. In: *Extremophiles in Eurasian Ecosystems: Ecology, Diversity, and Applications. Microorganisms for Sustainability*, vol. 8 (ed. D. Egamberdieva, N.-K. Birkeland, H. Panosyan, and W.-J. Li), 388–403. Singapore: Springer.

Al-Ani, L.K.T. (2019a). Recent patents on endophytic fungi and their international market. In: *Intellectual Property Issues in Microbiology* (ed. H.B. Singh, C. Keswani, and S.P. Singh), 271–284. Singapore: Springer.

Al-Ani, L.K.T. (2019b). The importance of endophytic fungi from the medicinal plant: diversity, natural bioactive compounds, and control of plant pathogens. In: *Medically Important Plant Biomes Source of Secondary Metabolites* (ed. D. Egamberdieva and A. Tiezzi), 189–238. Singapore: Springer.

Al-Ani, L.K.T. (2019c). A patent survey on *Trichoderma* spp. (from 2007–2017). In: *Intellectual Property Issues in Microbiology* (ed. H.B. Singh, C. Keswani, and S.P. Singh), 163–192. Singapore: Springer.

Al-Ani, L.K.T. (2019d). Entomopathogenic fungi in IP landscape. In: *Intellectual Property Issues in Microbiology* (ed. H.B. Singh, C. Keswani, and S.P. Singh), 223–238. Singapore: Springer.

Al-Ani, L.K.T. (2019e). Secondary *metabolites* of nonpathogenic Fusarium spp.; scope in agriculture. In: Singh, H.B., Keswani, C., Reddy, M.S., Royano, E.S. and Garcia-Estrada, C. (Eds.), Secondary metabolites of plant growth promoting rhizomicroorganisms -discovery and applications. Springer, Singapore, pp. 59–76. https://doi.org/10.1007/978-981-13-5862-3 3.

Al-Ani, L.K.T. (2019f). Bioactive Secondary *Metabolites* of Trichoderma spp. for Efficient Management of Phytopathogens. In: Singh, H.B., Keswani, C., Reddy, M.S., Royano, E.S. and Garcia-Estrada, C. (Eds.), Secondary metabolites of plant growth promoting rhizomicroorganisms - discovery and applications. Springer, Singapore, pp. 125–143. https://doi.org/10.1007/978-981-13-5862-3_7.

Al-Ani, L.K.T., Aguilar-Marcelino, L., Fiorotti, J., Sharma, V., Sarker, M.S., Raza, W., Furtado, E.L., Wijayawardene, N.N., and Herrera-Estrella, A. (2020 a). *Biological Control* Agents and

Their Importance for the Plant Health. In: Singh, J.S. and Vimal, S.R. (Eds.), Microbial services in ecological restoration. Elsevier, USA, pp. 13–36. https://doi.org/10.1016/B978-0-12-819978-7.00002-6.

Al-Ani, L.K.T. and Mohammed, A.M. (2020). Versatility of *Trichoderma* in plant disease management. In: *Molecular Aspects of Plant Beneficial Microbes in Agriculture* (ed. V. Sharma, R. Salwan, and L.K.T. Al-Ani), 159–168. Cambridge: Elsevier Science.

Al-Ani, L.K.T., et al. (2020). The effect of incompatible plant pathogen on the host plant. In: Sharma, V., Salwan, R. and Al-Ani, L.K.T. (Eds.), Molecular Aspects of Plant Beneficial Microbes in Agriculture. Elsevier Science, Cambridge, pp.47–57. In: https://doi.org/10.1016/B978-0-12-818469-1.00004-3 .

Al-Ani, L.K.T., Surono, Aguilar-Marcelino, L. et al. (2021). Role of useful fungi in agriculture sustainability. In: *Recent Trends in Mycological Research, Fungal Biology* (ed. A.N. Yadav), 1–44. Springer Nature Switzerland.

Al-Ani, L.K.T., Soares, F.E.D.F., Sharma, A. et al. (2022). Strategy of nematophagous fungi in determining the activity of plant parasitic nematodes and their prospective role in sustainable agriculture. *Frontiers in Fungal Biology* 3: 863198. https://doi.org/10.3389/ffunb.2022.863198.

Al-Ani, L.K.T., Mitrović, I., Zimowska, B. et al. (2024). Metabolites of fungi and applications in the agricultural. In: *Present Status and Prospects of Plant Associated Fungi* (ed. S. Siddiqui, V. Meshram, R.S. Patil, and M. Gupta), 351–397. Cambridge Scholars Publishing.

Al-Ani, L.K.T., Mitrović, I., Furtado, E.L., Bhattacharyya, P.N., Hajji-Hedfi, L., Sukmawati, D., Aguilar-Marcelino, L., Phookamsak, R., Garganese, F.Guimaraes, R.A., Roca, M.E.M., Carrasco, F., Poitevin, C.G., Dautt-Castro, M., Acheampong, M.A., and Surono, S. (2025). Trichoderma a Holistic Fungus with Biocontrol Potential for Agricultural Diseases. In: Shaaban, M. (Eds), Soils and Sustainable Agriculture. Frontier Studies in Soil Science. Springer, Cham., pp. 877–912. https://doi.org/10.1007/978-3-031-91114-9_37 .

Al-Ani, L.K.T., Aguilar-Marcelino, L., Santos-Villalobos, S.D.L. et al. (2026a). Identification and characterization of plant parasitic nematodes using molecular techniques. In: *Plant-Nematode Interactions and Their Management Strategies* (ed. M.S. Akhtar and K.R. Hakeem). CRC Press.

Al-Ani, L.K.T., Aguilar-Marcelino, L., Cruz-Arévalo, J. et al. (2026b). Progress and development of plant nematology and future consequences. In: *Plant-Nematode Interactions and Their Management Strategies* (ed. M.S. Akhtar and K.R. Hakeem). CRC Press.

Al-Ani, L.K.T., Acheampong, M.A., Thakur, R., Bhattacharyya, P.N., Kumari, R., Sukmawati, D., Raza, W. and Phookamsak R. (2027). Endophytic Microorganisms: Foes and Friends of Plants. In: Vimal, S.R., Singh, J.S., and Kumar, A. (Eds.), Botanical Microbiomes: Sustainable Practices in Agriculture and Ecosystem Management. Elsevier, pp. 447–468. http://dx.doi.org/10.1016/B978-0-443-27628-6.00018-8.

Aguilar-Marcelino, L., Al-Ani, L.K.T., Castaneda-Ramirez, G.S., Garcia-Rubio, V. and Ojeda-Carrasco, qJ.J. (2020 a). Microbial technologies to enhance crop production for future needs. In: Rastegari, A.A., Yadav, A.N., Awasthi, A.K. and Yadav, N. (Eds.), New and Future Developments in Microbial *Biotechnology and Bioengineering*. Elsevier, USA, pp. 29–47. https://doi.org/10.1016/B978-0-12-820526-6.00003-8.

Aguilar-Marcelino, L., Mendoza-de-Gives, P., Al-Ani, L.K.T., et al. (2020 b). Using molecular techniques applied to beneficial *microorganisms* as biotechnological tools for controlling agricultural plant pathogens and pest. In: Sharma, V., Salwan, R. and Al-Ani, L.K.T. (Eds.),

Molecular Aspects of Plant Beneficial Microbes in Agriculture. Elsevier Science, Cambridge, pp.333–349. https://doi.org/10.1016/B978-0-12-818469-1.00027-4.

Alghuthaymi, M.A., Almoammar, H., Rai, M. et al. (2015). Myconanoparticles: synthesis and their role in phytopathogens management. *Biotechnology and Biotechnological Equipment* 29 (2): 221–236. https://doi.org/10.1080/13102818.2015.1008194.

Ameyaw, D., Eziah, V.Y., Al-Ani, L.K., Nyahe, P.A., Coombes, C.A., Adom, M., Fening, K.O., Billah, M.K., Osae, M.Y., Rangel, D.E. and Sukmawati, D. (2026). Evaluation of the *biocontrol* potential of native Metarhizium (Hypocreales: Clavicipitaceae) species against the false codling moth (Thaumatotibia leucotreta Meyrick) on chilli peppers in Ghana. *African Entomology*, 34: 1–9. Doi:https://doi.org/10.17159/2254-8854/2026/a24601.

Arunthirumeni, M., Veerammal, V., and Shivakumar, M.S. (2022). Biocontrol efficacy of mycosynthesized selenium nanoparticle using *Trichoderma* sp. on insect pest *Spodoptera litura*. *Journal of Cluster Science* 33: 1645–1653.

Attitalla, I.H., Mansour, S.E., Mohamed, W.S. et al. (2010a). Influence of *Aspergillus flavus* and *Aspergillus terreus* on the protein value of the two varieties of peanut grains. *International Conference, International Mycotoxin Conference*, MycoRed, Penang–Malaysia (1–4 December 2010), p. 177.

Attitalla, I.H., Al-Ani, L.K.T., Nasib, M.A. et al. (2010b). Screening of fungi associated with commercial grains and animal feeds in Al-Bayda Governorate, Libya. *World Applied Sciences Journal* 9 (7): 746–756.

Avalos, J. and Limón, M.C. (2022). Fungal secondary metabolism. *Encyclopedia* 2: 1–13. https://doi.org/10.3390/encyclopedia2010001.

Aygün, A., Ozdemir, S., Gülcan, M. et al. (2020). Synthesis and characterization of Reishi mushroom-mediated green synthesis of silver nanoparticles for the biochemical applications. *Journal of Pharmaceutical and Biomedical Analysis* 178: 112970.

Aziz, B.N. and Abdel-Azeem, A.M. (2024). Antifungal activity of secondary metabolites produced by fungi. *Microbial Biosystems* 9 (1): 2024.371458. https://doi.org/10.21608/mb.2024.371458.

Azmath, P., Baker, S., Rakshith, D., and Satish, S. (2016). Mycosynthesis of silver nanoparticles bearing antibacterial activity. *Saudi Pharmaceutical Journal* 24: 140–146. https://doi.org/10.1016/j.jsps.2015.01.008.

Banu, A.N. and Balasubramanian, C. (2014). Optimization and synthesis of silver nanoparticles using *Isaria fumosorosea* against human vector mosquitoes. *Parasitology Research* 113: 3843–3851. https://doi.org/10.1007/s00436-014-4052-0.

Banu, A.N. and Balasubramanian, C. (2014). Myco-synthesis of silver nanoparticles using *Beauveria bassiana* against dengue vector, *Aedes aegypti* (Diptera: Culicidae). *Parasitology Research* 113: 2869–2877.

Barbosa, A.C.M.S., Silva, L.P.C., Ferraz, C.M. et al. (2019). Nematicidal activity of silver nanoparticles from the fungus *Duddingtonia flagrans*. *International Journal of Nanomedicine* 2341–2348.

Bashyal, B.P., Wellensiek, B.P., Ramakrishnan, R. et al. (2014). Altertoxins with potent anti-HIV activity from QUE1Se, a fungal endophyte of *Quercus emoryi*. *Bioorganic & Medicinal Chemistry* 22: 6112–6116.

Bhainsa, K.C. and D'Souza, S.F. (2006). Extracellular biosynthesis of silver nanoparticles using the fungus *Aspergillus fumigatus*. *Colloids and Surfaces. B, Biointerfaces* 47: 160–164.

Bhattacharyya, P.N., Nath, B.C., Sarma, B. et al. (2024). Interaction of *Metarhizium anisopliae* against emergent insect pest problems in the north-eastern tea industry. In: *Entomopathogenic Fungi: Prospects and Challenges* (ed. S.K. Deshmukh and K.R. Sridhar), 319–357. Springer Nature Singapore https://doi.org/10.1007/978-981-97-5991-0_13.

Birla, S.S., Gaikwad, S.C., Gade, A.K., and Rai, M.K. (2013). Rapid synthesis of silver nanoparticles from *Fusarium oxysporum* by optimizing physicocultural conditions. *The Scientific World Journal* 796018. https://doi.org/10.1155/2013/796018.

Blackwell, M. (2011). The fungi: 1, 2, 3 … 5.1 million species? *American Journal of Botany* 98 (3): 426–438. https://doi.org/10.3732/ajb.1000298.

Boroumand Moghaddam, A., Namvar, F., Moniri, M. et al. (2015). Nanoparticles biosynthesized by fungi and yeast: a review of their preparation, properties, and medical applications. *Molecules* 20 (9): 16540–16565. https://doi.org/10.3390/molecules200916540.

Brauer, V.S., Rezende, C.P., Pessoni, A.M. et al. (2019). Antifungal agents in agriculture: friends and foes of public health. *Biomolecules* 9 (10): 521. https://doi.org/10.3390/biom9100521.

Castañeda-Ramírez, G.S., Torres-Acosta, J.F.d.J., Sánchez, J.E. et al. (2020). The possible biotechnological use of edible mushroom bioproducts for controlling plant and animal parasitic nematodes. *BioMed Research International* 2020: 1–12. https://doi.org/10.1155/2020/6078917.

Castro-Longoria, E. (2016). Fungal biosynthesis of nanoparticles, a cleaner alternative. In: *Fungal Applications in Sustainable Environmental Biotechnology, Fungal Biology* (ed. D. Purchase), 323–351. Springer.

Castro-Longoria, E., Vilchis-Nestor, A.R., and Avalos-Borja, M. (2011). Biosynthesis of silver, gold and bimetallic nanoparticles using the filamentous fungus *Neurospora crassa*. *Colloids and Surfaces. B, Biointerfaces* 83 (1): 42–48.

Chinnaperumal, K., Govindasamy, B., Paramasivam, D. et al. (2018). Bio-pesticidal effects of *Trichoderma viride* formulated titanium dioxide nanoparticle and their physiological and biochemical changes on *Helicoverpa armigera* (Hub.). *Pesticide Biochemistry and Physiology* 149: 26–36.

Cimmino, A., Masi, M., Evidente, M. et al. (2015). Fungal phytotoxins with potential herbicidal activity: chemical and biological characterization. *Natural Product Reports* 32: 1629. https://doi.org/10.1039/c5np00081e.

Comans-Pérez, R.J., Sánchez, J.E., Al-Ani, L.K.T. et al. (2021). Biological control of sheep nematode *Haemonchus contortus* using edible mushrooms. *Biological Control* 152: 104420. https://doi.org/10.1016/j.biocontrol.2020.104420.

Dalimunthe, C.I., Soekarno, B.P.W., Al-Ani, L.K.T. et al. (2023). First report of inhibitory abilities of dark septate endophytic fungi against white root rot disease on *Hevea brasiliensis* seedlings in nursery conditions. *Egyptian Journal of Biological Pest Control* 33 (1): 1–12.

Daniel, J.J. Jr., Zabot, G.L., Tres, M.V. et al. (2018). *Fusarium fujikuroi*: A novel source of metabolites with herbicidal activity. *Biocatalysis and Agricultural Biotechnology* 14: 314–320. https://doi.org/10.1016/j.bcab.2018.04.001.

De Lucca, A.J. (2007). Harmful fungi in both agriculture and medicine. *Revista Iberoamericana de Micología* 24 (1): 3–13.

Demain, A.L. (2014). Valuable secondary metabolites from fungi. In: *Biosynthesis and Molecular Genetics of Fungal Secondary Metabolites, Fungal Biology* (ed. J.-F. Martín, C. García-Estrada, and S. Zeilinger), 1–15. New York: Springer.

Deshmukh, S.K., Gupta, M.K., Prakash, V., and Saxena, S. (2018). Endophytic fungi: a source of potential antifungal compounds. *Journal of Fungi* 4: 77. https://doi.org/10.3390/jof4030077.

Devi, L.S. and Joshi, S.R. (2012). Antimicrobial and synergistic effects of silver nanoparticles synthesized using soil fungi of high altitudes of Eastern Himalaya. *Mycobiology* 40 (1): 27–34.

Devi, R., Kaur, T., Guleria, G. et al. (2020). Fungal secondary metabolites and their biotechnological applications for human health. *New and Future Developments in Microbial Biotechnology and Bioengineering* 147–161. https://doi.org/10.1016/B978-0-12-820528-0.000.0-7.

Dhanasekaran, D., Latha, S., Saha, S. et al. (2013). Extracellular biosynthesis, characterisation and in-vitro antibacterial potential of silver nanoparticles using *Agaricus bisporus*. *Journal of Experimental Nanoscience* 8: 579–588. https://doi.org/10.1080/17458080.2011.577099.

Díaz, C.E., Andrés, M.F., Bolaños, P., and González-Coloma, A. (2024). Nematicidal and insecticidal compounds from the laurel forest endophytic fungus *Phyllosticta* sp. *Molecules* 29 (19): 4568. https://doi.org/10.3390/molecules29194568.

Diko, C.S., Zhang, H., Lian, S. et al. (2020). Optimal synthesis conditions and characterization of selenium nanoparticles in *Trichoderma* sp. WL-Go culture broth. *Materials Chemistry and Physics* 246: 122583.

Dixon, D.M. and Walsh, T.J. (1996). Antifungal agents. In: *Medical Microbiology*, 4the (ed. S. Baron). Galveston, TX: University of Texas Medical Branch at Galveston Chapter 76.

Elamawi, R.M., Al-Harbi, R.E., and Hendi, A.A. (2018). Biosynthesis and characterization of silver nanoparticles using *Trichoderma longibrachiatum* and their effect on phytopathogenic fungi. *Egyptian Journal of Biological Pest Control* 28: 28. https://doi.org/10.1186/s41938-018-0028-1.

Elamawi, R.M., Al-Harbi, R.E., and Hendi, A.A. (2018). Biosynthesis and characterization of silver nanoparticles using *Trichoderma longibrachiatum* and their effect on phytopathogenic fungi. *Egyptian Journal of Biological Pest Control* 28: 1–11.

Elechiguerra, J.L., Burt, J.L., Morones, J.R. et al. (2005). Interaction of silver nanoparticles with HIV-1. *Journal of Nanbiotechnology* 3: 6.

Feroz, S., Muhammad, N., Ranayake, J., and Dias, G. (2020). Keratin-based materials for biomedical applications. *Bioactive Materials* 5: 496–509. https://doi.org/10.1016/j.bioactmat.2020.04.007.

Fisher, M.C., Hawkins, N.J., Sanglard, D., and Gurr, S.J. (2018). Worldwide emergence of resistance to antifungal drugs challenges human health and food security. *Science* 360: 739–742. https://doi.org/10.1126/science.aap7999.

de Freitas Soares, F.E., Ferreira, J.M., Genier, H.L.A. et al. (2023). Biological control 2.0: use of nematophagous fungi enzymes for nematode control. *Journal of Natural Pesticide Research* 4: 100025.

Gade, A. and Rai, M. (2021). Mycofabrication of metal nanoparticles: a green approach. In: *Progress in Mycology* (ed. T. Satyanarayana, S.K. Deshmukh, and M.V. Deshpande). Singapore: Springer https://doi.org/10.1007/978-981-16-3307-2_9.

Gade, A.K., Bonde, P., Ingle, A.P. et al. (2008). Exploitation of *Aspergillus niger* for synthesis of silver nanoparticles. *Journal of Biobased Materials and Bioenergy* 2: 243–247. https://doi.org/10.1166/jbmb.2008.401.

Gade, A.K., Ingle, A.P., Whiteley, C., and Rai, M.K. (2010). Mycogenic metal nanoparticles: progress and applications. *Biotechnology Letters* 32: 593–600. https://doi.org/10.1007/s10529-009-0197-9.

Gajbhiye, M., Kesharwani, J., Ingle, A. et al. (2009). Fungus-mediated synthesis of silver nanoparticles and their activity against pathogenic fungi in combination with fluconazole. *Nanomedicine: Nanotechnology, Biology and Medicine* 5 (4): 382–386.

Grahovac, M., Balaž, J., Grahovac, J. et al. (2014). Screening of antagonistic activity of selected microorganisms against apple rot pathogens. *Romanian Biotechnological Letters* 19 (1): 1821–1827.

Greeff-Laubscher, M.R., Beukes, I., Marais, G.J., and Jacobs, K. (2019). Mycotoxin production by three different toxigenic fungi genera on formulated abalone feed and the effect of an aquatic environment on fumonisins. *Mycology* 11 (2): 105–117. https://doi.org/10.1080/21501203.2019.1604575.

Guilger-Casagrande, M. and Lima, R. (2019). Synthesis of silver nanoparticles mediated by fungi: a review. *Frontiers in Bioengineering and Biotechnology* 7: 287. https://doi.org/10.3389/fbioe.2019.00287.

Gupta, A.K., Cooper, E.A., Ryder, J.E. et al. (2004). Optimal management of fungal infections of the skin, hair, and nails. *American Journal of Clinical Dermatology* 5 (4): 225–237.

Hajji-Hedfi, L., ... Pollmann, S. ((2024)). Exploring endophytic fungal diversity: Review of traditional and molecular techniques. In: Abdel Azeem, A.M., Yadav, A.N. and Yadav, N. (Eds.), Developments in Applied Microbiology and Biotechnology, Endophytic Fungi. Academic Press, Pages 1–24. In: https://doi.org/10.1016/B978-0-323-99314-2.00005-X.

Hajji-Hedfi, L., Dali, S., Rhouma, A., Al-Ani, L.K.T., Chihani-Hammas, N., Khlif, A. and Bargougui O. (2025 a). Microbial inoculants for plant resilience performance: roles, *prospects* and challenges. *Journal of Plant Diseases and Protection*, 132 (2): 88. https://doi.org/10.1007/s41348-025-01080-7.

Hajji-Hedfi, L., Rhouma, A., Al-Ani, L.K.T., Bargougui, O., Tlahig, S., Jaouadi, R., Zaouali, Y., Degola, F., & Abdel-Azeem, A.M. (2025 b). The Second Life of Citrus: Phytochemical Characterization and Antifungal Activity Bioprospection of C. limon and C. sinensis Peel Extracts Against Potato Rot Disease. *Waste and Biomass Valorization* 16, 6863–6875. https://doi.org/10.1007/s12649-025-03095-5.

Hajji-Hedfi, L., Wannassi, T., Al-Ani, L.K.T., Balbool, B.A., Tissaoui, S., Mougou-Hamdane, A., Hamdi, W., Azeem, A.M.A., & Rebouh, N.Y. (2025 c) Investigating the potential role of beneficial rhizobacteria for protecting grapevine health and promoting growth. *Frontiers in Sustainable Food Systems*, 9, 1619801. 10.3389/fsufs.2025.1619801.

Hajji-Hedfi, L., Bargougui, O., Al-Ani, L.K.T., Oueslati, S., Sukmawati, D., Wannassi, T., & Triki, M. A. (2026a). Sustainable Integrated Management of Olive Verticillium Wilt (Verticillium dahliae) Using a Microbial Consortium of Pseudomonas yamanarum-Trichoderma longibrachiatum, and Compost Mill Waste. Journal of *Soil Science and Plant Nutrition*, 1–17. https://doi.org/10.1007/s42729-026-03297-3.

Hajji-Hedfi, L., Al-Ani, L. K. T., Wannassi, T., Khlif, A., L'taief, B., & Acheampong, M. A. (2026b). Estimating Efficacy of Indigenous Isolates of Three Trichoderma Species as Biocontrol Agents Against Alternaria alternata and Curvularia spicifera. *Journal of Fungi*, 12 (6), 421. https://doi.org/10.3390/jof12060421

Hamilton-Miller, J.M.T. (1973). Chemistry and biology of the polyene macrolide antibiotics. *Journal of the American Chemical Society* 37: 166–196.

Hartmann, T. (2007). From waste products to ecochemicals. Fifty years research of plant secondary metabolism. *Phytochemistry* 68: 2832–2846.

Hemashekhar, B., Chandrappa, C.P., Govindappa, M., and Chandrashekar, N. (2019). Endophytic fungus *Alternaria* spp isolated from *Rauvolfia tetraphylla* root arbitrate synthesis of gold nanoparticles and evaluation of their antibacterial, antioxidant and antimitotic activities. *Advances in Natural Sciences: Nanoscience and Nanotechnology* 10: 035010.

Ho, T.-Y., Wu, S.-L., Chen, J.-C. et al. (2007). Emodin blocks the SARS coronavirus spike protein and angiotensin-converting enzyme interaction. *Antiviral Research* 74: 92–101.

Honary, S., Barabadi, H., Gharaei, E., and Naghibi, F. (2012). Green synthesis of copper oxide nanoparticles using *Penicillium aurantiogriseum, Penicillium citrinum* and *Penicillium waksmani. Digest Journal of Nanomaterials and Biostructures* 7 (3): 999–1005.

Hussain, F.S., Abro, N.Q., Ahmed, N. et al. (2022). Nano-antivirals: a comprehensive review. *Frontiers in Nanotechnology* 4: 1064615. https://doi.org/10.3389/fnano.2022.1064615.

Husseiny, S.M., Salah, T.A., and Anter, H.A. (2015). Biosynthesis of size controlled silver nanoparticles by *Fusarium oxysporum*, their antibacterial and antitumoral activities. *Beni-Suef University Journal of Basic and Applied Sciences* 4: 225–231. https://doi.org/10.1016/j.bjbas.2015.07.004.

Ingle, A., Gade, A., Pierrat, S. et al. (2008). Mycosynthesis of silver nanoparticles using the fungus Fusarium acuminatum and its activity against some human pathogenic bacteria. *Current Nanoscience* 4: 141–144.

Isaka, M., Berkaew, P., Intereya, K. et al. (2007). Antiplasmodial and antiviral cyclohexadepsipeptides from the endophytic fungus *Pullularia* sp. BCC 8613. *Tetrahedron* 63: 6855–6860.

Jagtap, P., Nath, H., Kumari, P.B. et al. (2021). Mycogenic fabrication of nanoparticles and their applications in modern agricultural practices & food industries. *Fungi Bio-Prospects in Sustainable Agriculture, Environment and Nano-technology* 3: 475–488.

Jiang, H., Li, C., Xu, S., Bhat, J.D., Dawoud, T.M., Al-Ani, L.K.T., Xu, J. and Phookamsak, R. (2026). Querciphoma Phyllanthi-Emblicae, a Novel Leaf Endophyte in Amla (Phyllanthus Emblica L.) from Yunnan, China. *Mycobiology*, 54 (2): 217–225. https://doi.org/10.1080/12298093.2026.2632471.

Khan, N.T., Khan, M.J., Jameel, J. et al. (2017). An overview: biological organisms that serves as nanofactories for metallic nanoparticles synthesis and fungi being the most appropriate. *Bioceramics Development and Applications* 7: 101. https://doi.org/10.4172/2090-5025.1000101.

Khandel, P. and Shahi, S.K. (2018). Mycogenic nanoparticles and their bio-prospective applications: current status and future challenges. *Journal of Nanostructure in Chemistry* 8: 369–339. https://doi.org/10.1007/s40097-018-0285-2.

Khandel, P., Yadaw, R.K., Soni, D.K. et al. (2018). Biogenesis of metal nanoparticles and their pharmacological applications: present status and application prospects. *Journal of Nanostructure in Chemistry* 8 (3): 217–254.

Khooshe-Bast, Z., Sahebzadeh, N., Ghaffari-Moghaddam, M., and Mirshekar, A. (2016). Insecticidal effects of zinc oxide nanoparticles and Beauveria bassiana TS11 on Trialeurodes vaporariorum (Westwood, 1856) (Hemiptera: Aleyrodidae). *Acta agriculturae Slovenica* 107: 299–309.

Kremer, R.J. (2019). Bioherbicides and nanotechnology: current status and future trends. In: *Nano-Biopesticides Today and Future Perspectives* (ed. O. Koul). Burlington: Academic.

Li, E., Tian, R., Liu, S. et al. (2008). Pestalotheols A–D, bioactive metabolites from the plant endophytic fungus *Pestalotiopsis thea*. *Journal of Natural Products* 71: 664–668.

Li, G., He, D., Qian, Y. et al. (2012). Fungus-mediated green synthesis of silver nanoparticles using *Aspergillus terreus*. *International Journal of Molecular Sciences* 13: 466–476.

Li, Q., Liu, F., Li, M. et al. (2022). Nanoparticle and nanomineral production by fungi. *Fungal Biology Reviews* 41: 31e44. https://doi.org/10.1016/j.fbr.2021.07.003.

Linnakoski, R., Reshamwala, D., Veteli, P. et al. (2018). Antiviral agents from fungi: diversity, mechanisms and potential applications. *Frontiers in Microbiology* 9: 2325. https://doi.org/10.3389/fmicb.2018.02325.

Liu, S.-S., Jiang, J.-X., Huang, R. et al. (2019). A new antiviral 14-nordrimane sesquiterpenoid from an endophytic fungus *Phoma* sp. *Phytochemistry Letters* 29: 75–78.

Loshchinina, E.A., Vetchinkina, E.P., and Kupryashina, M.A. (2023). Diversity of biogenic nanoparticles obtained by the fungi-mediated synthesis: a review. *Biomimetics* 8: 1. https://doi.org/10.3390/biomimetics8010001.

Luceri, A., Francese, R., Lembo, D. et al. (2023). Silver nanoparticles: review of antiviral properties. *Mechanism of Action and Applications. Microorganisms* 11: 629. https://doi.org/10.3390/microorganisms11030629.

Luo, X., Yang, J., Chen, F. et al. (2018). Structurally diverse polyketides from the mangrovederived fungus *Diaporthe* sp. SCSIO 41011 with their antiinfluenza A virus activities. *Frontiers in Chemistry* 6: 1–10.

Ma, K.X., Shen, X.-T., Huang, R. et al. (2014). Bioactive metabolites produced by the endophytic fungus *Phomopsis* sp. YM355364. *Natural Product Communications* 9: 1–49.

Marin, S., Ramos, A.J., Cano-Sancho, G., and Sanchis, V. (2013). Mycotoxins: occurrence, toxicology, and exposure assessment. *Food and Chemical Toxicology* 60: 218–237.

Mehra, R.K. and Winged, I.L. (1991). Metal ion resistance in fungi: molecular mechanisms and their regulated expression. *Journal of Cellular Biochemistry* 45: 30–40.

Meng, J., Li, R., Huang, Q. et al. (2023). Survey and toxigenic abilities of *Aspergillus*, *Fusarium*, and *Alternaria* fungi from wheat and paddy grains in Shanghai, China. *Frontiers in Plant Science* 25 (14): 1202738. https://doi.org/10.3389/fpls.2023.1202738.

Metuku, R.P., Pabba, S., Burra, S. et al. (2014). Biosynthesis of silver nanoparticles from Schizophyllum radiatum HE 863742.1: their characterization and antimicrobial activity. *3 Biotech* 4 (3): 227–234. https://doi.org/10.1007/s13205-013-0138-0.

Mitra, D., Nayeri, F.D., Sansinenea, E. et al. (2023). Unraveling arbuscular mycorrhizal fungi interaction in rice for plant growth development and enhancing phosphorus use efficiency through recent development of regulatory genes. *Journal of Plant Nutrition* 46: 1–37.

Mitrović, I., Grahovac, J., Hrustić, J. et al. (2021). Utilization of waste glycerol for the production of biocontrol agents nigericin and niphimycin by *Streptomyces hygroscopicus*: bioprocess development. *Environmental Technology* https://doi.org/10.1080/09593330.2021.1913241.

Mitrović, I., Lukić, N., Grahovac, M. et al. (2021). Optimization of *Streptomyces hygroscopicus* cultivation parameters in lab scale bioreactor. *Chemical Engineering and Technology* 44: 349–358. https://doi.org/10.1002/ceat.202000380.

Mitrović, I., Vučurović, D., Al-Ani, L.K.T. et al. (2023). Production of *Trichoderma harzianum* K179 bioagent for maize diseases control: complete laboratory stage bioprocess development. *Journal of Applied Microbiology* 134: 1–13. https://doi.org/10.1093/jambio/lxad115.

Möbius, N. and Hertweck, C. (2009). Fungal phytotoxins as mediators of virulence. *Current Opinion in Plant Biology* 12 (4): 390–398.

Mohammed, A.M. and Al-Ani, L.K.T. (2021). Identification and production of Beauvericin by *Fusarium subglutinans* and *F. sacchari* from sugarcane. *Brazilian Archives of Biology and Technology* 64: e21200088.

Namasivayam, S.K.R., Tony, B., Bharani, R.S.A., and Russell, R.F. (2015). Herbicidal activity of soil isolate of *Fusarium oxysporum* free and chitosan nanoparticles coated metabolites against economic important weed *Ninidam theenjan*. *Asian Journal of Microbiology, Biotechnology & Environmental Sciences* 17 (4): 1015–1020.

Namasivayam, S.K.R., Bharani, R.A., and Karunamoorthy, K. (2018). Insecticidal fungal metabolites fabricated chitosan nanocomposite (IM-CNC) preparation for the enhanced larvicidal activity-An effective strategy for green pesticide against economic important insect pests. *International Journal of Biological Macromolecules* 120: 921–944.

Namasivayam, S.K.R., Pandian, U.K., Chava, V. et al. (2023). Chitosan nanocomposite as an effective carrier of potential herbicidal metabolites for noteworthy phytotoxic effect against major aquatic invasive weed water hyacinth (*Eichhornia crassipes*). *International Journal of Biological Macromolecules* 226: 1597–1610. https://doi.org/10.1016/j.ij.iomac.2022.11.272.

Numata, M., Hasegawa, T., Fujisawa, T. et al. (2004). B-1, 3-glucan (Schizophyllan) can act as a one-dimensional host for creation of novel poly (aniline) nanofiber structures. *Organic Letters* 6: 4447–4450. https://doi.org/10.1021/ol0483448.

Nyahe PAS, Eziah VY, Al-Ani LKT, Akumyoungta M, Coombes CA, Rangel DEN, Alder-Rangel A, Sukmawati D, Aidoo OF and Acheampong MA(2025). Extreme UV sensitivity of native Metarhizium spp. as potential biocontrol agent for False Codling Moth (Thaumatotibia leucotreta Meyrick) on chili pepper in Ghana. *Front. Fungal Biol.* 6: 1660692. doi: 10.3389/ffunb.2025.1660692

Patocka, J. (2016). Bioactive metabolites of entomopathogenic fungi *Beauveria bassiana*. *Military Medical Science Letters (Vojenske Zdravotnicke Listy)* 85 (2): 80–88. https://doi.org/10.31482/mmsl.2016.015.

Petersen, A.B., Rønnest, M.H., Larsen, T.O., and Clausen, M.H. (2014). The chemistry of griseofulvin. *Chemical Reviews* 114: 12088–12107. https://doi.org/10.1021/cr400368e.

Peyrat, L.-A., Eparvier, V., Eydoux, C. et al. (2017). Betulinic acid, the first lupane-type triterpenoid isolated from both a *Phomopsis* sp. and its host plant *Diospyros carbonaria* benoist. *Chemistry & Biodiversity* 14: 1–8.

Phanjom, P. and Ahmed, G. (2017). Effect of different physicochemical conditions on the synthesis of silver nanoparticles using fungal cell filtrate of *Aspergillus oryzae* (MTCC No. 1846) and their antibacterial effects. *Advances in Natural Sciences: Nanoscience and Nanotechnology* 8: 1–13. https://doi.org/10.1088/2043-6254/aa92bc.

Philip, D. (2009). Biosynthesis of Au, Ag and Au–Ag nanoparticles using edible mushroom extract. *Spectrochimica Acta, Part A: Molecular and Biomolecular Spectroscopy* 73: 374–381.

Pusztaheli, T., Holb, I.J., and Pócsi, I. (2015). Secondary metabolites in fungus-plant interactions. *Frontiers in Plant Science* 6: 573.

Qian, Y., Yu, H., He, D. et al. (2013). Biosynthesis of silver nanoparticles by the endophytic fungus *Epicoccum nigrum* and their activity against pathogenic fungi. *Bioprocess and Biosystems Engineering* 36: 1613–1619. https://doi.org/10.1007/s00449-013-0937-z.

Rai, M., Yadav, A., and Gade, A. (2009). Silver nanoparticles as a new generation of antimicrobials. *Biotechnology Advances* 27 (1): 76–83.

Rai, M., Ingle, A.P., Trzcinska-Wencel, J. et al. (2021a). Biogenic silver nanoparticles: what we know and what do we need to know? *Nanomaterials* 11: 2901.

Rai, M., Bonde, S., Golinska, P. et al. (2021b). *Fusarium* as a novel fungus for the synthesis of nanoparticles: mechanism and applications. *Journal of Fungi* 7: 139.

Raliya, R. and Biswas, P. (2015). Environmentally benign bio-inspired synthesis of Au nanoparticles, their self-assembly and agglomeration. *RSC Advances* 5: 42081.

Rhouma, A., Hajji-Hedfi, L., Bouqellah, N.A., Khaire, P.B., Dali, S., Bargougui, O., Khlif, A., & Al-Ani, L.K. (2024). Uniting the Role of Entomopathogenic Fungi against Rhizoctonia solani JG Kuhn, the Causal Agent of Cucumber Damping-Off and Root Rot Diseases. *Phyton*, 93 (11): 2857-2881. https://doi.org/10.32604/phyton.2024.057591.

Riddin, T.L., Gericke, M., and Whiteley, C.G. (2006). Analysis of the inter- and extracellular formation of platinum nanoparticles by *Fusarium oxysporum* f. sp. lycopersici using response surface methodology. *Nanotechnology* 17: 3482–3489.

Saitawadekar, A. and Kakde, U.B. (2020). Green synthesis of copper nanoparticles using *Aspergillus flavus*. *Journal of Critical Reviews* 7: 9.

Saitoh, N., Fujimori, R., Yoshimura, T. et al. (2020). Microbial recovery of palladium by baker's yeast through bioreductive deposition and biosorption. *Hydrometallurgy* 196: 105413.

Sanghi, R. and Verma, P. (2009). Biomimetic synthesis and characterisation of protein capped silver nanoparticles. *Bioresource Technology* 100 (1): 501–504. https://doi.org/10.1016/j.biortech.2008.05.048.

dos Santos Lacerda, I.C., Polonio, J.C., and Correia Golias, H. (2021). Endophytic fungi as a source of antiviral compounds—a review. *Chemistry & Biodiversity* 19: e202100971. https://doi.org/10.1002/cbdv.202100971.

Sarkar, J. and Acharya, K. (2017). *Alternaria alternata* culture filtrate mediated bioreduction of chloroplatinate to platinum nanoparticles. *Inorganic and Nano-Metal Chemistry* 47: 365–369.

Saxena, J., Sharma, P.K., Sharma, M.M., and Singh, A. (2016). Process optimization for green synthesis of silver nanoparticles by *Sclerotinia sclerotiorum* MTCC 8785 and evaluation of its antibacterial properties. *Springerplus* 5: 861. https://doi.org/10.1186/s40064-016-2558-x.

Senapati, U.S., Jha, D.K., and Sarkar, D. (2015). Structural, optical, thermal and electrical properties of fungus guided biosynthesized zinc sulphide nanoparticles. *Research Journal of Chemical Sciences* 2231: 606X.

Shahzad, A., Saeed, H., Iqtedar, M. et al. (2019). Size-controlled production of silver nanoparticles by *Aspergillus fumigatus* BTCB10: likely antibacterial and cytotoxic effects. *Journal of Nanomaterials* 5168698. https://doi.org/10.1155/2019/5168698.

Sharma, V., Salwan, R., and Al-Ani, L.K.T. (2020). *Molecular Aspects of Plant Beneficial Microbes in Agriculture*, 454. Cambridge: Elsevier Science.

Shelar, G. and Chavan, A. (2015). Myco-synthesis of silver nanoparticles from *Trichoderma harzianum* and its impact on germination status of oil seed. *Nanomaterials* 13 (4): 702. https://doi.org/10.1080/21501203.2024.2354302.

Shi, Y., Ji, M., Dong, J. et al. (2024). New bioactive secondary metabolites from fungi: 2023. *Mycology* 15 (3): 283–321. https://doi.org/10.1080/21501203.2024.2354302.

Shukla, G., Gaurav, S.S., Singh, A., and Rani, P. (2022). Synthesis of mycogenic silver nanoparticles by *Fusarium pallidoroseum* and evaluation of its larvicidal effect against white grubs (*Holotrichia* sp.). *Materials Today Proceedings* 49: 3517–3527.

Siddiqi, K.S. and Husen, A. (2016). Fabrication of metal nanoparticles from fungi and metal salts: scope and application. *Nanoscale Research Letters* 11: 98. https://doi.org/10.1186/s11671-016-1311-2.

Singh, S.B., Zink, D.L., Guan, Z. et al. (2003). Isolation, structure, and HIV-1 integrase inhibitory activity of Xanthoviridicatin E and F, two novel fungal metabolites produced by *Penicillium chrysogenum*. *Helvetica Chimica Acta* 86: 3380–3385.

Singh, S.B., Ondeyka, J.G., Tsipouras, N. et al. (2004). Hinnuliquinone, a C2-symmetric dimeric non-peptide fungal metabolite inhibitor of HIV-1 protease. *Biochemical and Biophysical Research Communications* 324: 108–113.

Singh, H., Du, J., Singh, P., and Yi, T.H. (2018). Extracellular synthesis of silver nanoparticles by *Pseudomonas* sp. THG-LS1.4 and their antimicrobial application. *Journal of Pharmaceutical Analysis* 8 (4): 258–264. https://doi.org/10.1016/J.JPHA.2018.04.004.

Singh, A., Singh, K., Sharma, A. et al. (2023). Recent advances in antifungal drug development targeting lanosterol 14α-demethylase (CYP51): a comprehensive review with structural and molecular insights. *Chemical Biology and Drug Design* 102: https://doi.org/10.1111/cbdd.14266.

Skinner, M., Parker, B.L., and Kim, J.S. (2014). Role of entomopathogenic fungi. In: *Integrated Pest Management* (ed. D.P. Abrol), 169–191. Cambridge, MA: Academic Press.

Soleimani, P., Mehrvar, A., Michaud, J.P., and Vaez, N. (2022). Optimization of silver nanoparticle biosynthesis by entomopathogenic fungi and assays of their antimicrobial and antifungal properties. *Journal of Invertebrate Pathology* 190: 107749. https://doi.org/10.1016/j.jip.2022.107749.

Soumare A, Diedhiou AG, Arora NK, Al-Ani LKT, Ngom M, Fall S, Hafidi M, Ouhdouch Y, Kouisni L and Sy MO (2021) Potential Role and Utilization of Plant Growth Promoting Microbes in Plant Tissue *Culture. Front. Microbiol.* 12: 649878. doi: 10.3389/fmicb.2021.649878.

Sousa, F., Ferreira, D., Reis, S., and Costa, P. (2020). Current insights on antifungal therapy: novel nanotechnology approaches for drug delivery systems and new drugs from natural sources. *Pharmaceuticals* 13: 248. https://doi.org/10.3390/ph13090248.

Strobel, G. and Daisy, B. (2003). Bioprospecting for microbial endophytes and their natural products. *Microbiology and Molecular Biology Reviews* 67: 491–502. https://doi.org/10.1128/MMBR.67.4.491-502.

Sukmawati, D., Nurmayanti, F., Supiyani, A., Kurniati, T.H., Adisyahputra, A., Tawfeeq Al-Ani, L.K., Widowati, R., Nida Al Husna, S., Hidayat, I., Bimo Setiarto, R.H., Sulistiani, S., & Abdul Rahman, R. (2024). Molecular Identification and Antifungal Activity of Candida chrysomelidarum and Rhodotorula toruloides Isolated from Indonesian Indigenous Fruit Flacourtia inermis Roxb. (Lobi-lobi) against Pathogenic Mold Aspergillus flavus. *Trends in Sciences*, 20 (10): 8321. https://doi.org/10.48048/tis.2024.8321.

Swapnil, G., Avinash, I., Aniket, G. et al. (2013). Antiviral activity of mycosynthesized silver nanoparticles against herpes simplex virus and human parainfluenza virus type. *International Journal of Nanomedicine* 4303–4314. https://doi.org/10.2147/IJN.S50070.

Ulrich, A., Lerin, L.A., Camargo, A.F. et al. (2021). Alternative bioherbicide based on *Trichoderma koningiopsis*: enzymatic characterization and its effect on cucumber plants and soil organism. *Biocatalysis and Agricultural Biotechnology* 36: 102127. https://doi.org/10.1016/j.bcab.2021.10.127.

Ulrich, A., Müller, C., Gasparetto, I.G. et al. (2023). Bioherbicide effects of *Trichoderma koningiopsis* associated with commercial formulations of glyphosate in weeds and soybean plants. *Crop Protection* 172: 106346. https://doi.org/10.1016/j.cropro.2023.106346.

Vigneshwaran, N., Kathe, A.A., Varadarajan, P.V. et al. (2007). Silver-protein (core-shell) nanoparticle production using spent mushroom substrate. *Langmuir* 23: 7113–7117.

Villar-Luna, E., Gomez-Rodriguez, O., Villar-Luna, H., Aguilar-Marcelino, L., Al-Ani, L.K.T., and Fernandez-Herrera, E. (2022). Eco-friendly Management of False Root-Knot Nematode Nacobbus aberrans: An Overview. In: Chaudhary, K.K. and Meghvansi, M.K. (Eds.), Sustainable Management of Nematodes in Agriculture, Vol.1: Organic Management. Springer *Nature* Switzerland, pp. 327–342. https://doi.org/10.1007/978-3-031-09943-4_13.

Wang, X., Xu, J., Wang, X. et al. (2019). *Isaria fumosorosea*-based zero-valent iron nanoparticles affect the growth and survival of sweet potato whitefly, *Bemisia tabaci* (Gennadius). *Pest Management Science* 75: 2174–2181.

Wang, D., Xue, B., Wang, L. et al. (2021). Fungus-mediated green synthesis of nano-silver using *Aspergillus sydowii* and its antifungal/antiproliferative activities. *Scientific Reports* 11: 10356. https://doi.org/10.1038/s41598-021-89854-5.

Wu, F., Groopman, J.D., and Pestka, J.J. (2014). Public health impacts of foodborne mycotoxins. *Annual Review of Food Science and Technology* 5: 351–372.

Yang, Z.-J., Zhang, Y.-F., Wu, K. et al. (2020). New azaphilones, phomopsones AC with biological activities from an endophytic fungus *Phomopsis* sp. CGMCC No. 5416. *Fitoterapia* 145: 1–6.

Yelikbayev, B., Correa, E.J.A., Al-Ani, L.K., Duarte, N.F. and Pagano, M.C. (2024). Revitalizing Degraded Rural Landscapes: Approaches and Strategies for Ecological Restoration. Environmental Reports;an International Journal. https://doi.org/10.51470/ER.2024.6.2.23.

Zhang, G., Sun, S., Zhu, T. et al. (2011a). Antiviral isoindolone derivatives from an endophytic fungus *Emericella* sp. associated with Aegiceras corniculatum. *Phytochemistry* 72: 1436–1442.

Zhang, X., Yan, S., Tyagi, R.D., and Surampalli, R.Y. (2011b). Synthesis of nanoparticles by microorganisms and their application in enhancing microbiological reaction rates. *Chemosphere* 82 (4): 489–494. https://doi.org/10.1016/j.chemosphere.2010.10.023.

Zhang, X., Li, G., Wu, D. et al. (2020). Recent advances on emerging nanomaterials for controlling the mycotoxin contamination: from detection to elimination. *Food Frontiers* 1–22.

5

Potential Applications of Cyanobacterial Nanoparticles in Agriculture

Rajeshwari Negi[1], Babita Sharma[2], Faezeh Parastesh[3], Tanvir Kaur[1], Samrendra Singh Thakur[4], Subhikshaa Bagavathiappan[5], Ashok Yadav[6], Sofia Sharief Khan[7], Neelam Yadav[8], Paridhi Puri[9], and Ajar Nath Yadav[10]

[1] Department of Biosciences, Graphic Era Deemed to Be University, Dehradun, Uttarakhand, India
[2] Department of Microbiology, Akal College of Basic Sciences, Eternal University, Sirmour, Himachal Pradesh, India
[3] Department of Soil Science, University of Manitoba, Winnipeg, Manitoba, Canada
[4] Department of Biotechnology, School of Biological Sciences, Dr Harisingh Gour Vishwavidyalaya (A Central University), Sagar, Madhya Pradesh, India
[5] Department of Biology, Concordia University, Montreal, Quebec, Canada
[6] Department of Botany, Banaras Hindu University, Varanasi, Uttar Pradesh, India
[7] Department of Biotechnology, Shri Mata Vaishno Devi University, Katra, Jammu and Kashmir, India
[8] Centre of Research Impact and Outcome, Chitkara University Institute of Engineering and Technology, Chitkara University, Rajpura, Punjab, India
[9] University Centre for Research and Development, Chandigarh University, Mohali, Punjab, India
[10] Department of Genetics, Plant Breeding and Biotechnology, Dr. Khem Singh Gill Akal College of Agriculture, Eternal University, Sirmour, Himachal Pradesh, India

5.1 Introduction

A remarkable progress in nanotechnology has been gained in recent years in the synthesis of nanomaterials with superior physicochemical characteristics (Khan et al. 2019). Nearly every scientific discipline has made use of nanotechnology to develop more innovative solutions for a range of research-related problems. With the goal of developing nanoparticles (NPs) with biological counterparts that affect their physical characteristics, the field of nano-biotechnology has developed (Omran and Omran 2020). The synthesis of NPs in a range of sizes and shapes sparked a lot of attention due to their distinct qualities when compared to their huge counterparts (Salem et al. 2022). The larger aspect ratio between surface area and volume at the nanoscale allows for significant variation in the stability of the NPs physicochemical, biological, and chemical properties (Altammar 2023). Significant variations in melting point, mechanical properties, thermal and electrical conductivities, optical absorbance, and catalytic and biological activity are thus encouraged. NPs serve as aqueducts connecting the bulk goods to the atomic or molecular structures (Li et al. 2021). The structure, dimensions, and origin of NPs determine their classification, and their sizes range from 1 to 100 nm (Bhatia and Bhatia 2016).

Nearly every scientific domain, including energy, space, communication, defence, biomedicine, and agriculture, uses NPs extensively due to their distinct physicochemical and morphological (charge, size, and shape) characteristics (Pal 2022). The use of NPs in ailment diagnostics, imaging, drug administration, treatment, cancer therapies, and tissue engineering has grown dramatically in recent years (Harish et al. 2022). However, it is preferable if the NPs used for biomedical applications come from green sources because their use in humans is quite crucial (Rónavári et al. 2021). Thus, the existing NP synthesis process needs to change to the safer option while taking into account many criteria such as the bioavailability, biocompatibility, biosafety, and biodistribution of NPs (Chandrakala et al. 2022). NPs are mostly created by physical and chemical processes such as electroexplosion, solgel, laser ablation, and mechanical milling, yet they are terrible for the environment and human health (Islam et al. 2022). Consequently, there is now a greater need for NPs synthesis using safer, non-hazardous, alternative, economical, and environmentally acceptable green technologies (Ahmed et al. 2022). Green synthesis technology for NPs, which are devoid of hazardous contaminants, can be employed in organisms such as bacteria, fungi, algae, and plants. In extracellular or intracellular NP production, these biological methods use several biological units acting as capping and reducing agents (Salem and Fouda 2021). The biological units functional groups (thiol, carbonyl, amine, and hydroxyl), peptides, and proteins are referred to as stabilizing or reducing agents since they convert the salt precursor into NPs (Sidhu et al. 2022). This type of synthesis is relatively stable, repeatable, and easy because they both stabilize and cap the NPs at the same time (Sajid and Płotka-Wasylka 2020).

The integration of algae has expanded for green synthesis methods because they are thought to be good sources of proteins, peptides, secondary metabolites, and pigments that can function as nanobiofactories (Nitnavare et al. 2022). They are, nevertheless, promising entities for NPs biosynthesis because of their fast growth rate, ease of harvesting, and low cost of scale-up (Bahrulolum et al. 2021). Algae are among the most ancient creatures that live in a variety of environments and are important photosynthetic organisms. One of the greatest options for environmentally friendly NP synthesis is algae, which have the ability to hyper-accumulate metals before turning them into NPs (El Messaoudi et al. 2024). A variety of algae, including brown algae (*Phaeophyceae*), blue-green algae (*Cyanobacteria*), red algae (*Rhodophyceae*), and green algae (*Chlorophyceae*), synthesize a large number of metal oxide and metallic NPs (Arya and Chundawat 2020).

Cyanobacteria, commonly known as blue-green algae, are the largest groups of photosynthetic prokaryotes found in all types of environments around the world (Saad and Atia 2014). Most cyanobacteria live in water as phytoplankton. Cyanobacteria represent one of the oldest life forms on the planet. They are morphologically and physiologically diverse prokaryotes and characterized as unusual prokaryotic microorganisms that can perform oxygenic photosynthesis (Hamida et al. 2020b). They have chlorophyll, like eukaryotic algae and plants, that plays an essential role in photosynthesis. They also use water as an electron donor for the production of oxygen during the process of photosynthesis (Patel et al. 2015). They play a crucial role in the formation of the geochemistry and biosphere of the earth. Cyanobacteria directly convert organic and inorganic molecules in value-added products as they have biocatalytic property (Bakku and Rakwal 2022). With the help of genetic engineering, the introduction of a few heterologous genes can endow cyanobacteria with the ability to transform specific central metabolites into many end products. This capability of cyanobacteria can be used in nanotechnology to synthesize nontoxic NPs that can be used in biomedical, agricultural, and other potential applications (Singh 2020).

5.2 Synthesis of Nanoparticles

Synthesis of NPs is one of the crucial steps in the field of nanotechnology. It determines certain properties such as size, shape, geometry, surface area, homogeneity, dispersion and the chemical composition of the nanoparticle Nayak et al. (2023) which can play a crucial role in the application. The synthesis can be broadly classified into physical, chemical, and biological methods. Generally, physical synthesis uses a top-down approach where the synthesis involves breaking down larger structures into smaller, nanoscale particles, while chemical and biological synthesis uses a bottom-up approach where materials are constructed starting from its smallest units, such as molecules, atoms, or clusters layer by layer (Kumari et al. 2023).

5.2.1 Physical Synthesis

Physical synthesis of NP relies on mechanical processes such as thermal energy, radiation, and mechanical pressure to produce nanoscale materials from bulk sources without chemical intervention (Al-Harbi and Abd-Elrahman 2024). This approach produces nanomaterials that are homogeneous and environment friendly (Kumari et al. 2023). However, in hindsight this method demands substantial energy input, which makes it well suited for large-scale industrial application. The physical synthesis encompasses techniques like ball milling, lithography fabrication, laser ablation, sputtering, thermal decomposition, and vapor condensation.

5.2.1.1 Ball Milling

The fundamental concept of this technique relies on utilizing ball milling to decrease particle dimensions through high-energy processes. The reduction in particle size alters the material's surface characteristics. Mechanical milling is categorized into low-energy and high-energy processes, with the outcome dependent on the milling powder properties and the applied mechanical energy. This method is frequently used to create metal NPs (Joy et al. 2022; Piras et al. 2019). In this procedure, the material intended for nanoparticle conversion is placed in a container with steel or carbide balls, maintaining a 2:1 ball-to-material ratio under an inert gas atmosphere. The container rotates rapidly around its axis, resulting in the raw materials being properly crushed between the container wall and balls. The rotation speed of the container and the time of milling affect the production rate and the size of NPs (Al-Harbi and Abd-Elrahman 2024). Various milling containers are employed, including attrition ball mills, high-energy and low-energy ball mills, planetary ball mills, and vibrating ball mills (Jamkhande et al. 2019). NPs produced through ball milling typically range from 3 to 25 nm in size. Multiple factors affect the morphological and physical properties of the resulting nanomaterials, such as material type, mill type, rotation speed, grinding duration, energy input, ball-to-powder ratio, and the inert gas used in the grinding environment (Nyabadza et al. 2023).

There are several benefits associated with ball milling methods, including the fact that this method is widely used to produce nanomaterials with enhanced physical properties like high purity and solubility. It is an eco-friendly process that does not generate pollution (Joy et al. 2022; Piras et al. 2019). Apart from benefits, some of the key drawbacks are also linked with this method, such as the production of nanomaterials requiring significant energy input. The grinding process is time-consuming. Contamination of nanomaterials may occur during grinding. The material's microstructure can be susceptible to the grinding process. Control over the size of the resulting nanomaterials is limited (de Oliveira et al. 2020).

5.2.1.2 Lithography Fabrication Method

Lithography technology offers a cost-effective approach to generating NPs with controllable dimensions, encompassing conventional and unconventional methods (Ijaz et al. 2020; Nyabadza et al. 2023). The conventional technique involves pattern transfer to a thin resist through interactions between the resist and particles, such as photons or electrons. This transfer occurs via direct or indirect writing. Direct writing employs a point-by-point mechanism, making it slower than indirect writing, which utilizes a mask for one-step pattern transfer. In contrast, the unconventional method involves imprinting a pattern onto a polymer using a mold and then transferring it to the substrate after separation (Aassime and Hamouda 2017). Various lithography techniques exist, including photolithography, electron beam, soft, focused ion, nano-imprint, and dip-pen lithography. Photolithography encompasses contact, proximity, and projection printing (Ijaz et al. 2020). In optical lithography, a light source and a pattern-bearing mask are typically utilized. The mask can be positioned relative to the light source, including in contact or non-contact modes with the substrate. The mask is considered binary, as light either passes through areas without metal or reflects off metal-covered areas. The pattern may function as an intermediary through a resist layer on the substrate surface, enabling specific molecular configurations in the resist via light-resist interactions (Nyabadza et al. 2023).

Electron beam lithography is a widespread technique in scientific literature due to its high precision and user-friendliness. This method can be divided into scanning electron microscopy (SEM) to mask writers, shaped spot systems, and electron projection beam machines. In modified SEM systems, pattern transfer from the computer is achieved by manipulating beam displacement and activating or deactivating it at specific locations in a pixel-by-pixel fashion (Aassime and Hamouda 2017). Nanosphere lithography is an unconventional method that combines top-down and bottom-up approaches. This process consists of two primary phases: mask preparation and nanostructure formation. The substrate's hydrophilic properties can be enhanced through chemical treatment. A monolayer or bilayer with an Hexagonal Close-Packed (HCP) crystal structure as the substrate dries forms a colloidal crystal mask (CCM). Selective material deposition allows for mask distribution across the substrate surface. The second phase involves mask removal and nanostructure creation. Mask removal can be accomplished through various methods, such as dispersion in an appropriate solvent or stripping. This technique can be considered both bottom-up (due to the self-organization of colloidal spheres into an HCP) and top-down (resulting in dot formation) (Colson et al. 2013; Nyabadza et al. 2023).

5.2.1.3 Laser Ablation Method

Laser ablation synthesis in a solution is a straightforward method for creating NPs using various solvents. When a laser beam irradiates different metals submerged in solution, it forms a plasma that generates NPs. This top-down approach differs from traditional chemical methods of reducing metals to NPs (Baig et al. 2021; Ijaz et al. 2020). The NPs produced through laser ablation are stable and require no stabilizing agents or chemicals. Scientists have employed laser ablation to create lead sulfide NPs on silicon substrates and silver NPs on nickel hydroxide nanosheets. This technique can also be applied to carbon nanotube production. The size and formation of NPs are significantly influenced by laser beam type, wavelength, pulse duration, and solvent type (Baig et al. 2021). The benefits of this method include efficiency in producing nanoparticle suspensions without surfactants, preparing silicon and Al_2O_3 nanoparticle coatings, and high nanoparticle purity of about 90%. Additionally, it is considered an environmentally friendly approach (Kumari et al. 2023). However, drawbacks include low productivity and high costs (Kim et al. 2017).

5.2.1.4 Sputtering

This method is also known as ion beam deposition (IBD), ionizing mass beam deposition (IMBD), and ion sputtering (IS). Ion sputtering is a technique developed in 1985 to produce high-quality single-crystalline nanofilms (Son et al. 2017). The process occurs in a vacuum chamber at 0.05–0.1 mbar pressure. A high electrical potential is applied to the target material, causing surface molecule evaporation. An electron beam is passed, with free electrons moving in a spiral path via magnetic poles, colliding with inert gas (argon) atoms and ionizing them. As the process continues, the accelerated electrical voltage creates a plasma, attracting positively charged gas ions that continuously collide with the targeted substance surface. This process repeats using energy exceeding the binding energy between substrate molecules. Ongoing collisions between gas molecules and substrate atoms result in atom dispersion and vapor cloud formation. The cloud of matter atoms then deposits as thin nanofilms (Al-Harbi and Abd-Elrahman 2024; Jamkhande et al. 2019).

5.2.1.5 Physical Vapor Depositions (PVD)

A physical technique for creating nanomaterials involves transferring materials at the atomic level through material evaporation. This process yields NPs in fragile layers and is considered eco-friendly as it produces no pollution (El-Khawaga et al. 2023). Also known as evaporative condensation, this method enhances metal hardness and resistance to oxidation or corrosion by depositing nanoparticle thin films on metal surfaces. It has been employed to produce silver and germanium NPs for electronic and optical applications. Additionally, this approach is used to deposit metal NPs and thin films onto carbon nanotubes (Escorcia-Díaz et al. 2023). Benefits of this method include relatively low temperatures in the PVD process, simplicity in producing thin metal nano-films, controllable nanoparticle size, and good crystal structure of the resulting NPs. However, drawbacks include high production costs and low productivity (Al-Harbi and Abd-Elrahman 2024).

5.2.2 Chemical Synthesis

The constructive method, called the bottom-up approach, is the opposite of the top-down method. This technique involves creating NPs from simpler components. The major chemical synthesis processes used in NPs, such as chemical vapor deposition (CVD), sol-gel, spinning, pyrolysis, solvothermal, and hydrothermal methods, reverse micelle methods, inert gas condensation/molecular condensation, and co-precipitation methods.

5.2.2.1 Chemical Vapor Deposition

CVD is a process that produces a thin layer on a substrate surface through a chemical reaction involving vapor-phase precursors (Escorcia-Díaz et al. 2023). Suitable precursors for CVD should be sufficiently volatile, chemically pure, stable during evaporation, affordable, safe, and have a long shelf life. Moreover, their decomposition should not leave residual contaminants. CVD has several variations, including atomic layer epitaxy, metal-organic CVD, plasma-enhanced CVD, and vapor phase epitaxy (Altammar 2023). Some of the key advantages of this technique are producing highly pure, rigid, uniform, and robust NPs. CVD is recognized as an excellent technique for synthesizing high-quality nanomaterials and is mainly known for creating two-dimensional NPs (Altammar 2023; Baig et al. 2021). Lee et al. (2011) used the CVD method to synthesize and coat TiO_2, utilizing titanium tetra iso-peroxide (TTIP) as the precursor.

5.2.2.2 Sol-Gel Method

The sol-gel technique is a widespread wet-chemical technique for producing high-grade metal-oxide NPs. The metal alkoxides typically act as a precursor to convert sol into a gel. Initially the metal oxide hydrolysis into water or alcohol to form sol is followed by condensation. During the condensation process a metal-polymer structure is formed, which is either formed by hydroxo- (M–OH–M) or oxo- (M–O–M) bridges. This process increases the viscosity and creates porous structures. As the process continues with aging, the porosity decreases, followed by removal of water and solvents making it dry. Finally, the calcination procedure is carried out to complete the formation of NPs (Baig et al. 2021). The key factors in this method include the type of precursor used, the rate of hydrolysis, aging duration, pH, and water-to-precursor ratio. Sol-gel method produces homogeneous materials, operates at low temperature and often considered ideal for creating composites and complex nanostructures (Altammar 2023; Al-Harbi and Abd-Elrahman 2024).

5.2.2.3 Spinning

NPs also synthesized using the spinning method. In this method a spinning disc reactor (SDR) is used to synthesize NPs. This instrument features a rotating disc with adjustable physical parameters, including temperature. The reactor is filled with nitrogen (N_2) or other inert gases to prevent unwanted chemical reactions and eliminate oxygen (O_2). The chamber or reactor is supplied with liquid, such as water and precursor, *via* pumping (Ijaz et al. 2020). Several factors influence the properties of NPs produced by SDR, including the disc surface, the ratio of liquid to precursor, disc rotation speed, liquid flow rate, and the location of feed input.

5.2.2.4 Pyrolysis

Pyrolysis is the widely used commercial method for NPs synthesis. This technique involves burning the precursor with a flame. The precursor can be in vapor or liquid state and is introduced into a furnace at high pressure to produce NPs. Lasers or plasma are sometimes used instead of flames to generate high temperatures, facilitating easier evaporation (Al-Harbi and Abd-Elrahman 2024). Advantages of pyrolysis include its efficiency, cost-effectiveness, simplicity, and continuous processing with high yield (Ijaz et al. 2020).

5.2.2.5 Solvothermal and Hydrothermal Techniques

The hydrothermal technique is a widely recognized and frequently employed method for creating nanostructured materials. This process entails a heterogeneous reaction occurring in an aqueous solution, conducted under conditions of high temperature and pressure approaching the critical point, within a confined, sealed apparatus (Mekuye and Abera 2023). The solvothermal approach is similar but uses a non-aqueous medium. Both methods typically occur in closed systems. Recently, the microwave-assisted hydrothermal technique has gained attention in nanomaterial engineering, combining the advantages of both hydrothermal and microwave methods. These approaches are valuable for producing various nanowires, nanorods, nanosheets, and nanospheres (Mekuye and Abera 2023).

5.2.2.6 Reverse Micelle Technique

An alternative effective method for creating nanomaterials with specific sizes and shapes is the reverse micelle technique. Unlike normal micelles that form in oil-in-water emulsions with hydrophobic tails facing inward, reverse micelles occur in water-in-oil emulsions with hydrophilic heads oriented toward a water-containing core. This core acts as a nanoreactor for NPs synthesis (Mekuye and Abera 2023). The dimensions of these nanoreactors can be modified by changing the water-to-surfactant ratio, which influences the resulting NPs size. A reduction in water concentration results in smaller water droplets and, subsequently, smaller NPs. This approach provides a simple way to produce uniform, precisely sized NPs that are extremely fine and monodispersed (Khan et al. 2023).

5.2.2.7 Inert Gas Condensation/Molecular Condensation

The inert gas compression method is utilized for large-scale production of metal NPs. This technique involves evaporating a metallic source in an inert gas to generate NPs. Metals vaporize at a suitable rate at achievable temperatures. For example, copper NPs are created by vaporizing copper in argon, helium, or neon. The vaporized atoms undergo rapid cooling by the inert gas, and liquid nitrogen is employed to further cool the gases, yielding NPs ranging from 2 to 100 nm in size (Altammar 2023).

5.2.2.8 Co-precipitation

The solvent displacement method is a liquid-based chemical process that utilizes various solvents including acetone, ethanol, hexane, and a non-solvent polymer. It involves combining a polymer solution, triggering rapid diffusion of the polymer solvent into the non-solvent polymer phase. NPs form as a result of interfacial tension between the two phases. The main benefit of this technique lies in its ability to generate substantial amounts of water-soluble NPs through a simple procedure. This approach is utilized for the production of various commercially available MRI contrast agents based on iron oxide NPs, such as Combidex, Feridex, and Reservist (Altammar 2023; Baig et al. 2021).

5.2.3 Biological Synthesis of NPs

The conventional methods were used to produce large quantities of NPs that are of specific size and shape; however, the method was complex, it involved outdated protocols and required significant financial investment. While the green-mediated nanoparticle synthesis utilizes natural substances like plants, microorganisms, and other organic agents, making it cost-effective and eco-friendly without toxic effects (Das et al. 2017). Every green-mediated approach varies in technique; however, the general use of biological material or their extracts that readily react with metallic salts remains the same. The metals transform into NPs by biological stabilization. The microbial-mediated synthesis is considered significant where it utilizes enzymes and secondary metabolites from fungi, bacteria, and viruses to synthesize and manipulate well-organized NPs (Nair et al. 2022). The plant-mediated methods have also proven beneficial for nanoparticle synthesis. Various components found in plant extracts, such as proteins, polysaccharides, amino acids, and phytochemicals like flavonoids, alkaloids, tannins, and polyphenols, have shown potential for stabilizing and reducing NPs. Here the plant extract is used, which has the advantage of reducing metallic ions and facilitating the formation and aids the stabilization of metallic NPs (Karunakaran et al. 2023; Naikoo et al. 2021). In contrast to chemical and physical methods, the green synthesis method has several advantages such as ease of development, simple, efficient, cost-effective, and minimal waste generation during nanoparticle production (Karunakaran et al. 2023). The green-mediated synthesis has become popular due to its affordability, nontoxicity, and high stability of NPs. It is believed that this eco-friendly method can avoid harmful impacts on both the ecosystem and human well-being.

5.2.3.1 Synthesis Using Plants

The bark, flowers, fruits, leaves, roots, shoots, and stems serve as primary sources for nanoparticle synthesis. Numerous plant leaf extracts, containing bioactive elements, are utilized in an environmentally friendly synthesis process (Hano and Abbasi 2021). While plant extracts have been traditionally used to produce metallic NPs, recent studies have revealed an innovative and effective method using carefully processed plant components (Ahmad et al. 2011). This approach has effectively enhanced the scale-up process related to downstream processing and reduced manipulation time. Research has indicated that extracted phenolic components, such as secondary metabolites, sugars, and proteins, can be employed to synthesize various metallic NPs. Plants and their extracts, particularly flavonoids, have been used for metal bioremediation. Flavonoids like quercetin have been applied to biosynthesize copper and silver NPs in micelle suspension, resulting in improved antibacterial activity and increased surface area. In an experiment with *Desmodium triflorum* plants, silver ions were introduced in an agar suspension, followed by shooting. This led to an increase in silver ion concentration within the tissue, demonstrating the plant's uptake of silver ions (Ahmad et al. 2011). The study observed the detoxification, capping of silver, and formation of AgNPs within the tissue.

NPs developed using *Chrysophyllum oliviforme* plants range from 20 to 50 nm in diameter and exhibit low polydispersity, indicating effective *in vivo* nanoparticle synthesis. These materials allow enhanced positioning of materials and synthesis possibilities (Anju Varghese et al. 2015). The extract of *Veronica amygdalina* was studied for its ability to bioconvert silver ions. This process achieved rapid development of silver NPs with an average size of 15 nm and controlled material sizes between 2 and 18 nm. Terpenoids found in geranium leaf extracts, such as geraniol, linalool, and citronellal, have been identified as effective reducing agents for silver ions (Aisida et al. 2019).

5.2.3.2 Synthesis Using Bacteria

NPs can be synthesized in both unicellular and multicellular organisms within the cells or external. In bacteria, the enzymatic processes are crucial for intracellular signaling mechanisms and play a key role in transforming the metal ions into NPs. The NP synthesis from bacteria often involves cellular extracts, followed by downstream processes namely collection, separation, and purification to ensure high-quality NPs for further applications. The bacterial approach to nanoparticle synthesis is simple and has enhanced control over the biosynthesis process (Karunakaran et al. 2023). Studies suggest sulfate-reducing bacteria, such as *Acinetobacter calcoaceticus* and *Desulfovibrio desulfuricans*, are frequently employed for controlled nanoparticle synthesis (He et al. 2008). These bacteria present a unique capability to convert the precursor solutions into NPs of specific sizes. Likewise, *Rhodopseudomonas capsulata* has also been shown to produce elemental NPs at neutral pH and 37 °C (He et al. 2008). This level of precision makes bacterial synthesis a highly valuable approach particularly in applications where precise sizes and shapes are critical. Additionally, genetic engineering techniques can also be employed to optimize and reduce the toxicity during bacterial synthesis of NPs. However, bacterial nanoparticle synthesis has its limitations, which include complex downstream processing, difficulties in maintaining cost-effectiveness, and challenges in maintaining nanoparticle precise size and shape. Despite these challenges, bacterial synthesis still remains the widely used method for the green synthesis of NPs (Karunakaran et al. 2023).

5.2.3.3 Synthesis Using Fungi

Fungi are gaining popularity in NP production, offering distinct advantages of generating higher quantities of protein over other microorganisms, which enhances the NP synthesis. They have the advantage of producing NPs with well-defined shapes and sizes, which is crucial for various applications (Dhillon et al. 2012). Additionally, fungi readily produce biomass, which contributes to their efficiency in nanoparticle formation. The fungal enzymes play a vital role in facilitating the reduction and stabilization processes, which are necessary for producing stable NPs. Extracellular nanoparticle production by the fungi *Fusarium oxysporum* produces extracellular NPs, typically at 37 °C, yielding NPs ranging from 15 to 30 nm in size. Notably, these extracellular NPs exhibit reduced toxicity (Soni and Prakash 2011). NP synthesis using fungi has been proven to be simple, efficient, scalable, and has trivial downstream processes. The scientific community has shown significant interest in fungal biosynthesis of NPs, focusing on their ability to efficiently manage nanoparticle dimensions and other properties. Compared to bacteria, fungi demonstrate higher nanoparticle output and secrete more proteins influencing the nanoparticle formation and stability, making the fungal system a promising option for nanoparticle synthesis (Rami et al. 2024).

5.2.3.4 Synthesis Using Actinomycetes

Actinomycetes are gram-positive bacteria having a filamentous structure similar to fungi. These bacteria serve as a unique source for NP synthesis. Studies have demonstrated actinomycetes exhibit unique ability to stabilize Ag and Au NPs through proteins, with free amino acids or cysteine attaching to Au NPs. There are specific proteins ranging from 10 to 80 kDa excreted by these microorganisms that play a crucial role in the reduction and stabilization of gold NPs (Ranjitha and Rai 2017). The actinomycetes have the ability to produce certain proteins that can interact with the crystallographic features of nanomaterials, enabling them to form gold nanocrystals with complex shapes and controllable sizes. Research has shown that alkali thermophilic actinomycetes can produce Au NPs as small as 8 nm in alkaline conditions. Species like *Rhodococcus* and *Thermomonospora* have shown to generate intracellular Au NPs ranging from 5 to 15 nm in size (Manimaran and Kannabiran 2017). Actinomycetes-mediated biogenic synthesis of metal and metal oxide NPs: progress and challenges. *Letters in Applied Microbiology*, 64(6), 401–408. It was found that the proportion of NPs formed on the cell membrane was notably more than within the cell wall, suggesting that membrane-associated enzymes are playing a crucial role in synthesizing NP. Furthermore, the synthesized NPs were found to be nontoxic to the cells, underscoring the potential of actinomycetes to be safe and efficient and can be used for various applications. In addition to NP synthesis, actinomycetes are well known for their secondary metabolite production, antibiotics, etc which enhance their versatility in the field of biotechnology (Karunakaran et al. 2023; Manimaran and Kannabiran 2017).

5.2.3.5 Synthesis Using Yeast

Yeasts, a group of ascomycetes, have the ability to aggregate heavy metals, which have shown significant potential in synthesizing NPs. They produce extracellular polysaccharides and peptides to manage metal toxicity (Liu et al. 2021). Additionally, glutathione, metallothioneins, and phytochelatins in yeast cells contribute to heavy metal detoxification and NP stabilization help yeasts to form nontoxic polymer complexes. *Schizosaccharomyces pombe* synthesizes semiconductor nanocrystals during its log phase, while the fungus *Verticillium luteoalbum* synthesizes intracellular gold nanoparticles (AuNPs). Researchers can optimize NP synthesis and material development by adjusting physical parameters like pH and temperature (Karunakaran et al. 2023).

Extracellular methods have also been employed to produce AgNPs. For example, *Cladosporium cladosporioides* synthesizes Ag NPs using proteins, organic acids, and polysaccharides, which controls spherical crystal formation (Bhargava et al. 2016). Similarly, *Penicillium fellutanum*, isolated from coastal mangrove sediment, has been used for the extracellular synthesis of Ag NPs. These examples highlight the potential of fungi and yeast for producing NPs with precise sizes and shapes through environmentally friendly processes (Gajbhiye et al. 2009). Biologically mediated NP synthesis is promising due to its efficiency, nontoxicity, and adaptability for large-scale production.

5.3 Cyanobacteria as a Source of Nanoparticles Synthesis

5.3.1 Synthesis of Gold Nanoparticles from Cyanobacteria

Recently, biosynthesized AuNPs have been found and studied because of their valuable application in the medical and biomedical industries (Khan et al. 2022). AuNPs can be made in different shapes such as spheres, rod-like, or core-shell by some specific surface chemistry techniques (Moreira et al. 2018). The defect of both size and shape can affect their behavior with light. Biosynthesized NPs have been found for their therapeutic applications in antimicrobial, anticancer, and labeling agents in designing immunoassays (Bamal et al. 2021). AuNPs have been used to target the delivery of anticancer medicines, such as methotrexate, doxorubicin (DOX), and paclitaxel. AuNPs absorption relies on its shape and size, which could be controlled by temperature, gold concentration, and pH (Eyu 2018; Gericke and Pinches 2006). It has been found that the main and dominant shape of AuNPs is spherical, but there are many other shapes. Hexagons, triangles, cells, nanospheres, octahedrons, stars, wells, and nanorods are different forms of AuNPs, which could be different based on their productions (Mukherjee et al. 2012). These particles' physical properties depend on their shape and size (Boisselier and Astruc 2009). There is dependence between the construction of extract and the form and shapes of AuNPs. Triangular and prismatic NPs are produced in higher quantities than hexagonal and spherical NPs under lower extraction concentrations. Chandran et al. (2006) and also more hexagonal and triangular AuNPs are produced when the reaction time is decreased (Usman et al. 2018).

Traditionally, AuNPs have been synthesized by physical method (UV irradiation, laser ablation, and plasma synthesis) and chemical methods (citrate synthesis, Turkish method, wet chemical synthesis, and chemical reduction method), physicochemical methods involve sonochemical and sonoelectrochemical methods. All chemical, physical, and physicochemical methods have some disadvantages such as the usage of dangerous chemicals, high energy demanding, and difficult synthesis conditions (Shedbalkar et al. 2014). AuNPs synthesis by cyanobacteria has been invented by Parial et al. (2016). They culture different cyanobacteria, add 15 mL hydrogen tetrachloroaurate under different pH levels, and incubate them for three days. They realized that the color of treatments with *Phormidium valderianum*, *P. tenue*, and *Microcoleus chthonoplastes* changed from green to purple indicating they had thallus colors which are indicative of intracellular gold NPs production (Bhardwaj and Naraian 2021; Parial et al. 2016). Cyanobacteria have been found suitable for the production of NPs. They produce large amounts of biomass, grow fast, and their cost of production is low.

Cyanobacteria are a group of prokaryotic that are oxygenic phototrophs. They are gram-negative and can be found in almost every habitat on the planet. They are a source of several bioactive compounds. The researchers reported that about 19 strains of cyanobacteria could produce 20 bioactive particles (Shedbalkar et al. 2014). The mechanism of AuNPs production and synthesis by cyanobacteria remains unclear. In addition, aerobic respiration during N_2 fixation and both oxygenic and anoxygenic photosynthesis in vegetative cells are possible through cyanobacterial thylakoids and intracytoplasmic membrane complexes. As a result, during NP production, cyanobacteria ought to effectively reduce gold ions (Mandhata et al. 2022). Two methods have been found for the synthesis of AuNPs. The first method is the top-down method, where bulk metal material is broken down into nanoscale particles using physical techniques such as grinding, milling, and stirring. Lithographic processes that assist in reducing bulk metal material to nanoscale size and are a subset of this method (Pan et al. 2020). The second method is bottom-up and in this method by assembling atoms into new nuclei, they can dissociate into NPs. Both chemical and biological approaches for metal synthesis can be part of this method. Chemical approaches such as spinning, pyrolysis, sol-gel process, atomic or molecular condensation, chemical precipitation, vapor deposition, and aerosol processes (Mandhata et al. 2022).

Bio-reduction by microorganisms such as cyanobacteria was adopted under the bottom-up method (Purohit et al. 2019). Activation, growth, and termination or, to put it more simply, trapping, reduction, and capping are the three main stages of the process. In the first stage, which is activation, the metal iron will reduce and then the atom nucleation will occur. After that, all particles will be merged and form a bigger-sized particle when the thermodynamics of the process stabilizes. In cyanobacteria, the polysaccharides and proteins function as reducing agents. Proteins, specifically amino groups ($-NH_2$) like cysteine (-cys) residues, are essential for the stability and covering of metal NPs in aqueous conditions (Singaravelu et al. 2007).

For instance, biosynthesis of AuNPs has been demonstrated using the cyanobacterium *S. platensis* to biosynthesize AuNPs under 37 °C with pH 5.6. Further analysis confirmed the reduction of Au to nanoform (Govindaraju et al. 2008). The absorbance of surface plasmon at 530 nm was used to characterize the Au nanoparticle dispersions. The absorption peak at 75, 50, and 25 mol concentrations was at 509, 486, and 464 nm respectively. High-resolution TEM confirmed the production of NPs in the 6–10 nm range. FTIR reveals the potential role of a biomolecule (protein) as a capping and reducing agent of the biosynthesized NPs. The brown solution, which signifies the reduction of gold ions, is formed with the assistance of the $HAuCl_4$ solution. Wherein NPs of 13 ± 2 nm were produced, as evidenced by the dark patches seen in the cytoplasm, plasma membrane, thylakoid membrane, and cell wall (Focsan et al. 2011). For AuNPs production, the cyanobacterial polysaccharides loaded with sulfate, their unique enzyme-reducing capacities, and stabilizing units remained beneficial (Younis et al. 2019) (Figure 5.1).

Gold is the most well-known new metal nanoparticle due to its strong hazardous impact and anti-proliferative action. Gold is an inert or non-reactive metal. Due to localized surface plasma resonance, it has strong optical properties. These are a few of the characteristics that make AuNPs a viable platform for the development of cancer theragnostic (Mandhata et al. 2022). AuNPs functionalized with the ability to target ligands, and anticancer medications are among the several nanomaterials used to treat cancer that show potential for cancer therapy and diagnosis. It was shown that the oral administration of AuNPs conjugated with the milk protein lactoferrin, and the biocompatible polymer

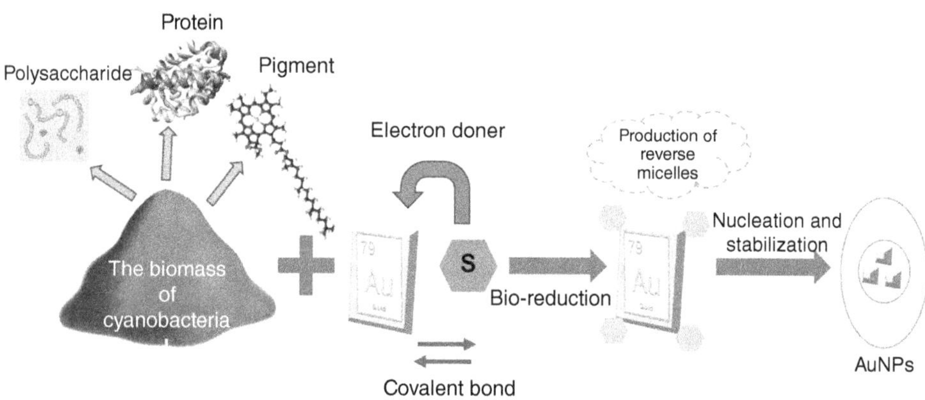

Figure 5.1 The process of synthesis of gold nanoparticles. *Source:* BlueBee/Adobe Stock Photos.

polyethylene glycol (PEG) may bind with highly expressed lactoferrin receptors in the brain and penetrate the blood-brain barrier (Kim et al. 2022). AuNPs have the capability to cause hyperthermia or deliver the medication in the targeted area or cell to have a synergistic impact, either alone or in conjunction with other treatment techniques like radiotherapy or chemotherapy. This ability helps to facilitate the treatment of cancer.

5.3.2 Synthesis of Silver Nanoparticles from Cyanobacteria

Silver, a transition metal with high thermal and electrical conductivity, has been used for centuries in various forms, such as coins, solutions, and colloids, to prevent and treat infections (Ahmed et al. 2016). Silver nanoparticles (AgNPs) are the most widely used type of metal nanoparticles (MNPs). Because of their appealing physicochemical characteristics, AgNPs are very useful in studying biology and medicine. AgNPs could be practical and found in various areas such as drug industries, dressings, surgical instruments and masks, detergents, and care products (Vijayaraghavan and Nalini 2010). AgNPs are believed to have antiviral, anti-inflammatory, antiplatelet, and antifungal characteristics (Shankar et al. 2005).

AgNPs are known to have been created for several physical, biological, and medicinal uses. These applications can be guided by a variety of methods, such as solution irradiation, electrochemical reduction, cryochemical synthesis, spark discharging, and control of roughly 20–15,000 silver atoms (Prabu and Johnson 2015). Like AuNPs, conventional methods of synthesis of AgNPs including chemical and physical approaches had lots of negativities such as high expenses, the requirement of high energy, and the production of toxic by-products. Producing AgNPs by green methods and using cyanobacteria is not only an environmentally friendly approach but also prevents the harmful effects of those techniques (Thakkar et al. 2010).

An eco-friendly method of synthesizing NPs without the use of harmful chemicals is becoming more and more important (Prabu and Johnson 2015). Biological microorganisms, including fungi, bacteria, cyanobacteria, and plant extracts, are utilized in green synthetic systems. There has recently been a lot of interest in the use of microbes as ecologically friendly precursors for the synthesis of NPs like gold or silver. Nowadays, organism's macro and micro play a crucial purpose in the remediation of toxic metals by the reduction of metal ions. Prokaryotes have been extensively studied as a potential source of metallic NPs (Das et al. 2017). This is because they are abundant and have adapted to harsh environments; they are also inexpensive to cultivate, develop quickly and have easily controllable growth parameters, including temperature, oxygenation, and incubation period (Li et al. 2011).

There are two methods for AgNPs production, i.e., intracellular synthesis and extracellular synthesis (Murillo-Rábago et al. 2022). The process of producing NPs inside of host cells via biological processes is known as intracellular production of NPs. Microorganisms are cultivated, living cells are interacted with, and NPs are purified during this procedure (Park et al. 2016). Centrifuging the mixture, cleaning it, and examining the pellets or lysing cells are the last steps. According to some research, cyanobacteria such as *Lyngbya majuscule*, *Anabaena* sp., *Plectonema, Phormidium*, and *Boryanum* may produce silver NPs in this method. In the second method, because extracellular manufacturing is external and depends on biological materials such as proteins, ions, enzymes, pigments, antioxidants, and hormones, it is a commonly utilized process to create NPs. By utilizing cell biomass

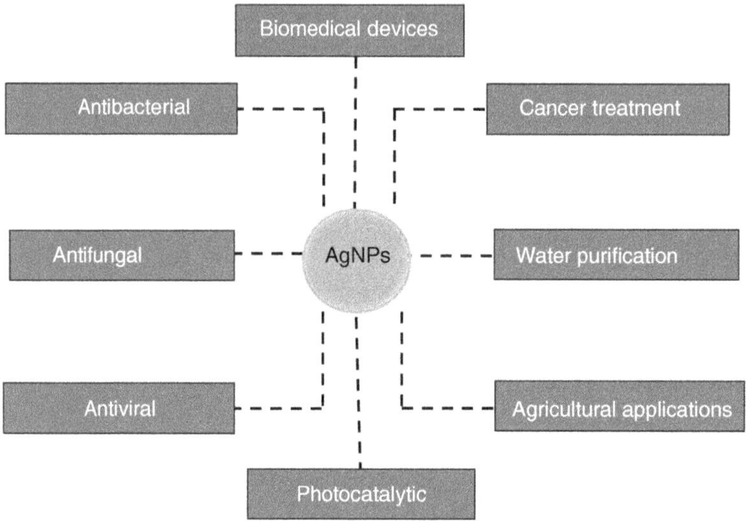

Figure 5.2 Application of silver nanoparticles.

extracts and cell-free culture medium for bio-fabrication, biomolecules act as capping and reducing agents. Many extracellular methods have been used to produce silver NPs from cyanobacteria strains including *Arthrospira*, *Anabaena*, *Aulosira*, *Calothrix*, *Chroococcus*, *Nostoc*, *Oscillatoria*, *Spirulina*, and *Tolypothrix tenuis* (Behera et al. 2024).

The chemical stability and catalytic activity of silver NPs for antibacterial, antiviral, anticancer, antifungal, and anti-inflammatory properties provide them an advantage over other metallic NPs (Ramachandran et al. 2017; Saifuddin et al. 2009; Veerasamy et al. 2011). Microbes prevalent in industrial and medicinal processes have long been known to be inhibited by silver. AgNPs are applied topically to burn and open wounds in medicine to prevent infection (Iravani et al. 2014). Because of their potential for use in antibacterial, anticancer, antifungal, anti-algal, and photocatalytic applications, cyanobacteria are being used more and more in applied science and technology to synthesize NPs. Numerous industries, including textiles, medicine, and agriculture, stand to gain from this safe, eco-friendly, and energy-efficient synthesis. Electronic materials, biological sciences, and environmental sciences may also profit (Figure 5.2).

5.4 Mechanism of Cyanobacteria Mediated Nanoparticles Biosynthesis

The biosynthesis of NPs varies within the organisms. The NPs can be synthesized via intracellular and extracellular mechanisms.

5.4.1 Intracellular Synthesis of Nanoparticles

Intracellular synthesis is achieved by reducing bulk material into NPs inside the cells of the cyanobacteria under controlled biological factors. In this type of synthesis, three steps are generally involved including culturing of cyanobacteria, interaction of the cells of cyanobacteria and precursor solution and separation and purification of NPs using physiochemical techniques (Lengke et al. 2007). Interaction between cells and bulk materials can be done by collecting cellular biomass of the cells of the organisms in the log phase by centrifugation method. Afterwards, the washing of biomass is done in order to remove any salt traces present in the medium. After the clean biomass has been collected, it is dissolved in water and combined with a suitable concentration of precursor solution. Another method for the interaction of cells and bulk materials can be achieved by culturing the cells along with bulk materials solution, and it is kept for two weeks under culture conditions (Jena et al. 2013). After the interactions, precipitation is used to purify the NPs, and the precipitated NPs are then centrifuged at 12,000g. Cells can be lysed using the sonication process to remove synthesized NPs that can be further examined, or the resulting pellets are repeatedly rinsed with distilled water for direct analysis (Hamida et al. 2020b).

5.4.2 Extracellular Synthesis of Nanoparticles

Extracellular synthesis of NPs occurs outside of the cell and involves biological substances such as ions, pigments, proteins, hormones, enzymes, and antioxidants that contribute to the process of NPs reduction. These biomolecules during the biofabrication process act as capping and reducing agents. The extracellular synthesis of NPs can be done by various methods that depend on the use of cell biomass extract, biomolecules, and free culture media. In cell-free media, the NPs synthesis is achieved by removing biomass from culture media using centrifugation technique. For the synthesis of NPs, the extracellular culture media are filtered and combined with metal precursor solutions under specific conditions (Hamida et al. 2020b). In cell biomass filtrate approach of NPs synthesis, cells of cyanobacteria are centrifuged, and the pellets are washed for more than three times with distilled water to remove the metals from the medium. Following the step, cleaned pallets are freeze dried using a lyophilizer then the dried pellets are crushed into fine powder. The obtained powder is mixed with distilled water, and then filtration is done. The collected filtrate is mixed with a metal precursor solution and incubated under certain conditions. After this the mixture is centrifuged, and the obtained supernatant is further used for synthesis of NPs (Bin-Meferij and Hamida 2019). In biomolecule-based synthesis, pigments like carotenoids, R-phycoerythrin, and C-phycocyanin, proteins, and enzymes can be used (El-Naggar et al. 2017). In these purified biomolecules and bulk material is mixed and incubated at certain conditions. After incubation, reduction is done to obtain NPs suspension, which is further centrifuged to recover NPs (Hamida et al. 2020b).

5.5 Characterization of Biosynthesized Nanoparticles

This biosynthesis provides a green and eco-friendly route towards the preparation of cyanobacteria nanoparticles (CNPs) with specific physicochemical properties. To fully characterize these properties, such as size, shape, stability, and chemical analysis, there is a need to apply various elaborated techniques (Menon et al. 2017; Shah et al. 2015). Such approaches ensure that NPs are adequately developed to suit their intended purposes. Different spectroscopic, X-ray-based, microscopic, and magnetic methods are used to confirm the biosynthesized CNPs. To determine wavelength ranges and the chemical makeup of biocoats covering biogenic NPs, validate NP formation, and characterize their charge and hydrodynamic diameter, a variety of spectroscopic techniques are employed, such as nuclear magnetic resonance spectroscopy, zeta potential measurement, Fourier transform infrared spectroscopy (FTIR), dynamic light scattering, and ultraviolet (UV)–visible spectroscopy (Hamida et al. 2020a; MubarakAli et al. 2012). Composition, structure, and crystal phase are evaluated using X-ray-based methods such as energy dispersive spectroscopy (EDAX or EDS), X-ray diffraction (XRD), and X-ray photoelectron spectroscopy (XPS) (Dahoumane et al. 2016; Shah et al. 2015). To determine the size and shape of NPs, microscopic techniques such as transmission electron microscopy (TEM), high-resolution TEM, SEM, and atomic force microscopy are used (Hamida et al. 2020b; Khan et al. 2019). Additionally, many approaches are employed to assess the magnetic characteristics of NPs, including ferromagnetic resonance (FMR) and vibrating sample magnetometry (VSM) (Loiola et al. 2022; Upadhyay et al. 2016).

These techniques are now a day's widely used for characterization of cyanobacteria-derived NPs. done the characterization of cyanobacteria-derived ZnO nanoparticles utilizing UV–vis, FTIR, XRD, SEM-EDAX, TEM, zeta potential, and DLS analysis (Asif et al. 2021). A study reported the characterization of AgNPs by TEM and SEM analysis. Their results indicated the presence of spherical crystals exhibiting face-centered cubic structures, with size ranges of 6.24–11.4 nm and 6.46–12.2 nm for AgNPs. Subsequently, using XRD and energy-dispersive X-ray spectroscopy (EDX) to validate the effective synthesis of AgNPs in either their oxide or chloride forms (Kota et al. 2017). Ultimately, FTIR spectral data were employed to verify the existence of hydroxyl and amide functional groups (Hanna et al. 2022). A study reported the use of UV–vis, FTIR, XRD, SEM, and zeta potential for characterizing biogenic CuONPs (Yadav et al. 2024). Therefore, it is necessary to employ multiple analytical techniques to determine the biosynthesized CNPs size, shape, surface charge, chemical composition, and stability. UV–vis, FTIR, and NMR demonstrate optical properties and chemical bonding. XRD and XPS can be used to differentiate the crystalline structures and elemental characteristics of samples (Mourdikoudis et al. 2018). SEM, TEM, and AFM are used to reveal the shape of the NPs and their surface topography. These methods enhance knowledge about NPs characterization and possible future applications of these materials in healthcare, farming, and the environmental remediation (Patil et al. 2022).

5.6 Molecular Insights into Nanoparticle Toxicity

NPs are known to inhibit the growth of insects, which helps in increasing the growth of plants and their productivity. NPs are potential insecticides, and they have mode of action against insects. These particles trigger the oxidative stress in the tissues of the arthropods (Mao et al. 2018). Moreover, NPs penetrate inside the insects through exoskeletons that bind to the sulfur of the proteins and phosphorus from the DNA, which results in rapid denaturation of enzymes and organelles. NPs act on the activity of Cu-dependent enzymes, which leads to mimicking starvation and to cuticular demelanization. It also affects the acetylcholinesterase activity, antioxidants, and detoxifying enzymes. The upregulation and downregulation of ecdysone receptor genes, CrGnRH1 and CrBR2.2 is up regulate are also done by the NPs, which leads to chronic toxicity which fitness reduction. The disruption of reproduction and development has also been reported in the insects. The binding of NPs with the cuticles has also been known to result in the sorbtion of cuticular lipids and waxes that result in insect dehydration. NPs lead to cortex damage, epithelial cells, midgut, shrinkage of the abdominal region, changes the thorax shape, accumulation of reactive oxygen species (ROS), progeny loss, and reduced invertase activity (Benelli 2018).

5.7 Applications of Cyanobacterial-based Nanoparticles in Agriculture

5.7.1 Crop Improvement

Crop protection and fertilization are two potential uses of nanotechnology in agriculture. Nevertheless, such advantages could come with risks for the environment: non-target plants, plant-beneficial soil microorganisms, and other living things could be impacted if NPs contaminate the environment (Dimkpa 2014). Cyanobacterial nanotechnology is a very promising way to achieve sustainable development in agriculture and boost crop yields by controlling weed and pest populations. For example, the high penetration strength, wide surface area, nontoxicity, and targeted delivery mechanism of nano-fertilizers, nanopesticides, and nanoherbicides allow for their application in agriculture (Arora et al. 2022; Prasad et al. 2017). However, the application of cyanobacterial NPs in agriculture to enhance food production and disease elimination is still in the early stages. Furthermore, the application of engineered NPs in agriculture appears to be minimal or nonexistent (Gupta et al. 2022). According to a study conducted by Yadav and Kumar (2024), the application of nano-formulations has significantly increased plant growth of rice plants. The study also examined the effect of cyanobacteria-based CuONPs on hydroponically produced paddy crops.

5.7.2 Crop Disease Management

The global food security is seriously threatened by plant diseases brought on by pathogenic organisms. The availability of food and the sustainability of agriculture depend on the development of efficient sensory applications and technologies for identifying plant diseases (Kumar et al. 2022). In response to the growing food demand, more advanced crop protection technologies are being developed to guarantee greater crop yields and help feed the world's expanding population. Among these cutting-edge technologies with enormous promise is nanotechnology (Yadav and Yadav 2018). Plant disease management can be made more affordable and environmentally friendly with the use of NPs produced by agriculturally important microbes (Delilah et al. 2022). Cyanobacterium-based NPs are regarded as a bio-stimulant substance for plant growth in addition to being an effective nano-nematicide. They might be used in an integrated program to control some plant diseases (Khan et al. 2021). The biological, physical, and chemical characteristics of cyanobacteria-mediated NPs offer a wide range of applications. Numerous investigations have demonstrated the potential of these NPs as photocatalytic, antibacterial, and anticancer agents. It may also be a very useful substitute for chemical-based bactericides in the pharmaceutical and agricultural sectors, where it can be used to manage bacterial infections in crops (Singh et al. 2020). Although the antibacterial potential of metal NPs and cyanobacteria has been investigated for the management of phytopathogens, their interactions or synergy in a combined manner has not been investigated.

In a report, Mahawar et al. (2019) studied the biocontrol potential of CuNPs, cyanobacterium *Calothrix elenkinii* and CuNPs augmented *C. elenkinii* (CuNPs—Ce). Plant vitality was enhanced, and the severity of the disease was lessened by the application of CuNPs—Ce. In another investigation (Hamed et al. 2019) reported the nematicidal potential of cyanobacteria-based AgNPs on egg hatching and larval mortality of *Meloidogyne javanica* in vitro and under greenhouse conditions, indicating their potential application in *M. javanica* biological control. Additionally, the study demonstrated the beneficial effects of AgNPs derived from *Nostoc* sp. on a few plant development metrics when grown in soil infected with nematodes. In a report, the biocontrol potential of a cyanobacterium *C. elenkinii* (Ce), AgNPs, and their augmented complex (AgNPs—Ce) was evaluated (Mahawar et al. 2020). Increased plant growth characteristics, reduced fungal burden, and elevated expression of triggered antioxidant enzymes demonstrate the biocontrol ability of AgNPs—Ce treatment in *Alternaria alternata*-infected plants. In diverse agroclimatic circumstances, this synergistic relationship may be investigated as a viable biocontrol alternative against tomato plants attacked by *A. alternata*.

5.7.3 Bioremediation of Heavy Metals

The term "bioremediation" refers to the use of biological resources, or their parts and extracts, in remediation operations. This is accomplished by breaking down harmful materials into less toxic ones (Jayaprakash et al. 2019). Nanotechnology has been used as advancement in the bioremediation process, in which NPs produced by using microorganisms and plants are employed. Nanomaterials are beneficial in remediation because of their advantages over microparticles, including better diffusion and penetration rates and reactivity to redox components. They were applied in the decomposition of herbicides, petroleum-based products, heavy metals, uranium cleanup, and groundwater cleanup (Ravi et al. 2023). Nano-bioremediation shows promise in effectively removing contaminants from polluted environments by combining NPs with bioremediation strategy (Karnwal et al. 2024). One of the most advantageous organisms in the green synthesis of NPs is cyanobacteria, because of its capacity to accumulate different contaminants from their surroundings (Ghosh et al. 2022). Cyanobacteria can degrade a broad range of pollutants by altering their metabolic activities and producing different metabolites into their surroundings. They can efficiently remediate polluted sites that are contaminated with huge amounts of phenolic substances, polycyclic aromatic hydrocarbons (PAHs), heavy metals, petroleum by-products, and dyes (Yadav et al. 2021).

5.8 Future Projections of Nanoparticle Synthesis

There is much debate on the future implication of NPs. NPs are frequently used in the development of consumer products and devices, especially in the fields of medicine, electronics, and energy production (Santos et al. 2015). This move toward the use of NPs also raised several issues related to the harmful and adverse effect of NPs on the environment and human health (Staggers et al. 2008). These concerns have accounted to make special statutory regulation for the application of NPs in biomedical and other uses. Despite some disputes, the NPs render huge expectation for the future. It may play a huge role in biomedical applications varying from drug delivery to bioimaging, biomarkers, and biosensors (Yao et al. 2014). The use of NPs in imaging applications is commercially available and in near future monitoring of cellular activities in tissues may also be possible (Padmanabhan et al. 2016). The highly specific and sensitive biosensing kits were already developed in various laboratories for detecting nucleic acids and proteins. Polymer-based nanoporous drug delivery systems were already developed and commercialized in the market (Radhakrishnan et al. 2022). Continuous research work in the field of nano-medicine and nano-drug delivery systems has opened new opportunities for the treatment of various diseases that are not curable in present days (Patra et al. 2018). By the use of NPs, disease diagnosis and detection of infected cells would be faster, and as a result, the disease may be cured before they spread into and affect other parts of the body. Also, a person suffering from trauma or impaired organ functions could benefit from NPs as antimicrobial agents and implant devices (Aminian et al. 2022).

The primary challenge of nanotechnology is to develop tools for assessing the exposure of NPs in the environment (Shatkin 2017). After the use of NPs, some amount of NPs spread in air and water contaminated the surrounding environment. These NPs are directly exposed to humans, animals, or the food chain, which causes serious issues related to the environment and human health (Sajid et al. 2015). Measurement of quantity and quality of NPs in the surrounding environment is crucial for proper risk assessment to human health and environment. Economic challenges of nanotechnology comprise ambiguity of effectiveness, scalability, funding for projects, availability of resources, and patience. However, companies recognize an immense possibility in the field of nanotechnology for the development of new commodities and the upgrading of existing products (Niosi and Reid 2007). Thus, raising fundamental questions and the need for new regulations are required for research and development in the field of nanotechnology. The government bodies and other authorities all around the world should assess the probable risks and appropriate regulatory response to the extensive use of nanotechnology (Larsson et al. 2019). While there are favorable aspects found in the application of nanotechnology and NPs in various fields of science and technology, there is still a pressing need for further investigation into the potential toxicity that can occur following nanoparticle exposure (Acharya and Pal 2020). Governments must thoroughly explore additional regulation of nanoparticle applications in the biomedical, agriculture, and food industries across the world to develop standards to avoid possible health risks to humans and the environment (Jain et al. 2018).

5.9 Conclusion

The science of tiny devices is becoming more and more tangible because of the nearly incompatible multidisciplinary field of nanotechnology. NPs have a wide range of applications in the food business, electronics, environmental remediation, agriculture, and biomedical fields. The synthesis of cyanobacterial NPs has demonstrated significant potential in recent decades for the advancement of novel techniques since it enables the economical, energy-efficient, and ecologically benign synthesis of NPs of various sizes and shapes. These NPs have distinctive biological, chemical, and physical characteristics that make them useful for photocatalysis, antibacterial, and anticancer applications. Recent remarkable developments in the field of cyanobacteria synthetic biology and the synthesis of nanomaterials from the species have given scientists believed that this innovation will be widely applied in industrial applications. A wealth of natural resources could be found in cyanobacteria due to their distinctive characteristics and growing variety. However, little is currently understood regarding the safety and toxicological assessment of NPs. It is necessary to adequately evaluate the risks and safety issues associated with using NPs for health and other objectives. There is a pressing need in the field of nanotechnology to find a dependable and environmentally benign method for creating metallic nanoparticles. Researchers, scientists, and entrepreneurs should cooperate to create the framework for additional cyanobacterial innovation research because these microorganisms have the potential to be employed in hundreds of fascinating applications.

References

Aassime, A. and Hamouda, F. (2017). Conventional and un-conventional lithography for fabricating thin film functional devices. In: *Modern Technologies for Creating the Thin-Film Systems and Coatings* (ed. N.N. Nikitenkov), 43–58. Intechopen.

Acharya, A. and Pal, P.K. (2020). Agriculture nanotechnology: translating research outcome to field applications by influencing environmental sustainability. *NanoImpact* 19: 100232.

Ahmad, N., Sharma, S., Singh, V. et al. (2011). Biosynthesis of silver nanoparticles from *Desmodium triflorum*: a novel approach towards weed utilization. *Biotechnology Research International* 2011: 454090.

Ahmed, S., Ahmad, M., Swami, B.L., and Ikram, S. (2016). A review on plants extract mediated synthesis of silver nanoparticles for antimicrobial applications: a green expertise. *Journal of Advanced Research* 7: 17–28.

Ahmed, S.F., Mofijur, M., Rafa, N. et al. (2022). Green approaches in synthesising nanomaterials for environmental nanobioremediation: technological advancements, applications, benefits and challenges. *Environmental Research* 204: 111967.

Aisida, S.O., Ugwu, K., Akpa, P.A. et al. (2019). Biosynthesis of silver nanoparticles using bitter leave (*Veronica amygdalina*) for antibacterial activities. *Surfaces and Interfaces* 17: 100359.

Al-Harbi, N. and Abd-Elrahman, N.K. (2024). Physical methods for preparation of nanomaterials, their characterization and applications: a review. *Journal of Umm Al-Qura University for Applied Sciences* https://doi.org/10.1007/s43994-024-00165-7.

Altammar, K.A. (2023). A review on nanoparticles: characteristics, synthesis, applications, and challenges. *Frontiers in Microbiology* 14: 1155622.

Aminian, A., Fathi, A., Gerami, M.H. et al. (2022). Nanoparticles to overcome bacterial resistance in orthopedic and dental implants. *Nanomedicine Research Journal* 7: 107–123.

Anju Varghese, R., Anandhi, P., Arunadevi, R. et al. (2015). Satin leaf (*Chrysophyllum oliviforme*) extract mediated green synthesis of silver nanoparticles: antioxidant and anticancer activities. *Journal of Pharmaceutical Sciences and Research* 7 (6): 266–273.

Arora, S., Murmu, G., Mukherjee, K. et al. (2022). A comprehensive overview of nanotechnology in sustainable agriculture. *Journal of Biotechnology* 355: 21–41.

Arya, A. and Chundawat, T.S. (2020). Metal nanoparticles from algae: a green approach for the synthesis, characterization and their biological activity. *Nanoscience and Nanotechnology—Asia* 10: 185–202.

Asif, N., Fatima, S., Aziz, M.N. et al. (2021). Biofabrication and characterization of cyanobacteria derived ZnO NPs for their bioactivity comparison with commercial chemically synthesized nanoparticles. *Bioorganic Chemistry* 113: 104999.

Bahrulolum, H., Nooraei, S., Javanshir, N. et al. (2021). Green synthesis of metal nanoparticles using microorganisms and their application in the agrifood sector. *Journal of Nanobiotechnology* 19: 1–26.

Baig, N., Kammakakam, I., and Falath, W. (2021). Nanomaterials: a review of synthesis methods, properties, recent progress, and challenges. *Materials Advances* 2: 1821–1871.

Bakku, R.K. and Rakwal, R. (2022). Applications of cyanobacterial compounds in the energy, health, value-added product, and agricultural sectors: a perspective. In: *Cyanobacterial Physiology* (ed. H. Kageyama and R. Waditee-Sirisattha), 149–164. Elsevier.

Bamal, D., Singh, A., Chaudhary, G. et al. (2021). Silver nanoparticles biosynthesis, characterization, antimicrobial activities, applications, cytotoxicity and safety issues: an updated review. *Nanomaterials* 11: 2086.

Behera, M., Behera, P.R., Sethi, G. et al. (2024). Cyanobacterial silver nanoparticles and their potential utility—recent progress and prospects: a review. *Journal of Basic Microbiology* 64: e2400256.

Benelli, G. (2018). Mode of action of nanoparticles against insects. *Environmental Science and Pollution Research* 25: 12329–12341. https://doi.org/10.1007/s11356-018-1850-4.

Bhardwaj, A.K. and Naraian, R. (2021). Cyanobacteria as biochemical energy source for the synthesis of inorganic nanoparticles, mechanism and potential applications: a review. *3 Biotech* 11: 445.

Bhargava, A., Jain, N., Khan, M.A. et al. (2016). Utilizing metal tolerance potential of soil fungus for efficient synthesis of gold nanoparticles with superior catalytic activity for degradation of rhodamine B. *Journal of Environmental Management* 183: 22–32.

Bhatia, S. and Bhatia, S. (2016). Nanoparticles types, classification, characterization, fabrication methods and drug delivery applications. In: *Natural Polymer Drug Delivery Systems: Nanoparticles, Plants, and Algae* (ed. S. Bhatia), 33–93. Springer.

Bin-Meferij, M.M. and Hamida, R.S. (2019). Biofabrication and antitumor activity of silver nanoparticles utilizing novel *Nostoc* sp. Bahar M. *International Journal of Nanomedicine* 14: 9019–9029.

Boisselier, E. and Astruc, D. (2009). Gold nanoparticles in nanomedicine: preparations, imaging, diagnostics, therapies and toxicity. *Chemical Society Reviews* 38: 1759–1782.

Chandrakala, V., Aruna, V., and Angajala, G. (2022). Review on metal nanoparticles as nanocarriers: current challenges and perspectives in drug delivery systems. *Emergent Materials* 5: 1593–1615.

Chandran, S.P., Chaudhary, M., Pasricha, R. et al. (2006). Synthesis of gold nanotriangles and silver nanoparticles using Aloevera plant extract. *Biotechnology Progress* 22: 577–583.

Colson, P., Henrist, C., and Cloots, R. (2013). Nanosphere lithography: a powerful method for the controlled manufacturing of nanomaterials. *Journal of Nanomaterials* 2013: 948510.

Dahoumane, S.A., Wujcik, E.K., and Jeffryes, C. (2016). Noble metal, oxide and chalcogenide-based nanomaterials from scalable phototrophic culture systems. *Enzyme and Microbial Technology* 95: 13–27.

Das, R.K., Pachapur, V.L., Lonappan, L. et al. (2017). Biological synthesis of metallic nanoparticles: plants, animals and microbial aspects. *Nanotechnology For Environmental Engineering* 2: 1–21.

Delilah, D., Mathew, N.M., Joseph, A. et al. (2022). Plant micronutrient nanoparticles from microalgae-biosynthesis and their applications–a review. *International Journal of Botany Studies* 7: 135–139.

Dhillon, G.S., Brar, S.K., Kaur, S., and Verma, M. (2012). Green approach for nanoparticle biosynthesis by fungi: current trends and applications. *Critical Reviews in Biotechnology* 32: 49–73.

Dimkpa, C.O. (2014). Can nanotechnology deliver the promised benefits without negatively impacting soil microbial life? *Journal of Basic Microbiology* 54: 889–904. https://doi.org/10.1002/jobm.201400298.

El Messaoudi, N., Ciğeroğlu, Z., Şenol, Z.M. et al. (2024). Green synthesis of nanoparticles for remediation organic pollutants in wastewater by adsorption. In: *Advances in Chemical Pollution, Environmental Management and Protection*, vol. 10 (ed. A. Kumar, M. Bilal, and L.F.R. Ferreira), 305–345. Elsevier.

El-Khawaga, A.M., Zidan, A., and El-Mageed, A.I.A.A. (2023). Preparation methods of different nanomaterials for various potential applications: a review. *Journal of Molecular Structure* 1281: 135148. https://doi.org/10.1016/j.molstruc.2023.135148.

El-Naggar, N.E.-A., Hussein, M.H., and El-Sawah, A.A. (2017). Bio-fabrication of silver nanoparticles by phycocyanin, characterization, in vitro anticancer activity against breast cancer cell line and in vivo cytotxicity. *Scientific Reports* 7: 10844. https://doi.org/10.1038/s41598-017-11121-3.

Escorcia-Díaz, D., García-Mora, S., Rendón-Castrillón, L. et al. (2023). Advancements in nanoparticle deposition techniques for diverse substrates: a review. *Nanomaterials* 13: 2586.

Eyu, B. (2018). Gold nanoparticles in the diagnosis and treatment of cancer. Bulletin of Russian State Medical University, pp. 79–85.

Focsan, M., Ardelean, I.I., Craciun, C., and Astilean, S. (2011). Interplay between gold nanoparticle biosynthesis and metabolic activity of cyanobacterium *Synechocystis* sp. PCC 6803. *Nanotechnology* 22: 485101.

Gajbhiye, M., Kesharwani, J., Ingle, A. et al. (2009). Fungus-mediated synthesis of silver nanoparticles and their activity against pathogenic fungi in combination with fluconazole. *Nanomedicine: Nanotechnology Biology and Medicine* 5: 382–386.

Gericke, M. and Pinches, A. (2006). Biological synthesis of metal nanoparticles. *Hydrometallurgy* 83: 132–140.

Ghosh, S., Nag, M., Lahiri, D. et al. (2022). 4—Biogenic nanoparticles from cyanobacteria and their applications. In: *Agricultural Nanobiotechnology* (ed. S. Ghosh, S. Thongmee, and A. Kumar), 79–103. Woodhead Publishing https://doi.org/10.1016/B978-0-323-91.08-1.00003-1.

Govindaraju, K., Basha, S.K., Kumar, V.G., and Singaravelu, G. (2008). Silver, gold and bimetallic nanoparticles production using single-cell protein (*Spirulina platensis*) Geitler. *Journal of Materials Science* 43: 5115–5122.

Gupta, V., Verma, N.K., and Choudhary, K.K. (2022). Cyanobacterial nanoparticles: application in agriculture and allied sectors. In: *Cyanobacterial Lifestyle and Its Applications in Biotechnology* (ed. P.K. Singh, M.F. Fillat, and A. Kumar), 23–40. Elsevier.

Hamed, S.M., Hagag, E.S., and El-Raouf, N.A. (2019). Green production of silver nanoparticles, evaluation of their nematicidal activity against *Meloidogyne javanica* and their impact on growth of faba bean. *Beni-Suef University Journal of Basic and Applied Sciences* 8: 9. https://doi.org/10.1186/s43088-019-0010-3.

Hamida, R.S., Abdelmeguid, N.E., Ali, M.A. et al. (2020a). Synthesis of silver nanoparticles using a novel cyanobacteria *Desertifilum* sp. extract: their antibacterial and cytotoxicity effects. *International Journal Nanomedicine* 15: 49–63.

Hamida, R.S., Ali, M.A., Redhwan, A., and Bin-Meferij, M.M. (2020b). Cyanobacteria—a promising platform in green nanotechnology: a review on nanoparticles fabrication and their prospective applications. *International Journal of Nanomedicine* 15: 6033–6066.

Hanna, A.L., Hamouda, H.M., Goda, H.A. et al. (2022). Biosynthesis and characterization of silver nanoparticles produced by *Phormidium ambiguum* and *Desertifilum tharense* cyanobacteria. *Bioinorganic Chemistry and Applications* 2022: 9072508.

Hano, C. and Abbasi, B.H. (2021). Plant-based green synthesis of nanoparticles: production, characterization and applications. *Biomolecules* 12: 31.

Harish, V., Tewari, D., Gaur, M. et al. (2022). Review on nanoparticles and nanostructured materials: bioimaging, biosensing, drug delivery, tissue engineering, antimicrobial, and agro-food applications. *Nanomaterials* 12: 457.

He, S., Zhang, Y., Guo, Z., and Gu, N. (2008). Biological synthesis of gold nanowires using extract of *Rhodopseudomonas capsulata*. *Biotechnology Progress* 24: 476–480.

Ijaz, I., Gilani, E., Nazir, A., and Bukhari, A. (2020). Detail review on chemical, physical and green synthesis, classification, characterizations and applications of nanoparticles. *Green Chemistry Letters and Reviews* 13: 223–245. https://doi.org/10.1080/17518253.2020.1802517.

Iravani, S., Korbekandi, H., Mirmohammadi, S.V., and Zolfaghari, B. (2014). Synthesis of silver nanoparticles: chemical, physical and biological methods. *Research in Pharmaceutical Sciences* 9: 385–406.

Islam, F., Shohag, S., Uddin, M.J. et al. (2022). Exploring the journey of zinc oxide nanoparticles (ZnO-NPs) toward biomedical applications. *Materials* 15: 2160.

Jain, A., Ranjan, S., Dasgupta, N., and Ramalingam, C. (2018). Nanomaterials in food and agriculture: an overview on their safety concerns and regulatory issues. *Critical Reviews in Food Science and Nutrition* 58: 297–317.

Jamkhande, P.G., Ghule, N.W., Bamer, A.H., and Kalaskar, M.G. (2019). Metal nanoparticles synthesis: an overview on methods of preparation, advantages and disadvantages, and applications. *Journal of Drug Delivery Science and Technology* 53: 101174. https://doi.org/10.1016/j.jddst.2019.10.174.

Jayaprakash, K., Govarthanan, M., Mythili, R. et al. (2019). Bioaugmentation and biostimulation remediation technologies for heavy metal lead contaminant. *Microbial Biodegradation of Xenobiotic Compounds* 24.

Jena, J., Pradhan, N., Dash, B.P. et al. (2013). Biosynthesis and characterization of silver nanoparticles using microalga *Chlorococcum humicola* and its antibacterial activity. *International Journal of Nanomaterials and Biostructures* 3: 1–8.

Joy, J., Krishnamoorthy, A., Tanna, A. et al. (2022). Recent developments on the synthesis of nanocomposite materials via ball milling approach for energy storage applications. *Applied Sciences* 12: 9312.

Karnwal, A., Martolia, S., Dohroo, A. et al. (2024). Exploring bioremediation strategies for heavy metals and POPs pollution: the role of microbes, plants, and nanotechnology. *Frontiers in Environmental Science* 12: https://doi.org/10.3389/fenvs.2024.1397850.

Karunakaran, G., Sudha, K.G., Ali, S., and Cho, E.-B. (2023). Biosynthesis of nanoparticles from various biological sources and its biomedical applications. *Molecules* 28: 4527.

Khan, I., Saeed, K., and Khan, I. (2019). Nanoparticles: properties, applications and toxicities. *Arabian Journal of Chemistry* 12: 908–931.

Khan, F., Ansari, T., Shariq, M., and Siddiqui, M.A. (2021). Nanotechnology: a new beginning to mitigate the effect of plant-parasitic nematodes. In: *Innovative Approaches in Diagnosis and Management of Crop Diseases* (ed. R.K. Singh and Gopala), 19–43. Apple Academic Press.

Khan, M.A.R., Al Mamun, M.S., Habib, M.A. et al. (2022). A review on gold nanoparticles: biological synthesis, characterizations, and analytical applications. *Results in Chemistry* 4: 100478.

Khan, T., Singh, B., and Manikandan, M. (2023). Synthesis, characteristics, and applications of nanomaterials. In: *Nanomaterials: The Building Blocks of Modern Technology: Synthesis, Properties and Applications* (ed. T. Khan, M. Jawaid, K.A. Ahmad, and B. Singh), 11–26. Springer.

Kim, M., Osone, S., Kim, T. et al. (2017). Synthesis of nanoparticles by laser ablation: a review. *Kona Powder and Particle Journal* 34: 80–90.

Kim, H.S., Lee, S.J., and Lee, D.Y. (2022). Milk protein-shelled gold nanoparticles with gastrointestinally active absorption for aurotherapy to brain tumor. *Bioactive Materials* 8: 35–48.

Kota, S., Dumpala, P., Anantha, R.K. et al. (2017). Evaluation of therapeutic potential of the silver/silver chloride nanoparticles synthesized with the aqueous leaf extract of *Rumex acetosa*. *Scientific Reports* 7: 11566.

Kumar, A., Choudhary, A., Kaur, H. et al. (2022). Potential applications of engineered nanoparticles in plant disease management: a critical update. *Chemosphere* 295: 133798. https://doi.org/10.1016/j.chemosphere.2022.133798.

Kumari, S., Raturi, S., Kulshrestha, S. et al. (2023). A comprehensive review on various techniques used for synthesizing nanoparticles. *Journal of Materials Research and Technology* 27: 1739–1763. https://doi.org/10.1016/j.jmrt.2023.09.291.

Larsson, S., Jansson, M., and Boholm, Å. (2019). Expert stakeholders' perception of nanotechnology: risk, benefit, knowledge, and regulation. *Journal of Nanoparticle Research* 21: 1–17.

Lee, H., Song, M.Y., Jurng, J., and Park, Y.-K. (2011). The synthesis and coating process of TiO_2 nanoparticles using CVD process. *Powder Technology* 214: 64–68. https://doi.org/10.1016/j.powtec.2011.07.036.

Lengke, M.F., Fleet, M.E., and Southam, G. (2007). Biosynthesis of silver nanoparticles by filamentous cyanobacteria from a silver (I) nitrate complex. *Langmuir* 23: 2694–2699.

Li, G., He, D., Qian, Y. et al. (2011). Fungus-mediated green synthesis of silver nanoparticles using *Aspergillus terreus*. *International Journal of Molecular Sciences* 13: 466–476.

Li, H., Chen, X., Shen, D. et al. (2021). Functionalized silica nanoparticles: classification, synthetic approaches and recent advances in adsorption applications. *Nanoscale* 13: 15998–16016.

Liu, X., Chen, J.-L., Yang, W.-Y. et al. (2021). Biosynthesis of silver nanoparticles with antimicrobial and anticancer properties using two novel yeasts. *Scientific Reports* 11: 15795.

Loiola, A.R., Bessa, R.A., Oliveira, C.P. et al. (2022). Magnetic zeolite composites: classification, synthesis routes, and technological applications. *Journal of Magnetism and Magnetic Materials* 560: 169651.

Mahawar, H., Prasanna, R., and Gogoi, R. (2019). Elucidating the disease alleviating potential of cyanobacteria, copper nanoparticles and their interactions in *Fusarium solani* challenged tomato plants. *Plant Physiology Reports* 24: 533–540. https://doi.org/10.1007/s40502-019-00490-8.

Mahawar, H., Prasanna, R., Gogoi, R. et al. (2020). Synergistic effects of silver nanoparticles augmented *Calothrix elenkinii* for enhanced biocontrol efficacy against Alternaria blight challenged tomato plants. *3 Biotech* 10: 102. https://doi.org/10.1007/s13205-020-2074-0.

Mandhata, C.P., Sahoo, C.R., and Padhy, R.N. (2022). Biomedical applications of biosynthesized gold nanoparticles from cyanobacteria: an overview. *Biological Trace Element Research* 200: 5307–5327.

Manimaran, M. and Kannabiran, K. (2017). Actinomycetes-mediated biogenic synthesis of metal and metal oxide nanoparticles: progress and challenges. *Letters in Applied Microbiology* 64: 401–408.

Mao, B.-H., Chen, Z.-Y., Wang, Y.-J., and Yan, S.-J. (2018). Silver nanoparticles have lethal and sublethal adverse effects on development and longevity by inducing ROS-mediated stress responses. *Scientific Reports* 8: 2445.

Mekuye, B. and Abera, B. (2023). Nanomaterials: an overview of synthesis, classification, characterization, and applications. *Nano Select* 4: 486–501.

Menon, S., Rajeshkumar, S., and Kumar, V. (2017). A review on biogenic synthesis of gold nanoparticles, characterization, and its applications. *Resource-Efficient Technologies* 3: 516–527.

Moreira, A.F., Rodrigues, C.F., Reis, C.A. et al. (2018). Gold-core silica shell nanoparticles application in imaging and therapy: a review. *Microporous and Mesoporous Materials* 270: 168–179.

Mourdikoudis, S., Pallares, R.M., and Thanh, N.T. (2018). Characterization techniques for nanoparticles: comparison and complementarity upon studying nanoparticle properties. *Nanoscale* 10: 12871–12934.

MubarakAli, D., Gopinath, V., Rameshbabu, N., and Thajuddin, N. (2012). Synthesis and characterization of CdS nanoparticles using C-phycoerythrin from the marine cyanobacteria. *Materials Letters* 74: 8–11.

Mukherjee, S., Sushma, V., Patra, S. et al. (2012). Green chemistry approach for the synthesis and stabilization of biocompatible gold nanoparticles and their potential applications in cancer therapy. *Nanotechnology* 23: 455103.

Murillo-Rábago, E.I., Vilchis-Nestor, A.R., Juarez-Moreno, K. et al. (2022). Optimized synthesis of small and stable silver nanoparticles using intracellular and extracellular components of fungi: an alternative for bacterial inhibition. *Antibiotics* 11: 800.

Naikoo, G.A., Mustaqeem, M., Hassan, I.U. et al. (2021). Bioinspired and green synthesis of nanoparticles from plant extracts with antiviral and antimicrobial properties: a critical review. *Journal of Saudi Chemical Society* 25: 101304.

Nair, G.M., Sajini, T., and Mathew, B. (2022). Advanced green approaches for metal and metal oxide nanoparticles synthesis and their environmental applications. *Talanta Open* 5: 100080.

Nayak, V., Rb Singh, K., Paliwal, R. et al. (2023). Chapter 15—Introduction to nanotechnological utility in the pharmaceutical industry. In: *Nanotechnology for Drug Delivery and Pharmaceuticals* (ed. R. Pratap Singh, K. Rb Singh, J. Singh, and C.O. Adetunji), 337–355. Academic Press https://doi.org/10.1016/B978-0-323-95325-2.00013-4.

Niosi, J. and Reid, S.E. (2007). Biotechnology and nanotechnology: science-based enabling technologies as windows of opportunity for LDCs? *World Development* 35: 426–438.

Nitnavare, R., Bhattacharya, J., Thongmee, S., and Ghosh, S. (2022). Photosynthetic microbes in nanobiotechnology: applications and perspectives. *Science of the Total Environment* 841: 156457.

Nyabadza, A., McCarthy, É., Makhesana, M. et al. (2023). A review of physical, chemical and biological synthesis methods of bimetallic nanoparticles and applications in sensing, water treatment, biomedicine, catalysis and hydrogen storage. *Advances in Colloid and Interface Science* 321: 103010. https://doi.org/10.1016/j.cis.2023.103010.

de Oliveira, P.F., Torresi, R.M., Emmerling, F., and Camargo, P.H. (2020). Challenges and opportunities in the bottom-up mechanochemical synthesis of noble metal nanoparticles. *Journal of Materials Chemistry A* 8: 16114–16141.

Omran, B.A. and Omran, B.A. (2020). Fundamentals of nanotechnology and nanobiotechnology. In: *Nanobiotechnology: A Multidisciplinary Field of Science* (ed. B.A. Omran), 1–36. Springer Nature.

Padmanabhan, P., Kumar, A., Kumar, S. et al. (2016). Nanoparticles in practice for molecular-imaging applications: an overview. *Acta Biomaterialia* 41: 1–16.

Pal, K. (2022). *Green Nanoarchitectonics: Smart Natural Materials*. CRC Press.

Pan, Z., Wen, Y., Wang, T. et al. (2020). One-step synthesis of hollow PEI-NaBiF4: Yb3+/Er3+ upconversion nanoparticles for water-responsive luminescent probe. *Journal of Rare Earths* 38: 362–368.

Parial, D., Gopal, P.K., Paul, S., and Pal, R. (2016). Gold (III) bioreduction by cyanobacteria with special reference to in vitro biosafety assay of gold nanoparticles. *Journal of Applied Phycology* 28: 3395–3406.

Park, T.J., Lee, K.G., and Lee, S.Y. (2016). Advances in microbial biosynthesis of metal nanoparticles. *Applied Microbiology and Biotechnology* 100: 521–534.

Patel, V., Berthold, D., Puranik, P., and Gantar, M. (2015). Screening of cyanobacteria and microalgae for their ability to synthesize silver nanoparticles with antibacterial activity. *Biotechnology Reports* 5: 112–119.

Patil, R.M., Deshpande, P.P., Aalhate, M. et al. (2022). An update on sophisticated and advanced analytical tools for surface characterization of nanoparticles. *Surfaces and Interfaces* 33: 102165.

Patra, J.K., Das, G., Fraceto, L.F. et al. (2018). Nano based drug delivery systems: recent developments and future prospects. *Journal of Nanobiotechnology* 16: 1–33.

Piras, C.C., Fernández-Prieto, S., and De Borggraeve, W.M. (2019). Ball milling: a green technology for the preparation and functionalisation of nanocellulose derivatives. *Nanoscale Advances* 1: 937–947.

Prabu, H.J. and Johnson, I. (2015). Plant-mediated biosynthesis and characterization of silver nanoparticles by leaf extracts of *Tragia involucrata*, *Cymbopogon citronella*, *Solanum verbascifolium* and *Tylophora ovata*. *Karbala International Journal of Modern Science* 1: 237–246.

Prasad, R., Bhattacharyya, A., and Nguyen, Q.D. (2017). Nanotechnology in sustainable agriculture: recent developments, challenges, and perspectives. *Frontiers in Microbiology* 8: https://doi.org/10.3389/fmicb.2017.01014.

Purohit, J., Chattopadhyay, A., and Singh, N.K. (2019). Green synthesis of microbial nanoparticle: approaches to application. In: *Microbial Nanobionics: Volume 2, Basic Research and Applications* (ed. R. Prasad), 35–60. Springer.

Radhakrishnan, D., Mohanan, S., Choi, G. et al. (2022). The emergence of nanoporous materials in lung cancer therapy. *Science and Technology of Advanced Materials* 23: 225–274.

Ramachandran, R., Krishnaraj, C., Sivakumar, A.S. et al. (2017). Anticancer activity of biologically synthesized silver and gold nanoparticles on mouse myoblast cancer cells and their toxicity against embryonic zebrafish. *Materials Science and Engineering: C* 73: 674–683.

Rami, M.R., Meskini, M., and Sharafabad, B.E. (2024). Fungal-mediated nanoparticles for industrial applications: synthesis and mechanism of action. *Journal of Infection and Public Health* 17: 102536.

Ranjitha, V. and Rai, V.R. (2017). Actinomycetes mediated synthesis of gold nanoparticles from the culture supernatant of *Streptomyces griseoruber* with special reference to catalytic activity. *3 Biotech* 7: 1–7.

Ravi, A., Hemapriya, J., Sudha, P. et al. (2023). Nanobiomaterials in bioremediation. In: *Nanobiomaterials* (ed. T. Gomathi, P.N. Sudha, and S. Thomas), 283–313. CRC Press.

Rónavári, A., Igaz, N., Adamecz, D.I. et al. (2021). Green silver and gold nanoparticles: biological synthesis approaches and potentials for biomedical applications. *Molecules* 26: 844.

Saad, A. and Atia, A. (2014). Review on freshwater blue-green algae (Cyanobacteria): occurrence, classification and toxicology. *Biosciences, Biotechnology Research Asia* 11: 1319–1325.

Saifuddin, N., Wong, C., and Yasumira, A.N. (2009). Rapid biosynthesis of silver nanoparticles using culture supernatant of bacteria with microwave irradiation. *Journal of Chemistry* 6: 61–70.

Sajid, M. and Płotka-Wasylka, J. (2020). Nanoparticles: synthesis, characteristics, and applications in analytical and other sciences. *Microchemical Journal* 154: 104623.

Sajid, M., Ilyas, M., Basheer, C. et al. (2015). Impact of nanoparticles on human and environment: review of toxicity factors, exposures, control strategies, and future prospects. *Environmental Science and Pollution Research* 22: 4122–4143.

Salem, S.S. and Fouda, A. (2021). Green synthesis of metallic nanoparticles and their prospective biotechnological applications: an overview. *Biological Trace Element Research* 199: 344–370.

Salem, S.S., Hammad, E.N., Mohamed, A.A., and El-Dougdoug, W. (2022). A comprehensive review of nanomaterials: types, synthesis, characterization, and applications. *Biointerface Research in Applied Chemistry* 13: 41.

Santos, C.S., Gabriel, B., Blanchy, M. et al. (2015). Industrial applications of nanoparticles–a prospective overview. *Materials Today: Proceedings* 2: 456–465.

Shah, M., Fawcett, D., Sharma, S. et al. (2015). Green synthesis of metallic nanoparticles via biological entities. *Materials* 8: 7278–7308.

Shankar, S.S., Rai, A., Ahmad, A., and Sastry, M. (2005). Controlling the optical properties of lemongrass extract synthesized gold nanotriangles and potential application in infrared-absorbing optical coatings. *Chemistry of Materials* 17: 566–572.

Shatkin, J.A. (2017). Introduction: assessing nanotechnology health and environmental risks. In: *Nanotechnology* (ed. J.A. Shatkin), 1–29. CRC Press.

Shedbalkar, U., Singh, R., Wadhwani, S. et al. (2014). Microbial synthesis of gold nanoparticles: current status and future prospects. *Advances in Colloid and Interface Science* 209: 40–48.

Sidhu, A.K., Verma, N., and Kaushal, P. (2022). Role of biogenic capping agents in the synthesis of metallic nanoparticles and evaluation of their therapeutic potential. *Frontiers in Nanotechnology* 3: 801620.

Singaravelu, G., Arockiamary, J., Kumar, V.G., and Govindaraju, K. (2007). A novel extracellular synthesis of monodisperse gold nanoparticles using marine alga, *Sargassum wightii* Greville. *Colloids and Surfaces B: Biointerfaces* 57: 97–101.

Singh, D. (2020). Cyanobacteria as a source of nanoparticle: application and future projections. In: *Advances in Cyanobacterial Biology* (ed. P.K. Singh, A. Kumar, V.K. Singh, and A.K. Shrivistava), 319–331. Elsevier.

Singh, Y., Kaushal, S., and Sodhi, R.S. (2020). Biogenic synthesis of silver nanoparticles using cyanobacterium *Leptolyngbya* sp. WUC 59 cell-free extract and their effects on bacterial growth and seed germination. *Nanoscale Advances* 2: 3972–3982.

Son, H.H., Seo, G.H., Jeong, U. et al. (2017). Capillary wicking effect of a Cr-sputtered superhydrophilic surface on enhancement of pool boiling critical heat flux. *International Journal of Heat and Mass Transfer* 113: 115–128. https://doi.org/10.1016/j.ijheatmasstransfer.2017.05.055.

Soni, N. and Prakash, S. (2011). Factors affecting the geometry of silver nanoparticles synthesis in *Chrysosporium tropicum* and *Fusarium oxysporum*. *American Journal of Nanotechnology & Nanomedicine* 2: 112–121.

Staggers, N., McCasky, T., Brazelton, N., and Kennedy, R. (2008). Nanotechnology: the coming revolution and its implications for consumers, clinicians, and informatics. *Nursing Outlook* 56: 268–274.

Thakkar, K.N., Mhatre, S.S., and Parikh, R.Y. (2010). Biological synthesis of metallic nanoparticles. *Nanomedicine: Nanotechnology, Biology and Medicine* 6: 257–262.

Upadhyay, S., Parekh, K., and Pandey, B. (2016). Influence of crystallite size on the magnetic properties of Fe_3O_4 nanoparticles. *Journal of Alloys and Compounds* 678: 478–485.

Usman, A.I., Aziz, A.A., and Noqta, O.A. (2018). Bio-synthesis of triangular and hexagonal gold nanoparticles using palm oil fronds' extracts at room temperature. *Materials Research Express* 5: 015042.

Veerasamy, R., Xin, T.Z., Gunasagaran, S. et al. (2011). Biosynthesis of silver nanoparticles using mangosteen leaf extract and evaluation of their antimicrobial activities. *Journal of Saudi Chemical Society* 15: 113–120.

Vijayaraghavan, K. and Nalini, S.K. (2010). Biotemplates in the green synthesis of silver nanoparticles. *Biotechnology Journal* 5: 1098–1110.

Yadav, A. and Yadav, K. (2018). Nanoparticle-based plant disease management: tools for sustainable agriculture. In: *Nanobiotechnology Applications in Plant Protection* (ed. K.A. Abd-Elsalam and R. Prasad), 29–61. Cham: Springer International Publishing https://doi.org/10.1007/978-3-319-91161-8_2.

Yadav, P., Gupta, R.K., Singh, R.P. et al. (2021). Chapter 9—Role of cyanobacteria in green remediation. In: *Sustainable Environmental Clean-up* (ed. V. Kumar Mishra and A. Kumar), 187–210. Elsevier https://doi.org/10.1016/B978.0-12-823828-8.00009-8.

Yadav, R., Kumar, M., and Tomar, R.S. (2024). Cyanobacteria based nanoformulation of biogenic CuO nanoparticles for plant growth promotion of rice under hydroponics conditions. *Current Microbiology* 81: 118. https://doi.org/10.1007/s00284-024-03619-7.

Yao, J., Yang, M., and Duan, Y. (2014). Chemistry, biology, and medicine of fluorescent nanomaterials and related systems: new insights into biosensing, bioimaging, genomics, diagnostics, and therapy. *Chemical Reviews* 114: 6130–6178.

Younis, N.S., Bakir, E.M., Mohamed, M.E., and El Semary, N.A. (2019). Cyanobacteria as nanogold factories II: chemical reactivity and anti-myocardial infraction properties of customized gold nanoparticles biosynthesized by *Cyanothece* sp. *Marine Drugs* 17: 402.

6

Role of Beneficial Microbes for Sustainable Agriculture and Environmental Nanotechnology

Chandra Kant Sharma[1], Harshit Chavda[1], Saleha Diwan[1], Vijay Upadhye[1], Nitin Mahendra Chauhan[2], and Sunil Tulshiram Hajare[2]

[1] *Faculty of Agriculture, Parul University, Vadodara, Gujarat, India*
[2] *Department of Biotechnology, College of Natural and Computational Sciences, Dilla University, Dilla, Ethiopia*

6.1 Introduction

The human population is increasing rapidly day by day and creating excessive pressure on the existing land area for food, fuel, and raw materials. It needs to intensify the existing agricultural production systems. Agriculture production is high due to excessive use of synthetic agrochemicals, i.e., pesticides, chemical fertilizers etc. which has resulted in soil and environmental deterioration, and loss of biodiversity. The beneficial microorganisms have a large number of benefits when they are applied as manure or spray to the field crops (Revillas et al. 2000). According to Cho and Koyama (1997), microorganisms are the smallest units of life that are too minute; they are found everywhere in nature and play a vital role in maintaining the ecological balance. Beneficial microbes (BMs) are mixed cultures of naturally occurring life forms that can be applied as inoculants to enhance the microbial diversity of the soil ecosystem.

A lot of scientific evidence is available to point out that intensive irrigation and high chemical use, including fertilizers and pesticides, add to the degradation of land. According to the estimates, about 141 million hectares of land out of 328.7 million hectors are used for agricultural crop production, due to excessive chemical usage and modern agronomical practices that significantly contribute to land degradation. Many technological inventions are being carried out for the improvement of crop productivity, but still, we are not in a position to fulfill the global demand for food. BMs associated with plants can be used to overcome the problems related to soil salinity, fertility, degradation, habitat loss, and plant protection. Soil contains a mixture of diverse group of organisms like bacteria, archaea, fungi, algae, insects, annelids, and other invertebrates, which show an innermost relationship to each other and with plants (Glick 2010). Among these relationships, microbial entities are found to be unique, as they are directly involved in increasing soil fertility, plant growth promotion and lowering biotic and abiotic stresses by different mechanisms. Among them, microbial entities are unique as they are directly involved in enhancing soil fertility, promoting plant growth, and reducing biotic and abiotic stresses by different mechanisms. BMs enhance the plant growth by nutrient uptake, form a complex soil matrix and help in plant's defence response by secretion of various metabolites. In addition, BMs can show tolerance toward adverse environmental conditions such as salt stress, drought stress, weed infestation, nutrient deficiency, and heavy metal contamination. In the past few years, researchers have observed that soil microbes play both beneficial and harmful roles in the soil ecosystem. However, BMs gained much importance not only for their plant growth promoting attributes but also for their role in the decomposition of organic wastes, detoxification of toxic substances such as pesticides and relieve of soil stresses (Aislabie and Deslippe 2013; Ma et al. 2016).

A major focus in the coming decades would be on safe and eco-friendly methods by exploiting the role of beneficial microorganisms in sustainable crop production (Nina et al. 2014). In general, those include a diverse group of naturally occurring microbes whose inoculation in the soil ecosystem improves soil physical and chemical properties, microbial biodiversity within soil, soil health conditions, growth and development of plants as well as crop production (Sahoo et al. 2013a). The most useful microbial populations in agricultural biotechnology include plant growth-promoting *Rhizobacteria*, N_2-fixing *Cyanobacteria*, *Mycorrhiza*, beneficial bacteria that are prone to suppress plant disease, endophytes that help plants in stress tolerance and bio-degrading active microbes (Sahoo et al. 2013b). Biofertilizers act as supplementary components to soil and crop cultivation practices, which helps in rotation of crop, organic maintenance, tillage conservation, reuse of crop residue, soil fertility improvement, and biological control of plant pathogens and insect pests. As such biofertilizers collectively can significantly contribute to maintaining the sustainability of various crop productions (Dogan et al. 2011). For example, *Azotobacter*, *Azospirillum*, *Rhizobium*, *Cyanobacteria*, phosphorus and potassium solubilizing microorganisms can significantly provide the desirable amount of nitrogen to *Helianthus annus* and thus enhance the height of plant, the number of leaves, stem thickness, seed filling percentage, and dry weight of seed (Choudhury and Kennedy 2004). Similarly, the addition of *Azotobacter*, *Azospirillum*, and *Rhizobium* promotes the physiology of plant and improves the root morphology in rice crops (Revillas et al. 2000).

6.2 Beneficial Microorganisms in Sustainable Agriculture

6.2.1 Plant Growth Promoting Rhizobacteria

Plant growth promoting rhizobacteria (PGPR) are potential microbes that facilitate plant growth either directly or indirectly by colonizing the plant root. These soil bacteria have the ability to colonize roots and stimulate plant growth. PGPR species belong to various different genera including *Azoarcus*, *Azospirillum*, *Rhizobium*, *Azotobacter*, *Arthrobacter*, *Bacillus*, *Clostridium*, *Enterobacter*, *Gluconoacetobacter*, *Pseudomonas*, and *Serratia*. Recently, a lot of research has been carried out to understand the mechanisms of plant–microbe interactions. Besides, PGPR may affect the growth of plants via direct or indirect means (Vadakattu 2012). Direct growth promotion involves the production of phytohormones, fixation of atmospheric nitrogen (N_2), synthesizing iron chelators known as siderophores, solubilizing inorganic minerals such as phosphorus (P), potassium (K) and zinc (Zn), making them readily available for plant growth in large quantity. An indirect role can be seen in the form of their growth inhibition activity against plant pathogens by one of several mechanisms, such as antibiotic or antifungal metabolite(s) production, remediation of iron from the rhizosphere, induced systemic resistance (ISR), fungal cell wall lytic enzyme production and competition toward binding site(s) on the roots. In addition, PGPR are also recognized as potential microbes that can protect the plant from various environmental stresses in normal as well as stressed environments (Mishra et al. 2017).

6.2.1.1 Biofertilizers

6.2.1.1.1 Rhizobium

This beneficial bacteria belongs to family Rhizobiaceae, is found symbiotic in nature, fixes atmospheric nitrogen at 50–100 kg/ha in association with leguminous crops. It is useful for pulse legumes like chickpea, red gram, pea, lentil, black gram, etc., oilseed legumes like soybean and groundnut and forage legumes like *Berseem* and *lucerne*. Successful nodulation of leguminous crops by *Rhizobium* bacteria mainly depends on the availability of compatible strain of particular legume crop. The bacteria colonizes the roots of specific legumes to develop tumor-like growth called root nodules, which serve as factories for ammonia production. *Rhizobium* bacteria possess important features to fix atmospheric nitrogen by making symbiotic association with legumes and certain nonlegumes like *Parasponia*. *Rhizobium* population in the soil depends on the presence of legume crops in the field (Patil and Medhane 1974).

6.2.1.1.2 Azospirillum

This organism belongs to the family Spirilaceae, and is found to be heterotrophic and associative in nature. In addition to their nitrogen-fixing property of about 20–40 kg/ha, they are also known for the production of growth-regulating substances. Even though there are many species that come under this genus, like *A. amazonense, A. halopraeferens, A. brasilense,* but, globally distribution and application of inoculation have been reported mainly with *A. lipoferum* and *A. brasilense*. Plants having the C4-dicarboxyliac pathway for photosynthesis are known to have associative symbiosis in the form of *Azospirillum* (Hatch and Slack pathway), as this organism is known to grow and fix nitrogen from salts of organic acids such as malic and aspartic acid. Therefore, it is mainly recommended for maize, sugarcane, sorghum, pearl millet, etc. (Dhanasekar and Dhandapani 2012).

6.2.1.1.3 Azotobacter

This class of bacteria belongs to the family Azotobacteriaceae and is aerobic, free-living, and heterotrophic in nature. Azotobacters are present in neutral or alkaline soil condition, and the most commonly occurring species in arable soils is *A. chroococcum*. *A. vinelandii, A. beijerinckii, A. insignis,* and *A. macrocyto* genes are other reported species belonging to the genera *Azotobacter*. The number of *Azotobacter* rarely exceeds a threshold level of 10^4 to 10^5 1/g in soil due to the absence of organic matter and availability of antagonistic microorganisms in soil. The bacterium produces antifungal antibiotics, which affect the growth of several pathogenic fungi in the root section, thereby preventing seedling mortality to a certain extent. The occurrence of this organism has been reported from the rhizosphere of a number of crop plants such as rice, maize, sugarcane, bajra, vegetables, and plantation crops.

6.2.1.1.4 Cyanobacteria and Azolla

These microbes belong to eight different classes of families, naturally phototrophic in nature and are known to produce auxin, indole acetic acid, and gibberellic acid. This organism can fix 20–30 kg N/ha in submerged rice fields, as they are found abundantly in paddy and are therefore commonly known as "paddy organisms." The 50–60% N requirement is accomplished through the mixture of mineralization of soil organic N and biological nitrogen fixation by free-living and rice plant-associated bacteria. For successful achievement of food security through sustainable agriculture, the requirement of fixed nitrogen must be increasingly replaced by BNF rather than by industrial nitrogen fixation. Fungi, liverworts, ferns, and flowering plants can form symbiotic associations with blue-green algae and are capable of fixing nitrogen. The most common symbiotic association has been reported between a free-floating aquatic fern, the *Azolla* and *Anabaena azollae* (BGA). *Azolla* contains 4–5% N on dry basis and 0.2–0.4% on wet basis and can be the potential source of organic manure and nitrogen in rice production. The important factor in using *Azolla*as as a biofertilizer for rice crops is its quick decomposition in the soil and efficient availability of its nitrogen to rice plants. Apart from N-fixation ability, the biofertilizers or biomanures significantly contribute to the amounts of P, K, S, Zn, Fe, Mb, and other micronutrients for plants. The fern developed a green mat over water along with a branched stem, deeply bi-layered leaves and roots. Within the central cavity of the dorsal fleshy lobe of the leaf are rich in algal symbiont. *Azolla*can be applied as green manure by incorporating it in the fields prior to rice planting. *A. pinnata* is the most common species in India, which can be propagated easily at commercial scale by vegetative means. On average, about 1.5 kg/m^2 can probably be yielded in a week at commercial level. Recently, India has also introduced some species of *Azolla* for their large biomass production, which includes *A. caroliniana, A. microphylla, A. filiculoides,* and *A. Mexicana.*

6.2.1.1.5 Phosphate Solubilizer

Tremendous data have reported the ability of various bacterial species to solubilize insoluble inorganic phosphate compounds such as tricalcium phosphate, dicalcium phosphate, hydroxyapatite, and rock phosphate. Among the bacterial genera with this capacity are *Pseudomonas, Bacillus, Rhizobium, Burkholderia, Achromobacter, Agrobacterium, Micrococcus, Aereobacter, Flavobacterium* and *Erwinia.* There are notable populations of phosphate-solubilizing bacteria present in soil and plant rhizospheres. These include both aerobic and anaerobic strains, with a prevalence of aerobic strains in submerged soils. A large number of higher concentrations of phosphate-solubilizing bacteria are commonly found in the rhizosphere when compared to non-rhizosphere soil. The soil bacteria belonging to the genera *Pseudomonas* and *Bacillus* and *Fungi* are more common (Azziz et al. 2012).

6.2.1.1.6 Mycorrhiza

The term *Mycorrhiza* mostly represents "fungus roots." It is a symbiotic association between host plants and certain groups of fungi within the root system, in which the fungal partner is benefited by obtaining its carbon requirements from the photosynthates of the host plant and in return the host plant is gifted by obtaining the much-needed nutrients, especially in the form of phosphorus, calcium, copper, zinc, etc. which are not easily accessible by plants with the help of the fine absorbing hyphae of the fungi. These organisms are mostly associated with the majority of agricultural crops, with an exemption of those crops that belong to families such as Chenopodiaceae, Amaranthaceae, Caryophyllaceae, Polygonaceae, Brassicaceae, Commelinaceae, Juncaceae, and Cyperaceae.

6.2.1.1.7 Zinc Solubilizers

Apart from above-mentioned microbes that can be used as biofertilizers, there lies a host of microorganism that can transform micronutrients in the form of zinc, iron, copper, etc. that can also be used as biofertilizers. This zinc-degrading microorganism includes *B. subtilis, Thiobacillus thioxidans,* and *Saccharomyces* sp. These microorganisms can be used as biofertilizers for solubilization of fixed micronutrients, mostly in the form of zinc. An example includes *Bacillus* spp. (Zn solubilizing bacteria), which can serve as biofertilizer in soils where native zinc is higher or in conjunction with insoluble cheaper zinc compounds in the form of zinc oxide (ZnO), zinc carbonate ($ZnCO_3$), and zinc sulphide (ZnS) rather than of costly zinc sulphate (Table 6.1).

Table 6.1 Characteristics of some BSM based biofertilizers.

Type of bio-fertilizers	Name of beneficial microbes used	Description	References
Nitrogen fixers	*Rhizobium* spp.	*Rhizobium* is known for its ability to fix atmospheric nitrogen in symbiotic association with plants forming nodules in roots.	Mia and Shamsuddin (2010)
	Azospirillum spp.	It is a free-living and nonsymbiotic nitrogen fixer, beneficial to a wide array of crops covering cereals, millets, vegetables, cotton, and sugarcane.	Bashan et al. (1993)
	Azotobacter spp.	Rapidly grow in various agricultural soils fixes high level of nitrogen.	Siddiqui et al. (2014)
	Cyanobacteria	Some cyanobacteria have terminally differentiated, specialized structures called heterocyst and considered as photodiazotrophs. Their presence in rice fields provided aid in nitrogen uptake.	Kaushik (2014)
	Azolla spp.	This water fern has asymbiotic relation with blue, green algae and can help rice or other crops cultivation	Carrapico et al. (2000)
Phosphate solubilizers	*Pseudomonas* spp. *Bacillus* spp., *Rhizobium* spp., *Burkholderia* spp. and Mycorrhiza,	PSB and Mycorrhiza can help in improving the uptake of an insoluble form of phosphate to plants in different ways	Duarah et al. (2011)

(*Continued*)

Table 6.1 (Continued)

Type of bio-fertilizers	Name of beneficial microbes used	Description	References
Zinc solubilizers	*Bacillus* spp., *Pseudomonas* spp.	Solubilise zinc compounds like zinc oxide (ZnO), zinc carbonate ($ZnCO_3$) and zinc sulphide (ZnS) and may reduce the rate of costly zinc sulphate	Ahemad and Kibret (2014)
Potassium solubilizers	*Bacillus* spp., *Burkholderia* spp., *Pseudomonas* spp.	Solubilise potassium, present in the insoluble form of rocks and silicate minerals	Shanware et al. (2014)
Phytohormones producers	*Pseudomonas, Bacillus, Rhizobium, Burkholderia* and *Mycorrhiza*	Produce phytohormones such as auxins, cytokinins, gibberellins and ethylene can affect cell proliferation in the root architecture by overproduction of lateral roots and root hairs with a subsequent increase in nutrient and water uptake. The enzyme 1-aminocyclopropane-1 carboxylic acid (ACC) is a pre-requisite for ethylene production, catalysed by ACC oxidase	Ahemad and Kibret (2014)

6.2.1.2 Biopesticides

About 40% of the potential global crop yield is destroyed by pests (invertebrates, plant pathogens and weeds) before it is harvested, and another 20% is destroyed post-harvest (Bailey et al. 2010). However, for sustainable agriculture, the use of chemical-based pesticides should be restricted, as they possess a negative impact on the environment. Workers around the world are now using biopesticides and trying to minimize the use of conventional chemical pesticides. Biopesticides are pesticides derived from natural materials such as animals, plants, bacteria, and minerals (USEPA 2008). Biocontrol agents can be grouped into three broad categories, namely bacterial, fungal, and viral. Among the most widely used bacterial pesticides are the strains of *Bacillus thuringiensis* (Bt). Several species of *Pseudomonas* showed biocontrol potential against plant pathogens; for example, biopesticides in the form of *Pseudomonas fluorescence* and *Pseudomonas syringae* have been used in large numbers nowadays (Usta 2013). Fungal biopesticides include several species of *Trichoderma* used against plant pathogens, which are antagonists of several soil-borne fungi such as *Rhizoctonia, Pythium, Fusarium* and other plant pathogens (Table 6.2).

6.2.1.2.1 Bacillus thuringiensis (Bt)

Bt is a sporulating bacterium, Gram-positive facultative aerobic in nature, and familiar as soil bacterium. Its principal characteristic is the synthesis of a crystalline inclusion containing proteins known as δ-endotoxins or Cry proteins during sporulation. This protein possesses an insecticidal property. The considerable diversity of these toxins, their efficacy

Table 6.2 Characteristics of some important biopesticide organism found in nature.

Type of biopesticide	Name of beneficial microbe used	Host range
Bacteria	*Bacillus thuringiensis* var. *kurstaki* (*Bt*)	Caterpillars (larvae of moths and butterflies)
	Bacillus thuringiensis var. israelensis (*Bt*)	Larvae of *Aedes* and *Psorophora* mosquitoes, black flies, and fungus gnats
	Bacillus thuringiensis var. *tenebrinos*	Larvae of Colorado potato beetle, elm leaf beetle adults
	Bacillus thuringiensis var. *aizawai*	Wax moth caterpillars
	Bacillus popilliae and *Bacillus lentimorbus*	Larvae (grubs) of Japanese beetle
	Bacillus sphaericus	Larvae of *Culex, Psorophora*, and *Culiseta* mosquitos, larvae of some *Aedes* spp.
	Pseudomonas chlororaphis	Soil and seed-borne fungi
	Pseudomonas syringae	Postharvest diseases
	Pseudomonas aureofaciens	Turf fungal diseases
	Pseudomonas fluorescens	Bacterial wilt, root rot
	Pseudomonas alcaligenes	Locusts, grasshoppers
	Pseudomonas aureofaciens	Turf fungal diseases
	Pseudomonas resinovorans	Insect pest control
	Pseudomonas syringae	Bacterial speck
	Agrobacterium radiobacter	Crown gall disease
	Pantoeaagglomerans	Fire blight
Fungi	*Beauveriabassiana*	Aphids, fungus gnats, mealy bugs, mites, thrips, whiteflies
	Lagenidiumgiganteum	Larvae of most pest mosquito species
	Metarhiziumanisopliae	Coleoptera and lepidopteran insect
	Trichodermaviride, Trichodermagamsii, Trichodermaharzianum, T. virens, T. fasciculatum, T. atroviride, T asperellum	Soil-borne pathogens
	Verticilliumlecanii	Whitefly, coffee green bug
	Pythium oligandrum	Root rots
	Trichodermapolysporum	Soil and foliar diseases
	Paecilomycesfumosoroseus	Greenhouse whiteflies
	Alternariadestruens	Herbicide—dodder
	Candida oleophila	Postharvest disease
	Aspergillus flavus	*Aspergillus flavus*-producing aflatoxins

and their relatively cheap production have made Bt the most widely used biopesticide globally. It is widely used in the fight against many agricultural crop pests, mostly lepidopteran and coleopteran larvae notably in the creation of new genetically modified plant varieties expressing Bt cry genes. Some of the BMs containing biopesticides are also used in the management of pests (Sanchis and Bourguet 2008).

6.2.1.2.2 *Pseudomonas fluorescens*

This bacterium belongs to PGPR and is an important group of bacteria that play a major role in the plant growth promotion, ISR, biological control of pathogens, etc. Many strains of *Pseudomonas fluorescens* are able to enhance plant growth promotion and decrease the severity of many diseases. The efficacy of bacterial antagonists in controlling fungal diseases was often better alone, and sometimes in combination with fungicides (Ganeshan and Manojkumar 2005).

6.2.1.2.3 *Agrobacterium radiobacter*

This microorganism is a gram negative bacillus found in soil that is enriched with organic material (rhizosphere). It is a saprophytic organism that mainly uses dead plant material as nutrients (Sanchis and Bourguet 2008). The bacterium namely *A. radiobacter* strain k84 is utilized commonly for its competitive behavior with the closely related *A. tumefaciens* the causative agent of crown gall disease in plants. The bacteria synthesizes and secretes a substance called agrocin 84, which has the functional ability to block *A. tumefaciens* from replicating its own DNA, thereby effectively halting its ability to reproduce (Sanchis and Bourguet 2008). The strain *A. radiobacter* J14 is useful in its ability to degrade atrazine, which has the power of an agricultural herbicide.

6.2.1.2.4 *Trichoderma*

This organism mostly belongs to a genus of fungi that is present in most types of soils, where they are the most prevalent cultural fungi in nature. *Trichoderma* spp. can be frequently isolated from forest or agricultural soils and from wood easily. Some species are also found growing on other fungi. There are almost 90 species that fall under *Trichoderma* genus. *Trichoderma* strains are familiar as among the most useful fungi used in industrial enzyme production, agriculture sector, and bioremediation.

6.3 Environmental Nanotechnology

Nanotechnology is a branch of engineering that deals with creating objects smaller than 100 nm in dimension. The field of nanotechnology is one of the most popular areas for advanced research and development nowadays basically in all technical disciplines. Environmental nanotechnology plays an important role in the discipline of environmental engineering and science due to its simulative advancement and cost-effective technology for remediation, pollution detection, and monitoring. Environmental protection and pollution problems are worldwide spread hot topics of debate since the pollution controlled by nanotechnology on one hand results in a reduction of the use of raw materials, water, or other resources and the elimination or reduction of waste, and on the other hand, it has a more energy-efficient production (Walls and Palmer 2001). One of the major challenges for the global advancement of the nanomaterials market is the environmental sustainability of nano-manufacturing processes. Indeed, traditional top-down or bottom-up chemical and physical Nano-manufacturing approaches have a greater energy intensity compared to manufacturing processes of bulk materials (Khanna et al. 2008; Prasad et al. 2016).

They are often characterized further by low process yields (using acidic/basic chemicals and organic solvents), greenhouse gas generation, and the need for desirable facilities and operative conditions (e.g., from moderate to high vacuum), and high purity-level quality starting materials (Sengül et al. 2008; Yuan and Zhang 2013). Green chemistry ("the principle processes to reduce or to eliminate the use and generation of hazardous substances") combined with white biotechnology ("biotechnology that uses living cells, yeasts, molds, bacteria, plants, and enzymes to synthesize products at industrial scale") can really contribute to the development of more sustainable industrial processes (Ribeiro et al. 2016), and for the manufacturing of nonmaterial's with environment value. The microbial-mediated biosynthesis of nanomaterials is a promising biotechnological-based nano manufacturing process that represents a "green" alternative approach to physical and chemical strategies of nanosynthesis (Li et al. 2011; Khan et al. 2018). The microbial-mediated biosynthesis of metallic (also as alloys), nonmicrobial strains of bacteria, yeast, molds, and microalgae (Hulkoti and Taranath 2014). In addition, some microorganisms have shown the capability to biosynthesize unique nanostructured materials, i.e., biomineralized nano-structures like silicified frustules (Kroger and Poulsen 2008), calcified coccoliths (Skeffngton and Scheffel 2018), magnetosomes (Yan et al. 2017), and organic nonmaterial like bacterial nano-cellulose (Gama et al. 2012). Exopolysaccharide nanoparticles (Nwodo et al. 2012) and bacterial nanowires (Malvankar and Lovely 2012).

6.3.1 Nanotechnology as a Tool for Environmental Remediation, Pollution Prevention, and Treatment

Environmental remediation often includes the degradation, sequestration, or other related approaches that reduce the risks to human and environmental receptors mostly imposed by chemical and radiological contaminants. Benefits that mostly arise from the application of nanomaterials for remediation include more rapid or cost-effective cleanup of wastes (Reyesa et al. 2014). The most common environmental remediation techniques traditionally used to treat large volumes of water intended for municipal water supplies include carbon adsorption, air stripping, oxidation through ozonation or chlorination and incineration, ultrafiltration and sedimentation (Carp et al. 2004). Some nanotechnology applications are near commercialization: nanosensors and nanoscale coatings to replace thicker, more wasteful polymer coatings that prevent corrosion, nanosensors for detection of aquatic toxins, nanoscale biopolymers for improved decontamination and recycling of heavy metals, nanostructured metals that break down hazardous organics at room temperature, smart particles for environmental monitoring and purification, nanoparticles as novel photocatalyst for environmental cleanup. The implementation of green chemistry principles for the production of nanoparticles and for nanotechnological applications in standard chemical engineering will lead to a great reduction in waste generation, less hazardous chemical syntheses, improved catalysis, and finally an inherently safer chemistry (Albrecht et al. 2006).

Environmental cleanup has enhanced the development of highly efficient photocatalysts, which directly convert pollutants into bio-friendly nontoxic products, and also contributes to the use of oxygen as oxidant and the elimination of expensive oxidizing chemicals; potential for using free and abundant solar energy; self-regeneration and recycling of photocatalyst. The progress in this area enables the catalyzed destruction of different organic chemicals and completes oxidation of organics to CO_2, dilution of mineral acids, lack of inherent toxicity and resistance to photodegradation at low cost render this process highly suitable for environmental remediation. The environmental remediation is very important factor for choosing a particular technology since the chosen approach is dependent on the treatment costs and the complexity of the contaminated media (soil, water, or air) (Bhawana and Fulekar 2012). Thus, by controlling and modifying the structure of significant substances for remediation of soils, sediment, and groundwater rich with heavy metals (mercury, lead, and cadmium) and organic compounds (benzene, chlorinated solvents, creosote, and toluene), it can be possible to improve their affinity, capacity, and selectivity toward certain pollutants (Agarwal and Joshi 2010).

6.3.1.1 Decontamination of Polluted Air and Soil

Air pollution is a potential area where nanotechnology has great promise. Filtration techniques similar to the water purification methods could be used in buildings to purify indoor air volumes. Nano filters can easily be applied to automobile tailpipes and factory smokestacks to filter out contaminants and stop them from entering into the atmosphere. Finally, nanosensors could be developed to detect toxic gas leaks at extremely low concentrations. The size of nanoscale materials enables them to be captured by narrow passages between soil particles and allows their higher adsorption capacity and aggregation compared to larger particles. Very few reports in the form of literature about the nanoparticles movement in soil are available. The form and chemical modification of nanoparticles, as well as mechanisms and properties that govern binding between colloidal particles, are very important to the mobility and sorption of nanoparticles (Schrick et al. 2004). Chemically modified fullerenes (C60) and carbon nanotubes have a great potential for dispersing in soil. The decontamination of polluted soil research area is constantly growing, and it is found that metallic iron in nanoparticle form is highly effective in catalyzing the treatment of soil polluted by chlorinated organic substances (Weisner et al. 2006). The large ratio between surface area and mass in nanoparticles can also be used to immobilize heavy metals. In these cases it means that relatively large quantities of nanoparticles are released into the soil and will reach the groundwater.

6.3.1.2 Water Treatment

Clean water is a prerequisite for all properly functioning societies globally, but is often available in limited form. The significance of new approaches such as using nanoparticles as adsorbents, nanoparticles used as catalysts for chemical or photochemical destruction of contaminants; nanosized zero-valent iron used for the removal of metals and organic compounds from water, and nanofiltration membranes are discussed. Removal of contaminated groundwater using nanoparticle remediation containing zero-valent iron is one of the most prominent examples of a rapidly emerging technology of bioremediation with considerable potential benefits. Nanopowder of zero-valent iron, due to their large ratio of surface area to mass, can be used for the removal of nitrate in water (Lockman et al. 2004). Many researchers are focused on TiO_2 nanoparticles as photocatalysts in water treatment due to their activation by light, and on ZnO that are able to remove organic contaminants from various media. The advantages of low-toxical nanoparticles are also their availability and low cost. The semiconducting property of TiO_2 is essential for the remediation of different organic pollutants via excitation of TiO_2 semiconductor with a light energy that is greater than its band gap, which can generate the required electron-hole pairs. These may be exploited in different reduction processes at the semiconductor/solution interface (Reyesa et al. 2014). FeS nanoparticles, synthesized by the wet chemical method and stabilized using a polymer from basidiomycetous, can degrade Lindane (γ-hexachloroccyclohexane, persistent organic pollutants) from drinking water (Zhang 2003). Cellulose nanobeads, applied as ion exchangers, are a promising adsorbent for heavy metal ions and proteins, and the carriers for immobilization of biocatalysts due to their special characteristics, including hydrophilic, porous, high surface area, and excellent mechanical and hydraulic properties (Makarova et al. 2000). Hydrated Fe(III)oxide, as an inexpensive and chemically stable compound, dispersed within a macroporous polymeric cation exchanger, forms a hybrid material suitable for arsenic removal. The resulting polymeric/inorganic hybrid adsorbent combines excellent mechanical properties of spherical polymeric beads (Guo and Chen 2005).

Nanocomposite consisting of iron oxide and silicate was also designed for degradation of azo-dye orange (II). Organo clays have also recently attracted lots of attention in a number of applications, such as dithiocarbamate-anchored polymer/organosmectite for the removal of heavy metal ions from aqueous media (Feng et al. 2003). Carbon nanotubes show adsorption capability and high adsorption efficiency for removal of heavy metals such as lead (DeMarco and SenGupta 2003). Nanofiltration membranes that are used in pressure-driven membrane separation processes have a pore size between 1 and 10 nm and are widely recommended for treatment of drinking water production or wastewater treatment (removing turbidity, microorganisms. and inorganic ions such as Ca and N) (DeMarco and SenGupta 2003). They are also used for softening of groundwater, for removal of dissolved organic matter and trace pollutants from surface water and for pretreatment in seawater desalination.

6.3.1.3 Environmental Risk Assessment

Understanding the knowledge of the environmental effects and risks associated with nanotechnology is very less and contradictory in nature. Risk assessment is an important component in the development of effective risk management strategies (Say et al. 2006; Dahl et al. 2006). Even though the technology has uncertain environmental, health, and safety impacts, nanotechnology has shown a great capability for "smart" multifunctional and effective performance products for innumerable commercial and industrial application. Society might accept the potential risks if the benefits of nanotechnology (cancer treatment and other areas of medicine and in more efficient energy systems) are clear. The downsides of nanotechnologies are rarely acknowledged, while "benefits" claimed are often emphasized. An approach for improving the understanding of the potential influences of nanotechnology is life cycle assessment (LCA). This comprehensive analysis tool can be used for the estimation of the effect of a product or material life cycle on ecosystems and human health. Nanotechnology delivers cleaner production, but nanomaterials production requires a lot of water and energy, and the used chemicals are often highly toxic. Early LCAs show certain doubts that nanotechnology applications can save energy in most industrial processes. It was estimated that nanomaterial fabrications have an unexpectedly large ecological footprint (Li et al. 2005) concerning the energy and water needs of processing, low product yields, high waste generation, and the production and use of greenhouse gases originating from toxic chemicals and solvents (Gajewicza et al. 2012; Khanna et al. 2008).

Also, carbon nanofiber production was found to have a much greater contribution to global warming, ozone layer depletion, environmental or human toxicity compared to conventional materials (aluminum, steel, and polypropylene) (Laurent et al. 2012). Nanoparticles (carbon black) can be also used for the reinforcement of polymer matrix, and literature data show that 90% of their use is for rubber reinforcement (for automotive tire production). Between 25% and 35% of the automotive tires consist of carbon black nanoparticles, which can be emitted in the atmosphere after long-term consumption of tires (Bauera et al. 2008).

6.4 Conclusions

Use of BSMs in sustainable agriculture and environmental management offers countless benefits. Their utilization in the form of bio-fertilizers and bio-pesticides is becoming popular and providing substantial aid to the agro-ecosystems. Advanced nanotechnology, going beyond the knowledge of natural system and cycles, has a great potential for the continued improvement of technologies regarding environmental protection. With developing different aspects of nanotechnology, the broader environmental impacts need to be discussed, including determination of potential benefits of reduction or prevention of pollutants from industrial sources, but also monitoring or controlling unintended negative effects by proper and comprehensive risk assessment and life cycle analysis. Although nanotechnology application has many important benefits for industry development and environmental sustainability and protection, the energy demands of producing nanoparticles, the significant quantities of potentially toxic waste generated from their production, and the ecotoxic behavior of many nanoparticles cause doubts in society that nanotechnology and new technological innovations offer cleaner solutions to the current ecological problems (energy consumption and air, water, and soil pollution).

References

Agarwal, A. and Joshi, H. (2010). Application of nanotechnology in the remediation of contaminated groundwater: a short review. *Recent Research in Science and Technology* 2: 51–57.

Ahemad, M. and Kibret, M. (2014). Mechanisms and applications of plant growth promoting rhizobacteria: current perspective. *Journal of King Saud University - Science* 26: 1–20.

Aislabie, J. and Deslippe, J.R. (2013). Soil microbes and their contribution to soil services. In: *Ecosystem Services in New Zealand-Conditions and Trends* (ed. J.R. Dymond), 143–161. New Zealand: Manaaki Whenua Press.

Albrecht, M.A., Evans, C.W., and Raston, C.L. (2006). Green chemistry and the health implications of nanoparticles. *Green Chemistry* 8: 417–432.

Azziz, G., Bajsa, N., Haghjou, T. et al. (2012). Abundance, diversity and prospecting of culturable phosphate solubilizing bacteria on soils under croppasture rotations in a no-tillage regime in Uruguay. *Applied Soil Ecology* 61: 320–326.

Bailey, K.L., Boyetchko, S.M., and Langle, T. (2010). Social and economic drivers shaping the future of biological control: a Canadian perspective on the factors affecting the development and use of microbial biopesticides. *Biological Control* 52: 221–229.

Bashan, Y., Holguin, G., and Lifshitz, R. (1993). Isolation and characterization of plant growthpromoting rhizobacteria. In: *Methods in Plant Molecular Biology and Biotechnology* (eds. B.R. Glick and J.E. Thompson), 331–345. Boca Raton, FL: CRC Press.

Bauera, C., Buchgeistera, J., Hischierb, R. et al. (2008). Towards a framework for life cycle thinking in the assessment of nanotechnology. *Journal of Cleaner Production* 16: 910–926.

Bhawana, P. and Fulekar, M.H. (2012). Nanotechnology: remediation Technologies to clean up the Environmental pollutants. *Research Journal of Chemical Sciences* 2: 90–96.

Carp, O., Huisman, C.L., and Reller, A. (2004). Photo induced reactivity of titanium dioxide. *Progress in Solid State Chemistry* 32: 33–177.

Carrapico, F., Teixeira, G., and Diniz, M.A. (2000). Azolla as a biofertilizer in Africa. *Revista de Ciencias Agrarias* XXIII: 120–138.

Cho, H.K. and Koyama, A. (1997). Korean natural farming: indigenous microorganisms and vital power of crop/livestock. Korean Natural Farming.

Choudhury, M.A. and Kennedy, I.R. (2004). Prospects and potentials for system of biological nitrogen fixation in sustainable rice production. *Biology and Fertility of Soils* 39 (4): 219–227.

Dahl, A., Gharibi, A., Swietlicki, E. et al. (2006). Traffic-generated emissions of ultrafine particles from pavement–tire interface. *Atmospheric Environment* 40: 1314–1323.

DeMarco, M.J. and SenGupta, A.K. (2003). Arsenic removal using a polymeric/inorganic hybrid sorbent, Greenleaf. *Water Research* 37: 164–176.

Dhanasekar, R. and Dhandapani, R. (2012). Effect of biofertilizers on the growth of *Helianthus annuus. International Journal of Plant, Animal and Environmental Sciences* 2 (4): 143–147.

Dogan, K., Celik, I., Gok, M., and Coskan, A. (2011). Effect of different soil tillage methods on rhizobial nodulation, biyomas and nitrogen content of second crop soybean. *African Journal of Microbiology Research* 5 (20): 3186–3194.

Duarah, I., Deka, M., Saikia, N., and Deka Boruah, H.P. (2011). Phosphate solubilizers enhance NPK fertilizer use efficiency in rice and legume cultivation. *3 Biotech* 1: 227–238.

Feng, J., Hu, X., Yue, P.L. et al. (2003). Degradation of azo-dye orange II by a photoassisted fenton reaction using a novel composite of iron oxide and silicate nanoparticles as a catalyst. *Industrial & Engineering Chemistry Research* 42: 2058–2066.

Gajewicza, A., Rasulevb, B., Dinadayalaneb, T.C. et al. (2012). Advancing risk assessment of engineered nanomaterials: application of computational approaches. *Advanced Drug Delivery Reviews* 64: 1663–1693.

Gama, M., Gatenholm, P., and Klemm, D. (ed.) (2012). *Bacterial Nanocellulose: A Sophisticated Multifunctional Material*. Boca Raton, FL: CRC Press.

Ganeshan, G. and Manojkumar, A. (2005). *Pseudomonas fluorescens*, a potential bacterial antagonist to control plant diseases. *Journal of Plant Interactions* 1 (3): 123–134.

Glick, B.R. (2010). Using soil bacteria to facilitate phytoremediation. *Biotechnology Advances* 2: 367–374.

Guo, X. and Chen, F. (2005). Removal of arsenic by bead cellulose loaded with iron oxyhydroxide from groundwater. *Environmental Science & Technology* 39: 6808–6818.

Hulkoti, N.I. and Taranath, T.C. (2014). Biosynthesis of nanoparticles using microbes—a review. *Colloids and Surfaces B: Biointerfaces* 121: 474–483.

Kaushik, B.D. (2014). Developments in cyanobacterial biofertilizer. *Proceedings of the Indian National Science Academy* 80: 379–388.

Khan, T., Abbas, S., Fariq, A., and Yasmin, A. (2018). Microbes: nature's cell factories of nanoparticles synthesis. In: *Exploring the Realms of Nature for Nanosynthesis*, 25–50. Cham, Switzerland: Springer.

Khanna, V., Bakshi, B., and Lee, L. (2008). Carbon nanofiber production life cycle energy consumption and environmental impact. *Journal of Industrial Ecology* 12: 394–410.

Kroger, N. and Poulsen, N. (2008). Diatoms-From cell wall biogenesis to nanotechnology. *Annual Review of Genetics* 42: 83–107.

Laurent, A., Olsen, S.I., and Hauschild, M.Z. (2012). Limitations of carbon footprint as indicator of environmental, sustainability. *Environmental Science & Technology* 46: 4100–4108.

Li, Y.H., Di, Z., Ding, J. et al. (2005). Adsorption thermodynamic, kinetic and desorption studies of Pb^{2+} on carbon nanotubes. *Water Research* 39: 605–609.

Li, X., Xu, H., Chen, Z.S., and Chen, G. (2011). Biosynthesis of nanoparticles by microorganisms and their applications. *Journal of Nanomaterials* 2011: 1–16.

Lockman, P.R., Koziara, J.M., Mumper, R.J., and Allen, D.D. (2004). Nanoparticle surface charges alter blood–brain barrier integrity and permeability. *Journal of Drug Targeting* 12: 635–641.

Ma Y, Oliveira RS, Freitas H and Zhang C (2016) Biochemical and molecular mechanisms of plant-microbe-metal interactions: relevance for phytoremediation. *Frontiers in Plant Science* https://doi.org/10.3389/fpls.2016.00918.

Makarova, O.V., Rajh, T., Thurnauer, M.C. et al. (2000). Surface modification of TiO_2 nanoparticles for photochemical reduction of nitrobenzene. *Environmental Science & Technology* 34: 4797–4803.

Malvankar, N.S. and Lovley, D.R. (2012). Microbial nanowires: a new paradigm for biological electron transfer and bioelectronics. *ChemSusChem* 5: 1039–1046.

Mia, M.A.B. and Shamsuddin, Z.H. (2010). Rhizobium as a crop enhancer and biofertilizer for increased cereal production. *African Journal of Biotechnology* 9: 6001–6009.

Mishra, J., Singh, R., and Arora, N.K. (2017). Alleviation of heavy metal stress in plants and remediation of soil by rhizosphere microorganisms. *Frontiers in Microbiology* 8: 1706.

Nina, K., Kuyper, T.W., and Bindraban, P.S. (2014). Beneficial organisms for nutrient uptake. VFRC Report 2014/1, Virtual Fertilizer Research Center. Wageningen Academic Publishers, Washington, DC, p. 63.

Nwodo, U., Green, E., and Okoh, A. (2012). Bacterial exopolysaccharides: functionality and prospects. *International Journal of Molecular Sciences* 13: 14002–14015.

Patil, P.L. and Medhane, N.S. (1974). Seed inoculation studies in groundnut with Rhizobium sp. *Journal of Maharashtra Agricultural Universities* 1: 46–47.

Prasad, R., Pandey, R., and Barman, I. (2016). Engineering tailored nanoparticles with microbes: Quo vadis? *Wiley Interdisciplinary Reviews. Nanomedicine and Nanobiotechnology* 8: 316–330.

Revillas, J.J., Rodelas, B., Pozo, C. et al. (2000). Production of B-group vitamins by two Azotobacter strains with phenolic compounds as sole carbon source under diazotrophic and adiazotrophic conditions. *Journal of Applied Microbiology* 89: 486–493.

Reyesa, J.T., Peralta-Videaa, J.R., and Gardea-Torresdeya, J.L. (2014). Supported and unsupported nanomaterials for water and soil remediation: are they a useful solution for worldwide pollution. *Journal of Hazardous Materials* 280: 487–503.

Ribeiro, B.D., Coelho, M.A.Z., and de Castro, A.M. (2016). Principles of green chemistry and white biotechnology. In: *White Biotechnology for Sustainable Chemistry* (ed. M.A.Z. Coelho and B.D. Ribeiro), 1–8. London: The Royal Society of Chemistry.

Sahoo, R.K., Ansari, M.W., Dangar, T.K. et al. (2013a). Phenotypic and molecular characterization of efficient nitrogen fixing Azotobacter strains of the rice fields. *Protoplasma* 251 (3): 511–523.

Sahoo RK., Tuteja N, Gill SS, (2013b). Biofertilizers: a sustainable eco-friendly agricultural approach to crop improvement. Plant Acclimation to Environmental Stress. Narendra Tuteja and Sarvajeet Singh Gill, New York: Springer Science plus Business Media b. pp.: 403–432.

Sanchis, V. and Bourguet, D. (2008). *Bacillus thuringiensis*: applications in agriculture and insect resistance management. A review. *Agronomy for Sustainable Development* 28: 11–20.

Say, R., Birlik, E., Denizli, A., and Ersöz, A. (2006). Removal of heavy metal ions by dithiocarbamate-anchored polymer/organosmectite composites. *Applied Clay Science* 31: 298–305.

Schrick, B., Hydutsky, B.W., Blough, J.L., and Mallouk, T.E. (2004). Delivery vehicles for zerovalent metal nanoparticles in soil and groundwater. *Chemistry of Materials* 16: 2187–2193.

Şengül, H., Theis, T., and Ghosh, S. (2008). Toward sustainable nanoproducts, an overview of nanomanufacturing methods. *Journal of Industrial Ecology* 12: 329–359.

Shanware, A.S., Kalkar, S.A., and Trivedi, M.M. (2014). Potassium solubilisers: occurrence, mechanism and their role as competent biofertilizers. *International Journal of Current Microbiology and Applied Sciences* 3: 622–629.

Siddiqui, A., Shivle, R., Magodiya, N., and Tiwari, K. (2014). Mixed effect of Rhizobium and Azotobacter as biofertilizer on nodulation and production of chickpea, Cicer arietinum L. International *Journal of Current Microbiology and Applied Sciences* 3: 531–536.

Skeffngton, A.W. and Scheffel, A. (2018). Exploiting algal mineralization for nanotechnology: bringing coccoliths to the fore. *Current Opinion in Biotechnology* 49: 57–63.

USEPA (2008). *What are Biopesticides*? Washington, DC: United States Environmental Protection Agency, Office of Pesticide Programs. https://www.epa.gov/ingredients-used-pesticide-products/what-are-biopesticides.

Usta, C. (2013). Microorganisms in biological pest control—a review (bacterial toxin application and effect of environmental factors). *Current Progress in Biological Research* 13: 287–316.

Vadakattu, V.S.R.G. (2012). Beneficial microorganisms for sustainable agriculture. *Microbiology Australia* 33 (3): 113–115.

Walls, M. and Palmer, K. (2001). Upstream pollution, downstream waste disposal, and the design of comprehensive environmental policies. *Journal of Environmental Economics and Management* 41: 94–108.

Weisner, M.R., Lowry, G.V., Alvarez, P. et al. (2006). Assessing the risks of manufactured nanomaterials. *Environmental Science & Technology* 40: 4336–4345.

Yan, L., Da, H., Zhang, S. et al. (2017). Bacterial magnetosome and its potential application. *Microbiological Research* 203: 19–28.

Yuan, C. and Zhang, T. (2013). Environmental implications of nano-manufacturing. In: *Green Manufacturing* (ed. D. Dornfeld), 179–202. Boston, MA: Springer.

Zhang, W. (2003). Nanoscale iron particles for environmental remediation: an overview. *Journal of Nanoparticle Research* 5: 323–332.

7

Microbial Nanotechnology in Plant Growth Promotion and Crop Protection

Neelam Thakur[1], Samiksha Jhamta[1], Simranjeet Kaur[1], Jaspreet Kaur Brar[1], Taniya Chauhan[1], Hanika Kapoor[1], Avtar Kaur Sidhu[2], and Ajar Nath Yadav[3]

[1] *Department of Zoology, Akal College of Basic Sciences, Eternal University, Sirmour, Himachal Pradesh, India*
[2] *Zoological Survey of India, High Altitude Regional Center, Solan, Himachal Pradesh, India*
[3] *Department of Genetics, Plant Breeding and Biotechnology, Dr. Khem Singh Gill Akal College of Agriculture, Eternal University, Sirmour, Himachal Pradesh, India*

7.1 Introduction

The rising global population's constant demand for nutrient-dense plant-based foods has become a matter of critical concern. Ranging in size from 1 to 100 nm, inorganic, organic, and hybrid nanoparticles (NPs) hold great promise for enhancing crop productivity and ensuring sustainability (Ahmed et al. 2024; Goyal et al. 2023). NPs boost nutritional content, enhance metabolism, and increase photosynthetic activity; they are crucial for plant development and quality improvement. The prudent use of nanotechnology can improve agricultural crop yield and quality by increasing input effectiveness and lowering associated losses. Food products free of pathogens are kept stable with the use of nanotechnology. Products that promote plant development, such as fertilizers and insecticides, have been developed using NMs. To increase soil fertility and increase agricultural yield, a vast number of different fertilizers are needed. However, the fertility and soil health are adversely affected by the frequent application of larger amounts of fertilizers. Large amounts of traditional fertilizer can either drain down or remain in the rhizosphere, contaminating the ecosystem and harming nearby plants and animals. As a result, applying fertilizer correctly is a challenging undertaking for the advancement of sustainable agriculture. In addition to reducing nutritional deficiencies, nano fertilizers also release nutrients based on the crop's needs. Therefore, farmers may increase agricultural yield without harming the environment by using nano fertilizers, which increase fertilizer usage efficiency.

Although there are many different methods for creating NPs, they can be roughly divided into two groups: top-down and bottom-up methods. Whereas bottom-up initiatives employ the building-up strategy, top-down strategies employ destructive measures. These methods are further separated into several subcategories, such as biological, chemical, and physical, according to their methods, reaction circumstances, and established protocols. NPs are created using chemical and physical methods that are challenging, costly, outdated, and hazardous to environmental and human health (Arshad et al. 2017; Khan et al. 2019). Furthermore, these methods have serious limitations due to the challenges of process scalability, the high surfactant consumption, and the extraction and purification of NPs from the microemulsion (Pilarska et al. 2013). The biological methods are better than the chemical and physical methods for creating NPs (Awad et al. 2021). Because it is feasible and employs less dangerous methods, green and biogenic bottom-up synthesis has piqued the curiosity of numerous scientists. These processes are less expensive, easier, faster, safer, good for the environment, and above all—nontoxic. Furthermore, it uses wasteless resources, less energy, and manages chemicals and components more efficiently. Another advantage is that this is a bottom-up approach (Malik et al. 2014; Mustafa et al. 2015; Schröfel et al. 2014).

Agriculture powered by nanotechnology is a new idea and is becoming a fascinating option for enhancing their qualitative qualities through the creation of nano-biofertilizers and nano-fertilizers. (Li et al. 2017). Nutrient formulations that can be applied to plants and enable consistent, long-term uptake make up nano-fertilizers. Nano-fertilizers improve the nutrient uptake capability of plants, also decrease soil toxicity, and alleviate the hostile effects of too much chemical fertilizer use and hasty fertilizer application (Deng et al. 2020; Raliya et al. 2017; Wang et al. 2021). These nano-fertilizer usages reduced the waste, saving farmers money and protecting our mother earth. An intentional combination of the biocompatible nanomaterial with a biological source-driven fertilizer results in a very effective nano-biofertilizer. They improve nutrient absorption and boost production as well as ensure sustainability. These particles are destined to boost constant nutrient discharge throughout the crop period (Thirugnanasambandan 2018). Studies in the laboratory and polyhouses as well as in the open fields are essential in developing nano-fertilizer safe dosages that are eco-friendly (Hu et al. 2016). Nano-based fertilizers and biofertilizers showed the potential to upsurge the crop yield to agricultural innovation. Due to their partial commercialization, these cutting-edge products frequently experience a low level of uptake. Providing farmers with scientific evidence of its effectiveness is required to solve this issue (Sharma et al. 2021).

The undeviating entrance of many NPs into the body of a human is enabled by their use in the agricultural fields, medicine, and pharmacology (Tsekhmistrenko et al. 2020). Creating a low-cost, environmentally friendly way of making NPs is necessary. Many species of plants, bacteria, and fungi have the capability to produce and use NPs (Das and Chatterjee 2019; Sanjay 2019). Controlled synthesis of NMs in the preferred size is needed, as key nanomaterial applications are generally size-dependent (Wang et al. 2020). Many reducing agents that are tremendously expensive and environmentally harmful or have a resilient reactivity and toxic consequence for human use are required for the chemical and physical synthesis of the NPs (Shah et al. 2019). Mono-dispersity and low stability are the difficulties associated with large-scale synthesis. The common tactics used to synthesize NPs by means of biological techniques are utilization of microbial flora, including fungi, bacteria, viruses, microalgae, nematodes, and yeast (Castillo-Henríquez et al. 2020; Kim et al. 2018; Vetchinkina et al. 2018). Furthermore, many microbial biopesticides are derived from such natural sources and serve as eco-friendly alternatives for managing agricultural pests and pathogens. These biopesticides leverage species-specific, biodegradable, and less toxic mechanisms, promoting sustainable crop production and sustainability while addressing environmental concerns associated with synthetic agrochemicals (Kaur et al. 2024; Thakur et al. 2020, 2021, 2022a, 2022b, 2023, 2024; Tomar and Thakur 2022; Tomar et al. 2022, 2023, 2024a, 2024b; Tomar et al. 2023; Tomar et al. 2024b; Tomar et al. 2022). Fungi, bacteria, viruses, plants, and algae are used in the green synthesis of many NPs (Pal et al. 2019; Shahid et al. 2017). Iron, gold, silver, nickel, zinc, and other significant NPs have been manufactured using numerous bacteria such as *Bacillus licheniformis, Pyrococcus furiosus, Pseudomonas deceptionensis, Pseudomonas aeruginosa, Pyrobaculum islandicum,* and many others (Sanaeimehr et al. 2018). As compared to the conventional chemical processes, the usage of biological constituents offers many benefits, such as biocompatibility, cost-effectiveness, and environmental friendliness (Alprol et al. 2023).

The increased plant's capacity to take in nutrients, use fertilizer well, and endure stress may be boosted by the PGPM. These PGPM reside in plant tissues near the rhizospheric soil and support the growth of plants through many processes, including enhanced nutrient accessibility and absorption, hormone synthesis, phytostimulation, and stress tolerance. To improve crop productivity and reduce reliance on synthetic agrochemicals, PGPM applications need to be investigated. Many studies conducted *in vitro* in optimum conditions showed a positive special effect of PGPM inoculation on numerous crops. However, more fieldwork seems to be essential to evaluate the effects of plant and microbe interaction to improve the yield of inoculated plants, mainly in harsh environmental circumstances. The relationship between plants, PGPM, and NPs is the focus of the study. The chosen manuscript emphasizes how crucial these interactions are for improving agricultural productivity, growth, and nutritional status. The study also emphasizes the necessity of additional transcriptomic, metabolomic, proteomic, and genomic research to determine advantageous and detrimental connections (Figure 7.1).

7.2 Microorganisms as Nano-factories

Microbes can be used as naturally occurring "nano-factories" to produce many NPs via biological mechanisms. One can synthesize NPs from bacteria, algae, fungi, and yeast by reducing the metal ions given environmental circumstances (Ghosh et al. 2021; Singh et al. 2023). This biogenic approach is environmentally benign, scalable, and cost-effective, making it a viable substitute for traditional physical and chemical synthesis approaches. Additionally, microbial production enables exact control over the nanoparticle shape, size, and functionality, offering potential applications in the domains of agriculture, medicine, and remediation of the environment (Sharma et al. 2019; Ying et al. 2022).

7.2.1 Synthesis of Nanoparticles

Different genera of bacteria synthesize NPs by reducing their metal ions through metabolic or enzymatic mechanisms (Lahiri et al. 2021). Gold and silver NPs are synthesized from *Bacillus subtilis* and *P. aeruginosa* which are valued for their antibacterial qualities (Alsamhary 2020). *Aspergillus niger* and *Fusarium oxysporum* are two examples of fungal nanofactories that generate numerous NPs with high stability and yield (Adebayo et al. 2021; Adeleke et al. 2024). The extracellular enzymes of these microorganisms facilitate the stabilization of formed particles (Costa et al. 2018). Whereas algae such as the

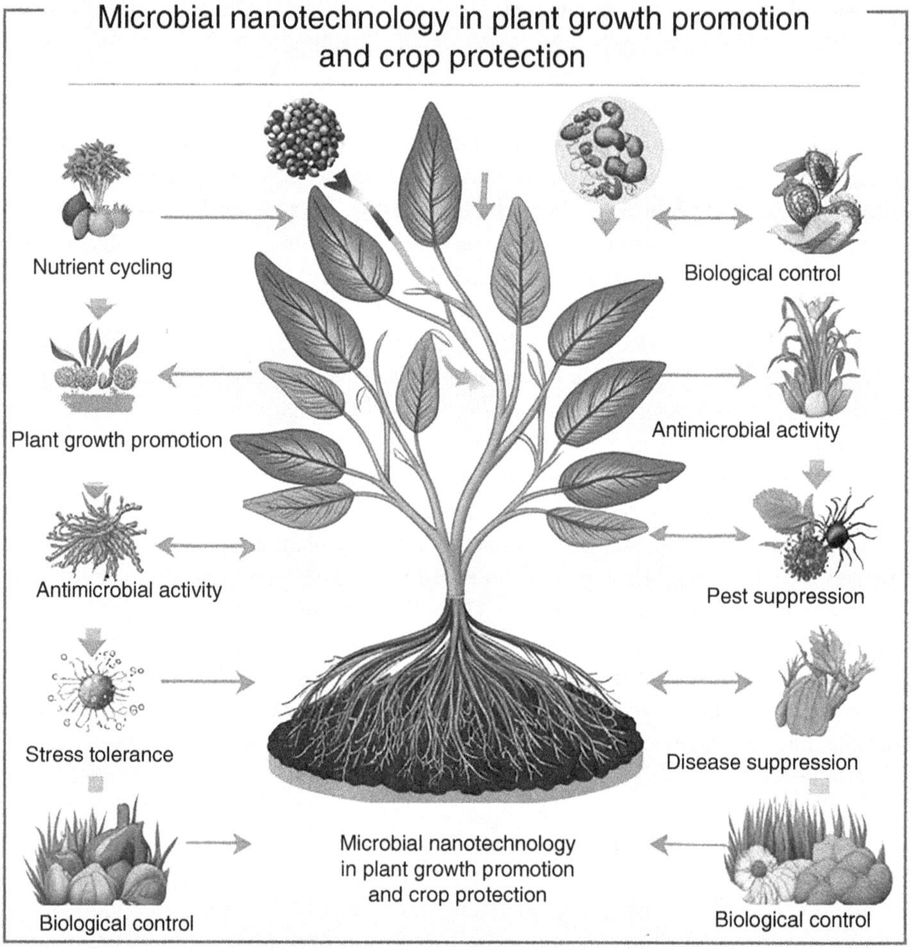

Figure 7.1 Microorganisms in combination with the nanotechnology contribute to enhancement of plant growth and crop protection.

genus *Spirulina* and *Chlorella*, biosynthesize NPs by employing biomolecules such as proteins and polysaccharides to reduce the metal ions (Chaudhary et al. 2020; El-Sheekh et al. 2022). These eco-friendly techniques are useful in the fields of agriculture and biomedicine (Table 7.1).

7.3 Microbial Nanoparticles in Plant Growth Promotion

NPs are being applied to soil and sprayed on foliage in agricultural production on a growing scale. By using NPs, agricultural chemical efficiency can be increased and pollution from chemical use can be decreased. Studies demonstrate that plants may absorb NPs and that they can affect the plant physiological mechanisms. Although crop production has

Table 7.1 Application and unique features of nanoparticles synthesized by different microorganisms.

Aspect	Bacterial synthesis	Fungal synthesis	Algal synthesis
Organisms used	*Bacillus subtilis*, *Pseudomonas* spp., and *Escherichia coli*	*Aspergillus* spp., *Penicillium* spp., and *Fusarium* spp.	*Spirulina* spp., *Chlorella vulgaris*, and *Diatoms* spp.
Mechanism	Metal ion reduction by the action of enzymes such as nitrate reductase	Secretions of extracellular enzymes and proteins	Photosynthesis and enzymatic reduction of ions
Nanoparticles made	Silver, gold, titanium dioxide and zinc oxide	Gold, silver, copper oxide and selenium	Gold, silver, silica and iron oxide
Advantages	Rapid synthesis and controlled particle size	High yield and synthesis of larger particles	Eco-friendly and simultaneous biomass production
Challenges	Limited scalability and requires strict aseptic conditions	Long processing times and potential contamination	Low production rate and seasonal availability
Applications	Antimicrobials, sensors and drug delivery	Bioremediation, antifungal agent and catalysts	Environmental monitoring, nanomedicine and coatings
Reference	Singh et al. (2024)	Ahmad et al. (2024)	Kumar et al. (2024)

been revolutionized by NPs, it is still unknown how plant-associated bacteria will react to nano-based amendments. Compared to bulk chemical pesticides or fertilizers, nano-based agricultural supplements synthesized from microorganisms less affect the soil microfauna. Consequently, additional research is necessary to guarantee sustainable precision agriculture methods regarding how plant-associated microorganisms interact and react to NPs. According to Sanyasi et al. (2016), microorganisms are critical to general soil health and plants. Numerous species of microorganisms have been found in soil and crops, and they are crucial for controlling the productivity of agroecosystems, the physicochemical properties of soil, and the health of plants. Many microorganisms linked with plants are in the rhizosphere. There, they perform vital ecological tasks such as stimulating plant growth (Ndaba et al. 2021).

PGPR are helpful bacteria found in the rhizosphere of plants, whereas endophytes are microorganisms that inhabit the endosphere, or the interior of plants. Endophytes and PGPR may both directly and indirectly enhance plant development; they can both be classified as Plant Development-Promoting Microorganisms (PGPMs). The breakdown of organic matter, nitrogen fixation, nutrient solubilization, phytohormone synthesis, and metabolite synthesis, which promote plant development and disease prevention, are among the vital metabolic processes carried out by PGPMs (Kalwani et al. 2022). Since the turn of the 20th century, researchers have examined the variety and quantity of Plant Growth Promoting Rhizobacteria (PGPR) in the rhizosphere (Fagorzi and Mengoni 2022). However, the endosphere was long thought to be sterile before the discovery of fungal endophytes in the middle of the 20th century. Fungal research initially dominated endosphere microbiology, but since then, it has changed. The scope has expanded with a greater emphasis on bacteria because of recent developments.

Every plant species that has been examined so far has a unique microbial community. As two of the most common and varied genera of PGPR, *Bacillus* and *Pseudomonas* are plant microorganisms that may be advantageous for these kinds of partnerships. By shielding plants from phytopathogens, they serve as crucial biocontrol agents (Santacruz-Márquez et al. 2020). The groundwork for this research was established by earlier studies on plant-microbe interactions, which described specific variables—sometimes called filters—that affect the connection. The host plant chemistry, the surrounding environment, and the interactions between microbes are the most noteworthy of these filters (Saunders 2012). The possible impact of other external elements introduced through anthropogenic means connected to technology growth is unclear, though. The usage of NMs in an agroecosystem is one system that could be impacted by the link between plants and microbes. The utilization of nanoparticles in applications is growing in popularity, particularly in precision and sustainable agriculture, where they are employed as nanopesticides and nanofertilizers. Regretfully, no guideline now in existence considers their possible effects on bacteria connected to plants.

Beneficial bacterial characteristics, including nitrogen fixing and the synthesis of secondary metabolites, can be favorably impacted by NMs. Dimkpa et al. (2015) reported that copper oxide NPs (CuO NPs) enhanced the synthesis of IAA in the plant growth-promoting bacterium *Pseudomonas chlororaphis*, while zinc oxide NPs (ZnO NPs) boosted the creation of siderophores, likely due to ion release within the bacteria. By upregulating the expression of the nitrification-related genes amoA1 and amoC2, silver nanoparticles (AgNPs) enhanced *Nitrosomonas europaea's* capacity to fix nitrogen (Yang et al. 2017). These investigations did, however, also demonstrate that NPs had detrimental effects on advantageous bacterial features, demonstrating the variability and interdependence of NPs' effects on bacteria. Although their exact methods are unknown, NMs do affect bacterial metabolism in several ways. These methods appear to involve modifications of cell membranes, alterations in gene expression, and ion-to-ion interactions.

Titanium dioxide (TiO_2) is a semiconductor that is frequently utilized for photocatalysis. It is an oxide that is produced from a transition metal (Tripathi et al. 2013). Sol-gel synthesis is a method that uses wet chemicals that enables enhanced stoichiometry control of doping and the creation of high-purity NPs at a very low temperature. The precursor is hydrolyzed or solubilized to form the solution (sol), which is made up of colloidal particles scattered throughout a liquid. Condensation produces a gel. In both the roots and the shoots, under the same conditions, the ability to absorb and accumulate Cd increased. However, the impact was inhibitory at 1,000 mg/kg TiO_2 (Daryabeigi Zand et al. 2020). Moreover, broad bean plants grown in salty soil benefited from TiO_2 NPs at a concentration of 0.01% by either improving the solubility of carbohydrates and amino acids or enzymatic activity. According to these results, TiO_2 nanoparticles can help cultivate plants in contaminated soil (Abdel Latef et al. 2018).

According to Wainwright (2011), silicon is found in large amounts in the soil and is the second most prevalent element on Earth after oxygen. Silica is naturally required by plants to help them respond to both natural and artificial stressors (Ma et al. 2012). Some of these responses include improving the plants' ability to photosynthesize and use water more efficiently, as well as strengthening and stiffening their leaves to prevent pathogen invasion and prevent plants from toppling over (de Moraes et al. 2021). By drawing soluble silica from groundwater and transferring it via the circulatory system to various tissues, plants create silica bodies. One of the most often utilized NPs is silicon oxide (SiO_2) nanoparticles, sometimes referred to as silica (de Moraes et al. 2021). They can be produced artificially or taken out from the environment. Upon extraction, they take on the form of crystalline minerals like quartz, as opposed to the amorphous nature of synthesized NPs. The polar heads arrange themselves to create reverse micelles, which are water-filled microcavities when they encounter water. After the silicon precursor is added, the NPs enter these microwells. Extraordinary expenses and challenges in eliminating the surfactant from the finished product are two drawbacks of this process (Pak and Yoo 2013). The degradation of metal-organic precursors by flame is the foundation of flame synthesis, sometimes referred to as chemical vapor condensation. The most widely used method is a reaction between silicon tetrachloride ($SiCl_4$), a precursor to silicon, and oxygen and hydrogen. The challenge of managing the NP size and morphology is one of this method's drawbacks (Shekar et al. 2012). Ultimately, the condensation and hydrolysis of an alkoxide is a silicon precursor that happens during the sol-gel process when a catalyst, either basic or acidic, is present. The Stöber method is the sol-gel approach that is most frequently utilized. Tetraethylorthosilicate, or Si ($(OC_2H_5)_4$), is the precursor that is employed. It is put into a solution that contains ethanol and water. Additionally, NP binding to the bacterial surface may change the size and form of cells, promoting growth (Phenrat et al. 2010).

According to Phenrat et al. (2010), ZnO is an n-type semiconductor possessing robust chemical stability, nontoxicity, photothermal stability, and biocompatibility. It is also rather inexpensive to create these NPs. Zinc oxide has a high absorption capacity of UV light, which makes it a promising photocatalyst. Due to its similar bandgap of 3.37 eV, this molecule is seen as an alternative to TiO_2 (Kołodziejczak-Radzimska and Jesionowski 2014). Because of their antibacterial, biodegradable, and biocompatible qualities, zinc oxide NPs are highly beneficial in the fields of medicine and food packaging (Akbar et al. 2019). ZnO is a suitable material for sensors because it also possesses piezoelectric qualities, which cause electrical tension to be generated under mechanical pressure (Bhatia et al. 2016). Based on their structure, these NPs can be categorized as one-, two-, or three-dimensional; examples of one-dimensional objects are wires, tubes, needles, and nanorods. Several impacts of these NPs' interactions with Plant Growth-Promoting Bacteria (PGPB) have been documented.

In 24 hours, the bacterium *P. chlororaphis* with 500 mg/L ZnO NPs produced more siderophores than the controls. This could be explained by the Zn^{2+} ions that NPs release when they attach to siderophores. At 48 hours, ZnO was found to suppress IAA synthesis. However, since IAA generation was demonstrated to be unaffected in a Zn^{2+} solution, this reduction cannot be explained by the release of cations (Dimkpa et al. 2015). The ZnS plant growth-promoting bacterial species *B. amyloliquefaciens, P. fluorescens,* and *P. aeruginosa* were also reduced in their ability to produce IAA by ZnO NPs. NPs were employed at concentrations ranging from 100 to 400 μg/mL; output fell with increasing concentration. The three bacteria's ability to solubilize phosphate was also reduced by the NPs, and this process once more demonstrated how zinc oxide concentration affects it. Nevertheless, when ZnO NPs were present, all the bacteria produced more siderophores than when the conditions were control; this chemical was created at a higher rate due to the increased NP concentration (Haris and Ahmad 2017). In a different study, ZnO NPs injected with PGPR induced higher grain weight, nodule counts, and plant height in soybean plants. With an increase in zinc oxide content, the number of grains per plant, the number of pods per plant, and the dry weight of nodules per plant all increased (Seyed and Khoramdel 2016). Similarly, $ZnSO_4$ NPs were shown to raise the levels of potassium, zinc, phosphorus, and nitrogen in rice plants when they were in conjunction with *Pseudomonas* spp. PGPB. This increased grain production and seed nutrient contents, according to de Moraes et al. (2021).

Plant diseases can be effectively controlled by AgNPs (Elmer and White 2018). According to Karuppanapandian et al. (2011), these particles function as growth promoters and help plants resist senescence brought on by reactive oxygen species (ROS) in stressful environments. Pallavi et al. (2016) examined the impact of silver NPs on the variety of bacteria in the rhizosphere and the growth of brassica (*Brassica juncea*), wheat (*Triticum aestivum*), and cowpea (*Vigna sinensis*) plants. Brassica displayed a favorable reaction for shoot parameters, cowpea responded favorably in terms of growth metrics and root nodulation, and wheat only demonstrated a negative effect. When it came to the diversity of bacteria found in cowpea and wheat, a 75 ppm concentration decreased the number of N fixers and siderophore makers, whereas a 50 ppm concentration had a beneficial effect on P solubilizers. Using either concentration did not alter the diversity in the case of Brassica. According to Chavan and Nadanathangam (2019) AgNPs have shown bactericidal activity against bacteria that fix nitrogen, solubilize phosphate, and form biofilms at low doses (2–22 μg/mL). The bacteria *Pseudomonas* and *Stenotrophomonas* species, which have a strong ability to stimulate plant development, were more prevalent (Chavan and Nadanathangam 2019). *Pseudomonas* spp., *Bacillus cereus* PGPR, and AgNPs were examined for their effects on maize growth. The length and area of the roots were enhanced by the NPs affixed to the bacteria. According to Khan and Bano (2016), nanosilver particles also enhanced the capacity of rhizobacteria to bioremediate Pb, Cd, and Ni. It has also been reported that cucumber (*Cucumis sativus* L.) plants benefit from AgNPs containing *Pseudomonas putida*. CuO NPs improved nitrogen fixation, lowered denitrification processes, and improved microbial community health in wheat (*T. aestivum*) soil, all of which promoted plant development. Pan et al. (2023) found an association between elevated nitrate concentrations in the rhizosphere and elevated gene expression linked to N fixing. These results imply that CuO NPs could be involved in the regulation of genes associated with N fixing.

This nanotechnology relies on microbial metabolites such as proteins, enzymes, and other reducing agents to facilitate the formation of NPs. We can say that this technology in which the synthesis of NPs occurs with the involvement of microorganisms is cost-effective and eco-friendly and develops particles with unique properties. However, the scalability of developed NPs and controlling their size and morphology remain key issues. Considering all the points mentioned above, the genetic engineering of the microorganisms should be done to enhance the specificity and efficiency of synthesised NPs (Table 7.2).

Table 7.2 Nanoparticles synthesized by micro-organisms their key features and applications.

Type of nanoparticle	Microorganism	Synthesis process	Key features	Application	References
Gold (Au)	*Bacillus subtilis*, *Fusarium oxysporum*	Enzymatic gold ions (Au^{3+}) reduction into gold nanoparticles (Au^0) by microbial enzymes in aqueous solutions.	High biocompatibility, tunable size, and shape.	Biomedicine (drug delivery, imaging), sensors, and electronics.	Gupta et al. (2020), Pourali et al. (2017)
Silver (Ag)	*Escherichia coli*, *Pseudomonas aeruginosa*	Reduction of Ag^+ ions to Ag nanoparticles by nitrate reductase or other reducing agents produced by microorganisms.	Antimicrobial properties, surface plasmon resonance.	Antibacterial agents, wound healing, textiles, and water purification	Ahmad et al. (2021), Singh et al. (2021a)
Zinc Oxide (ZnO)	*Aspergillus fumigatus*, *Pseudomonas aeruginosa*	Biochemical conversion of precursors like zinc nitrate to ZnO nanoparticles via fungal or bacterial metabolic pathways.	UV-blocking, antibacterial properties.	Sunscreens, biomedical applications, photocatalysis.	Abdelrahman et al. (2024), Singh et al. (2019)
Titanium Dioxide (TiO_2)	*Saccharomyces cerevisiae*, *Aspergillus niger*	Biosynthesis involving reduction of titanium precursors and deposition as nanoparticles.	Photocatalytic activity, high stability	Solar cells, water splitting, antimicrobial coatings	Li et al. (2015)
Iron Oxide (Fe_3O_4, Fe_2O_3)	*Magnetospirillum magnetotacticum*, *Bacillus subtilis*	Reduction of iron salts (e.g., ferric chloride) to magnetite nanoparticles under aerobic or anaerobic conditions	Magnetic properties, biocompatibility	Magnetic resonance imaging (MRI), drug delivery, and wastewater treatment.	Jain et al. (2018), Kumar et al. (2020)
Copper (Cu)	*Penicillium* sp., *Staphylococcus aureus*	Enzymatic reduction of Cu^{2+} ions to form copper nanoparticles in controlled conditions	Antimicrobial, conductive properties	Antimicrobial agents, catalysis, conductive inks	Ivanova et al. (2024), Ramesh et al. (2020)

(*Continued*)

Table 7.2 (Continued)

Type of nanoparticle	Microorganism	Synthesis process	Key features	Application	References
Cadmium Sulfide (CdS)	*Klebsiella pneumoniae*, *Shewanella oneidensis*	Precipitation of cadmium ions (Cd^{2+}) and sulfide ions (S^{2-}) to form semiconductor nanoparticles.	Quantum confinement, optoelectronic properties	Quantum dots, photovoltaics, bioimaging	Carrasco et al. (2021), Mehta et al. (2022)
Silica (SiO_2)	*Bacillus subtilis*, *Rhizopus oryzae*	Biogenic production using silica precursors such as sodium silicate under microbial enzymatic conditions	High surface area, thermal stability	Drug delivery, chromatography, and biosensors	Peréz et al. (2019)
Platinum (Pt)	*Desulfovibrio desulfuricans*, *Cladosporium* sp.	Reduction of Pt^{4+} ions to Pt nanoparticles by enzymatic and non-enzymatic pathways	Catalytic activity, stability	Catalysis, fuel cells, and anticancer applications	Park et al. (2021)
Selenium (Se)	*Pseudomonas* sp., *Bacillus cereus*	Reduction of selenium oxyanions (e.g., selenite or selenate) by microbial metabolic pathways.	Antioxidant, antibacterial properties	Nutraceuticals, biomedicine, and photocatalysis	Tugarova and Kamnev (2017), Ojeda et al. (2020)
Palladium (Pd)	*Shewanella oneidensis*, *Enterococcus faecalis*	Microbial reduction of Pd^{2+} ions to form nanoparticles on microbial surfaces or intracellularly	Catalytic efficiency, durability	Catalysis in hydrogenation reactions, biosensors	Windt et al. (2005)
Quantum Dots (QDs)	*Bacillus subtilis*, *Pseudomonas aeruginosa*	Synthesis of CdTe or CdSe dots using microbial enzymes as stabilizing agents.	Exhibit unique electronic and optical properties	Bioimaging, fluorescence labeling	Banerjee and Kar (2023)

7.4 Microbial Nanoparticles in Crop Protection

Plant pathogens confine crop productivity, and accurate detection is the key to disease control. Traditional methods like culture-based assays and Q-PCR are time-consuming and lack sensitivity. Nanobiotechnology improves pathogen detection by enhancing specificity, speed, and sensitivity, aiding disease control. Silver and copper nanoparticles act as antimicrobials, inhibiting fungal growth, while biological agents facilitate nanoparticle formation through enzyme secretion. Nanobiotechnology also reduces detection time and cost with biosensors and phage proteins, advancing integrated disease management and precise diagnostics in agriculture (Kaushal 2018).

Microorganisms that are good for the environment increase soil fertility and crop nutrient availability. The potential of NPs to increase microbial efficiency is being investigated by researchers. Nanofertilizers like ZnO nanoparticles and nanopyrite dressings improve nutrient uptake in plants, though the interaction between NPs and beneficial microbes is still under-researched. While silver and copper nanoparticles can inhibit microorganisms, the effects of sublethal doses on pathogens remain unclear. AgNPs, known for their antimicrobial properties, show potential in plant disease management (Table 7.3).

7.4.1 Nanoparticles in Pre-harvest Crop Management

Mesoporous silica nanoparticles (MSNPs) bid an innovative approach to increase agricultural productivity, an urgent requirement of our ever-increasing population. These NPs are biodegradable, biocompatible, can be encapsulated, and have antimicrobial properties. Essential oils provide a controlled targeted delivery reducing the use of chemicals and are environmentally safe. Cinnamaldehyde-loaded MSNPs can reduce bacterial counts to 99.9% and increase antimicrobial activity by 10-fold. By MSNPs' use in alginate seed coatings, they promoted faster seed germination and taller plants with symptomless plants against the *Pseudomonas syringae* pv. *Pisi* (Cadena et al. 2018). After that, Amer and Shoala (2020) explored the special effects of hydrogen peroxide (H_2O_2) and chitosan nanoparticles (CH NPs) on the production and quality of sweet marjoram (*Majorana hortensis*). They used three different concentrations of CH NPs, i.e. 0.13, 0.065, and 0.0325 mM, and H_2O_2 (10, 5, and 2.5 mM) alongside a control set. Data recorded from the two-year experiment showed that pre-harvest treatment of 10 mM H_2O_2 efficiently enhances the marjoram productivity as well as shelf life. Additionally, these minimize the decreases in chlorophyll content and unveil the lowest weight loss and essential oil percentage, along with abridged O_2 levels and increased CO_2 levels.

Zeyruk et al. (2022) conducted a study that focused on the management of potato pathogens through using metal NPs. Polymer film that contains zinc, iron, molybdenum, and copper NPs was developed to guard potatoes from black scurf (*Rhizoctonia solani*), early blight (*Alternaria solani*), stem canker, and late blight by *Phytophthora infestans*. Different concentrations of these metal NPs were used in potato tubers coated with a thin film of NPs before planting in field trials. Results indicated that the pre-planting treatment with metal NPs considerably abridged the incidence of *Alternaria* spp. and *P. infestans*. The predominance of *R. solani* also decreased, and additionally, this treatment increased the crop yield with rising commercial tuber yields, without affecting the starch content

Table 7.3 Summarize types, synthesis methods, properties, functions, applications, example, mechanism of action, advantages, challenges, and references for microbial nanoparticles used in agriculture.

Category	Details	References
Types of microbial nanoparticles (MNPs)	1) Bacterial nanoparticles (BNPs) 2) Fungal nanoparticles (FNPs) 3) Algal nanoparticles (ANPs) 4) Virus-based nanoparticles (VNPs)	Ahmad et al. (2012)
Synthesis methods	1) Biological synthesis: Microorganisms such as bacteria, fungi, and algae are used to synthesize nanoparticles through extracellular or intracellular mechanisms. 2) Green synthesis: Involves reducing metal ions to nanoparticles using plant extracts or microbial culture supernatants.	Khani et al. (2015)
Properties	1) Size: Typically ranging from 1 to 100 nm. 2) Shape: Spherical, rod-like, or irregular depending on the synthesis method. 3) Surface charge: Generally negatively charged, improving interaction with target pests or pathogens. 4) Stability: High stability in various environmental conditions	Jiang et al. (2018), Manivasagan et al. (2015)
Functions in crop protection	1) Antifungal: Inhibit the growth of plant pathogenic fungi (e.g., *Fusarium*, *Alternaria*). 2) Antibacterial: Used against bacterial diseases (e.g., *Xanthomonas*, *Pseudomonas*). 3) Insecticidal: Toxic to pests like aphids, weevils, and moths. 4) Growth promotion: Enhance plant growth by improving nutrient uptake or inducing systemic resistance.	Abbas et al. (2021), Singh et al. (2020)
Applications in agriculture	1) Pest control: Effective against insect pests through direct toxicity or by disrupting feeding and reproduction. 2) Disease management: Target pathogenic microbes responsible for crop diseases. 3) Plant growth promotion: Enhance seed germination, root development, and overall plant health. 4) Biological control: Serve as biocontrol agents, providing an eco-friendly alternative to chemical pesticides.	Bharathi et al. (2018), Narayanan and Park (2014)
Examples of microbial nanoparticles	1) Silver nanoparticles (AgNPs) synthesized by *Bacillus* spp. or *Trichoderma* spp. for antifungal and antibacterial properties. 2) Gold nanoparticles (AuNPs) synthesized by *Bacillus subtilis* with antibacterial and antifungal activity. 3) Chitosan nanoparticles from fungi such as *Aspergillus* spp., effective against fungal pathogens and as plant growth promoters. 4) 4. Magnetic nanoparticles synthesized by *Magnetospirillum* spp., useful in soil and plant disease management.	Bhattacharyya et al. (2015)

Table 7.3 (Continued)

Category	Details	References
Mechanism of action	1) Toxicity to pathogens/pests: Nanoparticles disrupt cell membranes, cause oxidative stress, and inhibit enzyme activity. 2) Induced resistance: Enhance plant defenses, such as by activating the plant immune system. 3) Gene expression modulation: Influence gene expression to promote growth or resist diseases.	Prasad et al. (2012)
Advantages	1) Environmentally friendly: Biodegradable and nontoxic to humans and animals. 2) Sustainable: Reduce reliance on synthetic chemicals and reduce environmental pollution. 3) Targeted action: High specificity to pests/pathogens without affecting beneficial organisms. 4) Enhanced efficacy: Higher surface area for interaction, leading to increased effectiveness at lower concentrations.	Kumar et al. (2017), Rajput et al. (2020)
Challenges	1) Toxicity to nontarget organisms: Potential effects on beneficial insects and microorganisms. 2) Stability issues: Nanoparticles can aggregate or degrade in certain environmental conditions. 3) Regulatory hurdles: Lack of standardized protocols for testing and approval. 4) Cost of production: Large-scale production of microbial nanoparticles can be expensive.	Bhattacharya and Samanta (2016), Singh et al. (2021b)

or the consumer qualities. Popletaeva and Arslanova (2021) used chitosan NPs, which are biocompatible, biodegradable as well as nontoxic in pre-harvest plant protection. These NPs stimulate plant immunity, also enhance crop yields, and provide better water solubility. The main advantages comprise their gradual release, providing a large surface area, being easily degradable, and can enter the plant cells easily. Yablokov et al. (2020) study explores the use of polymer-coated metal nanoparticles (PCMNPs) in the pre-sowing treatment of spring barley seeds. These NPs contain zinc, copper, and iron in a polymer coating. Their coating improved the germination, root mass of seedlings and fresh green mass, ultimately leading to an increase in the crop yields and reduced the moisture content (Figure 7.2).

7.4.2 Nanoparticles in Post-harvest Crop Management

Nanotechnology has now gained careful approval from the regulatory authorities for its uses in the fields of agriculture and food, as they directly impact human life. Compared to traditional farming nanotechnology can address many benefits also enhancing crop productivity as well as quality at different levels from storing seeds to managing them after harvest. Their key applications include: (i) Seed Priming: NPs enhance overall

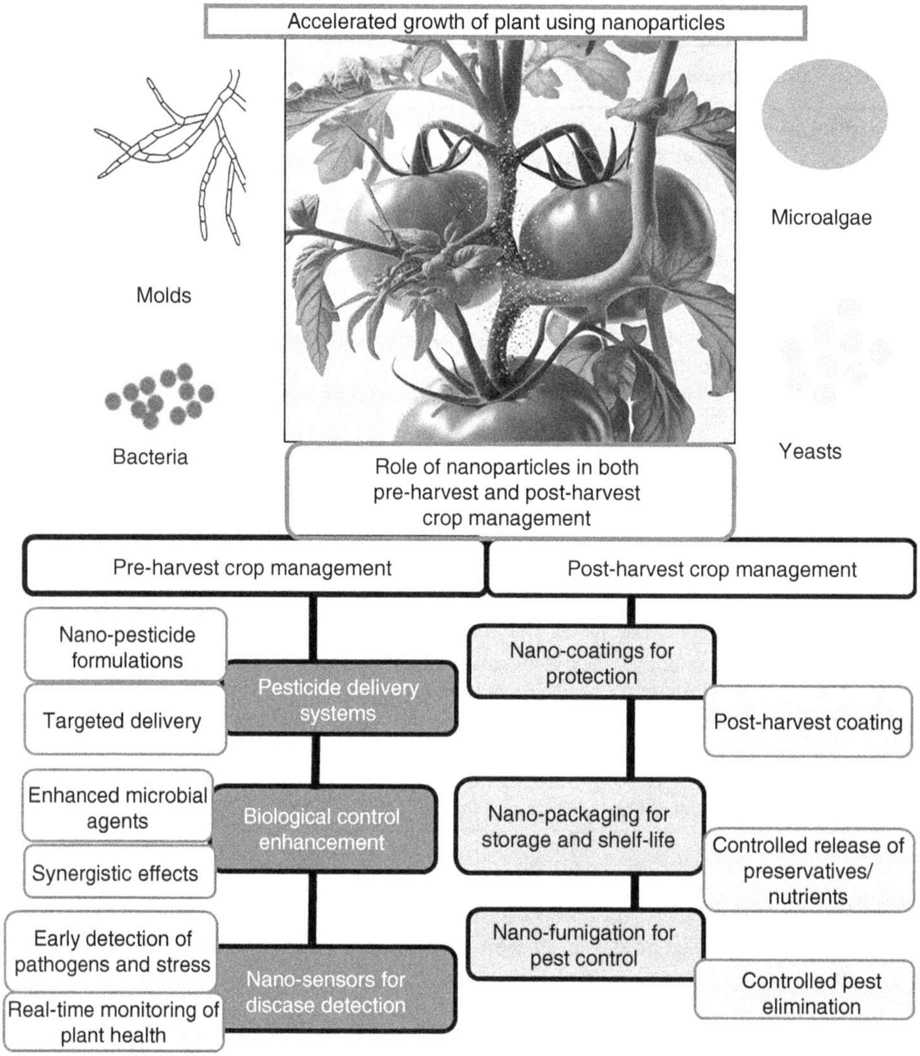

Figure 7.2 Microbial nanoparticles in crop protection which also accelerate plant growth and yield.

development, the shoot-to-root ratio, and seed germination. (ii) Controlled Release: Pesticides and fertilizers that are encapsulated enable for site-directed, controlled nutrient release, promoting soil health. (iii) Packaging: The development of cost-effective, smart, nanocomposite packaging materials with superior gas and water barrier properties increases shelf life (Shukla et al. 2019). Mir et al. (2023) suggested that NPs have great potential for post-harvest applications with their ability to decrease water consumption, increase crop yields, and minimize the use of fertilizers and pesticides in a sustainable way. These particles have special qualities for pest control, nutrient delivery, and plant growth stimulation, even though there are issues such as soil depletion and environmental contamination.

Sharma et al. (2024) reported that every year, post-harvest infections result in large crop losses due to climatic change, careless handling, lengthy shipping periods, and inadequate packaging. Zinc nanoparticles, especially ZnO, are needed for controlling many illnesses. These NPs assist as sensors for sensing uric acid, urea, proteins, DNA, glucose, phenols, and lactic acid and are also essential in biosensing and bioimaging, possessing antiviral and antibacterial properties. These can be used to create protective vegetation layers and regulate many post-harvest illnesses in plants. Substantial post-farming losses, low agricultural production, rapid population expansion, and resource degradation make it problematic to ensure food security in developing nations. By improving agricultural productivity by applying nano fertilizers, effective herbicides, soil control, pathogen identification, and wastewater treatment, nanotechnology can fill this gap (Neme et al. 2021). Industrial food processing is also upgraded, leading to perfections in food safety, nutritional value, antimicrobial protection, and quality (Yadollahi et al. 2009). In agriculture, nanotechnology is being used more to create sophisticated packaging films, slow down the growth of microorganisms, increase the shelf life of horticultural products, and employ nanobiosensors for automated storage control (Juárez-Maldonado et al. 2023).

7.5 The Use of Nanoscale Materials Produced by Microbes Carries Some Risks

Nanoscale materials have gained potential attention as a viable replacement for conventional fertilizers and insecticides, though they are also involved with possible risks. Research on nano-plant interactions is still in its early stages; it is important to consider both negative and positive viewpoints to weigh the advantages against the risks. Using NPs aided by microorganisms is a more eco-friendly method than using artificially produced NPs. Some important points still need to be considered.

7.5.1 Accumulation/Phytotoxicity

Many NPs derived from microbes are utilized for pesticidal purposes and are made from metals such as titanium dioxide, copper, zinc, and silver. They are encapsulated with some active component that kills plant disease-causing pathogens (Giri et al. 2022; Kumari et al. 2017c; Pandey et al. 2021). Nanoscale material has phytotoxicity and accumulating effects in different plant sections that require careful consideration. Information regarding the barrier, transfer, and absorption of NPs made by microorganisms during transportation may affect phytotoxicity (Kumari et al. 2017a; Schwab et al. 2016). According to Pandey et al. (2018), there are about 260,000–309,000 metric tonnes of engineered NMs accumulating in fields (63–91%) and soils (8–28%). It serves as a reminder to carefully consider the risk associated with nanoscale materials.

7.5.2 The Ideal Concentration of Nanoparticles for Growth-Promoting and Plant Pathogen Control

The ideal concentrations of NPs, particularly those of microbial origin, must be thoroughly checked before their application to plants in the field. Some of them are producing either phytotoxicity or plant protection depending on the number of NPs employed (Geisler-Lee et al. 2013;

Kumari et al. 2017b). Compared to their chemical equivalents, these NPs are known to cause less harm to the plant host because they are produced using environmentally friendly methods (Ma et al. 2015; Zhang et al. 2021). Additionally, how plants react to interactions with NMs depends greatly on the type of plant, soil, and NMs as well as the size, shape, and type of microbial production.

7.5.3 Effect on Biodiversity, Microflora, and Nontargeted Organisms

The microbially generated nanomaterial has an impact on biodiversity, microflora, and nontargeted organisms. The microbially produced NMs utilized for growth and plant protection have an impact that is associated with upper- and below-ground microbiota. According to Raliya et al. (2016), green synthetic NPs of ZnO assisted in releasing soil phosphorus for plant growth and development rather than having an impact on the microflora in the soil. Additionally, Lin et al. (2019) showed that cadmium-contaminated soil has improved soil microbial composition by green synthesized iron oxide nanoparticles. Although some researchers may have observed the detrimental effects during the interaction between NPs and the microbiome in their studies (Calder et al. 2012; Kumar et al. 2011). Agathokleous et al. (2020) stated that NMs at low concentrations have a neutral to favorable impact on native microflora and biodiversity.

7.5.4 Safe Application for Human Consumption

Both humans and animals directly consume plants. Not much research has been done on the buildup of microbially produced NMs inside the human body through plant intake or biomagnification (Hossain et al. 2021). While there may be certain dangers associated with using NMs in plants, most of them are the result of a lack of study, awareness, and optimization. Before using NMs, care should be taken to ensure that they don't have the same detrimental effects on the environment as their traditional counterparts. This is because microbially generated NMs may react differently to conventional pesticides and fertilizers in terms of the risks they bring. This is because they may react differently to danger than traditional pesticides and fertilizers (Giri and Kumari 2023).

7.6 Microbial Nanoparticle Formulations

NPs are generally made from synthetic or natural materials and have become useful tools in many biological and environmental applications. These developed formulations help improve the stability, efficacy, and bioavailability of many microbial agents used in different biocontrol, bioremediation, and agriculture sectors. The formulations are modified to regulate the active compounds' release preventing deterioration and enhancing their effectiveness. The key categories of microbial nanoparticle formulations are lipid-based, inorganic porous, clay-based and nonencapsulated (Table 7.4).

Table 7.4 Microbial nanoparticle formulations categorized by their type, characteristics, advantages, and applications.

Type	Characteristics	Advantages	Applications	Reference
Lipid-based	Nanoparticles composed of lipids include liposomes and solid lipid nanoparticles.	Biocompatible, nontoxic, enhance bioavailability, controlled drug release, and protect microbial agents from degradation.	Drug delivery, vaccines, agricultural pest management, antimicrobial delivery	Torchilin (2009)
Inorganic porous	Nanoparticles with a porous structure (e.g., silica, zeolites) provide high surface area for microbial adhesion.	High thermal and chemical stability, controlled release of agents, excellent adsorption properties.	Wastewater treatment, environmental remediation, slow release of fertilizers or biocontrol agents.	Huh et al. (2003)
Clay-based	Nanocomposites made from natural or synthetic clay minerals such as montmorillonite, kaolinite.	Low cost, environmentally friendly, enhances microbial stability, and modulate release.	Bioremediation, pesticide and herbicide carriers, soil improvement, and controlled delivery of beneficial microbes.	Mousa et al. (2018)
Nonencapsulated	Free nanoparticles or aggregates without a carrier matrix, directly formulated with microbial agents or bioactive compounds.	Simple production process, cost-effective, suitable for short-term applications.	Direct biocontrol applications, rapid response to agricultural pests or diseases, and immediate antimicrobial action.	Bhattacharya and Gupta (2005)

7.7 Conclusions

By increasing the effectiveness of microbial inputs, microbial nanotechnology improves sustainability, pest and disease management and aids plant growth promotion. It encourages ecological equilibrium and reduces the usage of chemical pesticides and fertilizers. Scalability, cost-effectiveness, and regulatory barriers are obstacles, nevertheless. To comprehend the long-term impacts of nano formulations, more scientific research is required. Precision farming and digital tools may be examples of integrated solutions that can maximize productivity and resource utilization, facilitating the transference to more resilient and sustainable agricultural systems.

Acknowledgement

We are grateful to the Vice Chancellor and Pro Vice Chancellor of Eternal University, Baru Sahib, India, for their unwavering support and the required facilities to carry out the study.

References

Abbas, Q., Yousaf, B., Munir, M.A.M. et al. (2021). Biochar-mediated transformation of titanium dioxide nanoparticles concerning TiO_2NPs-biochar interactions, plant traits and tissue accumulation to cell translocation. *Environmental Pollution* 270: 116077.

Abdel Latef, A.A.H., Srivastava, A.K., El-sadek, M.S.A. et al. (2018). Titanium dioxide nanoparticles improve growth and enhance tolerance of broad bean plants under saline soil conditions. *Land Degradation and Development* 29: 1065–1073.

Abdelrahman, S.E.S.A.H., El Hawary, S., Mohsen, E. et al. (2024). Bio-fabricated zinc oxide nanoparticles mediated by endophytic fungus *Aspergillus* sp. SA17 with antimicrobial and anticancer activities: *in vitro* supported by in silico studies. *Frontiers in Microbiology* 15: 1366614.

Adebayo, E.A., Azeez, M.A., Alao, M.B. et al. (2021). Fungi as veritable tool in current advances in nanobiotechnology. *Heliyon* 7: e08480.

Adeleke, B.S., Olowe, O.M., Ayilara, M.S. et al. (2024). Biosynthesis of nanoparticles using microorganisms: a focus on endophytic fungi. *Heliyon* 10: e39636.

Agathokleous, E., Feng, Z., Iavicoli, I., and Calabrese, E.J. (2020). Nano-pesticides: a great challenge for biodiversity? The need for a broader perspective. *Nano Today* 30: 100808.

Ahmad, J., Ahamed, M., Akhtar, M.J. et al. (2012). Apoptosis induction by silica nanoparticles mediated through reactive oxygen species in human liver cell line HepG2. *Toxicology and Applied Pharmacology* 259: 160–168.

Ahmad, J., Memon, A.G., Shaikh, A.A. et al. (2021). Insight into single-element nobel metal anisotropic silver nanoparticle shape-dependent selective ROS generation and quantification. *RSC Advances* 11: 8314–8322.

Ahmad, S., Ahmad, N., Islam, M.S. et al. (2024). Rice seeds biofortification using biogenic iron oxide nanoparticles synthesized by using *Glycyrrhiza glabra*: a study on growth and yield improvement. *Scientific Reports* 14: 12368.

Ahmed, B., Sharif, M., Hassan, S. et al. (2024). Evaluation the effects of copper nanoparticles dietary supplementation on growth performance, carcass characteristics, and serum antioxidant status in broilers. *Biological Trace Element Research* 1–8.

Akbar, A., Sadiq, M.B., Ali, I. et al. (2019). Synthesis and antimicrobial activity of zinc oxide nanoparticles against foodborne pathogens *Salmonella typhimurium* and *Staphylococcus aureus*. *Biocatalysis and Agricultural Biotechnology* 17: 36–42.

Alprol, A.E., Mansour, A.T., El-Beltagi, H.S., and Ashour, M. (2023). Algal extracts for green synthesis of zinc oxide nanoparticles: promising approach for algae bioremediation. *Materials* 16: 2819.

Alsamhary, K.I. (2020). Eco-friendly synthesis of silver nanoparticles by *Bacillus subtilis* and their antibacterial activity. *Saudi Journal of Biological Sciences* 27: 2185–2191.

Amer, A. and Shoala, T. (2020). Physiological and phenotypic characters of sweet marjoram in response to pre-harvest application of hydrogen peroxide or chitosan nanoparticles. *Scientia Horticulturae* 268: 109374.

Arshad, M.S., Joong-Ho, K., Faqir, M. et al. (2017). Wheat antioxidants, their role in bakery industry, and health perspective. In: *Wheat Improvement, Management and Utilization* (ed. R. Wanyera), 365–381. London: Intech Open.

Awad, M.A., Al Olayan, E.M., Siddiqui, M.I. et al. (2021). Antileishmanial effect of silver nanoparticles: green synthesis, characterization, *in vivo* and *in vitro* assessment. *Biomedicine & Pharmacotherapy* 137: 111294.

Banerjee, D. and Kar, A.K. (2023). Structural strategy to enhance the quantum and photocatalytic efficiency of ZnO quantum dots by incorporation of interface states. *Journal of Photochemistry and Photobiology A: Chemistry* 437: 114500.

Bharathi, D., Vasantharaj, S., and Bhuvaneshwari, V. (2018). Green synthesis of silver nanoparticles using *Cordia dichotoma* fruit extract and its enhanced antibacterial, anti-biofilm and photo catalytic activity. *Materials Research Express* 5: 055404.

Bhatia, S., Verma, N., and Bedi, R. (2016). Optical application of Er-doped ZnO nanoparticles for photodegradation of direct red-31 dye. *Optical Materials* 62: 392–398.

Bhattacharya, D. and Gupta, R.K. (2005). Nanotechnology and potential of microorganisms. *Critical Reviews in Biotechnology* 25: 199–204.

Bhattacharya, S. and Samanta, S.K. (2016). Soft nanocomposites of nanoparticles and nanocarbons with supramolecular and polymer gels and their applications. *Chemical Reviews* 116: 11967–12028.

Bhattacharyya, J., Bellucci, J.J., Weitzhandler, I. et al. (2015). A paclitaxel-loaded recombinant polypeptide nanoparticle outperforms *Abraxane* in multiple murine cancer models. *Nature Communications* 6: 7939.

Cadena, M.B., Preston, G.M., Van der Hoorn, R.A. et al. (2018). Enhancing cinnamon essential oil activity by nanoparticle encapsulation to control seed pathogens. *Industrial Crops and Products* 124: 755–764.

Calder, A.J., Dimkpa, C.O., McLean, J.E. et al. (2012). Soil components mitigate the antimicrobial effects of silver nanoparticles towards a beneficial soil bacterium, *Pseudomonas chlororaphis* O_6. *Science of the Total Environment* 429: 215–222.

Carrasco, V., Amarelle, V., Lagos-Moraga, S. et al. (2021). Production of cadmium sulfide quantum dots by the lithobiontic Antarctic strain *Pedobacter* sp. UYP1 and their application as photosensitizer in solar cells. *Microbial Cell Factories* 20: 1–10.

Castillo-Henríquez, L., Alfaro-Aguilar, K., Ugalde-Álvarez, J. et al. (2020). Green synthesis of gold and silver nanoparticles from plant extracts and their possible applications as antimicrobial agents in the agricultural area. *Nanomaterials* 10: 1763.

Chaudhary, R., Nawaz, K., Khan, A.K. et al. (2020). An overview of the algae-mediated biosynthesis of nanoparticles and their biomedical applications. *Biomolecules* 10: 1498.

Chavan, S. and Nadanathangam, V. (2019). Effects of nanoparticles on plant growth-promoting bacteria in Indian agricultural soil. *Agronomy* 9: 140.

Costa, A., Sarmento, B., and Seabra, V. (2018). Mannose-functionalized solid lipid nanoparticles are effective in targeting alveolar macrophages. *European Journal of Pharmaceutical Sciences* 114: 103–113.

Daryabeigi Zand, A., Tabrizi, A.M., and Heir, A.V. (2020). The influence of association of plant growth-promoting rhizobacteria and zero-valent iron nanoparticles on removal of antimony from soil by *Trifolium repens*. *Environmental Science and Pollution Research* 27: 42815–42829.

Das, M. and Chatterjee, S. (2019). Green synthesis of metal/metal oxide nanoparticles toward biomedical applications: Boon or bane. In: *Green Synthesis, Characterization and Applications of Nanoparticles* (ed. K.S. Shutosh and I. Siavash), 265–301. Elsevier.

Deng, C., Wang, Y., Cota-Ruiz, K. et al. (2020). Bok choy (*Brassica rapa*) grown in copper oxide nanoparticles-amended soils exhibits toxicity in a phenotype-dependent manner: translocation, biodistribution and nutritional disturbance. *Journal of Hazardous Materials* 398: 122978.

Dimkpa, C.O., McLean, J.E., Britt, D.W., and Anderson, A.J. (2015). Nano-CuO and interaction with nano-ZnO or soil bacterium provide evidence for the interference of nanoparticles in metal nutrition of plants. *Ecotoxicology* 24: 119–129.

Elmer, W. and White, J.C. (2018). The future of nanotechnology in plant pathology. *Annual Review of Phytopathology* 56: 111–133.

El-Sheekh, M.M., Deyab, M.A., Hassan, N.I., and Abu Ahmed, S.E. (2022). Green biosynthesis of silver nanoparticles using sodium alginate extracted from *Sargassum latifolium* and their antibacterial activity. *Rendiconti Lincei – Scienze Fisiche e Naturali* 33: 867–878.

Fagorzi, C. and Mengoni, A. (2022). Endophytes: improving plant performance. *Microorganisms* 10: 1777.

Geisler-Lee, J., Wang, Q., Yao, Y. et al. (2013). Phytotoxicity, accumulation and transport of silver nanoparticles by *Arabidopsis thaliana*. *Nanotoxicology* 7: 323–337.

Ghosh, S., Ahmad, R., Banerjee, K. et al. (2021). Mechanistic aspects of microbe-mediated nanoparticle synthesis. *Frontiers in Microbiology* 12: 638068.

Giri, V.P. and Kumari, M. (2023). Microbial approaches in fabrication of nanoscale materials effectively enhance the antimicrobial and crop protection potential—a review. *Plant Nano Biology* 3: 100027.

Giri, V.P., Pandey, S., Kumari, M. et al. (2022). Hybridization of chitosan and biosynthesized silver nanoparticles to enhance antimicrobial activity against phytopathogens in tomato (*Solanum lycopersicum*). *ACS Agricultural Science & Technology* 2: 719–733.

Goyal, V., Rani, D., Ritika, M.S. et al. (2023). Unlocking the potential of nano-enabled precision agriculture for efficient and sustainable farming. *Plants* 12: 3744.

Gupta, V., Kumar, K., Fatima, K. et al. (2020). Role of biocontrol agents in management of corm rot of saffron caused by *Fusarium oxysporum*. *Agronomy* 10: 1398.

Haris, Z. and Ahmad, I. (2017). Impact of metal oxide nanoparticles on beneficial soil microorganisms and their secondary metabolites. *International Journal of Life-Sciences Scientific Research* 3: 1020–1030.

Hossain, A., Skalicky, M., Brestic, M. et al. (2021). Application of nanomaterials to ensure quality and nutritional safety of food. *Journal of Nanomaterials* 2021: 9336082.

Hu, Z., Azmi, S.M., Raza, G. et al. (2016). Nanoparticle-assisted water-flooding in Berea sandstones. *Energy & Fuels* 30: 2791–2804.

Huh, S., Wiench, J.W., Yoo, J.-C. et al. (2003). Organic functionalization and morphology control of mesoporous silicas via a co-condensation synthesis method. *Chemistry of Materials* 15: 4247–4256.

Ivanova, I.A., Daskalova, D.S., Yordanova, L.P., and Pavlova, E.L. (2024). Copper and copper nanoparticles applications and their role against infections: a minireview. *Processes* 12: 352.

Jain, A., Singh, S.K., Arya, S.K. et al. (2018). Protein nanoparticles: promising platforms for drug delivery applications. *ACS Biomaterials Science & Engineering* 4: 3939–3961.

Jiang, J., Pi, J., and Cai, J. (2018). The advancing of zinc oxide nanoparticles for biomedical applications. *Bioinorganic Chemistry and Applications* 2018: 1062562.

Juárez-Maldonado, A., González-García, Y., and Pérez-Labrada, F. (2023). Nanomaterials and postharvest management of horticultural crops. In: *Engineered Nanoparticles in Agriculture: From Laboratory to Field* (ed. V. Fotopoulos and G. Gohari), 283. Walter de Gruyter GmbH & Co KG.

Kalwani, M., Chakdar, H., Srivastava, A. et al. (2022). Effects of nanofertilizers on soil and plant-associated microbial communities: emerging trends and perspectives. *Chemosphere* 287: 132107.

Karuppanapandian, T., Wang, H.W., Prabakaran, N. et al. (2011). 2, 4-dichlorophenoxyacetic acid-induced leaf senescence in mung bean (*Vigna radiata* L. Wilczek) and senescence inhibition by co-treatment with silver nanoparticles. *Plant Physiology and Biochemistry* 49: 168–177.

Kaur, S., Thakur, N., and Yadav, A.N. (2024). Evaluation of different mass production techniques to enhance the progeny of isolated indigenous entomopathogenic nematodes from Indian Himalayan Region. *National Academy Science Letters* 1-6.

Kaushal, M. (2018). Role of microbes in plant protection using intersection of nanotechnology and biology. In: *Nanobiotechnology Applications in Plant Protection. Nanotechnology in the Life Sciences* (ed. K. Abd-Elsalam and R. Prasad), 111–135. Cham: Springer.

Khan, N. and Bano, A. (2016). Modulation of phytoremediation and plant growth by the treatment with PGPR, Ag nanoparticle and untreated municipal wastewater. *International Journal of Phytoremediation* 18: 1258–1269.

Khan, I., Saeed, K., and Khan, I. (2019). Nanoparticles: properties, applications and toxicities. *Arabian Journal of Chemistry* 12: 908–931.

Khani, O., Shoushtari, M.Z., and Farbod, M. (2015). Excellent improvement in the static and dynamic magnetic properties of carbon coated iron nanoparticles for microwave absorption. *Physica B: Condensed Matter* 477: 33–39.

Kim, S.H., Kim, J.A., Kim, I.S. et al. (2018). Effects of *Rubus coreanus* byproducts on intestinal microbiota and the immune modulation. *Asian-Australasian Journal of Animal Sciences* 31: 429.

Kołodziejczak-Radzimska, A. and Jesionowski, T. (2014). Zinc oxide—from synthesis to application: a review. *Materials* 7: 2833–2881.

Kumar, N., Shah, V., and Walker, V.K. (2011). Perturbation of an arctic soil microbial community by metal nanoparticles. *Journal of Hazardous Materials* 190: 816–822.

Kumar, V., Sharma, N., and Maitra, S. (2017). *In vitro* and in vivo toxicity assessment of nanoparticles. *International Nano Letters* 7: 243–256.

Kumar, H., Bhardwaj, K., Nepovimova, E. et al. (2020). Antioxidant functionalized nanoparticles: a combat against oxidative stress. *Nanomaterials* 10: 1334.

Kumar, P., Thakur, N., Kumar, K. et al. (2024). Catalyzing innovation: exploring iron oxide nanoparticles-origins, advancements, and future application horizons. *Coordination Chemistry Reviews* 507: 215750.

Kumari, M., Pandey, S., Giri, V.P. et al. (2017a). Tailoring shape and size of biogenic silver nanoparticles to enhance antimicrobial efficacy against MDR bacteria. *Microbial Pathogenesis* 105: 346–355.

Kumari, M., Pandey, S., Mishra, S.K. et al. (2017b). Effect of biosynthesized silver nanoparticles on native soil microflora via plant transport during plant–pathogen–nanoparticles interaction. *3 Biotech* 7: 1–10.

Kumari, M., Shukla, S., Pandey, S. et al. (2017c). Enhanced cellular internalization: a bactericidal mechanism more relative to biogenic nanoparticles than chemical counterparts. *ACS Applied Materials & Interfaces* 9: 4519–4533.

Lahiri, D., Nag, M., Sheikh, H.I. et al. (2021). Microbiologically synthesized nanoparticles and their role in silencing the biofilm signaling cascade. *Frontiers in Microbiology* 12: 636588.

Li, R., Weng, Y., Zhou, X. et al. (2015). Achieving overall water splitting using titanium dioxide-based photocatalysts of different phases. *Energy & Environmental Science* 8: 2377–2382.

Li, Q., Lu, F., Zhou, G. et al. (2017). Silver inlaid with gold nanoparticle/chitosan wound dressing enhances antibacterial activity and porosity and promotes wound healing. *Biomacromolecules* 18: 3766–3775.

Lin, J., He, F., Su, B. et al. (2019). The stabilizing mechanism of cadmium in contaminated soil using green synthesized iron oxide nanoparticles under long-term incubation. *Journal of Hazardous Materials* 379: 120832.

Ma, M., Chen, H., Chen, Y. et al. (2012). Hyaluronic acid-conjugated mesoporous silica nanoparticles: excellent colloidal dispersity in physiological fluids and targeting efficacy. *Journal of Materials Chemistry* 22: 5615–5621.

Ma, C., Chhikara, S., Minocha, R. et al. (2015). Reduced silver nanoparticle phytotoxicity in *Crambe abyssinica* with enhanced glutathione production by overexpressing bacterial γ-glutamylcysteine synthase. *Environmental Science & Technology* 49: 10117–10126.

Malik, P., Shankar, R., Malik, V. et al. (2014). Green chemistry based benign routes for nanoparticle synthesis. *Journal of Nanomaterials* 2014: 302429.

Manivasagan, P., Alam, M.S., Kang, K.H. et al. (2015). Extracellular synthesis of gold bionanoparticles by *Nocardiopsis* sp. and evaluation of its antimicrobial, antioxidant and cytotoxic activities. *Bioprocess and Biosystems Engineering* 38: 1167–1177.

Mehta, J., Kumar, N., Singh, R.K. et al. (2022). Synthesis, characterization and impact of cadmium sulfide nanoparticles on the growth, pigment content and anti-oxidative defence system of *Pistia stratiotes*. *Journal of Medical, Pharmaceutical, and Allied Sciences* 12I4: 5991–5998.

Mir, M.A., Bansal, N., Sharma, S., and Negi, N.P. (2023). Nanoparticle application for sustainable agriculture and post-harvest technology: advances and future perspective. *Indian Journal of Agricultural Biochemistry* 36: 10–25.

de Moraes, A.C.P., Ribeiro, L.S., de Camargo, E.R., and Lacava, P.T. (2021). The potential of nanomaterials associated with plant growth-promoting bacteria in agriculture. *3 Biotech* 11: 318.

Mousa, M., Evans, N.D., Oreffo, R.O., and Dawson, J.I. (2018). Clay nanoparticles for regenerative medicine and biomaterial design: a review of clay bioactivity. *Biomaterials* 159: 204–214.

Mustafa, G., Sakata, K., and Komatsu, S. (2015). Proteomic analysis of flooded soybean root exposed to aluminum oxide nanoparticles. *Journal of Proteomics* 128: 280–297.

Narayanan, K.B. and Park, H.H. (2014). Antifungal activity of silver nanoparticles synthesized using turnip leaf extract (*Brassica rapa* L.) against wood rotting pathogens. *European Journal of Plant Pathology* 140: 185–192.

Ndaba, B., Bello-Akinosho, M., Roopnarain, A. et al. (2021). Nanobioremediation of contaminated agro-ecosystems applications, challenges and prospect. In: *Rhizomicrobiome Dynamics in Bioremediation* (ed. V. Kumar), 48–68. CRC Press.

Neme, K., Nafady, A., Uddin, S., and Tola, Y.B. (2021). Application of nanotechnology in agriculture, postharvest loss reduction and food processing: food security implication and challenges. *Heliyon* 7: e08539.

Ojeda, J.J., Merroun, M.L., Tugarova, A.V. et al. (2020). Developments in the study and applications of bacterial transformations of *Selenium* species. *Critical Reviews in Biotechnology* 40: 1250–1264.

Pak, J. and Yoo, H. (2013). Facile synthesis of spherical nanoparticles with a silica shell and multiple Au nanodots as the core. *Journal of Materials Chemistry A* 1: 5408–5413.

Pal, I., Bhattacharyya, D., Kar, R.K. et al. (2019). A peptide-nanoparticle system with improved efficacy against multidrug resistant bacteria. *Scientific Reports* 9: 4485.

Pallavi, M.C., Srivastava, R., Arora, S., and Sharma, A. (2016). Impact assessment of silver nanoparticles on plant growth and soil bacterial diversity. *3 Biotech* 6: 1–10.

Pan, X., Nie, D., Guo, X. et al. (2023). Effective control of the tomato wilt pathogen using TiO_2 nanoparticles as a green nanopesticide. *Environmental Science: Nano* 10: 1441–1452.

Pandey, S., Giri, K., Kumar, R. et al. (2018). Nanopesticides: opportunities in crop protection and associated environmental risks. *Proceedings of the National Academy of Sciences, India Section B: Biological Sciences* 88: 1287–1308.

Pandey, S., Giri, V.P., Tripathi, A. et al. (2021). Interaction, fate and risks associated with nanomaterials as fertilizers and pesticides. In: *Advances in Nano-Fertilizers and Nano-Pesticides in Agriculture* (ed. S. Jogaiah, H.B. Singh, L.F. Fraceto, and R. de Lima), 229–248. Elsevier.

Park, J.H., Jiang, Y., Zhou, J. et al. (2021). Genetically engineered cell membrane–coated nanoparticles for targeted delivery of dexamethasone to inflamed lungs. *Science Advances* 7: eabf7820.

Peréz, C.D., De La Torre, R.R., Zuverza-Mena, N. et al. (2019). Metalloid and metal oxide nanoparticles suppress sudden death syndrome of soybean. *Journal of Agricultural and Food Chemistry* 68: 77–87.

Phenrat, T., Kim, H.-J., Fagerlund, F. et al. (2010). Empirical correlations to estimate agglomerate size and deposition during injection of a polyelectrolyte-modified FeO nanoparticle at high particle concentration in saturated sand. *Journal of Contaminant Hydrology* 118: 152–164.

Pilarska, A., Wysokowski, M., Markiewicz, E., and Jesionowski, T. (2013). Synthesis of magnesium hydroxide and its calcinates by a precipitation method with the use of magnesium sulfate and poly (ethylene glycols). *Powder Technology* 235: 148–157.

Popletaeva, S. and Arslanova, L. (2021). Use of chitosan nanoparticles loaded with biologically active substances for pre-harvest plant protection from pathogens (a review). *Journal of Physics: Conference Series* 1942: 012077.

Pourali, P., Badiee, S.H., Manafi, S. et al. (2017). Biosynthesis of gold nanoparticles by two bacterial and fungal strains, *Bacillus cereus* and *Fusarium oxysporum*, and assessment and comparison of their nanotoxicity *in vitro* by direct and indirect assays. *Electronic Journal of Biotechnology* 29: 86–93.

Prasad, T., Sudhakar, P., Sreenivasulu, Y. et al. (2012). Effect of nanoscale zinc oxide particles on the germination, growth and yield of peanut. *Journal of Plant Nutrition* 35: 905–927.

Rajput, V., Minkina, T., Mazarji, M. et al. (2020). Accumulation of nanoparticles in the soil-plant systems and their effects on human health. *Annals of Agricultural Science* 65: 137–143.

Raliya, R., Tarafdar, J.C., and Biswas, P. (2016). Enhancing the mobilization of native phosphorus in the mung bean rhizosphere using ZnO nanoparticles synthesized by soil fungi. *Journal of Agricultural and Food Chemistry* 64: 3111–3118.

Raliya, R., Saharan, V., Dimkpa, C., and Biswas, P. (2017). Nanofertilizer for precision and sustainable agriculture: current state and future perspectives. *Journal of Agricultural and Food Chemistry* 66: 6487–6503.

Ramesh, G., Roopa, G., Shehzad, S., and Khan, S. (2020). Interaction of Al_2O_3–Ag and Al_2O_3–Cu hybrid nanoparticles with water on convectively heated moving material. *Multidiscipline Modeling in Materials and Structures* 16: 1651–1667.

Sanaeimehr, Z., Javadi, I., and Namvar, F. (2018). Antiangiogenic and antiapoptotic effects of green-synthesized zinc oxide nanoparticles using *Sargassum muticum* algae extraction. *Cancer Nanotechnology* 9: 1–16.

Sanjay, S.S. (2019). Safe nano is green nano. In: *Green Synthesis, Characterization and Applications of Nanoparticles* (ed. A.K. Shukla and S. Iravani), 27–36. Elsevier.

Santacruz-Márquez, R., Solorio-Rodríguez, A., González-Posos, S. et al. (2020). Comparative effects of TiO_2 and ZnO nanoparticles on growth and ultrastructure of ovarian antral follicles. *Reproductive Toxicology* 96: 399–412.

Sanyasi, S., Majhi, R.K., Kumar, S. et al. (2016). Polysaccharide-capped silver Nanoparticles inhibit biofilm formation and eliminate multi-drug-resistant bacteria by disrupting bacterial cytoskeleton with reduced cytotoxicity towards mammalian cells. *Scientific Reports* 6: 24929.

Saunders, J.A. (2012). Textural evidence of episodic introduction of metallic nanoparticles into bonanza epithermal ores. *Minerals* 2: 228–243.

Schröfel, A., Kratošová, G., Šafařík, I. et al. (2014). Applications of biosynthesized metallic nanoparticles—a review. *Acta Biomaterialia* 10: 4023–4042.

Schwab, F., Zhai, G., Kern, M. et al. (2016). Barriers, pathways and processes for uptake, translocation and accumulation of nanomaterials in plants—critical review. *Nanotoxicology* 10: 257–278.

Seyed, S.R. and Khoramdel, S. (2016). Effects of nano-zinc oxide and seed inoculation by plant growth promoting rhizobacteria (PGPR) on yield, yield components and grain filling period of soybean (*Glycine max* L.). *Iranian Journal of Field Crop Science* 13: 738–753.

Shah, A., Ahmad, K., Khalil, A.T. et al. (2019). *Monotheca buxifolia* mediated synthesis of gold nanoparticles and its diverse biomedical applications. *Materials Research Express* 6: 125416.

Shahid, M., Cheng, X.-W., Tang, R.-C., and Chen, G. (2017). Silk functionalization by caffeic acid assisted in-situ generation of silver nanoparticles. *Dyes and Pigments* 137: 277–283.

Sharma, V.K., Sayes, C.M., Guo, B. et al. (2019). Interactions between silver nanoparticles and other metal nanoparticles under environmentally relevant conditions: a review. *Science of the Total Environment* 653: 1042–1051.

Sharma, S., Parveen, R., and Chatterji, B.P. (2021). Toxicology of nanoparticles in drug delivery. *Current Pathobiology Reports* 9: 133–144.

Sharma, P., Aggarwal, K., Pal, A.K. et al. (2024). Role of zinc nanoparticles for the management of post-harvest diseases. In: *Metal and Metal-Oxide Based Nanomaterials. Smart Nanomaterials Technology* (ed. R.K. Bachheti, A. Bachheti, and A. Husen), 95–109. Singapore: Springer.

Shekar, S., Smith, A.J., Menz, W.J. et al. (2012). A multidimensional population balance model to describe the aerosol synthesis of silica nanoparticles. *Journal of Aerosol Science* 44: 83–98.

Shukla, P., Chaurasia, P., Younis, K. et al. (2019). Nanotechnology in sustainable agriculture: studies from seed priming to post-harvest management. *Nanotechnology For Environmental Engineering* 4: 1–15.

Singh, J., Kumar, S., Alok, A. et al. (2019). The potential of green synthesized zinc oxide nanoparticles as nutrient source for plant growth. *Journal of Cleaner Production* 214: 1061–1070.

Singh, K.R., Nayak, V., Sarkar, T., and Singh, R.P. (2020). Cerium oxide nanoparticles: properties, biosynthesis and biomedical application. *RSC Advances* 10: 27194–27214.

Singh, P., Pandit, S., Jers, C. et al. (2021a). Silver nanoparticles produced from *Cedecea* sp. exhibit antibiofilm activity and remarkable stability. *Scientific Reports* 11: 12619.

Singh, R.P., Handa, R., and Manchanda, G. (2021b). Nanoparticles in sustainable agriculture: an emerging opportunity. *Journal of Controlled Release* 329: 1234–1248.

Singh, S., Jaiswal, V., Singh, J.K. et al. (2023). Nanoparticle formulations: a smart era of advanced treatment with nanotoxicological imprints on the human body. *Chemico-Biological Interactions* 373: 110355.

Singh, G., Thakur, N., and Kumar, R. (2024). Nanoparticles in drinking water: Assessing health risks and regulatory challenges. *Science of the Total Environment* 949: 174940.

Thakur, N., Kaur, S., Tomar, P. et al. (2020). Microbial biopesticides: current status and advancement for sustainable agriculture and environment. In: *New and Future Developments in Microbial Biotechnology and Bioengineering* (ed. A.A. Rastegari, A.N. Yadav, and N. Yadav), 243–282. Amsterdam: Elsevier.

Thakur, N., Tomar, P., Kaur, S. et al. (2021). Entomopathogenic soil microbes for sustainable crop protection. In: *Soil Microbiomes for Sustainable Agriculture: Functional Annotation* (ed. A.N. Yadav), 529–571. Springer Nature Switzerland.

Thakur, N., Kaur, S., Kaur, T. et al. (2022a). Organic agriculture for agro-environmental sustainability. In: *Trends of Applied Microbiology for Sustainable Economy* (ed. R. Soni, D.C. Suyal, A.N. Yadav, and R. Goel), 699–735. Amsterdam: Elsevier.

Thakur, N., Tomar, P., Kaur, S., and Kumari, P. (2022b). Virulence of native entomopathogenic nematodes against major lepidopteran insect species of tomato (*Solanum lycopersicum* L.). *Journal of Applied Biology & Biotechnology* 10: 1–9.

Thakur, N., Tomar, P., Kaur, J. et al. (2023). Eco-friendly management of *Spodoptera litura* (Lepidoptera: Noctuidae) in tomato under polyhouse and field conditions using *Heterorhabditis bacteriophora* Poinar, their associated bacteria (*Photorhabdus luminescens*), and *Bacillus thuringiensis* var. kurstaki. *Egyptian Journal of Biological Pest Control* 33: 7.

Thakur, N., Sharma, A., Kaur, S. et al. (2024). Insect pest *Spodoptera litura* (Fabricius) and its resistance against the chemical insecticides: a review. *Plant Science Today* 11: 1–12.

Thirugnanasambandan, T. (2018). Advances and trends in nano biofertilizers. SSRN, p. 59. https://doi.org/10.2139/ssrn.3306998.

Tomar, P. and Thakur, N. (2022). Biocidal potential of indigenous isolates of Entomopathogenic Nematodes (EPNs) against tobacco cutworm, *Spodoptera litura* Fabricius (Lepidoptera: Noctuidae). *Egyptian Journal of Biological Pest Control* 32: 107.

Tomar, P., Thakur, N., and Yadav, A.N. (2022). Endosymbiotic microbes from entomopathogenic nematode (EPNs) and their applications as biocontrol agents for agro-environmental sustainability. *Egyptian Journal of Biological Pest Control* 32: 80.

Tomar P, Thakur N, Sidhu AK, Laskar BA, Hashem A, Avila-Quezada GD, Abd_Allah EF (2023) The isolation, identification, and insecticidal activities of indigenous entomopathogenic nematodes (*Steinernema carpocapsae*) and their symbiotic bacteria (*Xenorhabdus nematophila*) against the larvae of *Pieris brassicae*. *Horticulturae* 9:874.

Tomar, P., Thakur, N., Jhamta, S. et al. (2024a). Bacterial biopesticides: diversity, role in pest management and beneficial impact for agricultural and environmental sustainability. *Heliyon* 10: e31550.

Tomar, P., Thakur, N., Singh, S. et al. (2024b). Soil inhabiting bacto-helmith complex in insect pest management: current research and future challenges. *Heliyon* 10: e36365.

Torchilin, V.P. (2009). Nanotechnology for intracellular delivery and targeting. In: *Nanotechnology in Drug Delivery. Biotechnology: Pharmaceutical Aspects*, vol. X (ed. M.M. de Villiers, P. Aramwit, and G.S. Kwon), 313–346. New York: Springer.

Tripathi, R., Gupta, R.K., Shrivastav, A. et al. (2013). *Trichoderma koningii* assisted biogenic synthesis of silver nanoparticles and evaluation of their antibacterial activity. *Advances in Natural Sciences: Nanoscience and Nanotechnology* 4: 035005.

Tsekhmistrenko, S., Bityutskyy, V., Tsekhmistrenko, O. et al. (2020). Bacterial synthesis of nanoparticles: a green approach. *Biosystems Diversity* 28: 9–17.

Tugarova, A.V. and Kamnev, A.A. (2017). Proteins in microbial synthesis of selenium nanoparticles. *Talanta* 174: 539–547.

Vetchinkina, E., Loshchinina, E., Kupryashina, M. et al. (2018). Green synthesis of nanoparticles with extracellular and intracellular extracts of basidiomycetes. *PeerJ* 6: e5237.

Wainwright, S. (2011). *True Liquid Crystal Templating of Nanoparticle-Doped Mesoporous Silica*. University of York.

Wang, H., Liang, X., Wang, J. et al. (2020). Multifunctional inorganic nanomaterials for energy applications. *Nanoscale* 12: 14–42.

Wang, F., Li, K., and Shi, Z. (2021). Phosphorus fertilization and mycorrhizal colonization change silver nanoparticle impacts on maize. *Ecotoxicology* 30: 118–129.

Windt, W.D., Aelterman, P., and Verstraete, W. (2005). Bioreductive deposition of palladium (0) nanoparticles on *Shewanella oneidensis* with catalytic activity towards reductive dechlorination of polychlorinated biphenyls. *Environmental Microbiology* 7: 314–325.

Yablokov, A.G., Bogoslovskaya, O.A., Olkhovskaya, I.P., and Glushchenko, N.N. (2020). Application of metal nanoparticles for pre-sowing treatment of spring barley seeds. *Australian Journal of Crop Science* 14: 1140–1149.

Yadollahi, A., Arzani, K., and Khoshghalb, H. (2009). The role of nanotechnology in horticultural crops postharvest management. *Southeast Asia Symposium on Quality and Safety of Fresh and Fresh Cut Produce* 875: 49–45.

Yang, Y., Qin, Z., Zeng, W. et al. (2017). Toxicity assessment of nanoparticles in various systems and organs. *Nanotechnology Reviews* 6: 279–289.

Ying, S., Guan, Z., Ofoegbu, P.C. et al. (2022). Green synthesis of nanoparticles: current developments and limitations. *Environmental Technology and Innovation* 26: 102336.

Zeyruk, V., Vasilieva, S., Belov, G. et al. (2022). A boost to integrated management of certain potato diseases using metal nanoparticles. *Potato Research* 65: 273–288.

Zhang, H., Chen, S., Jia, X. et al. (2021). Comparation of the phytotoxicity between chemically and green synthesized silver nanoparticles. *Science of the Total Environment* 752: 142264.

8

Microbial Nanotechnology in Biocontrol of Insect Pest

Preety Tomar[1], Parul Sharma[1], Yankita Negi[1], Sakshi Sharma[1], Vijay Bharti Sharma[1], and Ajar Nath Yadav[2]

[1] *Department of Zoology, Akal College of Basic Sciences, Eternal University, Sirmour, Himachal Pradesh, India*
[2] *Department of Genetics, Plant Breeding and Biotechnology, Dr. Khem Singh Gill Akal College of Agriculture, Eternal University, Sirmour, Himachal Pradesh, India*

8.1 Introduction

The population of the planet has been rapidly growing, reaching 7.7 billion and continuously rising, which is projected to reach 9 billion by 2050 (FAO, 2018). In this scenario, providing food to the escalating population will be challenging (Fróna et al. 2019). Furthermore, the largest obstacles are the losses brought on by insect pests in the crops. It is estimated that biotic stresses, such as insect pests, account for roughly $40 billion in economic losses worldwide every year. The losses in the agricultural crops have been estimated to be between 20 and 40% (Jaiswal et al. 2022). Insects account for about 38% agricultural losses (Junaid and Gokce 2024; Oerke 2006). The plants and agricultural crops have been protected using various kinds of chemical, biological, and botanical-based insecticides and pesticides for the last several decades (Hazafa et al. 2021).

These chemical insecticides have negative impacts on all environmental matrices, including soil, water, air, and biota, in addition to exclusively targeting the pest organisms. Therefore, the development of environment-friendly insecticides is necessary (Bhattacharyya et al. 2016). The development of the biocontrol domain addresses various factors such as the adverse impact of chemical pesticides, pesticide residue issues, environmental hazards, emergence of resistance species, and the other hurdles faced by modern agricultural practices. The biocontrol approaches offered an effective, sustainable, and environmentally compatible substitute for chemical-based agriculture (Jaiswal et al. 2022; Kumar 2016; Sabarwal et al. 2018). Microbial bioagents provide long-lasting, sustainable, and efficient management of plant diseases (Kumar et al. 2021). Though they offers benefits over traditional crop protection methods, biocontrol agents cannot yet replace them (Sabarwal et al. 2018).

One potential solution in the area of biopesticides is the integration of nanotechnology. Currently, nanotechnology is revolutionizing agriculture, with the greatest potential for managing insect pests coming from nano-based insecticides (Shahid et al. 2021). Nanotechnology has become a very appealing substitute for chemical pesticides (Nivetha et al. 2021). Biologically derived compounds were integrated with a specific polymer to produce nanoparticles (NPs) that can be employed for insect-pest control (Kumar et al. 2021). The term "nano" has its roots in the Greek language, meaning "dwarf" and "small science." Nano is defined as 10^{-9} of something. In nature, particles employed in nanotechniques have a size between 1 and 100 nm (Bhattacharyya et al. 2010). Kannan et al. (2020) demonstrated that nanomaterials may be able to significantly reduce the damaging effects of insect pests that decimate crops. They used nanotechniques for the release of biocontrol agents over an extended period, providing a long-term crop protection strategy (Singh et al. 2024; Akhtar et al. 2020).

Varieties of NPs (organic and inorganic) have been formulated through various approaches including physical, chemical, and biological approaches. However, NPs produced by non-biological means are unstable, costly, and environmentally dangerous because harmful chemicals and costly energy-intensive procedures are used (Nivetha et al. 2021). Combining the principles of microbiology and nanotechnology offered an innovative solutions for pest management (Prasad et al. 2017). Microbe-based formulation of NPs has acquired popularity during the last few years, and microorganisms are now thought to be possible sources of bioactive NPs. Nanomaterials are widely synthesized using a variety of microorganisms, including yeasts like *Schizosaccharomyces pombe* and *Candida glabrata*, cyanobacteria like *Plectonema boryanum*, actinobacteria like *Thermomonospora*, and bacteria like *Pseudomonas aeruginosa, Bacillus subtilis, Serratia*, *Bacillus licheniformis,* and *Escherichia coli*, and fungi like *Verticillium* and *Fusarium* (Nivetha et al. 2021).

Under green agricultural practices, application of biological NPs in the control of insect pests is important, effective, and rapidly replacing chemical-based agriculture in many nations (Bhattacharyya et al. 2010). Metal NPs such as aluminum (Al), silver (Ag), gold (Au), zinc (Zn), silicon (Si), and zinc oxide (ZnO) have convincingly shown their toxic effects against a variety of insect pests. NPs significantly affect protein synthesis, gene regulation, detoxifying and antioxidant compounds in insects, creating oxidative stress, disruption of growth and reproduction, denaturation of enzymes, and cell death. NPs are being used in crop pest management, and their primary testing has been conducted against a variety of arthropod pests and vectors (Nivetha et al. 2021). Therefore, this chapter will give an overview of the microbe-mediated synthesis of NPs, providing insights into their potential as biopesticides, their insecticidal mechanism, possible hazards, regulatory rules and considerations, and future prospects.

8.2 Microbial Biosynthesis of Nanoparticles

Green production of NPs by using a variety of living organisms like plants and microorganisms can mitigate many of the unfavorable impacts that are associated with physical and chemical-based approaches. In this method, the biosynthesis of NPs occurs under mild pH, pressure, and temperature conditions. Additionally, there is no need for hazardous materials and other capping, reducing, and stabilizing agents (Usman et al. 2019). Microbes have the ability to transform inorganic elements into NPs through both intracellular and extracellular processes, and they can adapt to environments with elevated levels of metals (Singh et al. 2023). Microbial NPs can be formulated through intracellular and

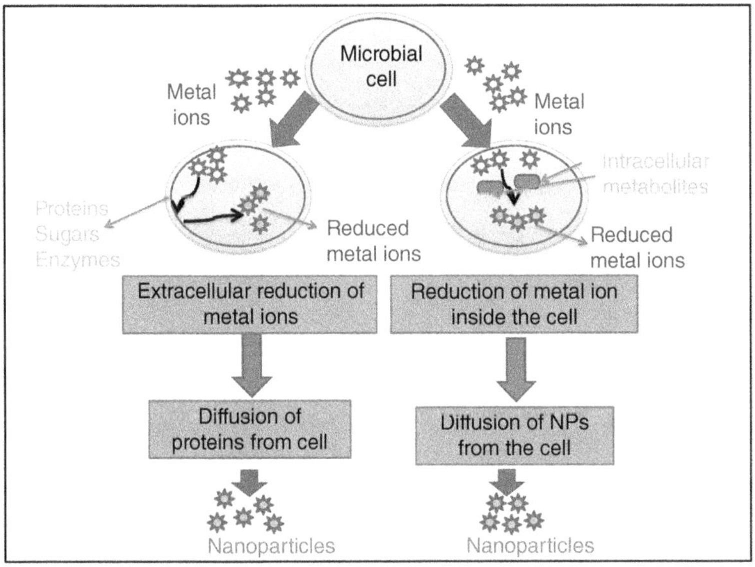

Figure 8.1 Mechanism of the synthesis of nanoparticles by microbes.

extracellular production approaches (Figure 8.1). In both methods, the choice of microorganisms, metal salts, and specific conditions such as temperature and pH is critical for efficient NP production and stabilization. In intracellular synthesis of NPs, the metal's salt concentration is optimized by the microbial utilization of the whole microbial cells. During intracellular synthesis, NPs are produced via microbes by reducing metal ions within the cellular environment at a desirable temperature and pH. On the other hand, in extracellular synthesis, NPs have been produced from the crude extract of the microbes generated during their culture. This crude extract contained microbial metabolites (cell-free filtrate) that are combined with a metal salt solution. After incubation, this mixture has been subjected to the reduction followed by stabilization that occurs outside the cells (Velusamy et al. 2016).

NPs can be produced via top-down and bottom-up approaches (Figure 8.2). The top-down method involves the use of lithographic methods such as mechanical grinding, wet etching, physical vapor deposition, and laser-assisted evaporation to break down the massive materials into fine particles. Through this approach, the material undergoes severe plastic deformation, producing nanoscale materials. Though this method is effective, however, it is not feasible for bulk production as it is lethargic, expensive, and tedious. Conversely, the bottom-up method involves the production of NPs by allowing intrinsic assembly of atoms and molecules into nuclei that eventually form a nanostructure. This method is cheap and effective, and the accumulation of atoms and molecules to synthesize NPs can be performed through biological and chemical-based approaches (Iqbal et al. 2012; Mazhar et al. 2017). During the synthesis of NPs through bottom-up method, initially atoms were reduced using reducing substances such as enzymes, protein, polysaccharides, or metabolites produced by microorganisms. Then, the atoms and molecules were nucleated to produce nanostructure and biotemplate (Purohit et al. 2019).

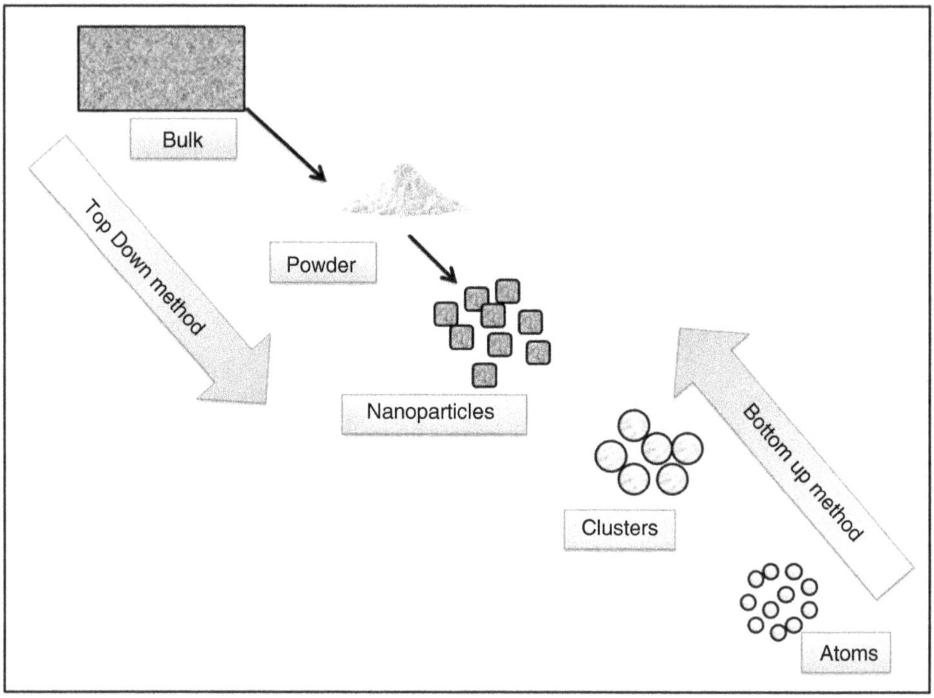

Figure 8.2 Schematic representation of the synthesis of nanoparticles. *Source:* ValentinValkov/ Adobe Stock Photos.

8.2.1 Bacterial Biosynthesis of Nanoparticles

In the biological synthesis of NPs, prokaryotic bacteria are the most attention-grabbing micro-organism. The main reason behind the succession of bacterial NPs is the easy handling and fast growth of the bacteria and their ability to be manipulated genetically. Bacteria have been considered efficient biotemplates for the production of NPs because they exhibit heterogeneity in their morphological forms and surface appearance. A considerable role has been played by the exo-polysaccharides present on the cell wall and the proteins on the surface layer. The aggregation and synthesis of NPs occurs as the metals bind with the enzymes, proteins, and sugars, containing various functional groups, synthesized from the microbial cell wall (Sanyasi et al. 2016; Sathiyanarayanan et al. 2017). In the extracellular process of NPs synthesis through bacteria, the metal ions are initially reduced and then allowed to bind with cell wall proteins through electrostatic interactions with functional groups (Banu and Balasubramanian 2014).

Klaus et al. (1999) demonstrated the production of silver nanoparticles (AgNPs) that were smaller than 299 nm in size. They reported that when the bacterium *Pseudomonas stutzeri* AG-259 was treated with the solution of silver nitrate (50 mM), the bacteria accumulated the silver atom into their periplasmic space. This accumulation produced the silver NPs of 200 nm size (Klaus et al. 1999). Moreover, several bacterial systems have been used for the production of the NPs by employing diverse biochemical mechanisms.

For example, *Alcaligenes faecalis* facilitates the production of AgNPs via a reduction process and utilizes NADH and NADH-dependent reductase (Divya et al. 2019).

Even the bacterial appendages such as flagella and pilli have also been used as a biotemplate for the production of NPs. These structures have been developed from flagellin and pilin proteins, that have the property of self-organization, and can be used for the production of particular-sized and shaped nanostructured materials (Sleytr et al. 1993; Purohit et al. 2019). Bacteria like *Geobacter sulfurreduces* have pilin protein that are favorable for the extracellular synthesis of the NPs (Singh et al. 2023).

Recently, several studies have reported the synthesis of silver oxide NPs using various microbes, including *Bacillus paramycoides*, *Bacillus thuringiensis* SSV1 *Lactobacillus mindensis*, *Nitrobacter* sp., and *Xanthomonas* sp. P5, *Nodularia haraviana* (Danish et al. 2022; Singh et al. 2022; Minhas et al. 2023). These microbe-based NPs showed remarkable insecticidal as well as antibacterial properties, along with better consistency (Dharmaraj et al. 2021). The silver oxide NPs formulated using *B. thuringiensis* also exhibit pesticidal and antifungal properties (Ge et al. 2024).

8.2.2 Fungal Biosynthesis of Nanoparticles

Fungi are also utilized for the formulation of NPs as the proteins and enzymes present in fungi poses the ability to reduce metal salts (Abdel-Aziz et al. 2018; Sastry et al. 2003). Biosynthesis of NPs through fungi is done by two methods: intracellular and extracellular (Vetchinkina et al. 2018). In the intracellular method, metal salts enter the fungal cells and reduction takes place inside the cell. In contrast, the extracellular methods involve the attachment of the metals on the cell wall, which are then reduced by the production of the enzymes and other metabolites produced by the fungi (Pantidos and Horsfall 2014; Vetchinkina et al. 2018). For the bulk production of the NPs, fungi are the most favored microorganisms compared with bacteria. Moreover, mycelia of fungi offer numerous binding sites for metal salts, and the synthesis of NPs occurs rapidly as the fungi grows more biomass (Aziz et al. 2019; Prasad et al. 2018).

8.2.3 Algal Biosynthesis of Nanoparticles

Algae have been utilized for the synthesis of NPs, and this process is known as phyconanotechnology. Algae have unique structural conformation and release a variety of biomolecules (Siddiqi and Husen 2016) such as proteins, sugars, and other secondary metabolites (SMs). The extracellular polysaccharides of algal membrane reduces the metal ions and stabilize the metal NPs. Furthermore, the SMs of alga acts as capping agents controlling the size and form of the NPs (Anwar 2018; Kannan et al. 2013; Purohit et al. 2019; Sharma et al. 2019; Vasquez et al. 2016). Biosynthesis of NPs using algae is advantageous as it is a low-cost approach and useful for the production of stable, biocompatible, and environmentally friendly NPs at a large scale. Toxic metal ions can be transformed into a non-toxic form through chelation by microalgae. These metals were then utilized for the production of NPs through chemical reaction. Several bio-molecules such as alkaloids, amides, amines, carbonyl groups, flavonones, phenolics, pigments, proteins, terpenoids, and other reducing compounds have been released by the algae for the production of NPs (Terra et al. 2019). The extract of *Spirogyra* has been employed for the production of Ag-NPs and the biomolecules released by *S. hyaline* acts as binding, stabilizing, and reducing agent during the NPs synthesis. The intrinsic cytotoxic effects of Ag-NPs enhance their ability to combat with bacteria, fungi, insects, oxidative stress, and even cancer (Al-Radadi et al. 2022).

8.2.4 Viral Biosynthesis of Nanoparticles

In natural nanoarchitecture, the viruses have been employed for the production of NPs due to their distinct nanoscale size and shapes. A virus is developed from the genetic material covered by the protein shell known as capsid. The capsid has a 3-D structure and contains protein subunits that bear carboxyl, amino, and thiol groups. The highly reactive groups bind with the metal ions and help in the synthesis of NPs over the virus surface (Flenniken et al. 2009; Shah et al. 2015). Moreover, the hollow cavity inside the capsid offers a site for the binding of metal ions with amino acids. Furthermore, it acts as a mini reactor for the generation of NPs with particular size and shape. Using both external surface and internal environment of the virus, various nanostructures such as nanorods and nanotubes have been developed (Bittner 2005; Wen et al. 2012).

8.3 Role of Nanotechnology in Agricultural Sustainability

A significant role has been played by nanotechnology in enhancing agricultural sustainability. Nanofertilizers and nano-based pesticides are intended to decrease the quantity of synthetic pesticides and minimize environmental damage by increasing the effectiveness of nutrient and pesticide application on plants. In addition to monitoring water quality and pesticides for agricultural sustainability, nanotechnology can improve crop productivity via nutrient regulation (Gruère 2012; Mukhopadhyay 2014; Prasad et al. 2014). It is difficult to give a broad judgment on the environmental impacts of nanomaterials because of their diverse properties and activities (Prasad et al. 2014). The factors affecting the toxicity of nanomaterials include surface structure and charges, chemical composition, shape of particle, behavior, and the extent of aggregation or separation (Ion et al. 2010).

The NPs with different sizes and shapes but similar composition showed varying degrees of toxicity. Research on nanotechnology is becoming increasingly important and even required for sustainable development in the agriculture sector. The formulation of nanoencapsulated pesticides has reduced the amount of pesticide use and human contact, thereby offering a sustainable way for crop defense (Nuruzzaman et al. 2016). It is now possible to enhance food productivity throughout the world by lowering detrimental effects on ecosystems and developing non-toxic and promising nanopesticides (De Oliveira et al. 2014; Grillo et al. 2016). The nanopesticides have been adopted by the agriculturist as they are easy to use, have mobility and handling ease, longer storage stability, and have maximum efficiency (Kim et al. 2018). Nanotechnology reduces the environmental impact, enhancing efficiency and improving agricultural practices by formulating innovative nanopesticides and disease management techniques (Hazarika et al. 2022). It is predicted that advances in nanoscale research and nanotechnology would transform agricultural and food systems. Soil quality and nutrient richness have been improved by nanopesticides and fertilizers, lowering the risk of overdosing, boosting nutrient usage efficiency, and decreasing the frequency of application (Kour et al. 2020; Naderi and Danesh-Shahraki 2013).

The availability of food can be enhanced and pesticide contamination can be reduced using NPs to protect plants (Adisa et al. 2018). Among the 17 sustainable development goals of the United Nations, one goal is to achieve zero hunger. Nanotechnology can help in achieving this goal to a greater extent (Usman et al. 2020).

8.4 Nanoparticles as Biopesticides

Nanotechnology offers a promising approach for sustainable agriculture by providing safer means for safe, stable, prolonged, and effective pest control. Koul (2019) demonstrated nanobiopesticides for the rapid and effective management of plant pathogens. There are different kinds of NPs, developed from various kinds of metals shows promising effectiveness in managing insects pests and pathogens.

8.4.1 Silver Nanoparticles

AgNPs are well documented for their insect-suppressing abilities and inducing other physiological and biochemical alterations in plants (Bapat et al. 2022). AgNPs are highly stable and formulated using toxic substances that reduce silver ions into nanosized particles. AgNPs hold great promise as a biopesticide due to their multifunctional properties. Furthermore, they showed their effect against diverse range of pests, including bacteria, fungi, and insects, which demonstrated their potential mechanism as pesticides. The pesticidal mechanism involves cell membrane disruption, alteration in cellular metabolism, and oxidative stress induction (Sharma 2024).

8.4.2 Silicon Nanoparticles

Silicon metal (SiNPs) are well recognized for attenuating biotic stresses from insect pest and abiotic stresses in plants. Due to the specific physical and chemical characteristics, SiNPs have greater advantages over synthetic pesticides in suppressing insect pests. SiNPs act as an insecticidal agent and also act as a delivery agent for the sustained release of insecticidal formulations. They also induce growth promotion and defense reaction against insect pests in plants. They are biocompatible in nature and reduce environmental pollution compared with synthetic pesticides (Saw et al. 2023).

8.4.3 Zinc Oxide Nanoparticles

Zinc oxide nanoparticles (ZnONPs) have sparked interest as biopesticides because of their considerable capacity to control various plant pathogen and pests. In the agricultural sector, ZnONPs have been known to enhance plant growth, nutrients, and show antimicrobial properties. ZnONPs play a major role in mitigating pest populations. They have been reported to manage *Tribolium castaneum* population in cereal crops. Significant mortality in *Bactericera cockerelli* has been observed after the application of ZnONPs. They were also reported to manage *Trialeurodes vaporariorum* and *Spodoptera litura* population. Even the combination of ZnO-NPs with pesticide thiamethoxam also showed promising results in insect control (Agredo-Gomez et al. 2024).

8.4.4 Iron Oxide Nanoparticles

Iron oxide nanoparticles (IONPs) are gaining much attention as biopesticides because of their sustainable effect in pest management practices. These NPs show several advantages:

- **Enhanced Efficacy:** IONPs can improve the effectiveness of plant growth-promoting bacteria (PGPB) used as biopesticides. For example, studies have shown that combining IONPs with PGPB can significantly boost seed quality and early seedling establishment in crops like soybeans. Use of PGPB has now been popularized in sustainable agricultural practice. The integration of PGPB with IONPs showed a synergistic way as biopesticides and biofertilizers and has been reported to improve the production in soybean plants (Benedetti et al. 2024).
- **Sustainable Pest Management:** Nanopesticides, especially IOs offer a green strategy for pest management. They minimized the dependency on chemical pesticides, thereby mitigating the potential ecological risk (Shekhar et al. 2021).
- **Antimicrobial Properties:** IONPs possess strong antimicrobial properties, making them effective in managing various plant pathogens. They release reactive oxygen species (ROS) production, which act as antibacterial. The specific characteristics of IONP such as size, appearance, charge on the surface, and super paramagnetism affect their antibacterial action (Zhang and Miao 2024).

8.4.5 Gold Nanoparticles

Gold nanoparticles (AuNPs) act as biopesticides because of their specific physical and chemical characteristics. AuNPs are effective in combating the issues related to pest resistance and also showed indirect effectiveness against bacteria responsible for causing disease in plants (Goyal et al. 2024).

8.4.6 Chitosan

Chitosan is prepared by recycling the shell waste of the shellfish, which is a commercially important derivative of *N*-acetyl chitin produced through *N*-deacetylation process. Chitosan has a very wide range of applications in agriculture as well as in industries. In agriculture, chitosan is used for plant growth promotion and in industry, especially in the food processing sector and bioengineering; it has been used for coating active food ingredients, controlled drug delivery carrier, and enzyme fixation. It acts as a plant defense activator and shows pathogen-inhibiting activities (Divya and Jisha 2018). The chitosan NPs are also known to exhibit bactericidal, fungicidal, and virucidal activities against multiple pathogens. They have been reported to induce local and systemic defense in plants, promote soil fertility, and also minimize pathogen incidence (Dutta et al. 2022). These nanochitosans cause changes in the membrane permeability, intracellular alteration, influence replication, activate defense genes, and cell disruption.

8.5 Applications Against Insect Pests

Bacterial biopesticides are the most preferred microbial pesticides that have shown remarkable potential in several aspects. Although they have been well recognized for their insect-killing properties, they also exhibit bactericidal and fungicidal activities. In order to achieve optimal efficacy, they need to come into contact with the possible pest, and consumption might be necessary beforehand. Insect-borne bacteria impede digestion by producing endotoxins. According to Kachhawa (2017), these biopesticides inhibit the bacterial and fungal pathogens responsible for causing diseases in plants. Integration of bacteria with nanotechnology enhances their efficacy against insect pests. According to Damalas and Koutroubas (2018), *B. thuringiensis* nanocomposites with *Bt* bioactives in different nanoformulations, like nanocapsules, nanosuspension, and nanoemulsion, have been used as nanopesticides.

During the last 20 years, a number of bacterial species, namely *Bacillus* sp., *Pseudomonas* sp., *Corynebacterium* sp., and *Shewanella* sp., have been integrated with several inorganic metals to synthesize antagonistic NPs. The NPs have been synthesized only because these bacteria can reduce the metal ions by actively absorbing them. These bacteria-driven NPs can improve agricultural productivity by reducing the post-harvest losses related to crop processing, transportation, and storage (Kumari et al. 2019). A very limited number of microbial NPs have been developed against insect pests. The list of the insecticidal NPs developed from the micro organism and their activities against various insect pests is mentioned in Table 8.1. Using *B. thuringiensis kurstaki*, Sayed et al. (2017) developed biocompatible AgNPs. Testing was done on the insecticidal effectiveness of bacterially synthesized AgNPs against the immature stage of *Agrotis ipsilon* and *Trichoplusia ni*, and it was observed that *Btk*-synthesized AgNPs were significantly more pathogenic towards *T. ni* larvae.

Table 8.1 List of microbial nanoparticles against insect pest.

Nanoparticles	Microbial agents	Target organisms
ZnO NPS	*Bacillus thuringiensis*	*Callosobruchus maculates*
AgNPs	*Cochliobolus lunatus*	*Aedes aegypti, Anopheles stephensi*
AgNPs and AuNPs	*Chrysosporium tropicum*	*Aedes aegypti*
AgNPs	*Trichoderma harzianum rifai*	*A. aegypti*
AgNPs	*Bacillus thuringiensis kurstaki*	*Trichoplusia ni, Agrotis ipsilon*
AgNPs	*Bacillus marisflavi*	*Aedes aegypti, Culex quinquefasciatus* and *Anopheles stephensi*
ZnO-NPs	*Pseudomonas aeruginosa*	*Culex pipiens*
MgO-NPs	*Penicillium chrysogenum*	*A. stephensi*
CuNPs	*Metarhizium robertsii*	*Artemia salina, Artemia nauplii, Eudrilus andrei, Eudrilus eugeniae, A. aegypti, A. stephensi, C. quinquefasciatus*, and *Tenebrio molitor*
AgNPs	*Cochliobolus lunatus*	*A. aegypti* and *A. stephensi*
AgNPs	*Metarhizium anisopliae*	*Anopheles culicifacies*
AgNPs	*Isaria fumosorosea*	*A. aegypti* and *C. quinquefasciatus*
AgNPs	*Trichoderma harzianum*	*A. aegypti*
AuNPs	*Parmelia sulcata*	*A. stephensi* and *A. aegypti*
AgNPs	*Beauveria bassiana* and *Metarhizium brunneum*	*Tetranychus urticae*
AuNPs	*Chrysosporium tropicum*	*A. aegypti*

Thelma and Balasubramanian (2021) synthesized AgNPs using *Bacillus* strain *Bacillus marisflavi,* and their effectiveness was assessed against eggs, larval instars, and pupal stages of *Culex quinquefasciatus, Anopheles stephensi*, and *Aedes aegypti*. Disease-transmitting mosquitoes are greatly harmed by the potent poisonous action of AgNPs. ZnO NPs synthesized using *Pseudomonas aeruginosa*, were tested for their larvicidal effectiveness against *Culex pipiens*, which exhibited a high fatality rate (Abdo et al. 2021). The metabolites released by *Penicillium chrysogenum* has been utilized to formulate magnesium oxide nanoparticles (MgO-NPs) that exhibit significant mortality rates against larvae and pupae of *A. stephensi* (Fouda et al. 2021). Cell-free filtrate of *Pseudomonas fluorescens* have been used to develop copper nanoparticles (CuNPs) that showed efficacy against *T. castaneum* (El-Saadony et al. 2020).

Several NPs developed from fungi that have been reported to exhibit insecticidal activities. The *Metarhizium robertsii* (aqueous extract) was used to develop CuNPs that have been tested for their insecticidal properties against various nontargeted and targeted organisms, including *Artemia nauplii*, *A. salina*, *Eudrilus andrei*, *E. eugeniae*, *A. stephensi*, *A. aegypti*, *C. quinquefasciatus*, and *Tenebrio molitor*. These were safer to non-target organisms than the targeted pests (Vivekanandhan et al. 2021). Filamentous fungus *Cochliobolus lunatus* has been used to synthesize AgNPs that demonstrated considerable insecticidal potential against larval stages *of A. aegypti* and *A. stephensi* (Salunkhe et al. 2011). AgNPs may have penetrated the larval membranes to cause the possible larvicidal action. Green synthesized AgNPs using *Metarhizium anisopliae* have been reported to manage the *Anopheles culicifacies* vector of malaria in rural areas (Amerasan et al. 2016).

AgNPs developed using *Isaria fumosorosea* were extremely potent against important vector mosquitoes, namely *A. aegypti* and *C. quinquefasciatus* (Banu and Balasubramanian 2014). Using an entomopathogenic fungus, *Trichoderma harzianum*, Sundaravadivelan and Padmanabhan (2014) synthesized AgNPs and assessed their insecticidal ability against the pre-adult stages of *A. aegypti*. The fungus *Parmelia sulcata* was used to develop the AuNPs. The mosquitocidal efficacy of these NPs against *A. stephensi* and *A. aegypti* was evaluated. They demonstrated that these NPs inhibited the development of larvae, pupae, adults, and hatching of eggs in mosquitoes (Gandhi et al. 2019). The CuNPs formulated by integrating *Metarhizium roberstii* also exhibit insect-killing properties and cause considerable mortality in *A. aegypti*, *A. stephensi*, C. *quinquefasciatus*, and *T. molitor* (Vivekanandhan et al. 2021).

The entomopathogenic fungi *Beauveria bassiana* (Bb) and *Metarhizium brunneum* (Mb) have been utilized to formulate AgNPs. The NPs was evaluated against *Tetranychus urticae* that demonstrate that these treatments were highly virulent against newly emerged adult *T. urticae* females (Sayed and Kim 2018). *Chrysosporium tropicum* was used by Soni and Prakash (2012) to develop the AgNPs and AuNPs. These NPs have been tested against *A. aegypti* larvae as a larvicide. Their findings suggested that use of NPs offers a quick, safer, and ecologically sound method for controlling mosquitoes. This may open up new avenues for a vector control approach.

8.6 Mechanism of Action of Nanoparticles Against Insects

Microbial NPs cause harm to insects by entering the body via food, respiratory pathway, or direct touch (Hazafa et al. 2021; Saranya et al. 2020). These NPs cause damage and denaturation to the cell membrane when entering the insect body, which allows them to enter

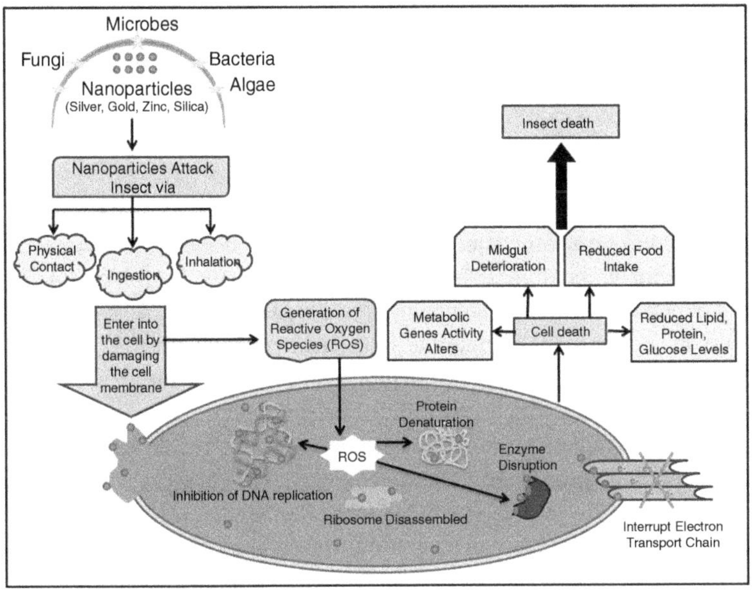

Figure 8.3 Mechanism of action of nanoparticles against insects.

the cells (Anandhi et al. 2020; Hazafa et al. 2021). These NPs induce the formation of ROS upon their entry into the cell (Fu et al. 2014; Canaparo et al. 2020) (Figure 8.3). Several studies indicated that ROS cause a number of cell dysfunctions, including protein denaturation, ribosome disassembly, disrupting enzyme activity, interrupting electron transport chain, and generate excessive free radicals that induce DNA damage through inflammatory processes, inhibiting DNA replication (Abdal Dayem et al. 2017; Qiu et al. 2016; Yu et al. 2020). These dysfunctions can ultimately lead to insect death. When NPs come into physical touch with one another, they pierce the exoskeleton and interact intercellularly by binding the nanomaterials to either phosphorus from DNA or sulfur from protein. This results in the denaturation of cellular organelles and enzymes which finally causes cellular fraction death (Abd-Elsalam and Prasad 2018).

The process for controlling stored grain pests with nano dust particles is based on physical disruption (Jasrotia et al. 2022; Malaikozhundan et al. 2019). Through attachment to the layer of cuticle wax, the NPs cause the insect's body to become dehydrated (Bihal et al. 2023; Shahzad and Manzoor 2021). The nanoclay, nanoalumina, and nanosilica absorbs water from the insect body (Asif et al. 2021). The hydrophobic nature of AgNPs results in DNA damage through the production of harmful Ag^{+} ions, and finally breaks the insect body (Asif et al. 2021; El-Samad et al. 2022). These changes in membrane characteristics affect permeability and respiration. AgNPs obstruct the production of melanin. When NPs are inhaled, they are internalized via phagocytosis, which degrades the midgut, changes the activity of genes involved in metabolism, and lowers glucose, protein, and fat levels. In the end, the insect dies from reduced food intake (Gustafson et al. 2015). Additionally, inhalation changed the membrane potential and functioning of nervous system enzymes (Feng et al. 2015; Wang et al. 2008). For example, ZnO NPs and TiO_2 NPs bind to β-carboxyl esterase and acetylcholine esterase, respectively, and modify their action

(Ragavendran et al. 2023). These NPs acts similarly to carbamate and organophosphorous groups of insecticides. Ag NPs have an impact on the enzyme glutathione S transferase (Bhat et al. 2019). Ag NPs function as an amylase inhibitor (Debnath et al. 2019, 2020).

8.7 Risk Concerns, Regulatory Policies, and Considerations

Numerous prospects for scientific advancement and innovation are presented by nanotechnology. However, because of its possible effects on the environment, health, and safety issues, it also brings up regulatory issues. The major regulatory issues includes risk evaluation and toxicity assessment as these nanomaterials have distinct characteristics, establishing standardized procedures for identifying and describing them, ensuring consumer safety, accurately labeling of nano-products, evaluating environmental effects, establishing global standards, protecting intellectual property rights, and considering ethical and societal implications (Ghosh and Kumar 2024). Various countries have been reviewing their established legal frameworks to check their feasibility for managing nanotechnologies efficiently (Amenta et al. 2015). It is important to take into account the ethical ramifications of nanotechnology, including issues with privacy, potential misuse, and equal access. Regulatory agencies must make sure that considerations such as social values and concerns are taken into account when making decisions about regulations (Fan and Alexeeff 2010). Global regulatory bodies are proactively tackling these concerns by means of joint ventures, research endeavors, and the formulation of particular protocols and laws pertaining to nanotechnology (Ghosh and Kumar 2024).

8.8 Conclusion and Future Perspective

Agriculture production loss and farmer incomes are significantly impacted by insect pests and diseases caused by them. To control these insect pests and increase agricultural yields, farmers mostly utilize synthetic pesticides that have the capacity to harm consumers, workers, and the environment. In order to create effective nanopesticide formulations, researchers have synthesized and assessed NPs and/or nanoemulsions, which include metal, agrochemical, essential oils, and bioactive agents. But they are expensive, hazardous, nonspecific, and labor-intensive. Combining the microbial systems with nanotechnology for controlling the insect population has evolved as an innovative and sustainable alternative. This technique utilizes the integration of microbes with nanomaterials to improve the effectiveness, reliability, and relevance of bioagents in the management of insect pests. These microbial bioactive agents-based nanopesticides have been developed to counter the adverse effects of synthetic pesticides. These are less hazardous, resolve pesticide residue issues, specific, environmentally compatible, and offer a sustainable means for insect pest management. Although utilization of microbial nanotechnology is effective, several challenges remain. The production of NPs is not economically efficient. Nobody knows the relationship between loading efficiency and pertinent NP properties. Nobody knows the correlation between the dose and their long-term effect against organisms. The effect of their residues on nontargeted organisms is still unsolved. Future research should be focused on extended experimental research to assess the optimal dosage, environmental safety, and effectiveness of microbial nanopesticides. Efforts should be made to formulate economical methods, knowing NPs and bioagent communications, and developing formulations that are nontoxic, effective, and sustainable for widespread agricultural deployments.

References

Abdal Dayem, A., Hossain, M.K., Lee, S.B. et al. (2017). The role of reactive oxygen species (ROS) in the biological activities of metallic nanoparticles. *International Journal of Molecular Sciences* 18: 120.

Abdel-Aziz, S.M., Prasad, R., Hamed, A.A., and Abdelraof, M. (2018). Fungal nanoparticles: a novel tool for a green biotechnology? In: *Fungal Nanobionics: Principles and Applications* (ed. R. Prasad, V. Kumar, M. Kumar, and S. Wang), 61–87. Springer.

Abd-Elsalam, K.A. and Prasad, R. (2018). *Nanobiotechnology Applications in Plant Protection*. Springer.

Abdo, A.M., Fouda, A., Eid, A.M. et al. (2021). Green synthesis of zinc oxide nanoparticles (ZnO-NPs) by *Pseudomonas aeruginosa* and their activity against pathogenic microbes and common house mosquito, *Culex pipiens*. *Materials* 14: 6983.

Adisa, I.O., Reddy Pullagurala, V.L., Rawat, S. et al. (2018). Role of cerium compounds in *Fusarium wilt* suppression and growth enhancement in tomato (*Solanum lycopersicum*). *Journal of Agricultural and Food Chemistry* 66: 5959–5970.

Agredo-Gomez, A., Molano-Molano, J., Portela-Patiño, M., and Rodríguez-Páez, J. (2024). Use of ZnO nanoparticles as a pesticide: in vitro evaluation of their effect on the phytophagous *Puto barberi* (mealybug). *Nano-Structures & Nano-Objects* 37: 101095.

Akhtar, I., Iqbal, Z., and Saddiqe, Z. (2020). Nanotechnology in pest management. *Nanoagronomy* 12: 69–83.

Al-Radadi, N.S., Hussain, T., Faisal, S., and Shah, S.A.R. (2022). Novel biosynthesis, characterization and bio-catalytic potential of green algae (*Spirogyra hyalina*) mediated silver nanomaterials. *Saudi Journal of Biological Sciences* 29: 411–419.

Amenta, V., Aschberger, K., Arena, M. et al. (2015). Regulatory aspects of nanotechnology in the agri/feed/food sector in Eu and non-Eu countries. *Regulatory Toxicology and Pharmacology* 73: 463–476.

Amerasan, D., Nataraj, T., Murugan, K. et al. (2016). Myco-synthesis of silver nanoparticles using *Metarhizium anisopliae* against the rural malaria vector *Anopheles culicifacies* Giles (Diptera: Culicidae). *Journal of Pest Science* 89: 249–256.

Anandhi, S., Saminathan, V., Yasotha, P. et al. (2020). Nano-pesticides in pest management. *Journal of Entomology and Zoology Studies* 8: 685–690.

Anwar, S.H. (2018). A brief review on nanoparticles: types of platforms, biological synthesis and applications. *Research & Reviews: Journal of Material Sciences* 6: 109–116.

Asif, M., Islam, S., Malik, M.A. et al. (2021). Nano pesticides application in agriculture and their impact on environment. In: *Agricultural Development in Asia-Potential Use of Nano-Materials and Nano-Technology* (ed. M. Asaduzzaman and M. Afroz), 1–11. Intechopen.

Aziz, N., Faraz, M., Sherwani, M.A. et al. (2019). Illuminating the anticancerous efficacy of a new fungal chassis for silver nanoparticle synthesis. *Frontiers in Chemistry* 7: 65.

Banu, A.N. and Balasubramanian, C. (2014). Optimization and synthesis of silver nanoparticles using Isaria Fumosorosea against human vector mosquitoes. *Parasitology Research* 113: 3843–3851.

Bapat, M.S., Singh, H., Shukla, S.K. et al. (2022). Evaluating green silver nanoparticles as prospective biopesticides: an environmental standpoint. *Chemosphere* 286: 131761.

Benedetti, T., Tamagno, W.A., Sordi, E., and Bortoluzzi, E.C. (2024). Iron oxide nanoparticles as enhancers of growth-promoting bacteria: a step towards developing nano-biofertilizers. *Environmental Science: Nano* 11 (1): https://doi.org/10.1039/D4EN00049H.

Bhat, R.S., Almusallam, J., Al Daihan, S., and Al-Dbass, A. (2019). Biosynthesis of silver nanoparticles using *Azadirachta indica* leaves: characterisation and impact on *Staphylococcus aureus* growth and glutathione-S-transferase activity. *IET Nanobiotechnology* 13: 498–502.

Bhattacharyya, A., Bhaumik, A., Rani, P.U. et al. (2010). Nano-particles – a recent approach to insect pest control. *African Journal of Biotechnology* 9: 3489–3493.

Bhattacharyya, A., Duraisamy, P., Govindarajan, M. et al. (2016). Nano-biofungicides: emerging trend in insect pest control. In: *Advances and Applications Through Fungal Nanobiotechnology* (ed. R. Prasad), 307–319. Springer.

Bihal, R., Al-Khayri, J.M., Banu, A.N. et al. (2023). Entomopathogenic fungi: an eco-friendly synthesis of sustainable nanoparticles and their nanopesticide properties. *Microorganisms* 11: 1617.

Bittner, A.M. (2005). Biomolecular rods and tubes in nanotechnology. *Naturwissenschaften* 92: 51–64.

Canaparo, R., Foglietta, F., Limongi, T., and Serpe, L. (2020). Biomedical applications of reactive oxygen species generation by metal nanoparticles. *Materials* 14: 53.

Damalas, C.A. and Koutroubas, S.D. (2018). Current status and recent developments in biopesticide use. *Agriculture* 8: 13.

Danish, M.S.S., Estrella-Pajulas, L.L., Alemaida, I.M. et al. (2022). Green synthesis of silver oxide nanoparticles for photocatalytic environmental remediation and biomedical applications. *Meta* 12: 769.

De Oliveira, J.L., Campos, E.V.R., Bakshi, M. et al. (2014). Application of nanotechnology for the encapsulation of botanical insecticides for sustainable agriculture: prospects and promises. *Biotechnology Advances* 32: 1550–1561.

Debnath, G., Das, P., and Saha, A.K. (2019). Green synthesis of silver nanoparticles using mushroom extract of *Pleurotus giganteus*: characterization, antimicrobial, and A-amylase inhibitory activity. *BioNanoScience* 9: 611–619.

Debnath, G., Das, P., and Saha, A.K. (2020). Characterization, antimicrobial and A-amylase inhibitory activity of silver nanoparticles synthesized by using mushroom extract of Lentinus tuber-Regium. *Proceedings of the National Academy of Sciences, India Section B: Biological Sciences* 90: 37–45.

Dharmaraj, D., Krishnamoorthy, M., Rajendran, K. et al. (2021). Antibacterial and cytotoxicity activities of biosynthesized silver oxide (Ag_2O) nanoparticles using *Bacillus paramycoides*. *Journal of Drug Delivery Science and Technology* 61: 102111.

Divya, K. and Jisha, M. (2018). Chitosan nanoparticles preparation and applications. *Environmental Chemistry Letters* 16: 101–112.

Divya, M., Kiran, G.S., Hassan, S., and Selvin, J. (2019). Biogenic synthesis and effect of silver nanoparticles (AgNPs) to combat catheter-related urinary tract infections. *Biocatalysis and Agricultural Biotechnology* 18: 101037.

Dutta, P., Kumari, A., and Mahanta, M. (2022). Chitosan nanoparticle: synthesis, characterization, and use as plant health materials. In: *Chitin and Chitosan-Isolation, Properties, and Applications* (ed. B. Kumar), 1–17. Intechopen.

El-Saadony, M.T., Abd El-Hack, M.E., Taha, A.E. et al. (2020). Ecofriendly synthesis and insecticidal application of copper nanoparticles against the storage pest *Tribolium castaneum*. *Nanomaterials* 10: 587.

El-Samad, L.M., Bakr, N.R., El-Ashram, S. et al. (2022). Silver nanoparticles instigate physiological, genotoxicity, and ultrastructural anomalies in midgut tissues of beetles. *Chemico-Biological Interactions* 367: 110166.

Fan, A.M. and Alexeeff, G. (2010). Nanotechnology and nanomaterials: toxicology, risk assessment, and regulations. *Journal of Nanoscience and Nanotechnology* 10: 8646–8657.

FAO (2018). World fertilizer trends and outlook to 2018. Food and Agriculture Organization of the United Nations—Rome 2015. http://www.fao.org/publication (accessed 12 May 2020).

Feng, X., Chen, A., Zhang, Y. et al. (2015). Central nervous system toxicity of metallic nanoparticles. *International Journal of Nanomedicine* 10: 4321–4340.

Flenniken, M. L., Uchida, M., Liepold, L. O., Kang, S., Young, M. J., and Douglas, T. (2009). A library of protein cage architectures as nanomaterials. In: *Viruses and nanotechnology* (ed. M. Manchester and N.F. Steinmetz), 71–93. Springer Nature.

Fouda, A., Awad, M.A., Eid, A.M. et al. (2021). An eco-friendly approach to the control of pathogenic microbes and *Anopheles stephensi* malarial vector using magnesium oxide nanoparticles (Mg-NPs) fabricated by *Penicillium chrysogenum*. *International Journal of Molecular Sciences* 22: 5096.

Fróna, D., Szenderák, J., and Harangi-Rákos, M. (2019). The challenge of feeding the world. *Sustainability* 11: 5816.

Fu, P.P., Xia, Q., Hwang, H.-M. et al. (2014). Mechanisms of nanotoxicity: generation of reactive oxygen species. *Journal of Food and Drug Analysis* 22: 64–75.

Gandhi, A.D., Murugan, K., Umamahesh, K. et al. (2019). Lichen parmelia sulcata mediated synthesis of gold nanoparticles: an eco-friendly tool against *Anopheles stephensi* and *Aedes aegypti*. *Environmental Science and Pollution Research* 26: 23886–23898.

Ge, J., Hu, J., Cui, S. et al. (2024). Biosynthesis of Bt-Ag_2O nanoparticles using *Bacillus thuringiensis* and their pesticidal and antimicrobial activities. *Applied Microbiology and Biotechnology* 108: 157.

Ghosh, M. and Kumar, R. (2024). Regulatory issues in nanotechnology. In: *Nanotechnology Theranostics in Livestock Diseases and Management* (ed. M. Prasad, R. Kumar, M. Ghosh, et al.), 765–788. Springer.

Goyal, U., Chaubey, K.K., Pandey, S. et al. (2024). Role of gold nanoparticles in plant protection against pathogen. In: *Metal and Metal-Oxide Based Nanomaterials: Synthesis, Agricultural, Biomedical and Environmental Interventions* (ed. R.K. Bachheti, A. Bachheti, and A. Husen), 201–211. Springer.

Grillo, R., Abhilash, P.C., and Fraceto, L.F. (2016). Nanotechnology applied to bio-encapsulation of pesticides. *Journal of Nanoscience and Nanotechnology* 16: 1231–1234.

Gruère, G.P. (2012). Implications of nanotechnology growth in food and agriculture in OECD countries. *Food Policy* 37: 191–198.

Gustafson, H.H., Holt-Casper, D., Grainger, D.W., and Ghandehari, H. (2015). Nanoparticle uptake: the phagocyte problem. *Nano Today* 10: 487–510.

Hazafa, A., Murad, M., Masood, M.U. et al. (2021). Nano-biopesticides as an emerging technology for pest management. In: *Insecticides: Impact and Benefits of Its Use for Humanity* (ed. R.E.R. Ranz). Intechopen https://doi.org/10.5772/intechopen.101285.

Hazarika, A., Yadav, M., Yadav, D.K., and Yadav, H.S. (2022). An overview of the role of nanoparticles in sustainable agriculture. *Biocatalysis and Agricultural Biotechnology* 43: 102399.

Ion, A.C., Ion, I., Culetu, A., and Gherase, D. (2010). *Carbon-Based Nanomaterials. Environmental Applications*. Politehnica University of Bucharest.

Iqbal, P., Preece, J. A. and Mendes, P. M. (2012). Nanotechnology: the "topdown" and "bottom-up" approaches. In: *Supramolecular chemistry: from molecules to nanomaterials*. (ed. P.A. Gale and J.W. Steed). Wiley Online Library. https://doi.org/10.1002/9780470661345.smc195.

Jaiswal, D.K., Gawande, S.J., Soumia, P. et al. (2022). Biocontrol strategies: an eco-smart tool for integrated pest and diseases management. *BMC Microbiology* 22: 324.

Jasrotia, P., Nagpal, M., Mishra, C.N. et al. (2022). Nanomaterials for postharvest management of insect pests: current state and future perspectives. *Frontiers in Nanotechnology* 3: 811056.

Junaid, M. and Gokce, A. (2024). Global agricultural losses and their causes. *Bulletin of Biological and Allied Sciences Research* 2024: 66–66.

Kachhawa, D. (2017). Microorganisms as a biopesticides. *Journal of Entomology and Zoology Studies* 5: 468–473.

Kannan, R.R.R., Arumugam, R., Ramya, D. et al. (2013). Green synthesis of silver nanoparticles using marine macroalga *Chaetomorpha linum*. *Applied Nanoscience* 3: 229–233.

Kannan, M., Elango, K., Tamilnayagan, T. et al. (2020). Impact of nanomaterials on beneficial insects in agricultural ecosystems. In: *Nanotechnology for Food, Agriculture, and Environment* (ed. D. Thangadurai, J. Sangeetha, and R. Prasad), 379–393. Springer.

Kim, D.Y., Kadam, A., Shinde, S. et al. (2018). Recent developments in nanotechnology transforming the agricultural sector: a transition replete with opportunities. *Journal of the Science of Food and Agriculture* 98: 849–864.

Klaus, T., Joerger, R., Olsson, E., and Granqvist, C.-G. (1999). Silver-based crystalline nanoparticles, microbially fabricated. *Proceedings of the National Academy of Sciences of the United States of America* 96: 13611–13614.

Koul, O. (2019). Nanobiopesticides: an introduction. In: *Nano-Biopesticides Today and Future Perspectives* (ed. O. Koul), 1–16. Elsevier.

Kour, D., Rana, K.L., Yadav, A.N. et al. (2020). Microbial biofertilizers: bioresources and eco-friendly technologies for agricultural and environmental sustainability. *Biocatalysis and Agricultural Biotechnology* 23: 101487.

Kumar, B. (2016). Biocontrol of insect pests. In: *Ecofriendly Pest Management for Food Security* (ed. Omkar), 25–62. Elsevier.

Kumar, P., Pandhi, S., Mahato, D.K. et al. (2021). Bacillus-based nano-bioformulations for phytopathogens and insect–pest management. *Egyptian Journal of Biological Pest Control* 31: 1–11.

Kumari, B., Mallick, M., Solanki, M.K. et al. (2019). Applying nanotechnology to bacteria: an emerging technology for sustainable agriculture. In: *Role of Plant Growth Promoting Microorganisms in Sustainable Agriculture and Nanotechnology* (ed. A. Kumar, A.K. Singh, and K.K. Choudhary), 108–120. Elsevier.

Malaikozhundan, B., Vinodhini, J., and Vaseeharan, B. (2019). Nanopesticides for the management of insect pests of stored grains. In: *Nanotechnology for Agriculture: Crop Production & Protection* (ed. D.G. Panpatte and Y.K. Jhala), 303–322. Springer.

Mazhar, T., Shrivastava, V., and Tomar, R.S. (2017). Green synthesis of bimetallic nanoparticles and its applications: a review. *Journal of Pharmaceutical Sciences and Research* 9: 102.

Minhas, L.A., Kaleem, M., Jabeen, A. et al. (2023). Synthesis of silver oxide nanoparticles: a novel approach for antimicrobial properties and biomedical performance, featuring *Nodularia haraviana* from the Cholistan Desert. *Microorganisms* 11: 2544.

Mukhopadhyay, S.S. (2014). Nanotechnology in agriculture: prospects and constraints. *Nanotechnology, Science and Applications* 7: 63–71.

Naderi, M. and Danesh-Shahraki, A. (2013). Nanofertilizers and their roles in sustainable agriculture. *International Journal of Agriculture and Crop Sciences* 5: 2229–2232.

Nivetha, N., Asha, A., Bandeppa, GS et al. 2021. Microbial-based nanoparticles as potential approach of insect pest management. *Microbes for Sustainable Lnsect Pest Management: Hydrolytic Enzyme & Secondary Metabolite, Volume* 2, Md. Aslam Khan, Wasim Ahmad, 135–157. Springer.

Nuruzzaman, M., Rahman, M.M., Liu, Y., and Naidu, R. (2016). Nanoencapsulation, Nano-guard for pesticides: a new window for safe application. *Journal of Agricultural and Food Chemistry* 64: 1447–1483.

Oerke, E.-C. (2006). Crop losses to pests. *The Journal of Agricultural Science* 144: 31–43.

Pantidos, N. and Horsfall, L.E. (2014). Biological synthesis of metallic nanoparticles by bacteria, fungi and plants. *Journal of Nanomedicine & Nanotechnology* 5: 1.

Prasad, R., Kumar, V., and Prasad, K.S. (2014). Nanotechnology in sustainable agriculture: present concerns and future aspects. *African Journal of Biotechnology* 13: 705–713.

Prasad, R., Bhattacharyya, A., and Nguyen, Q.D. (2017). Nanotechnology in sustainable agriculture: recent developments, challenges, and perspectives. *Frontiers In Microbiology* 8: 1014.

Prasad, R., Kumar, V., Kumar, M., and Wang, S. (2018). *Fungal Nanobionics: Principles and Applications*. Springer.

Purohit, J., Chattopadhyay, A., and Singh, N.K. (2019). Green synthesis of microbial nanoparticle: approaches to application. In: *Microbial Nanobionics: Volume 2, Basic Research and Applications* (ed. R. Prasad), 35–60. Springer.

Qiu, T.A., Gallagher, M.J., Hudson-Smith, N.V. et al. (2016). Research highlights: unveiling the mechanisms underlying nanoparticle-induced ROS generation and oxidative stress. *Environmental Science: Nano* 3: 940–945.

Ragavendran, C., Kamaraj, C., Jothimani, K. et al. (2023). Eco-friendly approach for ZnO nanoparticles synthesis and evaluation of its possible antimicrobial, Larvicidal and photocatalytic applications. *Sustainable Materials and Technologies* 36: E00597.

Sabarwal, A., Kumar, K., and Singh, R.P. (2018). Hazardous effects of chemical pesticides on human health–cancer and other associated disorders. *Environmental Toxicology and Pharmacology* 63: 103–114.

Salunkhe, R.B., Patil, S.V., Patil, C.D., and Salunke, B.K. (2011). Larvicidal potential of silver nanoparticles synthesized using fungus *Cochliobolus lunatus* against *Aedes aegypti* (Linnaeus, 1762) and *Anopheles stephensi* Liston (Diptera; Culicidae). *Parasitology Research* 109: 823–831.

Sanyasi, S., Majhi, R.K., Kumar, S. et al. (2016). Polysaccharide-capped silver nanoparticles inhibit biofilm formation and eliminate multi-drug-resistant bacteria by disrupting bacterial cytoskeleton with reduced cytotoxicity towards mammalian cells. *Scientific Reports* 6: 24929.

Saranya, S., Selvi, A., Babujanarthanam, R. et al. (2020). Insecticidal activity of nanoparticles and mechanism of action. In: *Model Organisms to Study Biological Activities and Toxicity of Nanoparticles* (ed. B. Siddhardha, M. Dyavaiah, and K. Kasinathan), 243–266. Springer.

Sastry, M., Ahmad, A., Khan, M.I., and Kumar, R. (2003). Biosynthesis of metal nanoparticles using fungi and actinomycete. *Current Science* 85: 162–170.

Sathiyanarayanan, G., Dineshkumar, K., and Yang, Y.-H. (2017). Microbial exopolysaccharide-mediated synthesis and stabilization of metal nanoparticles. *Critical Reviews in Microbiology* 43: 731–752.

Saw, G., Nagdev, P., Jeer, M., and Murali-Baskaran, R. (2023). Silica nanoparticles mediated insect pest management. *Pesticide Biochemistry and Physiology* 194: 105524.

Sayed, A. and Kim, S. (2018). Myco-silver nanoparticles synthesized using *Beauveria bassiana* and *Metarhizium brunneum* as a smart pest control. *Egyptian Journal of Plant Protection Research Institute* 1: 1–18.

Sayed, A.M., Kim, S., and Behle, R.W. (2017). Characterisation of silver nanoparticles synthesised by *Bacillus thuringiensis* as a nanobiopesticide for insect pest control. *Biocontrol Science and Technology* 27: 1308–1326.

Shah, M., Fawcett, D., Sharma, S. et al. (2015). Green synthesis of metallic nanoparticles via biological entities. *Materials* 8: 7278–7308.

Shahid, M., Naeem-Ullah, U., Khan, W. et al. (2021). Application of nanotechnology for insect pests management: a review. *Journal of Innovative Sciences* 7: 28–39.

Shahzad, K. and Manzoor, F. (2021). Nanoformulations and their mode of action in insects: a review of biological interactions. *Drug and Chemical Toxicology* 44: 1–11.

Sharma, G. (2024). Eco-friendly silver nanoparticles synthesized from plant extracts as biopesticides: a comprehensive review. *Bulletin of Pure and Applied Sciences-Zoology* 1235–1259.

Sharma, D., Kanchi, S., and Bisetty, K. (2019). Biogenic synthesis of nanoparticles: a review. *Arabian Journal of Chemistry* 12: 3576–3600.

Shekhar, S., Sharma, S., Kumar, A. et al. (2021). The framework of nanopesticides: a paradigm in biodiversity. *Materials Advances* 2: 6569–6588.

Siddiqi, K.S. and Husen, A. (2016). Fabrication of metal nanoparticles from fungi and metal salts: scope and application. *Nanoscale Research Letters* 11: 1–15.

Singh, M.P., Shabir, S., Deopa, A.S. et al. (2022). Synthesis of green engineered silver nanoparticles through *Urtica dioica*: an inhibition of microbes and alleviation of cellular and organismal toxicity in *Drosophila melanogaster*. *Antibiotics* 11: 1690.

Singh, N.A., Narang, J., Garg, D. et al. (2023). Nanoparticles synthesis via microorganisms and their prospective applications in agriculture. *Plant Nano Biology* 5: 100047.

Singh, A., Shraogi, N., Verma, R. et al. (2024). Challenges in current pest management practices: navigating problems and a way forward by integrating controlled release system approach. *Chemical Engineering Journal* 498: 154989.

Sleytr, U.B., Messner, P., Pum, D., and Sára, M. (1993). Crystalline bacterial cell surface layers. *Molecular Microbiology* 10: 911–916.

Soni, N. and Prakash, S. (2012). Efficacy of fungus mediated silver and gold nanoparticles against *Aedes aegypti* larvae. *Parasitology Research* 110: 175–184.

Sundaravadivelan, C. and Padmanabhan, M.N. (2014). Effect of mycosynthesized silver nanoparticles from filtrate of *Trichoderma harzianum* against larvae and pupa of dengue vector *Aedes aegypti* L. *Environmental Science and Pollution Research* 21: 4624–4633.

Terra, A.L.M., Kosinski, R.D.C., Moreira, J.B. et al. (2019). Microalgae biosynthesis of silver nanoparticles for application in the control of agricultural pathogens. *Journal of Environmental Science and Health, Part B* 54: 709–716.

Thelma, J. and Balasubramanian, C. (2021). Ovicidal, larvicidal and pupicidal efficacy of silver nanoparticles synthesized by *Bacillus marisflavi* against the chosen mosquito species. *PLoS One* 16: E0260253.

Usman, A.I., Aziz, A.A., and Noqta, O.A. (2019). Green sonochemical synthesis of gold nanoparticles using palm oil leaves extracts. *Materials Today: Proceedings* 7: 803–807.

Usman, M., Farooq, M., Wakeel, A. et al. (2020). Nanotechnology in agriculture: current status, challenges and future opportunities. *Science of the Total Environment* 721: 137778.

Vasquez, R.D., Apostol, J.G., De Leon, J.D. et al. (2016). Polysaccharide-mediated green synthesis of silver nanoparticles from *Sargassum siliquosum* Jg Agardh: assessment of toxicity and hepatoprotective activity. *Opennano* 1: 16–24.

Velusamy, P., Kumar, G.V., Jeyanthi, V. et al. (2016). Bio-inspired green nanoparticles: synthesis, mechanism, and antibacterial application. *Toxicological Research* 32: 95–102.

Vetchinkina, E., Loshchinina, E., Kupryashina, M. et al. (2018). Green synthesis of nanoparticles with extracellular and intracellular extracts of basidiomycetes. *PeerJ* 6: E5237.

Vivekanandhan, P., Swathy, K., Thomas, A. et al. (2021). Insecticidal efficacy of microbial-mediated synthesized copper Nano-pesticide against insect pests and non-target organisms. *International Journal of Environmental Research and Public Health* 18: 10536.

Wang, J., Liu, Y., Jiao, F. et al. (2008). Time-dependent translocation and potential impairment on central nervous system by intranasally instilled TiO_2 nanoparticles. *Toxicology* 254: 82–90.

Wen, A.M., Shukla, S., Saxena, P. et al. (2012). Interior engineering of a viral nanoparticle and its tumor homing properties. *Biomacromolecules* 13: 3990–4001.

Yu, Z., Li, Q., Wang, J. et al. (2020). Reactive oxygen species-related nanoparticle toxicity in the biomedical field. *Nanoscale Research Letters* 15: 115.

Zhang, T.-G. and Miao, C.-Y. (2024). Iron oxide nanoparticles as promising antibacterial agents of new generation. *Nanomaterials* 14: 1311.

9

Role of Actinomycetes in the Bioactive Synthesis of Nanoparticles

Dibyajit Lahiri[1], Moupriya Nag[1], Manoj Singh[2], Arunava Goswami[3], Ritwik Banerjee[1], Dipro Mukherjee[1], Sayantani Garai[1], Ankita Dey[2], and Rina Rani Ray[4]

[1] *Department of Biotechnology, University of Engineering & Management, Kolkata, West Bengal, India*
[2] *Department of Biotechnology, Adamas University, Kolkata, West Bengal, India*
[3] *Agricultural and Ecological Research Unit, Biological Sciences Division, Institute of National Importance, Indian Statistical Institute, West Bengal, India*
[4] *Department of Biotechnology, Maulana Abul Kalam Azad University of Technology, Kolkata, West Bengal, India*

9.1 Introduction

Nanoparticles (NPs), nanomaterials (NMs), and nanostructured materials (NSMs) represent an active area of research in the field of nanotechnology, with profoundly significant applications in a large arena of various biomedical and chemical fields. NPs or NMs are described as substances of dimension in the range of 1–100 nm or with length of 1–1,000 nm in at least one dimension (Jeevanandam et al. 2018). NPs offer enhanced performances and additive advantages over the bulk counterparts because of their tunable physicochemical characteristics such as increased surface area, high reactivity, surface plasmon resonance (SPR), high light absorption rates, Rayleigh scattering and surface-enhanced Raman scattering (SERS) (Daniel and Astruc 2004; Kim et al. 2007).

NPs can be synthesized using various physical, chemical, and biological methods. Traditional methods for the synthesis of NPs are primarily edging around nucleation-based chemical degradation methods involving reduction of salts in their solutions by various organic and inorganic reducing agents. However, the use of such chemical agents can destabilize the particles formed or cause various forms of cellular toxicity, which makes the NMs unfit for usage in biomedical and health care products, in food industry, drug delivery, nanoantibiotics, pharmaceutical nanoengineering, catalysis, cosmetics, and personal care products (Chau et al. 2007; Li et al. 2011). Furthermore, the physical and chemical methods involve costly appliances and chemicals (Ingle et al. 2009). Hence, biological methods for synthesizing NPs have been favored over the physico-chemical methods, and scientists are trying to study and optimize existing methods of biogenic nanoparticle production along with finding newer alternatives. Biosynthesis of NPs is used to obtain large-scale production with diverse shapes, sizes, and chemical composition, offering better advantages in the production of monodispersed particles compared to chemical alternatives. Biogenic NPs are mostly water soluble, less toxic, and biocompatible, as they involve the use of microorganisms, plant extracts and enzymes derived from various biotic sources to synthesize NPs (Song and Kim 2008). There have been various incidences about natural capping of the synthesized NPs by proteins present in the biological source, which often act in a synergistic way, thereby increasing their efficiency (Kumar et al. 2013).

Microorganisms like bacteria and fungi are considered as one of the most promising options for nanoparticle synthesis, owing to their diversity and controllable growth kinetics (Płaza et al. 2014). Bacterial growth kinetics can be easily handled and manipulated efficiently by designing proper reactor systems, which makes them a preferred choice for the production of NPs with specific shape, size, and application (Zonooz and Salouti 2011). Actinomycetes are prokaryotic organisms that are classified as bacteria, share important characteristics of fungi, and are known for their ability to produce various bioactive compounds, which can be useful for biosynthesis of metal NPs (Vimal et al. 2009). They are unicellular, Gram-positive bacteria that belong to the order Actinomycetales and are one of the most important microorganisms that produce a wide variety of useful secondary metabolites, many of which have potent biological activities. Actinomycetes are known to be good producers of antibacterial substances, and offer the most promising synthesizers of many industrially and commercially meaningful metabolites (Neelu et al. 2013). Hence, they are often preferred for the synthesis of NPs, which confer additional antibacterial properties and are extensively favored in clinical applications. Actinobacteria isolated from different ecosystems were recognized as potential synthesizers of gold (AuNPs) and silver nanoparticles (AgNPs).

Although more research works are needed to be performed, this chapter provides pertinent concepts on the mechanism of synthesis of biogenic NPs by the use of actinomycetes and their applications in various fields (Figure 9.1).

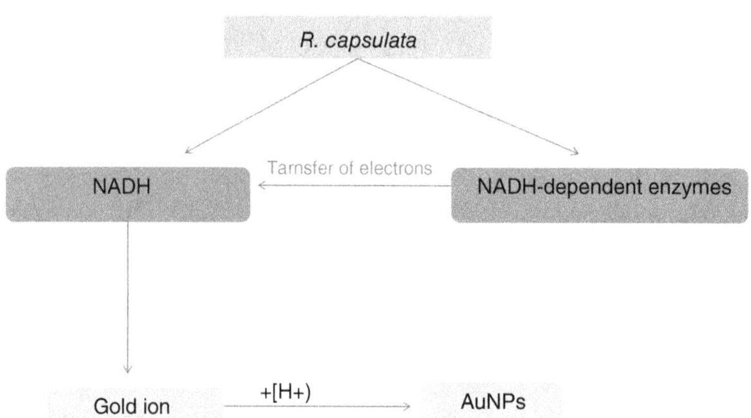

Figure 9.1 Synthesis of AuNPs by *R. capsulate* outside the cell.

9.2 Synthesis of Biogenic Nanoparticles from Actinomycetes

Actinomycetes are capable of both intracellular and extracellular nanoparticle synthesis. The intracellular synthesis of NPs in actinomycetes is mediated by cellular enzymes. These enzymes reduce the metal ions in the surrounding media or solution as they attach themselves on the surface of the mycelia by electrostatic binding to the negatively charged cell wall of mycelia, which arises due to the carboxylate groups present in the enzymes. There have been various accounts of intracellular production of AgNPs (Abdeen et al. 2013) and AuNPs by diverse groups of actinobacteria (Ahmad et al. 2003a). Although there are instances of other metal and metal oxide nanoparticle synthesis, researches are yet to flourish. On the other hand, the extracellular synthesis of NPs occurs mainly by the action of the enzymes that are involved in the nitrogen cycle, which are responsible for the electron shuttle enzymatic metal reduction process, which plays an important role in reducing the metal ions to their corresponding nanoforms (Karthik et al. 2013).

9.3 Nanoparticle (Biogenic) Production by Actinomycetes

Actinomycetes possess the capability of producing extracellular and intracellular types of NPs. The production of intracellular NPs takes place over the mycelia surface because of an electrostatic interaction of positively charged silver ions with the negatively charged molecules of carboxylate ions of the enzyme present within the mycelia cell wall. The positively charged silver ions are reduced to metallic silver (Ag) due to the action of these enzymes. The storage of the silver nuclei in the cell wall leads to the production of AgNPs (Abdeen et al. 2013). Moreover, some new and novel types of actinomycetes, *Rhodococcus* sp. are required for the synthesis of AuNPs. The transmission electron microscopic (TEM) image proves the existence of NPs on the cell walls of the actinomycetes, proving the presence of intracellular Au NPs (Ahmad et al. 2003a). Formation of intercellular AgNPs by the usage of *Rhodococcus* NCIM 2891 is also observed (Otari et al. 2012). *Streptomyces hygroscopicus*'s biomass in wet condition is usually exposed to $HAuCl_4$ and at 72 hours, it is seen that the yellow-colored biomass turns pink, which indicates the creation of extracellular AuNPs.

Streptomyces species HBUM171191 is observed for the production of NPs of Ag, Mn, and Zn, exposing the biomass, which is wet toward the surrounding metal solution. Differences in color in the biomass are seen to change from yellow to light brown, light yellow and dark yellow, and it also shows the production of Ag, Mn, and Zn nanoparticles (Waghmare et al. 2011). Extracellular nanoparticle synthesis is mediated via enzymes present in these regions, which also play critical roles in the nitrogen cycle. This is done by the enzymatic metal reduction process of the electron shuttle system (Karthik et al. 2014). Nitrate reductase is an NADH-dependent enzyme that possesses important roles in the reduction of Ag^+ to Ag^0. The extracellular production of NPs is done by actinomycetes by the usage of *Streptomyces glaucus* 71MD (Tsibakhashvili et al. 2011). The *Streptomyces* sp. ERI-3 pure culture alters $AgNO_3$ solution having no color to a brownish-red color within 12 hours (Zonooz and Salouti 2011), which dictates that the production of AgNPs is complete.

9.4 Development of Metal Oxides, Metal-based Nanomaterials, and Nano-antibiotics from Actinomycetes

Silver has been used as an antimicrobial agent for ages now. It is largely known for its properties like catalytic agent, providing stability to the chemicals and providing better conductivity. In a minimum quantity, silver is not toxic towards human health. The said characteristics make silver a good option for the production of AgNPs for different applications in medical biology. Actinomycetes are observed to possess the power of production of secondary metabolites or newer chemical groups that act as antibiotics. As actinomycetes are the first source for antibacterial components, there lie many instances of actinomycetes-mediated metal oxide and metal nanoparticle production and its subsequent antimicrobial activity. Ag-NPs formed from actinomycetes have antibacterial properties. Ag-NPs formed from the usage of *Streptomyces* sp. BDUKAS10 shows generally 13–16 mm inhibition zones in response to *Bacillus cereus, Pseudomonas aeruginosa,* and *Staphylococcus aureus* respectively (Sivalingam et al. 2012). AgNPs are produced extracellularly by *Streptomyces rochei* by the usage of $AgNO_3$ and possess antibacterial activities. The inhibition zones usually lie from 22 to 31 mm against *P. aeruginosa*, *Escherichia coli*, *Enterococcus faecalis*, *Klebsiella pneumoniae,* and *S. aureus* are observed through many researches (Selvakumar et al. 2012). The formation of AgNPs by the usage of *Rhodococcus* sp. and *Streptomyces* sp. possesses high antibacterial activities with inhibition zones of 25.43 mm against *P. aeruginosa* and 22.2 mm against *S. aureus*, 20.9 mm against *K. pneumoniae*, and 19 mm against *E. coli* are also been observed in many research journals conducted in the last decade (Sukanya et al. 2013). The extracellular AgNPs, which are produced by the *Streptomyces parvulus* SSNP11, show inhibition zones in the range of 21–27 mm against *Pseudomonas putida*, *Salmonella typhi*, *Bacillus subtilis,* and *K. pneumoniae*, respectively (Prakasham et al. 2014). Intracellular AuNPs, which are produced by *Streptomyces viridogens*, usually show 14–16 mm inhibition zones against *S. aureus* and *E. coli* (Balagurunathan et al. 2011) (Table 9.1).

9.5 Mechanism of Synthesis of NPs by Actinomycetes

Actinomycetes possess resistance to metals, and thus this aspect helps in the detoxification of chemicals along with the presence of chemiosmotic cations or protons or ATPase that brings about energy-dependent efflux from the cell. They help in the easy detoxification of the metal ions to nontoxic metallic NPs by the involvement of intracellular bioaccumulation, extracellular precipitation and biomineralization (Kumari et al. 2020). The mechanism of synthesis of NPs is associated with the involvement of various types of reductive species like cofactors, enzymes, and intracellular proteins. The positively charged metal ions bind with the negatively charged cell wall. The positively charged silver ions from the solution of silver nitrate solution undergo the mechanism of reduction to Ag^0, thereby resulting in the development of NPs.

Table 9.1 Anti-bacterial properties of NPs from *Actinomycetes*.

S. no.	*Actinomycetes*	Nanoparticles	References
1.	*Actinomycetes* sp.	Ag	Abdeen et al. (2013)
2.	*Streptomyces rochei* MHM13	Ag	Abd-Elnaby et al. (2016)
3.	*Streptacidiphilus durhamensis*	Ag	Buszewski et al. (2016)
4.	*Streptomyces* sp.VITPK1	Ag	Sanjenbam et al. (2014)
5.	*Streptomyces graminofaciens*	Ag	Kamel et al. (2016)
6.	*Streptomyces* sp. VDP-5	Ag	Singh et al. (2014)
7.	*Streptomyces* sp.	Ag	Abdeen et al. (2014)
8.	*Streptomyces*- MS 26	Ag	Zarina and Nanda (2014)
9.	*Streptomyces* sp. LK3	Ag	Karthik et al. (2014)
10.	*Streptomyces* sp. JAR1	Ag	Chauhan et al. (2013)
11.	*Nocardiopsis* sp. MBRC-1	Ag	Manivasagan et al. (2013)
12.	*Streptomyces parvuus* SSNP11	Ag	Prakasham et al. (2014)
13.	*Streptomyces* sp. I & II,	Ag	Sukanya et al. (2013)
14.	*Rhodococcus* sp.	Ag	Sukanya et al. (2013)
15.	*Streptomyces* sp. VITBT7	Ag	Subashini and Kannabiran (2013)
16.	*Thermoactinomyces* sp.	Ag	Deepa et al. (2013)
17.	*Streptomyces glaucus* 71MD	Ag	Tsibakhashvili et al. (2011)
18.	*Streptomyces rochei*	Ag	Selvakumar et al. (2012)
19.	*Streptomyces* sp. ERI-3	Ag	Zonooz and Salouti (2011)
20.	*Nocardia farcinica*	Au	Shah et al. (2012)
21.	*Rhodococcus* sp.	Ag	Otari et al. (2012)
22.	*Streptomyces* sp.	Zn/Cu	Usha et al. (2010)
23.	*Streptomycetes viridogens* HM10	Au	Balagurunathan et al. (2011)
24.	*Rhodococcus* sp.	Au	Ahmad et al. (2003a)
25.	*Streptomyces* sp. HBUM171191	Zn/Mn	Waghmare et al. (2011)

9.6 Intracellular and Extracellular Synthesis of NPs by Actinomycetes

Synthesis of NPs by actinomycetes occurs both intracellular and extracellular mechanisms. The type of metallic NPs synthesis is dependent on the location within the cell and the involvement of various types of reductive species like metabolites and enzymes. The mechanism of extracellular synthesis of NPs involves the reductants that are being produced from the cell wall, whereas the intracellular synthesis involves the various types of enzymes that are associated with the cells (Udaya Prakash et al. 2014).

9.7 Factors Controlling the Mechanism of Synthesis of NPs

Various types of physical factors play an important role in the synthesis of NPs. This mechanism controls the speed, size, shape, and stability of the NPs. One of the most critical parameters that controls the synthesis of NPs is the pH that helps in the reduction of the metal ions (Ahmad et al. 2019). Temperature also acts as an important factor that brings about its control in the rate of synthesis, size, and stability of the NPs (Elamawi et al. 2018). The concentration of the stabilizing and reducing agents also plays an important role in the mechanism of stabilizing the biosynthesis of NPs. The time of incubation also plays an effective role in maintaining the reaction between the metal and the plant extract.

9.8 Actinomycetes NPs and Their Antifungal Activities

AgNPs, which are produced biologically by the usage of *Streptomyces* sp. JAR1, show inhibition of *Aspergillus terreus* JAS1 and *Fusarium* sp. with inhibition areas lying nearly from 20.5 to 17.1 mm (Chauhan et al. 2013). AgNPs formed from *Streptomyces* sp. VITBT7 has been observed to show some particular inhibition centeres in response to *Aspergillus fumigatus* MTCC 3002 and *Aspergillus niger* MTCC 1344, having 21.3–21.9 mm inhibition zones (Subashini and Kannabiran 2013). AgNPs, which are produced by *Streptomyces* species VITSTK7, show antifungal properties. The antifungal index lies approximately 74% in response to *A. fumigatus*, 66% in response to *A. niger* and 63% against *Aspergillus flavus* (Thenmozhi et al. 2013). AgNPs, which are produced from *Streptomyces* sp. VITPK1, are observed to possess properties like optimum anti-candidal activity against *Candida tropicalis*, *Candida albicans,* and *Candida krusei,* having the inhibition areas lying from approximately 16–20 mm respectively (Sanjenbam et al. 2014). AuNPs are produced from *Streptomyces* sp. VITDDK3 showed antifungal properties in reply to *Tricophyton rubrum* and *Microphyton gypseum,* forming inhibition zones that lies from 11 to 14 mm, respectively. NPs are produced and have high efficacy against many fungi that are human pathogens, like *Saccharomyces cerevisiae*, *C. tropicalis*, *C. krusei*, *C. albicans,* and various groups of the genus *Aspergillus,* including *A. fumigatus*, *A. flavus*, *A. niger*, *Aspergillus brasiliensis,* and *A. terreus*. The synthesized NPs possessed properties against dermatophytes such as *Scedosporium* sp., *Trichophyton rubrum*, *Trichophyton tonsurans,* and *Ganoderma* sp. (Chauhan et al. 2013; Manivasagan et al. 2013).

9.9 Actinomycetes NPs with Antiparasitic Activity

Formation of AgNPs is done by *Actinomycetes* or *Streptomyces* sp., and it has been suggested to have the anti-parasitic property against *Haemaphysalis bispinosa* and *Rhipicephalus microplus* (Karthik et al. 2014) (Figure 9.2).

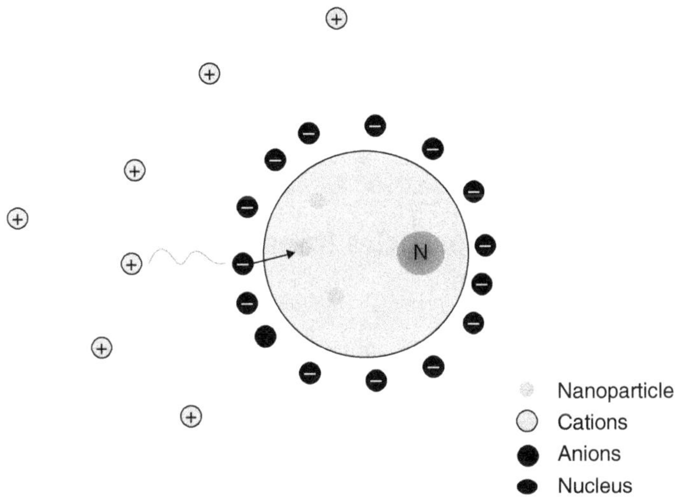

Figure 9.2 Diagrammatic representation of intracellular production of NPs.

9.10 Applications of Actinomycetes NPs Based on Biological Grounds

Antioxidants possess major functions in providing protection to cells against cell damage by neutralizing the free radicals. Some of the most effective free radical scavengers are the metal oxides and metals. These results have an important function in different medical arenas. For example, peroxidations of lipids are initiated by free radicals. This is basically decomposition of lipids oxidatively causing cell damage. AuNPs are produced from *Streptomyces* sp. NK52 showing nearly 46% inhibition in peroxidation of lipids. Production of AgNPs also takes place by *Nocardiopsis* species MBRC-1, and a very high cytotoxic property is observed in response to cervical cancer cell lines of humans. A further 49% viability of inhibition zones is seen at 198 log CFU/mL of the nanoparticle (Manivasagan et al. 2013). On the other hand, marine *Nocardiopsis* sp. MBRC-48 synthesizes AuNPs that are observed to give better reduction power than normal ascorbic acid samples. It again shows the 68% of free radical scavenging property or DPPH activity (Manivasagan et al. 2015).

Zinc and copper nanoparticles are produced by the usage of the *Streptomyces* species that are covered with about 98% cotton fabrication by pad-dry-cure process (Ghosh 2018). The prepared fabrication was kept on a bacteriostasis agar plate, which had undergone swabbing in the bacterial mediums. The zinc NPs were observed to provide a bacteriostatic zone of 28 and 23 mm after 24 hours for *S. aureus* and *E. coli* and for the copper NPs, it was observed that the bacteriostatic zone was 29 and 26 mm after 24 hours for *S. aureus* and *E. coli*. Produced fabrics may be utilized by hospitals for the prevention or minimization of infections (Usha et al. 2010). AgNPs, which are taken around 25 μg approximately, are produced from *Streptomyces* sp., and it is seen that a 21 mm inhibition zone is obtained in response to beta-lactamases having extended spectrums or ATCC 700603, which is the

ESBL pathogen *K. pneumoniae*. Production of AuNPs from *Streptomyces* sp. LK-3 was taken for the treatment of *Plasmodium berghei* PbA or ANKA-infected mouse and observed that percentage of survival has increased to about 84% when it is compared to the standard, which has a survival percentage of almost half of the total percentage in eight days post infection (Karthik et al. 2013).

9.11 Synthesis of Biogenic Nanoparticles from Actinomycetes-derived Compounds

The reports of biogenetically synthesized NPs using actinomycetes and their derived compounds are rare. Ultrafine metal particles having 10 to a few hundreds atoms and smaller than 100 nm are generally called NPs (Jeevanandam et al. 2018). Many bacteria and fungi have been investigated to successfully produce various metallic NPs (Cu, Ag, Zn, Au, Mg, and Ti) (Reza Ghorbani et al. 2011; Rai et al. 2009), and this biogenic synthesis has also been reported by many research papers with antimicrobial property (Ahmad et al. 2003b; Gade et al. 2008; Ingle et al. 2008; Kalimuthu et al. 2008). The biogenic synthesis of metal NPs gained popularity because of its striking physicochemical properties, as well as the formation of stable, size-administered, and scattered NPs (Kim et al. 2007; Morones et al. 2005; Salata 2004), caused by the depletion of metal ion concentration when using the biological systems (Golinska et al. 2014).

The formation of metal NPs can take place either inside the cell or outside. It has been reported that in the case of actinomycetes, the metal ion concentration gets reduced in the cytosol of the cell on the surface of mycelia along with the plasma membrane, which ultimately proceeds to the synthesis of NPs (Ahmad et al. 2003a). Another research concluded the entrapment of the silver (Ag^+) ions on the actinomycetes cell surfaces possibly by the coulombic forces between the metal ions and the negative carboxyl groups present in the enzymes of mycelium cell wall as the possible method for the synthesis (Sunitha et al. 2014). It said that silver nuclei are formed as the silver ions get reduced by the enzymes present in the cell wall and grow as they further accumulate in the cytosol. AgNPs were also synthesized using lactic acid bacteria (Sintubin et al. 2009).

A group of scientists have proposed the speculative production of AgNPs extracellularly from *Streptomyces* sp. (Karthik et al. 2014). This process, where the nitrate is reduced to nitrite by the nitrate reductase enzyme, which is also involved in the nitrogen cycle (Durán et al. 2005), is known for the vital NADH-dependent enzyme, which plays an important role in the bioreduction of the metal ions and subsequently forms NPs (Golinska et al. 2014). The silver ions present are reduced by the process of electron shuttle enzymatic reduction reaction. A similar mechanistic pathway is followed by the synthesis of AuNPs, where the gold ions are reduced from the aqueous solution of $AgCl_4$ utilizing NADH (Shah et al. 2012; Waghmare et al. 2011) (Figure 9.3).

A certain *Streptomyces coelicolor* produces a bluish actinorhodin pigment, which was used for the synthesis of AgNPs (after it was isolated from the cells by centrifugation) using the photo-irradiation method (Manimaran and Kannabiran 2017). These NPs were found to be resistant against the methicillin-resistant *Staphylococcus aureus* (MRSA) (Manikprabhu and Lingappa 2013). *Gordonia amicalis* was cultivated on a medium with

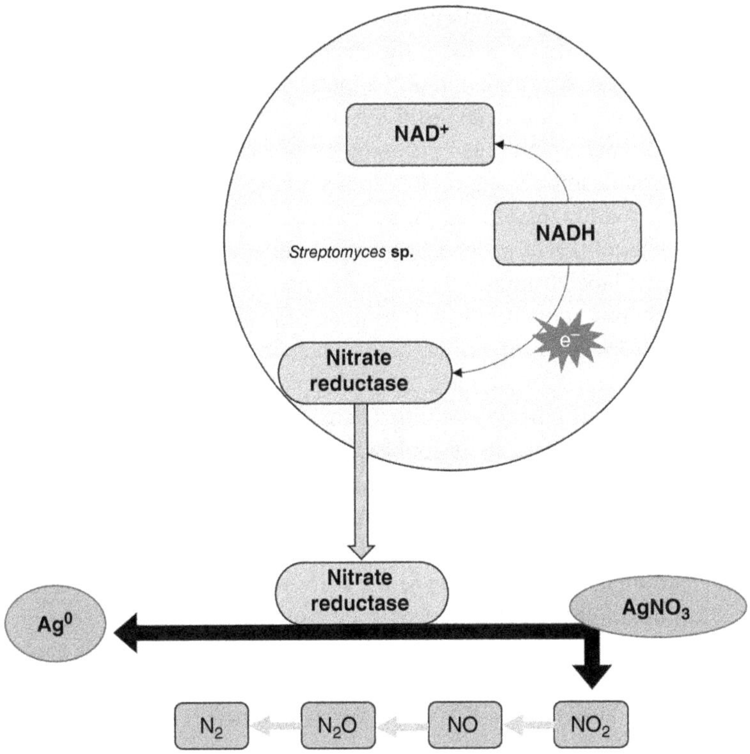

Figure 9.3 Hypothetical mechanism of the synthesis of AgNPs.

n-hexadecane as the primary source of carbon and esters of palmitate glycolipid and thermostable stearate along with myristate were extracted from it. These compounds were then used to produce metallic AuNPs and AgNPs, and this process of synthesis of the NPs showed approximately 90 and 95% hydroxyl radical scavenging activity (Ghosh et al. 2016; Manimaran and Kannabiran 2017) and approximately 75% and 68% nitric acid scavenging activity (Sowani et al. 2016), respectively. Thus, it is evident that the NPs synthesized using *Actinomycetes*-derived compounds have greater efficiency than the conventional ones. An individual strain of *Actinomycetes* has the capability to produce up to twenty secondary metabolites (Lam 2006), though its metal and metal oxide reducing potential for the synthesis of NPs and its possible pathway need to be studied.

9.12 Challenges of Using Metal and Metal Oxide Nanoparticles

The NPs synthesized by the biogenic method using *Actinomycetes* are more efficient than those produced using the conventional methods and have wide applications in various industries. However, there are some disadvantages of these biogenetically produced NPs that intercepts their use in certain applications. The NPs interact with the proteins present in our body in order to get distributed throughout the body by the process of opsonisation, which also has a role in altering the properties of the desired nanoparticle (Tsibakhashvili et al. 2011). Agglomeration is the accumulation of the metal particles due to the adhesion of particles to each other by weak forces, giving rise to micro-sized entities. The synthesized NPs have high surface energy and thus to prevent the process of agglomeration and protect the NPs, effective stabilization is required. Various methods have been adopted to eliminate the interactions of different cellular components like macrophages and other leukocytes present in the mononuclear phagocytic system, which tends to alter the nanoparticle properties during the biodistribution. The NPs are encased with a layer of polyethylene glycol (PEG), which has proven effective against protecting the NPs from the alterations by reducing the unwanted recognition and thus increasing half-life of circulation of the NPs (Manimaran and Kannabiran 2017; Owens and Peppas 2006). The NPs having greater solubility help in its biocompatibility. There are certain causes of toxicity of the NPs which also have a drastic effect on its applicability, though none of the NPs were extensively tested for their toxicity. In order to ensure a wide variety of the nanoparticle; they should not be toxic. Toxicity of the NPs can be caused by the internal properties of the particles, like the chemical structure, surface composition and stability, morphology, ionic charge, etc., or by the core metal or metal oxide or the surface coating. Also, the metallic NPs can secrete out harmful ions that make it toxic (Seabra and Duran 2015).

In order to ensure the industrial usage of the NPs, the process of their synthesis should be scaled up, but this is very difficult in practice. Moreover, the mechanistic pathway for the biogenic synthesis (green synthesis) of the NPs using the *Actinomycetes* is not clear to the full extent to date, and further improvements are required to make the synthesis reproducible. Another major disadvantage of these NPs is that there is much to be explored about the effect and result of the synthesized NPs on actual targets, having a grip on its biodistribution and bioavailability by improving the control over their size. Thus, there must be detailed studies, which are left to be done on the metal NPs synthesized from *Actinomycetes* and/or its derived compounds in terms of composition, toxicity to human and other living beings, and chemical stability, which are some of the major challenges we face today. These studies might make the nanoparticle synthesis easier and will help to make them safe and ideal to be used for a wide variety of purposes.

9.13 Conclusions

The synthesis of NPs by biogenic agents has been an eco-friendly, easily scalable, and high-productive technique; thus, it is playing an important role in replacing the conventional technique of synthesis. Developments of new therapeutics are important need of the hour to act against various microbial pathogens and various antibiotic-resistant groups of organisms. The importance of actinomycetes in the synthesis of NPs is increasing due to their ability to produce secondary metabolites having various types of biological applications. The NPs produced from actinobacteria possess high yield and significant purity. They are the potent biogenic source to develop nanoantibiotics and nanoantimicrobials that can be used as alternate therapeutic agents. Although there are researches in the field of NPs synthesis by the use of actinomycetes, but the development of metallic-NPs is still less, and more work needs to be accomplished in this field. The NPs developed by the use of actinomycetes are biocompatible and hence find their application in pharmaceutical and medical fields. In the last decades, there have been significant studies on the production of NPs by actinobacteria, but very few studies have been performed pertaining in scaling up the production. Future studies are needed for the purpose to scaling up the production of NPs. The NPs that are synthesized are less in comparison to the various physiochemical routes. The efficiency of the synthesis needs to be further improved. Along with this, various optimization studies need to be performed for achieving the enhancement in the production of NPs.

References

Abdeen, S., Isaac, R. R., Sornalekshmi, S., and Praseetha, P. K. (2013). Evaluation of antimicrobial activity of biosynthesized iron and silver nanoparticles using the fungi Fusarium Oxysporum and actinomycetes sp. on human pathogens. *Nano Biomedicine & Engineering* 5(1).

Abdeen, S., Geo, S., Sukanya, S., Praseetha, P. K., and Dhanya, R. P. (2014). Biosynthesis of silver nanoparticles from actinomycetes for therapeutic applications. *International Journal of Nano Dimension* 5 (2): 155–162.

Abd-Elnaby, H.M., Abo-Elala, G.M., Abdel-Raouf, U.M., and Hamed, M.M. (2016). Antibacterial and anticancer activity of extracellular synthesized silver nanoparticles from marine *Streptomyces rochei* MHM13. *Egyptian Journal of Aquatic Research* 42: 301–312.

Ahmad, A., Senapati, S., Khan, M.I. et al. (2003a). Intracellular synthesis of gold nanoparticles by a novel alkalotolerant actinomycete, *Rhodococcus* species. *Nanotechnology* 14: 824.

Ahmad, A., Senapati, S., Khan, M.I. et al. (2003b). Extracellular biosynthesis of monodisperse gold nanoparticles by a novel extremophilic actinomycete *Thermomonospora* sp. *Langmuir* 19: 3550–3553.

Ahmad, F., Ashraf, N., Ashraf, T. et al. (2019). Biological synthesis of metallic nanoparticles (MNPs) by plants and microbes: their cellular uptake, biocompatibility, and biomedical applications. *Applied Microbiology and Biotechnology* 103: 2913–2935.

Balagurunathan, R., Radhakrishnan, M., Rajendran, R.B., and Velmurugan, D. (2011). Biosynthesis of gold nanoparticles by actinomycete *Streptomyces viridogens* strain HM10. *Indian Journal of Biochemistry and Biophysics* 48: 331–335.

Buszewski, B., Railean-Plugaru, V., Pomastowski, P. et al. (2016). Antimicrobial activity of biosilver nanoparticles produced by a novel *Streptacidiphilus durhamensis* strain. *Journal of Microbiology, Immunology and Infection* 51 (1): 45–54.

Chau, C., Wu, S., and Yen, G. (2007). The development of regulations for food nanotechnology. *Trends in Food Science and Technology* 18: 269–280.

Chauhan, R., Kumar, A., and Abraham, J. (2013). A biological approach to the synthesis of silver nanoparticles with *Streptomyces* sp. JAR1 and its antimicrobial activity. *Scientia Pharmaceutica* 81: 607–621.

Daniel, M. and Astruc, D. (2004). Gold nanoparticles: assembly, supramolecular chemistry, quantum-size-related properties, and applications toward biology, catalysis, and nanotechnology. *Chemical Reviews* 104: 293–346.

Deepa, S., Kanimozhi, K., and Panneerselvam, A. (2013). Antimicrobial activity of extracellularly synthesized silver nanoparticles from marine derived actinomycetes. *International Journal of Current Microbiology and Applied Sciences* 2 (9): 223–230.

Durán, N., Marcato, P.D., Alves, O.L. et al. (2005). Mechanistic aspects of biosynthesis of silver nanoparticles by several *Fusarium oxysporum* strains. *Journal of Nanbiotechnology* 3: 8.

Elamawi, R.M., Al-Harbi, R.E., and Hendi, A.A. (2018). Biosynthesis and characterization of silver nanoparticles using *Trichoderma longibrachiatum* and their effect on phytopathogenic fungi. *Egyptian Journal of Biological Pest Control* 28: 1–11.

Gade, A.K., Bonde, P., Ingle, A.P. et al. (2008). Exploitation of *Aspergillus niger* for synthesis of silver nanoparticles. *Journal of Biobased Materials and Bioenergy* 2: 1–5.

Ghosh, S. (2018). Copper and palladium nanostructures: a bacteriogenic approach. *Applied Microbiology and Biotechnology* 101 (18): 7693–7701.

Ghosh, S., Patil, S., Chopade, N.B. et al. (2016). Gnidia glauca leaf and stem extract mediated synthesis of gold nanocatalysts with free radical scavenging potential. *Journal of Nanomedicine and Nanotechnology* 7: 358.

Golinska, P., Wypij, M., Ingle, A.P. et al. (2014). Biogenic synthesis of metal nanoparticles from actinomycetes: biomedical applications and cytotoxicity. *Applied Microbiology and Biotechnology* 98 (19): 8083–8097.

He, S., Guo, Z., Zhang, Y. et al. (2007). Biosynthesis of gold nanoparticles using the bacteria *Rhodopseudomonas capsulata*. *Materials Letters* 61: 3984–3987.

Ingle, A.P., Gade, A.K., Pierrat, S. et al. (2008). Mycosynthesis of silver nanoparticles using the fungus *Fusarium acuminatum* and its activity against some human pathogenic bacteria. *Current Nanoscience* 4: 141–144.

Ingle, A., Rai, M., Gade, A., and Bawaskar, M. (2009). *Fusarium solani*: a novel biological agent for the extracellular synthesis of silver nanoparticles. *Journal of Nanoparticle Research* 11: 2079–2085.

Jeevanandam, J., Barhoum, A., Chan, Y.S. et al. (2018). Review on nanoparticles and nanostructured materials: history, sources, toxicity and regulations. *Beilstein Journal of Nanotechnology* 9: 1050–1074.

Kalabegishvili, T., Kirkesali, E., Ginturi, E. et al. (2013). Synthesis of gold nanoparticles by new strains of thermophilic actinomycetes. *Nano Studies* 7: 255–260.

Kalimuthu, K., Babu, R.S., Venkataraman, D. et al. (2008). Biosynthesis of silver nanocrystals by *Bacillus licheniformis*. *Colloids and Surfaces B: Biointerfaces* 65: 150–153.

Kamel, Z., Saleh, M., and El Namoury, N. (2016). Biosynthesis, characterization, and antimicrobial activity of silver nanoparticles from actinomycetes. *Research Journal of Pharmaceutical Biological and Chemical Sciences* 7 (1): 119–127.

Karthik, L., Kumar, G., Keswani, T. et al. (2013). Marine actinobacterial mediated gold nanoparticles synthesis and their antimalarial activity. *Nanomedicine: Nanotechnology, Biology and Medicine* 9: 951–960.

Karthik, L., Kumar, G., Vishnu Kirthi, A. et al. (2014). *Streptomyces* sp. LK3 mediated synthesis of silver nanoparticles and its biomedical application. *Bioprocess and Biosystems Engineering* 37: 261–267.

Kim, J.S., Kuk, E., Yu, K.N. et al. (2007). Antimicrobial effects of silver nanoparticles. *Nanomedicine: Nanotechnology, Biology and Medicine* 3: 95–101.

Kiran, G.S., Sabu, A., and Selvin, J. (2010). Synthesis of silver nanoparticles by glycolipid biosurfactant produced from marine *Brevibacterium casei* MSA 19. *Journal of Biotechnology* 148: 221–225.

Klaus-Joerger, T., Joerger, R., Olsson, E., and Granqvist, C. (2001). Bacteria as workers in the living factory: metal accumulating bacteria and their potential for materials science. *Trends in Biotechnology* 19: 15–20.

Kumar, A., Kaur, K., and Sharma, S. (2013). Synthesis, characterization and antibacterial potential of silver nanoparticles by *Morus nigra* leaf extract. *Indian Journal of Pharmaceutical and Biological Research* 1: 16–24.

Kumari, S., Tehri, N., Gahlaut, A., and Hooda, V. (2020). Actinomycetes mediated synthesis, characterization, and applications of metallic nanoparticles. *Inorganic and Nano-Metal Chemistry* 1–10.

Lam, K.S. (2006). Discovery of novel metabolites from marine actinomycetes. *Current Opinion in Biotechnology* 9: 245–251.

Li, Z., Jin, R., Mirkin, C.A., and Letsinger, R.L. (2002). Multiple thiol-anchor capped DNA-gold nanoparticle conjugates. *Nucleic Acids Research* 30: 1558–1562.

Li, X., Xu, H., Chen, Z., and Chen, G. (2011). Biosynthesis of nanoparticles by microorganisms and their applications. *Journal of Nanomaterials* 2011: 270974.

Manikprabhu, D. and Lingappa, K. (2013). Antibacterial activity of silver nanoparticles against methicillin-resistant *Staphylococcus aureus* synthesized using model *Streptomyces* sp. pigment by photo-irradiation method. *Journal of Pharmacy Research* 6: 255–260.

Manimaran, M. and Kannabiran, K. (2017). Actinomycetes-mediated biogenic synthesis of metal and metal oxide nanoparticles: progress and challenges. *Letters in Applied Microbiology* 64 (6): 401–408. https://doi.org/10.1111/lam.12730.

Manivasagan, P., Venkatesan, J., Senthilkumar, K. et al. (2013). Biosynthesis, antimicrobial and cytotoxic effect of silver nanoparticles using a novel *Nocardiopsis* sp. MBRC-1. *BioMed Research International* 2013: 287638. (9 pages). https://doi.org/10.1155/2013/287638.

Manivasagan, P., Alam, M.S., Kang, K. et al. (2015). Extracellular synthesis of gold bionanoparticles by *Nocardiopsis* sp. and evaluation of its antimicrobial, antioxidant and cytotoxic activities. *Bioprocess and Biosystems Engineering* 38: 1167–1177.

Morones, J.R., Elechiguerra, J.L., Camacho, A., and Ramirez, J.T. (2005). The bactericidal effect of silver nanoparticles. *Nanotechnology* 16: 2346–2353.

Mukherjee, P., Ahmad, A., Mandal, D. et al. (2001). Fungus-mediated synthesis of silver nanoparticles and their immobilization in the mycelial matrix: a novel biological approach to nanoparticle synthesis. *Nano Letters* 1: 515–519.

Mukherjee, P., Senapati, S., Mandal, D. et al. (2002). Extracellular synthesis of gold nanoparticles by the fungus *Fusarium oxysporum*. *ChemBioChem* 3: 461–463.

Nair, B. and Pradeep, T. (2002). Coalescence of nanoclusters and formation of submicron crystallites assisted by *Lactobacillus strains*. *Crystal Growth & Design* 2: 293–298.

Narasimha, G., Janardhan, A., Alzohairy, M. et al. (2013). Extracellular synthesis, characterization and antibacterial activity of silver nanoparticles by actinomycetes isolative. *International Journal of Nano Dimension* 4: 77–83.

Narayanan, K.B. and Sakthivel, N. (2010). Biological synthesis of metal nanoparticles by microbes. *Advances in Colloid and Interface Science* 156: 1–13.

Naveen, K.S.H., Kumar, G., Karthik, L., and Rao, K.V.B. (2010). Extracellular biosynthesis of silver nanoparticles using the filamentous fungus *Penicillium* sp. *Archives of Applied Sciences* 6: 161–167.

Neelu, N., Aigle, B., Mandal, A. et al. (2013). Actinomycetes: role in biotechnology and medicine. *BioMed Research International* 2013: 687190, 1 page. https://doi.org/10.1155/2013/687190.

Otari, S.V., Patil, R.M., Nadaf, N.H. et al. (2012). Green biosynthesis of silver nanoparticles from an actinobacteria *Rhodococcus* sp. *Materials Letters* 72: 92–94.

Owens, D.E. III and Peppas, N.A. (2006). Opsonization, biodistribution, and pharmacokinetics of polymeric nanoparticles. *International Journal of Pharmaceutics* 307: 93–102.

Oza, G., Pandey, S., Gupta, A. et al. (2012). Biosynthetic reduction of gold ions to gold nanoparticles by *Nocardia farcinica*. *Journal of Microbiology and Biotechnology Research* 4: 511–515.

Pattekari, P., Zheng, Z., Zhang, X. et al. (2011). Top-down and bottom-up approaches in production of aqueous nanocolloids of low solubility drug paclitaxel. *Physical Chemistry Chemical Physics* 13: 9014–9019.

Płaza, G.A., Chojniak, J., and Banat, I.M. (2014). Biosurfactant mediated biosynthesis of selected metallic nanoparticles. *International Journal of Molecular Sciences* 15: 13720–13737.

Prakasham, R.S., Buddana, S.K., Yannam, S.K., and Guntuku, G.S. (2012). Characterization of silver nanoparticles synthesized by using marine isolate *Streptomyces albidoflavus*. *Journal of Microbiology and Biotechnology* 22: 614–621.

Prakasham, R.S., Kumar, B.S., Kumar, Y.S., and Kumar, P.K. (2014). Production and characterization of protein encapsulated silver nanoparticles by marine isolate *Streptomyces parvulus* SSNP11. *Indian Journal of Microbiology* 54: 329–336.

Rai, M., Yadav, A., and Gade, A. (2009). Silver nanoparticles as a new generation of antimicrobials. *Biotechnology Advances* 27: 76–83.

Reza Ghorbani, H., Akbar Safekordi, A., Attar, H., and Rezayat Sorkhabadib, S.M. (2011). Biological and non-biological methods for silver nanoparticles synthesis. *Chemical and Biochemical Engineering Quarterly* 25: 317–326.

Saifuddin, N., Wang, W.C., and Yasumira, A.A.N. (2009). Rapid biosynthesis of silver nanoparticles using culture supernatant of bacteria with microwave irradiation. *Journal of Chemistry* 6: 61–70.

Salata, O. (2004). Applications of nanoparticles in biology and medicine. *Journal of Nanbiotechnology* 2: 1–6.

Sanjenbam, P., Gopal, J.V., and Kannabiran, K. (2014). Anticandidal activity of silver nanoparticles synthesized using *Streptomyces* sp. VITPK1. *Journal of Medical Mycology* 24: 211–219.

Sanvicens, N. and Marco, M.P. (2008). Multifunctional nanoparticles-properties and prospects for their use in human medicine. *Trends in Biotechnology* 26: 425–433.

Sastry, M., Ahmed, A., Khan, M.I., and Kumar, R. (2003). Biosynthesis of metal nanoparticles using fungi and actinomycetes. *Current Nanoscience* 85 (2): 162–170.

Seabra, A.B. and Duran, N. (2015). Nanotoxicology of metal oxide nanoparticles. *Metals* 5: 934–975.

Selvakumar, P., Viveka, S., Prakash, S. et al. (2012). Antimicrobial activity of extracellularly synthesized silver nanoparticles from marine derived *Streptomyces rochei*. *International Journal of Pharma and Bio Sciences* 3: 188–197.

Shah, R., Oza, G., Pandey, S., and Sharon, M. (2012). Biogenic fabrication of gold nanoparticles using Halomonas salina. *J Microbiol Biotechnol Res* 2 (4): 485–492.

Singh, R., Pantarotto, D., Lacerda, L. et al. (2006). Tissue biodistribution and blood clearance rates of intravenously administered carbon nanotube radiotracers. *Proceedings of the National Academy of Sciences of the United States of America* 103: 3357–3362.

Singh, D., Rathod, V., Fatima, L. et al. (2014). Biologically reduced silver nanoparticles from Streptomyces sp. VDP-5 and its antibacterial efficacy. *International Journal of Pharmacy and Pharmaceutical Sciences* 4: 31–36.

Sintubin, L., De Windt, W., Dick, J. et al. (2009). Lactic acid bacteria as reducing and capping agent for the fast and efficient production of silver nanoparticles. *Applied Microbiology and Biotechnology* 84: 741–761.

Sivalingam, P., Antony, J.J., Siva, D. et al. (2012). Mangrove *Streptomyces* sp. BDUKAS10 as nanofactory for fabrication of bactericidal silver nanoparticles. *Colloids and Surfaces, B: Biointerfaces* 98: 12–17.

Składanowski, M., Wypij, M., Laskowski, D. et al. (2017). Silver and gold nanoparticles synthesized from *Streptomyces* sp. isolated from acid forest soil with special reference to its antibacterial activity against pathogens. *Journal of Cluster Science* 28: 59–79.

Song, J.Y. and Kim, B.S. (2008). Biological synthesis of bimetallic Au/Ag nanoparticles using persimmon (*Diopyros kaki*) leaf extract. *Korean Journal of Chemical Engineering* 25: 808–811.

Sowani, H., Mohite, P., Munot, H. et al. (2016). Green synthesis of gold and silver nanoparticles by an actinomycete *Gordonia amicalis* HS-11: mechanistic aspects and biological application. *Process Biochemistry* 51: 374–383.

Subashini, J. and Kannabiran, K. (2013). Antimicrobial activity of *Streptomyces* sp. VITBT7 and its synthesized silver nanoparticles against medically important fungal and bacterial pathogens. *Der Pharmacia Lettre* 5: 192–200.

Sukanya, M.K., Saju, K.A., Praseetha, P.K., and Sakthivel, G. (2013). Therapeutic potential of biologically reduced silver nanoparticles from actinomycete cultures. *Journal of Nanoscience* 2013: 940719.

Sundarmoorthi, E., Devarasu, S., and Vengadesh Prabhu, K. (2011). Antimicrobial and wound healing activity of silver nanoparticles synthesized from Streptomyces aureofaciens. *International Journal of Pharmaceutical Research and Development* 12: 69–75.

Sunitha, A., Geo, S., Sukanya, S. et al. (2014). Biosynthesis of silver nanoparticles from actinomycetes for therapeutic applications. *International Journal of Nano Dimension* 5: 155–162.

Thenmozhi, M., Kannabiran, K., Kumar, R., and Khanna, V.G. (2013). Antifungal activity of *Streptomyces* sp. VITSTK7 and its synthesized Ag_2O/Ag nanoparticles against medically important *Aspergillus* pathogens. *Journal of Medical Mycology* 23: 97–103.

Tsibakhashvili, N.Y., Kirkesali, E.I., Pataraya, D.T. et al. (2011). Microbial synthesis of silver nanoparticles by *Streptomyces glaucus* and *Spirulina platensis*. *Advanced Science Letters* 4: 3408–3417.

Udaya Prakash, N.K., Bhuvaneswari, S., Prabha, S.B. et al. (2014). Green synthesis of silver nanoparticles using airborne. *International Journal of ChemTech Research* 6: 4123–4127.

Usha, R., Prabu, E., Palaniswamy, M. et al. (2010). Synthesis of metal oxide nanoparticles by *Streptomyces* sp for development of antimicrobial textiles. *Global Journal of Biochemistry and Biotechnology* 5: 153–160.

Vidyasagar, G.M., Shankaravva, B., Begum, R., and Imrose, R.R.L. (2012). Antimicrobial activity of silver nanoparticles synthesized by *Streptomyces* species JF714876. *International Journal of Pharmaceutical Sciences and Nanotechnology* 5: 1638–1642.

Vimal, V., Rajan, B.M., and Kannabiran, K. (2009). Antimicrobial activity of marine actinomycete, *Nocardiopsis* sp. VITSVK 5 (FJ973467). *Asian Journal of Medical Sciences* 1: 57–63.

Waghmare, S.S., Deshmukh, A.M., Kulkarni, W., and Oswaldo, L.A. (2011). Biosynthesis and characterization of manganese and zinc nanoparticles. *Universal Journal of Environmental Research and Technology* 1: 64–69.

Zarina, A., and Nanda, A. (2014). Green approach for synthesis of silver nanoparticles from marine Streptomyces-MS 26 and their antibiotic efficacy. *Journal of Pharmaceutical Sciences and Research* 6 (10): 321.

Zonooz, N.F. and Salouti, M. (2011). Extracellular biosynthesis of silver nanoparticles using cell filtrate of *Streptomyces* sp. ERI-3. *Scientia Iranica* 18: 1631–1635.

10

Nanotechnology for Genetically Modified Organisms in Food Production

Rahul Nitnavare[1,2], *Joorie Bhattacharya*[3,4], *and Sougata Ghosh*[5]

[1] *Division of Plant and Crop Sciences, School of Biosciences, University of Nottingham, Sutton Bonington, Leicestershire, United Kingdom*
[2] *Department of Plant Sciences, Rothamsted Research, Harpenden, Hertfordshire, United Kingdom*
[3] *Genetic Gains, International Crops Research Institute for the Semi-Arid Tropics, Hyderabad, Telangana, India*
[4] *Department of Genetics, Osmania University, Hyderabad, Telangana, India*
[5] *Department of Microbiology, School of Science, RK University, Rajkot, Gujarat, India*

10.1 Introduction

The field of nanotechnology has gained immense popularity in various fields, holding economic significance and allowing for the manipulation, regulation, and deployment of nanomaterials. Nanomaterials are derived from nanotechnology and possess certain characteristics such as high surface-to-volume ratio, optical properties, and a size between 10 and 1,000 nm (Karmakar et al. 2020). They are usually processed in the form of nanosheets, nanotubes, nanofibers, nanofilms, and nanowires (Robkhob et al. 2020). In recent times, a significant increase in commercial nanomaterials has seen a rise in the form of carbon-based nanoparticles and metal nanoparticles in the field of pharmaceuticals, microelectronics, cosmetics, molecular catalysts and probes, pesticides, and wastewater remediation (Ghosh et al. 2016a–d; Rokade et al. 2017; Rokade et al. 2018). In the agriculture and food production industry, biogenic nanomaterials are being expansively produced for purposes such as pesticides and herbicides, soil sensors and monitors, and agrichemical delivery (Jamdade et al. 2019; Shende et al. 2017; Shende et al. 2018; Bhagwat et al. 2018; Shinde et al. 2018). Engineered nanomaterials (ENs) are primarily being utilized for the generation of genetically modified crops (GMCs).

Nanotechnology in agriculture is being used to enhance food production and yield and improve the quality of fuel. Nanomaterials interact at the genetic level in the plant cell, causing modifications at specific levels in the gene expression and interfering with biological pathways, subsequently regulating plant growth and yield. Such nanoparticles have varying effects on different crops at differing developmental stages and are also dependent on the shape, size, solubility, and composition of the nanomaterial used (Adersh et al. 2015; Ghosh 2018). The physical and chemical properties of nanomaterials are such that the particles are able to easily transmit through the cell wall of the plant system. ENs have the unique ability to modify their characteristics according to the physiochemical properties and needs of the plant system. For an efficient system, nanomaterials must be designed to have specific size, shape, and chemical properties, and most importantly, stability, specific targeting, and safe mode of delivery in the plant (Ghosh et al. 2015a–d; Kitture et al. 2015; Salunke et al. 2014; Sant et al. 2013). However, the technology has its share of drawbacks. Since they have a high contact surface ratio, they can often have deleterious effects upon accumulation (Joshi et al. 2019). Moreover, since they have a wider application, they often get introduced into the soil and aquatic ecosystems unknowingly interacting with the living species in them. The effects of nanomaterials on biological systems are not extensive; however, certain factors need to be taken into consideration during the commercialization of nanomaterials. It must be ensured that the concentration and quantity of nanomaterials are kept minimal up to a threshold to negate any possible deleterious effects.

Additionally, they must also be enclosed with any chemical or physical material to ensure the unavailability of free nanomaterials. ENs can remain suspended in air for days and are taken up by the plant's respiratory system readily. Due to this, they accumulate in several regions of plants, which might prove to be deleterious. Additionally, nanoparticles pose the issue of bioaccumulation, and with the lack of thorough research on the health risks of such particles, a generic assessment is of immense requirement (Aslani et al. 2014). The plant system differs from the animal system in terms of the presence of a cell wall, which does not allow easy transport of large particles across the membrane. However, owing to the small size of nanoparticles, they are often able to penetrate the cell wall and accumulate in various regions of the plant. Further, the nanoparticles, which are larger in size as compared to the pores on the plant cell wall, accumulate on the surface of the plant and cause obstruction in the movement of essential electrolytes and minerals (Figure 10.1).

(a) (b)

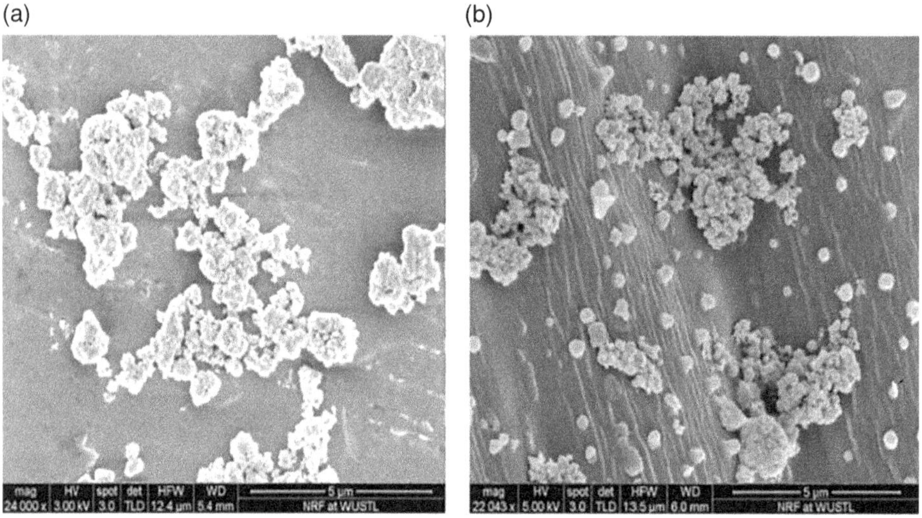

Figure 10.1 Scanning electron microscope images for NPs/lettuce seeds. In the aqueous phase, the SEM image shows that (a) TiO_2 NPs (1,000 mg L^{-1}) and (b) CuO NPs (1,000 mg L^{-1}) were adsorbed on the seed surface. Reprinted from Aslani, F., Bagheri, S., Muhd Julkapli, N. et al. (2014). Effects of engineered nanomaterials on plants growth: an overview. *Scientific World Journal* 2014: 641759

TiO_2 nanoparticles (25 nm size) were unable to cross the cell wall and therefore accumulated on the root surface, causing hindrance to hydraulic conductivity and water availability, leading to reduced transpiration rate and eventually affecting the plant growth. Apart from size, shape, and chemical composition, the uptake of engineered nanoparticles is also dependent on the plant species. For example, positively charged nanoparticles (NPs) can be taken up readily by the root system while negatively charged NPs are transported to the stem and leaves from roots. Due to the dynamic nature of NPs, they interact promptly with the environment and its biological components. Metal oxides of NPs release the metal ions and transform by reacting with organic and inorganic components present in various biological systems. Certain heavy metal ions also react with root exudates and chelate limiting their uptake by the roots. Hence, the different biological microenvironments have varying effects on the engineered nanoparticles and therefore will exhibit varying levels of toxicity in plants. In the soil system, NPs also undergo reduction, oxidation, and transformation by reagents present in plants. Compared to the extensive studies performed on the interaction between plants and metal nanoparticles, less data is available on carbon-based or polymer nanomaterials.

This is due to a lack of appropriate detection systems and also the high background of plant matrix. However, in a few studies, there are reports of degradation of carbon-based NPs by peroxides (Zhang et al. 2015). In plant tissue culture, there have been several positive outcomes from the usage of nanotechnology in seed germination, crop yield and growth, genetic modifications, bioactive compound production, and plant protection. It was also observed that ENs reduce microbial contamination in tissue culture media and enhance the potential of explants along with exhibiting somaclonal variation, callus induction, secondary metabolite production, and organogenesis. Various studies on numerous crop plants such as potato, tomato, and model crops such as *Arabidopsis thaliana* and *Nicotiana tabacum* have exhibited reduced microbial contamination upon treatment with metal NPs (Kim et al. 2017). Additionally, nanoparticles have been known to increase the abiotic stress tolerance by activating the plant's defense system. For plant genetic engineering, the major obstacle is the entry of the nanoparticle into the plant cell due to the presence of the plant cell wall. To overcome this, various delivery strategies have been implemented, which allow the easy introduction of the NPs such as *Agrobacterium*-mediated transformation or biolistic delivery. There have been several successful studies performed which portray the efficiency of delivery systems using NPs such as magnetic nanoparticles (MNPs) for stable transformation in cotton and nanotube scaffolds for gene expression in *Nicotiana benthamiana* and *Triticum aestivum*. This has helped researchers carry out expression in transgene systems by reducing the drawbacks associated with them. However, limitations to these approaches occur in the narrow range of host plants (Sanzari et al. 2019). Additionally, the biolistic method of gene delivery allows the penetration of exogenous DNA into a wide range of tissues and plants; however, it can be damaging to the plant tissues due to external physical force.

A significant breakthrough in this regard has been the development of mesoporous silica nanoparticles (MSNs), which behave like a carrier molecule for the transfer of DNA in the form of plasmid, dsRNA, or siRNA and proteins into plant cells with the help of biolistic. In this method, the DNA along with a chemical inducer is loaded into the MSN, which is capped with gold nanoparticles (AuNPs) at the end. Once the entire entity is transferred into the plant, the AuNPs are uncapped due to the change in physicochemical environment thus releasing the molecule of interest. This allows for a controlled release of the biogenic molecules into the plant system, which provides an added advantage. MSNs have also been used to deliver DNA into the root structure of *Arabidopsis* plant without using physical impact (Wang et al. 2019a).

The movement of nanoparticles within the plant system is facilitated by two major pathways, viz., apoplast and symplast tissues. Apoplastic transport occurs outside the plasma membrane through the extracellular spaces and xylem vessels, while the symplastic means of transport occurs within the cytoplasmic structures of cells through plasmodesmata. The symplastic pathways also include carrier proteins, such as aquaporins, which bind to NPs and facilitate the internalization and uptake of the NP inside the cell. Also, symplastic mechanisms such as endocytosis and pore formation in the plasma membrane help entry of the NP directly into the cell cytoplasm or specific cell organelles. The movement of the NPs is not restricted to a specific cell type and may occur between xylem and phloem as well. It is therefore extremely important to understand the translocation mechanism of NPs within plants to determine its potential to move toward specific tissues and bioaccumulate. Interestingly, the microorganisms residing in plants might also govern the plant uptake of NPs such as the presence of mycorrhizal fungi. The discourse associated with modern technologies such are often paired with distrust and skepticism. While nanotechnology has accelerated in various fields such as drug industries and medicine, it is still relatively unexplored in agriculture and food production. Research extrapolated into this field through existing studies offers an added complication as NPs tend to react differently with different plant systems, and therefore, a generalized view cannot be provided. The cost-effectiveness is another major aspect that must be taken into consideration. In spite of promising results in studies, the cost of generating large amounts of NPs for contained environments, such as greenhouses, and field trials, is unfeasible. For effective application in agriculture, nanomaterials should be low-priced, and this can be achieved by the use of relatively cheaper materials such as polymers from naturally occurring compounds like chitin from crustaceans and algae. Further, since these NPs would be released directly into the ecological system, the risk factor associated with them must be thoroughly assessed. This can be achieved by using nanomaterials that are nonhazardous in nature. Materials that are biodegradable, like natural polymers, are ideal for this purpose. Lastly, communication is key, and therefore, in order to ensure ease of commercial release of nanotechnology-based agriculture products, the masses should be made familiar with the technique and its associated pros and cons (Pérez-de-Luque 2017).

Nanoparticles in plant genetic engineering can be used for the transient expression of genes in genetically enhanced plants. Such forms of expression are not inherited by the offspring. Also, gene expression is controlled and not integrated directly into the plant genome exempting this technology from the stringent regulations surrounding genetically modified organisms (GMOs). The recent developments in nanotechnology can also be extrapolated into plant genetic studies by RNAi interference technology for studying the functions of various genes. More recently, nanotechnology-mediated CRISPR/Cas9 (clustered regularly interspersed palindromic repeats/CRISPR-associated proteins) cargo is being considered as another approach to generate genetically enhanced plants. MSNs have been successfully used for this purpose for introducing Cre recombinase carrying *loxP* integrated sites in *Zea mays* embryos (Sanzari et al. 2019; Wang et al. 2019a). Nanotechnology

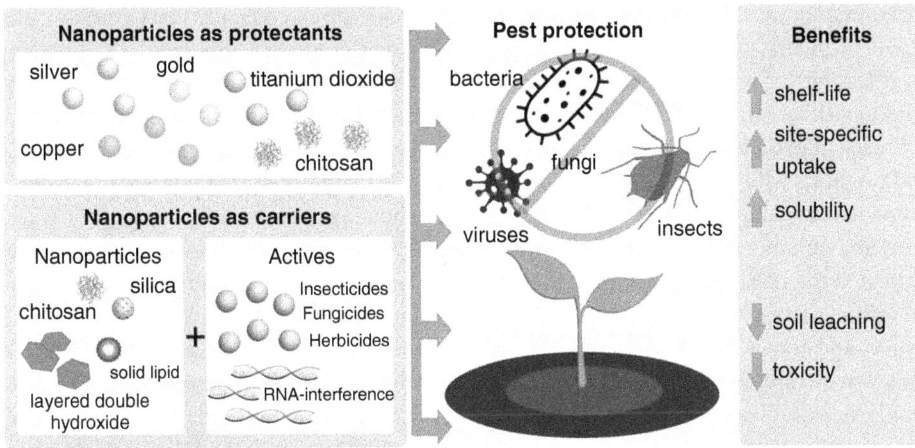

Figure 10.2 Nanomaterials as protectants or carriers crop protection. This schematic shows different nanomaterials as either protectants for actives such as insecticides, fungicides, herbicides, or RNA-interference molecules, targeting a wide range of pests and pathogens. Reprinted from Worrall, E. A., Hamid, A., Mody, K. T. et al. (2018). Nanotechnology for plant disease management. *Agronomy* 8: 285.

has tremendous potential in food production and agriculture. A multidisciplinary approach to the technology would provide a more wholesome and effective use (Figure 10.2). A collaborative initiative merging various disciplines such as genetics, biochemistry, engineering, and physics can deliver robust systems and a more extensive use of smart nanoparticles in the food industry (Giraldo et al. 2019; Sanzari et al. 2019).

10.2 Nanomaterials-Driven Genetic Engineering in Plants

The ENs used for plant-associated genetic engineering are made of various materials based on their biodegradation, accumulation, and uptake efficiency. Phytotoxicity studies have been carried out in various plants such as wheat, soybean, rice, tomato, lettuce, cucumber, corn, etc. The two primarily used nanomaterials are carbon and metal-based nanomaterials in plant systems (Aslani et al. 2014; Zhang et al. 2015). Apart from these, polymeric nanomaterials and hybrid nanomaterials have gained mileage in plant genetic engineering over the years. NPs require properties such as the ability to cross biological membranes including the cell wall, protection and release of appropriate cargoes, as well as specific target delivery (Cunningham et al. 2018).

10.2.1 Carbon-Based Nanomaterials

Carbon nanomaterials (carbon NMs) are being used for engineering in plant systems due to their unique physical properties. They are synthesized in various forms such as carbon nanotubes (CNTs), carbon nanofibers, graphite, fullerenes, and diamond. Due to their high physically tunable properties, CNTs have been used extensively for plant engineering purposes. CNTs are of two types, multiwalled carbon nanotubes (MWCNTs) and single-walled carbon nanotubes (SWCNTs). The MWCNTs are made up of coaxially placed hollow graphene cylinders, while SWCNTs are made up of a single graphene cylinder. CNTs are being used for delivery of nano-encapsulated slow-release fertilizers, which have been seen to have a positive impact on plant health and grain quality along with a reduced nutrient loss.

Encapsulated nano-fertilizers have exhibited enhanced efficiency as they are more stable and are not easily degraded (Majeed et al. 2020). CNTs and other carbon NMs have been found to invariably interact with the plant growth showing accumulation, apoplectic transport, and enhancement/inhibition of root growth (Aslani et al. 2014). A few other studies indicated the negative impact of carbon NMs while a few more proved that there was no inhibition of growth as such. In a study conducted on alfalfa and wheat, enhanced germination and root elongation were observed when exposed to industrial-grade MWCNTs. This was attributed to the easy water uptake by MWCNTs upon penetration of seed coat. Similarly, in a tobacco, cell growth was found to increase due to upregulation of genes associated with water channels such as aquaporins in the presence of MWCNTs (Miralles et al. 2012; Khodakovskaya et al. 2012; Khodakovskaya et al. 2013). In general, carbon-based nanomaterials have a mixed response toward plant systems and may have both positive and negative effect on them simultaneously. In a study by Anjum et al. (2013, 2014), the effect of graphene oxide (GO) was determined on germinating faba beans. A significant decrease in growth parameters and oxidative stress induction along with improvement in health status of the faba bean implies a more complex nature of these nanomaterials. Therefore, a more elaborate and in-depth study of these materials is vital for further application in agriculture (Zhang et al. 2015). Studies have proven that CNTs are able to penetrate the cell wall and enter cellular organelles such as plastids and nucleus. However, the penetration is limited in case of MWCNTs, which were able to traverse the cell membrane of only protoplast whereas SWCNTs showed mobility across cell wall as well as cell membranes of intact plant cells.

The CNTs also showed efficient transport and delivery of cargoes such as nucleic acids across plant tissues and also protect them from degrading enzymes allowing a stable transformation of plant systems (Jat et al. 2020). In spite of the application of carbon NMs, they possess certain drawbacks which hinder their application in plant systems. There has been limited research based on their phytotoxic properties at the cellular and phenotypic level. Further, the stability and leaching of attached metals are not well elucidated. Additionally, the accumulation and associated toxicological effects are one of the gray areas, which are yet to be explored. It is also imperative to deduce the subsequent effect on other biological systems upon consumption of these plants engineered with carbon NMs as excessive exposure has been known to cause oxidative stress and loss of cell viability. Finally, the consequential nutritional effects upon contact with such nanostructure materials are not well researched and calls for further study (Omar et al. 2019; Majeed et al. 2020).

10.2.2 Metal-Based Nanomaterials

Various metal-based nanomaterials have been fabricated for their applications in agriculture and food industry. Metals and their oxides for ENs exhibit size-dependent properties such as magnetism and fluorescence (Aslani et al. 2014) as well as opto-electrical properties due to their metal precursors (Zhang et al. 2015). They are synthesized through procedures such as thermal decomposition, coprecipitation, and microemulsions. Metal ENs can also be synthesized with specific functional groups that would allow them to bind with ligands (Omar et al. 2019). The nano-sized properties allow for enhanced stability of the NPs. The most frequently used metals for the generation of metal-based ENs are iron, magnesium, cobalt, cadmium, manganese, zinc, copper, nickel, titanium, gold, and silver, and their respective oxides. Similar to carbon NMs, metal-based ENs can demonstrate both positive and negative impacts on plant systems (Omar et al. 2019). The phytotoxicity was due to the physical properties of the nanoparticle as well as the nature of the experiment. In plants, aggregation of nanomaterial has been observed due to pH variation in the soil environment as well as high concentrations of the metal. Additionally, an excess of metal NPs would compete with enzyme cofactors and lead to the formation of intermediate metabolites. However, the effect of metal-based NMs on plants is greatly dependent on the nature and size of the NM along with the plant species and its developmental stage. Various research groups have studied the uptake mechanism and toxicity of various metal oxides in different plant systems.

Cellular damage in roots was reported due to the entrance of large AuNPs through small pores in *Medicago sativa* and *Brassica juncea*, and the deposition was also seen in the cell vacuoles (Niederberger 2007). Silver nanoparticles (AgNPs) are considered to be safe even in bulk concentrations, but their environmental risk factor is not known. At extensively higher concentrations of 3,000 μg mL^{-1}, toxic effects on seed germination, root, and shoot growth on crop species such as rice, cabbage, and mung bean are observed. In another study performed on *Cucurbita pepo*, the AgNPs caused 4–10 times reduction in the cell biomass as well as the transpiration rate at concentrations as low as 40 nm (Cheng et al. 2011).

In certain crops, such as *Sorghum bicolour*, AgNPs demonstrated concentration-dependent inhibitory effects on the plant uptake and seed germination. Therefore, it is vital to consider the physicochemical properties of NPs and their respective translocation and uptake efficiencies in various crops. Interestingly, cadmium nanoparticles (CdNPs) exhibited an inverse effect on the root growth of tomato, wherein a significant increase in root growth and reduction in root wilt was seen in carrot and cucumber. TiO_2NPs also have conflicting effects on various crop species. In *A. thaliana*, the TiO_2 forms reactive oxygen species (ROS) that result in the formation of a mucilage in the root system, which reduces the water uptake. On the other hand, these NPs showed an increased ability for water uptake and antioxidant production, nitrate reductase enzyme, and higher photosynthetic rate in spinach and soybean. Similar to these metal oxides, several other ENs such as Al_2O_3, Cu_2O, and ZnO have effects on the water channels, uptake mechanism, growth, nitrogen fixation, yield, and germination of different plants at both high and low concentrations (Aslani et al. 2014).

Other research indicates the toxicity of metal-based ENs due to the release of the metal ions, and therefore the dissolution mechanism needs to be explored thoroughly. MNPs such as FeCo- and FePt-based nanostructured cargo delivery system have been successfully implemented in canola, wherein the magnetic gold nanoparticles (mGNP) was bound to a functional group along with the plasmid DNA and delivered into the canola protoplast by applying a magnetic field. A delivery efficiency of 95% was observed (Hao et al. 2013). Functionalization of the MNP with polyethyleneimine (PEI) and DNA of concern was done and loaded onto a magnetic field. This resulted in the formation of an

MNP–DNA complex. This complex was then introduced into cotton pollen by the application of a magnetic field. These magnetofected pollen were used to fertilize plants circumventing the generation of transgenics with tissue culture. This method of gene delivery and transgenic generation was relatively quicker and simpler. However, magnetofection is dependent on the pollen size and aperture. The binding of the DNA to the NPs, mediated through electrostatic interaction and associated polymers, helps in the internalization of the DNA and has proven to be a promising technique (Jat et al. 2020).

10.2.3 Polymeric Nanomaterials

Polymer-based nanomaterials (NMs) include nanogels, nanocomposites, nanospheres, and nanocapsules and are based on naturally found materials such as chitosan and lignin (Omar et al. 2019). These NMs are the primary choice for agricultural purposes due to their antimicrobial properties as well as growth promotion. Due to certain disadvantages of using carbon or metal-based NPs in agriculture, researchers have opted for an alternative which would provide a more sustainable, nontoxic, and eco-friendly alternative to the existing ENs. Chitosan is highly tunable due to its unique physiochemical properties, presence of functional groups, ability to modulate size, and sensitivity to pH, and can be utilized for multiple applications.

Owing to these factors, chitosan-based ENs are being used for various agricultural purposes such as increasing yield and growth, inducing plant defense response and production of antioxidant enzymes. However, despite its high potential and application in fields other than agriculture, use of chitosans has been limited due to unidentified bioactivities of these ENs with microbial species and plants. Chitosans are soluble in aqueous solutions with low pH, due to which its application to plants proves to be toxic to the target organism. In order to enhance its solubility in neutral media, physicochemical modifications are being performed on chitosan. Synthetically derived forms of chitosan such as tri-ethylene diamine dithiocarbamate chitosan have greater solubility in aqueous solutions as well as an enhanced microbial activity. Chitosan-based ENs have been extensively used in plants; however, for agricultural purposes, ample amounts of NMs are required. Among the various methods used for synthesis of chitosan ENs such as emulsion-droplet coalescence, precipitation, sieving, spray drying, and ionic gelation, the latter two are appropriate for generating large amounts of chitosan-based ENs. They have been utilized in agriculture for a wide range of purposes, such as antimicrobial activities, growth promotion, increasing yield, immunity booster, and nutrition-use efficiency. The application of chitosan-based ENs and their derivatives for bioactivities like those above is based on several proposed mechanisms. The presence of surface amino acids, which confer them with positive charge, allows for the association with anionic targets and subsequently instigating antimicrobial activities and plant defense responses. Binding of the cationic components of chitosan to the polyanionic components on the microbial surface causes it to disintegrate and leak out the cellular content, eventually killing it.

Cations such as NH_3^+ can also bind to the DNA or proteins of microbial cells leading to inactivation of metabolic pathways and enzymes, disrupting cellular mechanism. Chitosan is also a known elicitor molecule and induces plant defense systems by interacting with the pattern-recognition receptor (PRR). This in turn elicits activation of various pathogenesis-related (PR) genes, ROS, nitric oxide (NO), and phytohormones. Chitosan-based NMs have been found to effectively trigger the generation of PRs, ROS, and NOs. Due to various regulations in nanotechnology-based materials in agriculture, field studies related to them are still at the preliminary stage (Kumaraswamy et al. 2018). Apart from chitosans, lignin is also being used as polymer-based NMs. The use of lignin-based ENs has been applied for therapeutic purposes and in nanomedicine. Such types of polymeric ENs can be extended in plants through transgenic approaches or through merging technologies like genome editing with nanotechnology to have a more fine-tuned and enhanced approach (Figure 10.3). Additionally, the complex structure of lignin poses an issue for incorporation into technologies like these. Therefore, genetic engineering of lignin to a simpler, linear form, C-lignin, allows for an adjustable base-material for synthesis of lignin-based nanomaterials (Grossman and Wilfred 2019).

10.2.4 Hybrid Nanomaterials

Hybrid nanomaterials are generated through the amalgamation of metal, carbon, and polymer-based nanomaterials imparting them with enhanced applicability and efficacy as compared to individual NMs. Hybrid NMs also provide a higher biocompatibility and can be used in various fields along with overcoming individual drawbacks of individual components. They have seen their use in plant protection and growth, pathogen diagnosis, food safety, and for introduction of genes in plants. Pollen magnetofection along with NMs has gained pace for generation of genetically modified plants. Hybrid exosomes are being employed to deliver small components such as miRNAs and siRNAs along with low-molecular-weight components like Cas9. The effectiveness of such deliveries is dependent mostly on the engineering strategy and nanoparticle synthesis. Therefore, the use of natural components produced in plant cells such as hybrid exosomes provides a more stable and functional nanodelivery system (Omar et al. 2019; Abd-Elsalam 2020).

10.3 Cargo Delivery Systems

The NP-based delivery systems are usually limited by the presence of the cell wall and the recalcitrance of certain plant species toward transformation. This can be overcome by employing noninvasive and non-damaging techniques for the introduction of genes and optimizing stable transformation of the explant material. Recently, embryos and calluses are being used in a wide range of plant species for successful transformation. Several physical, chemical, and biological biomolecule delivery methods are being taken up for the introduction of cargoes such as nucleic acids (DNA, siRNA, and miRNA) and ribonucleoproteins in plant tissues (Cunningham et al. 2018). The various strategies have been elaborated below.

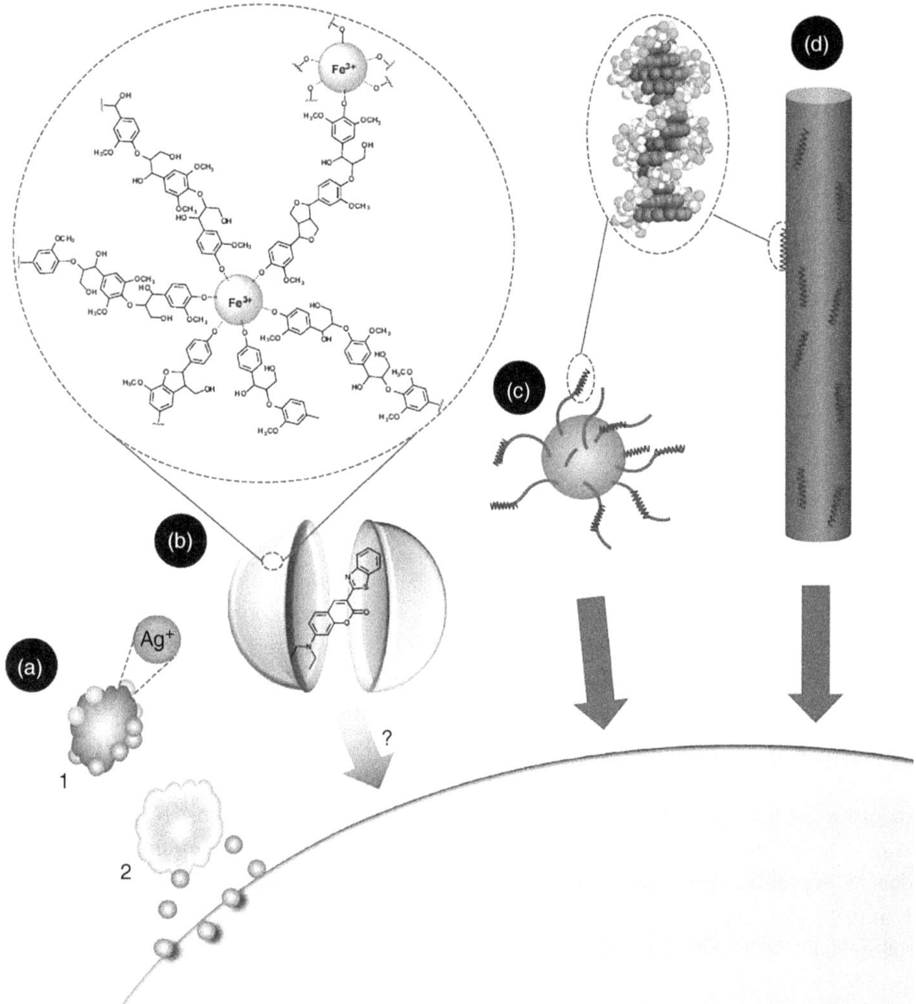

Figure 10.3 Lignin-based nanomaterials and their interactions with a generic cell (bacterial or mammalian), represented in the bottom right of the image. (a) Negatively charged kraft lignin nanoparticles (40–70 nm in diameter) infused with Ag^+ ions (1) and subsequently coated with a layer of cationic poly(diallyldimethylammonium chloride) (PDAC); (2) can be used to deliver Ag^+ ions, which interfere with essential bacterial cell functions. (b) Lignin nanocapsules (0.3–6 mm in diameter) containing the fluorescent model drug Coumarin-6 (shown inside a dissected nanocapsule) can be synthesized by low-power sonication of a mixture of oil, water, kraft lignin, $FeCl_3$, and Coumarin-6. The nanocapsules form and retain their shape as the result of complexation of lignin with Fe^{3+} via the phenolic hydroxyl moieties (see inset) and lignin's propensity to participate in π–π stacking. These lignin nanocapsules will release Coumarin-6 over time; delivery into live cells has yet to be demonstrated. (c) Nanoparticles (100–200 nm in diameter) consisting of a kraft lignin core (sphere) onto which cationic poly-(2-dimethylaminoethyl methacrylate) (PDMAEMA; represented by wires) was grafted via atom transfer radical polymerization (ATRP) are able to bind DNA (see inset) via electrostatic interactions and deliver the DNA to a variety of mammalian cells in culture. (d) Hollow lignin nanotubes (LNTs) or solid lignin nanowires with a diameter of 200 nm, prepared in a sacrificial alumina template, can be used to deliver DNA adsorbed onto the LNTs (inset) into human HeLa cells without the need for auxiliary transfection agents. Reprinted with permission from Grossman, A., and Wilfred, V. (2019). Lignin-based polymers and nanomaterials. *Current Opinion in Biotechnology* 56: 112–120. Copyright © 2018 Elsevier Ltd.

10.3.1 Physical Methods of Cargo Delivery

The physical methods of gene delivery include biolistic particle-mediated delivery and sonoporation and magnetofection (Cunningham et al. 2018). Biolistic method of gene delivery is the most prominent of the methods wherein gold microparticles containing biomolecules are bombarded with a gene gun into plant tissues. This technique overcomes the recalcitranct barrier and can deliver biomolecules into the plastid genome. However, it has a few drawbacks such as plant tissue damage due to high pressure, low efficiency and expression of gene, and the requirement for high concentrations of cargo biomolecules. Particle bombardment method also requires an appropriate and extensive setup for the biomolecule delivery, which might prove to be an expensive affair. The bottleneck in such approaches is crossing the cell wall barrier, which can be overcome by the use of nanoparticles. The cell wall, nucleus, and organelles in plants have a size exclusion limit of 300–500 nm, which is difficult to traverse using gold microparticles. Forceful bombardment, therefore, often causes irreversible tissue damage. Nanoparticles are hence a suitable option for delivery of particles as they have a size below 100 nm. Biolistic delivery using gold-capped MSN was successfully performed in *N. tabacum* cotyledons, which was the first of its kind in plants (Torney et al. 2007). Additionally, MSNs have also been used for direct protein delivery into plant cells. Cre recombinase protein was successfully delivered into *Zea mays* cells, wherein the protein was loaded into MSNs and the complex was bombarded using biolistic gun. These plant cells contained a *loxP* site flanking selection and reporter genes. Upon entry into the plant system, the recombinase caused recombination of the *loxP* site and removal of both the genes. This technology was revolutionary as it provided a non-transgenic genome-editing approach for gene delivery through biolistic approach (Martin-Ortigosa et al. 2014).

In another study by Jun et al. (2008), starch nanoparticles coated with poly-L-lysine were conjugated with plasmid DNA and treated with ultrasound. The DNA-nanoparticle complex was then cocultured with *Dioscorea zingiberensis CH Wright* calli cell suspension and subsequently treated with ultrasound. The nanoparticles had a size range of 50–100 nm, and fluorescence microscopy study revealed that they were able to traverse cell wall and membrane of the plant cell suspension. This technique of sonoporation demonstrated high efficiency of particle transport across cell wall, which usually posed as an obstacle with other transformation techniques. In addition to this, treatment of nanoparticle with magnetic field also has a high impact on the delivery efficiency. mGNPs were bound with fluorescein isothiocyanate (FITC) molecule, and the complex was delivered into plant cells by application of a magnetic field. FITC-labeled mGNPs demonstrated an efficiency of 95%. Further, these complexes were bound with plasmids and delivered into canola cells. The plasmid was successfully delivered into cells by examination through fluorescence microscopy. These mGNPs were also found to be nontoxic to the plant cells and therefore had an added advantage as compared to other components (Hao et al. 2013).

Apart from the existing physical means of cargo delivery, there have been studies based on passive internalization of cargoes into cell walls. These are MSN-mediated plasmid DNA delivery through coculture, topical application of layered double hydroxide (LDH) clay nanosheets, DNA origami structures, and silicon carbide whiskers (SCWs) (Wang et al. 2019b).

In passive delivery of plasmid DNA, MSN loaded with the DNA was cocultured with model crop system, *A. thaliana,* which showed efficient uptake through the roots. A 46.5% of transient expression was observed of the organically functionalized MSNs.

Incubation and vortexing of SCWs with plant cells and DNA facilitated the entry of the DNA into the undifferentiated plant tissues. However, due to vigorous vortexing, cell-wall rupture was often observed, which might cause reduction in transformation efficiency. A more reliable method in comparison was the use of LDH, which effectively demonstrated the delivery of RNAi molecules into model system, *N. tabacum*. However, this is yet to be implemented in delivery of nucleic acids (Demirer et al. 2019; Wang et al. 2019b). In addition to the above nonconventional means of nanoparticle delivery, SCW has been used widely owing to its various properties such as high tensile strength and resistance to degradation. SCW has been successfully used in transformation of groundnut with gene encoding for chitinase, and the generated transgenics showed tolerance to leaf-spot disease (Jat et al. 2020).

10.3.2 Chemical/Nonbiological Method of Cargo Delivery

In chemical-based nanoparticle delivery systems, the NPs are modified for cargo drafting via various interactive forces (Cunningham et al. 2018). Chemical vectors such as calcium phosphate nanoparticles (CaPNPs) have demonstrated a higher efficiency of delivery and incorporation in host organism. Nucleic acids encapsulated by CaPNPs are also protected from digestive enzymes such as nucleases. This method employed passive introduction of concerned genes through fluid-phase endocytosis of the CaPNP complex into *Brassica juncea* hypocotyls explants. An astounding 80.7% stable transformation efficiency was observed, which was due to factors such as cytosolic stability of the compact DNA, dissolution of the CaPNPs, and the endosomal escape of DNA due to osmotic imbalance of the released Ca^{2+} ions. Additionally, an increased uptake of DNA occurred due to the nuclear pore complex formation by calcium ions. The endoplasmically released plasmid DNA then integrates into the plant genome through nonhomologous recombination as was evident from the GUS assay performed. The significant efficacy of this method makes it a promising technique for transgene introduction into plant cells. Further, CaP is a relatively nontoxic and safe alternative, along with causing minimum damage to cellular tissues and allowing easy entry into other organelles apart from the nucleus (Naqvi et al. 2012).

Chemical cargo delivery for ENs is a relatively new platform in biotechnology. In this approach, biochemicals are utilized for the delivery of nanomaterial into plant organelles, such as chloroplasts, using guided peptide recognition motifs. This method avoids cell-wall-mediated delivery and induces target delivery as well. Chloroplasts play key roles in several plant metabolic activities including photosynthesis and as signaling organelles and are therefore important target organelles for genetic engineering studies in plants. Santana et al. (2020) developed a unique EN platform in *A. thaliana*, wherein quantum dots (QDs), hydrophilic in nature and coated with β-cyclodextrin, which allows functionalization of the QD, were generated. The guiding peptide, RbcS, was developed with high zeta potential leading to easy translocation across lipid membranes. The coated nanomaterial molecular baskets possess a size which facilitates the movement through leaf-cell-wall pores. The peptide recognition motif ensures an efficient delivery and QD along with chemical cargoes confirms specificity. This technology exhibited an efficacy of $74.6\% \pm 10.8\%$ and therefore can be applied to nanotechnology-based plant engineering in other crop species as well.

High aspect ratio nanomaterials such as CNTs have been widely studied and their surface chemistries imply toward their flexibility and biocompatibility. These structures are known to have a high tensile strength and biomolecules attached to CNTs display superior stability. Functionalized CNTs were used to deliver DNA into plant species such as *N. tabacum*, *Eruca sativa*, *T. aestivum*, and *Gossypium hirsutum* and generate robust protein expression through diffusion of the biomolecule into intact plant cells. The study proved the passive delivery of genetic material across different plant species through no tissue damage, and it is therefore a promising strategy for plant biotechnology applications (Demirer et al. 2019; Jat et al. 2020).

Cationic polymers are another aspect which are being gradually exploited. Such polymers protect the biomolecule from degradation and also confer stability while transportation into the plant cell. In a study by Finiuk et al. (2017) linear 2-(dimethylamino) ethyl methacrylate (DMAEM) was synthesized, and its potential as facilitators for exogenous nucleic acid transport using polyethylene glycol (PEG) was studied in tobacco and moss. A cationic polymer and plasmid DNA (pDNA) complex was formed and the efficiency of nanoparticle-mediated gene transfer was studied. The synthesized polymeric DMAEM (PDMAEM) carrier effectively transferred the gene into the protoplast of tobacco and moss and has scope for further development in nanotechnology-mediated plant biotechnology.

10.3.3 Biological Method of Cargo Delivery

Bionanotechnology is emerging as a method that involves a multidisciplinary approach for development of biohybrid EN by employing various fields of science including biology, chemistry, physics, and engineering in a coaction. It encompasses the usage of microbes such as viruses as templates for the production of nanomaterials (Soto and Ratna 2010). Viral-based delivery systems can deliver various cargoes in a target-specific manner and can also be regulated accordingly. Viruses possess several unique abilities such as biocompatibility and small size range, which makes them apt for nanoengineering purposes. Viruses also have an inherent mechanism of gene transfer, which can be explored for biomolecule delivery. Plant viral nanoparticles (PVNPs) are basically of two types, viz., virions and viral-like particles (VLP). VLPs are essentially empty capsids and therefore are non-virulent making them excellent materials as cargo delivery agents (Shahgolzari et al. 2020).

The active surface structures of PVNPs aid the easy loading of cargo materials onto them as well as the coat protein (CP). However, appropriate strategies are necessary for the proper loading of cargo. This can be achieved by direct interaction of PVNP with cargo or modification of PVNP to produce genetically modified PVNPs. In relation to this, PVNP loading of cargos can be of genetic and nongenetic nature. Nongenetic approach for cargo loading can be performed on all kinds of nanomaterials; however, genetic means are applicable only to protein-based NMs (Bhaskar and Lim 2017). The versatility of PVNPs allows the loading of a wide range of cargoes including proteins, enzymes, nucleic acids, as well as smaller nanoparticles. In spite of the several pros of PVNPs, the *in vivo* application is far from extensive due to off-target delivery and acceptance by the innate immune system of the species. Therefore, efforts are being made in order to modify PVNPs by addition of ligands and polymers for improved genetic engineering. PVNPs have been modified by using proteins or peptides which bind to certain cellular markers and enhance efficacy of genetic engineering. One such study implied toward improved cell growth and differentiation upon interaction of cellular mesenchymal cells (MSCs) and PVNPs. Further, arginine–glycine–aspartic (RGD) peptide-modified PVNPs enhance MSC differentiation and growth through release of NPs. PVNPs have also been modified for improving the cargo delivery by regulating shape and size for protection against phagocytic activity of cells and targeting ligands that have affinity for cellular markers to ensure accumulation in target cells. An ideal PVNP should be able to load and package the cargo effectively and deliver it to the appropriate target location with localization. PVNPs can thus be tailored according to the need through physical, chemical, and genetic manipulations along with a multidisciplinary approach, and can be further investigated for application in generating engineered plants (Shahgolzari et al. 2020).

10.4 RNAi and Nanotechnology

With the advent of modern biotechnological tools such as nanotechnology-mediated genetically engineered crops, there has been a tremendous economic leap in agricultural production. In spite of certain shortcomings, genetically modified has gained leverage. However, the strict regulations associated with them as well as its application in plant system being fairly recent makes this technology laborious. The future holds more promise due to the generation of nanomaterials such as MSNs, MNPs, and CNTs. These materials have been known to be able to deliver a wide range of cargoes and more recently RNA interference (RNAi) payload. RNAi technology in itself has been much exploited in various crop systems successfully, yet the commercialization is limited due to GM regulatory frameworks. RNAi is induced by double-stranded RNA (dsRNA) which are processed by *Dicer-like* (*DCL*) enzymes into small interfering RNA (siRNA). These siRNAs are then loaded by an RNA-induced silencing complex (RISC) and eventually targeted to pathogenic RNA, hindering its translation to proteins (Worrall et al. 2018). The topical application of double-stranded RNAi (dsRNAi) using spraying technique on plants is being adopted as an alternative to chemicals. However, these naked dsRNAi are not stable for longer periods of time. Hence, nanocarriers have emerged as effective delivery tools for transfer of these dsRNAi (Landry and Mitter 2019).

The triggering of RNAi-mediated silencing of genes is usually carried out by *Agrobacterium*-mediated or virus-mediated transformation, which is often tedious, time-consuming, and subject to regulations due to it being a transgenic technique. As a result of this, other carriers have been found. In a study by Jiang et al. (2014), cationic fluorescent nanoparticles (G1–G3) have been explored for possibility of gene transportation. The carrier consists of a central perylene-3,4,9,10-tetracarboxydiimide (PDI) chromophore surrounded by specific cationic group. The fluorescent nanoparticle moves into the plant cell with high efficacy. Out of the three gene carriers, G2 was seen to have the highest delivery efficiency, and therefore the G2/dsRNA mixture was exposed to different regions of root in *A. thaliana*. G2 was able to traverse the root cell wall and spread across the entire root tissue. Further, in order to examine the systemic silencing ability of the carrier, the expression of the G2/dsRNA was studied in the model crop plant. The two genes selected for synthesis of dsRNA were: *SHOOT MERISTEMLESS (STM)* and *WEREWOLF (WER)*. Expression profile analysis exhibited toward reduced levels of the genes compared to control, implying toward successful silencing. There was also an evident reduction of associated developmental genes and also apparent phenotypic changes in the plant. This was the first study of systemic silencing of genes through a nonviral means of gene nanocarrier.

In continuation to this study, nanomaterials were being gradually explored as a mode for dsRNA delivery into plants. The topical application of dsRNA was then done through loading the concerned dsRNA onto an LDH clay nanosheet. LDH is known to occur naturally due to precipitation in saline water and comes with the added advantages of being nontoxic and degradable. The LDH along with the dsRNA forms a complex that has been named Bioclay and can sustain on leaf surfaces for as long as 30 days as a result of protection from nuclease activities. The Bioclay was observed to enable the release of dsRNA on the leaf surfaces. Bioclay was also seen to have effects on newly emerged leaf tissues that were not sprayed and conferred them with systemic RNAi silencing even after over 20 days of spraying. Finally, the clay nanosheets in LDH are biodegradable in nature making them environmentally sustainable. The studies pertaining to nanoparticles for RNAi delivery have stayed confined to animal systems due to the presence of the cell wall in plants. LDH was able to deliver large amounts of dsRNA into plant system and facilitated the delivery of the dsRNA with the by-products degrading the nanosheets post reaction. The viruses taken up for this investigation were cucumber mosaic virus (CMV) and pepper mild mottle virus (PPMoV) and were studied in the plant systems, *N. tabacum* (tobacco) and *Vigna unguiculata* (cowpea). The study established conclusively that LDH provided stable systemic silencing for extended periods of time and also can be used for further study of mechanism associated with dsRNA uptake in various plant systems. Clay nanosheet-mediated topical application of RNAi would also overcome the hurdle faced with generation of GMOs (Mitter et al. 2017).

Li-Byarlay et al. (2013) used siRNA loaded in perfluorocarbon nanomaterials for studying the mechanism of methylation in honeybees. The siRNA was coupled with perfluorocarbon nanoparticles and nebulized for aerosol spraying. It was observed that the spraying of perfluorocarbon-loaded siRNA onto insect species caused higher efficiency of silencing of genes as compared to individual siRNA. This study laid the foundation for further studies pertaining to target-specific pest management studies in agriculture.

Owing to the controversies surrounding GMOs, researchers are continuously in search of strategies that are able to generate genetically improved crop plants without the integration of genes into the plant genome. Therefore, nanotechnology-mediated RNAi silencing and associated knockdown studies provide a robust, highly efficient, and species-independent alternative. It can also be used as a screening tool for reviewing phenotypic changes due to genotypic manipulations through silencing (Wang et al. 2019b).

As with any novel modern technology, nanocarriers and associated RNAi need to go through appropriate regulatory restrictions. The Office of the Gene Technology Regulator in Australia has categorized gene silencing using RNAi as non-GMO as long as it does not alter host genome and is not subsequently translated into proteins. Further, topical application of RNAi does not fall under GMO status and therefore forms scope for commercialization of such crop species. RNAi-based pesticides, for example, have already penetrated the global agricultural landscape in the forms of sprays, injections, and root/seed treatments.

However, the use of nanocarriers in this respect is not yet clear and subject to debate. Therefore, they are being taken up on a case-by-case basis (Landry and Mitter 2019). Considering the surge in application of nanotechnology and its effectiveness, it is high time that the global regulatory framework for this technology be established. This would also allow the opportunity for optimizing agricultural challenges which are yet to be solved.

10.5 Conclusion and Future Prospects

Genetic engineering has greatly influenced agricultural biotechnology and led to improvement in several crop varieties globally in terms of stress tolerance, disease/pest resistance, and trait improvement. However, the limitations in traversing the cell wall often render the technology unsettled. The synergy of genetic engineering with nanotechnology allows crossing such barriers. Nanoparticles, due to their physicochemical nature, size, and tunable aspect, offer scope for application in plant engineering. The applications in nanotechnology until now are mostly limited to animal cells; however, current research has also undertaken its applicability in plant systems. The various types of ENs that have been used are extensive and come in various forms such as carbon-based, metal-based, polymer-based, and hybrid NMs. Additionally, depending on the mode of delivery, NMs have seen their implementation through techniques such as biolistic gene guns and chemical aids. Interestingly, it has also been established that NPs developed with surface chemistries according to animal systems can also be administered in plant systems, both *in vivo* and *in vitro* with a few general modifications. It is essential to also keep in mind that these nanocarriers have to be designed such that they are able to move across a wide range of plant species efficiently along with being tunable to numerous cargo biomolecules for a transient as well as stable transformation (Cunningham et al. 2018). Apart from the physical methods employed for gene transfer, passive diffusion of these nanocarriers is a much preferred method.

Endocytosis is the major mechanism for cellular transportation of nanocarrier encapsulated transgene (Rai et al. 2015). The proper biodegradation and disposal of nanoparticles in the environment is an aspect which needs to be studied thoroughly in order to obtain a clear perspective of the risk assessment. Metal and metal oxide nanoparticles, for example, are one of the most popular and widely used forms of NPs. While there is no concrete data about the toxicity and relevant effects of such NPs to the plant and environment, research related to molecular mechanism associated with plant-NP interaction is still at the primary stage. On the other hand, biological approaches such as virus-mediated nanocarriers as well as plant polymer-based NPs like chitosans, starch, and lignin are more favored due to their nontoxicity, biocompatibility, and biodegradability (Chaudhary et al. 2019). The surface chemistry of NPs along with the physicochemical nature of plant surfaces imparts further complexity to NP designing. NPs should be able to successfully traverse the rigid cell wall, waxy cuticles, and other lipid membranes while delivering intact cargoes into the cell. The additional challenge for further research is the interactive mechanism of the engineered nanoparticles with plant cell and determination of threshold concentration of NPs beyond which it would impart stress to the plant system. Extensive research on toxic effects of NMs would help in establishing minimum toxic levels of NMs in plants as well as the ecosystem. Thus, the NPs need to be generated keeping these factors in mind. A multidisciplinary approach for synthesizing tunable NPs is required for efficient genetic engineering. CNTs enable transgene expression in plant cells through plasmid delivery targeting the nucleus. CNTs can also be used for transient expression in the chloroplast. This is guided by functionalization of the CNT by altering the surface chemistry and size of the particle. In order to attain nuclear targeting, the CNT is usually functionalized by PEI polymers, which facilitates both nuclear and chloroplast delivery. On the other hand, functionalization with chitosan polymers helps in delivery of the CNT complex into the chloroplast of the cell. Thus, it is evident that further studies on the surface chemistry of such NPs would assist in cellular localization and targeting to various organelles and destinations in the plant cell (Aslani et al. 2014; Wang et al. 2019b).

Considering the regulatory limitations of genetically engineered plants, a continuous alternative is always in search. While the regulations of GMOs and nanomaterials have individual regulatory frameworks, nanotechnology-based GMOs are highly debated and the associated risk factors are still unresolved. Therefore, such products are being taken up in a case-by-case basis (Landry and Mitter 2019). Although another alternative would be establishing a transgene-free approach and so, nanoparticles are being synthesized for delivery of gene-edited cargoes into plant cells including meristematic tissues (Sanzari et al. 2019; Wang et al. 2019a). These gene-edited plants can be further regenerated through tissue culture techniques. Similar to LDH topical spraying which was found to enhance the sustainability of dsRNA in leaf tissues (Mitter et al. 2017), the foliar application of Cas9/gRNAs-coated carbon dots exhibited targeted gene editing. This method a faster, simpler, and inexpensive approach to gene editing using NMs and therefore can be extrapolated to diverse range of crops. This provides a transient expression of the CRISPR-gene and is thus a transgene-free approach for non-sexually reproducing crops as well (He and Zhao 2020). The parallel development of nanotechnology and non-integrative DNA approaches such as genome editing and RNAi stands to improve the output by essentially bypassing GMO regulations. However, the current public outlook is toward a chemical-free agricultural approach in which GMOs are an excellent candidate technique (Landry and Mitter 2019). Along with this, nanotechnology, gene silencing, and genome editing also provide tremendous benefits and advantages over existing conventional techniques, and therefore it is necessary to assuage the current perspective of such technologies being harmful and intrusive.

References

Abd-Elsalam, K. (2020). *Multifunctional Hybrid Nanomaterials for Sustainable Agri-Food and Ecosystems*. Elsevier Amsterdam, Netherlands.

Adersh, A., Kulkarni, A.R., Ghosh, S. et al. (2015). Surface defect rich ZnO quantum dots as antioxidant inhibiting α-amylase and α-glucosidase: a potential anti-diabetic nanomedicine. *Journal of Materials Chemistry B* 3: 4597–4606.

Anjum, N.A., Singh, N., Singh, M.K. et al. (2013) Single-bilayer graphene oxide sheet tolerance and glutathione redox system significance assessment in faba bean (Vicia faba L.). *Journal of Nanoparticle Research* 15:1770.

Anjum, N.A., Singh, N., Singh, M.K. et al. (2014) Single-bilayer graphene oxide sheet impacts and underlying potential mechanism assessment in germinating faba bean (Vicia faba L.) *Science of the Total Environment* 472: 834–841.

Aslani, F., Bagheri, S., Muhd Julkapli, N. et al. (2014). Effects of engineered nanomaterials on plants growth: an overview. *Scientific World Journal* 2014: 641759.

Bhagwat, T.R., Joshi, K.A., Parihar, V.S. et al. (2018). Biogenic copper nanoparticles from medicinal plants as novel antidiabetic nanomedicine. *World Journal of Pharmaceutical Research* 7 (4): 183–196.

Bhaskar, S. and Lim, S. (2017). Engineering protein nanocages as carriers for biomedical applications. *NPG Asia Materials* 9 (4): e371.

Chaudhary, R.G., Bhusari, G.S., Tiple, A.D. et al. (2019). Metal/metal oxide nanoparticles: toxicity, applications, and future prospects. *Current Pharmaceutical Design* 25: 4013–4029.

Cheng, Y., Yin, L., Lin, S. et al. (2011). Toxicity reduction of polymer-stabilized silver nanoparticlesby sunlight. *Journal of Physical Chemistry C* 115 (11): 4425–4432.

Cunningham, F.J., Goh, N.S., Demirer, G.S. et al. (2018). Nanoparticle-mediated delivery towards advancing plant genetic engineering. *Trends in Biotechnology* 36: 882–897.

Demirer, G.S., Zhang, H., Matos, J.L. et al. (2019). High aspect ratio nanomaterials enable delivery of functional genetic material without DNA integration in mature plants. *Nature Nanotechnology* 14: 456–464.

Finiuk, N., Buziashvili, A., Burlaka, O. et al. (2017). Investigation of novel oligoelectrolyte polymer carriers for their capacity of DNA delivery into plant cells. *Plant Cell, Tissue and Organ Culture* 131: 27–39.

Ghosh, S. (2018). Copper and palladium nanostructures: a bacteriogenic approach. *Applied Microbiology and Biotechnology* 101 (18): 7693–7701.

Ghosh, S., Chacko, M.J., Harke, A.N. et al. (2016c). Barleria prionitis leaf mediated synthesis of silver and gold nanocatalysts. *Journal of Nanomedicine & Nanotechnology* 7: 4.

Ghosh, S., Gurav, S.P., Harke, A.N. et al. (2016d). Dioscorea oppositifolia mediated synthesis of gold and silver nanoparticles with catalytic activity. *Journal of Nanomedicine & Nanotechnology* 7: 5.

Ghosh, S., Harke, A.N., Chacko, M.J. et al. (2016b). Gloriosa superba mediated synthesis of silver and gold nanoparticles for anticancer applications. *Journal of Nanomedicine & Nanotechnology* 7: 4.

Ghosh, S., Jagtap, S., More, P. et al. (2015a). Dioscorea bulbifera mediated synthesis of novel Aucore Agshell nanoparticles with potent antibiofilm and antileishmanial activity. *Journal of Nanomaterials* 2015: 562938.

Ghosh, S., More, P., Derle, A. et al. (2015d). Diosgenin functionalized iron oxide nanoparticles as novel nanomaterial against breast cancer. *Journal of Nanoscience and Nanotechnology* 15 (12): 9464–9472.

Ghosh, S., More, P., Nitnavare, R. et al. (2015b). Antidiabetic and antioxidant properties of copper nanoparticles synthesized by medicinal plant Dioscorea bulbifera. *Journal of Nanomedicine & Nanotechnology* S6: 7.

Ghosh, S., Nitnavare, R., Dewle, A. et al. (2015c). Novel platinum-palladium bimetallic nanoparticles synthesized by Dioscorea bulbifera: anticancer and antioxidant activities. *International Journal of Nanomedicine* 10 (1): 7477–7490.

Ghosh, S., Patil, S., Chopade, N.B. et al. (2016a). Gnidia glauca leaf and stem extract mediated synthesis of gold nanocatalysts with free radical scavenging potential. *Journal of Nanomedicine & Nanotechnology* 7: 358.

Giraldo, J.P., Wu, H., Newkirk, G.M., and Kruss, S. (2019). Nanobiotechnology approaches for engineering smart plant sensors. *Nature Nanotechnology* 14: 541–553.

Grossman, A. and Wilfred, V. (2019). Lignin-based polymers and nanomaterials. *Current Opinion in Biotechnology* 56: 112–120.

Hao, Y., Yang, X., Shi, Y. et al. (2013). Magnetic gold nanoparticles as a vehicle for fluorescein isothiocyanate and DNA delivery into plant cells. *Botany* 91: 457–466.

He, Y. and Zhao, Y. (2020). Technological breakthroughs in generating transgene-free and genetically stable CRISPR-edited plants. *aBIOTECH* 1: 88–96.

Jamdade, D.A., Rajpali, D., Joshi, K.A. et al. (2019). Gnidia glauca and Plumbago zeylanica mediated synthesis of novel copper nanoparticles as promising antidiabetic agents. *Advances in Pharmacological Sciences* 2019: 9080279.

Jat, S.K., Bhattacharya, J., and Sharma, M.K. (2020). Nanomaterial based gene delivery: a promising method for plant genome engineering. *Journal of Materials Chemistry B* 8: 4165–4175.

Jiang, L., Ding, L., He, B. et al. (2014). Systemic gene silencing in plants triggered by fluorescent nanoparticle-delivered double-stranded RNA. *Nanoscale* 6: 9965–9969.

Joshi, K.A., Ghosh, S., and Dhepe, A. (2019). Green synthesis of antimicrobial nanosilver using in-vitro cultured Dioscorea bulbifera. *Asian Journal of Organic & Medicinal Chemistry* 4 (4): 222–227.

Jun, L., Wang, F., Wang, L. et al. (2008). Preparation of fluorescence starch-nanoparticle and its application as plant transgenic vehicle. *Journal of Central South University of Technology* 15 (768): 773.

Karmakar, S., Ghosh, S., and Kumbhakar, P. (2020). Enhanced sunlight driven photocatalytic and antibacterial activity of flower-like ZnO@MoS_2 nanocomposite. *Journal of Nanoparticle Research* 22: 11.

Khodakovskaya, M.V., De Silva, K., Biris, A.S. et al. (2012). Carbon nanotubes induce growth enhancement of tobacco cells. *ACS Nano* 6: 2128–2135.

Khodakovskaya, M.V., Kim, B.S., Kim, J.N. et al. (2013). Carbon nanotubes as plant growth regulators: effects on tomato growth, reproductive system, and soil microbial community. *Small* 9: 115–123.

Kim, D.H., Gopal, J., and Sivanesan, I. (2017). Nanomaterials in plant tissue culture: the disclosed and undisclosed. *RSC Advances* 7: 36492–36505.

Kitture, R., Chordiya, K., Gaware, S. et al. (2015). ZnO nanoparticles-red sandalwood conjugate: a promising anti-diabetic agent. *Journal of Nanoscience and Nanotechnology* 15: 4046–4051.

Kumaraswamy, R.V., Kumari, S., Choudhary, R.C. et al. (2018). Engineered chitosan based nanomaterials: bioactivities, mechanisms and perspectives in plant protection and growth. *International Journal of Biological Macromolecules* 113: 494–506.

Landry, M.P. and Mitter, N. (2019). How nanocarriers delivering cargos in plants can change the GMO landscape. *Nature Nanotechnology* 14: 512–514.

Li-Byarlay, H., Li, Y., Stroud, H. et al. (2013). RNA interference knockdown of DNA methyltransferase 3 affects gene alternative splicing in the honey bee. *Proceedings of the National Academy of Sciences of the United States of America* 110: 12750–12755.

Majeed, N., Panigrahi, K.C.S., Sukla, L.B. et al. (2020). Application of carbon nanomaterials in plant biotechnology. *Materials Today Proceedings* 30: 340–345.

Martin-Ortigosa, S., Peterson, D.J., Valenstein, J.S. et al. (2014). Mesoporous silica nanoparticle-mediated intracellular cre protein delivery for maize genome editing via lox P site excision. *Plant Physiology* 164: 537–547.

Miralles, P., Johnson, E., Church, T.L. et al. (2012). Multiwalled carbon nanotubes in alfalfa andwheat: toxicology and uptake. *Journal of the Royal Society, Interface* 9: 3514–3527.

Mitter, N., Worrall, E.A., Robinson, K.E. et al. (2017). Clay nanosheets for topical delivery of RNAi for sustained protection against plant viruses. *Nature Plants* 3: 16207.

Naqvi, S., Maitra, A.N., Abdin, M.Z. et al. (2012). Calcium phosphate nanoparticle mediated genetic transformation in plants. *Journal of Materials Chemistry* 22: 3500–3507.

Niederberger, M. (2007). Nonaqueous sol–gel routes to metal oxide nanoparticles. *Accounts of Chemical Research* 40 (9): 793–800.

Omar, R.A., Afreen, S., Talreja, N. et al. (2019). Impact of nanomaterials in plant systems. In: *Plant Nanobionics Volume 1, Advances in the Understanding of Nanomaterials Research and Applications* (ed. R. Prasad), 117–140. Springer Switzerland.

Pérez-de-Luque, A. (2017). Interaction of nanomaterials with plants: what do we need for real applications in agriculture? *Frontiers in Environmental Science* 5: 1–7.

Rai, M., Bansod, S., Bawaskar, M. et al. (2015). Nanoparticles-based delivery systems in plant genetic transformation. In: *Nanotechnologies in Food and Agriculture* (ed. M. Rai, C. Ribeiro, L. Mattoso, and N. Duran), 209–239. Cham: Springer.

Robkhob, P., Ghosh, S., Bellare, J. et al. (2020). Effect of silver doping on antidiabetic and antioxidant potential of ZnO nanorods. *Journal of Trace Elements in Medicine and Biology* 58: 126448.

Rokade, S., Joshi, K., Mahajan, K. et al. (2018). Gloriosa superba mediated synthesis of platinum and palladium nanoparticles for induction of apoptosis in breast cancer. *Bioinorganic Chemistry and Applications* 2018: 4924186.

Rokade, S.S., Joshi, K.A., Mahajan, K. et al. (2017). Novel anticancer platinum and palladium nanoparticles from *Barleria prionitis*. *Global Journal of Nanomedicine* 2 (5): 555600.

Salunke, G.R., Ghosh, S., Santosh, R.J. et al. (2014). Rapid efficient synthesis and characterization of AgNPs, AuNPs and AgAuNPs from a medicinal plant, Plumbago zeylanica and their application in biofilm control. *International Journal of Nanomedicine* 9: 2635–2653.

Sant, D.G., Gujarathi, T.R., Harne, S.R. et al. (2013). *Adiantum philippense* L. frond assisted rapid green synthesis of gold and silver nanoparticles. *Journal of Nanoparticles* 2013: 1–9.

Santana, I., Wu, H., Hu, P., and Giraldo, J.P. (2020). Targeted delivery of nanomaterials with chemical cargoes in plants enabled by a biorecognition motif. *Nature Communications* 11: 1–12.

Sanzari, I., Leone, A., and Ambrosone, A. (2019). Nanotechnology in plant science: to make a long story short. *Frontiers in Bioengineering and Biotechnology* 7: 1–12.

Shahgolzari, M., Pazhouhandeh, M., Milani, M. et al. (2020). Plant viral nanoparticles for packaging and in vivo delivery of bioactive cargos. *WIREs Nanomedicine and Nanobiotechnology* 12: e 1629.

Shende, S., Joshi, K.A., Kulkarni, A.S. et al. (2018). Platanus orientalis leaf mediated rapid synthesis of catalytic gold and silver nanoparticles. *Journal of Nanomedicine & Nanotechnology* 9: 2.

Shende, S., Joshi, K.A., Kulkarni, A.S. et al. (2017). Litchi chinensis peel: a novel source for synthesis of gold and silver nanocatalysts. *Glob. J. Nanomedicine* 3 (1): 555603.

Shinde, S.S., Joshi, K.A., Patil, S. et al. (2018). Green synthesis of silver nanoparticles using Gnidia glauca and computational evaluation of synergistic potential with antimicrobial drugs. *World Journal of Pharmaceutical Research* 7 (4): 156–171.

Soto, C.M. and Ratna, B.R. (2010). Virus hybrids as nanomaterials for biotechnology. *Current Opinion in Biotechnology* 21: 426–438.

Torney, F., Trewyn, B.G., Lin, V.S.Y., and Wang, K. (2007). Mesoporous silica nanoparticles deliver DNA and chemicals into plants. *Nature Nanotechnology* 2: 295–300.

Wang, J.W., Grandio, E.G., Newkirk, G.M. et al. (2019a). Nanoparticle-mediated genetic engineering of plants. *Molecular Plant* 12: 1037–1040.

Wang, P., Zhao, F.J., and Kopittke, P.M. (2019b). Engineering crops without genome integration using nanotechnology. *Trends in Plant Science* 24: 574–577.

Worrall, E.A., Hamid, A., Mody, K.T. et al. (2018). Nanotechnology for plant disease management. *Agronomy* 8: 285.

Zhang, P., Ma, Y., and Zhang, Z. (2015). Interactions between engineered nanomaterials and plants: phytotoxicity, uptake, translocation, and biotransformation. In: *Nanotechnology and Plant Sciences* (ed. M. Siddiqui, M. Al-Whaibi, and F. Mohammad), 77–99. Cham: Springer.

11

Bioremediation of Contaminants Using Microbial Nanotechnologies for Agricultural Sustainability

Farhan Zameer[1,2], Kounaina Khan[1,2], Pratheek Jain[1], Subrahmanya Padyana[1], and Ajar Nath Yadav[3]

[1] *PathoGutOmics Laboratory, Alva's Traditional Medicinal Archive (ATMA), Department of Ayurveda Pharmacology, Alva's Ayurveda Medical College, Dakshina Kannada, Karnataka, India*
[2] *TheOmic School Pvt. Ltd., Mysore, Karnataka, India*
[3] *Department of Genetics, Plant Breeding and Biotechnology, Dr. Khem Singh Gill Akal College of Agriculture, Eternal University, Sirmour, Himachal Pradesh, India*

11.1 Introduction

Soil is a finite and fragile resource that supports life on Earth (lithosphere), acting as the foundation for agriculture, ecosystems, and biodiversity. Soil acts as a platform for plant growth, and it provides essential nutrients and minerals necessary for the production of food, fiber, and fuel. Without healthy, fertile soil, the ability to cultivate crops and sustain human life would be severely compromised (Parikh and James 2012; Acharya et al. 2023a). One of the most important functions of soil is its ability to support plant life through nutrient cycling. It serves as a reservoir for essential elements like nitrogen, phosphorus, and potassium, in which plants absorb through their roots (Acharya et al. 2023b). These nutrients are replenished naturally through the decomposition of organic matter, including plant residues and animal waste. This process ensures the sustainability of ecosystems and agricultural systems, making soil an integral part of the global food supply (Robe and Barberon 2023). Beyond agriculture, soil contributes to water filtration and regulation. As rainwater percolates through the soil, it filters out pollutants and recharges underground aquifers, providing clean water for human consumption and wildlife. Additionally, soil plays a key role in climate regulation by acting as a carbon sink. It stores carbon dioxide (CO_2) from the atmosphere, helping to mitigate the effects of climate change (Certini and Scalenghe 2023).

Virgin soil refers to the land that has never been cultivated, disturbed, or altered by human activities. This type of soil is in its natural state, containing a rich and balanced composition of organic matter, minerals, and microorganisms. Virgin soil is more fertile and nutrient dense than soil that has undergone repeated agricultural use or industrial exploitation, making it highly valuable for supporting plant growth and biodiversity. The organic matter in virgin soil, accumulated over centuries from decomposing plants and animals, enhances its ability to retain moisture, cycle nutrients, and foster healthy ecosystems (Kaur and Kaur 2020). The presence of undisturbed soil layers allows for proper water infiltration and root growth, which supports vegetation. This kind of soil is also home to a diverse population of microorganisms, which play a key role in nutrient cycling, decomposition, and maintaining soil health. Virgin soils are often found in remote or protected natural areas, such as forests, grasslands, and wetlands. However, they are increasingly rare due to human activities like deforestation, agriculture, and urbanization. Preserving virgin soils is essential for maintaining ecosystem services, such as carbon sequestration, water filtration, and biodiversity, making them an important resource for sustainability and environmental conservation (Selmy et al. 2022). Due to deforestation, overgrazing, and industrial agriculture, urbanization leads to soil degradation, including erosion, nutrient depletion, and pollution. Sustainable soil management practices, such as crop rotation, cover cropping, and reduced tillage, are essential for maintaining soil health and ensuring long-term productivity. Soil preservation and careful management are essential for the survival and prosperity of future generations (Wang et al. 2018; Manohara et al. 2023).

11.2 Soil and Its Components Supporting Ecosystem Sustainability

Composition of the soil determines its fertility and usability. Few key components collectively contribute to the soil role in sustaining ecosystems, regulating climate, and supporting life on Earth. (i) Mineral Content: Soil is composed of minerals such as sand, silt, and clay, which provide structural support for plants and supply essential nutrients like nitrogen, phosphorus, and potassium for plant growth and agricultural productivity. (ii) Organic Matter: Decomposed organic material, such as dead plants and animals, contributes to soil fertility by enriching it with carbon-based compounds. This organic matter enhances water retention, nutrient cycling, and microbial activity, supporting sustainable ecosystems (Stenberg et al. 2010). (iii) Soil Microorganisms: Bacteria, fungi, and other microorganisms in soil play a key role in breaking down organic matter, fixing nitrogen, and forming symbiotic relationships with plant roots (e.g., mycorrhizae). These processes improve nutrient availability and promote sustainable plant growth. (iv) Soil Carbon Sequestration: Soil stores a significant amount of carbon through organic material decomposition, acting as a carbon sink (Yadav 2021). This helps to mitigate the effects of climate change by reducing atmospheric CO_2 levels, contributing to global carbon balance (Wang et al. 2024). (v) Nutrient Cycling: Soils facilitate nutrient cycling, where essential nutrients like nitrogen, phosphorus, and potassium are continuously replenished and made available to plants through microbial decomposition and mineral weathering, maintaining ecosystem productivity. (vi) Water Retention and Filtration: Soil structure and organic content help to retain moisture, allowing plants to access water during dry periods.

It also filters pollutants, improving water quality and recharging underground aquifers, which supports sustainability in water use (Deemy et al. 2022). (vii) Soil Structure and Aggregation: Stable soil aggregates (clumps of soil particles) improve soil aeration, water infiltration, and root penetration. A good soil structure reduces erosion and promotes sustainable land use by maintaining fertile land for agriculture. (viii) Soil pH Buffering: Soil regulates pH levels, affecting nutrient availability and microbial activity. Soils with appropriate pH levels maintain healthy ecosystems and crop growth, sustaining biodiversity and agricultural productivity (Dewangan et al. 2023). (ix) Cation Exchange Capacity (CEC): Soil ability to hold and exchange positively charged ions (nutrients) with plant roots is critical for nutrient retention and availability. High CEC allows soils to sustain nutrient levels, reducing the need for artificial fertilizers. (x). Erosion Control: Vegetation-covered soils reduce surface runoff and erosion, preventing the loss of fertile topsoil. Sustainable soil practices like contour farming and terracing help protect soil structure, ensuring long-term land viability for agriculture and ecosystems (Ciric et al. 2023) (Figure 11.1).

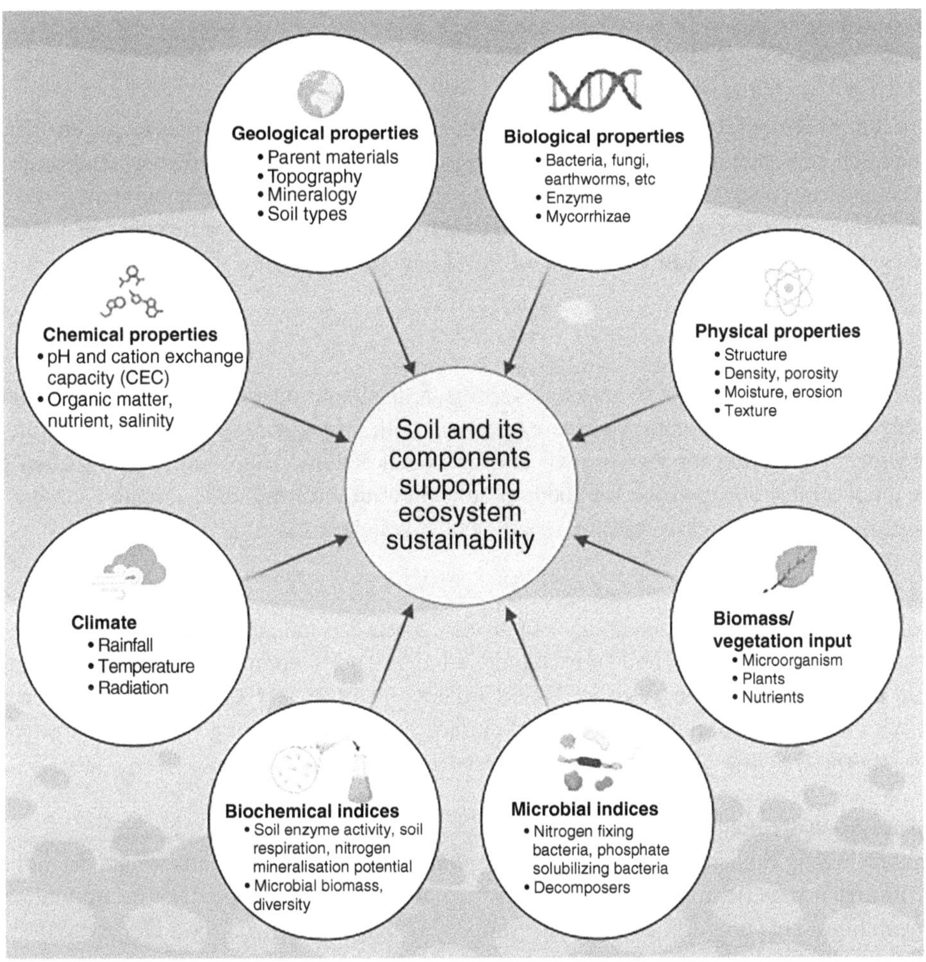

Figure 11.1 Components of soil supporting ecosystem sustainability.

11.2.1 Physical Indices Governing Soil Health

Soil health is determined by a variety of physical parameters that influence its ability to support plant growth, water retention, and ecosystem functions. These physical properties are fundamental to maintaining soil structure, nutrient availability, and ecosystem sustainability. The physical indices, in combination, dictate soil's ability to sustain life, maintain structure, and perform ecological functions critical to long-term soil health and productivity.

11.2.1.1 Soil Texture

Soil texture refers to the proportion of sand, silt, and clay particles, which govern many of its properties, including water retention, permeability, and aeration. Sandy soils have large particles and are well drained but poor at retaining nutrients and moisture. Clay soils, with smaller particles, have high water retention and nutrient-holding capacity but can be poorly drained. Loamy soils, a balanced mix of sand, silt, and clay, are generally considered the most fertile and healthy (Noble et al., 2023).

11.2.1.2 Soil Structure

The soil structure refers to the arrangement of soil particles into aggregates or clumps. The good soil structure allows for proper aeration, water infiltration, and root penetration. Well-aggregated soils are less prone to erosion and compaction, enhancing plant growth and microbial activity. Poor soil structure, such as in compacted or heavily tilled soils, reduces permeability and inhibits root development, negatively impacting soil health (Ghezzehei 2012).

11.2.1.3 Bulk Density

Bulk density is the mass of soil per unit volume, reflecting its degree of compaction. Lower bulk density indicates loose, well-structured soil with ample pore space for air and water movement, essential for root growth and microbial activity. High bulk density, often a result of compaction, restricts root penetration and water infiltration, leading to poor plant performance and increased erosion risk (Salehi Hikouei et al. 2021).

11.2.1.4 Porosity

Soil porosity refers to the percentage of pore space between soil particles, which influences water retention, drainage, and air exchange. A balance between macropores (for drainage) and micropores (for water retention) is essential for healthy soil. Higher porosity enhances water infiltration and reduces the risk of surface runoff, contributing to soil conservation and nutrient cycling (Zhou et al. 2021).

11.2.1.5 Water Holding Capacity

Water holding capacity is the soil ability to retain water against the force of gravity. Soils with higher organic matter and finer textures, such as clay, have greater water retention capabilities. This property is crucial for supporting plants during dry periods, particularly in agriculture, where consistent moisture levels are necessary for optimal crop growth (Lu et al. 2023).

11.2.2 Chemical Indices Governing Soil Health

Soil health is also profoundly influenced by various chemical properties that govern its fertility, nutrient availability, and overall capacity to support plant and microbial life. These chemical indices provide insight into how well soil can sustain growth, resist degradation, and cycle essential nutrients. These chemical indices interact dynamically, shaping the nutrient availability and biological activity in soils. Healthy soils exhibit balanced pH, high organic content, optimal CEC, and well-managed nutrient and salinity levels, supporting not just plant growth but the broader ecosystem.

11.2.2.1 Soil pH

Soil pH is a fundamental chemical indicator of soil health. It measures the acidity or alkalinity of soil, influencing the availability of essential nutrients like nitrogen, phosphorus, and potassium. Most crops thrive in a pH range of 6.0–7.5, where nutrients are optimally available. Soils outside this range can limit plant growth, as extreme acidity or alkalinity can lead to nutrient imbalances and toxicities (Wang and Kuzyakov 2024).

11.2.2.2 Cation Exchange Capacity

Cation exchange capacity refers to the soil ability to hold and exchange positively charged ions (cations) such as calcium (Ca^{2+}), magnesium (Mg^{2+}), and potassium (K^{+}). High CEC soils can retain and supply more nutrients to plants, making them more fertile and resilient. Soils with low CEC, like sandy soils, may require more frequent fertilization, as they struggle to retain nutrients (Zhang et al. 2021).

11.2.2.3 Organic Matter Content

Organic matter, primarily in the form of decomposed plant and animal residues, plays an important role in soil health. It improves nutrient retention, enhances microbial activity, and increases water-holding capacity. Higher organic matter content enriches soil fertility by releasing essential nutrients as it decompose, making soils more sustainable and productive over time (Sulthana et al. 2018; Sáez-Sandino et al. 2023).

11.2.2.4 Nutrient Content

The concentration of essential macronutrients like nitrogen, phosphorus, and potassium is a direct measure of soil fertility. Nitrogen is vital for plant growth and chlorophyll production, while phosphorus supports energy transfer and root development. Potassium regulates water uptake and disease resistance. Deficiencies or imbalances in these nutrients can significantly hinder plant growth and yield, making regular soil testing important (Chtouki et al. 2022).

11.2.2.5 Soil Salinity

Soil salinity measures the salt content in the soil, which can adversely affect plant growth by disrupting water absorption and causing toxicity. High salinity levels can occur in irrigated soils or areas with poor drainage. Managing soil salinity is critical to maintaining soil health, particularly in regions prone to salt buildup (Huang et al., 2023).

11.2.3 Biochemical Indices Governing Soil Health

Biochemical indices are components in assessing soil health, as they reflect the biological activity, nutrient cycling, and overall vitality of the soil. These indicators offer a deeper understanding of how soil sustains life by supporting plant growth, organic matter decomposition, and nutrient availability. Healthy soil is teeming with active microorganisms and biochemical processes, sustaining not just crops but the entire ecosystem.

11.2.3.1 Soil Enzyme Activity

Soil enzymes, produced by soil microorganisms and plant roots, catalyze biochemical reactions essential for nutrient cycling. Key enzymes like dehydrogenase, phosphatase, and urease are involved in breaking down organic matter, releasing essential nutrients like nitrogen and phosphorus into forms that plants can absorb. High enzyme activity is a strong indicator of active microbial populations and healthy soil nutrient dynamics (Daunoras et al. 2024).

11.2.3.2 Microbial Biomass

Microbial biomass refers to the total mass of living microorganisms in the soil, including bacteria, fungi, and protozoa. These microbes play a vital role in decomposing organic matter, fixing nitrogen, and recycling nutrients. A high microbial biomass indicates a rich and active soil community, promoting better nutrient availability and organic matter turnover, which directly impacts soil fertility and plant growth (Prommer et al. 2020).

11.2.3.3 Soil Respiration

Soil respiration measures the rate at which soil organisms respire, releasing CO_2 during the decomposition of organic matter. This process is a direct reflection of microbial activity and organic matter breakdown. High rates of soil respiration indicate robust microbial activity and suggest healthy carbon cycling, which is essential for maintaining soil organic matter levels and long-term fertility (Neogi et al. 2014).

11.2.3.4 Microbial Diversity

Microbial diversity refers to the variety of different microorganisms present in the soil, including bacteria, fungi, and archaea. Diverse microbial communities are more resilient to environmental changes and stressors such as drought or chemical contamination. A rich diversity of microorganisms enhances soil functions like nutrient cycling, disease suppression, and plant-microbe symbiosis, contributing to long-term soil health (Anthony et al. 2023).

11.2.3.5 Soil Organic Carbon

Soil organic carbon (SOC) is a key indicator of soil health, as it represents the organic matter content that fuels microbial activity. SOC serves as both an energy source for soil organisms and a reservoir for carbon sequestration, helping mitigate climate change. Soils rich in organic carbon have greater microbial activity, enhanced nutrient retention, and better soil structure, all of which contribute to sustainable plant growth and ecosystem stability (O'Rourke et al. 2015).

11.2.3.6 Nitrogen Mineralization Potential

Nitrogen mineralization refers to the conversion of organic nitrogen into inorganic forms (ammonium and nitrate) that plants can use. This process is driven by soil microbes and is a critical indicator of nutrient cycling efficiency. High nitrogen mineralization potential signals that soil microbes are actively decomposing organic matter and converting it into plant-available nutrients, which is essential for maintaining soil fertility (Li et al. 2019).

11.2.3.7 Microbial Quotient

The microbial quotient is the ratio of microbial biomass carbon to total SOC. It provides an indication of how efficiently SOC is being converted into microbial biomass. A higher microbial quotient suggests that the organic matter is actively supporting microbial growth and nutrient cycling, which is essential for maintaining soil fertility and long-term ecosystem productivity (Prommer et al. 2020).

11.2.4 Microbial Indices Governing Soil Health

Microbes play a essential role in governing soil health through their involvement in nutrient cycling, organic matter decomposition, and promoting plant growth. The microbial groups interact synergistically, driving essential soil processes like nutrient cycling, organic matter decomposition, and disease suppression.

11.2.4.1 Nitrogen-Fixing Bacteria

Nitrogen-fixing bacteria are essential for converting atmospheric nitrogen into forms that plants can utilize, such as ammonia (NH_3). Rhizobium species form symbiotic relationships with leguminous plants (e.g., peas and beans) by colonizing their root nodules. Through this process, they fix nitrogen, providing plants with a critical nutrient for growth while enhancing soil fertility. Free-living bacteria such as Azotobacter also contribute to nitrogen fixation in the rhizosphere, improving nitrogen availability in soils without the need for chemical fertilizers (Zameer et al. 2010; Aasfar et al. 2021).

11.2.4.2 Mycorrhizal Fungi

Mycorrhizal fungi form mutualistic associations with plant roots, which enhances nutrient and water uptake. Arbuscular mycorrhizal fungi (AMF), such as Glomus species, penetrate plant root cells and extend hyphal networks into the soil. This increases the surface area for nutrient absorption, particularly phosphorus, which is often limited in the soil. By improving the nutrient uptake and soil structure, these fungi contribute to plant health, soil aggregation, and long-term soil fertility (Gujre et al. 2021).

11.2.4.3 Decomposers

Decomposer microorganisms, including fungi and bacteria, break down organic matter such as dead plants and animal residues, releasing essential nutrients like carbon, nitrogen, and phosphorus back into the soil. Fungi like Trichoderma and bacteria such as Bacillus play an important role in this process. By decomposing complex organic compounds into simpler, plant-available nutrients, they enhance soil fertility, structure, and water-holding capacity (Zameer et al. 2016; Poveda and Eugui 2022).

11.2.4.4 Phosphate-Solubilizing Bacteria

Phosphate-solubilizing bacteria (PSB) make insoluble forms of phosphorus available to plants by secreting organic acids that dissolve phosphorus-bound minerals. Species like Pseudomonas and Bacillus produce enzymes and acids that release phosphorus from mineral-bound or organic sources in the soil. This process is essential for maintaining phosphorus levels in agricultural soils, reduces the need for chemical fertilizers and improves crop yields (Bakki et al. 2024).

11.2.4.5 Actinomycetes

Actinomycetes are a group of filamentous bacteria which helps in breaking down tough organic materials, such as cellulose and chitin. Streptomyces species are particularly important for decomposing complex organic matter and producing antibiotics that inhibit soil-borne pathogens. This dual role in decomposition and disease suppression helps to maintain soil health, enhances nutrient cycling, and promotes a balanced microbial community (Beulah et al. 2015; Abd-Elgawad et al. 2020).

11.3 Major Contaminants in Agricultural Soil

Agricultural soils are essential for food production, but they are often exposed to various contaminants that can severely affect soil health, crop yield, and even human and ecological well-being. These contaminants primarily originate from human activities such as industrial farming practices, use of agrochemicals, and improper waste disposal. Addressing these contaminants requires a multifaceted approach, including sustainable farming practices, better waste management, and stricter regulations, to protect both the environment and human health (Zameer and Gopal 2010).

11.3.1 Pesticides and Herbicides

Pesticides, herbicides, and fungicides are commonly used to control pests, weeds, and diseases in agricultural systems. They help to boost crop yields, but these chemicals can persist in the soil, leading to long-term contamination. Persistent organic pollutants (POPs) like dichlorodiphenyltrichloroethane (DDT) and atrazine accumulate in the soil, harming beneficial microorganisms and potentially entering the food chain, posing risks to both wildlife and human health (Manjunatha et al. 2013; Singh et al. 2016; Alengebawy et al. 2021).

11.3.2 Heavy Metals

Heavy metals such as lead (Pb), cadmium (Cd), arsenic (As), and mercury (Hg) often contaminate agricultural soils due to the use of fertilizers, sewage sludge, or industrial waste. These metals do not break down and can accumulate over time, reduce soil fertility, damage plant roots, and enter food crops, posing health risks when consumed. Even at low concentrations, heavy metals are toxic to humans and animals, affecting the nervous system, kidneys, and other vital organs (Raffa and Chiampo 2021).

11.3.3 Excessive Nutrients

Overapplication of synthetic fertilizers introduces excessive amounts of nutrients like nitrogen and phosphorus into the soil. While essential for plant growth, excess nitrogen can lead to soil acidification, and phosphorus runoff can cause eutrophication in nearby water bodies. This nutrient imbalance disrupts soil structure, reduces biodiversity, and contaminates water resources (Devlin and Brodie 2023).

11.3.4 Microplastics

Recent studies have shown the growing presence of microplastics in agricultural soils, often originating from plastic mulch, fertilizers, or wastewater irrigation. These tiny plastic particles can alter the soil structure, reduce water retention, and leach toxic chemicals that harm soil organisms and plants (Sajjad et al. 2022).

11.3.5 Industrial Pollutants

Soils near industrial areas may be contaminated with chemicals like polychlorinated biphenyls (PCBs), dioxins, and hydrocarbons from industrial emissions, vehicle exhaust, or wastewater irrigation. These pollutants are highly persistent and toxic, affecting soil microorganisms, disrupting ecosystems, and posing long-term health risks through bioaccumulation (Alloway 2013; FAO & UNEP 2021) (Figure 11.2).

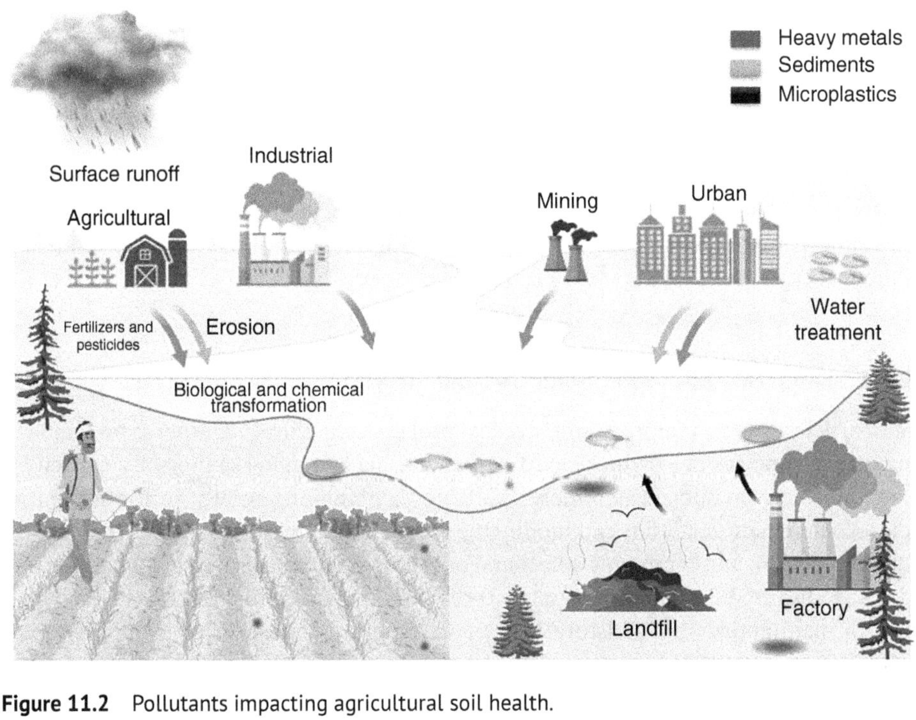

Figure 11.2 Pollutants impacting agricultural soil health.

11.4 Role of Nanotechnology in Soil Health

Nanotechnology, the science of manipulating matter at the atomic or molecular scale (1–100 nm), has shown immense potential in revolutionizing many fields, including agriculture and soil health. When applied to soil science, nanotechnology offers innovative solutions to enhance soil fertility, manage contaminants, and promote sustainable agricultural practices (Sharma and Kumar 2024).

11.4.1 Nanoscale Nutrient Delivery

One of the most promising applications of nanotechnology in soil health is the development of nanofertilizers. Traditional fertilizers result in nutrient loss due to leaching, volatilization, or poor uptake by plants. This inefficiency can cause soil degradation and contamination. Nanofertilizers are designed to deliver nutrients more efficiently, often with slow or controlled release mechanisms, ensuring that plants absorb a greater proportion of the nutrients. For example, nanoencapsulated nitrogen fertilizers can release nitrogen gradually, matching the plant's uptake needs, reducing wastage, and preventing soil acidification (Nongbet et al. 2022). This precision in nutrient delivery not only enhances soil fertility but also minimizes the environmental footprint of agriculture.

11.4.2 Remediation of Soil Contaminants

Nanotechnology offers novel approaches to soil remediation, especially for removing contaminants such as heavy metals, pesticides, and industrial pollutants. Nanoparticles, such as iron oxide (Fe_3O_4), TiO_2, and zero-valent iron (ZVI), have high surface reactivity, making them ideal for capturing, degrading, or immobilizing toxic substances in the soil. For instance, iron-based nanoparticles are used to reduce heavy metals like As and Cd by transforming them into less toxic forms or binding them, so they can no longer be absorbed by plants. Similarly, photocatalytic nanoparticles like TiO_2 can break down POPs such as pesticides or herbicides, thus reducing their harmful effects on soil health (Dhanapal et al. 2024).

11.4.3 Enhancing Soil Water Retention and Structure

Nanoparticles can also improve soil physical properties, such as water retention and soil structure. Nanoclays or polymer-based nanomaterials can increase the soil's capacity to retain moisture, making water more available to plants, especially in drought-prone regions. These nanomaterials can modify the soil structure by stabilizing soil aggregates, reducing erosion, and improving aeration. For example, hydrogels infused with nanoparticles can absorb and retain large amounts of water, releasing it slowly to plants, which helps in maintaining soil moisture levels and supports plant growth even during dry spells (Yadav et al. 2023).

11.4.4 Soil Microbial Health

Nanotechnology has been explored for its ability to enhance soil microbial health, which is essential for nutrient cycling, organic matter decomposition, and maintaining overall soil fertility. Certain nanoparticles, such as carbon-based nanomaterials (e.g., graphene oxide), can create a more favorable environment for beneficial microbes by improving soil aeration and reducing harmful contaminants (Mathew et al. 2020). However, the impact of nanomaterials on soil microbiota is complex. While some nanomaterials may support the microbial activity, others might negatively affect microbial communities depending on their concentration and chemical nature (Upadhayay et al. 2023). Thus, careful design and use of nanomaterials are necessary to avoid any unintended disruptions to microbial ecosystems.

11.4.5 Detection and Monitoring of Soil Health

Nanotechnology has advanced the development of sensitive tools for real-time soil health monitoring. Nanosensors can detect nutrient levels, soil pH, moisture content, and the presence of contaminants at extremely low concentrations. These sensors are capable of providing precise and timely data, helping farmers and soil scientists make informed decisions about soil management practices (Das et al. 2024). For example, nano-based sensors embedded in the soil can monitor the availability of critical nutrients such as nitrogen or phosphorus, sending alerts when nutrient levels are low and fertilizers are needed. This real-time feedback reduces the overuse of fertilizers, preventing nutrient imbalances and maintaining long-term soil health (Muthukrishnan 2022).

11.4.6 Mitigation of Soil Degradation

Nanotechnology contributes to combating soil degradation through soil restoration techniques. Nanomaterials can help to rehabilitate soils degraded by salinity, acidification, or heavy metal contamination. For example, nanoparticles can be used to neutralize excess salts in saline soils, making the land more suitable for crop growth. In acidic soils, nanolime or nanogypsum can be applied to increase pH and restore balance, making the soil more productive for agriculture (Huang et al. 2023).

11.4.7 Sustainable Pest Management

Nanotechnology also offers solutions for sustainable pest control, which indirectly benefits soil health. Nanopesticides or nanoherbicides are designed to target pests or weeds more effectively while reducing the amount of chemicals that enter the soil. Because these nanoformulations are more precise, they minimize the collateral damage to non-target organisms, including beneficial soil microbes, thus preserving the biological integrity of the soil (Yousef et al. 2023).

11.5 Microbial-Mediated Bioremediation of Soil Contaminants and Their Mode of Action

Bioremediation is a natural process that uses microorganisms to degrade or detoxify pollutants in the environment, particularly in contaminated soils. This method harnesses the capabilities of bacteria, fungi, and other microorganisms to break down hazardous substances into less harmful or non-toxic forms (Patil et al. 2021, 2022). This section delves into the mechanisms of microbial-mediated bioremediation in the management of soil health in agriculture (Pande et al. 2022). Microbial-mediated bioremediation can occur through several mechanisms, primarily classified into bioaccumulation, biotransformation, biosorption, and bioleaching.

11.5.1 Bioaccumulation

Bioaccumulation refers to the uptake and accumulation of contaminants by microorganisms. Certain bacteria and fungi can absorb heavy metals and other toxic substances from the soil and concentrate them within their cells. This process can reduce the bioavailability of contaminants, thus decreasing their toxic effects in the environment. The general mechanism involves contaminants enter microbial cells through passive or active transport mechanisms. Inside the cells, heavy metals can bind to intracellular proteins or be sequestered in vacuoles, reducing their toxic impact (Briffa et al. 2020).

11.5.2 Biotransformation

Biotransformation involves the conversion of contaminants into less harmful substances through enzymatic reactions. Microorganisms possess enzymes that can degrade complex organic compounds into simpler, less toxic forms. The mechanism involves (i) Oxidation-Reduction Reactions: Microbes can oxidize or reduce pollutants, converting them into non-toxic forms. (ii) Hydrolysis: Enzymes break chemical bonds, facilitating the breakdown of pollutants into simpler compounds. For instance, *Pseudomonas putida* can metabolize toluene into less toxic products through biotransformation processes (Pervaiz et al. 2013; Prasad et al. 2019; Satapathy et al. 2022).

11.5.3 Biosorption

Biosorption is a passive process where microorganisms adsorb and concentrate contaminants onto their cell surfaces. This method is particularly effective for heavy metals and some organic pollutants. The mode of action involves microbial cell walls, composed of polysaccharides and proteins, interact with contaminants through various forces, such as ionic bonding, van der Waals forces, and hydrogen bonding. This process can reduce the concentration of contaminants in the soil and prevent their uptake by plants (Joutey et al. 2015).

11.5.4 Bioleaching

Bioleaching is the process of using microorganisms to extract metals from their ores or contaminated soils. This mechanism is particularly useful for remediating heavy metal-contaminated soils. The microbes oxidize metal sulfides, releasing soluble metal ions into the solution. For example, *Acidithiobacillus ferrooxidans* oxidizes iron sulfide to release iron and copper (Cu) ions. The solubilized metals can then be removed or stabilized, effectively reducing soil contamination (Srichandan et al. 2014).

Microbial-mediated bioremediation offers a sustainable, eco-friendly approach to restoring contaminated soils. By employing the natural capabilities of microorganisms, hazardous substances can be effectively degraded, thus mitigating the risks associated with soil contamination (Zameer et al. 2010). Understanding the mechanisms behind microbial interactions with contaminants not only enhances our knowledge of soil health but also paves the way for innovative solutions to pressing environmental challenges (Atuchin et al. 2023). Continued research in this field will be essential for developing effective bioremediation strategies that promote soil sustainability and public health. While microbial bioremediation holds great promise, several challenges remain. Slow Process: Bioremediation can be slower than physical or chemical remediation methods. Microbial Community Dynamics: The effectiveness of bioremediation depends on the presence of suitable microbial species, which may not always be naturally available. Environmental Conditions: Extreme environmental conditions can limit microbial activity and effectiveness. Future research should focus on enhancing the efficiency of microbial bioremediation through genetic engineering, metabolic engineering, and the use of consortia of microorganisms to tackle complex contamination scenarios (Sahu et al. 2024).

11.6 Microbial Nanotechnology Bioremediation Strategies for Soil Contaminants

Soil contamination poses a significant threat to environmental health and agricultural productivity. Traditional methods for soil remediation often fall short in effectiveness and sustainability. However, the integration of nanotechnology with microbial bioremediation presents a promising solution for cleaning up contaminated soils.

11.6.1 Nanoparticle-Enhanced Microbial Growth

Nanoparticle-enhanced microbial growth represents a promising approach in sustainable agriculture, as nanoparticles can stimulate the metabolic activities of beneficial microorganisms in the soil. The high surface area and reactivity of nanoparticles facilitate their interaction with microbes, enhancing nutrient bioavailability and uptake. This interaction can lead to improved microbial growth, as nanoparticles often provide essential nutrients that are vital for microbial proliferation and enzymatic functions. Certain nanoparticles can serve as carriers for micronutrients, effectively delivering these nutrients to microbial cells and promoting their growth and metabolic processes. Additionally, nanoparticles can act as nanotools, enhancing electron transfer and supporting chemotaxis, which further optimizes microbial efficiency in nutrient cycling. These positive effects not only increase microbial populations but also improve soil health and fertility. These ultimately contribute to enhanced plant growth and resilience against biotic and abiotic stresses (Mansor and Xu 2020).

11.6.2 Nanofertilizers for Enhanced Microbial Activity

Nanofertilizers provide a controlled release of nutrients, fostering a favorable environment for microbial communities in contaminated soils. By delivering essential nutrients at the nanoscale, these fertilizers improve microbial metabolism and biodegradation processes. Nanoencapsulated nitrogen fertilizers have been employed in agricultural practices to support the growth of nitrogen-fixing bacteria such as Rhizobium in legume root nodules. This not only enhances plant growth but also contributes to soil health and the biodegradation of organic contaminants like herbicides (Jakhar et al. 2022).

11.6.3 Biosorption with Nanomaterials

Nanomaterials can significantly enhance the biosorption capacity of microbial cells, allowing for the effective removal of heavy metals and organic pollutants from contaminated soils. This method utilizes the high surface area and reactivity of nanomaterials to adsorb contaminants. Example: Magnetic nanoparticles (MNPs) coated with *Pseudomonas putida* have been used to remove Pb from contaminated soil. The MNPs effectively bind to Pb ions, which are then taken up by the bacteria through biosorption. The immobilization of these particles allows for easy recovery and reuse, making the process economically viable (Mandeep and Shukla 2020).

11.6.4 Biodegradation of Organic Pollutants Using Nanoscale Catalysts

Nanoscale catalysts can accelerate the biodegradation of organic pollutants by enhancing the microbial enzymatic activity. These catalysts promote the breakdown of complex organic compounds into less toxic forms. Example: TiO_2 nanoparticles are effective photocatalysts that facilitate the degradation of various organic pollutants, including pesticides. When combined with microbial strains such as *Deinococcus radiodurans*, TiO_2 nanoparticles can enhance the degradation rate of chlorinated solvents under UV light exposure (Remya et al. 2022).

11.6.5 Microbial Fuel Cells for Bioremediation

Microbial fuel cells (MFCs) utilize the metabolic activity of microorganisms to generate electricity while simultaneously treating contaminated soil. Nanotechnology can enhance the efficiency of these systems by improving electron transfer between microbes and the anode. Example: In a study, carbon nanotubes (CNTs) were incorporated into the anodes of MFCs using *Geobacter sulfurreducens*, a metal-reducing bacterium. The presence of CNTs improved electron transfer efficiency, leading to higher power output while degrading pollutants such as phenol and toluene (Abbas and Rafatullah 2021).

11.6.6 Bioleaching with Nanoparticles

Bioleaching employs microorganisms to extract metals from contaminated soils, and the addition of nanoparticles can enhance this process. Nanoparticles can facilitate the solubilization of metal ions and improve the overall efficiency of metal recovery. Example: Sulfur-oxidizing bacteria like *Acidithiobacillus ferrooxidans* can be combined with nano-zero-valent iron (nZVI) to bioleach metals from contaminated soils. The nZVI particles aid in the solubilization of heavy metals such as Cu and zinc (Zn) while the bacteria metabolize the iron, creating a synergistic effect that enhances metal recovery (Vera et al. 2022; Dinesh et al. 2024).

11.6.7 Controlled Release of Biostimulants Using Nanotechnology

Nanotechnology can be utilized to encapsulate biostimulants such as plant growth-promoting rhizobacteria (PGPR) within nanoparticles for controlled release into contaminated soils. This ensures that beneficial microbes are available over an extended period, promoting soil health and bioremediation. Example: Chitosan nanoparticles have been used to encapsulate *Bacillus subtilis*, a PGPR known for its ability to degrade pollutants (Akki et al. 2024). The slow release of these bacteria in contaminated soils not only improves the degradation of organic contaminants but also enhances soil structure and nutrient availability for plants.

11.7 Nanotechnology in Mycoremediation

Fungi play a crucial role in soil bioremediation, and nanotechnology can enhance their degradation capabilities. Nanomaterials can improve the growth and enzymatic activity of fungi, allowing for more effective breakdown of complex organic pollutants. Example: Nanosilica has been shown to stimulate the growth and ligninolytic enzyme production of the white-rot fungus *Phanerochaete chrysosporium*. This enhanced fungal activity leads to the effective degradation of POPs such as polycyclic aromatic hydrocarbons (PAHs) and chlorinated compounds (Sahu et al. 2023; Vaksmaa et al. 2023).

11.8 Nanoparticle-Enhanced Phytoremediation

Phytoremediation employs plants to remove contaminants from soil, and nanotechnology can enhance this process by improving plant-microbe interactions. Nanoparticles can be used to facilitate the uptake of heavy metals and organic pollutants by plants, as well as promote the beneficial microbial activity in the rhizosphere. Example: Zinc oxide (ZnO) nanoparticles can be applied to *Helianthus annuus* (sunflower) plants to improve their tolerance to heavy metals such as Cd. The presence of ZnO nanoparticles enhances microbial activity in the rhizosphere, facilitating the breakdown of pollutants and improving the overall phytoremediation process (Srivastav et al. 2019).

11.9 Smart Sensors for Monitoring Soil Remediation

Nanotechnology has also led to the development of smart sensors that can monitor microbial activity and soil health during bioremediation processes. These sensors can detect the presence of contaminants, nutrient levels, and microbial populations, providing valuable data for optimizing bioremediation strategies. Example: Graphene-based sensors can be employed to monitor soil pH, moisture, and microbial activity in real time. This data allows for timely adjustments in bioremediation efforts, ensuring maximum efficiency in pollutant degradation (Rajak et al. 2023). The integration of nanotechnology with microbial bioremediation offers a multifaceted approach to addressing soil contamination. Each of the strategies discussed highlights the potential of nanotechnology to enhance the effectiveness of microbial processes, improve pollutant degradation rates, and contribute to sustainable soil management. As research continues to advance, these innovative techniques will play a vital role in developing environmentally friendly solutions to combat soil pollution, ensuring the health and productivity of our soils for future generations. This approach not only mitigates the risks associated with soil contaminants but also promotes a healthier ecosystem, supporting agricultural productivity and environmental sustainability (Lloret et al. 2021). Future studies and applications will undoubtedly expand the possibilities of combining nanotechnology with microbial bioremediation, paving the way for a cleaner and greener planet (Figure 11.3).

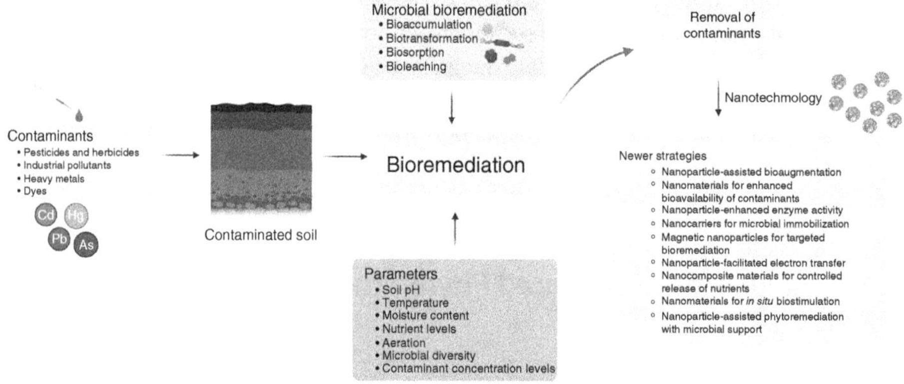

Figure 11.3 Advancements in microbial bioremediation of soil contaminants.

11.10 Opportunities in Microbial Nanotech Bioremediation

11.10.1 Enhanced Contaminant Degradation

One of the primary advantages of integrating nanotechnology with microbial bioremediation is the enhancement of contaminant degradation rates. Nanoparticles can serve as catalysts, providing a larger surface area for microbial interactions, which accelerates the breakdown of hazardous substances. Nano-sized TiO_2 is a potent photocatalyst that can facilitate the degradation of organic pollutants such as pesticides and dyes under UV light. When combined with specific bacteria, the degradation rate of these contaminants can be significantly increased.

11.10.2 Improved Nutrient Delivery

Nanotechnology can improve the delivery of nutrients to microorganisms, promoting their growth and metabolic activities. This is particularly beneficial in contaminated soils where nutrient deficiencies may inhibit microbial performance. Example: Nanofertilizers, such as nanoencapsulated nitrogen, provide controlled nutrient release, enhancing the growth of nitrogen-fixing bacteria. This not only supports the microbial community but also aids in the degradation of pollutants like hydrocarbons (Yang et al. 2023).

11.10.3 Targeted Delivery of Bioremediation Agents

Nanotechnology allows for the targeted delivery of microbial strains or bioremediation agents directly to contaminated sites. This precision minimizes the use of excess materials and reduces the impact on non-target organisms. Example: Encapsulation of beneficial microbes within nanocarriers can ensure their localized application in contaminated soils, maximizing their effectiveness while minimizing the ecological footprint (Maire et al. 2019; Avinash et al. 2023).

11.10.4 Enhanced Biosorption Capacities

Nanoparticles can improve the biosorption capabilities of microorganisms, enabling them to effectively remove heavy metals and organic pollutants from soils. The high surface area and reactivity of nanomaterials increase the binding sites available for contaminants. Example: MNPs can be functionalized with microbial cells to enhance biosorption of heavy metals like Pb. The MNPs bind to the contaminants, allowing for easy recovery of the bioremediation agents after treatment (Mandeep and Shukla 2020; More et al. 2021).

11.10.5 Real-Time Monitoring of Remediation Processes

Real-time monitoring of remediation processes using nanotechnology offers a newer approach to enhance the effectiveness and efficiency of soil decontamination efforts. By incorporating nanosensors and advanced nanomaterials, researchers can achieve high-resolution detection of contaminants at unique levels of sensitivity. These nanosensors can provide immediate feedback on the presence and concentration of pollutants, enabling adaptive remediation strategies that respond dynamically to changing environmental conditions. Moreover, nanotechnology facilitates the development of multifunctional materials that aid in contaminant degradation and improve the transport and bioavailability of essential nutrients, thereby supporting the recovery of microbial communities essential for bioremediation. The integration of real-time data collection and analysis through wireless communication technologies further enhances the ability to monitor remediation processes continuously, allowing for timely interventions and optimized treatment protocols. Ultimately, it minimizes ecological risks, which leads to more sustainable management of contaminated sites.

11.10.6 Potential for Phytoremediation Enhancement

Nanotechnology can be utilized to enhance phytoremediation (the use of plants to remove contaminants). Nanoparticles can improve plant-microbe interactions, facilitating the uptake of pollutants and enhancing the overall remediation process. Example: ZnO nanoparticles can enhance the growth of sunflowers (*Helianthus annuus*) and increase their ability to uptake heavy metals like Cd from contaminated soils (Tovar-Lopez 2023).

11.11 Challenges in Microbial Nanotech Bioremediation

11.11.1 Potential Toxicity of Nanomaterials

While nanoparticles offer numerous advantages, their potential toxicity to microorganisms and other soil organisms is a major concern. Some nanomaterials can exhibit harmful effects on microbial communities, which may hinder bioremediation efforts. Understanding the toxicity levels and mechanisms of various nanoparticles is important to ensure that their application does not adversely affect the very organisms needed for remediation (Bundschuh et al. 2018).

11.11.2 Environmental Persistence of Nanoparticles

The persistence of nanoparticles in the environment raises concerns about their long-term effects. Some nanomaterials may accumulate in soils or leach into groundwater, posing risks to human health and ecosystems. Extensive studies are needed to assess the environmental fate and transport of nanoparticles used in bioremediation to mitigate potential risks associated with their prolonged presence in the environment (Thu et al. 2023).

11.11.3 Complex Interactions with Soil Matrix

Nanoparticles interact with soil matrices through complex physicochemical processes, including adsorption, electrostatic interactions, and diffusion dynamics, which are influenced by their size, shape, surface charge, and functionalization. The adsorption affinity of nanoparticles for soil colloids is significantly affected by soil parameters such as pH, ionic strength, and organic matter content, all of which can alter the surface charge of both nanoparticles and soil particles, thus impacting their transport and retention. Additionally, soil microorganisms play a crucial role in modulating these interactions, microbial exudates can change soil properties, facilitating or inhibiting nanoparticle aggregation and enhancing or reducing their bioavailability. For example, exopolymeric substances produced by microbes can stabilize nanoparticles and promote their uptake, potentially enhancing bioremediation efficacy. To optimize the use of nanoparticles in bioremediation, it is essential to conduct in-depth studies utilizing advanced analytical techniques, such as high-resolution transmission electron microscopy (HR-TEM) and surface-enhanced Raman spectroscopy (SERS), to elucidate the dynamic behavior of nanoparticles in relation to soil components and microbial communities. Understanding these intricate relationships will enable the design of nanomaterials that effectively target contaminants while promoting soil health.

11.11.4 Cost and Scalability of Nanotechnology Applications

The production and application of nanomaterials can be expensive, potentially limiting the scalability of microbial-nanotech bioremediation technologies. While research demonstrates their effectiveness in laboratory settings, translating these methods to field applications poses economic challenges. Identifying cost-effective methods for synthesizing and applying nanomaterials at a larger scale is necessary to ensure the feasibility of these technologies for widespread use (Pandey and Jain 2020).

11.11.5 Regulatory Framework and Public Acceptance

The introduction of nanotechnology into bioremediation processes raises regulatory and public acceptance issues. There are currently limited guidelines and regulations surrounding the use of nanoparticles in environmental applications, which can hinder the deployment of these technologies. Establishing comprehensive regulatory frameworks that ensure safety and efficacy while addressing public concerns is essential for the successful integration of nanotechnology in bioremediation practices (Zainab et al. 2024).

11.11.6 Variability in Microbial Response

Microbial communities in contaminated soils can vary based on environmental conditions, contaminant types, and other factors. This variability can affect the performance of nanotechnology-enhanced bioremediation strategies. Developing adaptable bioremediation approaches that consider the diverse responses of microbial communities to nanoparticles is necessary to achieve optimal results.

11.11.7 Limited Research on Long-Term Effects

While there is substantial interest in the combination of nanotechnology and microbial bioremediation, there is still limited research on the long-term effects of nanoparticles on soil health, microbial communities, and ecosystems. Conducting long-term studies to evaluate the impacts of nanotechnology applications in bioremediation will provide valuable insights into their sustainability and safety (Köhler 2021).

11.12 Conclusions and Future Directions

To fully harness the opportunities presented by microbial-nanotech bioremediation, several future directions must be considered. Multidisciplinary Research: Collaborative research efforts that integrate microbiology, nanotechnology, environmental science, and toxicology will be essential for understanding the complex interactions between nanoparticles, microbes, and contaminants. Field Trials: Conducting field trials to evaluate the effectiveness and safety of nanotechnology-enhanced bioremediation strategies will help to bridge the gap between laboratory research and practical applications. Public Engagement: Engaging with communities and stakeholders to address concerns and provide education on the benefits and risks of nanotechnology in bioremediation will promote public acceptance and support for these innovative approaches. Regulatory Development: Establishing clear regulatory guidelines and safety assessments for the use of nanoparticles in environmental applications will be crucial for ensuring responsible implementation. Sustainable Nanomaterial Development: Research into the synthesis of environmentally friendly and biodegradable nanomaterials will help to mitigate potential risks while maximizing the benefits of nanotechnology in bioremediation. In conclusion, the integration of nanotechnology with microbial bioremediation holds great promise for effectively addressing soil contamination and promoting environmental sustainability. While numerous opportunities exist to enhance the degradation of pollutants and improve remediation efficiency, major challenges must be addressed to ensure the successful application of these technologies. By pursuing multidisciplinary research, engaging with the public, and developing robust regulatory frameworks, we can unlock the full potential of microbial-nanotech bioremediation, paving the way for cleaner soils and healthier ecosystems. As we move forward, it is essential to balance innovation with responsibility, ensuring that the benefits of these advanced techniques are realized without compromising environmental health.

Acknowledgments

FZ expresses gratitude toward Vision Group of Science and Technology (VGST); Karnataka Science and Technology Promotion Society (KSTePS) and Department of Science and Technology, Government of Karnataka (GoK), Bangalore, for K-FIST L1 grant with GRD No. 1051/2023-2025. All authors thank Dr. M. Mohan Alva, Chairman, Mr. Vivek Alva, Dr. Vinay Alva, Managing trustee, Alva's Education Foundation (AEF) and Dr. Sajith M., the Principal, Alva's Ayurveda Medical College for their constant support and motivation. All authors credit BioRender.com for design of the figures in this chapter.

References

Aasfar, A., Bargaz, A., Yaakoubi, K. et al. (2021). Nitrogen fixing *Azotobacter* species as potential soil biological enhancers for crop nutrition and yield stability. *Frontiers in Microbiology* 12: 628379. https://doi.org/10.3389/fmicb.2021.628379.

Abbas, S.Z. and Rafatullah, M. (2021). Recent advances in soil microbial fuel cells for soil contaminants remediation. *Chemosphere* 272: 129691. https://doi.org/10.1016/j.chemosphere.2021.129691.

AbdElgawad, H., Abuelsoud, W., Madany, M.M.Y. et al. (2020). Actinomycetes enrich soil rhizosphere and improve seed quality as well as productivity of legumes by boosting nitrogen availability and metabolism. *Biomolecules* 10 (12): 1675. https://doi.org/10.3390/biom10121675.

Acharya, M., Shobha, R., Raju Krishna, C., and Zameer, F. (2023a). Agrobotany – Chapter 1, Origin of cultivated plants. In: *Text Book in Accordance to NEP Scheme, Govt. of India*, 1e (vol. 1). Mangalore, Karnataka, India: United Agencies.

Acharya, M., Ramya Rai, P.D., Divakar, et al. (2023b). Basics of Gardening: Exploring Newer Paradigm in Plant-Field Techniques – Chapter 9, PrecisionGardening. In: *Text Book in Accordance to NEP Scheme, Govt. of India,* 1e (vol. 1). Bangalore, Karnataka, India: Chethan Book Publishers.

Akki, A.J., Jain, P., Kulkarni, R. et al. (2024). Microbial biotechnology alchemy: transforming bacterial cellulose into sensing disease—a review. *Sensors International* 100277. https://doi.org/10.1016/j.sintl.2023.100277.

Alengebawy, A., Abdelkhalek, S.T., Qureshi, S.R., and Wang, M.Q. (2021). Heavy metals and pesticides toxicity in agricultural soil and plants: ecological risks and human health implications. *Toxics* 9 (3): 42. https://doi.org/10.3390/toxics9030042.

Alloway, B.J. (ed.) (2013). *Heavy Metals in Soils. Environmental Pollution*. Springer Netherlands https://doi.org/10.1007/978-94-007-4470-7.

Anthony, M.A., Bender, S.F., and van der Heijden, M.G.A. (2023). Enumerating soil biodiversity. *Proceedings of the National Academy of Sciences of the United States of America* 120 (33): e2304663120. https://doi.org/10.1073/pnas.2304663120.

Atuchin, V.V., Asyakina, L.K., Serazetdinova, Y.R. et al. (2023). Microorganisms for bioremediation of soils contaminated with heavy metals. *Microorganisms* 11 (4): 864. https://doi.org/10.3390/microorganisms11040864.

Avinash, M.G., Aishwarya, S., Zameer, F., and Gopal, S. (2023). *Pseudomonas aeruginosa* biofilm and their molecular escape strategies. *Journal of Applied Biology and Biotechnology* https://doi.org/10.7324/jab.

Bakki, M., Banane, B., Marhane, O. et al. (2024). Phosphate solubilizing *Pseudomonas* and *Bacillus* combined with rock phosphates promoting tomato growth and reducing bacterial canker disease. *Frontiers in Microbiology* 15: 1289466. https://doi.org/10.3389/fmicb.2024.1289466.

Beulah, K.C., Aishwarya, T., Waseem, K. et al. (2015). Phyto-antiquorumones: an herbal approach for blocking bacterial trafficking and pathogenesis. *International Journal of Pharmacy and Pharmaceutical Science* 7 (1): 29–34. https://journals.innovareacademics.in/index.php/ijpps/article/view/8543.

Briffa, J., Sinagra, E., and Blundell, R. (2020). Heavy metal pollution in the environment and their toxicological effects on humans. *Heliyon* 6: e04691. https://doi.org/10.1016/j.heliyon.2020.e04691.

Bundschuh, M., Filser, J., Lüderwald, S. et al. (2018). Nanoparticles in the environment: where do we come from, where do we go to? *Environmental Sciences Europe* 30 (1): 6. https://doi.org/10.1186/s12302-018-0132-6.

Certini, G. and Scalenghe, R. (2023). The crucial interactions between climate and soil. *The Science of the Total Environment* 856 (Pt 2): 159169. https://doi.org/10.1016/j.scitotenv.2022.159169.

Chtouki, M., Laaziz, F., Naciri, R. et al. (2022). Interactive effect of soil moisture content and phosphorus fertilizer form on chickpea growth, photosynthesis, and nutrient uptake. *Scientific Reports* 12 (1): 6671. https://doi.org/10.1038/s41598-022-10703-0.

Ciric, V., Prekop, N., Seremesic, S. et al. (2023). The implication of cation exchange capacity (CEC) assessment for soil quality management and improvement. *Agriculture and Forestry* 69 (4): 89–102. https://doi.org/10.17707/AgricultForest.69.4.08.

Das, H., Singh, S.K., Priyadarshini, E. et al. (2024). Nanotechnology facilitated real-time soil monitoring for optimized crop production. *Asian Journal of Soil Science and Plant Nutrition* 10 (2): 582–607. https://doi.org/10.9734/ajsspn/2024/v10i2315.

Daunoras, J., Kačergius, A., and Gudiukaitė, R. (2024). Role of soil microbiota enzymes in soil health and activity changes depending on climate change and the type of soil ecosystem. *Biology* 13 (2): 85. https://doi.org/10.3390/biology13020085.

Deemy, J.B., Besterman, A.F., Hall, B.M. et al. (2022). Chapter 6—Nutrient cycling. In: *Fundamentals of Tropical Freshwater Wetlands* (ed. T. Dalu and R.J. Wasserman), 133–160. Elsevier https://doi.org/10.1016/B978.0-12-822362-8.00017-7.

Devlin, M. and Brodie, J. (2023). Nutrients and eutrophication. In: *Marine Pollution—Monitoring, Management and Mitigation. Springer Textbooks in Earth Sciences, Geography and Environment* (ed. A. Reichelt-Brushett). Cham: Springer https://doi.org/10.1007/978-3-031-10127-4_4.

Dewangan, S.K., Shrivastava, S., Kumari, L. et al. (2023). The effects of soil pH on soil health and environmental sustainability: a review. *Journal of Emerging Technologies and Innovative Research* 10: 611–616.

Dhanapal, A.R., Thiruvengadam, M., Vairavanathan, J. et al. (2024). Nanotechnology approaches for the remediation of agricultural polluted soils. *ACS Omega* 9 (12): 13522–13533. https://doi.org/10.1021/acsomega.3c09776.

Dinesh, B., Poyya, J., Zameer, F. et al. (2024). Effect of polyvinylpyrrolidone on antioxidant and antibacterial activity of silver metal nanoparticles: a comparative analysis. *Proceedings of the National Academy of Sciences, India Section A: Physical Sciences* 94: 359–368. https://doi.org/10.1007/s40010-024-00882-y.

FAO & UNEP (2021). Chapter 3—Sources of soil pollution. In: *Global Assessment of Soil Pollution: Report* (ed. FAO & UNEP). FAO & UNEP https://doi.org/10.4060/cb4894en.

Ghezzehei, T. A. (2012). Soil structure. In Handbook of Soil Sciences: Vol. 1 Properties and Processes (2nd ed., pp. 1–17). CRC Press. https://doi.org/10.1201/b16386-7

Gujre, N., Soni, A., Rangan, L. et al. (2021). Sustainable improvement of soil health utilizing biochar and arbuscular mycorrhizal fungi: a review. *Environmental Pollution (Barking, Essex: 1987)* 268 (Pt B): 115549. https://doi.org/10.1016/j.envpol.2020.115549.

Huang, K., Li, M., Li, R. et al. (2023). Soil acidification and salinity: the importance of biochar application to agricultural soils. *Frontiers in Plant Science* 14: 1206820. https://doi.org/10.3389/fpls.2023.1206820.

Jakhar, A.M., Aziz, I., Kaleri, A.R. et al. (2022). Nano-fertilizers: a sustainable technology for improving crop nutrition and food security. *NanoImpact* 27: 100411. https://doi.org/10.1016/j.impact.2022.100411.

Joutey, N.T., Sayel, H., Bahafid, W., and El Ghachtouli, N. (2015). Mechanisms of hexavalent chromium resistance and removal by microorganisms. *Reviews of Environmental Contamination and Toxicology* 233: 45–69. https://doi.org/10.1007/978-3-319-10479-9_2.

Kaur, R. and Kaur, S. (2020). Variation in the phosphate solubilizing bacteria from Virgin and the agricultural soils of Punjab. *Current Microbiology* 77 (9): 2118–2127. https://doi.org/10.1007/s00284-020-02080-6.

Köhler, J.M. (2021). Challenges for nanotechnology. *Encyclopedia* 1: 618–631. https://doi.org/10.3390/encyclopedia1030051.

Li, Z., Tian, D., Wang, B. et al. (2019). Microbes drive global soil nitrogen mineralization and availability. *Global Change Biology* 25 (3): 1078–1088. https://doi.org/10.1111/gcb.14557.

Lloret, J., Sendra, S., Garcia, L., and Jimenez, J.M. (2021). A wireless sensor network deployment for soil moisture monitoring in precision agriculture. *Sensors* 21 (21): 7243. https://doi.org/10.3390/s21217243.

Lu, C., Zhu, Q., Qiu, M. et al. (2023). Effects of different soil water holding capacities on vegetable residue return and its microbiological mechanism. *Frontiers in Microbiology* 14: 1257258. https://doi.org/10.3389/fmicb.2023.1257258.

Maire, J., Davarzani, H., Colombano, S., and Fatin-Rouge, N. (2019). Targeted delivery of hydrogen for the bioremediation of aquifers contaminated by dissolved chlorinated compounds. *Environmental Pollution* 249: 443–452. https://doi.org/10.1016/j.envpol.2019.03.033.

Mandeep and Shukla, P. (2020). Microbial nanotechnology for bioremediation of industrial wastewater. *Frontiers in Microbiology* 11: 590631. https://doi.org/10.3389/fmicb.2020.590631.

Manjunatha, H. P., Singh, H., Chauhan, J. B., et al. (2013). Induction of resistance against sorghum downy mildew by seed treatment with Duranta repens extracts. *IOSR Journal of Agriculture and Veterinary Science* 3: 37–44. https://scispace.com/pdf/induction-of-resistance-against-sorghum-downy-mildew-byseed-55fv805qp1.pdf

Manohara, A., Shobha, R., Raju Krishna, C., and Zameer, F. (2023). Agrobotany—Chapter 1—Origin of cultivated plants. In: *Text Book in Accordance to NEP Scheme*, vol. 1, 1–44. Govt. of India.

Mansor, M. and Xu, J. (2020). Benefits at the nanoscale: a review of nanoparticle-enabled processes favouring microbial growth and functionality. *Environmental Microbiology* 22: 3633–3649. https://doi.org/10.1111/1462-2920.15174.

Mathew, T., Sree, R. A., Aishwarya, S., et al. (2020). Graphene-based functional nanomaterials for biomedical and bioanalysis applications. *FlatChem*, 23: 100184. https://doi.org/10.1016/j.flatc.2020.100184

More, V.S. et al. (2021). Isolation and purification of microbial exopolysaccharides and their industrial application. In: *Microbial Polymers* (ed. A. Vaishnav and D.K. Choudhary). Springer, Singapore https://doi.org/10.1007/978-981-16-0045-6_3.

Muthukrishnan, L. (2022). An overview on the nanotechnological expansion, toxicity assessment and remediating approaches in agriculture and food industry. *Environmental Technology & Innovation* 25: 102136. https://doi.org/10.1016/j.eti.2021.10.136.

Neogi, S., Bhattacharyya, P., Roy, K.S. et al. (2014). Soil respiration, labile carbon pools, and enzyme activities as affected by tillage practices in a tropical rice-maize-cowpea cropping system. *Environmental Monitoring and Assessment* 186 (7): 4223–4236. https://doi.org/10.1007/s10661-014-3693-x.

Noble, D.T., MacDougall, A.S., and Levison, J. (2023). Impacts of soil, climate, and phenology on retention of dissolved agricultural nutrients by permanent-cover buffers. *The Science of the Total Environment* 860: 160532. https://doi.org/10.1016/j.scitotenv.2022.160532.

Nongbet, A., Mishra, A.K., Mohanta, Y.K. et al. (2022). Nanofertilizers: a smart and sustainable attribute to modern agriculture. *Plants (Basel, Switzerland)* 11 (19): 2587. https://doi.org/10.3390/plants11192587.

O'Rourke, S.M., Angers, D.A., Holden, N.M., and McBratney, A.B. (2015). Soil organic carbon across scales. *Global Change Biology* 21 (10): 3561–3574. https://doi.org/10.1111/gcb.12959.

Pande, V., Pandey, S.C., Sati, D. et al. (2022). Microbial interventions in bioremediation of heavy metal contaminants in agroecosystem. *Frontiers in Microbiology* 13: 824084. https://doi.org/10.3389/fmicb.2022.824084.

Pandey, G. and Jain, P. (2020). Assessing the nanotechnology on the grounds of costs, benefits, and risks. *Beni-Suef University Journal of Basic and Applied Sciences* 9: 63. https://doi.org/10.1186/s43088-020-00085-5.

Parikh, S.J. and James, B.R. (2012). Soil: the foundation of agriculture. *Nature Education Knowledge* 3 (10): 2.

Patil, A. G., Kounaina, K., Aishwarya, S., et al. (2021). Myco-nanotechnology for sustainable agriculture: challenges and opportunities. *Recent Trends in Mycological Research: Volume 1: Agricultural and Medical Perspective*, 457–479. https://link.springer.com/chapter/10.1007/978-3-030-60659-6_20

Patil, A. G., Khan, K., Aishwarya, S., et al. (2022). Fungal amylases and their industrial applications. In: *Industrially Important Fungi for Sustainable Development: Volume 2: Bioprospecting for Biomolecules* (pp. 407–434). Cham: Springer International Publishing. https://link.springer.com/chapter/10.1007/978-3-030-85603-8_11

Pervaiz, I., Ahmad, S., Madni, M.A. et al. (2013). Microbial biotransformation: a tool for drug designing. *Applied Biochemistry and Microbiology* 49: 437–450. https://doi.org/10.7868/s0555109913050097.

Poveda, J. and Eugui, D. (2022). Combined use of *Trichoderma* and beneficial bacteria (mainly *Bacillus* and *Pseudomonas*): development of microbial synergistic bio-inoculants in sustainable agriculture. *Biological Control* 176: 105100. https://doi.org/10.1016/j.biocontrol.2022.10.100.

Prasad, A., Baker, S., Nagendra Prasad, M. N., et al. (2019). Phytogenic synthesis of silver nanobactericides for anti-biofilm activity against human pathogen *H. pylori*. *SN Applied Sciences* 1 (4): 341. https://doi.org/10.1007//s42452-019-0362-2

Prommer, J., Walker, T.W.N., Wanek, W. et al. (2020). Increased microbial growth, biomass, and turnover drive soil organic carbon accumulation at higher plant diversity. *Global Change Biology* 26 (2): 669–681. https://doi.org/10.1111/gcb.14777.

Raffa, C.M. and Chiampo, F. (2021). Bioremediation of agricultural soils polluted with pesticides: a review. *Bioengineering (Basel, Switzerland)* 8 (7): 92. https://doi.org/10.3390/bioengineering8070092.

Rajak, P., Ganguly, A., Adhikary, S., and Bhattacharya, S. (2023). Internet of Things and smart sensors in agriculture: scopes and challenges. *Journal of Agriculture and Food Research* 14: 100776. https://doi.org/10.1016/j.jafr.2023.100776.

Remya, R.R., Julius, A., Suman, T.Y. et al. (2022). Role of nanoparticles in biodegradation and their importance in environmental and biomedical applications. *Journal of Nanomaterials* 2022: 6090846. https://doi.org/10.1155/2022/6090846.

Robe, K. and Barberon, M. (2023). Nutrient carriers at the heart of plant nutrition and sensing. *Current Opinion in Plant Biology* 74: 102376. https://doi.org/10.1016/j.pbi.2023.102376.

Sáez-Sandino, T., Delgado-Baquerizo, M., Egidi, E., and Singh, B.K. (2023). New microbial tools to boost restoration and soil organic matter. *Microbial Biotechnology* 16 (11): 2019–2025. https://doi.org/10.1111/1751-7915.14325.

Sahu, B., Sharma, D. D., Jayakumar, G. C., Madhan, B., & Zameer, F. (2023). A review on an imperative by-product: glycosaminoglycans—a holistic approach. *Carbohydrate Polymer Technologies and Applications*, 5, 100275. 10.1016/j.carpta.2022.100275

Sahu, B., Ramesh, A., and Zameer, F. (2024). Impact of nanomaterials on leather: a nano-Saga from processing to application. *Clean Technologies and Environmental Policy* 1-28: https://doi.org/10.1007/s10098-024-02912-0.

Sajjad, M., Huang, Q., Khan, S. et al. (2022). Microplastics in the soil environment: a critical review. *Environmental Technology & Innovation* 27: 102408. https://doi.org/10.1016/j.eti.2022.102408.

Salehi Hikouei, I., Kim, S.S., and Mishra, D.R. (2021). Machine-learning classification of soil bulk density in salt marsh environments. *Sensors (Basel, Switzerland)* 21 (13): 4408. https://doi.org/10.3390/s21134408.

Satapathy, P., Prakash, J. K., Gowda, V. C., et al. (2022). Targeting Imd pathway receptor in Drosophila melanogaster and repurposing of phytoinhibitors: structural modulation and molecular dynamics. *Journal of Biomolecular Structure and Dynamics* 40 (4): 1659–1670. https://doi.org/10.1080/07391102.2020.1831611

Selmy, S.A.H., Abd Al-Aziz, S.H., Ibrahim, A.G., and Jiménez-Ballesta, R. (2022). Impact of short-term cultivation on some selected properties of sandy soil in an arid environment. *Soil Systems* 6 (4): 82. https://doi.org/10.3390/soilsystems6040082.

Sharma, R. and Kumar, V. (2024). Nano enabled agriculture for sustainable soil. *Waste Management Bulletin* 2 (1): 152–161. https://doi.org/10.1016/j.wmb.2024.01.002.

Singh, H., Zameer, F., Khanum, S.A. et al. (2016). Durantol—A phytosterol antifungal contributor from *Duranta repens* Linn. For organic management of *Sorghum* Downy Mildew. *European Journal of Plant Pathology* 146: 671–682. https://doi.org/10.1007/s10658-016-0951-5.

Srichandan, H., Pathak, A., Singh, S. et al. (2014). Sequential leaching of metals from spent refinery catalyst in bioleaching–bioleaching and bioleaching–chemical leaching reactor: comparative study. *Hydrometallurgy* 150: 130–143. https://doi.org/10.1016/j.hydromet.2014.09.019.

Srivastav, A., Yadav, K., Yadav, S. et al. (2019). Nano-phytoremediation of pollutants from contaminated soil environment: current scenario and future prospects. *Management of Environmental Contaminants* 6: 10.1007/978-3-319-99651-6_16.

Stenberg, B., Viscarra Rossel, R.A., Mouazen, A.M., and Wetterlind, J. (2010). Chapter five—Visible and near infrared spectroscopy in soil science. In: *Advances in Agronomy*, vol. 107 (ed. D.L. Sparks), 163–215. Academic Press https://doi.org/10.1016/S0065-2113(10)07005-7.

Sulthana, R., Taqui, S.N., Zameer, F. et al. (2018). Adsorption of ethidium bromide from aqueous solution onto nutraceutical industrial fennel seed spent: kinetics and thermodynamics modeling studies. *International Journal of Phytoremediation* 20 (11): 1075–1086. https://doi.org/10.1080/15226514.2017.1365331.

Thu, H.E., Haider, M., Khan, S. et al. (2023). Nanotoxicity induced by nanomaterials: a review of factors affecting nanotoxicity and possible adaptations. *OpenNano* 14: 100190. https://doi.org/10.1016/j.onano.2023.10.190.

Tovar-Lopez, F.J. (2023). Recent progress in micro- and nanotechnology-enabled sensors for biomedical and environmental challenges. *Sensors (Basel, Switzerland)* 23 (12): 5406. https://doi.org/10.3390/s23125406.

Upadhayay, V.K., Chitara, M.K., Mishra, D. et al. (2023). Synergistic impact of nanomaterials and plant probiotics in agriculture: a tale of two-way strategy for long-term sustainability. *Frontiers in Microbiology* 14: 1133968. https://doi.org/10.3389/fmicb.2023.1133968.

Vaksmaa, A., Guerrero-Cruz, S., Ghosh, P. et al. (2023). Role of fungi in bioremediation of emerging pollutants. *Frontiers in Marine Science* 10: 1070905. https://doi.org/10.3389/fmars.2023.1070905.

Vera, M., Schippers, A., Hedrich, S., and Sand, W. (2022). Progress in bioleaching: fundamentals and mechanisms of microbial metal sulfide oxidation—Part A. *Applied Microbiology and Biotechnology* 106 (21): 6933–6952. https://doi.org/10.1007/s00253-022-12168-7.

Wang, C. and Kuzyakov, Y. (2024). Soil organic matter priming: the pH effects. *Global Change Biology* 30 (6): e17349. https://doi.org/10.1111/gcb.17349.

Wang, L.Y., Xiao, Y., Rao, E.M. et al. (2018). An assessment of the impact of urbanization on soil erosion in inner mongolia. *International Journal of Environmental Research and Public Health* 15 (3): 550. https://doi.org/10.3390/ijerph15030550.

Wang, X., Chi, Y., and Song, S. (2024). Important soil microbiota's effects on plants and soils: a comprehensive 30-year systematic literature review. *Frontiers in Microbiology* 15: 1347745. https://doi.org/10.3389/fmicb.2024.1347745.

Yadav, A.N. (2021). *Recent Trends in Mycological Research*. Springer International Publishing https://doi.org/10.1007/978-3-030-68260-6.

Yadav, N., Garg, V.K., Chhillar, A.K., and Rana, J.S. (2023). Recent advances in nanotechnology for the improvement of conventional agricultural systems: a review. *Plant Nano Biology* 4: 100032. https://doi.org/10.1016/j.plana.2023.100032.

Yang, K., Abu-Reesh, I.M., and He, Z. (2023). Enhancing organic contaminant degradation through integrating advanced oxidation processes with microbial electrochemical systems. *Journal of Hazardous Materials Letters* 4: 100075. https://doi.org/10.1016/j.hazl.2023.100075.

Yousef, H.A., Fahmy, H.M., Arafa, F.N. et al. (2023). Nanotechnology in pest management: advantages, applications, and challenges. *International Journal of Tropical Insect Science* 43: 1387–1399. https://doi.org/10.1007/s42690-023-01053-z.

Zainab, R., Hasnain, M., Ali, F. et al. (2024). Prospects and challenges of nanopesticides in advancing pest management for sustainable agricultural and environmental service. *Environmental Research* 261: 119722. https://doi.org/10.1016/j.envres.2024.119722.

Zameer, F. and Gopal, S. (2010). Evaluation of antibiotic susceptibility in mixed culture biofilms. *International Journal of Biotechnology and Biochemistry* 6 (1): 93–99. https://www.researchgate.net/publication/242100772.

Zameer, F., Kreft, J., and Gopal, S. (2010). Interaction of the dual species biofilms of *Listeria monocytogenes* and *Staphylococcus epidermidis*. *Journal of Food Safety* 30 (4): 954–968. https://doi.org/10.1111/j.1745-4565.2010.00254.x.

Zameer, F., Rukmangada, M.S., Chauhan, J.B. et al. (2016). Evaluation of adhesive and anti-adhesive properties of *Pseudomonas aeruginosa* biofilms and their inhibition by herbal plants. *Iranian Journal of Microbiology* 8 (2): 108. https://pubmed.ncbi.nlm.nih.gov/27307976/.

Zhang, L., Gadd, G.M., and Li, Z. (2021). Microbial biomodification of clay minerals. *Advances in Applied Microbiology* 114: 111–139. https://doi.org/10.1016/bs.aambs.2020.07.002.

Zhou, H., Whalley, W.R., Hawkesford, M.J. et al. (2021). The interaction between wheat roots and soil pores in structured field soil. *Journal of Experimental Botany* 72 (2): 747–756. https://doi.org/10.1093/jxb/eraa475.

12

Nanoparticles for Bioremediation of Heavy Metals from Agricultural Fields

Dalaq Aiysha[1] *and Salma Mukhtar*[2,3]

[1] *Department of Medical Laboratory Technology, Green International University, Lahore, Pakistan*
[2] *KAM School of Life Sciences, Forman Christian College (A Chartered University), Lahore, Pakistan*
[3] *Department of Plant Biology, Rutgers University, New Brunswick, NJ, USA*

12.1 Introduction

Soil forms the fundamental layer of the pedosphere and serves as a vital component sustaining life on Earth. Agricultural soils comprise a dynamic mixture of living organisms, mineral and organic fractions, water, and gases (Brevik et al. 2015). They support terrestrial ecosystems by providing habitat for diverse biota, regulating atmospheric exchange, filtering and storing water, and functioning as the growth substrate for plants. Because productive soils develop extremely slowly only a few centimeters over thousands of years preserving their quality is essential to meet global sustainability goals (Montgomery 2007; Wuepper et al. 2020).

Healthy soils are crucial for achieving multiple United Nations Sustainable Development Goals (SDGs), with at least eight of the seventeen SDGs directly linked to maintaining global soil health. Productive and uncontaminated soils are vital for mitigating climate change, as polluted soils cannot efficiently participate in the global carbon cycle (Hou et al. 2020; Le Quéré et al. 2019). Increasing recognition of soil degradation and contamination as pressing global issues has prompted several international efforts. At its 68th session, the United Nations General Assembly emphasized the need for soil conservation by designating December 5 as World Soil Day and declaring 2015 the International Year of Soils. In 2017, the United Nations Environment Assembly adopted a resolution urging the World Health Organization and the Food and Agriculture Organization to evaluate global soil pollution, its associated risks, and the consequences for sustainable development (FAO 2019; UNEP 2018).

Soil contamination occurs through the accumulation of toxic compounds, including organic, inorganic, and metallic substances, which are not naturally present in soils (Vázquez-Núñez et al. 2020). Metal elements characterized by properties such as cation stability, conductivity, ligand specificity, malleability, and ductility can become environmentally hazardous when present in excessive amounts, often referred to as heavy

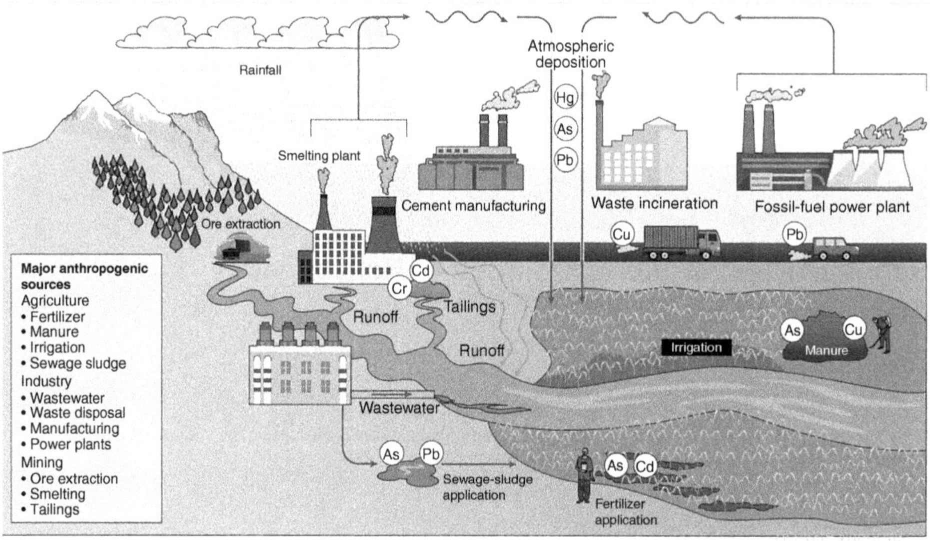

Figure 12.1 Sources of heavy metal pollution in agricultural fields. *Source:* Hou et al. (2020)/with permission of Springer Nature.

metals. Some, including zinc (Zn), vanadium (V), nickel (Ni), molybdenum (Mo), manganese (Mn), iron (Fe), copper (Cu), and cobalt (Co), are essential micronutrients critical for metabolic processes. Others, such as arsenic (As), mercury (Hg), cadmium (Cd), and lead (Pb), pose significant ecological and health risks (Pourret and Hursthouse 2019). Because heavy metals are nondegradable and tend to accumulate in the environment, their persistence in agricultural soils can lead to entry into food webs, threatening ecosystems and human health (Gall et al. 2015). Key anthropogenic sources of heavy metal contamination include fossil fuel combustion, tanning and smelting industries, mining and metal extraction, chemical manufacturing, and the overuse of synthetic fertilizers in agriculture (Tang et al. 2020) (Figure 12.1).

Integrating nanotechnology with biotechnology presents a promising strategy for the bioremediation of heavy metals in polluted environments. Nanoscale materials, characterized by their high surface area-to-volume ratios, superior reactivity, and catalytic properties, have significantly enhanced the efficiency of metal removal processes (Nikhil et al. 2020). In comparison with traditional physicochemical remediation techniques, nanoparticle-assisted bioremediation is increasingly valued for being environmentally friendly, cost-effective, and largely self-sustaining (Mohanty et al. 2020). Nanobioremediation generally follows two primary approaches. The first involves the direct deployment of nanoparticles at contaminated sites to promote in situ adsorption, reduction, or chemical transformation of heavy metals. The second, indirect method focuses on boosting microbial activity either by immobilizing beneficial microorganisms or by stimulating the production of metal-detoxifying enzymes, thereby enhancing the overall remediation efficiency (Benjamin et al. 2019). This chapter focuses on the major aspects of heavy metal contamination in agricultural soils, its impacts on crop productivity and human health, and emphasizes the potential of nanoparticles as sustainable, next-generation solutions for mitigating heavy metal pollution in terrestrial ecosystems.

12.2 Heavy Metal Contamination in Agricultural Fields

Humans have been utilizing heavy metals extracted from mineral ores for thousands of years, and their use remains widespread across agricultural, domestic, and industrial applications (Järup 2003). Despite the known health and environmental risks, the demand for these metals continues to grow, particularly in the renewable energy and technology sectors, including solar panel manufacturing and electric vehicle battery production (Lakherwal 2014; Li et al. 2016; Teske 2019). The extensive use of heavy metals, coupled with their tendency to precipitate, adsorb, and persist in the environment, has led to their accumulation as significant pollutants in agricultural soils.

Major sources of heavy metal buildup in cultivated lands include irrigation with contaminated water, application of sewage sludge or low-quality phosphate fertilizers, and runoff from mining and smelting activities (Atafar et al. 2010; Balkhair and Ashraf 2016; Boussen et al. 2013; Chaoua et al. 2019; Navarro et al. 2008; Srivastava et al. 1989; Walter et al. 2006). Atmospheric deposition also contributes substantially, largely through emissions from e-waste processing, cement production, vehicle exhaust, fossil fuel combustion, and metal extraction industries (Bi et al. 2006; Luo et al. 2011; Noli and Tsamos 2016; Ogunkunle and Fatoba 2013; Zhang et al. 2012).

Anthropogenic sources remain the primary drivers of heavy metal contamination in agricultural systems, particularly for elements such as zinc, copper, mercury, and lead (Hou et al. 2017). The historical use of lead (Pb) in gasoline and vehicle components has caused severe environmental accumulation of this metal (Salehi et al. 2016). Similarly, the application of manure and impure fertilizers has increased soil concentrations of zinc, copper, and cadmium, while irrigation with wastewater has significantly elevated heavy metal levels in soils (Alidokht et al. 2011; Lu et al. 2012; Nziguheba and Smolders 2008; Wang et al. 2018). Long-term studies have shown that three decades of wastewater reuse for irrigation resulted in substantial accumulation of chromium, cadmium, and lead in agricultural soils (Khan et al. 2008) (Table 12.1).

In addition to human-related sources, atmospheric deposition of heavy metals poses a significant challenge for remediation. This pathway is particularly problematic because it often goes unmonitored, yet it contributes substantially to metal accumulation in agricultural soils (Nicholson et al. 2003). Estimates suggest that atmospheric fallout is responsible for roughly 85% of mercury (Hg), 78% of lead (Pb), 60% of nickel (Ni), 56% of arsenic (As), and 53% of cadmium (Cd) present in farmland soils (Nicholson et al. 2003).

Geogenic sources represent another major contributor to heavy metal contamination (Marrugo-Negrete et al. 2017). For instance, cadmium (Cd) commonly occurs in sulfide minerals associated with copper, lead, and zinc ores (Mulligan et al. 2001). The presence of metals such as manganese, cobalt, and chromium in soils often reflects their lithogenic origin. Additionally, several studies have shown that chromium (Cr) contamination

Table 12.1 Global estimation of heavy metal polluted soils and their nanoparticles-based bioremediation.

Type of heavy metal pollution	Pollution intensity (mg/kg)	Nanoparticles applied	Removal efficacy	Reference
Arsenic-polluted soil	315	nZVI	40.0% decreased arsenic bio-accessibility	Zhang et al. (2010)
Arsenic-polluted soil	5800	nZVI	Increased arsenic residual fraction in soil	Gil-Diaz et al. (2016)
Lead-polluted soil	9	nZVI	Increased lead residual fraction in soil	Mar Gil-Diaz et al. (2014)
Zinc-polluted soil	20	nZVI	Increased zinc residual fraction in soil	Mar Gil-Diaz et al. (2014)
Lead-polluted soil	132	nZVI	83% lead removal efficacy achieved	Wang et al. (2014)
Chromium(VI)-polluted soil	120	nZVI	99% chromium(VI) reduction efficacy achieved at pH = 5	Zhu et al. (2016)
Chromium(VI)-polluted soil	15.84	nZVI	14.51–86.83% chromium reduction efficacy	Singh et al. (2012)
Chromium(VI)-polluted soil	320	nZVI	91.94–100% chromium reduction efficacy achieved	Su et al. (2016)
Chromium(VI)-polluted soil	400	nZVI	Reduced chromium bioavailability	Alidokht et al. (2011)
Lead, cadmium, chromium multimetal-polluted soil	Cd(360), Pb(600), Cr(80)	nZVI	Reduced cadmium, lead, chromium bioavailability, and toxicity	Chen et al. (2015)
Antimony-contaminated soil	0.685	nZVI	Reduced contamination concentration	Dorjee et al. (2014)
Uranium-polluted soil	50	nZVI	5–10 times increased uranium absorption capacity	Zhang et al. (2015)

originating from parent soil materials is frequently accompanied by elevated concentrations of arsenic (As), cadmium (Cd), and nickel (Ni) in agricultural fields (Facchinelli et al. 2001; Li et al. 2004; Shan et al. 2013; Zhou et al. 2016; Zhu et al. 2016).

12.3 Conventional Methods for the Removal of Heavy Metals from Agricultural Lands

Environmental remediation of heavy metals can generally be classified into four main categories: physical, chemical, biological, and nanotechnological methods, each offering distinct benefits and facing specific limitations. Physical techniques primarily rely on separation processes, including precipitation, ion exchange, electrowinning, electrocoagulation, cementation, membrane filtration, and electrodialysis. Chemical approaches typically involve reactions such as reduction, chemical washing, or chelate-assisted extraction to remove or transform metal contaminants. Biological strategies utilize living organisms—such as bacteria, fungi, algae, and plants to immobilize, absorb, or convert heavy metals through processes like bioremediation, biofiltration, biosorption, bioaugmentation, and bacterial sulfate reduction.

Recently, nano-bioremediation, which integrates nanotechnology with biological remediation, has emerged as a highly efficient and sustainable strategy for heavy metal removal from contaminated soils. By enhancing the reactivity, stability, and bioavailability of remediation agents, this approach significantly improves the effectiveness of detoxification processes (Tripathi et al. 2018).

12.4 Nanoparticles

In particle science, the study of particulate materials has long been a focus of scientific research, and nanoparticles have drawn particular attention. Nanoparticles are generally defined as particles with at least two or three dimensions ranging from 1 to 100 nm (Vázquez-Núñez et al. 2020). They occupy the scale between atomic or molecular structures and bulk materials, allowing the exploitation of unique properties not observed in larger-scale counterparts. Unlike bulk materials, whose physical properties remain largely consistent regardless of size or shape, nanoparticles exhibit highly tunable characteristics (Figure 12.2). Changes in their shape, size, aspect ratio, and morphology can substantially influence their physical, chemical, and biological behavior. These unique properties arise

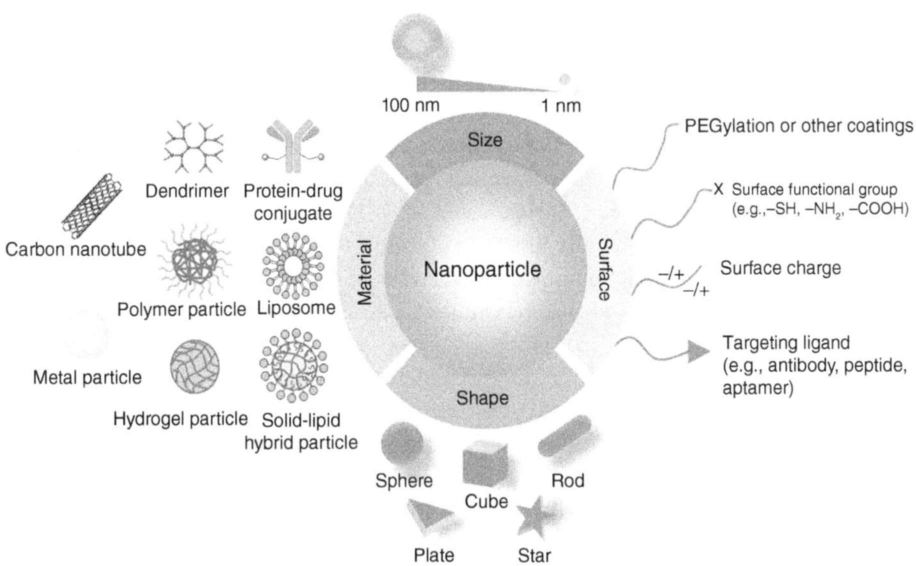

Figure 12.2 Characterization of nanoparticles according to their shape, material, size and surface. *Source:* Heinz et al. (2017)/Elsevier/CC BY 4.0.

from their nanoscale dimensions, where materials often behave differently compared to their bulk form. For instance, aluminum nanoparticles can become highly explosive (Vignes et al. 2019), gold nanoparticles show altered melting points (Essajai et al. 2018), and silver nanoparticles possess strong bactericidal and deodorizing effects (Abbasi et al. 2020; Weerasinghe et al. 2020).

The first use of nanoparticles in history was the making of "Lycurgus cup" in 4th century by Romans that was made using 50–100 nm nanoparticles of Ag:Au (silver-gold alloy) at 7:3 (Barber and Freestone 1990). The next use of nanoparticles in history was evident in the making of luster ceramic glazes in the 9th to 17th centuries by the Islamic world and Europeans (Pradell et al. 2007). In the 19th century, Michael Faraday used colloidal suspensions of Ruby gold and discovered that microparticles of gold are responsible for different color light scattering properties, which is now known as Faraday–Tyndall effect (Faraday 1857; Poole and Owens 2003). Such patterns of nanoparticle use and modification paved ways for future innovations and utilizations of nanoparticles with their underdefined intrinsic features (Figure 12.3).

Today, the remarkable potential of nanoparticles across diverse fields of science and technology is undeniable. Their core material composition, combined with the functional modifications of their surfaces, provides an exceptional range of chemical and physical versatility. Owing to these tunable properties, nanoparticles have found extensive applications in water and air purification systems, energy conversion and storage devices, catalysis, composite materials, targeted drug delivery, paints and coatings, medical imaging, and environmental remediation of toxic pollutants, as illustrated in Figure 12.4 (Heinz et al. 2017).

12.5 Nano-remediation of Soils and Groundwater with Nanoparticles

The application of nanoscale elemental or zero-valent metals, such as iron, nickel, and palladium has demonstrated significant potential in cleaning up environments polluted with hazardous substances. These metal nanoparticles are especially effective in immobilizing transition metals like chromium and arsenic, as well as promoting the dehalogenation of persistent organic contaminants (Murugesan et al. 2011; Thome et al. 2015). Because of their extremely small size (generally less than 100 nm), large surface area, and distinctive physicochemical characteristics, nanoparticles have become a major focus of research and are increasingly being employed in environmental remediation efforts (Sun et al. 2006; Tosco et al. 2014; Zhang et al. 2015).

Thome et al. (2015) noted that characteristics such as morphology, size distribution, specific surface area, surface charge, and crystallinity are crucial for anticipating nanoparticle performance. The morphology of nanoscale zero-valent iron (nZVI) is commonly characterized using scanning electron microscopy and transmission electron microscopy. Nanoparticles have proven especially effective in degrading stubborn organic contaminants, such as tetrachloroethylene (TCE) and polychlorinated biphenyls—offering faster remediation compared to conventional microbial degradation techniques (Eglal and Ramamurthy 2014; Nurmi et al. 2005; Tiraferri and Sethi 2009). Among the various

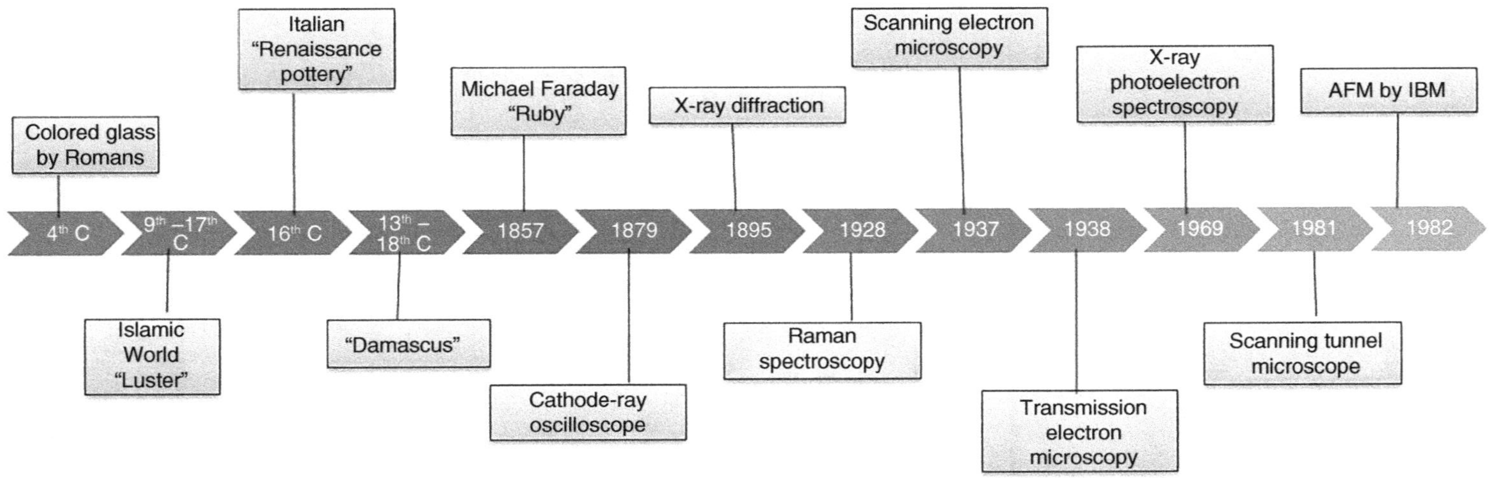

Figure 12.3 Timeline regarding the development of nanoparticles and their uses. *Source:* Adapted from Bayda et al. (2020).

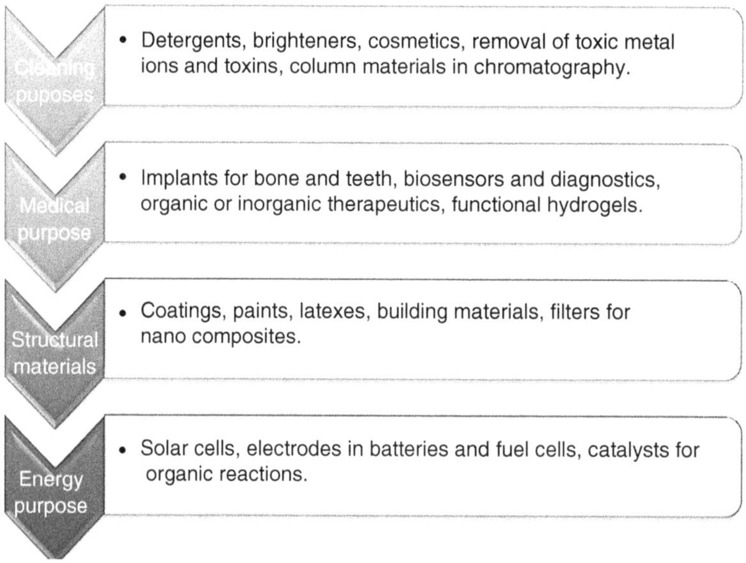

Figure 12.4 Applications of nanoparticles in purification systems, energy conversion devices, catalysis, composites, drug delivery, paints, imaging, and environment cleaning of toxic pollutants. *Source:* Adapted from Heinz et al. (2017).

nanomaterials explored, nZVI (often referred to as nanoiron) stands out for subsurface injection because of its relatively low toxicity, affordability, and extensive research backing it accounts for over 90% of related studies (Yan et al. 2012).

The success of nZVI largely depends on its mobility within porous media, allowing for even dispersion throughout contaminated zones. Over recent years, it has emerged as one of the most widely used in situ technologies for treating sites polluted with toxic chemicals (Table 12.1). Numerous studies have focused on its reactivity, transport mechanisms, and practical field applications (Liu and Zhang 2010).

Traditionally, ZVI in the form of iron filings has served as a reactive component in permeable reactive barriers (PRBs) (Gillham and O'Hannesin 1994). However, conventional ZVI systems encounter issues such as rapid depletion of reactivity, excessive gas production, buildup of precipitates, difficulties in deep excavation, and recycling requirements. To overcome these drawbacks, nanoscale ZVI was developed to improve the reductive dechlorination of TCE. Due to its much smaller particle size, nZVI can travel through soil pores, remain suspended in water, and reach contaminated areas more effectively (Berge and Ramsburg 2010; Tiraferri and Sethi 2009; Zhang et al. 2012).

ZVI exhibits high reactivity in soils and can be directly injected into contaminated sites. Factors influencing remediation efficiency include particle aggregation, mobility in porous media, reactivity, and long-term stability under field conditions (Tosco et al. 2012, 2014). Since the early 1990s, iron has been applied as a reducing agent to treat water contaminated with persistent organic pollutants such as chlorinated aliphatic hydrocarbons. The first field application involved using ZVI as a fill material in PRBs (Zanetti and Fiore 2005). Subsequent studies have employed ZVI to remediate sites contaminated with halogenated and polychlorinated compounds, pesticides, herbicides, polycyclic aromatic hydrocarbons, and metals (Chang et al. 2019; Liu and Zhang 2010). Notably, converting ZVI into its nanoscale oxide form (FeO) enhances both its specific surface area and reactivity (Sakulchaicharoen et al. 2010).

12.6 Nano-bioremediation Approaches for Heavy Metal-contaminated Agricultural Fields

The field of nano-bioremediation, a combination of three scientific disciplines (nanotechnology–biology–environmental science) documented encouraging environment cleaning solutions for removing heavy metals, organic and inorganic toxic pollutants from agricultural soils, water and air. This multidimensional process (nano-bioremediation) involves three fundamental categorical facts, that is, (i) preventing pollution, (ii) sensing and detection of pollutants, and (iii) treatment and remediation (Carata et al. 2017). The nanoparticles used in nano-bioremediation have intrinsic properties to either sequester or degrade heavy metals in the pollution sites.

12.6.1 Pretreatment of Soils

The cleaning of contaminated soil by chemical oxidation, reduction and sorption is best achieved by the application of nanoparticles due to their stable reactivity, higher surface area and smaller particle size (Guerra et al. 2018). Typically, the soil constituents affect the toxicity and activity of heavy metal pollutants due to their intrinsic sorption and desorption reactions (Hamid et al. 2020). Hence, modification of contaminated soil with remediation enhancers will increase heavy metal removal from soil by manipulating their toxicity and mobility in the soil (Hamid et al. 2019, 2020).

The most commonly adapted soil modification process was sorption of soil contaminants by surface precipitation, ion exchange, complexation with organic ligands and adsorption onto mineral surfaces for effective removal of heavy metals (Kumpiene et al. 2008). These modifications are characterized into two groups due to the alteration of contaminant's mobility into the soil: (i) immobilizing agents added to the contaminated soils decrease bioavailability and mobility of the heavy metals indirectly preventing their entrance into the food chain and (ii) while the mobilizing agents added to the contaminant soils enhance the mobility of the heavy metals for higher removal by either phytoaccumulation or soil washing (Robinson et al. 2009). For instance, the addition of nZVI (nanoscale ZVI) in heavy metal contaminated soil increased the removal of heavy metals from soil through soil washing after their adsorption on these NPs, which was later recovered by magnetic separation as shown in Table 12.1 (Boente et al. 2018).

12.6.2 Nanoparticle-assisted Whole-plant Nano-bioremediation

Nano-bioremediation of agricultural soils by using plants and nanoparticles in combination has three basic plant adaptations, that is, phytodegradation, phytoextraction, and phytostabilization, that enhanced heavy metals removal at an accelerated level in comparison to conventional phytoremediation practices (Ma and Wang 2010; Pillai and Kottekottil 2016; Souri et al. 2017). The underlying mechanisms of interactions between plants and applied nanoparticles used for heavy metals' removal from contaminated agricultural fields are germination protection (Anjum et al. 2014; Khodakovskaya et al. 2009; Kumar et al. 2013; Prasad et al. 2012), increased phytohormones production (Khan and Bano 2016), growth protection (Bao-Shan et al. 2004; Souri et al. 2017), increased biomass (Dimkpa et al. 2012; Rui et al. 2016), increased nitrogen uptake (Pradhan et al. 2013; Pradhan et al. 2014), and root elongation (Cañas et al. 2008; Kumar et al. 2013; Lee et al. 2010; Prasad et al. 2012; Wu et al. 2012).

Whole-plant-assisted nano-bioremediation significantly enhances heavy metals' removal from the contaminated agricultural fields. In a study, TiO_2 nanoparticles enhanced cadmium accumulation by soybean plant from cadmium-contaminated soils and reduced phytotoxicity of Cd to the plants. The reported underlying mechanisms were free radical scavenging produced due to cadmium toxicity and protection from oxidative damage by applied nanoparticles (Singh and Lee 2016). Another study reported the use of nano-carbon black (NCB) and nano-hydroxyapatite (NHAP) for enhanced nano-bioremediation of lead (Pb) from contaminated agricultural fields by *Lolium temulentum* L. (ryegrass) and enhanced phytoextraction by the host plant. The comparative analysis of applied nanoparticles (NHAP and NCB) showed that NHAP was more suitable for in situ nano-bioremediation than NCB (Liang et al. 2017). A number of studies explored the fundamental abilities of nanoparticles for altering heavy metal toxicity to the host plants, that is, modifying cell division, photosynthetic pathways, cell wall formation, water homeostasis, oxidative stress, and gene regulation related to heavy metal stress (Kaveh et al. 2013; Khodakovskaya et al. 2012; Nair and Chung 2014). Another investigation proposed that the increased transporter gene expression, co-transportation of heavy metals with nanoparticles and enhanced cell wall permeability as core mechanisms for increased accumulation of heavy metals during phytoremediation (Hu et al. 2014).

Among the various nanomaterials explored for soil amendment in nano-bioremediation, ZVI nanoparticles (nZVI) have been the most rapidly adopted due to their high efficiency in removing soil contaminants (Li et al. 2017), particularly heavy metals (Gil-Díaz et al. 2017; Němeček et al. 2014; Wang et al. 2014). However, the use of nZVI in agricultural fields presents both advantages and challenges. On the positive side, nZVI promotes immobilization of heavy metals, exhibits minimal adverse effects on soil biota, and can even enhance plant growth (Gil-Díaz et al. 2017). Conversely, concerns persist regarding its optimal application concentration and scalability under field conditions (Rónavári et al. 2016).

Several studies have demonstrated that low dosages of nZVI stimulate plant growth, while higher concentrations can exhibit phytotoxic effects (Guha et al. 2018; Li et al. 2015; Ma et al. 2013). The integration of nZVI-assisted phytoremediation is an emerging approach showing significant promise. For instance, low concentrations of nZVI have been reported to enhance cadmium (Cd) uptake during phytoremediation by ramie seedlings (Gong et al. 2017), while similar results were observed for lead (Pb) nanophytoremediation using *Lolium perenne* as the host plant (Huang et al. 2018).

12.6.3 Nanoparticle-assisted Arbuscular Mycorrhizal Fungi Bioremediation

Arbuscular mycorrhizal fungi (AMF) are found in nearly all soil ecosystems and form mutually beneficial relationships with approximately 80% of terrestrial plant species. In this symbiosis, plants provide carbon compounds to the fungi, while AMF enhance the plant's ability to absorb essential but poorly soluble nutrients such as nitrogen (N), phosphorus (P), and potassium (K) from the surrounding soil (Smith and Read 2010). These fungi are also known to ameliorate several abiotic stresses, including those caused by salinity, drought, and heavy metal toxicity (Chen et al. 2007; Evelin et al. 2009; Hutchinson et al. 2004; Li et al. 2014).

In soils contaminated with heavy metals, AMF contribute to stress tolerance through two complementary mechanisms. First, metals can be immobilized directly by binding to fungal cell wall components or by the secretion of chelating compounds. Second, AMF can modify the chemical environment of the mycorrhizosphere, altering metal speciation and reducing their mobility and bioavailability (Subramanian et al. 2009; Wu et al. 2015).

In one study, maize plants inoculated with AMF and grown in soils polluted with lead (Pb), zinc (Zn), and copper (Cu) were treated with 0.5% (w/w) nZVI. The AMF–nZVI combination markedly reduced heavy metal uptake and translocation in plant tissues and decreased metal availability in the rhizosphere, demonstrating that AMF can buffer plants from both heavy metal and nanoparticle toxicity (Wu et al. 2018).

A related experiment on wheat grown in lead-contaminated soil tested the combined effect of ZnO nanoparticles and the AMF species *Glomus macrocarpum*. While both treatments individually promoted plant growth and reduced Pb accumulation, their combined use produced a synergistic effect, increasing plant biomass by 30.66% and decreasing Pb concentrations in roots and shoots by 52.09% and 58.19%, respectively, at a contamination level of 100 mg/kg Pb (Raghib et al. 2020). These results underscore that integrating AMF with nanoparticles enhances the overall efficiency of nano-bioremediation strategies for heavy metal-polluted soils (Table 12.2).

12.6.4 Nanoparticle-assisted Plant Growth-promoting Rhizobacteria and AMF Bioremediation

Multiple studies have highlighted the effectiveness of integrated bioremediation strategies that combine nanoparticles with beneficial microorganisms. For example, the joint application of nanoscale nZVI, plant growth-promoting rhizobacteria (PGPR) such as

Table 12.2 Nanoparticle-assisted PGPRs bioremediation of heavy metal contaminated soils.

Plant growth-promoting rhizobacteria	Plant used for remediation	Nanoparticles	Heavy metals	References
Pseudomonas fluorescens	*Trifolim repens*	nZVI	Antimony	Zand et al. (2020)
Pseudomonas fluorescens	*Sorghum bicolor*	TiO_2	Lead	Daryabeigi et al. (2021)
Pseudomonas fluorescens	*Trifolim repens*	TiO_2	Cadmium	Zand et al. (2020)

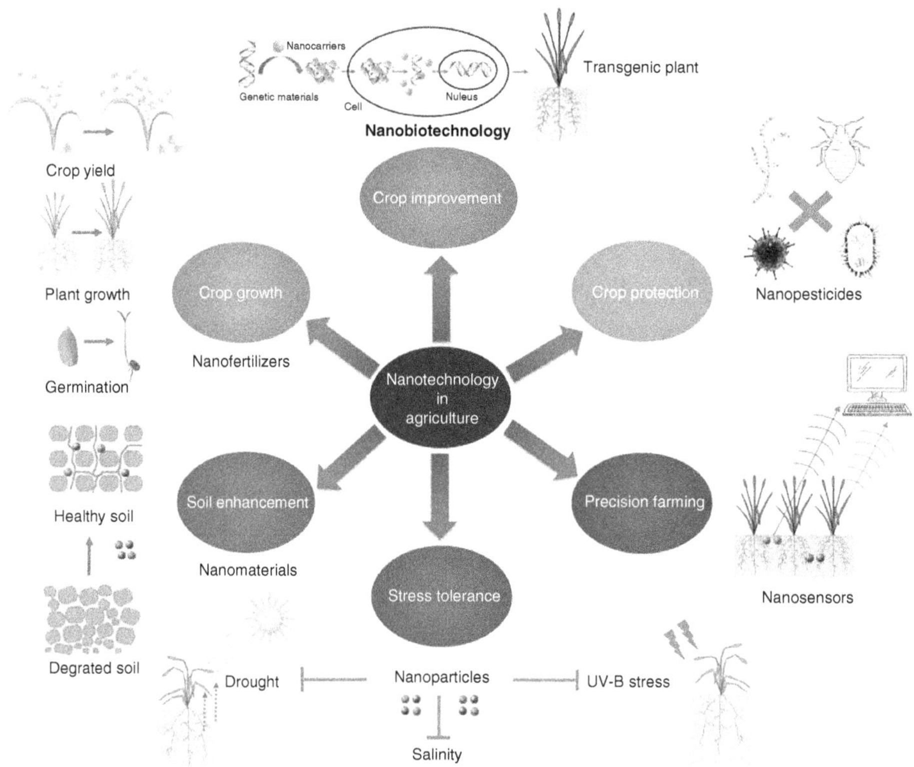

Figure 12.5 Different applications of nanoparticles in agriculture field. *Source:* Adapted from Shang et al. (2019).

Pseudomonas fluorescens, and the AMF *Rhizophagus irregularis* has been investigated for remediating soils contaminated with cadmium (Cd), copper (Cu), and lead (Pb) in *Salix alba* (white willow) plantations. In these experiments, soils were pretreated with nZVI at concentrations of 0, 150, and 300 mg/kg before planting. Willow plants were then inoculated with both PGPR and AMF and grown in pots for 160 days. The findings indicated that soils containing multiple heavy metals were most effectively remediated at lower nZVI concentrations, especially when both microbial inoculants were present, demonstrating the synergistic interactions between nanoparticles, bacteria, and fungi in enhancing heavy metal detoxification (Chen et al. 2015; Jiang et al. 2018; Mokarram-Kashtiban et al. 2019) (Figure 12.5).

12.6.5 Nanoparticle-assisted Plant Growth-promoting Rhizobacteria Bioremediation

PGPR have emerged as a sustainable and environmentally friendly strategy for the removal of heavy metals from contaminated agricultural soils (Yan et al. 2020). When used in combination with nanoparticles in nano-bioremediation approaches, PGPR can further improve the efficiency of soil decontamination. For example, certain PGPR strains, including *Pseudomonas fluorescens* and *Bacillus subtilis*, have been applied alongside TiO_2 nanoparticles to enhance the phytoremediation of lead-contaminated soils using *Sorghum bicolor* (sorghum). In this study, the most effective outcome was achieved with 500 mg/kg of TiO_2 nanoparticles combined with PGPR, which markedly increased lead removal from the soil (Daryabeigi et al. 2021).

In another investigation, the same bacterial strains were employed alongside nZVI to assist microbial remediation in combination with phytoremediation. Here, PGPR-inoculated *Trifolium repens* was cultivated in antimony-contaminated soil under controlled pot conditions, and the highest antimony removal was achieved with the combined use of PGPR and 500 mg/kg nZVI (Zand et al. 2020). Comparable outcomes have been observed for cadmium removal from soils, where various strains of *Bacillus, Pseudomonas,* and *Enterobacter* were used in conjunction with TiO_2 nanoparticles, as summarized in Table 12.2 (Zand et al. 2020) (Figure 12.6).

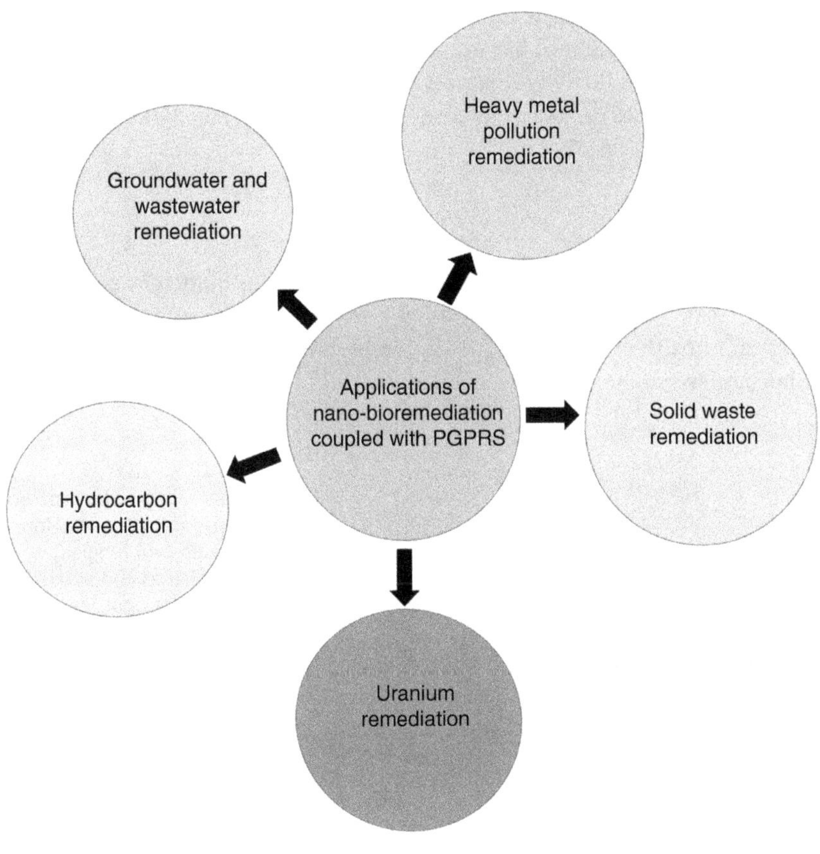

Figure 12.6 Applications of nano-bioremediation coupled with plant growth-promoting rhizobacteria microorganisms.

12.6.6 Plant Protection and Production

Nanoparticles can be utilized to create tiny pesticidal particles or structures with enhanced and efficient pesticidal activity (Ahmed et al. 2013; Bergeson 2010). These nanopesticides have the potential to improve soil quality, enhance the distribution of agricultural formulations, reduce runoff, and minimize the use of organic solvents as well as unintended pesticide dispersion (Figure 12.7). Various delivery methods can be employed to apply nanopesticides to plants, including nano-encapsulation, nano-emulsions, nano-cages, and nano-containers (El-Shetehy et al. 2020; Kookana et al. 2014) (Figure 12.7).

12.6.7 Detecting Plant Diseases

Plant diseases and environmental pollution can significantly reduce the production and yield of key crops. Various strategies have been employed to protect plants from fungal, bacterial, and viral pathogens. Recently, nanotechnology-based approaches have emerged as promising solutions. For example, different types of nanosensors can detect plant pathogens rapidly, often within six to seven hours, offering a simple and user-friendly option for agricultural applications (Kulkarni and Kulkarni 2012). Real-time field monitoring of soil conditions and crop performance has become increasingly feasible, providing valuable insights for improving productivity (Figure 12.7). Advances in nanosensor technology also enable timely responses to plant diseases and environmental changes (Sharon et al. 2010).

12.6.8 Micronutrient Supply

Nanoparticles can also enhance the availability of nutrients for plants. Micronutrient nano-formulations can be applied directly to the leaves or indirectly to the soil to improve overall soil health (Peteu et al. 2010). Research has shown that nanoparticle-based phytohormones, such as indole acetic acid and 1-naphthylacetic acid, can

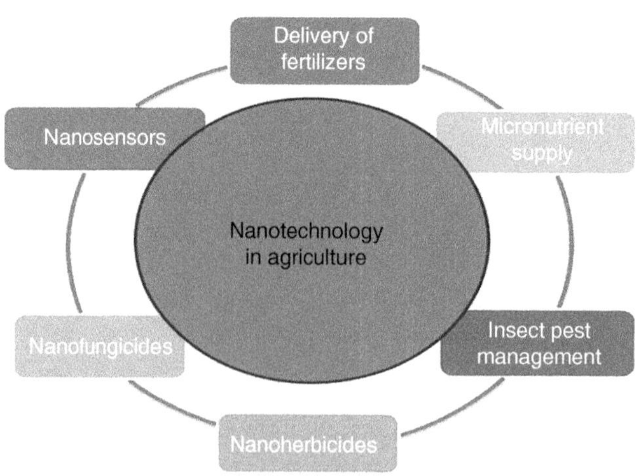

Figure 12.7 Uses of nanoparticles in plant growth promotion, plant protection, to develop new fungicides and herbicides and efficient supply of micronutrients. *Source:* Adapted from Rajwade et al. (2020).

promote plant growth under various environmental conditions (Figure 12.7). This approach can help reduce the use of chemical fertilizers while improving the productivity of important crops (Tao et al. 2012). Additionally, several studies have evaluated different nanoparticle-based micronutrients to assess their uptake and availability in plants (Wilson et al. 2008).

12.6.9 Nanofungicides

Certain nanoparticles can serve as effective antifungal agents, protecting plants from various fungal pathogens (Figure 12.7). For instance, zinc oxide nanoparticles (35–55 nm), titanium dioxide nanoparticles (80–90 nm), and silver nanoparticles (25–85 nm) have been tested against *Macrophomina phaseolina*, a soil fungus responsible for diseases in oilseed and pulse crops. Among these, silver nanoparticles demonstrated the strongest antifungal activity compared to titanium dioxide and zinc oxide nanoparticles (Shyla et al. 2014). Silica-based nanoparticles have also been applied to maize to control *Aspergillus niger* and *Fusarium oxysporum* (Sharon et al. 2010). The antifungal effectiveness of nanoparticles—such as causing reduced membrane permeability, cell lysis, and hyphal breakdown—depends on their size and shape. Research has indicated that nanoparticles with a truncated triangular shape exhibit greater antifungal activity than spherical ones (Pal et al. 2007).

12.7 Conclusion and Future Scope

Environmental pollution caused by heavy metals continues to increase due to rapid industrialization and various anthropogenic activities across the globe. The contamination of soil and groundwater with heavy metals has severely affected the production and yield of many economically important crops. Over the past few decades, environmental scientists have made significant efforts to mitigate soil and water pollution, with nanotechnology-based bioremediation emerging as a promising and widely adopted approach in recent years. Nanoparticles are increasingly utilized on a large scale owing to their unique structural, catalytic, electrical, surface, and sensing properties. The combined use of bentonite and kaolinite with nZVI nanoparticles and PGP microorganisms represents an effective strategy for both removing heavy metals and enhancing plant growth in contaminated agricultural fields.

Soil microorganisms possess the inherent ability to accumulate and detoxify heavy metals, thereby playing a crucial role in the bioremediation of polluted soils. PGP microorganisms can be applied on a larger scale through nanoparticle-based microbial films, integrating both conventional and modern techniques for ex situ and in situ remediation of contaminated environments. Looking ahead, nano-based biofertilizers hold great potential for diverse agricultural applications, including plant growth promotion, biological pest and disease control (as biopesticides and fungicides), and nutrient delivery. However, for the effective implementation of nano-bioremediation, the physical and chemical properties of nanoparticles, along with their sensitivity and stability, must be carefully optimized to ensure efficient and safe performance in contaminated soils.

References

Abbasi, S., Peerzada, M.H., Nizamuddin, S., and Mubarak, N.M. (2020). Functionalized nanomaterials for the aerospace, vehicle, and sports industries. In: *Handbook of Functionalized Nanomaterials for Industrial Applications* (ed. C.M. Hussain), 795–825. Elsevier.

Ahmed, F., Arshi, N., Kumar, S. et al. (2013). Nanobiotechnology: scope and potential for crop improvement. In: *Crop Improvement Under Adverse Conditions* (ed. N. Tuteja and S.S. Gill), 245–269. New York, NY: Springer.

Alidokht, L., Khataee, A.R., Reyhanitabar, A., and Oustan, S. (2011). Cr (VI) immobilization process in a Cr-spiked soil by zerovalent iron nanoparticles: optimization using response surface methodology. *Clean-Soil Air Water* 39: 633–640.

Anjum, N.A., Singh, N., Singh, M.K. et al. (2014). Single-bilayer graphene oxide sheet impacts and underlying potential mechanism assessment in germinating faba bean (*Vicia faba* L.). *Science of the Total Environment* 472: 834–841.

Atafar, Z., Mesdaghinia, A., Nouri, J. et al. (2010). Effect of fertilizer application on soil heavy metal concentration. *Environmental Monitoring and Assessment* 160: 83–89.

Balkhair, K.S. and Ashraf, M.A. (2016). Field accumulation risks of heavy metals in soil and vegetable crop irrigated with sewage water in western region of Saudi Arabia. *Saudi Journal of Biological Sciences* 23: S32–S44.

Bao-Shan, L., Chun-hui, L., Li-jun, F. et al. (2004). Effect of TMS (nanostructured silicon dioxide) on growth of *Changbai larch* seedlings. *Journal of Forest Research* 15: 138–140.

Barber, D.J. and Freestone, I.C. (1990). An investigation of the origin of the colour of the Lycurgus Cup by analytical transmission electron microscopy. *Archaeometry* 32: 33–45.

Bayda, S., Adeel, M., Tuccinardi, T. et al. (2020). The history of nanoscience and nanotechnology: from chemical–physical applications to nanomedicine. *Molecules* 25 (1): 1–12.

Benjamin, S.R., Lima, F.D., Florean, E.O.P.T., and Guedes, M.I.F. (2019). Current trends in nanotechnology for bioremediation. *International Journal of Environment and Pollution* 66: 19–40.

Berge, N.D. and Ramsburg, C.A. (2010). Iron-mediated trichloroethene reduction within nonaqueous phase liquid. *Journal of Contaminant Hydrology* 118: 105–116.

Bergeson, L.L. (2010). Nanosilver pesticide products: what does the future hold? *Environmental Quality Management* 19: 73–82.

Bi, X., Feng, X., Yang, Y. et al. (2006). Environmental contamination of heavy metals from zinc smelting areas in Hezhang County, western Guizhou, China. *Environment International* 32: 883–890.

Boente, C., Sierra, C., Martínez-Blanco, D. et al. (2018). Nanoscale zero-valent iron-assisted soil washing for the removal of potentially toxic elements. *Journal of Hazardous Materials* 350: 55–65.

Boussen, S., Soubrand, M., Bril, H. et al. (2013). Transfer of lead, zinc and cadmium from mine tailings to wheat (*Triticum aestivum)* in carbonated Mediterranean (Northern Tunisia) soils. *Geoderma* 192: 227–236.

Brevik, E.C., Cerdà, A., Mataix-Solera, J. et al. (2015). The interdisciplinary nature of soil. *The Soil* 1 (1): 117–129.

Cañas, J.E., Long, M., Nations, S. et al. (2008). Effects of functionalized and nonfunctionalized single-walled carbon nanotubes on root elongation of select crop species. *Environmental Toxicology and Chemistry: An International Journal* 27: 1922–1931.

Carata, E., Panzarini, E., and Dini, L. (2017). Environmental nanoremediation and electron microscopies. In: *Nanotechnologies for Environmental Remediation* (ed. G. Lofrano, G. Libralato, and J. Brown), 115–136. Cham: Springer.

Chang, Y., Hasan, M.K., Ahammed, G.J. et al. (2019). Applications of nanotechnology in plant growth and crop protection: a review. *Molecules* 24: 2558.

Chaoua, S., Boussaa, S., El Gharmali, A., and Boumezzough, A. (2019). Impact of irrigation with wastewater on accumulation of heavy metals in soil and crops in the region of Marrakech in Morocco. *Journal of the Saudi Society of Agricultural Sciences* 18: 429–436.

Chen, B.D., Zhu, Y.G., Duan, J. et al. (2007). Effects of the arbuscular mycorrhizal fungus *Glomus mosseae* on growth and metal uptake by four plant species in copper mine tailings. *Environmental Pollution* 147: 374–380.

Chen, C.F., Binh, N.T., Chen, C.W., and Dong, C.D. (2015). Removal of polycyclic aromatic hydrocarbons from sediments using sodium persulfate activated by temperature and nanoscale zero-valent iron. *Journal of the Air & Waste Management Association* 65: 375–383.

Daryabeigi Zand, A., Vaezi Heir, A., and Khodaei, H. (2021). Integrated remediation approach for metal polluted soils using plants, nanomaterials and root-associated bacteria. *Journal of Dispersion Science and Technology* https://doi.org/10.1080/01932691.2021.1878900.

Dimkpa, C.O., McLean, J.E., Latta, D.E. et al. (2012). CuO and ZnO nanoparticles: phytotoxicity, metal speciation, and induction of oxidative stress in sand-grown wheat. *Journal of Nanoparticle Research* 14: 1–15.

Dorjee, P., Amarasiriwardena, D., and Xing, B. (2014). Antimony adsorption by zero-valent iron nanoparticles (nZVI): ion chromatography–inductively coupled plasma mass spectrometry (IC–ICP-MS) study. *Microchemical Journal* 116: 15–23.

Eglal, M.M. and Ramamurthy, A.S. (2014). Nanofer ZVI: morphology, particle characteristics, kinetics, and applications. *Journal of Nanomaterials* 11: 152824.

El-Shetehy, M., Moradi, A., Maceroni, M. et al. (2020). Silica nanoparticles enhance disease resistance in Arabidopsis plants. *Nature Nanotechnology* https://doi.org/10.1038/s41565-020-00812-0.

Essajai, R., Rachadi, A., and Feddi, E. (2018). MD simulation-based study on the thermodynamic, structural and liquid properties of gold nanostructures. *Materials Chemistry and Physics* 218: 116–121.

Evelin, H., Kapoor, R., and Giri, B. (2009). Arbuscular mycorrhizal fungi in alleviation of salt stress: a review. *Annals of Botany* 104: 1263–1280.

Facchinelli, A., Sacchi, E., and Mallen, L. (2001). Multivariate statistical and GIS-based approach to identify heavy metal sources in soils. *Environmental Pollution* 114: 313–324.

FAO (2019). *Food and Agriculture Organization of the United Nations. The State of Food Security and Nutrition in the World*. FAO.

Faraday, M. (1857). LIX. Experimental relations of gold (and other metals) to light.—the bakerian lecture. *The London, Edinburgh, and Dublin Philosophical Magazine and Journal of Science*. 14: 512–539.

Gall, J.E., Boyd, R.S., and Rajakaruna, N. (2015). Transfer of heavy metals through terrestrial food webs: a review. *Environmental Monitoring and Assessment* 187 (4): 201.

Gil-Díaz, M., Diez-Pascual, S., González, A. et al. (2016). A nanoremediation strategy for the recovery of an As-polluted soil. *Chemosphere* 149: 137–145.

Gil-Díaz, M., Pinilla, P., Alonso, J., and Lobo, M.C. (2017). Viability of a nanoremediation process in single or multi-metal (loid) contaminated soils. *Journal of Hazardous Materials* 321: 812–819.

Gillham, R.W. and O'Hannesin, S.F. (1994). Enhanced degradation of halogenated aliphatics by zero-valent iron. *Ground Water* 32: 958–967.

Gong, X., Huang, D., Liu, Y. et al. (2017). Stabilized nanoscale zerovalent iron mediated cadmium accumulation and oxidative damage of *Boeheavy metaleria nivea* (L.) gaudich cultivated in cadmium contaminated sediments. *Environmental Science & Technology* 51: 11308–11316.

Guerra, F.D., Attia, M.F., Whitehead, D.C., and Alexis, F. (2018). Nanotechnology for environmental remediation: materials and applications. *Molecules* 23: 1760.

Guha, T., Ravikumar, K.V.G., Mukherjee, A. et al. (2018). Nanopriming with zero valent iron (nZVI) enhances germination and growth in aromatic rice cultivar (*Oryza sativa cv. Gobindabhog* L.). *Plant Physiology and Biochemistry* 127: 403–413.

Hamid, Y., Tang, L., Sohail, M.I. et al. (2019). An explanation of soil amendments to reduce cadmium phytoavailability and transfer to food chain. *Science of the Total Environment* 660: 80–96.

Hamid, Y., Tang, L., Hussain, B. et al. (2020). Efficiency of lime, biochar, Fe containing biochar and composite amendments for Cd and Pb immobilization in a co-contaminated alluvial soil. *Environmental Pollution* 257: 113609.

Hou, D., O'Connor, D., Nathanail, P. et al. (2017). Integrated GIS and multivariate statistical analysis for regional scale assessment of heavy metal soil contamination: a critical review. *Environmental Pollution* 23: 1188–1200.

Hou, D., O'Connor, D., Igalavithana, A.D. et al. (2020). Metal contamination and bioremediation of agricultural soils for food safety and sustainability. *Nature Reviews Earth and Environment* 1 (7): 366–381.

Hu, X., Kang, J., Lu, K. et al. (2014). Graphene oxide amplifies the phytotoxicity of arsenic in wheat. *Scientific Reports* 4: 1–10.

Huang, D., Qin, X., Peng, Z. et al. (2018). Nanoscale zero-valent iron assisted phytoremediation of Pb in sediment: impacts on metal accumulation and antioxidative system of Lolium perenne. *Ecotoxicology and Environmental Safety* 153: 229–237.

Hutchinson, J.J., Young, S.D., Black, C.R., and West, H.M. (2004). Determining uptake of radio-labile soil cadmium by arbuscular mycorrhizal hyphae using isotopic dilution in a compartmented-pot system. *The New Phytologist* 164: 477–484.

Järup, L. (2003). Hazards of heavy metal contamination. *British Medical Bulletin* 68: 167–182.

Jiang, D., Zeng, G., Huang, D. et al. (2018). Remediation of contaminated soils by enhanced nanoscale zero valent iron. *Environmental Research* 163: 217–227.

Kaveh, R., Li, Y.S., Ranjbar, S. et al. (2013). Changes in *Arabidopsis thaliana* gene expression in response to silver nanoparticles and silver ions. *Environmental Science & Technology* 47: 10637–10644.

Khan, N. and Bano, A. (2016). Modulation of phytoremediation and plant growth by the treatment with PGPR, Ag nanoparticle and untreated municipal wastewater. *International Journal of Phytoremediation* 8: 1258–1269.

Khan, S., Cao, Q., Zheng, Y.M. et al. (2008). Health risks of heavy metals in contaminated soils and food crops irrigated with wastewater in Beijing, China. *Environmental Pollution* 152: 686–692.

Khodakovskaya, M., Dervishi, E., Maheavy Metalood, M. et al. (2009). Carbon nanotubes are able to penetrate plant seed coat and dramatically affect seed germination and plant growth. *ACS Nano* 3: 3221–3227.

Khodakovskaya, M.V., De Silva, K., Biris, A.S. et al. (2012). Carbon nanotubes induce growth enhancement of tobacco cells. *ACS Nano* 6: 2128–2135.

Kookana, R.S., Boxall, A.B., Reeves, P.T. et al. (2014). Nanopesticides: guiding principles for regulatory evaluation of environmental risks. *Journal of Agricultural and Food Chemistry* 62: 4227–4240.

Kulkarni, V.D. and Kulkarni, P.S. (2012). Green synthesis of copper nanoparticles using *Ocimum sanctum* leaf extract. *International Journal of Chemical Studies* 1: 2321–4902.

Kumar, V., Guleria, P., Kumar, V., and Yadav, S.K. (2013). Gold nanoparticle exposure induces growth and yield enhancement in *Arabidopsis thaliana*. *Science of the Total Environment* 461: 462–468.

Kumpiene, J., Lagerkvist, A., and Maurice, C. (2008). Stabilization of As, Cr, Cu, Pb and Zn in soil using amendments—a review. *Waste Management* 28: 215–225.

Lakherwal, D. (2014). Adsorption of heavy metals: a review. *International Journal of Environmental Research and Development* 4: 41–48.

Le Quéré, C., Korsbakken, J.I., Wilson, C. et al. (2019). Drivers of declining CO_2 emissions in 18 developed economies. *Nature Climate Change* 9 (3): 213–217.

Lee, C.W., Mahendra, S., Zodrow, K. et al. (2010). Developmental phytotoxicity of metal oxide nanoparticles to *Arabidopsis thaliana*. *Environmental Toxicology and Chemistry: An International Journal* 29: 669–675.

Li, X., Lee, S.L., Wong, S.C. et al. (2004). The study of metal contamination in urban soils of Hong Kong using a GIS-based approach. *Environmental Pollution* 129: 113–124.

Li, T., Lin, G., Zhang, X. et al. (2014). Relative importance of an arbuscular mycorrhizal fungus (*Rhizophagus intraradices*) and root hairs in plant drought tolerance. *Mycorrhiza* 24: 595–602.

Li, X., Yang, Y., Gao, B., and Zhang, M. (2015). Stimulation of peanut seedling development and growth by zero-valent iron nanoparticles at low concentrations. *PLoS One* 10: e0122884.

Li, B., Zheng, M., Xue, H., and Pang, H. (2016). High performance electrochemical capacitor materials focusing on nickel based materials. *Inorganic Chemistry Frontiers* 3: 175–202.

Li, S., Wang, W., Liang, F., and Zhang, W.X. (2017). Heavy metal removal using nanoscale zero-valent iron (nZVI): theory and application. *Journal of Hazardous Materials* 322: 163–171.

Liang, S.X., Jin, Y., Liu, W. et al. (2017). Feasibility of Pb phytoextraction using nano-materials assisted ryegrass: results of a one-year field-scale experiment. *Journal of Environmental Management* 190: 170–175.

Liu, Z.G. and Zhang, F.S. (2010). Nano-zerovalent iron contained porous carbons developedfrom waste biomass for the adsorption and dechlorination of PCBs. *Bioresource Technology* 101: 2562–2564.

Lu, A., Wang, J., Qin, X. et al. (2012). Multivariate and geostatistical analyses of the spatial distribution and origin of heavy metals in the agricultural soils in Shunyi, Beijing, China. *Science of the Total Environment* 425: 66–74.

Luo, C., Liu, C., Wang, Y. et al. (2011). Heavy metal contamination in soils and vegetables near an e-waste processing site, south China. *Journal of Hazardous Materials* 186: 481–490.

Ma, X. and Wang, C. (2010). Fullerene nanoparticles affect the fate and uptake of trichloroethylene in phytoremediation systems. *Environmental Engineering Science* 27: 989–992.

Ma, X., Gurung, A., and Deng, Y. (2013). Phytotoxicity and uptake of nanoscale zero-valent iron (nZVI) by two plant species. *Science of the Total Environment* 443: 844–849.

Mar Gil-Díaz, M., Pérez-Sanz, A., Angeles Vicente, M., and Carmen Lobo, M. (2014). Immobilisation of Pb and Zn in soils using stabilised zero-valent iron nanoparticles: effects on soil properties. *Clean-Soil Air Water* 42: 1776–1784.

Marrugo-Negrete, J., Pinedo-Hernández, J., and Díez, S. (2017). Assessment of heavy metal pollution, spatial distribution and origin in agricultural soils along the Sinú River Basin, Colombia. *Environmental Research* 154: 380–388.

Mohanty, C., Satpathy, S.S., and Mohanty, S. (2020). An eco-friendly approach for the eradication of heavy metal contaminants by nano-bioremediation. In: *Recent Advancements in Bioremediation of Metal Contaminants* (ed. S. Dey and B. Acharya), 220–236. IGI Global.

Mokarram-Kashtiban, S., Hosseini, S.M., Kouchaksaraei, M.T., and Younesi, H. (2019). The impact of nanoparticles zero-valent iron (nZVI) and rhizosphere microorganisms on the phytoremediation ability of white willow and its response. *Environmental Science and Pollution Research* 26: 10776–10789.

Montgomery, D.R. (2007). Soil erosion and agricultural sustainability. *Proceedings of the National Academy of Sciences* 104 (33): 13268–13272.

Mulligan, C.N., Yong, R.N., and Gibbs, B.F. (2001). Remediation technologies for metal-contaminated soils and groundwater: an evaluation. *Engineering Geology* 60: 193–207.

Murugesan, K., Bokare, V., Jeon, J.R. et al. (2011). Effect of Fe-Pd bimetallic nanoparticles on phingomonas sp. PH-07 and a nano-bio hybrid process for triclosan degradation. *Bioresource Technology* 102: 6019–6025.

Nair, P.M.G. and Chung, I.M. (2014). Cell cycle and mismatch repair genes as potential biomarkers in *Arabidopsis thaliana* seedlings exposed to silver nanoparticles. *Bulletin of Environmental Contamination and Toxicology* 92: 719–725.

Navarro, M.C., Pérez-Sirvent, C., Martínez-Sánchez, M.J. et al. (2008). Abandoned mine sites as a source of contamination by heavy metals: a case study in a semi-arid zone. *Journal of Geochemical Exploration* 96: 183–193.

Němeček, J., Lhotský, O., and Cajthaml, T. (2014). Nanoscale zero-valent iron application for in situ reduction of hexavalent chromium and its effects on indigenous microorganism populations. *Science of the Total Environment* 485: 739–747.

Nicholson, F.A., Smith, S.R., Alloway, B.J. et al. (2003). An inventory of heavy metals inputs to agricultural soils in England and Wales. *Science of the Total Environment* 311: 05–219.

Nikhil, P., Sudheer, P., Joy, P.J. et al. (2020). Synthesis and characterization of nanoparticles using egg shell. *Materials Today Proceedings* 26: 3246–3249.

Noli, F. and Tsamos, P. (2016). Concentration of heavy metals and trace elements in soils, waters and vegetables and assessment of health risk in the vicinity of a lignite-fired power plant. *Science of the Total Environment* 563: 377–385.

Nurmi, J.T., Tratnyek, P.G., Sarathy, V. et al. (2005). Characterization and properties of metallic iron nanoparticles: spectroscopy, electrochemistry, and kinetics. *Environmental Science & Technology* 39 (5): 1221–1230.

Nziguheba, G. and Smolders, E. (2008). Inputs of trace elements in agricultural soils via phosphate fertilizers in European countries. *Science of the Total Environment* 390: 53–57.

Ogunkunle, C.O. and Fatoba, P.O. (2013). Pollution loads and the ecological risk assessment of soil heavy metals around a mega cement factory in southwest Nigeria. *Polish Journal of Environmental Studies* 22 (2).

Pal, S., Tak, Y.K., and Song, J.M. (2007). Does the antibacterial activity of silver nanoparticles depend on the shape of the nanoparticle? A study of the gram-negative bacterium Escherichia coli. *Applied and Environmental Microbiology* 73 (6): 1712–1720.

Peteu, S.F., Oancea, F., Sicuia, O.A. et al. (2010). Responsive polymers for crop protection. *Polymers* 2 (3): 229–251.

Pillai, H.P. and Kottekottil, J. (2016). Nano-phytotechnological remediation of endosulfan using zero valent iron nanoparticles. *Journal of Environmental Protection* 7: 734.

Poole, C.P. Jr. and Owens, F.J. (2003). *Introduction to Nanotechnology*. Wiley.

Pourret, O. and Hursthouse, A. (2019). It's time to replace the term "heavy metals" with "potentially toxic elements" when reporting environmental research. *International Journal of Environmental Research and Public Health* 16 (22): 4446.

Pradell, T., Climent-Font, A., Molera, J. et al. (2007). Metallic and nonmetallic shine in luster: an elastic ion backscattering study. *Journal of Applied Physiology* 101: 103518.

Pradhan, S., Patra, P., Das, S. et al. (2013). Photochemical modulation of biosafe manganese nanoparticles on *Vigna radiata*: a detailed molecular, biochemical, and biophysical study. *Environmental Science & Technology* 47: 13122–13131.

Pradhan, S., Patra, P., Mitra, S. et al. (2014). Manganese nanoparticles: impact on non-nodulated plant as a potent enhancer in nitrogen metabolism and toxicity study both in vivo and in vitro. *Journal of Agricultural and Food Chemistry* 62: 8777–8785.

Prasad, T.N.V.K.V., Sudhakar, P., Sreenivasulu, Y. et al. (2012). Effect of nanoscale zinc oxide particles on the germination, growth and yield of peanut. *Journal of Plant Nutrition* 35: 905–927.

Raghib, F., Naikoo, M.I., Khan, F.A. et al. (2020). Interaction of ZnO nanoparticle and AM fungi mitigates Pb toxicity in wheat by upregulating antioxidants and restricted uptake of Pb. *Journal of Biotechnology* 323: 254–263.

Rajwade, J.M., Chikte, R.G., and Paknikar, K.M. (2020). Nanomaterials: new weapons in a crusade against phytopathogens. *Applied Microbiology and Biotechnology* 104 (4): 1437–1461.

Robinson, B.H., Bañuelos, G., Conesa, H.M. et al. (2009). The phytomanagement of trace elements in soil. *Critical Reviews in Plant Sciences* 28: 240–266.

Rónavári, A., Balázs, M., Tolmacsov, P. et al. (2016). Impact of the morphology and reactivity of nanoscale zero-valent iron (NZVI) on dechlorinating bacteria. *Water Research* 95: 165–173.

Rui, M., Ma, C., Hao, Y. et al. (2016). Iron oxide nanoparticles as a potential iron fertilizer for peanut (*Arachis hypogaea*). *Frontiers in Plant Science* 7: 815.

Sakulchaicharoen, N., O'Carroll, D.M., and Herrera, J.E. (2010). Enhanced stability and dechlorination activity of presynthesis stabilized nanoscale FePd particles. *Journal of Contaminant Hydrology* 118: 117–127.

Salehi, F.M., Khaemba, D.N., Morina, A., and Neville, A. (2016). Corrosive-abrasive wear induced by soot in boundary lubrication regime. *Tribology Letters* 63: 1–11.

Shan, Y., Tysklind, M., Hao, F. et al. (2013). Identification of sources of heavy metals in agricultural soils using multivariate analysis and GIS. *Journal of Soils and Sediments* 13: 720–729.

Shang, Y., Hasan, M. K., Ahammed, G. J., Li, M., Yin, H., & Zhou, J. (2019). Applications of nanotechnology in plant growth and crop protection: A review. *Molecules*, 24 (14), 2558. https://doi.org/10.3390/molecules24142558

Sharon, M., Choudhary, A.K., and Kumar, R. (2010). Nanotechnology in agricultural diseases and food safety. *Journal of Phytology* 2: 83–92.

Shyla, K.K., Natarajan, N., and Nakkeeran, S. (2014). Antifungal activity of zinc oxide, silver and titanium dioxide nanoparticles against Macrophomina phaseolina. *Journal of Mycology and Plant Pathology* 44 (3): 269–274.

Singh, J. and Lee, B.K. (2016). Influence of nano-TiO2 particles on the bioaccumulation of Cd in soybean plants (*Glycine max*): a possible mechanism for the removal of Cd from the contaminated soil. *Journal of Environmental Management* 170: 88–96.

Singh, R., Misra, V., and Singh, R.P. (2012). Removal of Cr (VI) by nanoscale zero-valent iron (nZVI) from soil contaminated with tannery wastes. *Bulletin of Environmental Contamination and Toxicology* 88: 210–214.

Smith, S.E. and Read, D.J. (2010). *Mycorrhizal Symbiosis*. Academic Press.

Souri, Z., Karimi, N., Sarmadi, M., and Rostami, E. (2017). Salicylic acid nanoparticles (SANPs) improve growth and phytoremediation efficiency of *Isatis cappadocica* Desv., under As stress. *IET Nanobiotechnology* 11: 650–655.

Srivastava, S.K., Tyagi, R., and Pant, N. (1989). Adsorption of heavy metal ions on carbonaceous material developed from the waste slurry generated in local fertilizer plants. *Water Research* 23: 1161–1165.

Su, H., Fang, Z., Tsang, P.E. et al. (2016). Remediation of hexavalent chromium contaminated soil by biochar-supported zero-valent iron nanoparticles. *Journal of Hazardous Materials* 318: 533–540.

Subramanian, K.S., Tenshia, V., Jayalaksheavy Metali, K., and Ramachandran, V. (2009). Biochemical changes and zinc fractions in arbuscular mycorrhizal fungus (*Glomus intraradices*) inoculated and uninoculated soils under differential zinc fertilization. *Applied Soil Ecology* 43: 32–39.

Sun, Y.P., Li, X.Q., Cao, J. et al. (2006). Characterization of zerovalent iron nanoparticles. *Advances in Colloid and Interface Science* 120: 47–56.

Tang, J., Zhang, L., Zhang, J. et al. (2020). Physicochemical features, metal availability and enzyme activity in heavy metal-polluted soil remediated by biochar and compost. *Science of the Total Environment* 701: 134751.

Tao, S., Pang, R., Chen, C. et al. (2012). Synthesis, characterization and slow release properties of O-naphthylacetyl chitosan. *Carbohydrate Polymers* 88 (4): 1189–1194.

Teske, S. (2019). *Achieving the Paris Climate Agreement Goals: Global and Regional 100% Renewable Energy Scenarios with Non-energy GHG Pathways for+ 1.5 C and+ 2 C*, 491. Springer Nature.

Thomé, A., Reddy, K.R., Reginatto, C. et al. (2015). Review of nanotechnology for soil and groundwater remediation: Brazilian perspectives. *Water, Air, and Soil Pollution* 226: 121.

Tiraferri, A. and Sethi, R. (2009). Enhanced transport of zerovalent iron nanoparticles in saturated porous media by guar gum. *Journal of Nanoparticle Research* 11 (3): 635–645.

Tosco, T., Coisson, M., Xue, D., and Sethi, R. (2012). Zerovalent iron nanoparticles for groundwater remediation: surface and magnetic properties, colloidal stability, and perspectives for field application. In: *Nanoparticles Featuring Electromagnetic Properties: Research* (ed. A.P. Chiolerio), 2012–2223. Kerala: Signpost.

Tosco, T., Papini, M.P., Viggi, C.C., and Sethi, R. (2014). Nanoscale zerovalent iron particles for groundwater remediation: a review. *Journal of Cleaner Production* 77: 10–21.

Tripathi, S., Sanjeevi, R., Anuradha, J. et al. (2018). Nano-bioremediation: nanotechnology and bioremediation. In: *Biostimulation Remediation Technologies for Groundwater Contaminants* (ed. A.K. Rathoure), 202–219. IGI Global.

UNEP (2018). *United Nations Environment Assembly of the United Nations Environment Programme. Managing soil Pollution to Achieve Sustainable Development*. UNEP.

Vázquez-Núñez, E., Molina-Guerrero, C.E., Peña-Castro, J.M. et al. (2020). Use of nanotechnology for the bioremediation of contaminants: a review. *Processes* 8 (7): 826.

Vignes, A., Krietsch, A., Dufaud, O. et al. (2019). Course of explosion behaviour of metallic powders–from micron to nanosize. *Journal of Hazardous Materials* 379: 120767.

Walter, I., Martinez, F., and Cala, V. (2006). Heavy metal speciation and phytotoxic effects of three representative sewage sludges for agricultural uses. *Environmental Pollution* 139: 507–514.

Wang, Y., Fang, Z., Kang, Y., and Tsang, E.P. (2014). Immobilization and phytotoxicity of chromium in contaminated soil remediated by CMC-stabilized nZVI. *Journal of Hazardous Materials* 275: 230–237.

Wang, X., Daigger, G., Lee, D.J. et al. (2018). Evolving wastewater infrastructure paradigm to enhance harmony with nature. *Science Advances* 4: eaaq0210.

Weerasinghe, J., Li, W., Zhou, R. et al. (2020). Bactericidal silver nanoparticles by atmospheric pressure solution plasma processing. *Nanomaterials* 10: 874.

Wilson, M.A., Tran, N.H., Milev, A.S. et al. (2008). Nanomaterials in soils. *Geoderma* 46: 291–302.

Wu, S.G., Huang, L., Head, J. et al. (2012). Phytotoxicity of metal oxide nanoparticles is related to both dissolved metals ions and adsorption of particles on seed surfaces. *Journal of Petroleum & Environmental Biotechnology* 3 (4): 126.

Wu, S., Zhang, X., Sun, Y. et al. (2015). Transformation and immobilization of chromium by arbuscular mycorrhizal fungi as revealed by SEM–EDS, TEM–EDS, and XAFS. *Environmental Science & Technology* 49: 14036–14047.

Wu, S., Vosátka, M., Vogel-Mikus, K. et al. (2018). Nano zero-valent iron mediated metal (loid) uptake and translocation by arbuscular mycorrhizal symbioses. *Environmental Science & Technology* 52: 7640–7651.

Wuepper, D., Borrelli, P., and Finger, R. (2020). Countries and the global rate of soil erosion. *Nature Sustainability* 3 (1): 51–55.

Yan, W.L., Ramos, M.A.V., Koel, B.E., and Zhang, W.X. (2012). Sequestration by iron nanoparticles: study of solid-phase redox transformations with X-ray photoelectron spectroscopy. *Journal of Physical Chemistry C* 116: 5303–5311.

Yan, A., Wang, Y., Tan, S.N. et al. (2020). Phytoremediation: a promising approach for revegetation of heavy metal-polluted land. *Frontiers in Plant Science* 11: 513099.

Zand, A.D., Mikaeili Tabrizi, A., and Vaezi Heir, A. (2020). Application of titanium dioxide nanoparticles to promote phytoremediation of Cd-polluted soil: contribution of PGPR inoculation. *Bioremediation Journal* 24: 171–189.

Zanetti, M.C. and Fiore, S. (2005). Evaluation of mutual connections between zero-valent iron reactivity and groundwater composition in trichloroethylene degradation. *Annali di Chimica* 95: 779–789.

Zhang, M., Wang, Y., Zhao, D., and Pan, G. (2010). Immobilization of arsenic in soils by stabilized nanoscale zero-valent iron, iron sulfide (FeS), and magnetite (Fe_3O_4) particles. *Chinese Science Bulletin* 55: 365–372.

Zhang, F., Yan, X., Zeng, C. et al. (2012). Influence of traffic activity on heavy metal concentrations of roadside farmland soil in mountainous areas. *International Journal of Environmental Research and Public Health* 9: 1715–1731.

Zhang, Z., Liu, J., Cao, X. et al. (2015). Comparison of U (VI) adsorption onto nanoscale zero-valent iron and red soil in the presence of U (VI)–CO3/Ca–U (VI)–CO3 complexes. *Journal of Hazardous Materials* 300: 633–642.

Zhou, J., Feng, K., Pei, Z. et al. (2016). Multivariate analysis combined with GIS to source identification of heavy metals in soils around an abandoned industrial area, Eastern China. *Ecotoxicology* 25: 380–388.

Zhu, F., Li, L., Ma, S., and Shang, Z. (2016). Effect factors, kinetics and thermodynamics of remediation in the chromium contaminated soils by nanoscale zero valent Fe/Cu bimetallic particles. *Chemical Engineering Journal* 302: 663–669.

13

Biologically Synthesized Nanomaterials and their Antimicrobial Potentials for Crop Protection

Urvashi Mittal[1], Sarvjeet Kukreja[2], Nidhi Selwal[1], Nitu Gautam[1], and Umesh Goutam[1]

[1] *Department of Molecular Biology & Genetic Engineering, Lovely Professional University, Phagwara, Punjab, India*
[2] *Department of Agronomy, School of Agriculture, Lovely Professional University, Phagwara, Punjab, India*

13.1 Introduction

Nanotechnology has rapidly evolved into a cross-disciplinary science that provides innovative tools to address persistent challenges in global agriculture, particularly those related to crop protection, soil fertility, and sustainable food production. Nanoparticles (NPs) are molecules possessing dimension range between 1 and 100 nm, exhibiting physical and chemical characteristics arising from their large surface-to-volume ratio and quantum-scale behavior (Yang et al. 2025). These properties enable intimate interactions with biological macromolecules and cellular membranes, allowing NPs to modulate plant growth, nutrient acquisition, and defense signaling at the molecular level. Consequently, nanotechnology has emerged as a next-generation platform for designing precise nutrient delivery systems, improving pesticide functionality, and enhanced tolerance toward abiotic and biotic stressors (Zaman et al. 2025).

The diversity of nanomaterials directly influences their biological reactivity. According to composition and structural configuration, agricultural nanomaterials are commonly classified as metallic (Ag, Au, Cu, and Zn), metal-oxide (ZnO, TiO_2, Fe_2O_3, and CuO), carbon-based (graphene and carbon nanotubes), polymeric, and silica-based categories (Kapoor et al. 2025). Variations in morphology spherical, rod-like, cubic, hexagonal, or flower-shaped govern their catalytic activity and affinity toward microbial or plant cell surfaces. For example, spherical NPs generally demonstrate better colloidal stability, whereas anisotropic geometries with high-energy facets exhibit greater catalytic efficiency (Vinukonda et al. 2025).

Synthesis strategies for nanomaterials have traditionally relied on physical, chemical, and biological (green) routes, each differing in energy demand, cost, and environmental footprint. Physical or "top-down" techniques such as evaporation-condensation, laser ablation, and high-energy ball milling yield uniform and crystalline products with precise compositional control but are resource-intensive and difficult to scale for field applications (Setia et al. 2025). Chemical or "bottom-up" methods such as sol-gel processing, coprecipitation, and microemulsion synthesis allow fine control over particle size and dispersion stability but often employ toxic reductants or capping agents such as hydrazine and sodium borohydride, leaving hazardous residues in the environment (Zadehnazari, 2025).

Growing awareness of ecological risk has redirected attention toward biological or green or biogenic synthesis, which uses plant extracts, microbes, and algal biomass as natural reducing and stabilizing systems. These biogenic routes operate under mild conditions and exploit biomolecules flavonoids, phenolics, terpenoids, and enzymes that facilitate metal ion reduction and NP capping, producing biocompatible and stable structures (Kirubakaran et al. 2025). For instance, *Azadirachta indica* and *Moringa oleifera* extracts have successfully generated Ag- and ZnO-NPs with pronounced antimicrobial activity, while fungi such as *Aspergillus niger* and *Fusarium oxysporum* mediate enzyme-assisted reduction of metal salts (Eker et al. 2025).

Beyond environmental safety, biologically derived NPs exhibit functional advantages. Their organic surface corona enhances adhesion to pathogen membranes, stimulates reactive oxygen species (ROS) formation, and triggers plant defense pathways—collectively resulting in potent antimicrobial effects (Kapoor et al. 2025). Studies have demonstrated that biogenic Ag-NPs suppress *Xanthomonas oryzae* and *Acidovorax oryzae*, whereas ZnO-NPs inhibit fungal pathogens such as *Fusarium equiseti* (Subba et al. 2024). Integration of these nanomaterials into seed coatings, foliar sprays, or post-harvest treatments improves seed germination, growth, and yield (Kirubakaran et al. 2025).

Overall, green nanotechnology merges nanoscience with plant biochemistry and microbial ecology to deliver a sustainable framework for crop protection. It mitigates the toxicological and energy burdens associated with conventional synthesis while supporting the transition toward climate-smart and resource-efficient agriculture. In alignment with the United Nations Sustainable Development Goals (SDGs), biologically synthesized nanomaterials offer a low-cost, renewable, and eco-safe strategy for achieving food security and environmental stewardship (Yang et al. 2025; Ying et al. 2022).

13.2 Biologically (or Green) Synthesized Nanoparticles

Green nanotechnology has emerged as a cornerstone of sustainable scientific innovation, emphasizing environmentally responsible design, synthesis, and application of nanomaterials. The foundation of this discipline lies in the principles of green chemistry, which promote reduction of hazardous reagents, minimal waste generation, and energy-efficient synthesis routes (Khan et al. 2022). Within this framework, biogenic NP synthesis often referred to as green synthesis has gained global attention as an alternative to traditional physical and chemical approaches. Unlike conventional methods that employ toxic precursors and high energy inputs, the biological route utilizes natural reducing and stabilizing agents derived from plants, algae, fungi, and bacteria, enabling low-cost, scalable, and eco-safe production (Kirubakaran et al. 2025).

In biological systems, the synthesis of NPs (Figure 13.1) is primarily driven by redox-active biomolecules, including polyphenols, flavonoids, terpenoids, proteins, and enzymes such as NADH-dependent reductases, which mediate the transformation of metal ions into stable nanostructures (Ying et al. 2022). These molecules not only facilitate reduction but

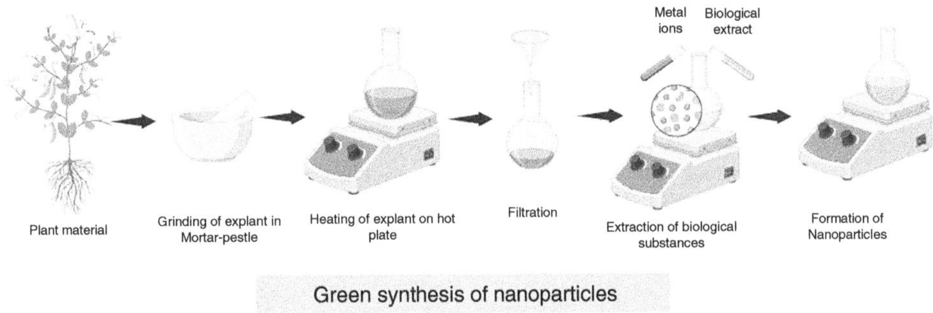

Figure 13.1 General procedure for the synthesis of green nanoparticles. *Source:* Created in Biorender by Urvashi Mittal, https://BioRender.com/u94i611, 2024.

also stabilize the growing NP by forming a capping layer that dictates its shape, size, and dispersion stability (Malik and Basu, 2025). For instance, phytochemical-rich samples of *Moringa oleifera* and *Azadirachta indica* have yielded silver and zinc oxide (ZnO) NPs with uniform morphologies and high antimicrobial efficacy, highlighting the versatility of botanical systems as nanofactories (Eker et al. 2025).

The mechanistic basis of green synthesis involves nucleation, growth, and capping stages, all influenced by the redox potential of the metal precursor and the nature of the biological extract. Metals with higher positive reduction potentials—such as gold (0.99 eV for $AuCl_4^-/Au$)—undergo faster reduction compared to those with lower potentials, such as palladium (0.59 eV for $PdCl_4^{2}-/Pd$), leading to distinct nanostructural outcomes (Kirubakaran et al. 2025). Slow reduction rates favor controlled nucleation, enabling the formation of monodisperse and thermodynamically stable NPs. Environmental conditions such as temperature, pH, and light exposure also modulate reaction kinetics and crystallinity (Kirubakaran et al. 2025; Ying et al. 2022).

Among the biological systems, plant-mediated synthesis is most prevalent due to its reproducibility, simplicity, and wide phytochemical diversity. Extracts of plants serve as reducing agents or reductants as well as contain metabolites that enhance stability through electrostatic and steric hindrance effects. Conversely, microbial synthesis employing bacteria (*Pseudomonas, Bacillus*) and fungi (*Fusarium, Aspergillus*) relies on enzymatic electron transfer pathways to convert metal ions to elemental forms within or outside the cell (Mallikarjuna et al. 2021). Although slower than plant-based systems, microbial routes permit precise control over particle morphology and surface chemistry, valuable for biomedical and agricultural applications (Zaman et al. 2025).

Figure 13.2 (Created in BioRender, Mittal 2024) summarizes the key merits of green nanoparticles in sustainable agriculture, including eco-friendliness, cost efficiency, simple operation, and reliance on renewable secondary metabolites. These attributes collectively make biological synthesis not only a cleaner alternative but also a strategic pathway for achieving circular economy goals in nanotechnology (Assad et al. 2025; Fahim et al. 2024; Sarwar et al. 2025).

The growing acceptance of plant- and microbe-derived NPs across scientific and industrial domains ranging from crop protection and food packaging to environmental remediation demonstrates their broad applicability and safety. Nonetheless, challenges remain in

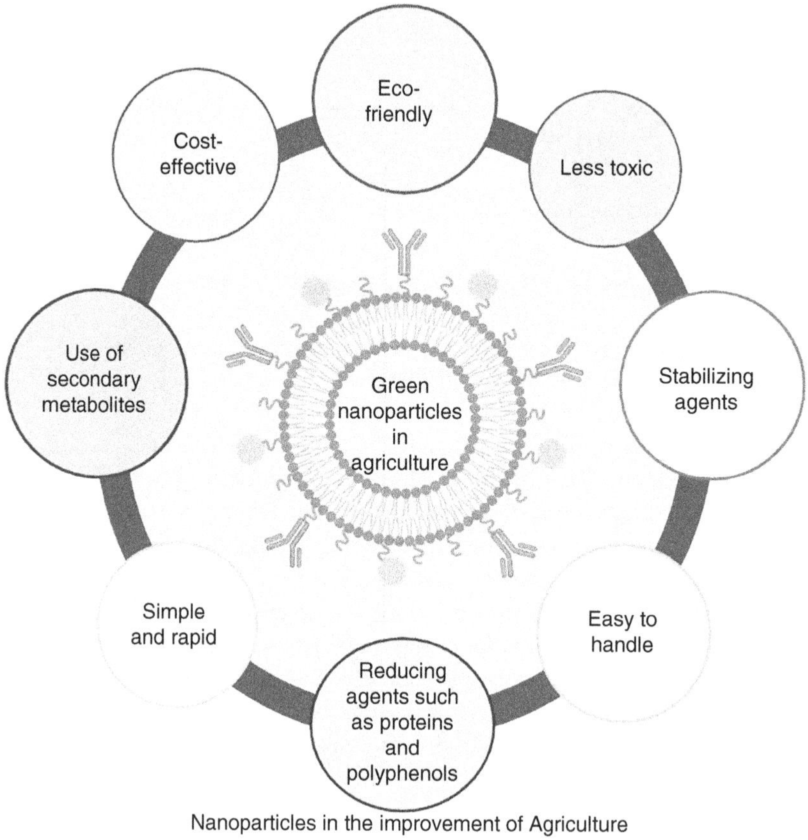

Figure 13.2 Merits of green nanoparticles in the sustainable agriculture development. *Source:* Created in Biorender by Urvashi Mittal, https://BioRender.com/u94i611, 2024.

scaling production, achieving consistent particle size distribution, and elucidating the complete molecular mechanisms underlying biomolecule metal interactions. Continued exploration into genomic and metabolomic regulation of biological synthesis will enhance yield predictability and open new prospects for precision nanofabrication under green paradigms (Assad et al. 2025; Fahim et al. 2024; Kirubakaran et al. 2025).

13.3 Interaction and Uptake of Biogenic Nanoparticles with Plants and Microbes

The interaction between biologically synthesized NPs and plant systems represents a complex and dynamic process shaped by physicochemical characteristics of NPs, plant anatomy, and associated microbial communities. Upon application, NPs encounter plant surfaces primarily roots and leaves where electrostatic forces, surface charge, and hydrophobicity determine their adhesion and subsequent uptake (Ballikaya et al. 2023). At the rhizosphere interface, root exudates containing organic acids and mucilage impart a net negative charge to the root surface, favoring the attraction of positively charged NPs and enhancing their retention. Root tips and young root hairs, due to their incompletely developed cell walls, serve as preferential entry sites (Lv et al. 2019).

Nanoparticles can penetrate plant tissues via apoplastic, symplastic, and endocytic pathways. The apoplastic route involves diffusion through the cell wall matrix, whereas the symplastic route depends on plasmodesmata-mediated transport (Li et al. 2020). The endocytosis-mediated pathway, wherein the plasma membrane invaginates to internalize NPs into vesicles, has been identified as a major mechanism for particles under 100 nm, independent of their charge polarity (Ali et al. 2021). Once internalized, NPs are translocated to vascular tissues and redistributed across organs via the xylem and phloem, driven by transpiration and assimilate flow (Ying et al. 2022; Zhang et al. 2024).

Surface charge polarity plays a pivotal role in determining mobility. Positively charged NPs tend to adhere to negatively charged cell walls, resulting in localized accumulation, while negatively charged or neutral particles exhibit higher tissue penetration and long-distance translocation. Confocal microscopy studies have shown that uncharged graphene quantum dots (22 nm) can traverse from epidermal to vascular tissues, while positively charged analogues remain confined near root surfaces (Parkinson et al. 2022; Spielman-Sun et al. 2017).

Beyond physical interaction, biogenic NPs influence plant physiology at multiple scales. They enhance nutrient assimilation, stimulate photosynthetic efficiency, and modulate redox homeostasis by upregulating antioxidant enzymes such as catalase (CAT) and superoxide dismutase (SOD; Zhang et al. 2024). The biologically synthesized surface corona—comprising flavonoids, phenolics, and proteins—acts as a signaling interface that activates defense genes and hormonal pathways, thereby enhancing resilience to biotic and abiotic stress (Sodhi et al. 2025). For example, Ag- and ZnO-based NPs have been shown to induce systemic resistance against *Xanthomonas oryzae* and *Fusarium oxysporum* by triggering ROS bursts and secondary metabolite synthesis (Sati et al. 2025; Subba et al. 2024).

Equally significant is the interaction between NPs and the rhizospheric microbiome, which governs nutrient cycling, pathogen suppression, and plant growth promotion. Beneficial microbes (*Bacillus subtilis*) synergize with biogenic NPs to enhance nitrogen fixation, phosphate solubilization, and phytohormone production (Zaman et al. 2025). Conversely, excessive NP accumulation may disrupt microbial equilibrium, underscoring the need for concentration- and size-specific optimization. Microbial-assisted NP transformations also modify redox states, influencing bioavailability and potential ecotoxicity (Fatima et al. 2021).

Although current research underscores the beneficial roles of biogenic nanomaterials in improving crop performance, potential bioaccumulation and trophic transfer risks demand careful ecotoxicological assessment. Long-term studies on NP degradation kinetics, persistence in soil matrices, and effects on non-target organisms are essential to ensure environmental safety. Hence, integrating molecular, physiological, and ecological investigations will facilitate the rational design of nanomaterials optimized for safe and efficient agricultural use, aligning with SDGs and regulatory frameworks (Sodhi et al. 2025).

13.4 Mechanisms of Antimicrobial Activity of Biogenic Nanoparticles

The antimicrobial potential of biologically synthesized NPs originates from a combination of physicochemical, biochemical, and molecular mechanisms that collectively disrupt microbial integrity and viability. Unlike conventional pesticides that target specific pathways, biogenic NPs exert multitargeted actions, minimizing the risk of resistance development among pathogens (Sati et al. 2025; Singh et al. 2024a). The first level of antimicrobial action involves direct interaction with the microbial cell envelope. The nanoscale dimensions and charge distribution of NPs facilitate electrochemical attraction to anionic walls of bacteria, leading to seepage of intracellular components (Ying et al. 2022). Scanning electron microscopy studies have revealed severe membrane deformation, pore formation, and cytoplasmic disintegration in bacterial cells treated with Ag- and ZnO-based NPs (Fatima et al. 2021). The cell wall damage compromises structural stability, resulting in osmotic imbalance and cell death.

A second critical mechanism is the generation of ROS, including superoxide anions (O_2^-), hydroxyl radicals (·OH), and hydrogen peroxide (H_2O_2). The overaccumulation of ROS induces oxidative stress, damaging proteins, nucleic acids, and lipids, and interfering with essential metabolic reactions (Subba et al. 2024). Biogenic NPs synthesized via plant or microbial routes often possess organic surface coatings rich in phenolic and flavonoid moieties that catalyze redox cycling, intensifying ROS production under physiological conditions (Kirubakaran et al. 2025). Additionally, metal ion release contributes to antimicrobial efficacy. Ag^+, Cu^{2+}, and Zn^{2+} ions liberated from NPs bind to thiol groups of enzymes, thereby inactivating respiratory and replication machinery (Singh et al. 2024a). This ion-mediated interference extends to nucleic acid synthesis and adenosine triphosphate (ATP) generation, leading to metabolic collapse. The extent of ion release and thus toxicity is influenced by particle size, morphology, and crystallinity; smaller or anisotropic particles display enhanced dissolution kinetics and antimicrobial potential (Jiang et al. 2025; Ojha et al. 2018). Another important aspect is the interaction of NPs with the microbial genetic material. Positively charged NPs can bind electrostatically to negatively charged DNA phosphate backbones, causing structural distortion, condensation, and inhibition of replication (Kapoor et al. 2025). Studies have shown that biogenic Ag-NPs derived from *Ocimum sanctum* and *Azadirachta indica* induce DNA fragmentation and suppress virulence gene expression in *Xanthomonas oryzae* and *Erwinia carotovora* (Sati et al. 2025). This genomic interference is complemented by protein denaturation through covalent modification of amino and sulfhydryl groups, altering enzymatic conformation and function (Asimuddin et al. 2020).

The antimicrobial mechanism also includes photocatalytic activation—particularly in metal-oxide NPs such as ZnO and TiO_2. Upon ultraviolet exposure, these particles create photoexcited charge pairs that react with dissolved oxygen or water to form highly reactive radicals, leading to oxidative degradation of microbial components (Sirelkhatim et al. 2015; Zaman et al. 2025). This process enhances the antimicrobial spectrum and can be further tuned by modifying particle size, defect density, and doping concentration. Biogenic NPs often display synergistic antimicrobial activity due to the retained biomolecular corona from plant or microbial metabolites. This organic layer not only stabilizes particles but also adds biofunctional properties that augment their interaction with microbial receptors and enzymes (Jiang et al. 2025). Moreover, their biocompatibility and controlled reactivity make them safer alternatives to chemically synthesized nanomaterials, reducing cytotoxicity toward non-target organisms.

In agricultural contexts, these antimicrobial pathways collectively underpin the success of NPs in controlling plant pathogens such as *Fusarium equiseti*, *Colletotrichum falcatum*, and *Staphylococcus aureus* while minimizing chemical load on ecosystems (Subba et al. 2024). The integration of biologically derived nanomaterials into crop management thus represents a dual advantage providing pathogen suppression and environmental protection through a naturally inspired mechanism of action (Yang et al. 2025).

13.5 Plant-Based Nanoparticles

13.5.1 Silver Nanoparticles (AgNPs)

Silver nanoparticles are extensively investigated nanomaterials owing to broad-spectrum antimicrobial and antifungal efficacy, stability, and biocompatibility. In agricultural biotechnology, AgNPs have demonstrated remarkable potential for managing phytopathogens and reducing the dependence on synthetic agrochemicals (Sati et al. 2025; Singh et al. 2024b). Typically 5–100 nm, AgNPs exhibit a distinct surface plasmon resonance phenomenon that enhances their electron transfer and oxidative activity (Shahzadi et al. 2025). This physicochemical property, coupled with their high surface-area-to-volume ratio, allows them to interact intimately with microbial membranes and intracellular components, culminating in metabolic disruption and cell death. Plant-mediated or biogenic AgNPs have gained priority over chemically synthesized counterparts because they employ phytochemicals such as flavonoids, terpenoids, alkaloids, and phenolic acids as natural reducing and capping agents. These molecules not only control nucleation and growth of Ag^0 nuclei but also impart stability and functional bioactivity to the final NPs (Saxena et al. 2025; Shahriary et al. 2018). Extracts from *Azadirachta indica*, *Moringa oleifera, Ocimum sanctum, Carissa carandas,* and *Calendula officinalis* have been used successfully to synthesize AgNPs with uniform morphology and enhanced antimicrobial properties (Santos et al. 2022; Sarwar et al. 2025; Singh et al. 2021).

The mechanistic basis of AgNP-mediated antimicrobial action involves three synergistic processes:

1) Electrostatic attachment to negatively charged cell membranes causing structural damage and permeability loss;
2) Generation of ROS such as $\cdot OH$ and O_2^- that oxidize proteins and lipids; and
3) Release of Ag^+ ions that interact with thiol-containing enzymes and nucleic acids, leading to metabolic inhibition (Fatima et al. 2021; Sati et al. 2025). Biogenic AgNPs, owing to their organic surface corona, sustain prolonged ROS generation at lower concentrations, achieving potent antimicrobial efficacy with reduced phytotoxicity.

Experimental studies illustrate that AgNPs synthesized using *A. indica* leaf extract exhibited pronounced inhibition of *Sclerotinia sclerotiorum, Colletotrichum falcatum, Escherichia coli*, and *Staphylococcus aureus* (Asimuddin et al. 2020; Singh et al. 2024b). Similarly,

O. sanctum-derived AgNPs demonstrated strong antibacterial activity against *Xanthomonas oryzae* and *Erwinia carotovora*, major pathogens affecting rice and vegetables (Singh et al. 2024b). The biomolecular capping confers improved colloidal stability and minimizes cytotoxicity, rendering biogenic AgNPs a safer and more sustainable option for crop protection (Jiang et al. 2025).

At the physiological level, low-dose AgNP applications stimulate plant defense responses, including elevated phenylalanine ammonia-lyase activity, accumulation of phytoalexins, and activation of pathogenesis-related (PR) genes. These responses enhance plant resistance to fungal and bacterial invasion (Ying et al. 2022). When used as seed coatings or foliar sprays, AgNPs have been shown to promote seed germination, root elongation, and chlorophyll biosynthesis, further improving crop vigor (Yang et al. 2025).

In conclusion, plant-derived AgNPs exemplify a multifunctional nanobiotechnological tool combining antimicrobial, elicitor, and growth-promoting attributes within a green synthesis framework. Continued optimization of synthesis parameters (pH, precursor concentration, and reaction time) alongside molecular-level investigations of NP–plant–pathogen interactions will be pivotal for their controlled, large-scale deployment in sustainable agriculture (Sati et al. 2025; Shahzadi et al. 2025).

13.5.2 Zinc Oxide Nanoparticles

Zinc oxide nanoparticles (ZnO-NPs) have become a focal point in agricultural nanobiotechnology owing to their role as an antimicrobial agent as well as a bioavailable zinc micronutrient. Their nanoscale size (10–50 nm) endows them with distinctive optical and catalytic properties, including a pronounced band-gap-driven photocatalytic and photo-oxidative potential that allows interaction with microbial surfaces and light-induced generation of ROS (Sirelkhatim et al. 2015). The antimicrobial mechanism of ZnO-NPs involves multiple complementary pathways. Contact between positively charged NPs and negatively charged bacterial membranes causes structural deformation, leakage of cytoplasmic contents, and disruption of transmembrane potential (Fatima et al. 2021). Concurrently, ZnO-NPs produce ROS superoxide (O_2^-) that oxidize cellular proteins and lipids, culminating in oxidative stress and cell lysis (Subba et al. 2024). Under UV or visible irradiation, electron-holes formation further enhances ROS production, while the Zn^{2+} ions which are released bind to thiol and phosphate atoms of biomolecules, inhibiting key metabolic enzymes (Xie et al. 2011). Collectively, these effects confer a broad-spectrum antibacterial and antifungal activity against both Gram-positive and Gram-negative microbes.

Physicochemical attributes strongly influence antimicrobial efficiency. Smaller particles dissolve more readily, generating higher Zn^{2+} ion concentrations and thereby greater toxicity to pathogens (Ojha et al. 2018). Morphological tuning also matters rod-shaped ZnO-NPs with exposed (111) facets exhibit more oxygen vacancies and thus greater ROS generation than spherical counterparts dominated by (100) facets (Sirelkhatim et al. 2015). Reaction conditions such as precursor concentration, solvent polarity, and pH determine crystallinity, defect density, and surface energy, enabling purposeful tailoring for agricultural applications (Adamu et al. 2024). Green or biogenic synthesis approaches have further advanced the eco-safety of ZnO-NPs. Phytochemical-mediated synthesis using plant extracts such as

Moringa oleifera and *Lupinus albus* provides natural reductants (e.g., phenolic acids, flavonoids, and terpenoids) that direct nucleation and passivation of NPs (Adamu et al. 2024; Pal et al. 2018). These biogenic ZnO-NPs display enhanced stability and uniformity, reduced ecotoxicity, and strong antimicrobial performance against *Bacillus subtilis*, *Escherichia coli*, and *Fusarium equiseti* (Subba et al. 2024). In addition, the retained organic corona from plant metabolites improves compatibility with soil microbiota and minimizes unintended environmental persistence (Sodhi et al. 2025). At the physiological level, low-dose ZnO-NPs stimulate plant defense and stress-response pathways, elevating the expression of antioxidant enzymes (CAT, peroxidase [POD]) and stress-related genes (dnaK, ahpC, and katA), thereby enhancing redox homeostasis and pathogen resistance. When applied as nanofertilizers or foliar sprays, they increase chlorophyll content, improve nitrogen and phosphorus assimilation, and strengthen cellular antioxidant capacity, ultimately promoting plant vigor and yield (Yang et al. 2025; Zaman et al. 2025).

In summary, ZnO-NPs derived through green routes offer a multifunctional, sustainable platform combining micronutrient delivery, photocatalytic sterilization, and stress mitigation. Future research should focus on elucidating long-term soil–microbe–NP interactions, optimizing synthesis for controlled ion release, and developing nanocomposite formulations that enhance performance while maintaining ecological safety (Yang et al. 2025).

13.5.3 Gold Nanoparticles

Gold nanoparticles (AuNPs) have attracted attention in green nanotechnology because of their distinctive optical, chemical, and catalytic properties, which arise from high surface plasmon resonance and exceptional electronic conductivity. These attributes make AuNPs particularly suitable for agricultural and biomedical applications that require precise biocompatibility, surface functionalization, and long-term stability (Kumalasari et al. 2024). Unlike bulk gold, AuNPs exhibit size-dependent quantum confinement effects, enabling them to interact efficiently with biomolecules and microbial membranes, thereby influencing metabolic and defense pathways (Hano and Abbasi 2021). Among available synthesis methods, biogenic or plant-mediated production of AuNPs has emerged as the most sustainable approach. Phytochemicals, including flavonoids, terpenoids, phenolic acids, and reducing sugars, act as natural electron donors and stabilizing ligands, driving the reduction of Au^{3+} ions to elemental Au^0 while simultaneously forming an organic coating that enhances colloidal stability (Vijayan et al. 2018). Plant-derived systems such as *Azadirachta indica*, *Equisetum diffusum*, *Indigofera tinctoria*, *Cucurbita pepo*, and *Malva crispa* have been successfully employed to generate monodisperse AuNPs ranging from 10 to 80 nm in diameter, exhibiting uniform morphology and prolonged shelf-life (Abalaka et al. 2014; Kumalasari et al. 2024).

The antimicrobial mechanism of AuNPs differs from that of other metallic NPs because gold ions themselves are relatively inert. Their biological activity is largely mediated through surface interactions and localized surface plasmon resonance (LSPR) effects, which enhance electron transfer to microbial membranes and promote oxidative stress (Ying et al. 2022). AuNPs can attach to bacterial cell walls and disrupt membrane potential, leading to leakage of intracellular components. Moreover, they induce structural modification of key proteins and interfere with DNA replication by binding to phosphorus and sulfur atoms within nucleic acids (Sati et al. 2025). The mild ROS generation associated with AuNP exposure results in oxidative imbalance in microbial cells, contributing to bacteriostatic or bactericidal effects while maintaining minimal cytotoxicity to plant tissues (Singh et al. 2024a).

Beyond the antimicrobial activity, AuNPs exhibit stimulatory effects on plant physiology. Several studies report that nanogold formulations enhance germination rates, root elongation, and chlorophyll biosynthesis in species such as wheat, mung bean, carrot, oat, and tobacco (Mahakham et al. 2016). These improvements are associated with modulation of redox-signaling cascades and activation of antioxidant defense enzymes, which collectively improve stress resilience and nutrient assimilation (Venzhik et al. 2021). Due to their optical responsiveness, AuNPs also find application in biosensing platforms for pathogen detection and in nanofertilizer coatings that enable controlled nutrient release (Hano and Abbasi 2021). The shape and size of AuNPs strongly influence their bioactivity. Triangular or decahedral structures demonstrate higher LSPR intensity and electron-transfer efficiency than spherical forms, while smaller (10–20 nm) particles tend to exhibit greater cellular uptake (Vijayan et al. 2018). These morphological parameters can be tuned during the biogenic synthesis by adjusting pH, reaction temperature, and the concentration of gold precursors and plant metabolites. Such fine control permits optimization of catalytic and biological functions without resorting to toxic reductants (Kumalasari et al. 2024).

In conclusion, plant-mediated AuNPs offer an eco-sustainable and multifunctional nanoplatform, combining antimicrobial, growth-promoting, and diagnostic potentials. Future directions should emphasize the integration of genomic and metabolomic tools to elucidate gold-ion reduction pathways, the development of AuNP-based nanobiosensors for rapid phytopathogen detection, and the assessment of long-term ecological behavior of AuNPs in soil–plant systems (Kumalasari et al. 2024; Yang et al. 2025).

13.5.4 Copper Nanoparticles (CuNPs)

Copper nanoparticles (CuNPs) have emerged as promising antimicrobial and micronutrient agents owing to their redox-active surface and strong catalytic potential. At the nanoscale (typically 10–60 nm), CuNPs display size-dependent plasmonic behavior, high electrical conductivity, and the ability to undergo Cu^0 to Cu^{2+} redox transitions that promote biocidal activity and nutrient cycling (Sirelkhatim et al. 2015). Unlike silver or gold, copper offers the advantage of low cost and natural abundance, making it a feasible candidate for large-scale agricultural deployment under sustainable frameworks (Noman et al. 2025; Patil and Suryawanshi 2025). The antimicrobial mechanism of CuNPs operates through several interconnected routes. Upon contact with microbial membranes, surface-bound Cu^+/Cu^{2+} ions initiate electron transfer reactions that destabilize the lipid bilayer, creating pores and causing efflux of cytoplasmic constituents. Concurrently, redox cycling on the CuNP surface generates ROS which oxidize proteins and nucleic acids, leading to cell death (Patil and Suryawanshi 2025). These oxidative reactions also inactivate critical respiratory enzymes by targeting thiol (—SH) groups, impairing energy metabolism (Ahmad et al. 2025; Noman et al. 2025). In fungi, CuNP exposure disrupts ergosterol synthesis and hyphal elongation, resulting in inhibition of spore germination and mycelial growth (Sati et al. 2025).

The biogenic synthesis of CuNPs using plant extracts, microbial filtrates, or algal biomass has addressed the environmental concerns associated with chemical reduction methods. Plant metabolites including polyphenols, ascorbic acid, terpenoids, and reducing sugars serve as natural reductants that convert Cu^{2+} to Cu^0 under mild conditions while providing organic capping layers that prevent oxidation and aggregation (Taghavi Fardood and Ramazani 2016). Extracts from *Moringa oleifera, Azadirachta indica*, and Aloe vera have successfully produced spherical or quasi-spherical CuNPs with enhanced colloidal stability and superior antimicrobial activity. Microbial routes, employing *Bacillus subtilis* or *Fusarium oxysporum,* exploit enzyme-mediated electron transfer to achieve controlled nucleation and fine-tuned particle size distribution (Ahmad et al. 2025, Noman et al. 2025).

The agricultural relevance of CuNPs extends beyond pathogen control. Copper is an essential cofactor in photosynthetic and antioxidant enzymes such as plastocyanin, SOD, and cytochrome c oxidase; thus, nano-enabled supplementation can correct micronutrient deficiencies and improve photosynthetic efficiency (Ahmad et al. 2025). Low-dose CuNP applications have been shown to enhance seed vigor, root elongation, and chlorophyll content in cereals and legumes, while excessive accumulation may induce oxidative stress and chlorosis. Hence, determining concentration thresholds that differentiate stimulatory from phytotoxic effects is critical for field-level adoption. From an ecological perspective, CuNPs demonstrate biodegradability and moderate persistence compared with silver analogues. Their gradual dissolution releases traceable Cu^{2+} ions, which can be assimilated by plants or immobilized in soil complexes, reducing long-term accumulation (Sibisi et al. 2025). However, NP size, surface coating, and oxidation state govern dissolution kinetics and bioavailability, necessitating systematic evaluation through soil-microbe-plant interaction studies (Zaman et al. 2025). Recent reports indicate that CuNPs synthesized via biogenic routes not only suppress phytopathogens such as *Fusarium oxysporum* and *Xanthomonas campestris* but also exhibit synergistic effects when combined with biocontrol agents like Pseudomonas fluorescens (Ahmad et al. 2025, Noman et al. 2025).

In conclusion, plant- and microbe-derived CuNPs represent a cost-effective and ecologically responsible nanomaterial class for integrated crop protection and micronutrient management. Future research should prioritize optimizing particle surface chemistry to balance antimicrobial potency with soil compatibility, investigating nano-Cu-induced transcriptional responses in plants, and assessing trophic transfer risks to ensure safe deployment under sustainable agricultural systems (Patil and Suryawanshi 2025).

13.5.5 Iron Nanoparticles

Iron nanoparticles (FeNPs) have garnered substantial attention for agricultural nanotechnology owing to their multifunctional properties as micronutrient sources, redox catalysts, and bioremediation agents. Their capacity to exist in multiple oxidation states (Fe^0, Fe^{2+}, and Fe^{3+}) facilitates electron transfer reactions critical for both antimicrobial and soil restoration processes (Azam et al. 2025; Razzaq and Zhou 2025). In the agricultural domain, FeNPs are employed to correct iron deficiency chlorosis, enhance photosynthesis, and improve plant growth, while their antimicrobial potential enables suppression of soilborne pathogens (Vinodhini et al. 2022). At the nanoscale (typically 20–80 nm), FeNPs exhibit high surface reactivity and magnetic susceptibility, which allow them to interact dynamically with plant tissues and microbial cells. Their antimicrobial mechanism primarily involves oxidative stress induction, wherein FeNPs catalyze the Fenton-like reaction to generate hydroxyl radicals (·OH) from hydrogen peroxide (H_2O_2), resulting in the oxidation of lipids, proteins, and DNA in microbial cells (Sati et al. 2025). These radicals disrupt membrane integrity and lead to enzyme inactivation, culminating in microbial death (Azam et al. 2025; Razzaq and Zhou 2025). In addition, Fe^{2+} ions released from FeNPs bind to phosphate and carboxyl groups in microbial membranes, disturbing ionic homeostasis and triggering apoptosis-like responses. Biogenic synthesis of FeNPs provides an eco-compatible route that circumvents the toxicity and instability associated with conventional chemical reduction. Phytochemicals such as tannins, polyphenols, and sugars act as natural reducing and capping agents, enabling rapid conversion of Fe^{3+} or Fe^{2+} precursors to Fe^0 while stabilizing particles against oxidation (Soliemanzadeh et al. 2016). For instance, the transformation efficiency of iron to iron-NPs was reported to be below 50% when using extracts from *Camellia sinensis, Mentha spicata,* and *Syzygium aromaticum* (Haider et al. 2024). These biologically derived FeNPs show greater colloidal stability and reduced aggregation due to biomolecular surface coatings rich in hydroxyl and carboxyl groups (Vinodhini et al. 2022).

Beyond antimicrobial applications, FeNPs contribute to soil fertility enhancement and pollutant removal. Their high adsorption capacity enables sequestration of toxic heavy metals such as arsenic, cadmium, and chromium, thereby reducing soil toxicity and improving plant nutrient availability. In hydroponic and soil systems, FeNPs can serve as a controlled iron delivery system, facilitating gradual release of Fe^{2+} ions that are readily assimilated by roots, improving chlorophyll biosynthesis and electron transport in photosystem I (Azam et al. 2025). Field studies indicate that FeNP-amended soils enhance root vigor, seed germination, and antioxidant enzyme activities in crops such as maize, mung bean, and tomato (Azam et al. 2025). At the molecular level, FeNPs influence plant signaling and gene expression, upregulating metal transporter genes (IRT1, FRO2) and antioxidant enzymes such as CAT, POD, and SOD (Mitra et al. 2023). These effects collectively mitigate oxidative stress under abiotic challenges like salinity and drought, promoting sustainable crop productivity. Nevertheless, excessive accumulation of FeNPs may cause oxidative imbalance and membrane lipid peroxidation, emphasizing the importance of dose optimization and environmental risk evaluation. From a sustainability perspective, biogenic FeNPs outperform chemically synthesized variants in terms of biodegradability, surface passivation, and compatibility with soil microbiota. When used in combination with beneficial microorganisms such as *Rhizobium* and *Pseudomonas fluorescens*, FeNPs can synergistically enhance nitrogen fixation and root nodulation while maintaining pathogen suppression (Sodhi et al. 2025). Such multifunctional integration aligns with the goals of precision and regenerative agriculture.

In conclusion, FeNPs represent a versatile nanomaterial class that bridges micronutrient biofortification and pathogen control. Green synthesis strategies have transformed their application potential, offering scalable, low-cost, and environmentally benign solutions for soil remediation and plant nutrition. Future work should focus on understanding nano-iron interactions at the transcriptomic and metabolomic levels, as well as their long-term environmental dynamics within soil–water–plant systems (Azam et al. 2025; Razzaq and Zhou 2025).

13.6 Applications of Biogenic Nanoparticles in Agriculture

The integration of biogenic NPs into agricultural systems represents one of the most promising strategies for achieving sustainable intensification increasing productivity while minimizing chemical input and ecological burden (Table 13.1). Owing to their small size, high reactivity, and surface tunability, biologically synthesized NPs have found applications across diverse agricultural sectors including crop protection, nutrient delivery, soil health management, post-harvest preservation, and environmental remediation (Kirubakaran et al. 2025; Ying et al. 2022).

13.6.1 Crop Protection and Pest Management

One of the most significant applications of biogenic NPs lies in their role as biocompatible antimicrobial agents capable of replacing or reducing conventional pesticides. Metallic NPs such as Ag, Cu, and ZnO derived from plant or microbial systems inhibit bacterial and fungal phytopathogens including *Fusarium oxysporum*, *Alternaria alternata*, and *Xanthomonas oryzae* by disrupting cell membranes, generating ROS, and interfering with the gene expression (Sati et al. 2025; Subba et al. 2024). These mechanisms exhibit multi-targeted action, limiting the development of microbial resistance a major concern with synthetic agrochemicals. Biogenic AgNPs and CuNPs have shown consistent success in

Table 13.1 Shows the effects of NPs on the growth of plants.

NPs	Plant species	Effect on growth	Reference
Silver	*Triticum aestivum* (wheat)	Enhanced seed germination, chlorophyll content, and root elongation	Lahuta et al. (2022)
Gold	*Oryza sativa* (rice)	Salinity stress alleviation	Song et al. (2022)
ZnO	*Oryza sativa* (rice)	Protection from abiotic stresses	Qiu et al. (2023)
TiO_2	*Solanum lycopersicum* (tomato)	Enhanced germination and photosynthesis	Tighe-Neira et al. (2024)
FeO	*Glycine max* (soybean)	Promoted root nodulation and chlorophyll synthesis	Burke et al. (2015)
CuO_2	*Oryza sativa* (rice)	At low concentration, increased nutrition and enzymatic activity	Da Costa and Sharma (2016)
Silicon dioxide	*Zea mays* (maize)	Enhanced salt tolerance	De Sousa et al. (2019)
NiO	Soyabean	Yield and nutrition increased	Zhou et al. (2023)
MgO	Green onion	Improved chlorophyll, and content of salts	El-Mogy et al. (2025)

controlling seedborne and soilborne pathogens when applied as seed coatings or foliar sprays, thereby improving germination, vigor, and yield.

Additionally, NP formulations have been utilized in nanoencapsulated biopesticides. Incorporation of essential oils, chitosan, and plant alkaloids into nanocarriers improves their stability, prolongs release time, and enhances adhesion to plant surfaces (Eker et al. 2025). The slow-release properties of such systems ensure prolonged pest suppression and reduce the frequency of chemical applications, contributing to safer food chains and environmental protection (Fatima et al. 2021).

13.6.2 Nanofertilizers and Nutrient Management

Biogenic NPs serve as efficient vectors for targeted nutrient delivery. Metal and metal-oxide NPs (ZnO, Fe, Mn, and Cu) synthesized using plant extracts or microbial filtrates provide slow and sustained nutrient release, improving fertilizer use efficiency while reducing nutrient leaching and volatilization (Zaman et al. 2025). For example, ZnO-NPs enhance zinc bioavailability and regulate enzymatic activity related to photosynthesis and nitrogen metabolism, while FeNPs correct iron-deficiency chlorosis and promote chlorophyll synthesis. Foliar application of nanofertilizers has been shown to increase biomass accumulation, grain protein content, and yield in cereals and legumes under both soil and hydroponic systems (Yang et al. 2025).

In addition, nanochelated formulations comprising amino acid-stabilized NPs enable enhanced root uptake and transport of micronutrients even under stress conditions. Biogenic nanofertilizers, due to their biodegradable surface coatings, demonstrate better compatibility with rhizospheric microorganisms, maintaining soil fertility and microbial diversity (Sodhi et al. 2025).

13.6.3 Stress Mitigation and Plant Growth Regulation

Biogenic NPs also play a pivotal role in alleviating abiotic stress induced by drought, salinity, or heavy metals. Controlled doses of AgNPs, ZnO-NPs, and FeNPs activate antioxidant defense mechanisms by upregulating SOD, CAT, and POD, which collectively scavenge ROS and prevent oxidative damage (Subba et al. 2024; Vinodhini et al. 2022). CuNPs and FeNPs have further been reported to enhance proline and phytohormone accumulation, stabilizing osmotic balance under salinity stress. By modulating signaling molecules such as nitric oxide and jasmonic acid, biogenic NPs strengthen the plant's adaptive response at molecular and physiological levels (Setia et al. 2025; Vinukonda et al. 2025).

13.6.4 Post-harvest Preservation and Packaging

In the post-harvest sector, biogenic NPs are being utilized to extend the shelf life of fruits, vegetables, and grains. AgNP- or ZnO-NP-infused biodegradable films exhibit strong antimicrobial and ethylene-scavenging properties, thereby delaying ripening and reducing spoilage (Kirubakaran et al. 2025). Integration of these nanomaterials into starch, cellulose, or chitosan matrices yields active packaging systems that maintain food quality without introducing harmful residues (Vinukonda et al. 2025). Such nanocomposites also provide UV protection and moisture regulation, crucial for preserving nutraceutical and sensory attributes of stored commodities.

13.6.5 Environmental Remediation and Soil Restoration

Biogenic FeNPs and CuNPs are increasingly used for removal of pollutants such as nitrates, phosphates, dyes, and heavy metals from soil and water. Their strong redox potential facilitates degradation and immobilization of contaminants, promoting soil detoxification and nutrient cycling (Vinukonda et al. 2025). Moreover, microbial-synthesized NPs coupled with biochar or humic substances can restore degraded soils by improving porosity, cation exchange capacity, and microbial colonization (Sodhi et al. 2025). These interventions align with circular bioeconomy principles, enabling waste valorization and resource recovery.

13.6.6 Biosensing and Precision Agriculture

Gold NPs and AgNPs, owing to their surface plasmon resonance and conductivity, are being incorporated into biosensing platforms for rapid detection of plant pathogens, pesticide residues, and heavy metals. AuNP-based sensors functionalized with antibodies or DNA probes enable real-time monitoring of crop health and environmental pollutants (Hano and Abbasi 2021; Kumalasari et al. 2024). The miniaturization of such nanodevices holds potential for precision agriculture, allowing site-specific input application and early disease detection, thereby reducing input cost and ecological footprint.

In summary, biogenic NPs offer a multifunctional technological framework for the transition toward sustainable agriculture. Their applications from pathogen suppression to nutrient fortification and biosensing integrate seamlessly with ecological principles, minimizing the environmental costs of conventional practices. The next generation of agricultural nanotechnology must focus on field-level validation, biosafety assessment, and regulatory standardization to ensure safe, efficient, and ethically responsible deployment (Kirubakaran et al. 2025; Yang et al. 2025).

13.7 Challenges and Future Prospects

Although biogenic NPs have demonstrated remarkable potential in sustainable agriculture, their practical deployment remains constrained by a combination of technical, toxicological, regulatory, and socio-economic challenges. These constraints underscore the need for a holistic approach that integrates nanoscience, ecology, and policy frameworks to ensure their safe and effective utilization.

13.7.1 Reproducibility and Scalability Issues

A primary bottleneck in the field of green nanotechnology lies in achieving reproducible and scalable synthesis. Biological extracts derived from plants or microbes vary significantly in composition due to factors such as growth stage, extraction method, pH, and seasonal variation (Eker et al. 2025). Such heterogeneity leads to inconsistent particle morphology, crystallinity, and yield (Sati et al. 2025). Moreover, transitioning from laboratory-scale batch reactions to industrial-scale bioreactors introduces difficulties in maintaining uniformity, sterility, and controlled reaction kinetics. Development of standardized protocols and process-analytical technologies is essential to ensure batch-to-batch consistency and quality assurance (Zaman et al. 2025).

13.7.2 Mechanistic and Molecular Knowledge Gaps

Despite extensive research on NP synthesis and characterization, the molecular mechanisms underlying plant–NP and microbe–NP interactions remain poorly understood. The specific biomolecular pathways involved in metal ion reduction, NP capping, and stress signaling are not yet fully elucidated (Vinodhini et al. 2022). Advanced omics-based tools including transcriptomics, metabolomics, and proteomics are required to map these interactions and identify the genes or enzymes responsible for NP biosynthesis and functionalization (Kirubakaran et al. 2025). A deeper mechanistic understanding will also enable rational design of biogenic NPs with predictable size, reactivity, and biological compatibility.

13.7.3 Environmental and Toxicological Concerns

Although biogenic synthesis reduces the use of hazardous chemicals, the ecotoxicological implications of NP accumulation in soil and aquatic ecosystems are still a major concern. Excessive or uncontrolled application may alter microbial community composition, enzyme activity, and nutrient cycling (Fatima et al. 2021). Metal oxide NPs such as ZnO and CuO can persist in soil matrices, where they may undergo transformation into less reactive or insoluble forms, thereby influencing bioavailability. Moreover, trophic transfer of NPs through plant-animal-human food chains remains insufficiently characterized. Long-term mesocosm and field studies are needed to evaluate persistence, dissolution kinetics, and potential biomagnification risks (Kirubakaran et al. 2025, Ying et al. 2022).

13.7.4 Regulatory and Policy Limitations

Currently, regulatory frameworks governing the agricultural use of NPs vary widely across countries. Most existing nanotechnology guidelines are designed for industrial or biomedical sectors and fail to address the unique ecological dynamics of agricultural systems (Ying et al. 2022). There is an urgent need for internationally harmonized protocols that define permissible exposure levels, toxicity thresholds, and environmental release standards. Implementation of transparent risk–benefit analyses, standardized labeling, and farmer education programs will foster responsible adoption of nano-enabled agricultural products.

13.7.5 Socio-economic Barriers

The successful integration of nanotechnology into agriculture depends not only on technical progress but also on public trust and socio-economic feasibility. Farmers and consumers often associate the term “nano” with artificial or hazardous technologies

(Kirubakaran et al. 2025). Lack of awareness, inadequate extension services, and limited affordability of NP formulations hinder large-scale adoption, particularly in developing nations. Establishing public–private partnerships, demonstration projects, and inclusive innovation platforms can bridge this gap and promote equitable access to green nanotechnologies (Yingyang et al. 2025).

13.7.6 Future Research and Technological Outlook

The future of biogenic NPs in agriculture lies in interdisciplinary convergence—combining nanotechnology with synthetic biology, materials engineering, and artificial intelligence. Precision agriculture can be revolutionized through the integration of nanobiosensors, smart nanofertilizers, and machine-learning-based models that predict optimal NP dosage and application timing (Hano and Abbasi 2021; Kumalasari et al. 2024). Emerging research should emphasize: development of hybrid nanocomposites combining biogenic NPs with biopolymers for enhanced stability and controlled release; use of bioinformatics and quantum modeling to simulate NP–biomolecule interactions; long-term life-cycle and carbon-footprint assessments to evaluate sustainability metrics; and establishment of global nanoregistries for safety data exchange and traceability. Ultimately, the success of biogenic NPs in transforming agriculture will depend on achieving a delicate balance between innovation and safety. Integrating green synthesis principles, regulatory foresight, and stakeholder participation can ensure that nanotechnology evolves into a truly sustainable pillar of next-generation agriculture (Yang et al. 2025; Zaman et al. 2025).

13.8 Conclusion

The advent of nanotechnology has opened transformative pathways for sustainable agriculture, enabling precision control over nutrient delivery, disease management, and environmental remediation. Among various approaches, biogenic or green synthesis of NPs found to be eco-compatible, low-cost, and safe strategy that aligns with the global imperative of reducing chemical dependence and environmental degradation. Phytochemical- and microbe-mediated NPs synthesis leverages biological reductants and stabilizers, providing NPs with enhanced biocompatibility, functional stability, and multitarget antimicrobial potential (Kirubakaran et al. 2025; Ying et al. 2022).

The cumulative evidence presented in this review demonstrates that biogenic NPs such as AgNPs, ZnO-NPs, CuNPs, AuNPs, and FeNPs are multifunctional: they act as antimicrobial agents, micronutrient sources, elicitors of plant defense, and biosensors in smart farming systems. Their ability to trigger ROS, modulate antioxidant enzymes, and activate defense-related genes underpins their efficacy antagonist to a broad class of phytopathogens (Sati et al. 2025; Subba et al. 2024). Furthermore, the incorporation of these NPs into seed coatings, foliar sprays, nanofertilizers, and post-harvest packaging has demonstrated measurable improvements in crop yield, quality, and resilience, making them indispensable components of next-generation precision agriculture (Zaman et al. 2025).

However, while the promise of green nanotechnology is substantial, its responsible translation from laboratory to field remains dependent on addressing critical gaps.

Standardization of biosynthesis protocols, mechanistic elucidation of NP–biomolecule interactions, and assessment of long-term ecotoxicological impacts are essential to ensure environmental and food safety. Regulatory clarity, risk assessment frameworks, and public awareness are equally crucial to prevent misuse and foster acceptance among farming communities (Malik and Basu, 2025; Ying et al. 2022).

The future trajectory of agricultural nanotechnology will likely converge with artificial intelligence, bioinformatics, and molecular biology to create smart, responsive, and sustainable systems capable of real-time monitoring and adaptive input management. Integrating multiomics tools with nanobiosensor platforms can deepen our understanding of plant–NP interactions and optimize nanoformulation design for targeted delivery. Collaborative efforts across disciplines—ranging from material science to agronomy and environmental ethics—will be pivotal in ensuring that biogenic NPs evolve as enablers of circular bioeconomy and food security rather than contributors to new ecological risks (Kirubakaran et al. 2025; Yang et al. 2025).

In conclusion, biologically synthesized nanomaterials represent a paradigm shift in agricultural innovation. Their potential to unify productivity enhancement, environmental stewardship, and economic feasibility makes them vital to achieving the SDGs, particularly those concerning zero hunger (SDG-2), responsible consumption and production (SDG-12), and climate action (SDG-13). The path forward lies in coupling scientific innovation with regulatory prudence and societal inclusivity to realize a green nanotechnological revolution that sustains both crops and the planet.

References

Abalaka, M.E., Daniyan, S.Y., Adeyemo, S.O., and Damisa, D. (2014). The antibacterial efficacy of gold nanoparticles derived from Gomphrena celosioides and Prunus amygdalus (Almond) leaves on selected bacterial pathogens. *International Journal of Bioengineering and Life Sciences* 8 (4): 348–351.

Adamu, T.B., Mengesha, A.M., Kebede, M.A. et al. (2024). Facile biosynthesis of zinc oxide nanoparticles (ZnO NPs) using Lupinus albus L (Gibto) seed extract for antibacterial and photocatalytic applications. *Results in Chemistry 10*: 101724.

Ahmad, W., Atiq, M., Rajput, N.A. et al. (2025). Multimodal impact of copper-silicon hybrid nanotools towards bacterial leaf streak, wheat biochemistry and productivity parameters. *PLoS One 20* (7): e0327021.

Ali, S., Mehmood, A., and Khan, N. (2021). Uptake, translocation, and consequences of nanomaterials on plant growth and stress adaptation. *Journal of Nanomaterials 2021* (1): 6677616.

Asimuddin, M., Shaik, M.R., Adil, S.F. et al. (2020). Azadirachta indica based biosynthesis of silver nanoparticles and evaluation of their antibacterial and cytotoxic effects. *Journal of King Saud University, Science 32* (1): 648–656.

Assad, N., Laila, M.B., Hassan, M.N.U. et al. (2025). Eco-friendly synthesis of gold nanoparticles using Equisetum diffusum D. Don. with broad-spectrum antibacterial, anticancer, antidiabetic, and antioxidant potentials. *Scientific Reports 15* (1): 19246.

Azam, R., Malik, K., Sultana, T. et al. (2025). Plant-mediated bioinspired iron nanoparticles as an alternative to enhance crop resistance against biotic and abiotic stress; a review. *Physiological and Molecular Plant Pathology* 136: 102586.

Ballikaya, P., Brunner, I., Cocozza, C. et al. (2023). First evidence of nanoparticle uptake through leaves and roots in beech (Fagus sylvatica L.) and pine (Pinus sylvestris L.). *Tree Physiology 43* (2): 262–276.

Burke, D.J., Pietrasiak, N., Situ, S.F. et al. (2015). Iron oxide and titanium dioxide nanoparticle effects on plant performance and root associated microbes. *International Journal of Molecular Sciences 16* (10): 23630–23650.

Da Costa, M.V.J. and Sharma, P.K. (2016). Effect of copper oxide nanoparticles on growth, morphology, photosynthesis, and antioxidant response in Oryza sativa. *Photosynthetica 54*(1): 110–119.

De Sousa, A., Saleh, A.M., Habeeb, T.H. et al. (2019). Silicon dioxide nanoparticles ameliorate the phytotoxic hazards of aluminum in maize grown on acidic soil. *Science of the Total Environment 693*: 133636.

Eker, F., Akdaşçi, E., Duman, H. et al. (2025). Green synthesis of silver nanoparticles using plant extracts: a comprehensive review of physicochemical properties and multifunctional applications. *International Journal of Molecular Sciences 26* (13): 6222.

El-Mogy, M.M., Hassan, H.A., Aboelghar, M. et al. (2025). Magnesium oxide nanoparticles enhance growth performance, spectral vegetation indices, yield quantity, and quality of green onion under Egyptian conditions. *Russian Journal of Plant Physiology 72* (4): 124.

Fahim, M., Shahzaib, A., Nishat, N. et al. (2024). Green synthesis of silver nanoparticles: a comprehensive review of methods, influencing factors, and applications. *JCIS Open 16*: 100125.

Fatima, F., Siddiqui, S., and Khan, W.A. (2021). Nanoparticles as novel emerging therapeutic antibacterial agents in the antibiotics resistant era. *Biological Trace Element Research 199*: 2552–2564. https://doi.org/10.1007/s12011-020-02394-3

Haider, F.U., Zulfiqar, U., ul Ain, N. et al. (2024). Harnessing plant extracts for eco-friendly synthesis of iron nanoparticle (Fe-NPs): characterization and their potential applications for ameliorating environmental pollutants. *Ecotoxicology and Environmental Safety 281*: 116620.

Hano, C. and Abbasi, B.H. (2021). Plant-based green synthesis of nanoparticles: production, characterization and applications. *Biomolecules* 12 (1): 31.

Jiang, X., Khan, S., Dykes, A. et al. (2025). Biogenic synthesis of silver nanoparticles and their diverse biomedical applications. *Molecules 30* (15): 3104.

Kapoor, D.U., Gandhi, S.M., Swarn, S. et al. (2025). Polymeric nanoparticles for targeted lung cancer treatment: review and perspectives. *Pharmaceutics 17* (9): 1091.

Khan, F., Shariq, M., Asif, M. et al. (2022). Green nanotechnology: plant-mediated nanoparticle synthesis and application. *Nanomaterials* 12 (4): 673.

Kirubakaran, D., Wahid, J.B.A., Karmegam, N. et al. (2025). A comprehensive review on the green synthesis of nanoparticles: advancements in biomedical and environmental applications. *Biomedical Materials & Devices*, 1–26.

Kumalasari, M.R., Alfanaar, R., and Andreani, A.S. (2024). Gold nanoparticles (AuNPs): a versatile material for biosensor application. *Talanta Open* 9: 100327.

Lahuta, L.B., Szablińska-Piernik, J., Stałanowska, K. et al. (2022). The size-dependent effects of silver nanoparticles on germination, early seedling development and polar metabolite profile of wheat (Triticum aestivum L.). *International Journal of Molecular Sciences 23*(21): 13255.

Li, L., Luo, Y., Li, R. et al. (2020). Effective uptake of submicrometre plastics by crop plants via a crack-entry mode. *Nature Sustainability* 3 (11): 929–937.

Lv, J., Christie, P., and Zhang, S. (2019). Uptake, translocation, and transformation of metal-based nanoparticles in plants: recent advances and methodological challenges. *Environmental Science: Nano* 6 (1): 41–59.

Mahakham, W., Theerakulpisut, P., Maensiri, S. et al. (2016). Environmentally benign synthesis of phytochemicals-capped gold nanoparticles as nanopriming agent for promoting maize seed germination. *Science of the Total Environment* 573: 1089–1102.

Malik, M., & Basu, S. (2025). Utilization of ayurveda and other indigenous medicinal systems for chronic diseases among older adults in India: evidence from LASI. *Discover Health Systems* 4 (1): 134.

Mallikarjuna, K., Nasif, O., Ali Alharbi, S. et al. (2021). Phytogenic synthesis of Pd-Ag/rGO nanostructures using stevia leaf extract for photocatalytic H_2 production and antibacterial studies. *Biomolecules* 11 (2): 190.

Mitra, M., Agarwal, P., and Roy, S. (2023). Plant response to heavy metal stress: an insight into the molecular mechanism of transcriptional regulation. In: *Plant Transcription Factors*, 337–367. Academic Press.

Noman, M., Ahmed, T., Islam, M.S. et al. (2025). Bacterial extracellular biomolecules-derived multimodal manganese nanoparticles control watermelon Fusarium wilt by dysregulating fusaric acid biosynthesis pathway and precise tuning of rhizosphere metabolome. *Journal of Nanobiotechnology 23* (1): 452.

Ojha, S., Singh, D., Sett, A. et al. (2018). Nanotechnology in crop protection. In: *Nanomaterials in Plants, Algae, and Microorganisms*, 345–391. Academic Press.

Pal, S., Mondal, S., Maity, J., and Mukherjee, R. (2018). Synthesis and characterization of ZnO nanoparticles using Moringa oleifera leaf extract: investigation of photocatalytic andantibacterial activity. *International Journal of Nanoscience and Nanotechnology 14* (2): 111–119.

Parkinson, S.J., Tungsirisurp, S., Joshi, C. et al. (2022). Polymer nanoparticles pass the plant interface. *Nature Communications* 13 (1): 7385.

Patil, V.W. and Suryawanshi, N.S. (2025). Copper oxide nanoparticles in agri-nanotechnology: green synthesis, pathogen targeting, and regulatory outlook. *International Journal of Scientific Research in Science and Technology 12* (3): 890–921.

Qiu, J., Chen, Y., Liu, Z. et al. (2023). The application of zinc oxide nanoparticles: an effective strategy to protect rice from rice blast and abiotic stresses. *Environmental Pollution 331*:121925.

Razzaq, S. and Zhou, B. (2025). Revolutionizing crop production with iron nanoparticles for controlled release of plant growth regulators and abiotic stress resistance. *Plant Nano Biology* 13: 100172.

Santos, A.P., Gonçalves, M.M., Justus, B. et al. (2022). Calendula officinalis L. flower extract-mediated green synthesis of silver nanoparticles under LED light. *Brazilian Journal of Pharmaceutical Sciences* 58: e19519.

Sarwar, K., Nazli, Z.I.H., Munir, H. et al. (2025). Biosynthesis of zinc oxide nanoparticles using Moringa oleifera leaf extract, probing antibacterial and antioxidant activities. *Scientific Reports 15* (1): 20413.

Sati, A., Ranade, T.N., Mali, S.N. et al. (2025). Silver nanoparticles (AgNPs): comprehensive insights into bio/synthesis, key influencing factors, multifaceted applications, and toxicity—a 2024 update. *ACS Omega 10* (8): 7549–7582.

Saxena, R., Kotnala, S., Bhatt, S.C. et al. (2025). A review on green synthesis of nanoparticles toward sustainable environment. *Sustainable Chemistry for Climate Action* 6: 100071.

Setia, A., Mehata, A.K., Malik, A.K., and Muthu, M.S. (2025). Synthesis of gold nanoparticles using physical and chemical methods: Strategies for stabilization, controlling shape, size and morphology. In: *Gold Nanoparticles, Nanomaterials and Nanocomposites*, 41–79. Elsevier.

Shahriary, M., Veisi, H., Hekmati. M., and Hemmati, S. (2018). In situ green synthesis of Ag nanoparticles on herbal tea extract (Stachys lavandulifolia)-modified magnetic iron oxide nanoparticles as antibacterial agent and their 4-nitrophenol catalytic reduction activity. *Materials Science & Engineering. C, Materials for Biological Applications* 90: 57–66.

Shahzadi, S., Fatima, S., Shafiq, Z., and Janjua, M.R.S.A. (2025). A review on green synthesis of silver nanoparticles (SNPs) using plant extracts: a multifaceted approach in photocatalysis, environmental remediation, and biomedicine. *RSC Advances 15* (5): 3858–3903.

Sibisi, M. P., Basson, A. K., Ntombela, Z. G., & Pullabhotla, R. V. (2025). Eco-friendly synthesis and optimization of CuNPs using a non-pathogenic bioflocculant from Kytococcus sedentarius. *Biotechnology Notes* 6: 89–99.

Singh, J., Kumar, A., Nayal, A.S. et al. (2024a). Comprehensive antifungal investigation of green synthesized silver nanoformulation against four agriculturally significant fungi and its cytotoxic applications. *Scientific Reports 14* (1): 5934.

Singh, K., Shukla, M., Singh, A. et al. (2024b). Bio-synthesis and biopotential assessment of silver nanoparticles (AgNPs) by using leaf extract of Ocimum sanctum. *International Journal of Plant Pathology and Microbiology* 4 (1): 98–105.

Sirelkhatim, A., Mahmud, S., Seeni, A. et al. (2015). Review on zinc oxide nanoparticles: antibacterial activity and toxicity mechanism. *Nano Micro Letters* 7 (3): 219–242.

Sodhi, G.K., Wijesekara, T., Kumawat, K.C. et al. (2025). Nanomaterials–plants–microbes interaction: plant growth promotion and stress mitigation. *Frontiers in Microbiology 15*:1516794.

Soliemanzadeh, A. et al. (2016). Biosynthesis of iron nanoparticles and their application in removing phosphorus from aqueous solutions. *Chemistry and Ecology* 32 (3): 286–300.

Song, Y., Zheng, C., Basnet, R. et al. (2022). Astaxanthin synthesized gold nanoparticles enhance salt stress tolerance in rice by enhancing tetrapyrrole biosynthesis and scavenging reactive oxygen species in vitro. *Plant Stress 6*: 100122.

Spielman-Sun, E., Lombi, E., Donner, E. et al. (2017). Impact of surface charge on cerium oxide nanoparticle uptake and translocation by wheat (Triticum aestivum). *Environmental Science & Technology* 51 (13): 7361–7368.

Subba, B., Rai, G.B., Bhandary, R. et al. (2024). Antifungal activity of zinc oxide nanoparticles (ZnO NPs) on Fusarium equiseti phytopathogen isolated from tomato plant in Nepal. *Heliyon 10* (22): e40198.

Taghavi Fardood, S. and Ramazani, A. (2016). Green synthesis and characterization of copper oxide nanoparticles using coffee powder extract. *Journal of Nanostructures* 6 (2): 167–171.

Tighe-Neira, R., Reyes-Díaz, M., Nunes-Nesi, A. et al. (2024). Physiological and agronomical traits effects of titanium dioxide nanoparticles in seedlings of Solanum lycopersicum L. *BMC Plant Biology 24* (1): 146.

Venzhik, Y.V., Moshkov, I.E., and Dykman, L.A. (2021). Gold nanoparticles in plant physiology: principal effects and prospects of application. *Russian Journal of Plant Physiology* 68 (3): 401–412.

Vijayan, R., Joseph, S., and Mathew, B. (2018). Indigofera tinctoria leaf extract mediated green synthesis of silver and gold nanoparticles and assessment of their anticancer, antimicrobial, antioxidant and catalytic properties. *Artificial Cells, Nanomedicine, and Biotechnology* 46 (4): 861–871.

Vinodhini, S., Vithiya, B.S.M., and Prasad, T.A.A. (2022). Green synthesis of palladium nanoparticles using aqueous plant extracts and its biomedical applications. *Journal of King Saud University, Science* 34 (4): 102017.

Vinukonda, A., Bolledla, N., Jadi, R.K. et al. (2025). Synthesis of nanoparticles using advanced techniques. *Next Nanotechnology 8*: 100169.

Xie, Y., He, Y., Irwin, P.L. et al. (2011). Antibacterial activity and mechanism of action of zinc oxide nanoparticles against Campylobacter jejuni. *Applied and Environmental Microbiology* 77 (7): 2325–2331.

Yang, Y., Ye, C., Zhao, M. et al. (2025). Nanoparticles in sustainable agriculture: enhancing nutrient use efficiency and abiotic stress resilience under climate change. *Plant Stress* 17: 100982.

Ying, S., Guan, Z., Ofoegbu, P.C. et al. (2022). Green synthesis of nanoparticles: current developments and limitations. *Environmental Technology & Innovation* 26: 102336.

YingYing, J., Balasubramanian, B., Park, S. et al. (2025). Green nanoparticles in agriculture: enhancing crop growth and stress tolerance. *Plant Stress* 18: 101017.

Zadehnazari, A. (2025). Chemical synthesis strategies for metal oxide nanoparticles: a comprehensive review. *Inorganic and Nano-Metal Chemistry 55* (6): 734–773.

Zaman, W., Ayaz, A., and Park, S. (2025). Nanomaterials in agriculture: a pathway to enhanced plant growth and abiotic stress resistance. *Plants 14* (5): 716.

Zhang, H., Zheng, T., Wang, Y. et al. (2024). Multifaceted impacts of nanoparticles on plant nutrient absorption and soil microbial communities. *Frontiers in Plant Science 15*: 1497006.

Zhou, P., Jiang, Y., Adeel, M. et al. (2023). Nickel oxide nanoparticles improve soybean yield and enhance nitrogen assimilation. *Environmental Science & Technology 57* (19): 7547–7558.

14

Nanostructures for Immobilization of Agriculturally Used Cells and Enzymes

Nafiseh Sadat Naghavi, Fahimeh Sarami, and Golnoosh Rezaeizadeh

Department of Microbiology, Fal.C, Islamic Azad University, Isfahan, Iran

14.1 Introduction

Biotechnology helps to solve several problems all over the world, especially those in accordance with health, agriculture, environment, and general business competitiveness (Demicheli 1997; Ranjkesh et al. 2021). Whole cells and their enzymes as biocatalysts have several advantages over chemical catalysts, including high specialty and catalytic activity, simplest operational conditions, and low side effects on the environment that make them suitable choices for manufacturing several industrial products (Anteneh and Franco 2019). Recently, nanotechnology, the science of nanoscale-sized particles, has altered the world of science (Desai 2012). Different nanoparticles such as nanovesicles, nanosuspensions, nanoemulsions, nanospheres, nanotubes, nanoshells, and lipid nanoparticles have been employed to advance diverse kinds of pharmaceuticals (Shah et al. 2010; El-Ridy et al. 2012; Abousamra and Mohsen 2016; Asfour and Mohsen 2018; El-Ridy et al. 2018).

The process of cell or protein immobilization onto a defined matrices while maintaining their biological activity is of special importance to many industrial productive or screening techniques (Niemeyer and Blohm 1999; MacBeath and Schreiber 2000; Wilson and Nock 2002; Wilson and Nock 2003; Chen et al. 2016). Methods used to immobilize a biocatalyst are divided into two main categories, including reversible methods (adsorption and ionic, disulfide, and affinity binding) and irreversible methods (covalent binding, crosslinking, and physical entrapment such as matrix entrapment and encapsulation) (Figure 14.1). In a different categorization, immobilization methods can also be classified into surface binding and entrapment methods by means of the type of chemical reaction involved in the immobilization process (Homaei et al. 2013). In this way, surface attachment techniques are considered a main protocol. Three main types of surface immobilization of biocatalysts onto surfaces include physical adsorption (electrostatic and hydrophobic interactions), covalent attachment (such as amide or imine bonds, and gold–thiol interactions), and biological interactions (including biotin–streptavidin binding, antibody–antigen interaction, and DNA hybridizations) (Liu and Amro 2002; Wadu-Renault et al. 2003; Homaei et al. 2013).

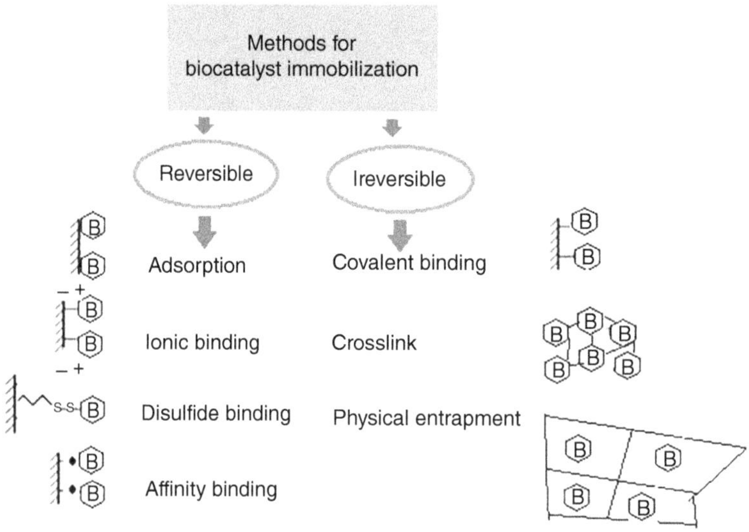

Figure 14.1 Classification of biocatalysts (B) immobilization methods.

Immobilization by electrostatic interactions is a practical simple route that attracted for attachment of biocatalysts onto solid surfaces. These reversible interactions can support the biological function of biocatalysts (Wadu-Mesthrige et al. 2001), although some contrary examples have also been seen (Ladam et al. 2001; Topoglidis et al. 2001; Lee et al. 2002). Early methods applied for immobilization of biocatalysts on nanoparticles were conducted by simple adsorption and covalent binding. Recently, other methods have been attracted for surface binding of biocatalysts on a variety of nanostructured materials, such as immobilization on porous nanomaterials, electroconductive nanofibers, carbon quantum dots (CQD), graphene oxide (GO), carbon nanoonions (CNO), and magnetic nanoparticles. This large field maintains better protection of the enzyme mobility in contrast to enzyme immobilization systems on conventional supports. One of the notable advantages of recent nanostructured materials is the porous size of the nanostructures that can be controlled at the nanometer scale (Homaei et al. 2013; Hetmann et al. 2018).

14.2 Immobilized Whole Cell Microorganisms Used in Agriculture

Chemical and biological reactions that lead to soil regeneration through underlying rock are done in long geological time (Buscot and Varma 2005; Sathya et al. 2016). Therefore, protection of soil health is important for sustainable agricultural production. Bacterial biofilm is a prevalent lifestyle of these organisms in ecosystems that is formed by the attachment of bacteria to surfaces. Mobile biofilms in the form of flocs and sludge can also be seen in the absence of any attachment surface by means of micro-sized particle aggregation (Flemming et al. 2016). Biofilm formation is often the cause of the possible successful survival and persistence of bacteria in different nutritional and environmental conditions for long times (Jacoby and Archer 1991; Rendueles and Ghigo 2015; Nocelli et al. 2016). Bacteria may use different natural (animate or inanimate) or artificial surfaces, including organic materials, soil particles, metals, glass, plastic, rubber, and plants (Sheng et al. 2008; Kokare et al. 2009; Tuson and Weibel 2013). Bacterial aggregations have evolutionary benefits over time for them and other organisms, although these properties can have harmful effects on living and nonliving hosts. The bacterial biofilm has also been observed in the soil in the last decades, especially in the rhizosphere (the area surrounding the roots) and rhizoplane (root surface) (Rudrappa et al. 2008). Biofilms are involved in the interactions between microbes and plants, which can cause benefits to both organisms, such as stimulation of growth and induction of resistance to unfavorable environmental conditions, and protection from pathogen or pest attacks (Rafique et al. 2015; Nocelli et al. 2016; Van Duijkeren et al. 2018). Therefore, recently scientists have been largely interested in the possible utilization of biofilms for sufficient crop cultivations (Galelli et al. 2015). On the other hand, the resistance of biofilms has been greatly concerned for industrial production of agriculturally useful biomolecules and in bioremediation of soil pollutants (Chen et al. 2011; Das and Adholeya 2012; Ercan and Demirci 2015).

New strategies have been used to induce microbes' favorites for surface attachment and biofilm formation. Different industries and stakeholders are interested in using microbial biofilms to improve their efficiency and to overcome the problems faced in the production processes involving microorganisms. For example, the concern for renewable energy sources led to the development of microbial fuel cells (MFCs) in which bacteria develop biofilms around electrodes (Chen et al. 2011). On the other hand, the extensive usage of microorganisms for remediation of organic and inorganic contaminants in ecosystems (Das and Adholeya 2012) is recently conducted for the usage of biofilms (Asri et al. 2018). The adhesion of microorganisms to the surfaces is a crucial strategy for biofilm formation. Therefore, strategies to induce biofilm formation have been typically aimed at modifying surface topography and coating depositions to facilitate adhesion of bacteria to the defined surfaces.

14.2.1 Bacteria

Bacterial species in association with the plant rhizosphere include beneficial effects on plant growth. These bacteria are highly concerned for their use in agriculture improvement according to their fertilization or pathogen control effects. This valuable group of bacteria has been named as "Plant Growth Promoting Rhizobacteria" (PGPR) (Glick 1995). PGPR strains mainly belong to the taxa Cyanobacteria, Firmicutes, Actinobacteria, Bacteroidetes, and Proteobacteria (Tilak et al. 2005). Immobilizing the living cells of these microorganisms has gained a wide range of attention. Encapsulation of soil-applicable microbial cells has provided various advantages over free cells, such as simple application to the soil, reduction of drifting probability, protection against severe environmental conditions, high cell loading capacity, establishment of cell viability, and elevated production rate. Also, the immobilizing matrix acts as a reservoir that can reserve slow release of bacteria at a constant rate (Cassidy et al. 1996).

Actinomycetes are Gram-positive filamentous bacteria that grow in aerobic conditions. These organisms are involved in nutrient cycling, and produce valuable secondary metabolites such as over half of ordinary clinical antibiotics, different antitumor and immunosuppressive drugs, anthelmintics, antifungal, herbicidal and insecticidal compounds, as well as plant growth inducers. Although actinomycetes are widespread in different habitats, including water, soil, and marsh, these organisms are mainly found in dry, humic, or calcareous soils. Actinomycetes have a great advantage in agriculture by secreting a wide variety of antimicrobial products in the rhizosphere, which are able to prevent the growth of root microbial pathogens. These bacteria have also been attracted because of their ability to solubilize phosphate, organic acid, and siderophores (iron-chelating compounds) production, and secretion of different enzymes, which have direct or indirect effects on the induction of plant growth (Doumbou et al. 2001; Bérdy 2005; Al-Aksar 2012; Sadeghi et al. 2012; Barka et al. 2016).

Enzymes such as cellulase, chitinase, phosphatase, protease, and phytase have critical effects on the maintenance of soil ecology, fertility, and health (Sinsabaugh et al. 1991). The enzyme productivity in soil is affected by different agents like organic and inorganic matter composition, and the macro- and micro-organisms that live in the soil (Kiss et al. 1978). Actinomycetes are able to produce these enzymes under submerged (SmF) as well as preferred solid-state fermentation (SSF) systems (Suneetha and Zaved 2011). In the SSF process that is free of water and has a sufficient moisture content, insoluble substrate is fermented (Chahal 1985). The substrates mainly consist of a low-cost agricultural residue like wheat bran, wheat straw, bagasse, or soya hulls (Mohana et al. 2008). SmF is a liquid fermentation system that involves free-flow liquid substrates like molasses. The biological products are released into the fermentation broth. The rapid utilization of substrates leads to the constant replacement of nutrients. This fermentation system is mainly preferred for bioactive compounds production by using microorganisms such as bacteria that require high moisture conditions for growth (Subramaniyam and Vimala 2012).

Homogeneous mixing in fermenters requires high-speed striation at to ensure sufficient oxygen and nutrient accessibility to the growing biomass. This vigorous mixing results in high rates of shear force, which may cause fragmentation of the mycelial organisms such as actinomycetes (Large et al. 1998; Roubos et al. 2001; Mehmood et al. 2010). This event can reduce the productivity of the microorganism as well as the release of intracellular contents into the culture medium as a concomitant that complicates purification of the product (Ohta et al. 1995). Immobilization of the cells, especially by microencapsulation, can physically separate a large volume of mycelia from the broth environment and protect them from shear force. Also the encapsulated mycelia are less permeable to small molecules than other immobilizing matrices such as calcium alginate and protect cells from extrinsic stress (Gaserod et al. 1999). Other polymers, such as polyurethane, have recently been proposed as cost-effective entrapping polymers in which the cells have maintained their activity for long times. Microbial cell consortia immobilized in polyurethane have retained 44.31% of their bioactivity after more than 6 months (Kazemzadeh et al. 2020b).

Some studies have been conducted to produce valuable agricultural compounds by immobilized actinomycetes. Higher yields and lower downstream-processing costs can be achieved by encapsulation of actinomycetes in suitable matrices. Zacchetti et al. (2018) applied *Streptomyces lividans* immobilized by micro-encapsulation in semi-solid alginate for extracellular tyrosinase production. The production rate was increased by extended viability of the mycelia along with the decrease in the intracellular proteins released into the culture medium. Shrivastava et al. (2008) encapsulated an indole-3-acetic acid (IAA)-producing isolate of *Kitasatospora* in the beads formed in calcium alginate beads. Growth pattern analysis by optical and scanning electron microscopy showed that the hyphae formed initially, and sporulation was seen at the seventh day of entrapment. Higher production of IAA was shown by the immobilized cells than free cell over a long period of time by inoculation of cells to sterile soil. Also, higher tolerance to severe environmental conditions, including pH, temperature, salinity and heavy metals, was shown by encapsulated *Kitasatospora* sp. than its free cells. Kar et al. (2009) produced alpha-amylase by immobilized *Streptomyces erumpens* in different matrices, including polyacrylamide, agar-agar, and luffa sponge, in batch fermentation. Enzyme yield in Luffa sponge was 21% and 51% more than agar-agar and polyacrylamide, respectively. The repeated experiments showed that the physiological activity of *S. erumpens* cells was almost equal on the supports after three cycles of fermentation.

The extracellular enzyme, cyclodextrin glucanotransferase (CGTase, EC.2.1.), which is in the α-amylase superfamily, catalyzis the hydrolysis of carbohydrates and noncarbohydrate moieties by the cleavage of glycoside bonds between them to produce alpha, beta, or gamma cyclodextrin (CDs) from starch (Costa et al. 2012; Eş et al. 2016). CDs are a significant material in industrial agricultural, food, and pharmaceutical processes (Ibrahim et al. 2014). Afzali et al. (2021) immobilized a CGTase-producing isolate of *Streptomyces griseobrunneus* in manufactured composites. FT-IR analyses for detection of physicochemical characteristics of the composites indicated an interaction between all of the elements. Enzyme activity measurement by phenolphthalein assay showed the activity of 1109 by using alginate/chitosan/microcrystalline cellulose (MCC) sheet, and 948 by using agar/gelatin/MCC beads as immobilization matrices in contrast to 606 U/mL enzyme activity by free cells. It concluded that the improved enzyme production by the immobilized cells may be the result of the protection of bacteria against produced inhibitors and toxins in the culture medium. In another study, *Bacillus cereus* and *Bacillus amyloliquefaciens,* which were immobilized in calcium alginate, showed a maximal CGTase activity of 47 U/mL and 110 U/mL, respectively, which was higher than that of the free cells (Vassileva et al. 2003).

Phosphorus (P), which is considered the second most limiting environmental nutrient, is produced primarily by weathering of minerals in rocks. Fertilization of agricultural fields may elevate the levels of P, especially in the most productive areas in which high P fertilizers are input. Therefore, P fertilizers have globally key roles in modern agriculture. On the other hand, potassium (K) also has a crucial different effect on plant life. Potassium acts as a constituent of enzymes, which regulates drought tolerance and water supply. Feldspars and micas are the main sources of K in soil, which release this element during weathering. Three forms of K are found in soils, including the exchangeable ionic form, which adsorbs to the clay particle surface; in circulating solution; and in organic matter. The widely distributed K-containing minerals prevent its leaching by cation exchange. Therefore, depletion of this element from the soil is not common (Ballabio et al. 2019).

Matthews (2019) evaluates the efficiency of immobilized potential P and K solubilizing bacteria, including *Klebsiella* sp., *Enterobacter* sp., and *Serratia* sp., to release the fixed P and K in soils from insoluble minerals. The bacteria were isolated from the rhizosphere and screened using insoluble P and K medium. The immobilized bacteria strongly increased total N, cation exchange capacity, root, and soil surface biomass compared to commercial biofertilizer and showed a higher increase in total P and exchangeable K compared to commercial biofertilizer. The immobilized solubilizing bacteria showed elevation of soil fertility as well as the root and shoot dry weight increase.

14.2.2 Fungi

Water activity (aw) is an important factor in microbial physiological activities. Although fungi require low aw levels and are suitable to be cultivated on SSF, aw is a critical factor in their growth and metabolism (Kredics et al. 2000; Parra et al. 2004). Therefore, aw is considered as an important parameter for fungal mass transfer (Gervais and Molin 2003). Fungal spores adsorb on solid supports and form biofilms (Gutiérrez-Correa and Villena 2003). Filamentous growth is the advantage of fungal cells that make them suitable for immobilization on different surfaces. Cellulase enzymes are highly used by several industries, such as food production, and fruit and textile processing (Bhat 2000; Kirk et al. 2002). Finding active microorganisms with high levels of cellulase production and employing biological techniques for detection of cellulase-coding genes suitable for transforming to fast-growing microorganisms are mainly concerned (Ghadiri et al. 2020). Cellulases like other industrial enzymes are produced by submerged fermentation (SmF) and to a lesser extent, by SSF. SSF has some advantages over SmF, such as low technological requirement and high productivity, that resulted in decreased downstream processing costs (Hölker et al. 2004). Cell adhesion leading to biofilm formation is another characteristic of SSF that is less investigated, although it,may enhance physiological advantages of SSF. Villena and Gutiérrez-Correa (2007) conducted a study in order to compare the lignocellulosic enzyme production (filter paper activity as well as endoglucanase and xylanase activities) by *Aspergillus niger* in SmF and immobilized biofilm fermentation (BF) at different aw levels. Enzyme activities were much higher in BF than in SF systems, but biomass production was higher in SF than in BF. At low aw content the enzymatic activity was decreased in both systems, but BF still had higher levels of enzyme production than SF cultures, and in the case of xylanase enzyme, the specific activity was increased more than twofold at low aw content, suggesting modifications in regulatory mechanisms in biofilm.

Plant hormones are important in several steps in plant growth and development and are required in plant defense (Costacurta and Vanderleyden 1995). Auxins, cytokinins, gibberellins, abscisic acid, and ethylene are considered the main plant growth hormones. Gibberellins (GAs) are in a large family of plant growth hormones that are structurally related to diterpenoid acids. These hormones are found in green plants and some microorganisms. GA controls many plant developmental mechanisms such as the induction of hydrolytic enzyme production during plant activities such as seed germination, stem formation, flowering, induction of crop productivity, dwarfism and dormancy control, sex expression, and fruit and leaf senescence (Kumar and Lonsane 1989; Bruckner and Blechschmidt 1991; Rios-Iribe et al. 2011; Rangaswamy 2012). It has been observed that GA production began earlier by cultivated immobilized microbial cells than suspended-cell cultures (Saucedo et al. 1989; Kim et al. 2006). Meleigy and Khalaf (2009) immobilized *Fusarium moniliforme* cells in loofa sponge for GA production and observed the best production of GA with high conversion and productivity rates during six days by immobilized cells. Saudagar et al. (2008) reported higher GA productivity by immobilized cells of

Gibberella fujikuroi in a copolymer of hydrophilic hydroxyl ethyl acrylate and hydrophobic trimethylpropane than the productivity of free cells.

Ivanova et al. (2011) immobilized ethanol-producing yeast in alginate, magnetic nanoparticles and covalently on magnetite-containing chitosan and cellulose-coated magnetic nanoparticles. The results showed that the ethanol productivity was affected by the type of support for immobilization, the initial seed, and the concentration of the sugar source. The magnetic particles containing immobilized yeast cells had high ethanol productivity and persisted in saline for more than one month at 4 °C.

14.2.3 Nanostructure Substrates for Immobilization of Cells Used in Agriculture

A few studies were recently conducted on the capability of synthetic micro- and nanoscale motifs that facilitate bacterial attachment to the smooth surfaces (Leonard et al. 2017; Truong et al. 2009). However, in most of these studies the interactions at the early stages, before biofilm formation, have been observed (Truong et al. 2009). Other studies conducted on the act of nanofibers as substrates for growing microorganisms until biofilm formation for some applications such as energy-productive reactions by nanofibrous electrodes in MFCs (Cavaliere 2015) or pollutant remediation (Balusamy et al. 2019). These studies did not provide the detailed information about biofilm progress mechanisms. It has been reported that some rod-shaped bacterial cells prefer to adhere to medium-sized nanofibers (Sarioglu et al. 2017) and small-sized nanofibers (De Cesare et al. 2019) compared to bacterial cell diameter. Coating nanofibers by the conditioning films released from the bacterial cells resulted in the rapid attachment of bacteria along the fibers (De Cesare et al. 2019).

Recently, microorganisms have started to be added in some agricultural products such as biofertilizers, biostimulants, and biopesticides to decrease the side effects of agrochemicals on health and the environment (Woo and Pepe 2018; Singh et al. 2019; Kour et al. 2020). On the other hand, microorganisms have been used to eliminate contaminants from soil and wastewater by means of bioremediation processes (Das 2014). It should be concerned that for both agriculture and bioremediation purposes, the microbe-involved products, such as plant growth-promoting rhizobacteria (PGPR) that are commercially available to promote crop yield or remove pollutants show ineffective activity, especially in the case of non-spore-forming species (Kaminsky et al. 2019). This is because of the fact that the microbes inoculated to the soil have to undergo several diverse conditions, such as nutrient deficiencies, harsh environment, and the presence of competing organisms that are resident in the soil and are more adapted to the adverse conditions. Hence, there is an increasing request for new efficient and effective microbial products for usage in both agricultural and environmental fields.

14.3 Immobilized Microbial Enzymes Used in Agriculture

Immobilized biocatalysts are more stable and more resistant to adverse environmental conditions such as the presence of heavy metals and other pollutants, as well as non-favorite pH values, temperature, nutrition, or salinity than free-dissolved enzymes. Most of the reported immobilized biocatalysts provide beneficial properties such as sufficient recovery of biocatalyst and product, which limits the number of processing steps, recyclability of biocatalyst, the advantage of continuous operation of biological reactions, high levels of productivity, the possibility to terminate the reaction in each point, and controlling environmental pollution and the costs of production (Sheldon 2007; Defaei et al. 2018). Despite the mentioned advantages, using immobilized biocatalysts faces some disadvantages including lowered activity and reaction rate compared to free biocatalysts, susceptibility to deactivation of biocatalysts, the cost of supports, and the requirement to dispose of immobilized biocatalysts.

Tyrosinase (EC 1.14.18.1) is a widespread enzyme found in animal and plant tissues and microorganism cells. It catalyzis the production of melanin from tyrosine by 3,4-dihydroxy phenylalanine (DOPA) and dopaquinone (Saruno et al. 1979). Harir et al. (2018) isolated different actinomycetes from Algerian Sahara soil to extract novel biotechnologically applicable enzymes. The extracted tyrosinase that was immobilized on nylon nanofiber layers displayed interesting properties suitable for biotechnological processes. The produced tyrosinase enzyme showed the highest activity and stability in different environmental conditions, including neutral and alkaline pH, $T = 55\,°C$, and in the presence of different cations. Furthermore, more specificity was observed by the enzyme for diphenols than monophenols, showing higher preferred activity for diphenolase than monophenolase.

14.3.1 Nanostructure Substrates for Immobilization Enzymes and their Usage in Agriculture

The usage of different nanoparticles and nanomaterials started since 2014 for enzyme immobilization in biological applications such as biosensors. In the early stages, conventional methods such as simple adsorption and covalent linkage were used for enzyme immobilization on different nanostructured materials. Then, the method that used porous nanomaterials, electroconductive nanofibers, and magnetic nanoparticles was more attractive than conventional matrices. These methods provide a larger area and better enzyme loading, which results in the improvement of enzyme mobility in pore-sized nanostructures (Homaei et al. 2013). An extensive attention is observed to metal nanoparticles, carbon nanomaterials, and magnetic nanoparticles among various nanoparticles and nanomaterials because of their fascinating advantages (Chen et al. 2017). Hung et al. (2011) used an electrospun polyacrylonitrile nanofibrous membrane for immobilizing cellulase enzyme. The immobilized enzyme that showed increased stability against pH and heating was obtained for hydrolyzing microalgae cell wall and the release of reducing sugars. Optimization of the activity condition by response surface methodology (RSM) showed that the hydrolysis efficacy was under the effect of temperature, pH, and substrate/enzyme mass ratio, and reached the hydrolysis efficacy of 60.86% by the optimization process. Lupoi and Smith (2011) compared the efficiency of the immobilized cellulose adsorbed on the 40 nm silica NPs with free enzyme in the hydrolysis of microcrystalline cellulose. The immobilized enzyme released 1.6 times ($p = 0.01$) more glucose than the free enzyme during 96 hours at pH = 4.8 and the temperature of 35 °C.

Chang et al. (2011) constructed two mesoporous silica (MS) nanoparticles with different particle and pore size and different surface area. They used the nanosupports to immobilize cellulase enzyme by two methods of physical adsorption and chemical binding. The cellulase, which attached covalently to MSNPs with large pore sizes containing carboxyl groups, was highly effective in the hydrolysis of cellulose into glucose. The immobilized cellulase on magnetic beads was highly effective in a wide range of temperatures and pH. The immobilized enzymes were protected from damage during long storage and were efficiently reused after being recovered from the reaction mixture. Hegedüs et al. (2012) demonstrated that the stability of cellulase enzyme coated with nanoparticles was highly increased in contrast to that of free enzyme. Also, the immobilized enzyme showed a higher activity of about 10 times in a temperature range between 20 °C and 37 °C than soluble enzyme, and a broad range of pH (1.5–12).

Among different nanoparticles, magnetic nanoparticles (MNPs) are subjected to vigorous attention due to their advantages features like low toxicity and high surface area, which facilitate loading of a large number of enzymatic molecules on their surfaces (Amirbandeh and Taheri-Kafrani 2016; Amin et al. 2018; Mehnati-Najafabadi et al. 2018). Despite these advantages, MNPs have some disadvantages especially the inhibition in acidic and oxidative conditions. For this reason, coating their outer surface is important to protect the consistency of MNPs (Landarani-Isfahani et al. 2015). In a recent report, Defaei et al. (2018) immobilized α-amylase (EC 3.2.1.1) on a naringin-functionalized MNP via ionic reactions. Then the surface of MNPs was coated with biocompatible flavonoid, naringin. The physical properties of functionalized MNPs and the immobilization efficiency of the nanostructure were tested by thermogravimetry (TGA), vibrating sample magnetometer (VSM), Fourier transform infrared spectroscopy (FTIR), scanning electron microscopy energy dispersive X-ray (SEM-EDX), and transmission electron microscopy (TEM). The best conditions for α-amylase immobilization on the manufactured nanocarrier were at pH = 6.5 and the temperature of 55 °C. The immobilized enzyme activity was maintained after 10 times of activity repetition. Furthermore, the consistency of the immobilized enzyme was 60% of its initial activity after six weeks of storage at 4 °C, which was better than that of the free enzyme. Improving the catalytic activities of enzymes by immobilization has made the synthesized nanobiocatalyst a confidence tool in bio-industrial procedures. A schematic comparison between the immobilization of enzymes by entrapment in polymers and adsorption on nanocomposites is shown in Figure 14.2. Different methods can be used for

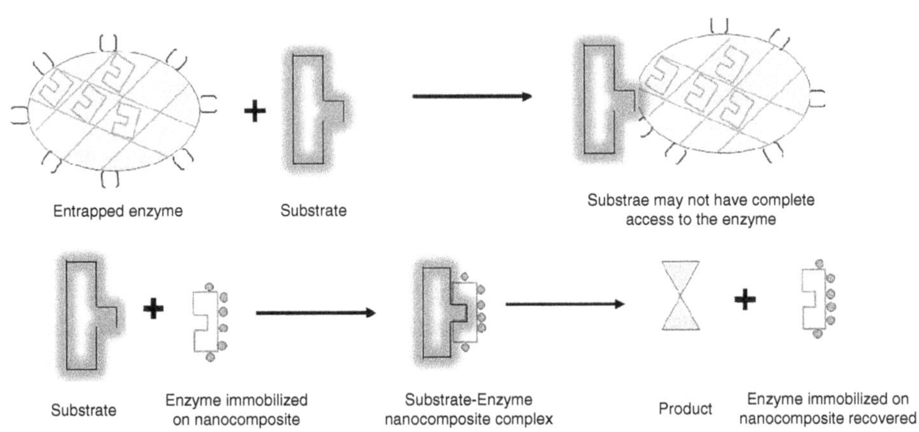

Figure 14.2 Immobilization of enzymes by adsorption on nanocomposites has the advantage of further access to the substrate compared to entrapment of enzymes in polymers.

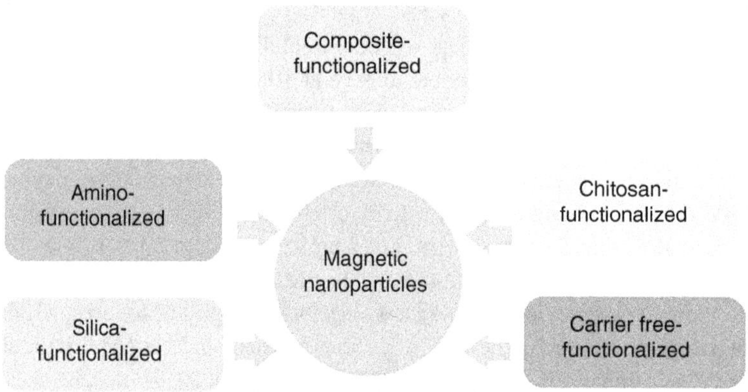

Figure 14.3 A view of of magnetic nanoparticles (MNPs) functionalization by different methods.

functionalization of MNP for enzyme immobilization (Figure 14.3). Further description of these methods as well as their advantages is discussed.

14.3.1.1 Silica-functionalized MNPs

Silica-functionalized MNPs provide valuable supports with low aggregation in liquids. These supports are biocompatible, and thermal and chemical stable (Bilal and Iqbal 2019). Co-immobilization of the enzymes, including cellulase, pectinase, and xylanase, has been studied by using MNP supports for industrial applications. As a result, the kinetic parameters (Vmax and Km) of the tri-enzyme were not changed, and the thermal and chemical stability of them was enhanced throughout the immobilization process (Perwez et al. 2017).

14.3.1.2 Carrier Free Enzyme Immobilization Strategy

Immobilization by using carrier-free methods such as co-immobilization via cross-linked enzyme crystals (CLECs) and cross-linking enzyme aggregates (CLEAs) are suitable methods to maintain enzyme stability (Cui and Jia 2015; Jafari Khorshidi et al. 2016; Bilal et al. 2017; Xu et al. 2018; Gaikwad et al. 2019). Muley et al. (2018) co-immobilized tri-enzymes on MNPs by cross-linking with glutaraldehyde to improve their reusability. As a result, co-immobilized MNPs maintained their stability for more than one month at 5 °C and also were active for up to four cycles. In another similar study, Kumari et al. (2018) immobilized bacterial derived enzymes (xylenes, cellulases, and amylases) on MNPs to enhance their stability and reusability. Tri-enzyme immobilized on MNPs is proposed for plant biomass production because of their excellent stability and recovery.

14.3.1.3 Immobilization on Amino-functionalized MNPs

Functionalization of MNPs is a common process for multi-enzyme immobilization via methods such as covalent bonding. Amines are one of the most commonly used groups that can be bound to proteins via cross-linking. Nadar and Rathod (2019) co-immobilized cellulase from *Trichoderma reesei* and pectinase from *Aspergillus aculeatus* on amino-functionalized MNPs (AMNPs) for antioxidant production from fruit peel wastes. This immobilization strategy resulted in the improved thermal stability, half-life, and Vmax for both enzymes; however, the enzymes activity was slightly decreased. Immobilization of the enzymes on MNPs resulted in an increased enzyme reusability. In another study, pectinase and cellulase were immobilized on AMNPs and used for extraction of lycopene from tomato peel. Although the enzymes' activity decreased by the immobilization process, their stability was increased after the process. The enzyme activity in AMNPs was increased by using ultrasonic irradiation. The time needed for biocatalyst extraction was also decreased in immobilized form compared to free enzyme forms (Ladole et al. 2018). In another similar study, application of ultrasound irradiation increased the activity of cellulase, which was immobilized on AMNPs, up to 3.6-fold, as well as an increase in the enzyme Vmax and a decrease in the enzyme Km. Also, the factors including MNP/enzyme ratio, concentration of cross-linking agent, and cross-linking time were effective on the enzyme activity (Ladole et al. 2017).

14.3.1.4 Immobilization on Composite-functionalized MNPs

Various nanomaterials can be used for coating MNPs that are used as supports for enzyme immobilization. MNP coating can prevent the oxidation of nanoparticles, improve the efficiency of enzyme immobilization, and decrease toxic effects of support materials. The process of coating should not change the essential magnetic characteristics of nanoparticles. Recently, metallic compounds such as metal oxides and carbon-related compounds were widely used in different copolymers to construct new nanocomposites that are used for enzyme immobilization. Gold, MgO, and GO are the most used compounds mixed in different polymers such as polyethylene glycol, glutamic acid, and poly(methyl methacrylate) to increase the enzyme persistence in a wide range of environmental conditions (Pei et al. 2018). MgO-coated MNPs are the examples that were used as supports for immobilization of cellulase from *Chlorella* sp. CYB2 by covalent binding. MgO showed a significant effect on increasing immobilization yield, activity enhancement, and the rate of substrate hydrolysis (Velmurugan and Incharoensakdi 2017).

14.3.1.5 Immobilization on Chitosan-functionalized MNPs

One of the most biocompatible coating agents used in enzyme immobilization processes is chitosan. Chitosan is a positively charged coating agent that can reduce the toxicity of immobilizing agent and facilitate immobilization processes for in vivo applications. Sanchez-Ramirez et al. (2017) immobilized cellulase enzyme on chitosan-coated MNPs by covalent binding. Although the enzyme activity was decreased, thermal stability, the Km value, and the number of enzyme reuses were increased following the immobilization process. The immobilization process resulted in efficient hydrolysis of lignocellulosic materials from *Agave atrovirens* leaves.

14.4 Transforming Agricultural Wastes by Immobilized Cells and Enzymes

Agriculture-related industries produced large amounts of waste every year. The entrance of these wastes to the environment without sufficient treating and reutilizing processes may cause environmental pollution that is harmful to human and animal health. Most of the agricultural/industrial wastes are disposed by burning, dumping, or unplanned landfilling without any treatment, which causes various problems such as climate changes by increasing the amounts of greenhouse gases (Bos and Hamelinck 2014). Juice and coffee industries as well as fiber processing industries are examples of industries that produce huge amounts of waste as peels, pulps, husks, and straw residues, respectively (Graminha et al. 2008; Belewu and Babalola 2009).

Most available methods for immobilization of microorganisms have been used to convert agricultural wastes, especially lignocellulose materials, into usable products (Ropars et al. 1992; Béguin and Aubert 1994). Cellulose biodegradation by using immobilized microorganisms in matrices such as polymerized hydrogels includes complex interactions between biotic agents including cell wall type, the morphogenetic characteristics of microorganisms, and the degrading acellulosic composition such as its co-polymerization with hemicellulose and lignin, and the abiotic agents, including physicochemical characteristics of hydrogel support surface and their porosity, the conditional pH, and ionic strength of nutritive solutions that are added to the substrata. By this method, whole cells are viable and form natural biofilms in which bacterial and fungal cells do their cellular cycles inside polymerized hydrogels. The porosity degree of hydrogel-forming polymer usually defines the immobilization efficacy by means of the degree of enzyme contact with cellulose particles (Ropars et al. 1992; Ge et al. 2018).

The extensive interest in further biotechnological usage of immobilized cells and the levels of current research in this area are going to lead to much more progressive commercial applications of immobilized cells for conventional fermentation systems (Wainwright 1992). Agricultural wastes, especially cellulose wastes are biologically composted. This is a concerned type of biodegradation process that can be performed using immobilized microorganisms in order to produce low molecular weight oligosaccharides that may be further transformed into usable compounds by other microorganisms (Anderson and Blain 1980; Cocker 1980). In a study conducted by Jewett et al. (1995), the immobilized cells of *Bacillus subtilis* and *Bacillus licheniformis* converted cellulose depolymerized compounds that were already produced by the fungus *Trichoderma viride* into compounds with lower molecular weight that are useful in agriculture as nutritional organic fertilizers.

Food and agriculture also contain lipid, carbohydrate, and protein compounds that cause high biological oxygen demand (BOD) in the waste. At the same time, these compounds have good potential to be transformed into valuable products that are among potential recoverable sources. Examples of such transforming reactions are carbohydrate hydrolysis, oxidation, phosphorylation, and acylation, as well as amino acid glycosylation and deamination, and lipid hydrogenation and esterification. For instance, esterification of waste lipids is extensively used for production of the different compounds that are valuable as additives in food and agricultural processes, and other usable products such as biofuels, which can be produced by trans-esterification of animal or industrial waste lipids with alcohol (Koller et al. 2018). Esterified sugars and esterified starch can be used as surfactants and biodegradable plastics or adhesives, respectively. Also, esterification of flavonoids may elongate their lifetime and induce their health and acceptance properties (Walle 2009). Traditional procedures for these conversions involved energy-consumed reactions that needed high amounts of chemical catalysts and formed by-products, especially when performed on complex substrates such as liquid food waste (Fang et al. 2002; Alissandratos and Halling 2012).

Pectin is a special polysaccharide that is a main structural compound in cell walls. Gelatinized and concentrated pectin is used in various industries (Sriamornsak 2003). Pectinases, which are pectin hydrolysis enzymes, play important roles in different biological procedures all over life (Naderi et al. 2012) and are used in different conventional industrial procedures like textile, plant fiber transformation, tea and coffee processing, oil extraction, industrial wastewater treatment, paper manufacturing industries, and purification of viruses (Jayani et al. 2005). Naghavi et al. (2012) isolated the fungus, *Monilia* sp. from putrefied tangerine and studied its exopectinase activity in submerged fermenting system. The best exopectinase production was seen at pH = 7, 50 °C, and 1:1 ratio of the enzyme to the substrate. In optimum condition, the activity of exopectinase, which was immobilized in agarose, was 20 units more than the fungus *Aspergillus niger*. Mukhopadhyay et al. (2015) used a nonomaterial-based immobilization technique that induced the activity of a psychrophilic pectate lyase (PL) enzyme. For this purpose, pectate lyase was substituted for Ca in calcium hydroxyapatite nanoparticles (NP-PL) and encapsulated in a self-assembly single-walled nanotube (SWNT). The PL was active and stable at a temperature range between 4 °C and 80 °C in the presence of NP and SWNT. The possibility of repeated release and re-trap of PL in SWNT while retaining significant activity was unique.

14.5 Biodegradation of Contaminants from Soil by Immobilized Cells and Enzymes

Fossil fuels cause increasing environmental pollutions that along with the expensive costs of traditional energy sources, encouraged researchers to establish novel methods toward environmentally friendly and biodegradable sources of energy as well as find new approaches for better degradation of fuel residues in contaminated areas. On the other hand, pesticides are other pollutants that are naturally degraded by microorganisms (Elgueta et al. 2016; Coseri 2017; Khoshnevisan et al. 2017; Kazemzadeh et al. 2019; Kazemzadeh et al. 2020a).

Immobilized biocatalysts are one of the attractive tools for biodegradation. Elgueta et al. (2016) formulated a novel support for immobilization of the white-rot fungus *Anthracophyllum discolor* in order to induce its ability for degradation of the herbicide atrazine. Different concentrations of starch, corn meal, sawdust, and flaxseed were used to construct three coated and uncoated pelletized supports. The results indicated that the half-life of atrazine was reached 6 days following the best immobilization process of *A. discolor* in contrast to 14 days for free enzyme. The treatment increased atrazine degradation and decreased the movement of the pollutant to water sources including surface and groundwater.

Organophosphates (OPs) potentially inhibit the activity of enzymes such as acetylcholinesterase (AChE). Their acute toxicity is associated with the exclusive cholinergic induction caused by the inhibitory effect of this enzyme in the central nervous system and on neuromuscular junctions (Jia et al. 2003; Wang et al. 2010). Wide usage of synthetic OPs in agriculture and as chemical warfare agents (Bajgar 2005; Wu et al. 2009; Saifuddin et al. 2013) has been caused numerous cases of poisonings worldwide (Wang et al. 2010), mainly because of the entrance of these toxic materials into ground and surface waters (Mulchandani et al. 2001). Many methods, including oxidation (Zhang and Pehkonen 1999; Wang and Lemley 2002), reverse osmosis (Chian et al. 1975), and activated carbon adsorption (Foo and Hameed 2010) have been used for OPs remediation. Biotechnological methods for OPs biodegradation such as hydrolysis by the enzyme organophosphate hydrolase (OPH), have drawn specific attention. Mechrez et al. (2014) proposed a novel "one-pot" method for rapid and sufficient fabrication of carbon nanotube (CNT) paper, which was activated enzymatically, for OPs bioremediation. Carboxylated CNTs were released in an aqueous surfactant solution by a microfiltration process to produce a paper-like membrane, which was assembled from entangled nanotubes. Cryo-transmission electron microscopy and infrared spectroscopy were used for confirmation of successful covalent binding of the enzyme onto the nanotubes surface.

Methyl paraoxon solution was used to study the potential of this method for OPs bioremediation. The immobilized OPH caused a significant decrease in methyl paraoxon content because of its in situ hydrolysis during the filtration procedure. The enzymatic activity was retained while owing to high OPH loading because of the unique composition of the mechanically robust CNT scaffold. In that study, a novel approach was presented for the designation of bioactive paper-like scaffolds for usage in different industrial approaches. Briceño et al. (2015) isolated 30 actinobacteria from organophosphorus pesticides (such as diazinon) involved in agricultural soil. Two isolates that were identified as *Streptomyces* spp. were selected because of their diazinon tolerance and removal properties. The removal efficiency of the isolates was 40–50% and 70–90% of the initial diazinon during 24 and 96 h incubation time, respectively. The comparison of diazinon removal by the free and immobilized actinobacteria on alginate beads showed a 60% higher diazinon removal by immobilized cells compared with free cells.

Lignocellulose is a natural compound that can be converted to valuable chemicals. Lignin is the major carrier of lignocellulose biomass in nature that has a key role in biomass digestibility. Biological degradation of lignin has been attracted as sustainable and environmentally friendly substitute for traditional degradation techniques (Amin et al. 2018). The immobilized enzymes were used for COD removal from pulp manufacturing waste. The endoglucanase activity was lost following immobilization in both polymers. But exoglucanase was completely active during 2 stages of usage for decontamination in the waste. Nam–Seok et al. (2008) purified an extracellular laccase from the white-rot fungus, *Cerrena unicolor*. Sephadex G-25 and ion-exchange chromatography on DEAE-Toyopearl column were used for enzyme purification, and the enzyme was immobilized on different supports.

The activity of purified laccase was increased two times after the desalting process and elevated 3.7 to 13.4 times after DEAE chromatography. During the immobilizing procedure, 93.8% of protein content and 100% of laccase activity were retained in association with the supports. The optimum pH for enzyme activity was changed to a wider range (5.0–6) following immobilization compared to its value for free enzyme (5). Both free and immobilized enzymes showed the most activity at 60°C; however, the range of immobilized enzyme activity was wider than the free enzyme. The immobilized laccase retained 80% of its activity at 9 °C and 95–100% of its initial activity after 10 days of storage at 4 °C. Amin et al. (2018) used porous magnetic nanocomposite of Fe3O4@SiO2@KIT-6 to immobilize laccase. The nanocomposite support enhanced delignification of olive pomace (77.3%) and degradation of its phenolic extracts (70%). Also, more than 70% of laccase activity was retained following immobilization.

14.6 Conclusion

Soil health is important for the protection of crop productivity in sustainable agriculture. In this context, the functions like nitrogen fixation, phosphorus solubility as well as biogeochemical cycle's homeostasis, soil structure protection, pollutants degradation, and suppression of plant pathogen control are crucial. Inefficiency of these activities causes the soil minerals and chemicals being insufficient. Various microbial cell and enzyme biocatalysts play important roles in different industrial areas. Immobilization of biocatalysts is an attractive tool for sustainable agriculture. Nanostructure substrates have significant advantages for immobilization of biocatalysts that are used in agriculture, although more researches are needed to improve and optimize nanomaterials for usage in sustainable development of agriculture.

References

Abousamra, M.M. and Mohsen, A.M. (2016). Solid lipid nanoparticles and nanostructured lipid carriers of tolnaftate: design, optimization and in-vitro evaluation. *Measurement 10*: 100.

Afzali, E., Eslaminejad, T., Amirheidari, B., and Ansari, M. (2021). Cell immobilization of Streptomyces griseobrunneus by microcrystalline cellulose for production of cyclodextrin glucanotransferase enzyme. *Cogent Engineering 8* (1): 1868145.

Al-Aksar, A.A. (2012). Microbiological studies on the *in vitro* inhibitory effect of *Streptomyces collinus albescens* against some phytopathogenic fungi. *African Journal of Microbiology Research* 6: 3277–3283.

Alissandratos, A. and Halling, P.J. (2012). Enzymatic acylation of starch. *Bioresource Technology 115*: 41–47.

Amin, R., Khorshidi, A., Shojaei, A.F. et al. (2018). Immobilization of laccase on modified Fe3O4@ SiO2@ Kit-6 magnetite nanoparticles for enhanced delignification of olive pomace bio-waste. *International Journal of Biological Macromolecules 114*: 106–113.

Amirbandeh, M. and Taheri-Kafrani, A. (2016). Immobilization of glucoamylase on triazine-functionalized Fe3O4/graphene oxide nanocomposite: improved stability and reusability. *International Journal of Biological Macromolecules 93*: 1183–1191.

Anderson, J.G. and Blain, J.A. (1980). Novel developments in microbial film reactors. In: *Fungal Biotechnology* (ed. J.E. Smith, D.R. Berry, and B. Kristiansen), 125–152. London: Academic Press.

Anteneh, Y.S. and Franco, C.M.M. (2019). Whole cell actinobacteria as biocatalysts. *Frontiers in Microbiology 10*: 77.

Asfour, M.H. and Mohsen, A.M. (2018). Formulation and evaluation of pH-sensitive rutin nanospheres against colon carcinoma using HCT-116 cell line. *Journal of Advanced Research 9*: 17–26.

Asri, M., Elabed, S., Ibnsouda, S., and El Ghachtouli, N. (2018). Biofilm-based systems for industrial wastewater treatment. In: *Handbook of Environmental Materials Management* (ed. C. Hussain), 1–21. Switzerland: Springer Nature.

Bajgar, J. (2005). Complex view on poisoning with nerve agents and organophosphates. *Acta Medica-Hradec Kralove 48* (1): 3.

Ballabio, C., Lugato, E., Fernández-Ugalde, O. et al. (2019). Mapping LUCAS topsoil chemical properties at European scale using Gaussian process regression. *Geoderma* 355: 113912.

Balusamy, B., Sarioglu, O.F., Senthamizhan, A., and Uyar, T. (2019). Rational design and development of electrospun nanofibrous biohybrid composites. *ACS Applied Bio Materials 2* (8): 3128–3143.

Barka, E.A., Vatsa, P., Sanchez, L. et al. (2016). Taxonomy, physiology, and natural products of the Actinobacteria. *Microbiology and Molecular Biology Reviews 80* (1): 1–43.

Béguin, P. and Aubert, J.P. (1994). The biological degradation of cellulose. *FEMS Microbiology Reviews 13* (1): 25–58.

Belewu, M.A. and Babalola, F.T. (2009). Nutrient enrichment of waste agricultural residues after solid state fermentation using Rhizopus oligosporus. *Journal of Applied Bioscience 13*: 695–699.

Bérdy, J. (2005). Bioactive microbial metabolites. *The Journal of Antibiotics (Tokyo) 58* (1): 1–26.

Bhat, M. (2000). Cellulases and related enzymes in biotechnology. *Biotechnology Advances 18* (5): 355–383.

Bilal, M. and Iqbal, H.M. (2019). Chemical, physical, and biological coordination: an interplay between materials and enzymes as potential platforms for immobilization. *Coordination Chemistry Reviews 388*: 1–23.

Bilal, M., Asgher, M., Parra-Saldivar, R. et al. (2017). Immobilized ligninolytic enzymes: an innovative and environmental responsive technology to tackle dye-based industrial pollutants—a review. *Science of the Total Environment 576*: 646–659.

Bos, A. and Hamelinck, C. (2014). Greenhouse gas impact of marginal fossil fuel use. *Project number: BIENL14773, 158.*

Briceño, G., Fuentes, M.S., Rubilar, O. et al. (2015). Removal of the insecticide diazinon from liquid media by free and immobilized *Streptomyces* sp. isolated from agricultural soil. *Journal of Basic Microbiology 55* (3): 293–302.

Brückner, B. and Blechschmidt, D. (1991). The gibberellin fermentation. *Critical Reviews in Biotechnology 11* (2): 163–192.

Buscot, F. and Varma, A. (2005). *Microorganisms in Soils: Roles in Genesis and Functions* (ed. A. Varma), 139–153. Germany: Springer.

Cassidy, M.B., Lee, H., and Trevors, J.T. (1996). Environmental applications of immobilized microbial cells: a review. *Journal of Industrial Microbiology 16* (2): 79–101.

Cavaliere, S. (ed.) (2015). *Electrospinning for Advanced Energy and Environmental Applications.* CRC Press.

Chahal, D.S. (1985). Solid state fermentation with *Trichoderma reesei* for cellulase production. *Applied and Environmental Microbiology* 49: 205–210.

Chang, R.H.Y., Jang, J., and Wu, K.C.W. (2011). Cellulase immobilized mesoporous silica nanocatalysts for efficient cellulose-to-glucose conversion. *Green Chemistry 13* (10): 2844–2850.

Chen, S., Hou, H., Harnisch, F. et al. (2011). Electrospun and solution blown three-dimensional carbon fiber nonwovens for application as electrodes in microbial fuel cells. *Energy & Environmental Science 4* (4): 1417–1421.

Chen, Y., Yu, B., Lin, J. et al. (2016). Simultaneous adsorption and biodegradation (SAB) of diesel oil using immobilized *Acinetobacter venetianus* on porous material. *Chemical Engineering Journal 289*: 463–470.

Chen, M., Zeng, G., Xu, P. et al. (2017). How do enzymes 'meet' nanoparticles and nanomaterials? *Trends in Biochemical Sciences 42* (11): 914–930.

Chian, E.S., Bruce, W.N., and Fang, H.H. (1975). Removal of pesticides by reverse osmosis. *Environmental Science & Technology 9* (1): 52–59.

Cocker, R. (1980). Interactions between fermenter and microorganisms: tower fermenter. In: Smith J. E. , Berry, D. R., Kristiansen B (eds.), *Fungal Biotechnology*, Academic Press, London, 111–124.

Coseri, S. (2017). Cellulose: to depolymerize ... or not to? *Biotechnology Advances 35* (2): 251–266.

Costa, H., Distefano, A.J., Marino-Buslje, C. et al. (2012). The residue 179 is involved in product specificity of the *Bacillus circulans* DF 9R cyclodextrin glycosyltransferase. *Applied Microbiology and Biotechnology 94* (1): 123–130.

Costacurta, A. and Vanderleyden, J. (1995). Synthesis of phytohormones by plant-associated bacteria. *Critical Reviews in Microbiology 21* (1): 1–18.

Cui, J.D. and Jia, S.R. (2015). Optimization protocols and improved strategies of cross-linked enzyme aggregates technology: current development and future challenges. *Critical Reviews in Biotechnology 35* (1): 15–28.

Das, S. (ed.) (2014). *Microbial Biodegradation and Bioremediation*. Elsevier.

Das, M. and Adholeya, A. (2012). Role of microorganisms in remediation of contaminated soil. In: Satyanarayana, T., Johri, B. (eds), *Microorganisms in Environmental Management*, 81–111. Dordrecht: Springer.

De Cesare, F., Di Mattia, E., Zussman, E., and Macagnano, A. (2019). A study on the dependence of bacteria adhesion on the polymer nanofibre diameter. *Environmental Science: Nano 6* (3): 778–797.

Defaei, M., Taheri-Kafrani, A., Miroliaei, M., and Yaghmaei, P. (2018). Improvement of stability and reusability of α-amylase immobilized on naringin functionalized magnetic nanoparticles: a robust nanobiocatalyst. *International Journal of Biological Macromolecules 113*: 354–360.

Demicheli, M.C. (1997). Realizing the potential of agricultural biotechnology in Europe. *Outlook on Agriculture 26* (2): 79–85.

Desai, N. (2012). Challenges in development of nanoparticle-based therapeutics. *The AAPS Journal 14* (2): 282–295.

Doumbou, C.L., Salove, M.K.H., Crawford, L.D., and Beaulieu, C. (2001). Actinomycetes, promising tools to control plant diseases and to promote plant growth. *Phytoprotection* 82: 85–102.

Elgueta, S., Santos, C., Lima, N., and Diez, M.C. (2016). Immobilization of the white-rot fungus Anthracophyllum discolor to degrade the herbicide atrazine. *AMB Express 6* (1): 1–11.

El-Ridy, M.S., Badawi, A.A., Safar, M.M., and Mohsen, A.M. (2012). Niosomes as a novel pharmaceutical formulation encapsulating the hepatoprotective drug silymarin. *International Journal of Pharmacy and Pharmaceutical Sciences 4* (1): 549–559.

El-Ridy, M.S., Yehia, S.A., Mohsen, A.M. et al. (2018). Formulation of Niosomal gel for enhanced transdermal lornoxicam delivery: in-vitro and in-vivo evaluation. *Current Drug Delivery 15* (1): 122–133.

Ercan, D. and Demirci, A. (2015). Current and future trends for biofilm reactors for fermentation processes. *Critical Reviews in Biotechnology 35* (1): 1–14.

Eş, I., Ribeiro, M.C., Dos Santos Junior, S.R. et al. (2016). Production of cyclodextrin glycosyltransferase by immobilized *Bacillus* sp. on chitosan matrix. *Bioprocess and Biosystems Engineering 39* (10): 1487–1500.

Fang, J.M., Fowler, P.A., Tomkinson, J., and Hill, C.A.S. (2002). The preparation and characterization of a series of chemically modified potato starches. *Carbohydrate Polymers 47* (3): 245–252.

Flemming, H.C., Wingender, J., Szewzyk, U. et al. (2016). Biofilms: an emergent form of bacterial life. *Nature Reviews Microbiology 14* (9): 563.

Foo, K.Y. and Hameed, B.H. (2010). Detoxification of pesticide waste via activated carbon adsorption process. *Journal of Hazardous Materials 175* (1–3): 1–11.

Gaikwad, S., Ingle, A.P., da Silva, S.S., and Rai, M. (2019). Immobilized nanoparticles-mediated enzymatic hydrolysis of cellulose for clean sugar production: a novel approach. *Current Nanoscience 15* (3): 296–303.

Galelli, M.E., Sarti, G.C., and Miyazaki, S.S. (2015). Lactuca sativa biofertilization using biofilm from Bacillus with PGPR activity. *Journal of Applied Horticulture 17* (3): 186.

Gaserod, O., Sannes, A., and Skjak-Braek, G. (1999). Microcapsules of alginate-chitosan. II. A study of capsule stability and permeability. *Biomaterials 20* (8): 773–783.

Ge, S., Chen, X., Li, D. et al. (2018). Hemicellulose structural changes during steam pretreatment and biogradation of Lentinus edodes. *Arabian Journal of Chemistry 11* (6): 771–781.

Gervais, P. and Molin, P. (2003). The role of water in solid-state fermentation. *Biochemical Engineering Journal 13* (2–3): 85–101.

Ghadiri, E., Naghavi, N.S., and Ghaedi, K. (2020). Detection of endoglucanase gene in cellulose degrading bacteria screened from forest soils of Mazandaran, Iran. *Water and Soil 34* (1): 73–83.

Glick, B. (1995). The enhancement of plant growth by free living bacteria. *Canadian Journal of Microbiology* 41: 109–117.

Graminha, E.B.N., Gonçalves, A.Z.L., Pirota, R.D.P.B. et al. (2008). Enzyme production by solid-state fermentation: application to animal nutrition. *Animal Feed Science and Technology 144* (1–2): 1–22.

Gutierrez-Correa, M. and Villena, G.K. (2003). Surface adhesion fermentation: a new fermentation category. *Revista Peruana de Biología 10* (2): 113–124.

Harir, M., Bellahcene, M., Baratto, M.C. et al. (2018). Isolation and characterization of a novel tyrosinase produced by Sahara soil actinobacteria and immobilization on nylon nanofiber membranes. *Journal of Biotechnology 265*: 54–64.

Hegedüs, I., Hancsók, J., and Nagy, E. (2012). Stabilization of the cellulase enzyme complex as enzyme nanoparticle. *Applied Biochemistry and Biotechnology 168* (6): 1372–1383.

Hetmann, A., Wujak, M., Bolibok, P. et al. (2018). Novel biocatalytic systems for maintaining the nucleotide balance based on adenylate kinase immobilized on carbon nanostructures. *Materials Science and Engineering: C 88*: 130–139.

Hölker, U., Höfer, M., and Lenz, J. (2004). Biotechnological advantages of laboratory-scale solid-state fermentation with fungi. *Applied Microbiology and Biotechnology 64* (2): 175–186.

Homaei, A.A., Sariri, R., Vianello, F., and Stevanato, R. (2013). Enzyme immobilization: an update. *Journal of Chemical Biology 6* (4): 185–205.

Hung, T.C., Fu, C.C., Su, C.H. et al. (2011). Immobilization of cellulase onto electrospun polyacrylonitrile (PAN) nanofibrous membranes and its application to the reducing sugar production from microalgae. *Enzyme and Microbial Technology 49* (1): 30–37.

Ibrahim, A.S.S., Al-Salamah, A.A., El-Toni, A.M. et al. (2014). Cyclodextrin glucanotransferase immobilization onto functionalized magnetic double mesoporous core–shell silica nanospheres. *Electronic Journal of Biotechnology 17* (2): 55–64.

Ivanova, V., Petrova, P., and Hristov, J. (2011). Application in the ethanol fermentation of immobilized yeast cells in matrix of alginate/magnetic nanoparticles, on chitosan-magnetite microparticles and cellulosecoated magnetic nanoparticles. *arXiv Preprint arXiv:1105.0619*.

Jacoby, G.A. and Archer, G.L. (1991). New mechanisms of bacterial resistance to antimicrobial agents. *New England Journal of Medicine 324* (9): 601–612.

Jafari Khorshidi, K., Lenjannezhadian, H., Jamalan, M., and Zeinali, M. (2016). Preparation and characterization of nanomagnetic cross-linked cellulase aggregates for cellulose bioconversion. *Journal of Chemical Technology & Biotechnology 91* (2): 539–546.

Jayani, R.S., Saxena, S., and Gupta, R. (2005). Microbial pectinolytic enzymes: a review. *Process Biochemistry 40* (9): 2931–2944.

Jewett, D.G., Hilbert, T.A., Logan, B.E. et al. (1995). Bacterial transport in laboratory columns and filters: influence of ionic strength and pH on collision efficiency. *Water Research 29* (7): 1673–1680.

Jia, H., Zhu, G., and Wang, P. (2003). Catalytic behaviors of enzymes attached to nanoparticles: the effect of particle mobility. *Biotechnology and Bioengineering 84* (4): 406–414.

Kaminsky, L.M., Trexler, R.V., Malik, R.J. et al. (2019). The inherent conflicts in developing soil microbial inoculants. *Trends in Biotechnology 37* (2): 140–151.

Kar, S., Swain, M.R., and Ray, R.C. (2009). Statistical optimization of alpha-amylase production with immobilized cells of Streptomyces erumpens MTCC 7317 in Luffa cylindrica L. sponge discs. *Applied Biochemistry and Biotechnology 152* (2): 177–188.

Kazemzadeh, S., Naghavy, N.S., Emami-Karvani, Z., and Fouladgar, M. (2019). Study of crude oil degradation by gas chromatography-mass spectrometry and methanoic acid and biosurfactant production by *Pseudomonas* sp. *Biological Journal of Microorganism 8* (31): 81–92.

Kazemzadeh, S., Naghavi, N.S., Emami-Karvani, Z. et al. (2020a). Production of glycolipid biosurfactant during crude oil degradation by the novel indigenous isolated *Achromobacter kerstersii* LMG3441. *Water Science and Technology 82* (10): 2134–2147.

Kazemzadeh, S., Naghavi, N.S., Emami-Karvani, Z. et al. (2020b). Gas chromatography-mass spectrometry analyses of crude oil bioremediation by the novel *Klebsiella variicola* SKV2 immobilized in polyurethane polymer scaffold and two-layer microcapsulation. *Bioremediation Journal 24* (2–3): 129–149.

Khoshnevisan, K., Vakhshiteh, F., Barkhi, M. et al. (2017). Immobilization of cellulase enzyme onto magnetic nanoparticles: applications and recent advances. *Molecular Catalysis 442*: 66–73.

Kim, C.J., Lee, S.J., Chang, Y.K. et al. (2006). Repeated-batch culture of immobilized Gibberella fujikuroi B9 for gibberellic acid production: an optimization study. *Biotechnology and Bioprocess Engineering 11* (6): 544–549.

Kirk, O., Borchert, T.V., and Fuglsang, C.C. (2002). Industrial enzyme applications. *Current Opinion in Biotechnology 13* (4): 345–351.

Kiss, S., Dragan-Bularda, M., and Radulescu, D. (1978). Soil polysaccharidases: activity and agricultural importance. In: *Soil Enzymes* (ed. R.G. Burns), 117–147. London: Academic Press.

Kokare, C.R., Chakraborty, S., Khopade, A.N., and Mahadik, K.R. (2009). Biofilm: importance and applications.

Koller, M., Shahzad, K., and Braunegg, G. (2018). Waste streams of the animal-processing industry as feedstocks to produce polyhydroxyalkanoate biopolyesters. *Applied Food Biotechnology 5* (4): 193–203.

Kour, D., Rana, K.L., Yadav, A.N. et al. (2020). Microbial biofertilizers: bioresources and eco-friendly technologies for agricultural and environmental sustainability. *Biocatalysis and Agricultural Biotechnology 23*: 101487.

Kredics, L., Antal, Z., and Manczinger, L. (2000). Influence of water potential on growth, enzyme secretion and in vitro enzyme activities of Trichoderma harzianum at different temperatures. *Current Microbiology 40* (5): 310–314.

Kumar, P.K.R. and Lonsane, B.K. (1989). Microbial production of gibberellins: state of the art. *Advances in Applied Microbiology 34*: 29–139.

Kumari, A., Kaila, P., Tiwari, P. et al. (2018). Multiple thermostable enzyme hydrolases on magnetic nanoparticles: an immobilized enzyme-mediated approach to saccharification through simultaneous xylanase, cellulase and amylolytic glucanotransferase action. *International Journal of Biological Macromolecules 120*: 1650–1658.

Ladam, G., Schaaf, P., Cuisinier, F.J. et al. (2001). Protein adsorption onto auto-assembled polyelectrolyte films. *Langmuir 17* (3): 878–882.

Ladole, M.R., Mevada, J.S., and Pandit, A.B. (2017). Ultrasonic hyperactivation of cellulase immobilized on magnetic nanoparticles. *Bioresource Technology 239*: 117–126.

Ladole, M.R., Nair, R.R., Bhutada, Y.D. et al. (2018). Synergistic effect of ultrasonication and co-immobilized enzymes on tomato peels for lycopene extraction. *Ultrasonics Sonochemistry 48*: 453–462.

Landarani-Isfahani, A., Taheri-Kafrani, A., Amini, M. et al. (2015). Xylanase immobilized on novel multifunctional hyperbranched polyglycerol-grafted magnetic nanoparticles: an efficient and robust biocatalyst. *Langmuir 31* (33): 9219–9227.

Large, K.P., Ison, A.P., and Williams, D.J. (1998). The effect of agitation rate on lipid utilization and clavulanic acid production in *Streptomyces clavuligerus*. *Journal of Biotechnology 63* (2): 111–119.

Lee, K.B., Park, S.J., Mirkin, C.A. et al. (2002). Protein nanoarrays generated by dip-pen nanolithography. *Science 295* (5560): 1702–1705.

Leonard, H., Halachmi, S., Ben-Dov, N. et al. (2017). Unraveling antimicrobial susceptibility of bacterial networks on micropillar architectures using intrinsic phase-shift spectroscopy. *ACS Nano 11* (6): 6167–6177.

Liu, G.Y. and Amro, N.A. (2002). Positioning protein molecules on surfaces: a nanoengineering approach to supramolecular chemistry. *Proceedings of the National Academy of Sciences 99* (8): 5165–5170.

Lupoi, J.S. and Smith, E.A. (2011). Evaluation of nanoparticle-immobilized cellulase for improved ethanol yield in simultaneous saccharification and fermentation reactions. *Biotechnology and Bioengineering 108* (12): 2835–2843.

MacBeath, G. and Schreiber, S.L. (2000). Printing proteins as microarrays for high-throughput function determination. *Science 289* (5485): 1760–1763.

Matthews, S. (2019). Immobilized microorganisms to improve the availability of phosphorus and potassium in soil. *Asian Journal of Microbiology, Biotechnology & Environmental Sciences 20* (1): 1–11.

Mechrez, G., Krepker, M.A., Harel, Y. et al. (2014). Biocatalytic carbon nanotube paper: a 'one-pot'route for fabrication of enzyme-immobilized membranes for organophosphate bioremediation. *Journal of Materials Chemistry B 2* (7): 915–922.

Mehmood, N., Olmos, E., Marchal, P. et al. (2010). Relation between pristinamycins production by Streptomyces pristinaespiralis, power dissipation and volumetric gas–liquid mass transfer coefficient, kLa. *Process Biochemistry 45* (11): 1779–1786.

Mehnati-Najafabadi, V., Taheri-Kafrani, A., and Bordbar, A.K. (2018). Xylanase immobilization on modified superparamagnetic graphene oxide nanocomposite: effect of PEGylation on activity and stability. *International Journal of Biological Macromolecules 107*: 418–425.

Meleigy, S.A. and Khalaf, M.A. (2009). Biosynthesis of gibberellic acid from milk permeate in repeated batch operation by a mutant *Fusarium moniliforme* cells immobilized on loofa sponge. *Bioresource Technology 100* (1): 374–379.

Mohana, S., Shah, A., Divecha, J., and Madamwar, D. (2008). Xylanase production by Burkholderia sp. DMAX strain under solid state fermentation using distillery spent wash. *Bioresource Technology* 99: 7553–7564.

Mukhopadhyay, A., Bhattacharyya, T., Dasgupta, A.K., and Chakrabarti, K. (2015). Nanotechnology based activation-immobilization of psychrophilic pectate lyase: a novel approach towards enzyme stabilization and enhanced activity. *Journal of Molecular Catalysis B: Enzymatic 119*: 54–63.

Mulchandani, A., Chen, W., Mulchandani, P. et al. (2001). Biosensors for direct determination of organophosphate pesticides. *Biosensors and Bioelectronics 16* (4–5): 225–230.

Muley, A.B., Thorat, A.S., Singhal, R.S., and Babu, K.H. (2018). A tri-enzyme co-immobilized magnetic complex: process details, kinetics, thermodynamics and applications. *International Journal of Biological Macromolecules 118*: 1781–1795.

Nadar, S.S. and Rathod, V.K. (2019). A co-immobilization of pectinase and cellulase onto magnetic nanoparticles for antioxidant extraction from waste fruit peels. *Biocatalysis and Agricultural Biotechnology 17*: 470–479.

Naderi, S., Naghavi, N.S., and Shahanipoor, K. (2012). Pectinolytic activity of aspergillus niger on pectic agricultural substrates. *The 1st International and the 4th National Congress on Recycling of Organic Waste in Agriculture*, Isfahan, Iran (26–27 April 2012), pp.1–6.

Naghavi, N.S., Naderi, S., Shahanipoor, K. et al. (2012). Optimization of exopectinase activity of the fungus Monilia isolated from tangerine in submerged fermentation. *Biological Journal of Microorganism 1* (1): 23–30.

Nam–Seok, C.H.O., Hee–Yeon, C.H.O., Soo–Jeong Shin, Y.J.C. et al. (2008). Production of fungal laccase and its immobilization and stability. *Journal of the Faculty of Agriculture, Kyushu University 53* (1): 13–18.

Niemeyer, C.M. and Blohm, D. (1999). DNA microarrays. *Angewandte Chemie International Edition 38* (19): 2865–2869.

Nocelli, N., Bogino, P.C., Banchio, E., and Giordano, W. (2016). Roles of extracellular polysaccharides and biofilm formation in heavy metal resistance of rhizobia. *Materials 9* (6): 418.

Ohta, N., Park, Y.S., Yahiro, K., and Okabe, M. (1995). Comparison of neomycin production from *Streptomyces fradiae* cultivation using soybean oil as the sole carbon source in an air-lift bioreactor and a stirred-tank reactor. *Journal of Fermentation and Bioengineering 79* (5): 443–448.

Parra, R., Aldred, D., Archer, D.B., and Magan, N. (2004). Water activity, solute and temperature modify growth and spore production of wild type and genetically engineered *Aspergillus niger* strains. *Enzyme and Microbial Technology 35* (2–3): 232–237.

Pei, J., Huang, Y., Yang, Y. et al. (2018). A novel layered anchoring structure immobilized cellulase via covalent binding of cellulase on MNPs anchored by LDHs. *Journal of Inorganic and Organometallic Polymers and Materials 28* (4): 1624–1635.

Perwez, M., Ahmad, R., and Sardar, M. (2017). A reusable multipurpose magnetic nanobiocatalyst for industrial applications. *International Journal of Biological Macromolecules 103*: 16–24.

Rafique, M., Hayat, K., Mukhtar, T. et al. (2015). Bacterial biofilm formation and its role against agricultural pathogens. In: *The Battle Against Microbial Pathogens: Basic Science, Technological advances and educational programs. Spain: Formatex Research Center*, 373–382.

Rangaswamy, V. (2012). Improved production of gibberellic acid by *Fusarium moniliforme*. *Journal of Microbiology Research 2* (3): 51–55.

Ranjkesh, F., Naghavi, N.S., and Mohammadi-Sichani, M. (2021). The in vitro survey of the effect of protease enzyme on reduction of biofilm producted by *Streptococcus mutans* isolated from dental plaque. *Journal of Isfahan Dental School 11*: 437–434.

Renault, J.P., Bernard, A., Bietsch, A. et al. (2003). Fabricating arrays of single protein molecules on glass using microcontact printing. *The Journal of Physical Chemistry B 107* (3): 703–711.

Rendueles, O. and Ghigo, J.M. (2015). Mechanisms of competition in biofilm communities. *Microbiology Spectrum 3*: MB-0009-2014.

Rios-Iribe, E.Y., Flores-Cotera, L.B., Chávira, M.M.G. et al. (2011). Inductive effect produced by a mixture of carbon source in the production of gibberellic acid by *Gibberella fujikuroi*. *World Journal of Microbiology and Biotechnology 27* (6): 1499–1505.

Ropars, M., Marchal, R., Pourquie, J., and Vandecasteele, J.P. (1992). Large-scale enzymatic hydrolysis of agricultural lignocellulosic biomass. Part 1: pretreatment procedures. *Bioresource Technology 42* (3): 197–204.

Roubos, J.A., Krabben, P., Luiten, R.G.M. et al. (2001). A quantitative approach to characterizing cell lysis caused by mechanical agitation of *Streptomyces clavuligerus*. *Biotechnology Progress 17* (2): 336–347.

Rudrappa, T., Biedrzycki, M.L., and Bais, H.P. (2008). Causes and consequences of plant-associated biofilms. *FEMS Microbiology Ecology 64* (2): 153–166.

Sadeghi, A., Karimi, E., Dahaji, P.A. et al. (2012). Plant growth promoting ability of an auxin and siderophore producing isolate of *Streptomyces* under saline salt conditions. *World Journal of Microbiology and Biotechnology* 28: 1503–1509.

Saifuddin, N., Raziah, A.Z., and Junizah, A.R. (2013). Carbon nanotubes: a review on structure and their interaction with proteins. *Journal of Chemistry 2013*.

Sanchez-Ramirez, J., Martinez-Hernandez, J.L., Segura-Ceniceros, P. et al. (2017). Cellulases immobilization on chitosan-coated magnetic nanoparticles: application for Agave Atrovirens lignocellulosic biomass hydrolysis. *Bioprocess and Biosystems Engineering 40* (1): 9–22.

Sarioglu, O.F., Celebioglu, A., Tekinay, T., and Uyar, T. (2017). Evaluation of fiber diameter and morphology differences for electrospun fibers on bacterial immobilization and bioremediation performance. *International Biodeterioration & Biodegradation 120*: 66–70.

Saruno, R., Kato, F., and Ikeno, T. (1979). Kojic acid, a tyrosinase inhibitor from *Aspergillus albus. Agricultural and Biological Chemistry 43* (6): 1337–1338.

Sathya, A., Vijayabharathi, R., and Gopalakrishnan, S. (2016). Soil microbes: the invisible managers of soil fertility. In: *Microbial Inoculants in Sustainable Agricultural Productivity*, 1–16. New Delhi: Springer.

Saucedo, J.E.N., Barbotin, J.N., and Thomas, D. (1989). Continuous production of gibberellic acid in a fixed-bed reactor by immobilized mycelia of *Gibberella fujikuroi* in calcium alginate beads. *Applied Microbiology and Biotechnology 30* (3): 226–233.

Saudagar, P.S., Shaligram, N.S., and Singhal, R.S. (2008). Immobilization of Streptomyces clavuligerus on loofah sponge for the production of clavulanic acid. *Bioresource Technology 99* (7): 2250–2253.

Shah, P., Bhalodia, D., and Shelat, P. (2010). Nanoemulsion: a pharmaceutical review. *Systematic Reviews in Pharmacy 1*: 1.

Sheldon, R.A. (2007). Cross-linked enzyme aggregates (CLEA® s): stable and recyclable biocatalysts. *Biochemical Society Transactions 35* (6): 1583–1587.

Sheng, X., Ting, Y.P., and Pehkonen, S.O. (2008). The influence of ionic strength, nutrients and pH on bacterial adhesion to metals. *Journal of Colloid and Interface Science 321* (2): 256–264.

Shrivastava, S., D'Souza, S.F., and Desai, P.D. (2008). Production of indole-3-acetic acid by immobilized actinomycete (Kitasatospora sp.) for soil applications. In: *Current Science*, 1595–1604.

Singh, V.K., Das, S., Dwivedy, A.K. et al. (2019). Assessment of chemically characterized nanoencapuslated Ocimum sanctum essential oil against aflatoxigenic fungi contaminating herbal raw materials and its novel mode of action as methyglyoxal inhibitor. *Postharvest Biology and Technology 153*: 87–95.

Sinsabaugh, R.L., Antibus, R.K., and Linkins, A.E. (1991). An enzymic approach to the analysis of microbial activity during plant litter decomposition. *Agriculture, Ecosystems and Environment* 34: 43–54.

Sriamornsak, P. (2003). Chemistry of pectin and its pharmaceutical uses: a review. *Silpakorn University International Journal 3* (1–2): 206–228.

Subramaniyam, R. and Vimala, R. (2012). Solid state and submerged fermentation for the production of bioactive substances: a comparative study. *International Journal of Natural Sciences 3* (3): 480–486.

Suneetha, V. and Zaved, A.K. (2011). Actinomycetes: sources for soil enzymes. In: *Soil Enzymology, Soil Biology* (ed. G. Shukla and A. Varma), 329–337. Berlin, Heidelberg: Springer-Verlag.

Tilak, K.V.B.R., Ranganayaki, N., Pal, K.K. et al. (2005). Diversity of plant growth and soil health supporting bacteria. *Current Science* 89: 136–150.

Topoglidis, E., Cass, A.E., O'Regan, B., and Durrant, J.R. (2001). Immobilization and bioelectrochemistry of proteins on nanoporous TiO2 and ZnO films. *Journal of Electroanalytical Chemistry 517* (1–2): 20–27.

Truong, V.K., Rundell, S., Lapovok, R. et al. (2009). Effect of ultrafine-grained titanium surfaces on adhesion of bacteria. *Applied Microbiology and Biotechnology 83* (5): 925–937.

Tuson, H.H. and Weibel, D.B. (2013). Bacteria–surface interactions. *Soft Matter 9* (17): 4368–4380.

Van Duijkeren, E., Schink, A.K., Roberts, M.C. et al. (2018). Mechanisms of bacterial resistance to antimicrobial agents. In: *Antimicrobial Resistance in Bacteria from Livestock and Companion Animals*, 51–82.

Vassileva, A., Burhan, N., Beschkov, V. et al. (2003). Cyclodextrin glucanotransferase production by free and agar gel immobilized cells of Bacillus circulans ATCC 21783. *Process Biochemistry 38* (11): 1585–1591.

Velmurugan, R. and Incharoensakdi, A. (2017). MgO-Fe3O4 linked cellulase enzyme complex improves the hydrolysis of cellulose from Chlorella sp. CYB2. *Biochemical Engineering Journal 122*: 22–30.

Villena, G.K. and Gutiérrez-Correa, M. (2007). Production of lignocellulolytic enzymes by Aspergillus niger biofilms at variable water activities. *Electronic Journal of Biotechnology 10* (1): 124–140.

Wadu-Mesthrige, K., Amro, N.A., Garno, J.C. et al. (2001). Fabrication of nanometer-sized protein patterns using atomic force microscopy and selective immobilization. *Biophysical Journal 80* (4): 1891–1899.

Wainwright, M. (1992). *An Introduction to Fungal Biotechnology*. Wiley.

Walle, T. (2009). Methylation of dietary flavones increases their metabolic stability and chemopreventive effects. *International Journal of Molecular Sciences 10* (11): 5002–5019.

Wang, Q. and Lemley, A.T. (2002). Oxidation of diazinon by anodic Fenton treatment. *Water Research 36* (13): 3237–3244.

Wang, L., Wei, L., Chen, Y., and Jiang, R. (2010). Specific and reversible immobilization of NADH oxidase on functionalized carbon nanotubes. *Journal of Biotechnology 150* (1): 57–63.

Wilson, D.S. and Nock, S. (2002). Functional protein biochips. *Current Opinion in Chemical Biology 6*: 81–85.

Wilson, D.S. and Nock, S. (2003). Recent developments in protein microarray technology. *Angewandte Chemie International Edition 42* (5): 494–500.

Woo, S.L. and Pepe, O. (2018). Microbial consortia: promising probiotics as plant biostimulants for sustainable agriculture. *Frontiers in Plant Science 9*: 1801.

Wu, J., Lan, C., and Chan, G.Y.S. (2009). Organophosphorus pesticide ozonation and formation of oxon intermediates. *Chemosphere 76* (9): 1308–1314.

Xu, M.Q., Wang, S.S., Li, L.N. et al. (2018). Combined cross-linked enzyme aggregates as biocatalysts. *Catalysts 8* (10): 460.

Zacchetti, B., Andrianos, A., van Dissel, D. et al. (2018). Microencapsulation extends mycelial viability of *Streptomyces lividans* 66 and increases enzyme production. *BMC Biotechnology 18* (1): 1–8.

Zhang, Q. and Pehkonen, S.O. (1999). Oxidation of diazinon by aqueous chlorine: kinetics, mechanisms, and product studies. *Journal of Agricultural and Food Chemistry 47* (4): 1760–1766.

15

Nanopesticides and Their Potential Applications for Crop Protection

Kounaina Khan[1,2], *Farhan Zameer*[1,2], *Pratheek Jain*[1], *Subrahmanya Padyana*[1], *and Ajar Nath Yadav*[3]

[1] *PathoGutOmics Laboratory, Alva's Traditional Medicinal Archive (ATMA), Department of Ayurveda Pharmacology, Alva's Ayurveda Medical College, Vidyagiri, Dakshina Kannada, Karnataka, India*
[2] *TheOmic School Pvt. Ltd., Mysore, Karnataka, India*
[3] *Department of Genetics, Plant Breeding and Biotechnology, Dr. Khem Singh Gill Akal College of Agriculture, Eternal University, Sirmour, Himachal Pradesh, India*

15.1 Introduction

In today's world, where the strain on natural resources is reaching intense levels, sustainable agriculture is no longer optional; it is essential. With a growing global population, the demand for food is surging, yet the environmental costs of conventional farming are becoming unsustainable (Shelef et al. 2018). Intensive agriculture has driven deforestation, water shortages, soil degradation, and contributed heavily to climate change. As we face increasing climate instability, resource scarcity, and biodiversity loss, sustainable agricultural practices offer a path forward. They integrate ecological principles to maintain soil health, conserve water, and protect ecosystems, while still meeting human needs for food (Abbass et al. 2022). By aligning farming practices with the planet's natural limits, sustainable agriculture is key to ensuring long-term food security and resilience against the environmental challenges of the Anthropocene. Agriculture stands at the intersection of human survival and planetary stability, serving as both the primary driver of global environmental change and the most affected by its consequences (Datta et al. 2023). In the Anthropocene, where human activities have become the dominant force shaping Earth's systems, the necessity of a fundamental shift in agricultural practices is undeniable. Sustainable agriculture must no longer be an afterthought to productivity; it must become the core strategy. This approach requires a deep integration of ecological processes and natural capital management to increase yields while maintaining biophysical stability. The paradigm of sustainable intensification emphasizes a dual goal: producing more food to meet the needs of a growing global population with minimizing environmental impacts and building adaptability across landscapes and ecosystems (Yang et al. 2024).

Crop protection and plant-pest interactions are fundamental factors to maintain global food security, as pests and pathogens (P&Ps) can significantly reduce crop yield and quality (Singh et al. 2016). P&Ps encompass a diverse array of organisms, including viruses, bacteria, fungi, oomycetes, nematodes, arthropods, and parasitic plants, all of which interact with crops in damaging ways (Yadav 2021; Patil et al. 2021a; More et al. 2021). These interactions range from feeding on plant tissues to infecting their cellular structures, leading to various plant diseases that severely affect crop productivity (Patil et al. 2021b). Modern agricultural practices, such as monoculture and extensive use of chemical pesticides, have disrupted the natural co-evolutionary balance between crops and their pests, often favoring the rapid adaptation of P&Ps. This disruption increases the likelihood of widespread outbreaks, as pests exploit the genetic uniformity in modern crops, leading to challenges in crop protection. These widespread outbreaks have led to the emergence of more aggressive pest strains through selective sweeps, such as *Puccinia striiformis* (yellow rust on wheat) and *Fusarium graminearum* (head blight on wheat and barley), demonstrating the ongoing arms race between plant defense mechanisms and pest adaptations (Bouvet et al. 2022).

The impact of P&Ps on food security is profound, with an estimated global crop loss of 21.5% for wheat, 30% for rice, and 22.6% for maize (Ristaino et al. 2021). These losses contribute to reduced food availability, higher costs, and increased vulnerability, especially in regions already facing food insecurity, like Sub-Saharan Africa and Southeast Asia (Wudil et al. 2022). Despite these staggering figures, accurately quantifying the impact of P&Ps remains a complex task. The sheer diversity of pests, their dynamic nature, and the variability in agroecosystems all contribute to the challenge of assessing crop losses. Additionally, data gaps, particularly in developing regions where infrastructure for pest monitoring is often inadequate, further complicate efforts to measure the real-world impact. Pest populations evolve continuously, often in response to global trade, climate change, and shifts in farming practices, which can lead to new pest strains emerging in previously unaffected regions. These factors, combined with the need for more reliable and region-specific data, emphasize the importance of developing comprehensive, sustainable crop protection strategies that consider both the immediate and long-term ecological dynamics between crops and pests (Finger et al. 2024).

Pesticides play an important role in modern agriculture by protecting crops from pests, diseases, and weeds, thereby enhancing food production and quality. These chemical agents, which include insecticides, herbicides, fungicides, and bactericides, help farmers manage and mitigate threats that could otherwise lead to substantial crop losses. However, their use raises concerns about environmental impact, human health risks from exposure, and the development of pesticide-resistant pests (Zameer and Gopal. 2010). To address these challenges, many farmers are adopting integrated pest management (IPM) practices, which combine chemical applications with sustainable agricultural techniques, promoting a balanced approach to pest control while minimizing negative consequences (Bajda and Grigoraki 2020).

Nanotechnology is revolutionizing agriculture by providing innovative solutions that enhance productivity, sustainability, and environmental safety. By manipulating materials at the nanoscale, researchers can develop advanced tools and techniques to improve farming practices. One of the key applications of nanotechnology in agriculture is the emergence of nanopesticides, which represent a transformative approach to pest management. Nanopesticides are designed to enhance the efficacy and safety of traditional pesticides by employing nanoscale materials that allow for improved targeting and delivery. Unlike conventional pesticides, which often affect nontarget organisms and beneficial insects, nanopesticides can be engineered to specifically target harmful pests, thereby minimizing ecological disruption.

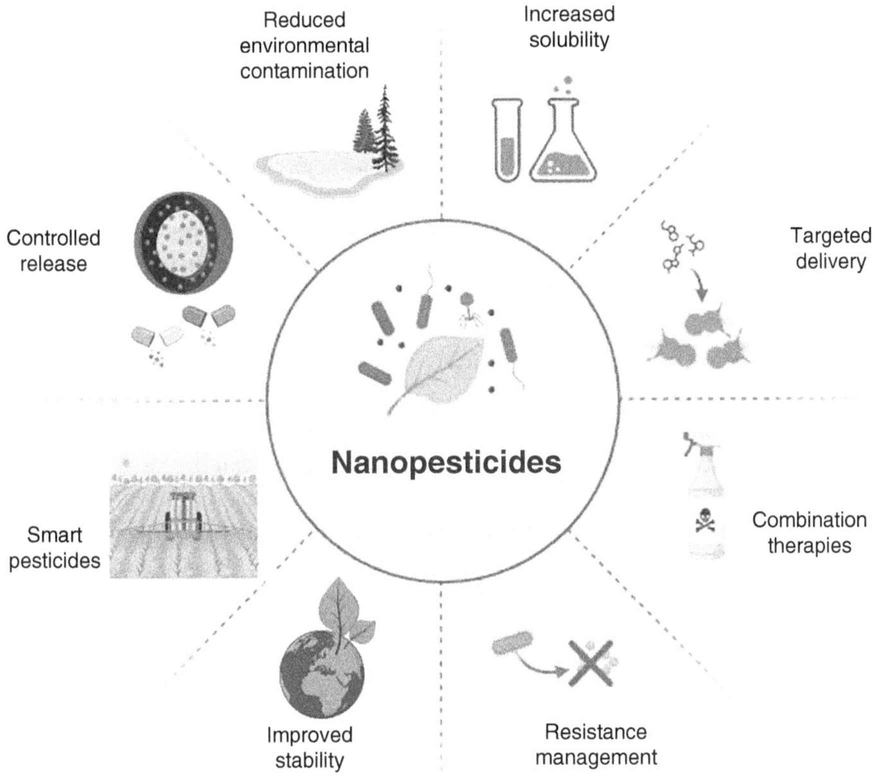

Figure 15.1 Applications of nanopesticides in crop protection. *Source:* BioRender.com.

Their increased surface area and reactivity facilitate better penetration into plant tissues, resulting in effective pest control with lower application rates (Manjunatha et al. 2013; Yin et al. 2023). Moreover, nanopesticides can be formulated for controlled or slow release, extending their effectiveness and reducing the need for frequent applications. This not only lowers costs for farmers but also lessens the environmental impact associated with chemical residues (Figure 15.1). While the potential of nanopesticides is substantial, challenges such as regulatory frameworks, public perception, and ongoing research are essential to address. By overcoming these hurdles, nanotechnology can significantly contribute to sustainable agricultural practices, ensuring food security while safeguarding environmental health (Xu et al. 2022).

15.2 Synthesis of Nanopesticides

Nanopesticides represent a cutting-edge solution in agricultural technology, designed to enhance pesticide efficacy, reduce environmental impact, and provide targeted pest control. The synthesis of nanomaterials (NMs) for pesticide delivery is important because the physical and chemical characteristics of these nanoparticles directly influence their behavior in biological systems, particularly their interaction with plant cells, pests, and the environment (Sahu et al. 2024). The mode of synthesis determines the factors like size, morphology, and surface properties, which affect the release, bioavailability, and stability of nanopesticides. Nanopesticides can be produced using both bottom-up and top-down synthesis approaches through the integration of nanoparticles and nanocarriers, nanoemulsions, and nanoencapsulation techniques (Satapathy et al. 2020; Abid et al. 2022).

The bottom-up approach to nanopesticide synthesis involves assembling smaller units such as atoms and molecules into larger NMs. This method is favored due to its ability to produce nanoparticles with minimal defects and uniform chemical composition, which are essential for applications in pesticide delivery. Chemical methods, in particular, offer versatile and efficient pathways for nanoparticle synthesis, allowing precise control over factors like size, shape, and surface chemistry by adjusting various synthesis parameters such as precursors, reducing agents, and surfactants. One common technique is co-precipitation, where controlled chemical reactions between pesticides and metal precursors lead to the formation of nanoparticles (Dinesh et al. 2024).

Another method is solvothermal decomposition, where specific solvents and temperature conditions are used to produce NMs with desired characteristics. Additionally, electrochemical synthesis could be used to generate nanopesticides at the electrode-electrolyte interface by manipulating electrochemical conditions, allowing fine-tuning of nanoparticle properties such as shape and size (Kumar et al. 2017). This method is highly adaptable for creating nanopesticides with controlled release and targeted activity. Furthermore, advanced techniques like chemical vapor deposition (CVD) and atomic or molecular condensation enable precise control over nanoparticle morphology for optimizing the delivery and efficacy of pesticides. CVD involves the deposition of nanoparticles from vapor-phase precursors onto a substrate, while atomic or molecular condensation relies on cooling vapor-phase particles to form solid nanoparticles. These methods provide flexibility in designing nanopesticides with tailored attributes that enhance their interaction with pests and plants, improve stability, and reduce the environmental impact of pesticide use (Zhao et al. 2019).

In response to the environmental challenges posed by chemical synthesis, green synthesis, or biological methods, provides an eco-friendly alternative for nanopesticide production. This approach leverages natural plant extracts and microorganisms, such as bacteria, fungi, and algae, to reduce metal precursors and produce nanoparticles without the toxicity associated with chemical reducing agents. Plant extracts, rich in natural reducing agents like polyphenols and flavonoids, facilitate nanoparticle formation, acting as capping agents to prevent agglomeration and ensure stability (Castillo-Henríquez et al. 2020). Similarly, microorganisms like bacteria and fungi internalize metal precursors and detoxify, which enables nanoparticle synthesis through intracellular or extracellular mechanisms. These biological methods are cost-effective and result in biocompatible nanocarriers that can efficiently deliver pesticides to their intended targets, for a sustainable alternative in nanotechnology-driven pest control (Prasad et al. 2019).

Further, the top-down approach for nanopesticide synthesis involves reducing larger materials into nanoscale particles through mechanical or physical processes, making it ideal for creating solid nanoparticle formulations. One of the most common techniques in this category is mechanical milling, where bulk pesticide materials are ground into nanoparticles using equipment with tungsten carbide, iron, or steel balls. This process effectively reduces the particle size to the nanometer scale, enhancing the solubility, bioavailability, and controlled release of the pesticide, making it a simple and widely used method for nanopesticide production. Other advanced techniques such as lithography and laser ablation are employed for precise cutting and shaping of materials into nanoparticles (Krishnaswamy and Orsat 2017). These methods are particularly useful for designing nanocarriers with specific shapes and sizes, which can be optimized for more effective pesticide delivery and targeted pest control. Additionally, sputtering and electrosputtering are physical methods used to coat surfaces with nanoparticles, ideal for encapsulating pesticides within nanocarriers like lipid-based nanoparticles or polymer matrices (An et al. 2022). These coatings improve the controlled release of active ingredients, offering more efficient and sustained pest control and reducing environmental contamination (Figure 15.2).

The application techniques of nanopesticides are important for maximizing their effectiveness in pest management. One of the most common methods is foliar spraying, where a solution containing nanoparticles is sprayed directly onto the leaves and stems of plants, allowing for better coverage and penetration into plant tissues, enhancing absorption against foliar pests. Soil application is another technique involving the incorporation of nanopesticides into the soil to target root-feeding pests or improve soil

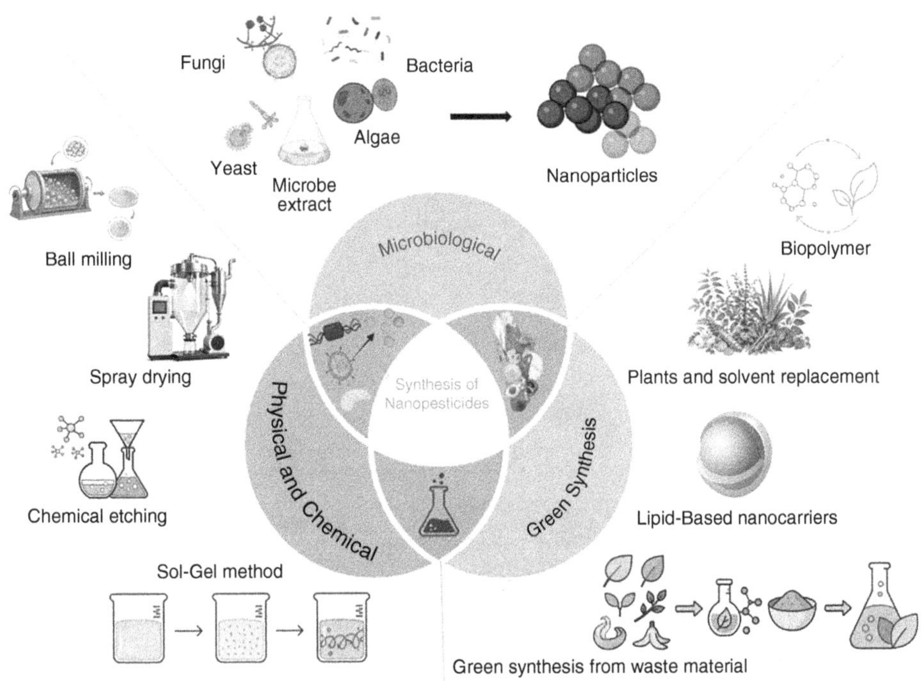

Figure 15.2 Methods on synthesis of nanopesticides. *Sources:* BioRender.com.

health. This allows for controlled release of active ingredients, making them available to the plant's root system over time and minimizing runoff. Trunk injection is particularly useful for woody plants, where nanopesticides are injected directly into the trunk, ensuring rapid delivery to the plant circulatory system, thus minimizing pesticide exposure to nontarget organisms.

Drenching involves applying a nanopesticide solution directly to the soil or plant base, allowing absorption through roots, which is effective against soil-borne pests and diseases. For large-scale operations, aerial application techniques like drone spraying are becoming popular, as they provide uniform coverage over extensive fields, which reduces labor costs and improves efficiency. Additionally, smart delivery systems utilizing nano-encapsulation can release active ingredients in response to specific environmental triggers, enhancing efficacy and minimizing overall usage. The combination of nanopesticides with beneficial microbes can also enhance effectiveness, resulting in synergistic pest management outcomes. Lastly, some nanopesticides can be formulated for fumigation, allowing uniform distribution of active ingredients in enclosed environments (Figure 15.3). These diverse application techniques offer flexibility in utilizing nanopesticides for effective pest management (Paradva and Kalla 2023; Shangguan et al. 2024a,b).

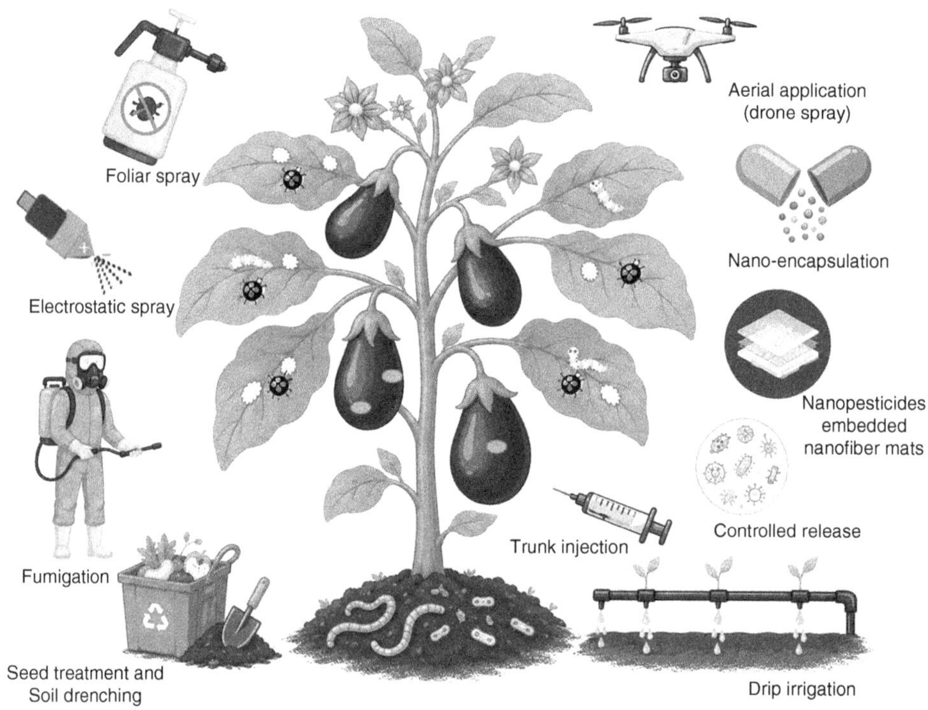

Figure 15.3 Various techniques of nanopesticides and their usage forms. *Sources:* BioRender.com.

15.3 Types of Nanopesticides

Nanopesticides can be categorized into two main types based on the materials used in their formulation. Type 1, where NMs serve as active ingredients (AIs), and Type 2, where NMs function as nanocarriers for conventional pesticide AI (Table 15.1).

15.3.1 Type 1 Nanopesticides (Nanomaterials as Active Ingredients)

In this category, the NMs themselves act as the active agents against pests. Metal-based nanoparticles dominate this type, and it includes materials such as silver (Ag), titanium (Ti), copper (Cu), zinc (Zn), iron (Fe), and aluminum (Al). These metal-based NMs are employed due to their inherent antimicrobial, antifungal, and insecticidal properties (Sahu et al. 2023; Akki et al. 2024). The Ag nanoparticles are widely used as nanobactericides, nanofungicides, and nanoinsecticides, while Ti-based NMs are applied as nanobactericides and nanofungicides. Cu-based NMs function primarily as fungicides and bactericides, offering high efficacy against various crop pathogens. Other metals like Zn, Fe, and Al are also used, though they are less common (Deka et al. 2021; Wang et al. 2022). These NMs are directly applied to target pests, pathogens (Zameer et al. 2010a; Beulah et al. 2015; Avinash et al. 2023), or weeds, leveraging their nanoscale

Table 15.1 Classification of nanopesticides and their implications.

Type of nanopesticide	Nanoparticles used	Synthesis method	Particle size/shape	Mode of application	Effects/impacts
Type 1	Ag, Cu, Zn, Ti, Fe nanoparticles	Chemical synthesis, green synthesis	1–100 nm, spherical	Foliar spray, soil application	Effective against pests and pathogens, reduced environmental impact, enhanced crop protection with lower dosages
Type 2	Polymers (e.g., chitosan, cellulose), clay nanoparticles (silica, montmorillonite), carbon nanotubes, liposomes	Solvent evaporation, coacervation, electrospinning	Varies (10–500 nm), diverse shapes	Encapsulation, controlled release systems	Targeted delivery of pesticides, reduced pesticide waste, improved bioavailability, enhanced effectiveness of AIs

Table 15.2 List of non-encapsulated nanopesticides.

Name of ingredients	Nature of synthesized nanoparticles	Target organisms
Nanosilica	Synthetic	*An. stephensi, Cx. quinquefasciatus, Aedes aegypti*
Silver, Aluminium Oxide, Zinc Oxide, and Titanium Dioxide Nanoparticles	Synthetic	Insect pests and pathogens
Nanoalumina	Synthetic	*Sitophilus oryzae (L.), Rhyzopertha dominica (F.)*
Silver and Gold Nanoparticles	Synthetic	*Aphis nerii*
Jatropha gossypifolia, Euphorbia tirucalli, Pedilanthus tithymaloides, Alstonia macrophylla	Plant	*Aedes aegypti*
Mukia maderaspatana	Plant	*Culex quinquefasciatus, Aedes aegypti*
Aloe vera	Plant	*Anopheles stephensi*
Plumeria rubra	Plant	*Aedes aegypti, Anopheles stephensi*
Nelumbo nucifera	Plant	*Anopheles stephensi, Culex quinquefasciatus*
Emblica officinalis	Plant	—
Phytophthora infestans	Microbial	—
Trichoderma reesei	Microbial	—
Aspergillus	Microbial	—
Aspergillus niger	Microbial	*An. stephensi, Cx. quinquefasciatus, Aedes aegypti*
Chrysosporium tropicum	Microbial	*Anopheles stephensi, Culex quinquefasciatus*
Epicoccum nigrum	Microbial	Pathogenic fungi
Rhizopus spps.	Microbial	—

properties to increase surface interaction (Sulthana et al. 2018), improve solubility, and enhance pest control efficiency (Table 15.2).

15.3.2 Type 2 Nanopesticides (Nanocarriers for Active Ingredients)

In type 2 NMs are used as carriers to encapsulate and deliver conventional pesticides, aiming for controlled, targeted, and synchronized release of AIs such as atrazine, avermectin, and glyphosate. These nanocarriers enhance the efficiency of pesticide delivery by ensuring the precise release of AIs at the right time, location, and dosage, which ultimately reduces environmental impact. A diverse range of nanocarrier materials is employed in this category. Polymers, in which chitosan, cellulose, and polyethylene can be fabricated into various structures such as nanocapsules, nanospheres, nano(hydro)gels, and nanomicelles, all of which provide exceptional control over pesticide stability and release dynamics. Clay NMs like silica, montmorillonite, and kaolinite serve as excellent carriers that facilitate prolonged release of the encapsulated pesticides and enhance their effectiveness. Additionally, nanocomposites and carbon-based materials like carbon nanotubes (CNTs) and graphene exhibit unique properties that improve the delivery and interaction of the pesticides with target organisms and increase their bioavailability (Bhan et al. 2018).

Other effective carriers include nanoliposomes, which encapsulate pesticides to protect them from environmental degradation while controlling their release, and dendrimers and nanozeolites, which are known for their high surface area and tunable properties, which makes them suitable for various applications in pest control. Solid lipid nanoparticles (SLNs) and layered double hydroxides (LDHs) are particularly noted for their biodegradability and ability to provide slow-release mechanisms, further contributing to the sustainability of pesticide applications. Moreover, polymersomes, composed of block copolymers, offer versatile hollow structures that can efficiently encapsulate a variety of pesticide molecules, allowing for customized delivery systems (Martirosyan and Schneider 2014; Maruyama et al. 2016; Prasad et al. 2017). The innovative design and utilization of these Type 2 nanopesticides can enhance targeting capabilities along with minimal use of pesticide by optimizing release strategies in crop protection (Table 15.3).

Table 15.3 List of encapsulated nanopesticides.

Name of ingredients	Nature of synthesized nanoparticles	Purpose/target organisms
Avermectin	Synthetic	*Martianus dermestoides*
Azadirachta indica (Oil)	Plant	*Bemisia abaci, Bemisia tabaci, Culex quinquefasciatus*
Atrazine	Polyhydroxybutyrate-co-hydroxyvalerate	Decreased genotoxicity, increased biodegradability
Bacillus sphaericus	Microbial	*Culex quinquefasciatus*
Aspergillus flavus	Microbial	*Anopheles stephensi*
Aspergillus flavus, Cuscuta reflexa	Microbial, Plant	*Anopheles stephensi, Culex quinquefasciatus*
Artemisia arborescens (Oil)	Plant	*Bemisia tabaci*
Azadirachtin	Alginate	Slower release
Chitosan, saponin, $CuSO_4$	Chitosan/tripolyphosphate, Chitosan-saponin, Chitosan-Cu	Antifungal activity
Copaifera sp. Oleoresin	Plant	*Aedes aegypti*
Cuscuta reflexa	Plant	*Culex quinquefasciatus*

(*Continued*)

Table 15.3 (Continued)

Name of ingredients	Nature of synthesized nanoparticles	Purpose/target organisms
Diethylphenylacetamide	Synthetic	*Culex tritaeniorhynchus*
Ethofumesate	Wheat gluten	Reduce diffusivity
Garlic Oil	Plant	*Tribolium castaneum*
Glyphosate	Surfactants/oil/water	Increase in bio-efficacy, reducing environmental toxicity
Imidacloprid	Alginate	Cytotoxicity, sucking pest (leafhoppers)
Imidacloprid	Polyacetic acid-polyethylene glycol-polyacetic acid	Decrease the lethal concentration
Imidacloprid	Synthetic	*Martianus dermestoides*
Imazapic and Imazapyr	Chitosan	Cytotoxicity assays
Lippia sidiodes oil	Plant	*Aedes aegypti*
Methomyl	Carboxymethylchitosan	Control release for longer time period
Moringa oleifera seeds	Plant	*Stegomya aegypti*
Nanopermethrin	Synthetic	*Aedes aegypti, Culex quinquefasciatus*
Nanohexaconazole, Nanosulfur	Synthetic	—
Paraquat	Alginate/chitosan	Increased period of action, reducing ecological toxicity
Paraquat	Chitosan/tripolyphosphate	Lower cyto- and genotoxicity
Piracetam, pentoxifylline, pyridoxine	Silica	Perfused brain tissue
Temephos	Synthetic	*Anopheles stephensi*
Temephos, *Cuscuta reflexa*	Synthetic, Plant	*Anopheles stephensi, Culex quinquefasciatus*
Temephos, Imidacloprid	Synthetic	*Culex quinquefasciatus*
Tropicamide	Xyloglucan/poloxamer	Higher corneal permeation, less toxic and non-irritant
2,4-Dichlorophenoxyacetate	Organic–inorganic nanohybrid	Control release

15.4 Mechanism of Nanopesticides

Nanopesticides represent an innovative approach to pest management by employing NMs to improve the efficacy of conventional pesticides or by acting as active agents themselves. The mode of action of nanopesticides is multifaceted, combining physical and chemical

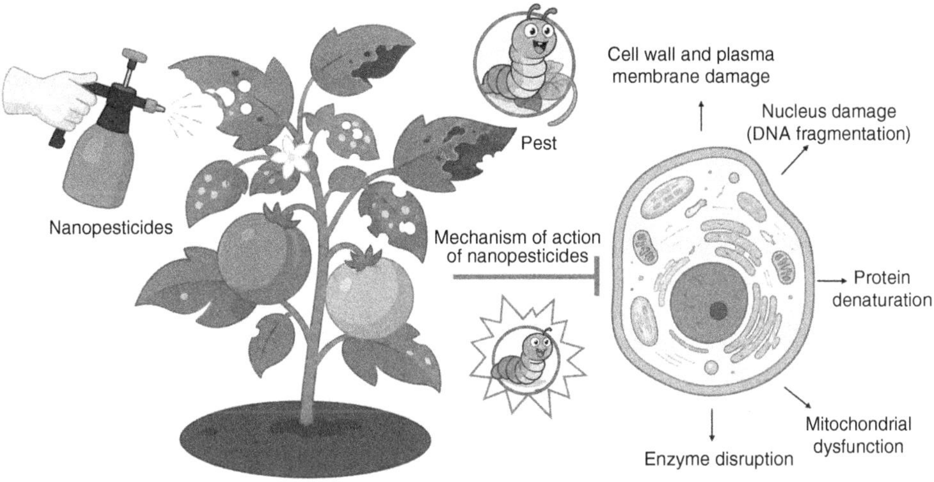

Figure 15.4 Nanopesticides and their mechanism of action. *Sources:* BioRender.com.

mechanisms that disrupt vital biological processes in pests and pathogens (Zameer et al. 2010a, 2016). The delivery of nanopesticides can be finely controlled, and their interactions at the cellular level in pests offer a more efficient, targeted, and environmentally friendly pest control strategy. Nanopesticides contribute to crop protection by providing sustained, targeted action against pests, fungi, bacteria, and other pathogens (Pontes et al. 2023). Their small size and surface properties enhance the penetration of the AIs into the pest's body and plant tissues. Nanoparticles can be engineered to attach to specific plant surfaces for localized delivery, which can reduce the need for frequent application, controlled release and stimuli-responsive release, such as pH, temperature, or light exposure (Figure 15.4). This minimizes pesticide loss and ensures prolonged action, continuous crop protection, and better crop yields (Hu et al. 2023).

Nanopesticides offer a novel approach to insect pest management by employing the unique properties of NMs to improve pesticide effectiveness and reduce environmental impact. These formulations include nano-sized active ingredients or nanocarriers that encapsulate conventional pesticides, allowing for precise, targeted delivery and controlled release. The mechanisms of action, such as diffusion, biodegradation, and osmotic pressure, enable a slow, sustained release of the pesticide, ensuring optimal dosage over time. This reduces the need for frequent applications and minimizes the risks of pesticide overuse. Their precision in targeting pests while protecting beneficial organisms makes them an eco-friendly and efficient alternative, with significant potential to revolutionize sustainable agricultural practices (Kannan et al. 2023).

15.4.1 Cell Wall and Plasma Membrane Damage

Nanopesticides, particularly metal-based ones like silver (Ag), copper (Cu), and zinc oxide (ZnO), interact with the cell wall and plasma membrane of pest or pathogen cells, compromising these protective barriers. The cell wall in bacteria, composed of peptidoglycans, or in fungi, made of chitin, serves as a first line of defense, but nanoparticles bind to these structural components, and weaken their integrity by making the wall more permeable. This disruption allows nanoparticles to penetrate further, to reach the lipid bilayer of the plasma membrane, where they embed and create pores. These perforations weaken the selective permeability of membrane, leading to uncontrolled leakage of vital cellular contents and allowing the entry of harmful substances into the cell. Moreover, the interaction of nanoparticles with the plasma membrane disturbs ion gradients essential for maintaining membrane potential, which is important for nutrient transport and energy production. The loss of this membrane potential severely impairs cellular function, ultimately leading to cell death due to an inability to regulate the internal environment. Additionally, metal-based nanopesticides generate reactive oxygen species (ROS), which induce oxidative stress, further damaging both the cell wall and plasma membrane and accelerating cellular breakdown (Kumar et al. 2019; Mathew et al. 2020).

15.4.2 Nucleus Damage

Nanoparticles can penetrate deep into the nucleus, where they disrupt cellular processes such as DNA replication and repair due to oxidative stress. This oxidative stress leads to direct damage to the DNA, causing single- or double-strand breaks that can result in mutations or prevent the proper functioning of the cell, which will trigger apoptosis, and lead to cell death to prevent the propagation of damaged or mutated cells. In addition to ROS-induced damage, nanoparticles, like Ag, can directly interfere with nuclear processes. By attaching to sulfur groups in DNA or interacting with replication enzymes, it causes denaturation of DNA strands or inhibits the activity of enzymes necessary for DNA replication and transcription. This interference prevents the normal copying of DNA during cell division and the transcription of DNA into RNA, ultimately leading to disrupted cellular function and cell death in pest and pathogen control (Bahrulolum et al. 2021).

15.4.3 Protein Denaturation and Enzyme Disruption

Nanoparticles exhibit a strong affinity for proteins, including enzymes, due to their large surface area and the ability to interact with protein structures through electrostatic interactions and van der Waals forces. When nanopesticides come into contact with proteins in pest cells, they often lead to disruption of the native three-dimensional structure of protein, causing it to unfold or misfold. This structural disruption renders the protein inactive and prevents from performing its essential biological functions. Further nanoparticles can bind to the enzyme's active site, blocking substrate access or altering the enzyme shape, which inhibits its catalytic activity (Rahman et al. 2019). For example, ZnO nanoparticles are known to displace essential metal ions that act as enzyme cofactors, or they can bind directly to the enzyme's active site, further impeding enzyme functionality. In addition, the oxidative stress induced by nanoparticles, particularly those made of metals Ag, Cu, generates ROS, which oxidize amino acids in the enzyme structure. This oxidation modifies the enzyme active site and destabilizes its structure, leading to permanent loss of function. The combination of structural protein denaturation and enzyme inactivation, through both direct nanoparticle interaction and oxidative stress, results in severe metabolic disruption in pest cells, effectively regulating their ability to survive and propagate (Fatima et al. 2021).

15.4.4 Soil Health and Environmental Interactions

Nanopesticides play a dual role in influencing soil health, with effects varying based on their composition and concentration. Biodegradable nanoparticles, derived from natural polymers like chitosan, can enhance soil quality by facilitating nutrient delivery and improving microbial activity. These biocompatible nanoparticles promote beneficial soil organisms, thereby fostering a healthy ecosystem that supports plant growth. The prolonged use of metal-based nanoparticles raises concerns regarding soil contamination and disruption of microbial communities (Verma and Saroop 2024). These heavy metals may accumulate in the soil over time, leading to toxic effects on soil biota, diminishing microbial diversity, and affecting the overall soil health and fertility. This can disrupt decomposition, nutrient availability, and crop productivity. Furthermore, the degradation of natural nanocarriers results in benign byproducts that can enrich the soil. Hence, the environmental interactions of nanopesticides emphasize the necessity for careful management to harness their benefits by mitigating adverse effects on soil health (Ahmed et al. 2023).

15.5 Applications of Nanopesticides in Crop Protection

The deployment of nanopesticides for pest control offers a promising advancement in agriculture, particularly in field crops, horticultural crops, and cash crops. These crops face unique challenges, including different pest species, growth environments, and economic implications. Nanopesticides, with their precision-targeting abilities and controlled release mechanisms, are emerging as an alternative to traditional chemical pesticides, offering enhanced efficacy (Figure 15.5).

15.5.1 Field Crops

Field crops such as rice, wheat, maize, and soybeans are susceptible to a wide variety of pests, including insects, fungi, nematodes, and weeds. Traditional pest management in field crops involves extensive use of chemical pesticides, which not only contributes to environmental degradation but also results in the development of pest resistance. Nanopesticides can offer enhanced pest control through mechanisms such as ROS generation, binding to peptidoglycans in bacteria and chitin in fungi, weakening their structural integrity and leading to cell death (Zameer et al. 2010b; Acharya et al. 2023a). In rice and maize fields, nanopesticides can be utilized to combat fungal pathogens like Fusarium spp. and Candida spp. through the sustained release of metal-based nanoparticles. Additionally, nanocarriers such as chitosan, which are biodegradable, are being explored for use in field crops to facilitate controlled release of AIs, reducing the need for repeated applications and lowering pesticide concentrations in the environment. For example, chitosan nanoparticles can encapsulate conventional herbicides, delivering them in a more efficient and

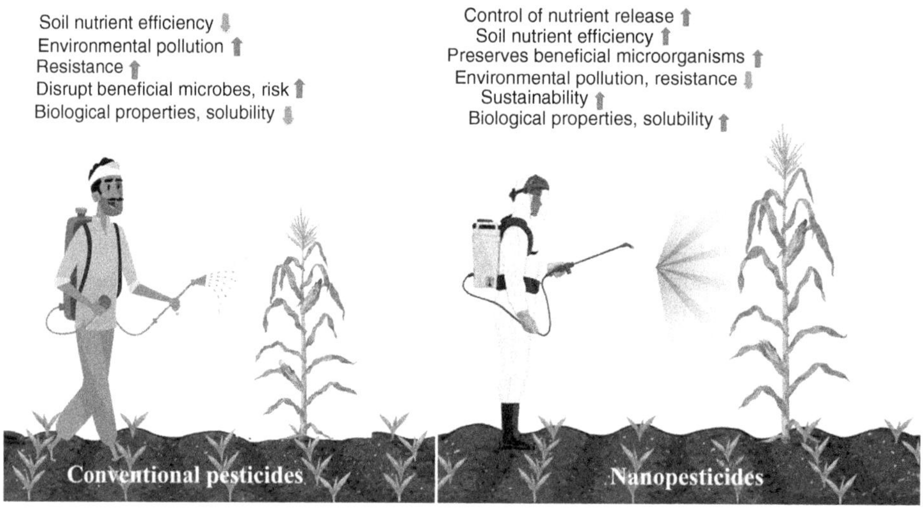

Figure 15.5 Differential modes of pesticide applications and their outcomes. *Sources:* BioRender.com.

environmentally friendly manner to target weed species in wheat fields, reducing herbicide runoff and soil leaching (Sarmah et al. 2023).

15.5.2 Horticultural Crops

Horticultural crops, which include fruits, vegetables, and ornamentals, require a higher degree of precision in pest control due to their high market value and the need to limit pesticide residues on consumable parts. Pests such as aphids, whiteflies, bacterial blights, and fungal pathogens like powdery mildew present many challenges. Nanopesticides formulated with Type 2 nanocarriers, such as silica or zinc nanoparticles, can be employed to deliver fungicides or insecticides with greater specificity and reduced off-target effects. Like, copper hydroxide ($Cu\ (OH)_2$)-based nanopesticides have shown efficacy against fungal pathogens like Alternaria and Botrytis in tomato and cucumber crops (Acharya et al. 2023b). In horticulture, the use of stimuli-responsive nanopesticides, which release their AI in response to environmental factors such as temperature, pH, or light, holds significant promise. For example, UV-responsive nanopesticides could be designed to release insecticides in outdoor crops like lettuce only during daylight hours, when pest activity is highest. Precise targeting will reduce the overall pesticide load, with higher crop protection efficiency without compromising soil health. Additionally, natural polymer-based nanocarriers have shown beneficial effects in promoting plant growth and enhancing resistance to pathogens, further supporting sustainable pest control in horticultural systems (Manzoor et al. 2024).

15.5.3 Cash Crops

Cash crops, such as cotton, tobacco, coffee, and sugarcane, are highly susceptible to pest infestations that can drastically affect their yield and economic value. For example, cotton bollworm, sugarcane borers, and coffee rust (a fungal pathogen) are some of the most common pests that threaten cash crops. Nanopesticides exhibit antimicrobial activity and are effective in controlling such pests through mitochondrial dysfunction and protein denaturation. This mode of action is particularly useful in cash crops where resistance to conventional pesticides is common (Acharya et al. 2023a).

In cotton cultivation, nanoparticle-encapsulated insecticides offer an efficient alternative to conventional sprays, as they provide prolonged release and enhanced stability, targeting pests like bollworms with minimal off-target effects. Similarly, in sugarcane fields, nanopesticides formulated with nanoclays or graphene oxide can be used to deliver herbicides that target specific weed species, ensuring that nontarget plants and beneficial organisms in the soil are less affected. Moreover, the adoption of nanotechnology in coffee plantations could help address fungal pathogens like *Hemileia vastatrix* by utilizing CuO nanoparticles to enhance pest control while reducing the risk of soil contamination (Kah et al. 2019).

The commercialization of nano-enabled products in agriculture is paving the way for enhanced productivity and sustainable farming practices. One notable product is AZteroid FC, a nanopolymer-based fungicide formulation that incorporates azoxystrobin to stabilize the active ingredient in high-salt environments, such as those created by fertilizers. This formulation effectively controls fungal pathogens in field crops like sugar beet and maize. Similarly, Bifender FC, developed by Vive Crop Protection, leverages nanotechnology for IPM, allowing for the simultaneous application of fungicides and fertilizers without the risk of clogging agricultural equipment, thus improving operational efficiency (Shangguan et al. 2024).

The Nualgi Foliar Spray represents another significant advancement; this nanoscale nutrient spray is designed to complement conventional fertilizers, effectively boosting crop yields in high-value horticultural and cash crops like tomatoes, strawberries, and tea. Lastly, Dedalo Elite, a nanotechnology-based solution from Red Surcos, aims to offer effective pest control; however, comprehensive field performance data for this product are still limited. Collectively, these examples exemplify the potential of nanotechnology in agriculture to provide targeted, efficient, and environmentally friendly solutions to various challenges faced by modern farming (Kah et al. 2019; Gilbertson et al. 2020).

15.5.4 Nanopesticides in Common Pest- and Vector-borne Diseases

Nanopesticides, leveraging the advancements of nanotechnology, have emerged as a promising solution for managing insect pests and vector-borne diseases in major crops such as rice, maize, wheat, soybean, and cotton. In rice (*Oryza sativa*), common pests like the brown planthopper (*Nilaparvata lugens*) and yellow stem borer (*Scirpophaga incertulas*) are effectively managed using silica nanoparticles, which enhance physical barriers and can act as carriers for insecticides, as well as zinc oxide nanoparticles that boost plant immunity. Maize (*Zea mays*) faces threats from the fall armyworm (*Spodoptera frugiperda*) and European corn borer (*Ostrinia nubilalis*), with silver nanoparticles proving effective against the fall armyworm by disrupting its biological processes, while chitosan nanoparticles allow for targeted pesticide delivery, minimizing environmental toxicity (Huang et al. 2020).

Wheat (*Triticum aestivum*) contends with pests like the wheat aphid (*Sitobion avenae*) and armyworm (*Pseudaletia unipuncta*), where titanium dioxide nanoparticles enhance the efficacy of conventional pesticides against the aphids, and nanostructured fungicide formulations improve control of wheat yellow rust caused by *Puccinia striiformis*. Soybean (*Glycine max*) crops, threatened by pests such as the soybean looper (*Chrysodeixis includens*) and Japanese beetle (*Popillia japonica*), benefit from copper nanoparticles that manage the looper and natural plant-based nanoparticles from neem for eco-friendly pest control (Nagoshi et al. 2023). In cotton (*Gossypium spp.*), the cotton bollworm (*Helicoverpa armigera*) and whiteflies (*Bemisia tabaci*) are tackled using gold nanoparticles to enhance the effectiveness of insecticides against the bollworm, while alginate-based nanoparticles provide controlled release for targeting whiteflies effectively (Mahmood et al. 2022). By integrating these innovative nanopesticides, farmers can enhance pest management strategies, reduce environmental impacts, and improve crop yields, paving the way for sustainable agricultural practices in a world facing increasing pest resistance and environmental challenges.

15.5.5 Weed Management Using Nano-herbicides

Nanoherbicides represent an innovative strategy for weed management, utilizing nanotechnology to enhance the efficacy, selectivity, and sustainability of herbicides. By encapsulating herbicidal agents within nanocarriers, such as silica, chitosan, or polymer-based nanoparticles, the controlled release of active compounds can be achieved, allowing for targeted application directly to weed species by minimizing off-target effects. For example, formulations utilizing nanostructured glyphosate or 2,4-dichlorophenoxyacetic acid (2,4-D) have demonstrated superior absorption and translocation in problematic weed species, including glyphosate-resistant *Amaranthus palmeri* (Palmer amaranth), *Echinochloa crus-galli* (barnyard grass), *Cynodon dactylon* (Bermuda grass), *Sorghum halepense* (Johnson grass), *Convolvulus arvensis* (field bindweed), and *Solanum nigrum* (black nightshade) (Suer and Tursun 2024).

The use of nanoparticles not only enhances the bioavailability of these herbicides but also prolongs their residual activity in the soil, leading to improved weed control efficacy. Moreover, the incorporation of biodegradable nanocarriers can mitigate environmental concerns associated with traditional herbicides, as these materials can degrade into non-toxic byproducts, thereby reducing chemical residues in the ecosystem. Studies have indicated that nanoherbicides can effectively suppress the growth of persistent weeds in common crops, contributing to integrated weed management strategies (Mishra and Gautam 2021).

Overall nanopesticides offer significant environmental and health benefits in agriculture, primarily by addressing the limitations associated with traditional pesticide use. One notable advantage is their ability to reduce environmental impacts related to crop protection. By allowing for more precise and efficient delivery of active ingredients, nanopesticides enable targeted applications that minimize off-target effects and reduce the overall quantity of chemicals released into the environment. This precision mitigates pesticide drift and runoff into water bodies, decreasing the risk of harming nontarget species, including beneficial insects, birds, and aquatic organisms. Additionally, nanopesticides help minimize human and animal toxicity. Traditional pesticides leave harmful residues on crops, posing risks to farmworkers and consumers. However, the controlled release mechanisms and targeted delivery systems in many nanopesticide formulations will reduce exposure to toxic substances (Mubeen et al. 2023). By ensuring that active ingredients are delivered specifically to intended pests, nanopesticides can lower chemical residue levels in food products, enhancing food safety and protecting public health. Furthermore, nanopesticides have the potential to reduce pest resistance, a common issue with conventional pesticide applications that often lead to rapid resistance development among pest populations, necessitating higher doses or more toxic alternatives.

In contrast, nanopesticides can incorporate multiple active ingredients or employ novel modes of action, reducing the likelihood that pests will develop resistance and prolonging the effectiveness of pest control strategies. Moreover, nanopesticides support soil health improvement, as excessive use of traditional pesticides can disrupt soil microbial communities, leading to decreased soil fertility and compromised plant health. Nanopesticides are often formulated to be more biodegradable and less harmful to beneficial soil organisms, promoting a balanced soil ecosystem that enhances nutrient cycling and improves soil structure. Ultimately, this contributes to healthier crops and better agricultural yields (Müller and Peter 1998; Yang et al. 2017).

15.6 Regulatory Challenges and Future Direction

The regulatory challenges surrounding nanopesticides are multifaceted, primarily due to the complexities introduced by nanotechnology in agriculture. As engineered nanoparticles enter the agrochemical market, they necessitate rigorous safety evaluations to assess their risks to human health and the environment. However, knowledge remains limited regarding the diverse range of NMs and the specific conditions relevant for agricultural applications. Current regulatory frameworks in developed countries mandate that new agrochemicals undergo stringent human and ecological risk assessments before market approval. Nevertheless, existing guidance often does not adequately capture the unique behavior of nanoparticles, suggesting that adaptations are required at various stages of the risk assessment process (Zainab et al. 2024).

A major regulatory challenge is the absence of standardized definitions and criteria for evaluating nanopesticides. While nanoformulations are assessed on a case-by-case basis, with comprehensive set of safety criteria that products must meet for approval by regulatory agencies. The registration process could be enhanced by establishing clear guidelines to direct the development and validation of new nanoformulations. Recently proposed guidelines, such as those from the European Food Safety Authority (EFSA) for risk assessment in the food and feed chain, offer frameworks for testing soil leaching and toxicity of NMs (Schoonjans et al. 2023). Current standard toxicity tests may not be sufficient to detect unintended consequences associated with nanotechnology in agriculture. Advanced testing methods that examine chronic or subtle effects are necessary to ensure long-term safety. This includes exploring potential trophic transfer of nanoparticles, conducting omics studies, and investigating the impact of nanoparticles on microbiomes, which can provide insights into soil, plant, or gut health (Kookana et al. 2014).

Public perception plays an important role in the acceptance of nanopesticides. The association of nanotechnology with previous controversies surrounding genetically modified organisms (GMOs) can fuel skepticism and resistance in certain regions. Therefore, transparent communication highlighting the clear societal benefits of nanopesticides is vital for building public trust. Engaging with agricultural producers during the development process is essential to tailor solutions to their needs, and effective public education can alleviate misconceptions regarding the safety of nanotechnology. The future of nanotechnology holds great promise for enhancing agricultural productivity and sustainability (Liu et al. 2022). However, to harness its full potential, a comprehensive life-cycle analysis is necessary, encompassing not only the application of agrochemicals but also their production, transportation, and disposal. This analysis will aid in comparing the safety of new nano strategies with existing practices, providing a critical assessment of the associated risks and benefits. Moreover, collaboration and interdisciplinary research, and the convergence of expertise across various scientific disciplines such as agronomy, chemistry, toxicology, social sciences, inter-, intra-, and trans-disciplinary sciences will be essential for developing viable nano-enabled solutions. Which will facilitate the identification of promising strategies to improve the efficiency of pesticides in crop protection (Tang et al. 2024).

15.7 Limitations

The implementation of nanopesticides in agricultural practices is subject to several limitations. Like, the environmental and ecological risks associated with nanopesticides stem from the insufficient characterization of their behavior in various environmental matrices, such as soil and aquatic systems. The bioaccumulation and toxicity in nontarget organisms, including beneficial insects and microbial communities, necessitates comprehensive risk assessments that are currently lacking. Further, regulatory frameworks for the approval and monitoring of nanopesticides need to be well-established to address challenges in ensuring safety and efficacy while fostering innovation in nanotechnology. Furthermore, the economic feasibility of nanopesticide production is constrained by the high costs associated with the synthesis and formulation of nanomaterials, which may limit their accessibility for farmers, particularly in low-resource settings. Public perception of nanotechnology also plays an important role, as prevailing concerns about health risks and ethical implications can impede acceptance and adoption. Finally, the long-term efficacy of nanopesticides is a concern; similar to conventional pesticides, there exists a risk of toxicity and pest resistance over time. Thus, long-term ecological studies are essential to evaluate the sustained effectiveness of nanopesticides and to develop strategies.

15.8 Conclusions

Nanopesticides represent a huge advancement in crop protection strategies, enhancing the effectiveness of pest management through improved targeting and controlled release of active ingredients. By employing the unique properties of nanomaterials, these formulations can reduce the environmental impact of traditional agrochemicals while maintaining efficacy against pests, pathogens, and weeds. The ability to minimize off-target effects not only benefits agricultural productivity but also contributes to environmental conservation and biodiversity. Furthermore, nanopesticides can support sustainable farming practices by decreasing the frequency of applications and overall chemical inputs, ultimately enhancing food security. As research continues to address the challenges associated with their implementation, including regulatory approval and public acceptance, nanopesticides hold the potential to revolutionize crop protection, aligning agricultural practices with the principles of sustainability and ecological balance.

Acknowledgments

FZ express gratitude towards Vision Group of Science and Technology (VGST); Karnataka Science and Technology Promotion Society (KSTePS) and Department of Science and Technology, Government of Karnataka (GoK), Bangalore for K-FIST L1 grant with GRD No. 1051/2023-2025. All authors thank the Dr. M. Mohan Alva, Chairman, Mr. Vivek Alva, Dr. Vinay Alva, Managing trustee, Alva's Education Foundation (AEF) and Dr. Sajith M., the Principal, Alva's Ayurveda Medical College for their constant support and motivation. All authors credit BioRender.com for design of the figures in this chapter.

References

Abbass, K., Qasim, M.Z., Song, H. et al. (2022). A review of the global climate change impacts, adaptation, and sustainable mitigation measures. *Environmental Science and Pollution Research International* 29 (28): 42539–42559. https://doi.org/10.1007/s11356-022-19718-6.

Abid, N., Khan, A.M., Shujait, S. et al. (2022). Synthesis of nanomaterials using various top-down and bottom-up approaches, influencing factors, advantages, and disadvantages: a review. *Advances in Colloid and Interface Science* 300: 102597. https://doi.org/10.1016/j.cis.2021.102597.

Acharya, M., Shobha Ramya Krishna C.R., and Zameer, F. (2023a). Chapter 1: origin of cultivated plants. In Agrobotany (Vol. 1, pp. 1–44). Text Book in Accordance to NEP Scheme, Govt. of India. United Agencies, Karnataka, India.

Acharya, M., Shobha Ramya Krishna C.R., and Zameer, F. (2023b). Chapter 8: propagation of specific plant types. In Basics of gardening (Vol. 1, pp. 378–435). Text Book in Accordance to NEP Scheme, Govt. of India. Chethan Books, Karnataka, India.

Ahmed, A., He, P., He, P. et al. (2023). Environmental effect of agriculture-related manufactured nano-objects on soil microbial communities. *Environment International* 173: 107819. https://doi.org/10.1016/j.envint.2023.107819.

Akki, A.J., Jain, P., Kulkarni, R. et al. (2024). Microbial biotechnology alchemy: transforming bacterial cellulose into sensing disease—a review. *Sensors International* 100277: https://doi.org/10.1016/j.sintl.2023.100277.

An, C., Sun, C., Li, N. et al. (2022). Nanomaterials and nanotechnology for the delivery of agrochemicals: strategies towards sustainable agriculture. *Journal of Nanobiotechnology 20* (11): https://doi.org/10.1186/s12951-021-01214-7.

Avinash, M. G., Aishwarya, S., Zameer, F., & Gopal, S. (2023). *Pseudomonas aeruginosa* biofilm and their molecular escape strategies. *Journal Of Applied Biology & Biotechnology* https://doi.org/10.7324/jab

Bahrulolum, H., Nooraei, S., Javanshir, N. et al. (2021). Green synthesis of metal nanoparticles using microorganisms and their application in the agrifood sector. *Journal of Nanobiotechnology* 19: 10. https://doi.org/10.1186/s12951-021-00834-3.

Bajda, S. and Grigoraki, L. (2020). Integrated pest management: novel tools, remaining challenges, and intriguing non-target effects. *Current Opinion in Insect Science* 39: iii–v. https://doi.org/10.1016/j.cois.2020.05.009.

Beulah, K.C., Aishwarya, T., Waseem, K. et al. (2015). Phyto-antiquorumones: an herbal approach for blocking bacterial trafficking and pathogenesis. *International Journal of Pharmacy and Pharmaceutical Sciences* 7 (1): 29–34. https://journals.innovareacademics.in/index.php/ijpps/article/view/8543.

Bhan, S., Mohan, L., and Srivastava, C.N. (2018). Nanopesticides: a recent novel ecofriendly approach in insect pest management. *Journal of Entomological Research* 42 (2): 263–270. https://doi.org/10.5958/0974-4576.2018.00044.0.

Bouvet, L., Holdgate, S., James, L. et al. (2022). The evolving battle between yellow rust and wheat: implications for global food security. *TAG. Theoretical and applied genetics. Theoretische und angewandte Genetik* 135 (3): 741–753. https://doi.org/10.1007/s00122-021-03983-z.

Castillo-Henríquez, L., Alfaro-Aguilar, K., Ugalde-Álvarez, J. et al. (2020). Green synthesis of gold and silver nanoparticles from plant extracts and their possible applications as antimicrobial agents in the agricultural area. *Nanomaterials (Basel, Switzerland)* 10 (9): 1763. https://doi.org/10.3390/nano10091763.

Datta, S., Hamim, I., Jaiswal, D.K., and Sungthong, R. (2023). Sustainable agriculture. *BMC Plant Biology* 23 (1): 588. https://doi.org/10.1186/s12870-023-04626-9.

Deka, B., Babu, A., Baruah, C., and Barthakur, M. (2021). Nanopesticides: a systematic review of their prospects with special reference to tea pest management. *Frontiers in Nutrition* 8:686131. https://doi.org/10.3389/fnut.2021.686131.

Dinesh, B., Poyya, J., Zameer, F. et al. (2024). Effect of polyvinylpyrrolidone on antioxidant and antibacterial activity of silver metal nanoparticles: a comparative analysis. *Proceedings of the National Academy of Sciences, India Section A: Physical Sciences* 94: 359–368. https://doi.org/10.1007/s40010-024-00882-y.

Fatima, F., Siddiqui, S., and Ahmad, W. (2021). Nanoparticles as novel emerging therapeutic antibacterial agents in the antibiotics resistant era. *Biological Trace Element Research* 199: 10. https://doi.org/10.1007/s12011-020-02394-3.

Finger, R., Sok, J., Ahovi, E. et al. (2024). Towards sustainable crop protection in agriculture: a framework for research and policy. *Agricultural Systems* 219: 104037. https://doi.org/10.1016/j.agsy.2024.104037.

Gilbertson, L.M., Pourzahedi, L., Laughton, S. et al. (2020). Guiding the design space for nanotechnology to advance sustainable crop production. *Nature Nanotechnology 15* (10): 801–810. https://doi.org/10.1038/s41565-020-0706-5.

Hu, S., Yan, C., Fei, Q. et al. (2023). MOF-based stimuli-responsive controlled release nanopesticide: mini review. *Frontiers in Chemistry* 11: 1272725. https://doi.org/10.3389/fchem.2023.1272725.

Huang, W., Reyes-Caldas, P., Mann, M. et al. (2020). Bacterial vector-borne plant diseases: unanswered questions and future directions. *Molecular Plant* 13 (10): 1379–1393. https://doi.org/10.1016/j.molp.2020.08.010.

Kah, M., Tufenkji, N., and White, J.C. (2019). Nano-enabled strategies to enhance crop nutrition and protection. *Nature Nanotechnology* 14 (6): 532–540. https://doi.org/10.1038/s41565-019-0439-5.

Kannan, M., Bojan, N., Swaminathan, J., and Prasad, A.G. (2023). Nanopesticides in agricultural pest management and their environmental risks: a review. *International journal of Environmental Science and Technology* 20 (10): 10507–10532. https://doi.org/10.1007/s13762-023-04795-y.

Kookana, R.S., Boxall, A.B., Reeves, P.T. et al. (2014). Nanopesticides: guiding principles for regulatory evaluation of environmental risks. *Journal of Agricultural and Food Chemistry* 62 (19): 4227–4240. https://doi.org/10.1021/jf500232f.

Krishnaswamy, K. and Orsat, V. (2017). Sustainable delivery systems through green nanotechnology. In: *Sustainable Food Systems from Agriculture to Industry* (Chapter 2) (ed. A.M. Grumezescu). Elsevier. https://doi.org/10.1016/B978-0-323-52727-9.00002-9.

Kumar, S., Bhushan, P., and Bhattacharya, S. (2017). Fabrication of nanostructures with bottom-up approach and their utility in diagnostics, therapeutics, and others. *Environmental, Chemical and Medical Sensors* 167–198: https://doi.org/10.1007/978-981-10-7751-7_8.

Kumar, S., Nehra, M., Dilbaghi, N. et al. (2019). Nano-based smart pesticide formulations: emerging opportunities for agriculture. *Journal of Controlled Release* 294: 131–153. https://doi.org/10.1016/j.jconrel.2018.12.012.

Liu, P., Zheng, X., Shangguan, S. et al. (2022). Public perceptions and willingness-to-pay for nanopesticides. *Nanomaterials (Basel, Switzerland)* 12 (8): 1292. https://doi.org/10.3390/nano12081292.

Mahmood, M.A., Ahmed, N., Hussain, S. et al. (2022). Dominance of Asia II 1 species of *Bemisia tabaci* in Pakistan and beyond. *Scientific Reports* 12 (1): 1528. https://doi.org/10.1038/s41598-022-05612-1.

Manjunatha, H.P., Singh, H., Chauhan, J.B. et al. (2013). Induction of resistance against sorghum downy mildew by seed treatment with *Duranta repens* extracts. *IOSR J Agric Vet Sci* 3 (6): 37–44.

Manzoor, M.A., Xu, Y., Lv, Z. et al. (2024). Nanotechnology-based approaches for promoting horticulture crop growth, antioxidant response and abiotic stresses tolerance. Plant. *Stress* 11: 100337. https://doi.org/10.1016/j.stress.2023.100337.

Martirosyan, A. and Schneider, Y.J. (2014). Engineered nanomaterials in food: implications for food safety and consumer health. *International Journal of Environmental Research and Public Health* 11: 5720–5750. https://doi.org/10.3390/ijerph110605720.

Maruyama, C.R., Guilger, M., Pascoli, M. et al. (2016). Nanoparticles based on chitosan as carriers for the combined herbicides imazapic and imazapyr. *Scientific Reports* 6: 23854. https://doi.org/10.1038/srep23854.

Mathew, T., Aishwarya Sree, R., Aishwarya, S. et al. (2020). Graphene-based functional nanomaterials for biomedical and bioanalysis applications. *FlatChem* 23: 100184. https://doi.org/10.1016/j.flatc.2020.10.184.

Mishra, Anand & Gautam, Vinay (2021). Weed species identification in different crops using precision weed management: a review. *ISIC'21:International Semantic Intelligence Conference, CEUR Workshop Proceedings*, New Delhi, India (25–27 February 2021), pp. 180–195. https://ceur-ws.org/Vol-2786/Paper25.pdf

More, V.S., Ebinesar, A., Prakruthi, A. et al. (2021). Isolation and purification of microbial exopolysaccharides and their industrial application. In: *Microbial Polymers: Applications and Ecological Perspectives* (pp. 69–86). Singapore: Springer Singapore. https://doi.org/10.1007/978-981-16-0045-6_3.

Mubeen, I., Bani Mfarrej, M.F., Razaq, Z. et al. (2023). Nanopesticides in comparison with agrochemicals: outlook and future prospects for sustainable agriculture. *Plant Physiology and Biochemistry 198*: 107670. https://doi.org/10.1016/j.plaphy.2023.107670.

Müller, R.H. and Peters, K. (1998). Nanosuspensions for the formulation of poorly soluble drugs: I. Preparation by a size-reduction technique. *International Journal of Pharmaceutics* 160: 229–237. https://doi.org/10.1016/S0378-5173(97)00311-6.

Nagoshi, R.N., Davis, J.A., Meagher, R.L. et al. (2023). Investigating the migratory behavior of soybean looper, a major pest of soybean, through comparisons with the corn pest fall armyworm using mitochondrial haplotypes and a sex-linked marker. *Genes* 14 (7): 1495. https://doi.org/10.3390/genes14071495.

Paradva, K. and Kalla, S. (2023). Nanopesticides: a review on current research and future perspective. *Chemistry Select* 8: https://doi.org/10.1002/slct.202300756.

Patil, A.G. et al. (2021a). Myco-nanotechnology for sustainable agriculture: challenges and opportunities. In: *Recent Trends in Mycological Research (Fungal Biology)* (ed. A.N. Yadav). Cham: Springer https://doi.org/10.1007/978-3-030-60659-6_20.

Patil, A.G. et al. (2021b). Fungal amylases and their industrial applications. In: *Industrially Important Fungi for Sustainable Development. Fungal Biology* (ed. A.M. Abdel-Azeem, A.N. Yadav, N. Yadav, and M. Sharma). Cham: Springer https://doi.org/10.1007/978-3-030-85603-8_11.

da Pontes, M.S., de Forini, M.M.L., Caires, A.R.L. et al. (2023). Nanopesticides-modes of action and impacts. In: *Nano-enabled Sustainable and Precision Agriculture* (ed. P. Zhang, I. Lynch, J.C. White, and R.D. Handy), 79–103. Academic Press https://doi.org/10.1016/B978-0-323-91233-4.00001-6.

Prasad, R., Bhattacharyya, A., and Nguyen, Q.D. (2017). Nanotechnology in sustainable agriculture: recent developments, challenges, and perspectives. *Frontiers in Microbiology* 8: 1014. https://doi.org/10.3389/fmicb.2017.01014.

Prasad, A., Baker, S., Nagendra Prasad, M.N. et al. (2019). Phytogenic synthesis of silver nanobactericides for anti-biofilm activity against human pathogen *H. pylori*. SN. *Applied Sciences* 1: 341. https://doi.org/10.1007/s42452-019-0362-2.

Rahman, S., Rahman, L., Khalil, A. et al. (2019). Endophyte-mediated synthesis of silver nanoparticles and their biological applications. *Applied Microbiology and Biotechnology* 103: 10. https://doi.org/10.1007/s00253-019-09661-x.

Ristaino, J.B., Anderson, P.K., Bebber, D.P. et al. (2021). The persistent threat of emerging plant disease pandemics to global food security. *Proceedings of the National Academy of Sciences* 118 (23): e2022239118. https://doi.org/10.1073/pnas.2022239118.

Sahu, B., Sharma, D.D., Jayakumar, G.C. et al. (2023). A review on an imperative by-product: glycosaminoglycans—a holistic approach. *Carbohydrate Polymer Technologies and Applications 5*: 100275. https://doi.org/10.1016/j.carpta.2022.100275.

Sahu, B., Ramesh, A., and Zameer, F. (2024). Impact of nanomaterials on leather: a nano-Saga from processing to application. *Clean Technologies and Environmental Policy* 1-28: https://doi.org/10.1007/s10098-024-02912-0.

Sarmah, K., Borah, R., Boruah, S., and Sarmah, D. (2023). Nanopesticides: revolutionizing pest management with nanotechnology. *Biological Forum—An International Journal* 15 (7): 210–218.

Satapathy, P., Aishwarya, S., Shetty, M.R. et al. (2020). Phyto-nano-antimicrobials: synthesis, characterization, discovery, and advances. *Frontiers in Anti-Infective Drug Discovery* 8: 196. https://doi.org/10.2174/9789811412387119080l.

Schoonjans, R., Castenmiller, J., Chaudhry, Q. et al. (2023). Regulatory safety assessment of nanoparticles for the food chain in Europe. *Trends in Food Science & Technology* 134: 98–111. https://doi.org/10.1016/j.tifs.2023.01.017.

Shangguan, W., Chen, H., Zhao, P. et al. (2024a). Scenario-oriented nanopesticides: shaping nanopesticides for future agriculture. *Advanced Agrochem* https://doi.org/10.1016/j.aac.2024.07.002.

Shangguan, W., Huang, Q., Chen, H. et al. (2024b). Making the complicated simple: a minimizing carrier strategy on innovative nanopesticides. *Nano-Micro Letters* 16: 193. https://doi.org/10.1007/s40820-024-01413-5.

Shelef, O., Fernández-Bayo, J.D., Sher, Y. et al. (2018). Elucidating local food production to identify the principles and challenges of sustainable agriculture. In: *Sustainable Food Systems from Agriculture to Industry* (ed. C.M. Galanakis), 47–81. Academic Press https://doi.org/10.1016/B978.0-12-811935-8.00002-0.

Singh, H., Zameer, F., Khanum, S.A. et al. (2016). Durantol—a phytosterol antifungal contributor from *Duranta repens* Linn. For organic management of Sorghum Downy Mildew. *European Journal of Plant Pathology* 146: 671–682. https://doi.org/10.1007/s10658-016-0951-5.

Suer, I. and Tursun, N. (2024). Weeds in the cotton growing areas in the Southeastern Anatolia Region. *Harran Tarım ve Gıda Bilimleri Dergisi* 28: 209–221. https://doi.org/10.29050/harranziraat.1406951.

Sulthana, R., Taqui, S.N., Zameer, F. et al. (2018). Adsorption of ethidium bromide from aqueous solution onto nutraceutical industrial fennel seed spent: kinetics and thermodynamics modeling studies. *International Journal of Phytoremediation* 20 (11): 1075–1086. https://doi.org/10.1080/15226514.2017.1365331.

Tang, Y., Zhao, W., Zhu, G. et al. (2024). Nano-pesticides and fertilizers: solutions for global food security. *Nanomaterials* 14 (1): 90. https://doi.org/10.3390/nano14010090.

Verma, R. and Saroop, S. (2024). Impact of nanopesticides in the environment: solutions, threats, and opportunities. In: *Pesticides in the Environment* (ed. A. Sharma, V. Kumar, and B. Zheng), 251–292. Elsevier. https://doi.org/10.1016/B978-0-323-99427-9.00010-0.

Wang, D., Saleh, N., Byro, A. et al. (2022). Nano-enabled pesticides for sustainable agriculture and global food security. *Nature Nanotechnology* 17: 1–14. https://doi.org/10.1038/s41565-022-01082-8.

Wudil, A.H., Usman, M., Rosak-Szyrocka, J. et al. (2022). Reversing years for global food security: a review of the food security situation in Sub-Saharan Africa (SSA). *International Journal of Environmental Research and Public Health* 19 (22): 14836. https://doi.org/10.3390/ijerph192214836.

Xu, Z., Tang, T., Lin, Q. et al. (2022). Environmental risks and the potential benefits of nanopesticides: a review. *Environmental Chemistry Letters 20*: 2097–2108. https://doi.org/10.1007/s10.11-021-01338-0.

Yadav, A.N. (2021). *Recent Trends in Mycological Research*. Springer International Publishing. https://doi.org/10.1007/978-3-030-68260-6.

Yang, D.S., Cui, B., Wang, C.X. et al. (2017). Preparation and characterization of emamectin benzoate solid nanodispersion. *Journal of Nanomaterials* 2017: 6560780. https://doi.org/10.1155/2017/6560780.

Yang, Y., Tilman, D., Jin, Z. et al. (2024). Climate change exacerbates the environmental impacts of agriculture. *Science (New York, N.Y.)* 385 (6713): eadn3747. https://doi.org/10.1126/science.adn3747.

Yin, J., Su, X., Yan, S., and Shen, J. (2023). Multifunctional nanoparticles and nanopesticides in agricultural application. *Nanomaterials (Basel, Switzerland)* 13 (7): 1255. https://doi.org/10.3390/nano13071255.

Zainab, R., Hasnain, M., Ali, F. et al. (2024). Prospects and challenges of nanopesticides in advancing pest management for sustainable agricultural and environmental service. *Environmental Research* 261: 119722. https://doi.org/10.1016/j.envres.2024.119722.

Zameer, F. and Gopal, S. (2010). Evaluation of antibiotic susceptibility in mixed culture biofilms. *International Journal of Biotechnology and Biochemistry* 6 (1): 93–99. https://www.researchgate.net/publication/242100772.

Zameer, F., Kreft, J., and Gopal, S. (2010a). Interaction of the dual species biofilms of *Listeria monocytogenes* and *Staphylococcus epidermidis*. *Journal of Food Safety 30* (4): 954–968. https://doi.org/10.1111/j.1745-4565.2010.00254.x.

Zameer, F., Meghashri, S., Gopal, S., and Rao, B.R. (2010b). Chemical and microbial dynamics during composting of herbal pharmaceutical industrial waste. *Journal of Chemistry 7*: 645978. https://doi.org/10.1155/2010/645978.

Zameer, F., Rukmangada, M.S., Chauhan, J.B. et al. (2016). Evaluation of adhesive and anti-adhesive properties of *Pseudomonas aeruginosa* biofilms and their inhibition by herbal plants. *Iranian Journal of Microbiology* 8 (2): 108. https://pubmed.ncbi.nlm.nih.gov/27307976/.

Zhao, X., Wei, C., Gai, Z. et al. (2019). Chemical vapor deposition and its application in surface modification of nanoparticles. *Chemical Papers 74*: 00963. https://doi.org/10.1007/s11696-019-00963-y.

16

Nanobiosensors in Agricultural and Environmental Sustainability

Monika Kundu[1], Shiv Prasad[2], Shipra Singh[3], Prameela Krishnan[1], and Ajar Nath Yadav[4]

[1] Division of Agricultural Physics, Indian Agricultural Research Institute, Delhi, New Delhi, India
[2] Division of Environmental Sciences, ICAR-Indian Agricultural Research Institute, Delhi, New Delhi, India
[3] Department of Chemistry, University of Delhi, Delhi, New Delhi, India
[4] Department of Genetics, Plant Breeding and Biotechnology, Dr. Khem Singh Gill, Akal College of Agriculture, Eternal University, Sirmaur, Himachal Pradesh, India

16.1 Introduction

Ensuring an adequate level of food quality and safety is necessary to protect consumers' health and the nation. Quality and security of the product are the prime issues in today's agriculture sector. This can be achieved by monitoring and implementing quality assurance measures down the entire food chain, wherever it is required. Every individual involved in the food chain, from the farm to the fork, has diverse responsibilities to maintain the food supply safe by taking the necessary measures to keep food secluded from hazards, hence increasing human health risks (Kundu et al. 2019a). The significant challenges are achieving sustainable agricultural production along with environmental security. The current existing problems in agriculture include: the rise in demand for food supply, changing dietary choices, water quality, land use and soil preservation and its conservation, climate change, and enhaned greenhouse gases. In agriculture, we have significantly improved the production levels by managing various biotic and abiotic measures. Worldwide, the number of people involved in agriculture is more than 1 billion (FAOSTAT 2017). The raw and processed food industry is a prime industry today, where a huge gap in demand and supply exists. With the introduction of the green revolution, food production was observed a great soar during that time. The emphasis has been laid upon increasing production through integrated approaches of diversified fields like plant genetics and breeding, chemical sciences, soil science, and agricultural engineering, including farm machinery and electronics and instrumentation, to narrow this demand and food supply gap.

With advancements in the instrumental sector aiming at faster analysis time, greater efficiency, and a high degree of resolution than HPLC, capillary electrophoresis is more cost-effective to develop and run due to relatively less expensive capillaries. It requires a minimum amount of sample as compared to other techniques. As an alternative to HPLC, the fluorimetry technique is also being used in agriculture. These analytical techniques have played a crucial role in food product authentication and characterization (Silva et al. 2015) to determine the adulteration in food products and antioxidants in foods. Based on antigen–antibody interactions, flow immune analysis has popularized due to rapidity, low cost of the assays, little sample pretreatment, and the efficiency for screening a large number of samples. The technique is widely used in agricultural applications to determine preservatives in food and beverages (Sajid et al. 2015), determine antioxidants and sweeteners (Pisoschi et al. 2011; Norfun et al. 2016), determine monitoring macro- and micronutrients in plants (Estela and Cerda 2005) and detect pesticide residue (Liu et al. 2013; Silva et al. 2015).

To assure compliance with international standards and evaluation of agricultural produce before they result in severe damage, applications of non-destructive techniques are very significant. Various spectroscopic techniques (infrared spectroscopy, Raman spectroscopy, hyperspectral imaging systems), magnetic resonance imaging, nuclear magnetic resonance, electronic nose, impedance tongue, ultrasound, and various sensors such as biosensors fall under this category of instruments. The spectroscopic techniques such as near-infrared reflectance spectroscopy (NIRS) are among the most promising strategies for monitoring and quality control of agri-produce. NIR is a non-destructive technique for pesticide residue monitoring in vegetables like pepper (Sanchez et al. 2010; Jamshidi et al. 2016). Emerging technologies such as Raman spectroscopy, laser-based detection, real-time PCR, sensors, and lab-on-chip (LOC) methods are evolving. With the advance of technology and precision agriculture becoming popular for smart and automated machines and devices, the need to invent more precision spectrometers for a more expansive range of applications led to the use of Raman spectrometer in the area of pesticide residue detection in fruits and vegetables (Liu et al. 2013).

UV, FTIR, and NMR have also been widely used to characterize and detect residues of insecticides being sprayed on fruits and vegetables (Upadhay et al. 2013). FTIR has been used to measure different types of materials in solid, liquid. or gas. It has recently been used to analyze pesticide residues on fruit peels (Xiao and Pignatello 2015). The spectrometric techniques require little or no sample preparation. But most of the spectroscopic techniques are carried out at the lab level and are not portable. Moreover, to develop a spectrometer for a particular application is very costly and complicated. Thus, in agriculture, spectroscopy-based instruments are limited to researchers' laboratory scale and do not find wide acceptance among the end-users, i.e., farmers or consumers.

However, existing conventional methods and techniques to evaluate quality and safety suffer from subjectivity, time consumption, lack of skilled personnel, high production cost, and labor-intensive nature. So there is a need to develop rapid, low-cost, portable sensors and devices to carry out the quality assessment in field conditions by even a layperson. The development of quick, simple, cheap, and non-destructive techniques may be regarded as essential for the agricultural sector and meet farmers' or consumers' requirements. The use of biosensors in different fields has drawn the attention of scientists and researchers worldwide. This chapter provides an in-depth understanding of biosensors' concept based on nanomaterials, biosensing components, characteristic features, their different types, fabrication process, and the nanobiosensor interventions in attaining agriculture and environmental sustainability.

16.2 Biosensor

The invention of the first biosensor was reported in 1967, and the innovations in this field continue today. Biosensors' history can be categorized into three broad areas: first generation, second generation, and third generation. The first generation-based biosensors are electrical response-based biosensors. Second-generation biosensors have mediators between the reaction and the transducer. The third-generation biosensors involve direct communication between the redox center of the biomolecules such as enzymes and the electrodes for electron transfer. Along with various types of innovative advancements for the development of nanobiosensors, nanomaterials' role is very significant.

A biosensor is an integrated pioneer approach towards sensing and characterization of materials. Nowadays, the farmers are ready to try the innovative leading technologies that provide clear benefits, hence opening a new gateway for the end-users to soar into smart and precision agriculture. In this chapter, we are exploring biosensors for applications in the agriculture and food industry. There is the utmost need for biosensors in agriculture due to unsolved problems and environmental sustainability issues. Nanosensors are similar to macro-level sensors but have at least one dimension in nanoscale and can measure signals available at that scale. The applications of nanotechnology are now widespread in all sectors. Various electronics industries have fed on miniaturization, and the biosensor field has got advantage of these advancements. An essential biosensor characteristic is that due to the minimal sample quantity requirements and fast response time (Table 16.1).

Table 16.1 Characteristic features of a biosensor.

S. no.	Characteristic	Measurement for a biosensor
1.	Sensitivity	σ = Change in the output response/change in the input variable
2.	Selectivity	It is a unique characteristic feature of a biosensor. Biosensor selectivity is defined as detecting the target analyte from a complex sample consisting of various types of interfering species.
3.	Resolution	Resolution is defined as the smallest measurable change in the target analyte's value that the biosensor can detect easily.
4.	Response time	Response time determines how fast is a biosensor device. It is measured by the time a biosensor takes to reach 63% of the final measured analyte value.
5.	Calibration	It is a plot of biosensor peak responses measured to the concentration of the analyte.
6.	Linearity	When the calibration curve of a biosensor is obtained, the linearity is measured in terms of matching the line of best fit to the calibration curve.
7.	Repeatability	Repeatability is also called as reproducibility of the biosensor response obtained for a fixed analyte concentration. It also represents the reproducibility of the set of experimental results when repeated under similar conditions.

(Continued)

Table 16.1 (Continued)

S. no.	Characteristic	Measurement for a biosensor
8.	Stability	It denotes the storage stability of a biosensor. It is the period for which there is negligible change in the response of a biosensor. It means that during this period, the calibration curve of the biosensor doesn't change.
9.	Drift	With the passage of storage time, there occurs a decrease in the response of the biosensor. This shift or decrease in the response ultimately reflected in the biosensor's calibration curve is called drift.
10.	Allowed ambient parameters	The allowed parameters include temperature, pressure, relative humidity, light or illumination, etc. The maximum permissible values of these parameters are defined below, which the device can operate satisfactorily.

16.3 The Fusion of Nanotechnology with the Biosensor: Nanobiosensor

The role of nanotechnology in the enhancement of productivity is well known to everyone today. Nanotechnology has played a significant role in nutrient management, monitoring water quality, detecting agricultural chemicals and target delivery, etc., implying a prominent part of sustainable agricultural development. In the development of nanotechnology, the material properties are studied at the nanoscale. There are various biomolecules that have sizes in the range of nanoscale (Figure 16.1).

The exploration of materials, processes, and phenomena at nanometer scale causes entirely new characteristics, and the material behaves differently (Kundu et al. 2019c). This

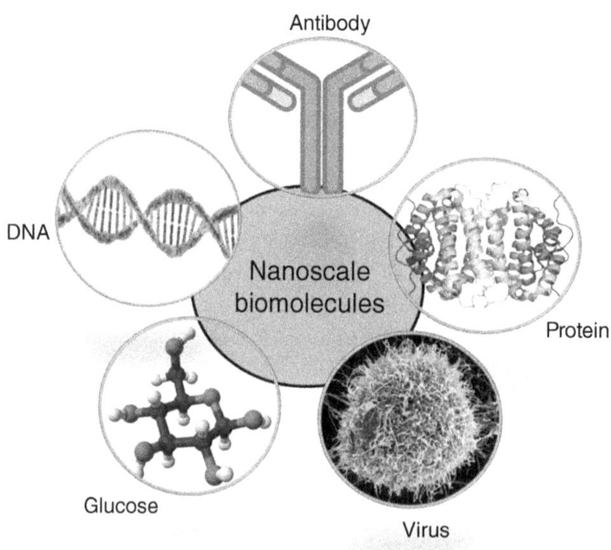

Figure 16.1 Various types of biomolecules at nanoscale.

innovative technology has significant consequences in agriculture. Among the various reported studies, nanoencapsulation is a very familiar and widely used technological intervention in agriculture. Nanoencapsulation (in the form of micelles, liposomes, or biopolymers) has been used for insecticide spray in plants (Khoobdel et al. 2017), the supply of nutrients and essential supplements for the growth and promotion of root along with target delivery of additives or supplements in foods (Syu et al. 2014). It is reported that herbicides' nanocapsules ensure precise target delivery leading to enhanced specificity and efficiency through slow and controlled release (Oliveira et al. 2015). In the food and agriculture industry, nanofilms of edible coatings have been developed for packaging raw, processed food items (Ribeiro 2014), and nanofibers possessing antibacterial properties have been designed for durable active food packaging (Saini et al. 2016). Such edible nanocoatings improve the shelf life and quality of food products.

Another important application area of nanotechnology is the early diagnosis of plant diseases. Nanoparticles with ultra-small dimensions compatible with most microorganisms interact better with bacteria and viruses, thus laying the path for detecting the pathogenic attack on agricultural produce (Ruttkay-Nedecky et al. 2017). Therefore, nanotechnology can offer a better way to provide very high-quality foods, simultaneously ensuring better bioavailability of nutrients. The introduction of nanoemulsions now benefits the food industry. In this technology, the food gets unique rheological and textural properties as desired. Nanoemulsions cause less fat use and ensure healthy food. More such nanotechnology applications in the food and agri-sector are currently at the research and development stage for coming in the near future. In agriculture, crop productivity is greatly influenced by plant health. The plants at this stage are very prone to the attack of diseases. Among the various diseases-causing agents, those by viruses are found to be the most destructive and hard to control.

In most cases after the disease, even applying pesticide doesn't find any use. In such cases, the early prediction of disease occurrence is the only preventive measure to avoid considerable losses to crops. Another problem is caused by lodging also. Lodging of crops can be due to the occurrence of various biotic and abiotic stresses. Various studies have been conducted to detect cereal crops' lodging tendencies (Monika et al. 2016a). Now nanomaterials and biosensors can play an essential role in the detection of lodging tendencies of various crops. Nanosurfaces or nanoparticles possessing fluorescent properties are reported for developing a sensing matrix for the biosensor to further detect bacteria through resulting electrical or chemical change as a response signal (Kahveci et al. 2016). Nanomaterials in various forms, such as nanofibers, nanodots, nanorods, nanoflowers, etc., have contributed to developing specific and sensitive devices. Among such devices, a biosensor is increasingly used for analyzing the environment, clinical and agricultural samples.

A biosensor is a monitoring and measurement system involving a nanomaterial-based matrix and bio-transducer to detect a specific chemical concentration sensitively. In research studies, it has been reported that the biosensors are advantageous for agricultural diagnosis due to high sensitivity, specificity, and wide linearity as a result of enhancement of active surface area (Wang et al. 2016). Nanomaterials of various shapes and dimensions affect the performance of biosensors. The compact size of biosensors is an added advantage for on-the-spot and real-time analysis. With the advancements in integrated technology, nanobiosensors fabricated with a compact integrated unit equipped with power, a sensing unit, a signal detector, and a display unit can be used as an autonomous device even in very harsh environments in different application areas. Such systems are advantageous due to increased life, ruggedness, and lower costs (Kuswandi et al. 2011; Sassolas et al. 2012).

Nanotechnology has contributed a lot to biosensor development due to its unique electronic, magnetic, and optical properties of the nanomaterials as substrate. The unique characteristics like nano-size and large surface-to-volume ratio enhance the nanosensor's sensitivity for the target analyte detection. Due to their unique and novel characteristic properties at the nanoscale, nanomaterials are now widely used in sensing technology. A nano-biosensor is defined as a device used to detect analytes quantitatively with a biologically active element attached to a suitable nano-structured transducer. Unlike the nanosensor, a biosensor has a biologically sensitive molecule immobilized onto the physiochemical transducer's surface. It is an interdisciplinary science involving electronics, biology, material chemistry, and physics. A nano biosensor is different from a sensor, as it has enhanced sensitivity, specificity, quick response time, and reliability. Among all the properties, specificity is unique as it enables a nano biosensor to detect and quantify the target analyte out of the other present potential interferents.

For good performance of a nano biosensor, analyte and biosensor interaction should not be affected by temperature and pH. An ideal nanobiosensor is stable under normal storage conditions. It should be inexpensive, manageable, and simple to handle by a layperson. The operational performance of nanobiosensors is not hidden at the atomic levels. High precision, accuracy, linearity, and increased sensitivity are considered error-prone devices. While fabricating a nano biosensor, three main essential components are the recognition element, the transducer, and the detector (Figure 16.2).

Among the various bioreceptor probes used are antibodies, nucleic acids, pathogens, and metabolites. The working of a nano biosensor is based on the binding of targeted bio-analytes with the baroreceptors. Due to this binding, a physicochemical signal is generated. A transducer measures electrochemical, mass, optical, or thermal changes occurring due to reaction at the interaction site of baroreceptor and analyte. Thus transducer converted the signal into measurable electrical output. The detector then traps this electrical signal and further amplifies it to be captured by the microprocessor. Optical nano biosensors utilize light to sense the effect of chemical activity on a biological system. Piezoelectric biosensors use piezoelectric devices like quartz, and work in an electric field medium. The electrochemical biosensors are based on the applied current. After applying a potential between two electrodes, the charge distribution changes have been detected through ion-selective electrodes (Mohsin et al. 2020).

16.3.1 Advantages of Nanobiosensor

Nanosensors based on the interfacing of nanomaterials with biomolecules are called nanobiosensors. The advantages of nanobiosensor are discussed below with the following sub-headings.

16.3.2 Nano Biosensors are Highly Sensitive than Macroscale Sensors

They provide more accurate and single-molecule detection. Since most of these nanosensors are based on target analytes' adsorption and adsorption depends on the surface area and surface chemistry. Using a nanoporous and nanocrystalline type of materials further

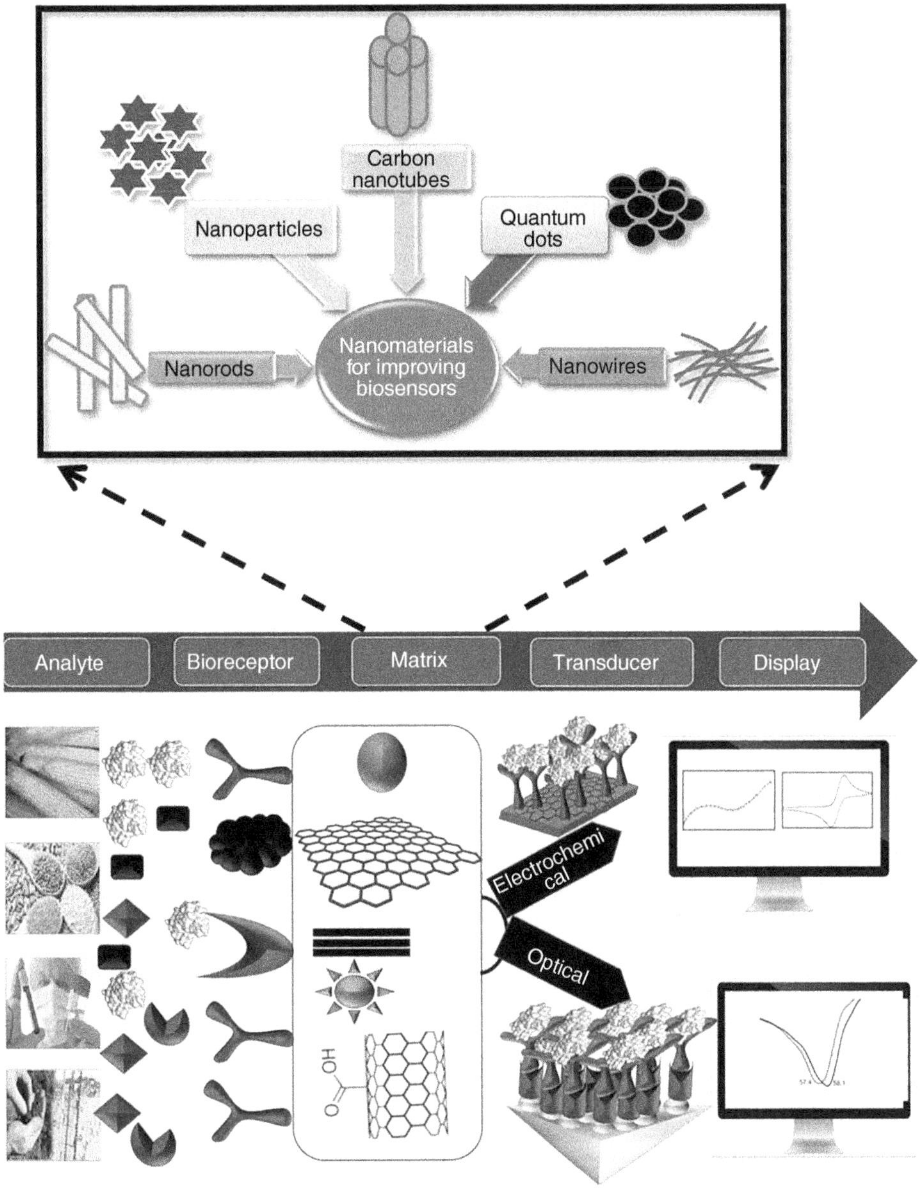

Figure 16.2 Schematic representation of different components of a nanobiosensor. *Source:* Modified from Bhardwaj et al. (2021).

increases the surface area to a considerable extent. In addition to this, various surface modifications of the nanomaterials are also essential techniques for enhancing sensitivity. For example, carbon nanotubes themselves have low sensitivity towards most of the analytes. Still, by surface modifications such as coating the nanotubes with metal atoms or metal oxides, they become selective towards particular species.

16.3.3 Miniaturized, Allowing them to Fit for Portable Applications

Besides, they consume less power, require less sample volume and cause minimum disturbance to the sensed phenomenon. Conventional sensors were bulky and power-hungry. Now in nanosensors, the improvement in the design and fabrication of sensor components has led to power saving. Most of such sensors hardly consume milliwatts of power.

16.3.4 They have a Low Response Time in the Analysis of the Agricultural Samples

Nanobiosensors are very helpful in rapid and real-time analysis of samples. Due to these sensors' miniature size, the time needed to attain the equilibrium will be reduced, and the speed will be enhanced as the time required to cover the shorter distances is less. Therefore, these sensors can implement the desired operations at faster speeds. Due to the rapid rate, the measurements are performed in real-time, and data are immediately analyzed for necessary action.

16.3.5 Nanobiosensors are Comparatively Low-cost Devices

Due to reduced nanomaterial-based biosensors' reduced size, the final fabrication and development costs in terms of components size and cost are less compared to their bulky counterparts. Also, the final integrated technology should be highly reproducible at a low cost. All such necessary requirements are satisfied by nanosensors so that the research on nanobiosensors will be directed to the development of cheaper devices. Availability of disposable nanosensors in abundant supply at throwaway prices is the need of the hour.

16.3.6 Multi Analyte Detection: Nanobiosensors are Multifunctional

Nanobiosensor is reported as a multifunctional and multi-analyte detection device. For example, if one has to detect several pollutant gases in the environment, then an array of gas sensors is required. To fabricate such an array consisting of hundreds or thousands of sensors will lead to an enormously bulky and costly sensor. It is virtually impractical to construct and use. The skill of multiplexing present in nano-bio sensors is a considerable level of improvement for real-time monitoring of gas composition. For such analysis, thousands of nanobiosensors integrated together to make a single assembly and coated with a different functional group are synchronized to detect a target analyte. Such an ultra-miniaturized device consuming significantly less power and equipped with signal processing and pattern recognition algorithms shows functional detection capabilities among target analytes. Such an array of biosensors will be producing a unique signature (or fingerprint) on exposure to a sample containing a mixture of chemical and biological characters required for environmental pollution control.

16.3.7 Less Sample Preparation/Pretreatment: Nanobiosensors are Easy to Use

Building a new class of intelligent sensors: nanobiosensors allow for creating a new level of devices that provide the elemental base for smart sensors capable of data processing, storage, and analysis. These sensors give high accuracy, ultra-high sensitivity, extreme specificity, real-time in vivo information with more incredible speed, having multi-analyte options, requiring a smaller quantity of sample and minimal sample preparation, durable, safe, and portable.

16.3.8 Immuno Nanobiosensor

Immunosensor technology has been increasing over the last decade in agricultural applications. There are significant advances in the areas, chiefly in food analyses and immuno-sensors for quality and safety analysis, but the challenges still exist to make their commercial uses feasible. These challenges are biomolecule specificity, immobilization matrix stability, transduction mode, and pretreatment of complex biological samples. There are three main components, i.e., bioreceptor, the matrix assembled, and the transducer type employed for the development of immunosensors. A biosensor's performance is directly dependent on the immobilization matrix used and the orientation and density of affinity biomolecules on the electrode surfaces. At present times, there are various techniques for the immobilization of the biomolecules. Conventionally, there are noncovalent and covalent techniques employed to immobilize the biomolecules based on adsorption, encapsulation, and entrapment in polymers, covalent bio-binding, and cross-linking of antibody aggregates.

16.3.9 Nanomanufacturing/Nanofabrication

It is defined as manufacturing the nanomaterials or various structures at the nanoscale for different applications. The main aim in the nanofabrication or manufacturing process remains to obtain smaller sizes. Now a microchip of the size of our fingertip can hold a considerable stock of data. Three different types of approaches are followed for the nanomanufacturing of materials for nanosensors. These are the top-down approach, the bottom-up approach, and molecular assembly.

16.3.10 Top-down Approach

In this nanomanufacturing process, we start with a large lump of material and start removing a little by little till the final shape is attained. Nanolithography and transfer of the pattern are two steps in this process. In nanolithography, we create the desired design on a particular type of sacrificial layer (called a resist). Among various nanolithography techniques are photolithography, electron beam lithography, X-ray lithography, soft lithography, and so forth. The underlying principle is that, first of all, a layer of resistance is applied to the substrate. Then using a pattern, the photoresist is exposed to some source of energy. As a result of such patterned exposure, there occurs a chemical reaction at the resist site. Then a part of the resist is removed, and a pattern is created. After etching, the design created by the resist is removed mechanically or chemically.

16.3.11 Bottom-up Approach

This approach is equivalent to building up a house using bricks. Thus, it is based on developing a final structure using small units or components. Among various bottom-up approaches are physical or chemical vapor deposition, contact printing, imprinting, assembly, joining, and coating.

16.3.12 Molecular Assembly

In this novel approach, the essential components assemble in the desired fashion to produce a nano object without an outside force's direction. This is still a growing field. The different properties of the molecule involved include shape, surface properties, charge, polarizability, and a magnetic dipole for assembling them to form a particular structure. The nanomanufacturing techniques for nano biosensors are electron beam lithography, focused ion beam lithography, and X-ray lithography.

16.3.13 Electron Beam Lithography

In this process, the patterns are transferred from one medium to another with the help of a material called to resist. This is a widespread technique in sensor fabrication as it possesses various characteristic features of precious pattern-making capability. Among various electron beam lithography technologies, direct writing electron beam lithography technology is simple and often used. In this method, a focused Gaussian round beam moves with the wafer, and a single pixel of a wafer is exposed at a time. The beam creates a desired pattern on the wafer, and, supported with the etching and deposition process, a highly complex nanostructure can be produced. Although this technique is cheap and famous, its main limitation is the large time requirement.

16.3.14 Focused Ion Beam Lithography

Similar to electron beam lithography, but different in the sense that instead of electrons here, the ions are used to perform the lithography. Since an ion is heavier than an electron, focused ion beam lithography is better than electron–electron beam lithography. In this technique, instead of resistance, a varying distance of the wafer is used to control ions' dose. Due to which the trench of different depths occurs on a wafer. In this technique, the heavy-ion species such as Ga^+ and Au^+ can also be used in this lithography for producing a more substantial effect. When a beam is passed over the water, a trench having a Gaussian shape of inverse nature is obtained. With the increase in strength, the trench becomes more sharp, narrow, and V-shaped.

16.3.15 X-ray Lithography

In X-ray lithography, shorter wavelengths are used, making it an advanced version of optical lithography. In this technique, a particular mask type is used. On this mask, different local X-ray absorption areas are present for defining the pattern. The patterns are then replicated on an X-ray-sensitive material (resist) that was earlier deposited on a substrate. When X-rays pass through, the pattern falls on the resist. It may cause cross-linking or bond-breaking based on the type of resists. After exposure, everything is dipped in a specific solvent. Depending on the resist's chemical nature, either the exposed area resist dissolves and creates a pattern or vice versa. The other part of the resist stays intact. This way, X-ray

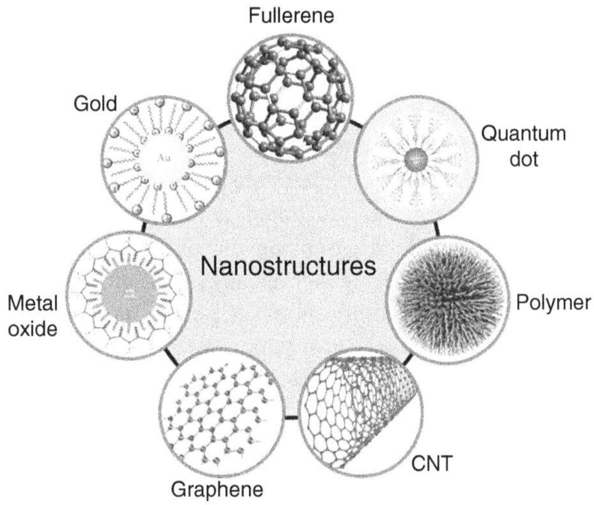

Figure 16.3 Various types of nanostructures for the fabrication of nanobiosensors.

lithography created nanopatterns on the substrate (Pal 2020). These days, various types of nanostructures are being used for the fabrication of nanomaterial substrate for biomolecule immobilization (Figure 16.3).

16.4 Nanobiosensors to Promote Agricultural Sustainability

Sustainable agriculture means food for all in all available conditions. Novel innovations can achieve this objective in agriculture by the introduction of sensors and instrumentation. Nanotechnology has a vast number of applications in agriculture. A nano fertilizer transports nutrients to crops via encapsulation technique using nanoparticles. Different nutrients can be encapsulated inside nanomaterials (such as nanotubes or nanoporous). The material can be coated with a thin protective polymer film, and the material can be delivered as particle emulsions of nanoscale dimensions.

Nano-fertilizers are very useful for decreasing nitrogen loss due to leaching, emissions, and long-term assimilation. Recently, researchers have reported that carbon nanotubes (CNTs) penetrate tomato seeds, while zinc oxide nanoparticles enter ryegrass's root tissue (Terry et al. 2005). It highlights that there is enormous scope for the development of new nutrient delivery systems. The role of nano fertilizers is preventing the early conversion of nutrients into improper forms that plants cannot absorb. Rather, it governs the release of nutrients as per the demand of the plant. In order to attain this, integration of biosensors and nanofertilizers will allow the selective release of nutrients (say nitrate) depending upon time and soil conditions.

Zeolite is crystalline aluminum silicates that are found naturally and are helpful for better plant growth. Due to the use of zeolites in agriculture, there is an improvement in fertilizer efficiency, water infiltration and retention, yield, retention of nutrients in plants, long-term soil quality, and reduced nutrient losses (Nakhli et al. 2017). Application of zeolite holds N and K fertilizers in the plant root zone and increases efficient use of fertilizer and higher crop yield. It also improves nutrient and water retention capacity. With successive applications, the zeolite can enhance soil's ability to retain nutrients and ultimately lead to improved yields. Zeolites linked to a nano biosensor can revolutionize agriculture because the biosensor can detect the deficiency in either plant or soil. It can further control the water release and nutrients retention in the zeolite (Rai et al. 2012). Pesticides-based nanoparticles are being developed to improve environmental quality.

A biosensor is an integrated device for applications in various fields. The major components include a sensing matrix incorporating the biomolecule, transducer, and signal processing for further display (Malhotra and Kaushik 2009). An ideal biosensor for widespread applicability should meet specific criteria. For a biosensor, the magnitude of electrode response per unit target concentration should be high. The linearity range of a biosensor should be wide for the detection of target analyte concentration. The biosensor should be highly selective, and the interference from possible chemical species should be minimal. Besides, the biosensor response should not decay with the storage period. Biosensor development involves nanomaterial synthesis, biomolecule selection, processing of the resulting signal, software development, and biosensing device fabrication (Mutlu 2016). Biosensors' development is a multidisciplinary area involving contributions from electrochemistry, optics, biochemistry, and electronic engineering. Various types of biosensors based on transduction mechanisms are shown in Figure 16.4.

The generation of ions due to different biochemical processes resulting in changes in the analyte sample's electrical properties is the underlying principle of electrochemical biosensors (Kundu et al. 2019a). With the difference in the analyte concentration, the magnitude of the resulting electrical signal also changes proportionately. The electrochemical biosensors are non-destructive tools as they provide sensing results without damaging the original system or sample. Electrochemical biosensing studies are conducted in electrochemical cells. Among

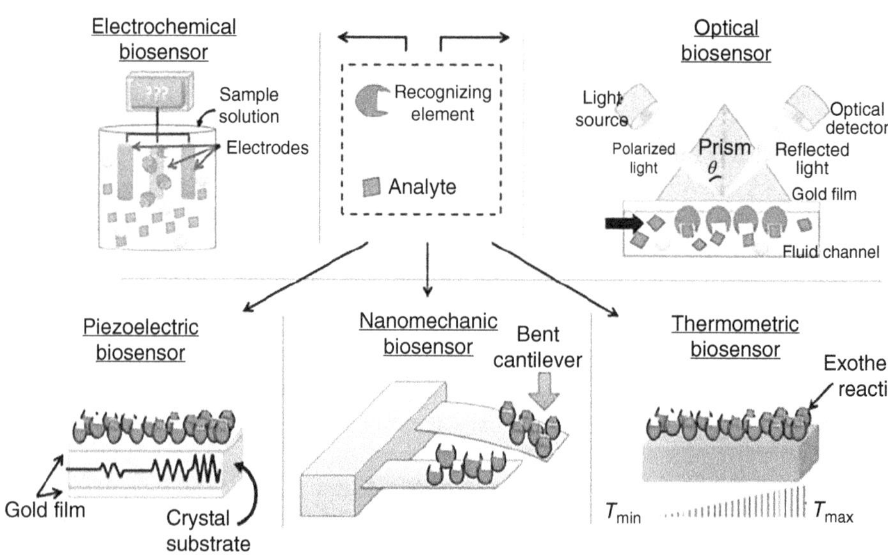

Figure 16.4 Different types of biosensors for application in agriculture and environment. *Source:* Colomer-Farrarons et al. (2011).

various electrochemical biosensors, potentiometric, amperometric, and conductometric are commonly used techniques. Another type of biosensors based on optical characteristics such as absorbance, scattering, fluorescence, etc., are called optical biosensors (Koyun et al. 2012). Due to the chemical reaction taking place in the analyte solution, a change in optical property is measured. The optical biosensors transduction section primarily consists of enzymes or antibodies. The biosensors based on the principle of sound waves are called acoustic biosensors or piezoelectric biosensors. Based on the piezoelectric effect, an electrical signal is produced due to mechanical force on the material. Quartz crystal microbalance (QCM) is based on the piezoelectric biosensor principle (Grumezescu 2016). The role of biosensors in the agriculture and food industry is a novel area of research.

Agriculture is a complex system involving many processes, methods, and techniques involved from farm to fork. Each specific stage is significant and influences the next one. The assessment of various parameters during these stages or processes is a very challenging task. The introduction of biosensors in agriculture has opened new avenues for exploring the hidden causes and establishing the correlations among various factors. The biosensors in agriculture are now utilized for the in-field assessment of crop health regarding measurements of nutrients, phytohormones, pollutants, and pathogens. In livestock rearing, wearable biosensors are now an integrated part of modern dairy management (Scott 1998). In the diversified agricultural fields, some problems and challenges need immediate attention and technology intervention for providing rapid, low-cost and on-the-spot testing solutions. Biosensors have a role in every sector of agriculture (Table 16.2).

Table 16.2 Nanobiosensors intervention in the agri-food sector.

Analyte	Sample	Nanobiosensor	Biosensing parameters	Electrode	References
Mercury, Cadmium, Copper	Water	Transgenic Zebrafish biosensor	Detection limit: 0.1 mg/L(for Hg), 0.5 mg/L (for Cd and Cu)	Transgenic Zebra fish (*Daniorerio*)	Pawar et al. (2016)
Cadmium	Milk	Whole-cell biosensor	Detection limit: 1.0 μg/L, shelf life stability: 65 days, Relative standard deviation <2%	Carbon paste electrode	Verma et al. (2011)
Arsenic	Water	Electrochemical biosensor	Sensitivity: 4.91 μA/ppb Detection limit: 8×10^{-4} ppb, Recovery rates: 95.7%–103.6%	The Fe_3O_4-RTIL composite modified electrode	Gao et al. (2013)
Lead	Milk	The colorimetric and potentiometric biosensor	Detection limit: 38.6 μm Shelf life stability >2 months, Relative standard deviation: 111%–119%	Nylon membrane	Kaur et al. (2014)
Lead and Mercury	Water	Amperometric biosensor	Sensitivity: 89.2 nA/μM (for Pb^{2+} ions) and 94.1 nA μM^{-1}(for Hg^{2+} ions	Pt/CeO_2 coated electrode	Gumpu et al. (2017)

Source: Kundu et al. (2019a)/with permission of Elsevier.

16.4.1 Nanobiosensors for Crop Improvement and Plant Protection

The Green Revolution led to increased wheat and rice yield, especially in developing countries, by improved varieties and exhaustive use of fertilizers and other agricultural chemical inputs. However, owing to the discovery of new high-yield varieties, the green revolution has resulted in many fewer varieties of crops being grown. In the green revolution, crop growth efficiency, longevity, and appearance are valued more than nutritional value. Due to this, many of the ancestral varieties of crops are extinct and not grown by farmers. Among the various agricultural techniques, the development of chemical pesticides and herbicides led to further improvements in crop yields by efficient weed control and eradication of insect pest infestation (Khan and Rizvi 2017).

Agricultural chemicals and pesticides are being used extensively to fight against pest infestations, increase nutrient intake efficiency, increase crop health, and increase production. Due to the dependence on agrochemicals, there are incredible gains in cereal production globally. But simultaneously, the fertilizers and pesticide's blind usage has caused severe health hazards and environmental problems. The pesticides and the residues are found to be very harmful and enter the food chain and water cycle (Kundu et al. 2019a). Water is also getting polluted by major chemical pollutants (fertilizer, pesticide, and herbicide runoff). Due to the after-effects of many earlier cancer-causing pesticides like 2, 4, 5-T, and DDT, their further use is nowadays discouraged, although the residual effects are still persistent. The day-to-day reported cases of human and animal food poisonings and deaths are the costs being paid for blind usage of agricultural chemicals (Pimentel 1996).

According to World Health Organization estimates, there are around 3 million pesticide poisoning cases each year and up to 220 thousand deaths, primarily in developing countries. During plants' growth, the chemical is used to protect the plant against the harmful impact of pests, pathogens, and weeds. Pesticides are used for horticultural crops (fruits, vegetables), cereals (wheat, rice), oilseeds (olives, canola), and others (cotton, grass, and flowers). However, since these pesticides act through inhibition acetylcholine esterase (AChE), they are lethal to human health (Malvano et al. 2017). Rapid detection of pesticides is more and more necessary due to their toxicity, persistence, and stability. Some pesticides have long-term residual effects. They are Chlordecone, DDT, Lindane, and Organophosphates. Chlordecone causes abnormal sperm and improper mobility of sperms. Lindane is caused by breast and fertility problems. DDT is linked to cancer and male infertility after it was shown to block male hormones' action (Nedellec et al. 2016).

The organophosphate pesticides malathion and chlorpyrifos are commonly used on all fruits, vegetables, and wheat. Pesticide detection at the levels established by the Environmental Protection Agency (EPA) remains a challenge. Various analytical methods like gas chromatography-mass spectrometry (GC–MS) and liquid chromatography-mass spectrometry (LC-MS), capillary electrophoresis (CE), flow injection immune analysis, and fluorimetry have been developed for pesticide residues detection (Kundu et al. 2019a). During the 1940s and 1950s, pesticides were used intensively to increase crop productivity. During that period, only thin-layer chromatography was the prevalent method of pesticide residue analysis. In the next decade, with commercial instruments and quantitative detectors' development advancements, many refined gas chromatography versions were commercialized. The 1970s brought reformatory improvements in separation powers, and HPLC came into the market.

The existing chromatographic techniques are advantageous due to being highly sensitive, reliable, and efficient. These are considered to be standardized techniques to compare any pesticide residue analysis. However, these are time-consuming, laborious, expensive, complicated, and bulky and require skilled persons to operate. But gas chromatography (GC) technique provides unsatisfactory results for ionic liquid samples. As GC separations are carried out at higher temperatures, a highly sophisticated instrument is required. Gas chromatography separations are mainly carried out on compounds ranging in molecular weights up to a few hundred. These limitations laid the way to search for better analytical techniques providing more accuracy and sensitivity. With the tremendous increase in applications of biodegradable pesticides, HPLC has been the popular analytical technique. It gives room temperature analysis involving two interacting phases. Thus, it is comparatively quick and efficient than the previous methods. HPLC is extremely precise for identifying and quantifying chemical components. This technique does have low sensitivity for certain compounds, and some cannot be detected. Moreover, it is more complex and costlier than other methods.

The ever-increasing population and simultaneously decreasing farming acreage have enhanced the agrochemicals usage in modern agricultural practice to improve yield. However, there is regulatory legislation for these agrochemical usages for the restrictions and ban on blindly using them (Handford et al. 2015). These conventional techniques for trace-level detection of these agrochemicals are not suitable, as they are costly and labor-intensive. Biosensor intervention in all the fields has now opened new avenues of application in the agriculture and environment sector. Recently biosensor development and instrumentation for the environment have been extensively studied by researchers all over the world. Biosensors for cereal crops such as corn have been developed to detect various quality attributes (Kundu et al. 2021). This chapter includes nanobiosensors and their applications in the field of agriculture and the environment. Among the various environmental components, plant pesticides and their detection using nanobiosensors are essential parts of the chapter.

16.4.2 Biosensors for Horticultural Food Quality and Safety Evaluation

Foods are materials that are raw, processed, formulated, or consumed by humans or animals for growth, health, general well-being, satisfaction, contentment, and quenching social needs. Food preservation is another desired area level of properties or nature for their maximum benefits. In general, each step of handling produce, processing of food, storage, and then distribution affects quality attributes, desirable or undesirable (Monika et al. 2016b). Biosensor devices are evolving as the foremost relevant diagnostic techniques for various applications such as food, clinical, and environmental monitoring, as they possess specificity, ease of mass fabrication, sensitivity, and field portability. Nano biosensing techniques and methods are now widespread in terms of applications such as:

- **Vitamin Analysis**: Various biosensors can be developed to monitor protein interactions with the vitamins on a sensor chip.
- **Antibiotics Detection**: Many of the medicines in the form of antibiotics leave their residual effect in animals. The presence of such antibiotics can be detected in food. Biosensors can analyze the presence of such antibiotics reliably, precisely, and in a short time.

- **Detection of Food Spoilage**: Biosensors based on amperometric technique can be developed for the determining and monitoring of the spoilage indicators such as histamine levels in food, and many other biosensors based on potentiometric and electrochemical processes could analyze additional spoilage-indicating compounds in different types of foods using selective electrodes and enzymes.
- **Detection of Microbial Contamination**: These days, immunobiosensors can detect microbial contamination in food caused by different sources such as growing, harvesting, storage, transportation, etc., in the whole food supply chain. Such biosensors are based on specific antibodies' surface immobilization onto selective electrodes based on a microorganism to be detected.

In today's world, people are exposed to many hazardous materials emanating from anthropogenic and natural sources. The environmental components (the air, water, and soil) and various products can be unknowingly entered into the body through various means (skin contact, inhalation, or ingestion). These toxic chemicals that enter into our body through the food chain can cause severe acute and chronic health effects. Since soil health is essential for nutrient recycling, the soil's chemical contamination should be continuously monitored (Prasad et al. 2021b). Alertness towards food quality has drawn focus due to the continually arising need for sustainable food availability for the estimated high global population presently and in coming times in the next few years. This global amplifying case poses noticeable dares to the agri-food sector.

The intensive agriculture and animal farming industries related to food handling, processing, and distribution play an important role in food quality and safety. Due to ignorance of quality attributes and increasing production, human health is severely affected (Ahmed et al. 2015). Innovation and development in the agriculture sector and the recent globalization of agro-industrial markets direct to a prior conviction for food safety and quality that is of utmost concern for human and environmental health. In order to combat this, various efforts have been steadfast to guarantee food quality. It has multiple parameters (such as appearance, taste, smell, nutritional content, ingredients, freshness, flavor, texture, and chemicals). The examination of food composition allows us to characterize food, and based on this, the quality of food can be adjudged if it contains desired constituents, natural components, and additives.

Furthermore, food composition assessment enables a broad estimation of freshness, revealing the presence of microorganisms' populations and various toxins produced due to damage or spoilage of food. The food-safety concept is beneficial in the production and the commercialization of food and does not embody a consumer's risk. Testing and quality evaluation of food help judge the presence of various unwanted harmful chemicals and components (allergens, pesticides, fertilizers, heavy metals, chemical compounds, pathogens, and toxins) (Kundu et al. 2019b). If present in food, these contaminants could be fatal for human health and general well-being, giving rise to the spread of foodborne diseases and severe consequences for the health-care system and loss to a country's economic productivity. Thus, there is a great demand to identify and establish an ensemble of rules and regulations, inspections, control systems, etc., to minimize dangers that may cause unhygienic or unacceptable-quality end products. The agricultural diagnostics industry is crucial to developing analytical methodologies and technologies equipped with high sensitivity, speed, and portability. Biosensors present an auspicious performance (speed, sensitivity, and fast response time) with enhanced technical knowledge and capacity, such as modularity, miniaturization, and automation. The international market of food laboratory analysis needs reliable, cheap, and fast-responsive methods for determining food quality and safety.

Biosensors propose the opportunity to satisfy this demand. They are ideal devices for improving and pioneering food diagnostics in quality control and evaluating genetically modified foods, authenticity and traceability, organicity, freshness, and residual chemicals and contaminants. Worldwide, various biosensors have been created and assembled for abundant target compounds of agrifood and environmental, from food (raw and processed), soil, plants, and water pollutants. Biosensors thus could be considered a leading technology and pose a potential alternative to conventional slow and time-consuming analytical methods. Biosensors have the potential for real-time monitoring, online measurements, and high-performance analysis along the production chain. In addition, there are various bioremediation strategies adopted for maintaining good soil and plant health (Prasad et al. 2021a).

Moreover, taking advantage of nanotechnology and developments in materials science, new opportunities for recuperating current technology have been explored. The integration of biological recognition techniques, optical and electrochemical methods, and nanomaterials provides novel innovations in biosensors' structures and techniques by improving the sensitivity, robustness, and performance of current practices. Bioinformatics and integration with synthetic biology have paved the way to transform biological systems and engineer them to perform novel tasks. Thus, the significant factors that will revolutionize the food industry due to advantages with biosensors,

- **Extreme Miniaturization not Necessary**: The biosensor's size depends on the fabrication process and signals also. Since the sample required is significantly less, the system developed is small, and extreme miniaturization is unnecessary.
- **Destructive Sampling Techniques are Acceptable**: Biosensor facilitates the use of techniques such as flow-injection analysis, as destructive sampling is also acceptable.
- **Enormous Work Done in the Medical Diagnostics Industry**: Since the medical diagnostic industry has a very well-developed biosensor technique, some of the fundamental techniques learned in this field can be applied directly to food industry requirements.
- **An instant or Continuous Measurement**: Biosensors are very useful for real-time monitoring and data generation in agriculture fields under constant study.

Fermentation is an area in food production that appears more elaborative and receptive for biosensors than other introduced new technologies. There is a vast range of fermented food products in the market, and the sophisticated equipment available for them is also extensive. Such systems have relied on many ways of detection. Biosensors have been widely deployed based on detection techniques such as amperometry, potentiometry, optical methods, and thermal.

16.4.3 Biosensors for Biochemical Analysis and Plant Physiology

There are some technical and social aspects to be taken care of before actual biosensor commercialization. First of all, the samples in daily life are very complex in composition and may vary in terms of their constituents. Thus, the possibility of lots of interferents co-exist. Before the commercialization of biosensors, there is a need for reliability testing of these devices, and also validation studies should be done with conventional analytical methods and techniques. An intricate matrix can affect biosensors' usage in actual samples and induce deviation from theoretically detected parameters. Pretreatment should be made as easy, rapid, and straightforward as possible on the factual matrix.

For this reason, the target motto in the development of a biosensor is linked to their application to real-life environmental samples. However, despite the significant number of newly developed biosensing devices, most of the literature references report applications of the biosensor in buffer solutions. Agricultural produce quality involves nutritional and organoleptic characteristics important for consumers, such as freshness, appearance, taste, and texture. In most of the food products that are stored due to perishability, some chemicals and compounds are produced that produce aroma, and abnormal flavors due to microbial growth and insufficient food safety need to be detected for the safety of the consumers (Kundu et al. 2017; Kundu et al. 2018). Food freshness is badly affected by storage time, improper packaging design, inadequate temperature, and oxygen management during the handling of fruits and vegetables in modified atmosphere storage conditions. Besides, among various other biosensors' uses are for monitoring various alcohols during controlled-atmosphere storage of fruits, rot development in stored vegetables, etc.

Moreover, microbial contamination increases the biogenic amine content of fruits and vegetables. Electrochemical biosensors for determining the amine content in fruits have been assembled using diamine oxidase (DAO) and polyamine oxidase (PAO) covalently immobilized onto polymeric membranes. Tempering packaged food, package design defect, or temperature fluctuations are the crucial causative agents for the deterioration of perishable food commodities such as fruits and vegetables. This is due to injurious oxygen levels, loss of quality, and ultimately the final product breakdown. Deficient package oxygen also may promote the growth of dangerous pathogens. The detection of ethanol would provide a sensitive technique for low-oxygen injury identification. The concentration of lactic acid is an essential parameter for the meat industry as it characterizes the state of fresh meat. Lactic acid is caused by anaerobic glycolysis from glycogen post mortem in muscles. The lactic acid concentration leads to conclusions concerning the pre-mortem metabolic situation, physical stress, and deficiency in the meat quality. The day-by-day increasing demand for online quality evaluation of milk focuses our research efforts and industry attention on practical solutions. The biosensor's characteristics recommended its use for the lactose measurement with an online detection facility in the milking parlor. Some biosensors can also be used for fat determination in milk.

16.5 Nanobiosensors for Natural Resource Management and Environmental Sustainability

The environment is vital for the sustainable development of humans. In this regard, concerns are continuously raised globally for the conservation of the environment. The need for conservation of the environment is rising as lots of harmful gases, emissions, and chemical vapors are being added to our environment carelessly in industrial growth and development. Thus, there is a need for rapid, sensitive, and economical technologies to accurately detect such harmful chemicals from time to time check over the causative agents. In this regard, the U.S. Environmental Protection Agency (USEPA) stringent regulation for the control of the release and levels of certain chemicals has created a need for reliable monitoring of these substances. All over the world, different agencies have established some limits for the levels of these harmful chemicals in the environment. In Europe, the European Parliament's regulation and the council in plant and animal origin products define a new fully harmonized set of rules for pesticide residues in the food and products proposed for children and young people. For instance, EU Baby Food Directives fixes at a low level that of the maximum acceptable daily intake of various pesticide residues in cereal-based processed foods.

The herbicides also represent the most common pollutants found in surface and ground waters. Permissible concentrations are 0.1 μg/L, for a single pollutant and 0.5 μg/L, for the total amount of pesticides (Chau et al. 2015). Pesticides are causing over three million poisoning death cases annually. Exposure to such pesticides or agricultural chemicals leads to several health problems ranging from breathing problems such as asthma attacks, skin allergies, and chronic ailments like cancer, sterility in male, female infertility, and neurological diseases (Parkinson's disease). The surge in the number of human sterility cases depends on the wide use of pesticides and pesticides in food. In view of the importance of controlling the environmental concentration of pesticides, even as extremely low concentration residues, more sensitive, stable, and versatile analytical tools have become urgent. In this context, biosensors can represent a concrete solution for analyzing real samples.

16.5.1 Biosensors in Hazard Control and Analysis

The quality assessment sector is a significant part of the agriculture and food processing industries. Biosensors are playing a very crucial role in a quality management system. The quality itself consists of lots of attributes, such as quality control, quality assurance, and a continuous improvement system. In all the quality management processes of agri-food products, hazard analysis, and critical control points (HACCP) are considered the most effective strategy to guarantee food safety.

Biosensors' implications in food safety facilitate fulfilling international standards of quality and safety, permitting the effective, efficient, safe, and reliable detection and quantification of pathogens involved in foodborne diseases and chemical contaminants that pose threats to consumers' health (Pisoschi et al. 2011). Still, there is a need to detect trace concentrations of chemical and biological polluting chemicals and materials in food products for humans and animals. Biosensors' transduction and reception systems determine the selectivity, specificity, precision, and rapid as they work based on recording reactions generated by physical, chemical, or immunological changes in the substance under study.

Along with the ease of handling, biosensors provide on-time detection and online process control. The complex metabolism of pathogenic bacteria and the guest organisms' defense reactions are used to build biosensors. Apart from this, antigen–antibody interactions are used because of efficiency and selectivity. The detection speed further depends on the transduction system's sensitivity and the signal amplification method (Table 16.3).

Table 16.3 Various types of a biosensor for environmental applications.

Analyte	Biosensor	Limit of detection/ stability/sensitivity	References
Urease and alkaline phosphatase enzyme	Cantilever	Detection limit: parts per trillion	Rego et al. (2020a)
Tyrosinase enzyme	Cantilever	Sensitivity: 0.49 and 0.101 v/μg; Detection limit: 0.1and 0.4 ng/L; Stability: 30 days	de Cezaro et al. (2020)
Aptamers, singled stranded DNA or RNA	Aptamer based biosensors	Selectivity: 1015–1016 sequences	Zhang et al. (2018)
Phosphatase alkaline, SAM technique	Cantilever	Stability: 15 days	Rigo et al. (2020b)
Gama-Al_2O_3 Nanorods, Graphene oxide (GO) nanosheets coupled with DNA/sulfide fluorophore (SF)	Gama Al_2O_3 -GO/ DNA/SF	Sensitivity: 75–2.5 μM	Samak et al. (2020)
Quantum dots	Fluorescent Hybrid biosensor	Detection limit:5–8 μM	Hajipour et al. (2021)
Acetylcholinesterase enzyme	Liposome based nanobiosensor	Stability(concentration): 10–10 M	Vamvakaki and Chaniotakis, (2007)
CdTe quantum dots	Bienzymes of acetylcholinesterase and oxidase biosensor	Detection limit: 10–12—10–6 M	Zheng et al. (2011)
Cholinesterase and sol-gel entrapped bromcresol purple	Fiber optic biosensor	Lifetime: 3 weeks Detection limit:108–5.2 μg/L	Andreou and Clonis (2002)
Indolyl acetate	Piezoelectric biosensor	Detection limit:5.0×10^{-8} and 1.0×10^{-7}	Abad et al. (1998)
Cytochrome P450 reductase and NADPH	Microsomal cytochrome P450-3A4 (CYP3A4) biosensor	Detection limit: −0.043 μg/L−1	Hendricks et al. (2009)

16.6 Future Prospects and Challenges

Biosensors are now familiar in every corner of the science field. There is no single aspect left where this novel field cannot contribute. Despite widespread applications and benefits, the commercialization of biosensor technology is still lagging behind. In agriculture, this technique has produced a proven significant role. But its application and actual usage are limited to laboratory scale only. This has to be popularized among the end-users, i.e., farmers and the general public. Although many biosensor-based patents are filed each year, very few play a prominent role in the agrifood industry, environmental agricultural, or veterinary applications. Some of the reasons for the slow technology transfer from the research laboratories to the marketplace include cost factors, shelf life or stability, sensitivity, lack of knowledge, etc. Many of the leading barriers include sensor calibration, fabrication, technology producing economical and highly reliable sensors, sensor stabilization and life, and total sensor system integration. Conceptually, the mechanism of biomolecule's adsorption and immobilization on surfaces during the fabrication and handling of biosensors needs to be better explored. There has been noteworthy progress in surface chemical analysis and scanning force electron microscopy (AFM) and scanning tunneling microscopy (STM) to characterize nonmaterial-based thin films.

Moreover, the proteins (enzymes, antibodies) used as essential components in biosensors have limited stability, thus hindering the development process. In various biosensor applications, the early detection of diseases in animals and plants will be the top priority of the areas highlighted for great attention to future challenges. The development of DNA probes using biosensors is a promising area now. Deoxyribonucleic acid sensors will be very useful to detect polymorphism or mutations in genes within flora and fauna. Another revolutionary system in biosensor development is biomimetic systems. Some researchers are trying to remove biological molecules' inefficiency or poor performance by replacing them with artificial molecular recognition systems with predetermined selectivity for target analytes. So far, molecularly imprinted polymers have been prepared with affinities for proteins, amino acid derivatives, sugars, vitamins, pesticides, and pharmaceuticals. Molecular imprinting is advantageous over biological receptors with low cost, ease of synthesis, and relative long-term stability and life. To effectively transfer technology to the market users, it has to be produced on a mass scale. The development of automated manufacturing technologies is significant in the commercial mass production of biosensors.

Deposition techniques such as screen-printing and ink-jet printing allow printing materials at very high precision and speed, thus producing large numbers of inexpensive and reproducible biosensors. In the fabrication of biosensors, various thin film deposition techniques are proving very helpful. Some of the leading thin-film deposition technology can create layers even less than 1 mm thick, suitable for micro biosensors' development. Electronic advancements with silicon fabrication technology have accelerated lab-on-chip device developments. The idea of lab-on-a-chip is brilliant and effective.

The fabrication of card-sized micro laboratories relies on advanced microfabrication, and micromachining technologies are now technically feasible. They can be miniaturized and integrate complex components such as pumps, valves, mixers, and flow cells in a single chip along with a biosensor. They were commercially accelerated by intelligent instrumentation, electronics, and multivariate signal processing methods such as chemometrics and artificial neural networks. The worldwide prime focus is being given to the design of both the necessary components and the device as a whole. In this sector, agricultural engineers have a significant role to play. Their expertise can be harnessed to improve the calibration, system analysis, and development of devices for farmers, consumers, and processors. Further nowadays multi-analyte biosensor technology is gaining popularity due to its adaptation to one-time multiple analyte detection, thus reducing overall costs of biosensor fabrication. Various future improvements in the design and development part will produce more competitive and rugged devices that will operate under even severe field conditions.

16.7 Conclusions

In recent times, nanoscience and nanotechnology for biosensor development have been growing areas of research. Integration of the techniques based on optical, mechanical, electrical, and electronics is essential for real-time monitoring of bio-recognition processes occurring on the surface of solids and in the solution phase. It is a niche research area where dedicated efforts need to be put into developing a single biosensor unit for multi-analyte detection in laboratory and field applications. In the future, advances in biosensor technology will be governed by embedded systems, microfluidics, telecommunication capability, and wearable facility, etc. There is an urgent need to develop biosensors for the farmers and the end-users to use in real-time and on-the-spot testing. Sensors available in the market are inadequate to tackle the present and future challenges in agriculture. Significant problems that arise during biosensors' development are (i) limited stability of immobilized biomolecules and (ii) interference-free transduction. A lot of scopes still exist for improvements in detection limit, linearity, sensitivity, specificity, design of electrode, and matrix fabrication in biosensing. In agriculture, there is emerging demand to use nanobiosensors as offline, online, or in vivo devices. As the agricultural samples are complex, nanobiosensors are prone to interferences from the biological and chemical components. These are mainly due to the electrochemical disturbances created by molecule metabolites, proteins, macromolecules, etc. Present time demands urgent research focus towards designing suitable strategies for enhancing the analytical performance of nanobiosensors for agricultural applications. In future researches, the need to develop smart portable gadgets based on nanobiosensing techniques flood the market if electronics, instrumentation, and electrochemistry together contribute collaboratively in this direction. The stability of the nanobiosensor is now dependent on the buffer, pH, ionic concentration, and biomolecule. The immobilization techniques, such as covalent bonding, used crosslinkers to efficiently bind the biomolecule, resulting in enhanced shelf life and the biosensor's robustness. More explorative studies in search of more biomolecule binding mechanisms open new avenues of research for improving nano biosensing performance. A remarkable achievement in nanobiosensors will be recorded in the future due to the wide-scale implementation of microfluidic technology. The novel technological intervention has the advantage of less sample and reagents, improves sensitivity, and provides a common platform for both sampling and detection. The present nanobiosensors integrated with microfluidics have progressively steered the research and exhibit vast potential in the future for areas such as field portability, real-time, and on-the-spot detection. Further in this direction, the introduction of micro-channels and novel techniques of miniaturizations are revolutionizing the field.

References

Abad, J.M., Pariente, F., Hernandez, L. et al. (1998). Determination of organophosphorus and carbamate pesticides using a piezoelectric biosensor. *Analytical Chemistry* 70 (14): 2848–2855.

Ahmed, K.M.F., Hafez, R.S., Morgan, S.D., and Awad, A.A. (2015). Detection of some chemical hazards in milk and some dairy products. *African Journal of Food Science* 9 (4): 187–193.

Andreou, V.G. and Clonis, Y.D. (2002). A portable fiber-optic pesticide biosensor based on immobilized cholinesterase and sol-gel entrapped bromcresol purple for in-field use. *Biosensors and Bioelectronics* 17 (1–2): 61–69.

Bhardwaj, H., Rajesh, & Sumana, G. (2021). Recent advances in nanomaterials integrated immunosensors for food toxin detection. *Journal of Food Science and Technology*, 59 (1), 12–33. https://doi.org/10.1007/s13197-021-04999-5

de Cezaro, A.M., Rigo, A.A., Martinazzo, J. et al. (2020). Cantilever nanobiosensor functionalized with tyrosinase for detection of estrone and B-estradiol in water. *Applied Biochemistry and Biotechnology* 190 (4): 1512–1524.

Chau, N.D.G., Sebesvari, Z., Amelung, W., and Renaud, F.G. (2015). Pesticide pollution of multiple drinking water sources in the Mekong Delta, Vietnam: evidence from two provinces. *Environmental Science and Pollution Research 22* (12): 9042–9058.

Colomer-Farrarons, J., Miribel-Catala, P.L., Rodriguez-Villarreal, A.I.N., and Samitier, J. (2011). Portable bio-devices: design of electrochemical instruments from miniaturized to implantable devices. In: *Serra P.A. (ed) New Perspectives in Biosensors Technology and Applications.* IntechOpen pp. 373–400.

Estela, J.M. and Cerda, V. (2005). Flow analysis techniques for phosphorus: an overview. *Talanta* 66 (2): 307–331.

FAOSTAT, F. (2017). Avalaible at http://www.fao.org/faostat/en/data.

Gao, C., Yu, X.-Y., Xiong, S.-Q. et al. (2013). Electrochemical detection of arsenic (III) completely free from noble metal: Fe_3O_4 microspheres-room temperature ionic liquid composite showing better performance than gold. *Analytical Chemistry* 85 (5): 2673–2680.

Grumezescu, A. (2016). *Nanobiosensors.* Academic Press.

Gumpu, M.B., Krishnan, U.M., and Rayappan, J.B.B. (2017). Design and development of amperometric biosensor for the detection of lead and mercury ions in water matrix–a permeability approach. *Analytical and Bioanalytical Chemistry* 409 (17): 4257–4266.

Hajipour, F., Asad, S., Amoozegar, M.A. et al. (2021). Developing a fluorescent hybrid nanobiosensor based on quantum dots and azoreductase enzyme for methyl red monitoring. *Iranian Biomedical Journal* 25 (1): 8.

Handford, C.E., Elliott, C.T., and Campbell, K. (2015). A review of the global pesticide legislation and the scale of challenge in reaching the global harmonization of food safety standards. *Integrated Environmental Assessment and Management 11* (4): 525–536.

Hendricks, N.R., Waryo, T.T., Arotiba, O. et al. (2009). Microsomal cytochrome P450-3A4 (CYP3A4) nanobiosensor for the determination of 2, 4-dichlorophenol-An endocrine disruptor compound. *Electrochimica Acta* 54 (7): 1925–1931.

Jamshidi, B., Mohajerani, E., and Jamshidi, J. (2016). Developing a Vis/NIR spectroscopic system for fast and non-destructive pesticide residue monitoring in agricultural product. *Measurement* 89: 1–6.

Kahveci, Z., Martinez-Tome, M.J., Mallavia, R., and Mateo, C.R. (2016). Fluorescent biosensor for phosphate determination based on immobilized polyfluorene-liposomal nanoparticles coupled with Alkaline Phosphatase. *ACS Applied Materials & Interfaces* 9: 136.

Kaur, H., Kumar, S., and Verma, N. (2014). Enzyme-based colorimetric and potentiometric biosensor for detecting Pb (II) ions in milk. *Brazilian Archives of Biology and Technology* 57 (4): 613–619.

Khan, M.R. and Rizvi, T.F. (2017). Application of nanofertilizer and nanopesticides for improvements in crop production and protection. In: Ghorbanpour, M., Manika, K., Varma, A. (eds) *Nanoscience and Plant–Soil Systems*, 405–427. Cham: Springer.

Khoobdel, M., Ahsaei, S.M., and Farzaneh, M. (2017). Insecticidal activity of polycaprolactone nanocapsules loaded with Rosmarinus officinalis essential oil in Tribolium castaneum (Herbst). *Entomological Research* 47: 175.

Koyun, A., Ahlatcioglu, E., and Ipek, Y.K. (2012). Biosensors and their principles. In: *A Roadmap of Biomedical Engineers and Milestones*. IntechOpen.

Kundu, M., Narsaiah, K., Muzaddadi, A.U. et al. (2017). Study of electrical impedance parameters of potassium chloride solution for potential applications in food and agriculture processing. *Journal of Agricultural Physics* 17 (2): 255–260.

Kundu, M., Bansal, S., and Krishnan, P. (2018). Assessment of maturity of mango (Mangifera indices L.) based on electrical and physiochemical properties during ambient storage. *Journal of Agricultural Physics* 18 (1): 107–113.

Kundu, M., Krishnan, P., Kotnala, R.K., and Sumana, G. (2019a). Recent developments in biosensors to combat agricultural challenges and their future prospects. *Trends in Food Science and Technology* 88: 157.

Kundu, M., Bhardwaj, H., Pandey, M.K. et al. (2019b). Development of electrochemical biosensor based on CNT-Fe_3O_4 nanocomposite to determine formaldehyde adulteration in orange juice. *Journal of Food Science and Technology* 56 (4): 1829–1840.

Kundu, M., Pramanik, P., Maity, A. et al. (2019c). Engineered nanomaterials: classification and strategies for physicochemical characterization and advanced analytical techniques for the measurement of nanomaterials in plant samples. In: *Advances in Phytonanotechnology*, 17–43. Elsevier.

Kundu, M., Krishnan, P., and Gajjala, S. (2021). Comparative studies of screen-printed electrode based electrochemical biosensor with the optical biosensor for formaldehyde detection in corn. *Food and Bioprocess Technology* 14: 1–13.

Kuswandi, B., Wicaksono, Y., Abdullah, A. et al. (2011). Smart packaging: sensors for monitoring of food quality and safety. *Sensing and Instrumentation for Food Quality and Safety* 5 (3–4): 137–146.

Liu, B., Zhou, P., Liu, X. et al. (2013). Detection of pesticides in fruits by surface-enhanced Raman spectroscopy coupled with gold nanostructures. *Food and Bioprocess Technology* 6 (3): 710–718.

Malhotra, B.D. and Kaushik, A. (2009). Metal oxide-chitosan based nanocomposite for cholesterol biosensor. *Thin Solid Films* 518 (2): 614–620.

Malvano, F., Albanese, D., Pilloton, R. et al. (2017). A new label-free impedimetric affinity sensor based on cholinesterases for detection of organophosphorous and carbamic pesticides in food samples: impedimetric versus amperometric detection. *Food and Bioprocess Technology* 10 (10): 1834–1843.

Mohsin, M., Naz, R., and Ahmad, A. (2020). *Nanobiosensors for Agricultural, Medical and Environmental Applications*. Springer.

Monika, K., Krishnan, P., and Singh, A.M. (2016a). Analysis of shoot shear strength to determine the lodging tendency in wheat. *International Journal of Tropical Agriculture* 34 (4): 919–924.

Monika, K., Arvind, J., and Sangita, B. (2016b). Analysis of physio-chemical properties of pear cv.' Patharnakh'(Pyrus pyrifolia Burm. F. Nakai) during storage. *International Journal of Tropical Agriculture* 34 (4): 925–930.

Mutlu, M. (2016). *Biosensors in Food Processing, Safety, and Quality Control*. CRC Press.

Nakhli, S.A.A., Delkash, M., Bakhshayesh, B.E., and Kazemian, H. (2017). Application of zeolites for sustainable agriculture: a review on water and nutrient retention. *Water, Air, & Soil Pollution 228* (12): 1–34.

Nedellec, V., Rabl, A., and Dab, W. (2016). Public health and chronic low chlordecone exposure in Guadeloupe, Part 1: hazards, exposure-response functions, and exposures. *Environmental Health* 15 (1): 75.

Norfun, P., Arqueropanyo, O., Liawruangrath, S., and Ounnunkad, K. (2016). Electrochemical flow injection determination of ascorbic acid in fruit samples employing a graphene-polyaniline electrode. *International Journal of Chemical Engineering and Applications* 7 (2): 142.

Oliveira, H.C., Stolf-Moreira, R., Martinez, C.B.R. et al. (2015). Nanoencapsulation enhances the post-emergence herbicidal activity of atrazine against mustard plants. *PLoS One* 10 (7): e0132971.

Pal, K. (2020). *Nanofabrication for Smart Nanosensor Applications*. Elsevier.

Pawar, N., Gireesh-Babu, P., Sivasubbu, S., and Chaudhari, A. (2016). Transgenic zebrafish biosensor for the detection of cadmium and zinc toxicity. *Current Science* 111: 1697–1701.

Pimentel, D. (1996). Green revolution agriculture and chemical hazards. *Science of the Total Environment 188*: S86–S98.

Pisoschi, A.M., Pop, A., Negulescu, G.P., and Pisoschi, A. (2011). Determination of ascorbic acid content of some fruit juices and wine by voltammetry performed at Pt and carbon paste electrodes. *Molecules* 16 (2): 1349–1365.

Prasad, S., Kannojiya, S., Kumar, S. et al. (2021a). Integrative approaches for understanding and designing strategies of bioremediation. In: *Rakshit. A., Parihar, M, Sarkar, B., et al. (eds) Bioremediation Science: From Theory to Practice*, 37.

Prasad, S., Malav, L.C., Choudhary, J. et al. (2021b). Soil microbiomes for healthy nutrient recycling. In: *Current Trends in Microbial Biotechnology for Sustainable Agriculture*, 1–21. Springer.

Rai, V., Acharya, S., and Dey, N. (2012). Implications of nanobiosensors in agriculture. J Biomater Nanobiotchnol 3: 315–324.

Ribeiro, J.M.P. (2014). Development and characterization of novel edible films with antimicrobial properties for food packaging. Universidade do Minho (Portugal).

Rigo, A.A., Cezaro, A.M.D., Muenchen, D.K. et al. (2020a). Heavy metals detection in river water with cantilever nanobiosensor. *Journal of Environmental Science and Health, Part B* 55 (3): 239–249.

Rigo, A.A., de Cezaro, A.M., Martinazzo, J. et al. (2020b). Detection of lead in river water samples applying cantilever nanobiosensor. *Water, Air, & Soil Pollution* 231 (4): 1–11.

Ruttkay-Nedecky, B., Krystofova, O., Nejdl, L., and Adam, V. (2017). Nanoparticles based on essential metals and their phytotoxicity. *Journal of Nanobiotechnology* 15 (1): 33.

Saini, S., Sillard, C., Belgacem, M.N., and Bras, J. (2016). Nisin anchored cellulose nanofibers for long-term antimicrobial active food packaging. *RSC Advances* 6 (15): 12422–12430.

Sajid, M., Kawde, A.N., and Daud, M. (2015). Designs, formats and applications of lateral flow assay: a literature review. *Journal of Saudi Chemical Society* 19 (6): 689–705.

Samak, N.A., Selim, M.S., Hao, Z., and Xing, J. (2020). Controlled-synthesis of alumina-graphene oxide nanocomposite coupled with DNA/sulfide fluorophore for eco-friendly "Turn off/on" H_2S nanobiosensor. *Talanta* 211: 120655.

Sanchez, M.T., Flores-Rojas, K., Guerrero, J.E. et al. (2010). Measurement of pesticide residues in peppers by near-infrared reflectance spectroscopy. *Pest Management Science: Formerly Pesticide Science* 66 (6): 580–586.

Sassolas, A., Prieto-Simon, B., and Marty, J.-L. (2012). Biosensors for pesticide detection: new trends. *American Journal of Analytical Chemistry* 3 (3): 210–232.

Scott, A.O. (1998). *Biosensors for Food Analysis*. Elsevier.

Silva, E., Daam, M.A., and Cerejeira, M.J. (2015). Aquatic risk assessment of priority and other river basin-specific pesticides in surface waters of Mediterranean river basins. *Chemosphere* 135: 394–402.

Syu, Y.-Y., Hung, J.-H., Chen, J.-C., and Chuang, H.-W. (2014). Impacts of size and shape of silver nanoparticles on Arabidopsis plant growth and gene expression. *Plant Physiology and Biochemistry* 83: 57–64.

Terry, L.A., White, S.F., and Tigwell, L.J. (2005). The application of biosensors to fresh produce and the wider food industry. *Journal of Agricultural and Food Chemistry 53* (5): 1309–1316.

Upadhay, N., Kumar, V., and Kumar, V. (2013). Extraction, UV-visible, FTIR, NMR spectroscopic study of Acephate and effect of pH. *International Journal of Environmental Sciences* 3 (6): 1849.

Vamvakaki, V. and Chaniotakis, N.A. (2007). Pesticide detection with a liposome-based nano-biosensor. *Biosensors and Bioelectronics* 22 (12): 2848–2853.

Verma, N., Kaur, H., and Kumar, S. (2011). Whole cell based electrochemical biosensor for monitoring lead ions in milk. *Biotechnology* 10 (3): 259–266.

Wang, D., Ha, Y., Gu, J. et al. (2016). 2D protein supramolecular nanofilm with exceptionally large area and emergent functions. *Advanced Materials* 28 (34): 7414–7423.

Xiao, F. and Pignatello, J.J. (2015). Interactions of triazine herbicides with biochar: steric and electronic effects. *Water Research* 80: 179–188.

Zhang, W., Liu, Q.X., Guo, Z.H., and Lin, J.S. (2018). Practical application of aptamer-based biosensors in detection of low molecular weight pollutants in water sources. *Molecules* 23 (2): 344.

Zheng, Z., Li, X., Dai, Z. et al. (2011). Detection of mixed organophosphorus pesticides in real samples using quantum dots/bi-enzyme assembly multilayers. *Journal of Materials Chemistry* 21 (42): 16955–16962.

17

Nanotechnology and Genetically Modified Organisms for Agricultural Sustainability

Guillermo Fernandez-Bunster[1,2]

[1] *Laboratorio de Biotecnología (BioTecLab), School of Medical Technology, Faculty of Medicine, Universidad de Valparaíso, Campus San Felipe, Camino la Troya S/N & El Convento, Circunvalación Oriente, San Felipe, Chile*
[2] *Centro de Investigación Interdisciplinario en Salud Territorial del Valle de Aconcagua (CIISTe), San Felipe, Chile*

17.1 Introduction

Anthropogenic activity directly changed the need and requirements for our sustenance and living necessities. The apparition of global climate changes has sped up in the last 200 years (Velásquez et al. 2018), increasing dioxide carbon (CO_2) concentration in the atmosphere, global temperature (increasing phytopathogens expansion and aggressiveness (Bebber et al. 2013) ocean levels, and the surge of intense extreme weather events (Fischer and Knutti 2015).

Food and sustainable agriculture are essential to keep the population fed, but numbers are daunting: According to statistics, the world population will be 9–10 billion people in 2050, requiring the increase of 25–70% more food than the current capabilities of food production (Hunter et al. 2017), projecting that global food demand must double by 2050 (Elmer and White 2018). Also, even if we get to that high numbers of production, there is a rough unbalance in how these resources are delivered: We still have 800 million people undernourished and lack essential nutrients, but ironically other two billion are obese and/or have excess weight. An incomplete food source, i.e., lacking or with insufficient micronutrients affects almost three billion of the world's population (Haddad et al. 2015) may result in incomplete physical or cognitive development.

Currently, more than three billion metric tons of crops are produced yearly worldwide, using 187 million metric tons of fertilizer, 4 million tons of pesticides (De et al. 2014), and 2.7 trillion cubic meters of water (Lowry et al. 2019). This demonstrates a deficient utilization of agrochemicals (Raliya et al. 2018), by trying to avoid crop losses by pathogens, which may also generate resistance to pesticides, insecticides, and herbicides, as well may decrease biodiversity and affect the full ecosystem (Kah 2015), producing air pollution, water pollution, and eutrophication and soil quality degradation (Congreves and Van Eerd 2015).

Also, because of climate change, agriculture has been greatly affected by depending on the weather to obtain adequate crop growth and productivity (Fu et al. 2020). But also, changes have promoted the apparition/existence of novel phytopathogens that, according to Kashyap (2017), produce an estimated crop loss of up to 30% per year, influencing directly the yield required numbers for population feeding (Kashyap et al. 2017).

Additionally, the lack of available water and drought cause enormous losses in crops, as well as indirectly contaminating drinking water sources (Dubey and Mailapalli 2016), with the consequent risk in the human population. Currently, crops provide between 7 and 15% of all greenhouse gas emissions (Solanki et al. 2015).

Not only are productivity and crop health taken into consideration, but the system must also obey certain requirements to keep it ongoing: High production in short land space, minimum negative impacts on the environment, as well as reduced emission of greenhouse gases, water, and energy demand, among others, requiring the most up-to-date technologies and advances (Scott et al. 2018).

17.2 Concept of Sustainability and Others

Kettenburg et al. defined sustainability as "*an ideal of human well-being within planetary boundaries across generations*" (Kettenburg et al. 2018). Espinoza-Orias et al. (2014) defined sustainable nutrition as "*The physical and economic access to sufficient, safe and nutritious food and water to fulfill dietary and cultural needs to enable an active and healthy lifestyle ... without compromising the ability of future generations to meet these needs*" (Espinoza-Orias et al. 2014).

According to United Nations, sustainability is based on three pillars that must be considered: Environment, social perception, and economy (Figure 17.1), focusing to give enough and the right nutrients to the population, considering all environmental contexts, from the soil of the crops, via animals and plants, to the food and consumer (Tilman and Clark 2014), linking directly the food nutritional quality to environmental sustainability and economic viability (Herforth et al. 2014).

To work with a unified concept, The United Nations World Commission on Environment and Development, in its 1987 report, defined sustainability as, "*meeting the needs of the*

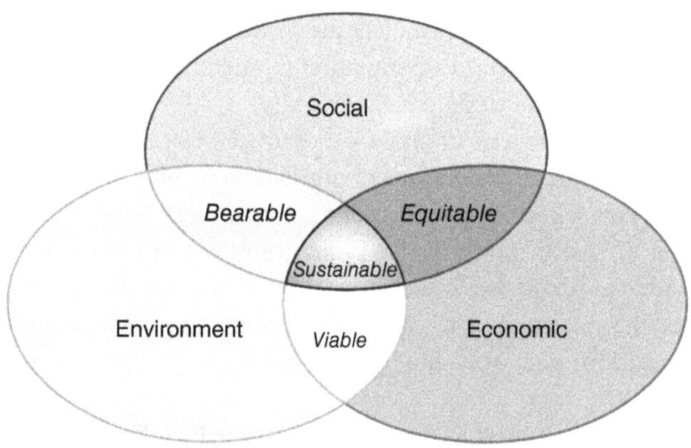

Figure 17.1 The three pillars of sustainability. *Source:* Based on "sustainable development" from http://Wikimedia.org under Creative Commons licensing.

present without compromising the ability of future generations to meet their own needs" (Cassen 1987). In the report, they also defined four goals to define sustainable agriculture (National Research Council 2010):

- "satisfy human food, feed, and fiber needs and contribute to biofuel needs
- enhance environmental quality and the resource base
- sustain the economic viability of agriculture
- enhance the quality of life for farmers, farmworkers, and society as a whole"

Scott (2018) goes beyond the definition, by proposing the concept as a "*process of change in which the direction of investment, the orientation of technology, the allocation of resources, the development and functioning of institutions and advancement of human and community well-being meets present needs and aspirations without compromising the ability of future generations to meet their own needs and aspirations*" (Scott et al. 2018).

These definitions can be summarized as the need to increase efficiency and productivity of crops and the food system, to feed a rapidly increasing population (Willett et al. 2019), by increasing food productivity and reduction of food waste, accompanied by an equitable food distribution, without damaging the environment in the short and long term, in a sustainable way for future generations (McClements 2020). Agrochemicals, such as fertilizers, pesticides, fungicides, antimicrobials, as well as nanomaterials and genetically modified organisms (GMOs), offer several benefits to achieve agricultural sustainability, but also their limitations and environmental consequences must be known (Scott et al. 2018).

This last point is well taken by Kaput (2015), who mentions that the green revolution increased dramatically the production of grain and rice (Kaput et al. 2015), up to feed one billion people, increasing production yields for wheat, rice, potato, and cassava (Herforth et al. 2014), but with grave consequences for the environment: Wastes, soil degradation, over- and misuse of water, and overuse of pesticides and agrochemicals ultimately diminished crop yields, affecting the whole ecosystem (de Fraiture et al. 2010). As can be seen in Figure 17.2, the yield is not the only parameter that should be taken into account for sustainability (Figure 17.2).

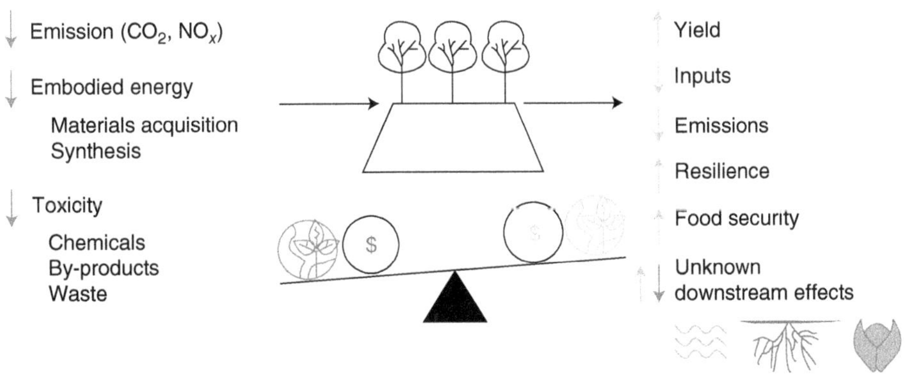

Figure 17.2 System trade-off. Represents the necessity of the necessities and requirements of any novel system to ensure agricultural sustainability. *Source:* Lowry et al. (2019).

Currently, crop agriculture remains mainly invariant which produces a bigger amount of crop in a particular area, namely the high-yield-variant (HYV), but has led to an estimated 75% loss in genetic diversity of agricultural plants between 1,900 and 2,000 (Kaput et al. 2015). Historically, humans used 7 thousand different plants for food, but now only 150 (2%) plants are used for nutrition, in which 12 species are the main energy suppliers (Esquinas-Alcázar 2005). This dependence of just 12 species, besides the lack of nutrient diversity, makes crops very vulnerable to pests, diseases, and extreme weather. Some species adapt more successfully to changes (term named "phenotypic plasticity" to refer to changes in phenotype from just one genotype), being able to grow in different soils, climate, and/or weather, but with a difference in their nutritional content (di Silvestro et al. 2012). As a consequence of using mono-crops, current food from cops may have less nutritional value than before (Kaput et al. 2015).

Green biotechnology must be changed, in a way to regulate it by policies, incentives, and structures that support sustainable agriculture, considering the socioeconomical system, networks, ethics, and societal needs required to keep this type of agriculture ongoing. Because of this, a concerted effort between agriculture practitioners, governments, academia, and stakeholders must be done to understand where to focus, in terms of sustainability (van Hove and Gillund 2017).

17.3 Nanotechnology and Sustainable Agriculture

Nanotechnology is considered a technology with the application of materials and/or particles at a length scale from 1 to 100 nm, corresponding a 1×10^{-9} m (Fincheira et al. 2020). The U.S. Environmental Protection Agency (EPA) considers it a science, and when it is associated with plants, it is named "Phytonanotechnology," which may allow modulating genetic and metabolic changes in plants (Wang et al. 2016).

The first mention of the "nanotechnology" concept was in 1974 by professor Norio Taniguchi from the Tokyo University of Science (Khan and Rizvi 2014), and since then, several advances in the assembly of nanomaterials have allowed to synthesize particles of different sizes and shapes (Gogos et al. 2012).

Agrochemicals are usually applied to crops by spraying or broadcasting, resulting in low efficiency in reaching target sites and/or not obtaining the minimum effective concentration to stimulate plant growth. Also, these chemicals may be light- and/or water degraded, as well as affected by microbial degradation (Nair et al. 2010). Several parameters must be considered, such as bioavailability of nutrients, fertilizer's effect on the soil, evaporation, hydrolysis, and degradation by microorganisms, among others (Gogos et al. 2012).

To enhance efficiency and delivery of the agrochemical, nanotechnology is a key technology that may be used to improve traditional agricultural practices for conservation and waste reduction (Jampílek and KráL'Ová 2015). The multiple uses of nanotechnology (Figure 17.3) offer a real alternative for treatment or management of plant diseases, improving agrichemical efficacy and low ecotoxicity (Elmer and White 2018), and despite the doubts about their potential accumulative toxicity, a correct dose does effectively suppress pathogen infections, deliver micronutrients to plants, and trigger plant immune responses (Servin et al. 2015) and crop growth improvement (Hao et al. 2018). Also, it may control the

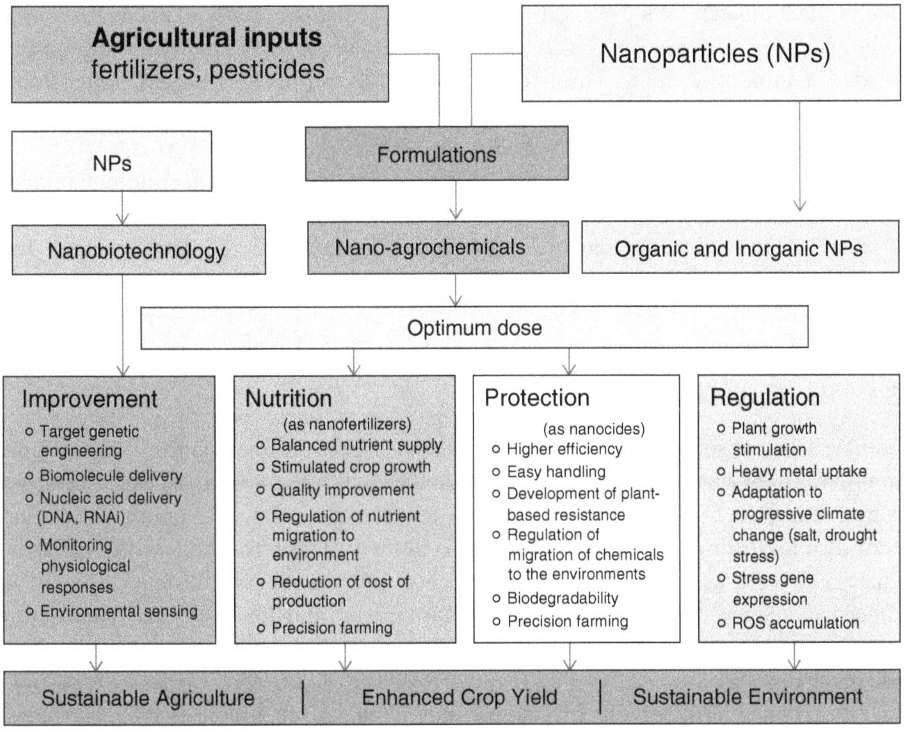

Figure 17.3 Overview of potential application of nanomaterials in sustainable agriculture production. *Source:* Shang et al. (2019).

release of nutrients and monitoring of water quality and pesticides to manage the sustainable progress of agriculture (Prasad et al. 2017). Other uses are the catalysis and degradation of waste and toxic materials, as well as improving how microorganisms can effectively degrade different toxins (namely, bioremediation; Dixit et al. 2015), restoring the natural situation of the soil.

Fu et al. (2020) classify the uses of nanotechnology, according to their use (Fu et al. 2020):

- Nanomaterials to provide protection and to fight against plant pathogens, such as fungi and bacteria
- Nanomaterials as biostimulants, boosting plant innate immunity
- Nanomaterials as carriers for agrochemicals, such as pesticides and fungicides, as well as micronutrients
- Nanosensors, which can facilitate the diagnosis of pathogen epidemics in crops, and also forecast the apparition of phytopathogens.

Also, different uses of nanotechnology have been developed toward sustainable agriculture, by avoiding toxicities and adverse effects in the ecosystem (Pretty 2008) but also looking for the most beneficial outcomes per nanomaterial application. For example, using silicon dioxide enhances crop production by measuring fresh/dry weight in plants and increased chlorophyll content in tomato and squash under high salt stress (Haghighi et al. 2012).

One of the nanotechnology properties is its ductility and the ability to make changes in the nanoparticles or nano vehicles according to their use, such as nanotubes, biosensors, fullerenes, nanofiltration, etc. (Sabir et al. 2014). For example, chemical composition, shape, surface structure and charge, and particle aggregation, among other properties, may be changed depending on the goal (Prasad et al. 2017).

He et al. (2019) didactically summarized the applications of nanotechnology in agriculture, as nanofertilizers, pesticides, herbicides, and sensors, but there must be a consideration of the consequences of nanotech applications, i.e., toxicity and risk assessments, public acceptance and awareness, regulations, and legislation, among others (He et al. 2019).

17.3.1 Nanomaterials

Nanomaterials, per se, can protect the agrochemicals directed to crops from degradation, by improving their stability and their release into the environment, increasing the effectiveness and reduction of agrochemicals, hence their bioaccumulation. Nanomaterials may provide also protection against phytopathogens as an antimicrobial and antifungal agent but also may work as a delivery system for active ingredients and nutrients. Having an enhanced efficiency, fewer concentration requirements and lower toxicity against the plant/environment makes them a very interesting tool to achieve sustainable agriculture goals (Fu et al. 2020).

Encapsulation of fertilizers, such as the use of graphene oxide films to encapsulate potassium nitrate, allows to control and prolong the release of the fertilizer, at a low cost for large-scale production (Zhang et al. 2014). Abdel-Aziz et al. (2016) described that the life cycle of wheat plants with nano fertilizers applied was almost 24% shorter for yield production when compared to plants with fertilizers in conventional applications, accelerating the whole process of plant growth and productivity, allowing for greater production in less time. The researchers recommend this practice on drought-prone soils (and even in flash flood-prone areas), where early crop maturity is a significant aspect of sustainable crop production (Abdel-Aziz et al. 2016).

For example, the use of multiwalled carbon-based nanotubes (MWCNTs) influences positively seed germination in barley, corn, wheat, maize, and even peanuts (Lahiani et al. 2013; Srivastava and Rao 2014). Similar results were obtained for silica and titanium nanoparticles (Lu et al. 2002; Manjaiah et al. 2018). Several nanomaterials, like ZnFeCu-oxide and hydroxyfullerenes, also are associated with increased crop growth and development in peanuts, wheat, onion, and potato, among others (Dubey and Mailapalli 2016; Shojaei et al. 2018).

As shown in Table 17.1, several studies have demonstrated the application of nanomaterials as an effective tool to enhance plant development and growth.

Other positive effects by nanomaterials are shown by Tripathi and Sarkar (2015), who reported an enhanced root and shoot growth after 10 days of exposure to water-soluble carbon nano-dots, compared to controls (Tripathi and Sarkar 2015). But negative effects have been also reported, Kerfahi et al. (2015) informed that, after two weeks, the use of multiwalled carbon nanotubes affected the soil bacterial community. After eight weeks, no changes were detected in the bacterial diversity in comparison to the control, stating that

Table 17.1 Examples of uses of nanomaterials in plants.

Nanomaterials	Crop species	Common name	Mode of application	Duration of treatments	Responses	Source
MWCNTs	*Hordeum vulgare, Zea mays*	Barley, maize	Seed priming	24 h	Enhanced germination and growth of seedlings	(Shang et al. 2019)
MWCNTs	*Triticum aestivum, Z. mays, Arachis hypogea, Allium sativum*	Bread wheat, maize, peanut, garlic	Seed priming	Overnight	Improved and rapid germination, increased biomass accumulation	
ZnO	*Coffea arabica*	Arabian coffee	Foliar spray	45 d	Enhanced growth, biomass accumulation, and net photosynthesis	
ZnO	*Triticum aestivum*	Bread wheat	Mixed with growth sustrate	Growth cycle	Increased grain yield and biomass accumulation	
ZnO	*Cyamopsis tetragonoloba*	Cluster bean	Foliar spray	6 wk	Improved plant growth, biomass accumulation, and nutrient content	
FeS2	*Brassica juncea, Sesamum indicum*	Mustard, sesame	Seed priming	12–14 h	Increased germination and crop yield	
CuO	*Spinacia oleracea*	Spinach	Mixed with soils	60 d	Improved photosynthesis and biomass production	
ZnO	*Nicotiana tabacum*	Cultivated tobacco	Hydroponics	21 d	Positively affected growth physiology, increased metabolites, enzymatic activities, and anatomical properties of plants	
Fe/SiO_2	*Arachis hypogea*	Peanut	As fertilizers	3 d	Enhanced plants growth and biomass accumulation	
Al	*Raphanus sativus, Brassica napus*	Radish, rapeseed	N.I.	N.I.	Improved root growth	(Duhan et al. 2017)

(Continued)

Table 17.1 (Continued)

Nanomaterials	Crop species	Common name	Mode of application	Duration of treatments	Responses	Source
Au	*Cucumis sativus, Lactuca sativa*	Cucumber, lettuce	N.I.	N.I.	Postive effect on germination index	
CeO_2	*Cucumis sativus, Z. mays*	Cucumber, maize	N.I.	N.I.	Increased root and stem growth	
CeO_2	*Coriandrum sativum*	Coriander	N.I.	N.I.	Increased shoot and rood length, biomass, catalase activity	
TiO_2	*Triticum aestivum*	Bread wheat	N.I.	N.I.	Increased shoot and seedling lengths	
Ag	*Trigonella foenum-graecum*	Fenugreek	N.I.	N.I.	Enhanced plant growth and diosgen in synthesis	
ZnO	*Arachis hypogea*	Peanut	N.I.	N.I.		
ZnO	*C. tetragonoloba*	Cluster bean	N.I.	N.I.	Improved shoot-root growth, chlorophyl, total soluble leaf protein	
SiO_2	*Lycopersicum esculentum Mill*	Tomato	N.I.	N.I.	Improved seed germination	

MWCNTs, Multiwalled carbon-based nanotubes.
Source: Information adapted from Duhan et al. (2017), Shang et al. (2019).

the initial changes after two weeks perhaps were produced by the acidic natures of the nanotube, decreasing soil pH and ultimately changing the microorganism's diversity, stimulating the growth of acidophilic bacteria (Kerfahi et al. 2015).

17.3.2 Nanoparticles

A nanoparticle (NP), according to the European Community, is defined as "*A natural, incidental or manufactured material containing particles, in an unbound state or as an aggregate or agglomerate and where, 50% or more of the particles in the number size distribution, one or more external dimension is in the size range of 1 nm – 100 nm.*" (Fincheira et al. 2020).

The requirements for plant growth inducers based on NPs, according to Fincheira et al., must comply with the following points, which are also very relatable to the point of sustainable agriculture: A better relation cost/efficiency, the possibility to be applied in any country, no matter the degree of development, to diminish product toxicity, and to comply with the regulatory measures to lessen the potential risks to the human population (Aschberger et al. 2015).

One of the most known nanoparticles is silver nanoparticles (AgNPs) with antibacterial activities that rely on the soil pH, organic material contents, and cation exchange capacity to determine their toxicity and bioavailability (Shoults-Wilson et al. 2011), because lower pH soil ranges and certain organic content may disrupt silver sorption by the plant, resulting in an over-accumulation of the NPs in the soil and, possibly, becoming toxic to the environment (Mishra et al. 2017).

Several examples can be found in nanotechnology supporting sustainability: Strayer-Scherer et al. (2018) designed a nanocomposite based on copper (Cu) that delivered 80% less Cu into the environment when compared with Cu-based commercial products (Strayer-Scherer et al. 2018). Giannousi et al. (2013) also described that Cu-nano-based particles were more effective against the tomato pathogen *Phytophthora infestans*, a fungoid protist when a lower copper weight content was used (Giannousi et al. 2013). Sustainably, NPs can promote plant growth, stress tolerance, and plant disease management (Mishra et al. 2017), shown by the use of titanium-NPs using *Aspergillus flavus* to stimulate plant growth of mung bean (*Vigna radiata*) (Raliya et al. 2015), as well as the fungicide activity of AgNPs against the wheat pathogen *Bipolaris sorokiniana* (Mishra et al. 2014).

Zn-based nanoparticles can inhibit the growth of bacterial and fungal strains, such as *Aspergillus niger*, *flavus*, and *fumigatus* (He et al. 2019). Application of AgNPs reduces the incidence of the fungus genera *Fusarium* in chickpea, without toxicity or affecting seed germination and soil microbiota (Kaur et al. 2018). Several manuscripts have focused on evaluating the impact of NPs on the soil microbial community (Simonin et al. 2015), stating that zinc (Zn) and titanium (Ti) nanoparticles effectively affect the bacterial community and the ecosystem (Shah et al. 2014). Additionally, an apparent connection was described between soil properties and the toxic behavior of NPs on soil microbiota (Fu et al. 2020). For example, Frenk et al. (2013) evaluated the effect of NPs based on copper oxide (CuO) and magnetite (Fe_3O_4) on the soil microbiota in two different types of soils: They found that microbial community was more affected in sandy loam soil using any of the tested NPs, but the Cu-NP influenced directly bacterial activity and composition, especially *Rhizobiales*, that belong to the alpha-proteobacteria class, and *Sphingobacteriaceae*, a family of environmental bacteria (Frenk et al. 2013).

NPs' phytotoxicity is dependent on the size of the particle and the concentration used, mainly because NPs of size less than 5 nm may go through the cell wall pores, and those from 5 to 10 nm may possess higher toxicity because of an easier uptake and posterior translocation within the plant system (Wang et al. 2013), while NPs up to 20 nm may move through the plasmodesmata, membrane-lines channels found in plants that connect cytoplasms between cells, enabling intercellular communication. Mishra (2017) states that the most common metallic NPs, such as AgNPs, size is pivotal to define their toxicity in plants (Mishra et al. 2017).

17.3.3 Pesticides

EPA defines pesticides, according to the pesticide law, as "*any substance or mixture of substances intended to prevent, destroy, repel or mitigate any pest, or as a chemical to be used as a plant growth regulator, defoliant, or desiccant. Also, any nitrogen stabilizer may be defined as a pesticide*" (US Environmental Protection Agency 2007).

The relevance of pesticides on sustainability relies on a considerable proportion of crops was lost due to pest damage. To avoid this situation, pesticides are used to reduce crop losses, but the misuse may lead to the accumulation of the agrochemical in the water, air, and land resources, as well as to bioaccumulate in the crop. Some known pesticides, such as arsenic, ethylene oxide, and lindane, belong to group 1 of carcinogens by the International Agency for Research on Cancer (IARC), corresponding as a conclusive cause of cancer, similarly for 2,3,7,8-tetrachlorodibenzo-p-dioxin (TCDD), which can be found in some pesticides (Siemiatycki et al. 2004). Exposure to a high level of some pesticides may promote chronic diseases, such as obesity, asthma, and aforementioned cancer. Also, the excess of pesticide may induce resistance to the agrochemical by the pest, diminishing the efficiency of the pesticide or requiring more chemicals to obtain the desired effect.

One way to solve this is by creating more potent pesticides to use less concentration of them, hence avoiding bioaccumulation in the environment. Another solution is to create "smart" pesticides, which are released upon a specific trigger that only activates in a certain situation, such as pH, temperature, and/or humidity, reducing the uncontrolled release of the pesticide, while the pesticide release rate is given by the pore size of the nanoparticle's carrier, such as polymeric hydrogels (McClements 2020). "Smart" pesticides may be obtained by encapsulation with nanomaterials, enhancing their potency and specificity (McClements 2019). For example, Cu-based NPs protect watermelon crops more efficiently than a conventional Cu-based pesticide, being the Cu-NP effectiveness dependent on size and composition (Servin et al. 2015). Zinc, silicon, silver, and aluminum derivatives, among others, are also used as effective pesticides (Shang et al. 2019).

NPs sizes make them better suited to penetrate plants through the pores of leaves, roots, and stems. Also, NPs can be engineered to better adhere to plant surfaces, enhancing the specificity for certain plant organ. For example, NPs may be synthesized as cationic or anionic, depending on if they are positively or negatively charged, respectively, and this decision is done depending on the target organ.

17.3.4 Nanofertilizers

Fertilizers, as different from pesticides, are used to ensure high yields of healthy plants and to reduce losses, giving the necessary nutrients for balanced growth. Fertilizers may be composed of macronutrients, such as phosphorus (P), potassium (K), and nitrogen (N), as well as micronutrients, like copper (Cu), manganese (Mn), and zinc (Zn). Chemical fertilizers, such as diammonium phosphate (DAP-$(NH_4)_2HPO_4$), urea (NH_3), and single superphosphate (SSP – $CaH_6O_8P_2^{+2}$ – SSP), are used to supplement the plant and soil with nitrogen, phosphorus, and potassium, but a very low amount reaches the target site, where a high proportion of the agrochemical is lost as runoff, volatilization, drift, hydrolysis, evaporation, or even microbial degradation (Sabir et al. 2014). Trenkel estimated that 40–70% of nitrogen, 80–90% of phosphorus, and 50–70% of potassium from the conventional fertilizers are lost and deposited into the soil, without any absorption by the crops (Trenkel 1997). As with any agrochemical, misuse may lead to polluting water resources and soils (McClements 2020); for example, sources of water, such as rivers and water reservoirs, can be contaminated due to the leaching of toxic materials, and in the worst cases, drinking water can be polluted (Solanki et al. 2015).

The concept of nanofertilizers is very similar to nanopesticides: In both of them, they can be trapped into small nano-sized particles that can be modified in composition, morphology, and dimensions. Because of their size and surfaces, nanofertilizers can adhere (preventing leaching and environmental pollution) and enter through the pores of the plant, reaching the interior of the plants where nutrients are needed. Additionally, a combination of fertilizers and/or pesticides may be used in a single nanoparticle, diminishing the number of NPs required (McClements 2020).

The encapsulation, or nanocoating, reduces the rate of dissolution and delivery of the fertilizer, allowing for a sustained release of the agrochemical to improve the absorption by plant roots, minimizing environmental accumulation (Wu and Liu 2008). For example, nanocoated nutrients may be delivered over 40–50 days in a slow-release mode, instead of the 4–10 days by conventional fertilizers (Scott et al. 2018).

Another example of agricultural sustainability by using nanofertilizers was described by Kale et al. demonstrated that the use of ZnO-nanoparticles with fertilizer inside promoted nutrient use and barley productivity by 91% when compared to the control, while the non-encapsulated $ZnSO_4$ enhanced productivity just by 31% (Kale and Gawade 2016). Liu used a microcapsule delivery system encapsulating the fertilizer chlorantraniliprole, obtaining a sustained and controlled release, avoiding leaching (Shang et al. 2019).

Raliva et al. described the relevance of nanofertilizers in the present and future by using ZnO-NPs as a cofactor for enzymes to enhance phosphorus and zinc uptake in seeds and plants (Raliya and Tarafdar 2013). In sugarcane plants, Prasad et al. evaluated the uptake and translocation of zein nanoparticles (ZNPs), confirming that these types of NPs can be found directly in the root systems, proposing that ZNPs would be a good delivery system for agrochemical in these type of plants (Prasad et al. 2017).

Shang et al. (2019) mentions the term "*control loss fertilizer*" to be used in sustainable agriculture: These types of agrochemicals are engineered to reduce soil pollution by forming a nano-network that, when in contact with water and once the network is formed, the nutrients from the fertilizer enter through hydrogen bonds, surface tension, and molecular/viscous forces, but they are big enough to be retained in the soil where the crop roots are located, facilitating nutrient absorption by the plant and reducing transfer rate (Shang et al. 2019); for example, this approach has been previously used by Cai et al. (2015) to reduce nitrogen transfer rate in the environment (Cai et al. 2015), as well as by Liu et al. who applied control loss fertilizer, diminishing nitrogen leaching and runoff by 24.5% and 21.6%, respectively, even increasing in almost 10% the mineral nitrogen and 6% of wheat production when compared to conventional fertilizers (Liu et al. 2016).

17.3.5 Biosensors

Biosensors are used to check and monitor the nutritional needs, maturity, and health status of the agricultural drops, aiming to improve yields and nutritional qualities of the plants (McClements 2019). This monitoring may help agricultural workers to apply an exact watering, as well as pesticide and fertilizer concentration. Also, nanosensors may be used to detect the presence of phytopathogens or chemicals of potential health risk a fast pathogen detection will guide the farmers to apply pesticides before any damage occurs, avoiding crop loss by damaged plants or roots, increasing sustainability. In general terms, the development of nanosensors for monitoring nutrients, toxic substances, and diseases would assure soil and plant health but also safety to the environment, supporting sustainable agriculture and ecosystem (Marchiol 2018).

These biosensors may be done at nano-scale, looking to improve their sensitivity and sensibility: The preferred way to use them is by taking a soil or plant sample, mix it with water and place it in a small chamber with sensors to detect target substances that may be indicative (or biomarker) of nutrition, maturity, and/or health status (McClements 2020). But, referring to sustainability, nanosensors must be affordable, robust, and safe. Wong et al. (2017) demonstrated that plants can internalize sensors based on carbon nanotubes that transmit signals to a smartphone when any pesticide is detected (Wong et al. 2017).

Biological organisms can sense environmental conditions, but with the support of nanomaterials, specificity, and sensitivity may improve sensor efficiency (Dubey and Mailapalli 2016). For example, in cultivated fields, global positioning systems (GPS) based on nanomaterials are used to monitor the release system via nanoscale carriers that use wireless signals positioned within the agricultural crop, assuring real-time monitoring to avoid the over- or misuse of agrochemicals (Shang et al. 2019). Atzberger (2013) describes the creation of nanosensors within the soil or plant root, designed to send data continuously to a remote computer with alarms if any parameters are low (Atzberger 2013). Not only agrochemicals can be controlled, but also water and phytopathogens can be detected; for example, Afsharinejad et al. developed a nanosystem to sense insect attacks by the identification of characteristic volatile organic compounds released by the plant species infected by the phytopathogen (Afsharinejad et al. 2016).

Singh et al. (2010) used a nanogold immunosensor to detect the karnal bunt disease in wheat, produced by the fungus *Tilletia indica* infecting wheat grains at flowering, reducing grain quality by the production of powdery spores that bleach the grain and grain-derived products (Singh et al. 2010). Pesticides may also be detected by biosensors, using a photosystem containing biosensors to bind different functional groups from pesticides, as well as other agrochemicals, providing a low-cost and easy technique to improve sustainability avoiding bioaccumulation in the ecosystem (Shang et al. 2019).

About stress responses, plants may respond to different stimuli produced by changes in physiological states mediated by stress hormones, for example, methyl jasmonate and jasmonic acid, as well as salicylic acid. The latter was used by Wang et al., who used a gold electrode nanosensor containing Cu-NPs to detect phytogenic fungus infestation through detecting variation levels on salicylic acid in oilseed (Wang et al. 2010). Multiwalled carbon nanotubes, or MWCNTs, may also be used to study plant growth to detect hormones like auxin to elucidate how plant roots can adapt to the environment, even if they are in harsh soils (McLamore et al. 2010). Ganeshkumar et al. used nanofibers of potassium niobate ($KNbO_3$) to sense humidity by evaluating the conductance belonging to the relative humidity, from humidity values ranging from 15 to 95% at room temperature (25 °C) (Ganeshkumar et al. 2016).

Detection of contaminants, such as allergens, toxins, and pesticides, is crucial to assure food-based safety requirements, one of the pillars of agricultural sustainability, namely avoiding negative impacts on human health. For example, silicon-based nanowires, based on biocompatibility and modifiable electrical properties, as well as gold NPs and carbon nanotubes are all used to detect the previously mentioned contaminants (Das et al. 2019). Fang et al. developed a platform based on AuNPs to detect the presence of the toxic element cadmium (Cd^{2+}) in water samples (Fang et al. 2017).

About allergen detection for food safety, nanosensors based on surface plasmon resonance have been used to detect the Arah1, one of the most important peanut protein-based allergens in chocolate candy bars (Pollet et al. 2011); ovalbumin as a whole milk contaminant (He et al. 2011) and tainted soy protein in soy products (Godoy-Navajas et al. 2011) may also be detected by nanosensors.

Other functional uses of nanosensors include moisture levels, soil pH, temperature, fertility crop nutrient levels, and concentrations, where all of them will support the decision of agro workers to determine what are the best fertilization strategies to increment crop production, with the reduction of inputs for not to waste key resources like water, fertilizers, and pesticides, accompanied by a better time and resources management sustainably for not producing any negative impacts in the future (Iavicoli et al. 2017). Mastronardi et al. (2015) proposed the detection of rhizosphere biomarkers, i.e., plant exudates and microbial secondary metabolites by using polymers, antibodies, and aptamers to provide information about the real status of crops and soils to design a specific agrichemical delivery method that is better suited to the target plant/soil (Mastronardi et al. 2015).

17.3.6 Contaminant Removal

Agriculture must be considered as a full process, from the crops to the final products, and one of the key points for sustainable agriculture is contaminant removal, where environmental pollution is one of the biggest challenges to solve. Soils and water sources could be contaminated by natural and/or anthropogenic (produced by the man) pollutants that may result in a serious human health risk or to the environment. Nanotechnology can also provide solutions by providing low-cost and effective methods to remove soil contaminants like heavy metals, industrial dyes, and organic compounds.

Because of the properties of nanomaterials, they may be used for separation processes like centrifugation or filtration (Bacchin et al. 2011), enabling the isolation of valuable compounds from waste streams generated by food and agro-industry, ultimately being transformed into functional ingredients used in other industries, diminishing toxic wastes and pollution. For example, McClements describes that the control of colloidal interactions between macromolecules/particles at a nanometer scale may be used to manipulate the aggregation states to further isolation. These strategies may be used to obtain valuable functional compounds from residual and waste streams, as minerals, lipids, proteins, polysaccharides, and polyphenol, as well as vitamins (McClements 2020).

17.3.7 Plant Properties Modified by Nanotech

Nutrient delivery, enhanced nutrient absorption, toxic compound detections are some of the multiple uses of nanotechnology. Additionally, nanomaterials can help to improve accumulation and enzyme activities within the plant. For example, the application of nano-SiO_2 or ZnO nanomaterials improves the uptake of amino acids, such as proline, nutrients, and water, while an activity enhancement of defense antioxidant enzymes, like superoxide dismutase (SOD), catalase (CAT), and glutathione reductase, among others, improves plant tolerance to harsh weather events (Wang et al. 2015). SOD, as well as other related enzymes, is part of the physiological defense strategies in plants and animals against free radicals and reactive oxygen species (ROS) produced by biotic (organic matter, fungi, bacteria, etc.) and abiotic stress (light, temperature, water, pH, etc.). ROS are necessary for redox homeostasis and participates in the immune and cardiovascular system, but an overproduction of ROS may provoke damages to cellular processes and organelles, disrupting the normal physiology of the plant (Shang et al. 2019).

Detection of hormones or molecules derived from stress has been largely evaluated, but also nanomaterial may help the regulation of gene expression associated with stress. To evaluate which genes are up- or downregulated by a certain stimulus, such as stress, presence of nutrients, or toxic elements, molecular biology techniques, such as microarrays, may be used. By applying this method, it can be evaluated which genes were affected by the application of AgNPs in *Arabidopsis*, a weed plant native to Europe, Asia, and Africa (Shang et al. 2019). It is not surprising that this approach to understanding gene expression was made in *Arabidopsis*: It is considered as the plant model organism, because it was the first genome sequenced from plants, and is currently used in molecular biology to understand plant traits, such as flower development and heterologous gene expression.

Banerjee and Cole (2016) found that the application of AgNPs induced the expression of genes associated with metal and oxidative stress response, such as SOD, peroxidases, and cytochrome p450-dependent oxidases but also repressed (or down-regulated) the expression of genes related to pathogen response and hormone stimulation, like auxin-regulated (hormone) genes involved in growth and organ size, as well as ethylene signaling (Banerjee and Kole 2016). This research demonstrated that nanomaterials are directly associated with stress responses for plant protection. But the response to nanomaterials and nano fertilizer will vary depending on the plant species, growth stage, and type of nanomaterial used (Abdel-Aziz et al. 2016). For that reason, Shang et al. (2019) stated consciously that further work is required to identify and understand signaling cascades and genes affected by nanomaterials in different crops before their use in the field (Shang et al. 2019).

Despite the next topic will be covered in more detail in the second part of the chapter, referring to genetic engineering and gene modifications in the plant, the combination of nanotechnology with molecular biology techniques has given new tools to transport gene-editing tools to modify specific target genes (and even produce new organisms) to enhance plant properties or giving them novel properties to help them resist pests and phytopathogens (Torney et al. 2007).

Nanobiotechnology may use nanoparticles, nanocapsules, and/or nanofibers to carry and deliver foreign deoxyribonucleotide acid (DNA) and chemicals required to modify the target gene. Before the use of nanomaterials, gene delivery was made by viral vectors that presented several restraints, such as limited host range, restricted size of the inserted gene material, and being able to transport itself across the cell membrane and to reach the nucleus (Ghormade et al. 2011). NPs may help to overcome these challenges.

Silicon dioxide NPs are being used to deliver DNA fragments or gene sequences to tobacco and corn plants without side effects (Galbraith 2007). Also, NPs coated in DNA may be used as bullets in gene-gun technology (also called biolistics) to deliver the desired genes to target plants (Vijayakumar et al. 2010). Chitosan-based NPs covering siRNA (or small-interfering ribonucleic acid) are used to control specific insect pests by interfering (hence the name) the expression of essential genes of the pathogen (Zhang et al. 2010). The most advanced gene-editing tool at the moment, the CRISPR/Cas9 system, has been used to successfully genome editing in plants (Miller et al. 2017), but with a low delivery efficiency. The combination of the genome editor with nanoparticles may increase the efficiency and specificity of the CRISPR/Cas9 system. For example, Mout et al. (2017) used cationic arginine gold nanoparticles (or Arg-NPs) to deliver ribonucleoproteins, showing 30% of efficiency in cytoplasmic and nuclear gene editing in cell lines, which could be used in plant future research (Mout et al. 2017).

17.3.8 Photosynthesis Improvement

One way to enhance sustainability is to achieve better use of the resources, and light is crucial for plant growth and development. Nanomaterials applied to leaves have presented an increased plant photosynthesis efficiency (Giraldo et al. 2014), and according to Lowry et al. (2019), these could be through the following ways (Lowry et al. 2019):

- By the protection of the photosynthetic processes from abiotic stress, such as pH and temperatures
- By increasing the chlorophyll content, hence increasing light absorption
- By enhancing light absorption and photon use
- By scavenging ROS, protecting the plant from oxygen-induced damages
- By assisting electron transfer

It must be considered that faster photosynthesis may result in a lower nutritional content of the agricultural crop (Myers et al. 2014).

Nanotechnology has very interesting applications in plants because nanoparticles can be inserted into the chloroplasts and cells of plants to sense compounds or changes in the environment, as well as to communicate as infrared devices, and even transform plants as light sources (Ghorbanpour and Fahimirad 2017). Giraldo et al., for example, informed that the insertion of SWNTs increased the electron transfer rate by 49% in chloroplasts by light absorption. Also, they demonstrate that SWNTs can contribute to the light-harvesting capacity by impeding ROS generation in chloroplasts, resulting in increased photosynthetic efficiency (Giraldo et al. 2014).

17.4 Consequences of the Use of Nanotechnology in Sustainability

17.4.1 Ecotoxicological Implications of the Nanoparticles

Despite all the promising features of nanomaterials and derived products, there is a lack of understanding about what happens with nanomaterials in the environment (Hao et al. 2018), and also about the potential bioaccumulation in trophic and food chains (Rui et al. 2017), making it necessary to have to regulate measures and to be carefully evaluated before using these types of technology in the fields. To minimize the possible environmental damage, the modification of size, shape, and functionality should optimize the efficacy of nanomaterials.

As shown before, a variety of examples of uses of nanotechnology in agricultural sustainability can be found, but to protect human health and plant crops, several research groups have focused on the current and future toxicity of the nanoparticles in the industry (Prasad et al. 2017). The toxicity of NPs based on metals, described as antibacterial and antifungal, will depend on factors like solubility, biological site, functional groups, etc., as well as the structure of the targeted cell wall (Abdel-Aziz et al. 2016).

Other photoactive nanoparticles, such as TiO_2, ZnO, and fullerenes, when exposed to light, generate excited electrons that will subsequently convert into free radicals in the form of superoxide, producing oxidative stress, being an excellent parameter to evaluate toxicity and ecotoxicity (Vannini et al. 2014). Plant cells have the protective enzymes to overcome them, such as SOD and catalases. In this precise point, we can find two very different situations: Sayes et al. (2004) confirmed ROD generation by TiO_2 and fullerenes, while others, like Tripathi (2016) and Venkatachalam (2017), described that fullerenes and silicon NPs may protect against oxidative stress, instead of producing it. More studies and ecotoxicological research are required to establish the real effect of nanomaterials in the environment (Tripathi et al. 2016; Venkatachalam et al. 2017). To achieve this, the structure of the nanomaterial and the complexity of the ecosystem must be fully understood to determine and predict the behavior of the nanomaterial in the environment (Cox et al. 2016). Research must be done to establish the bioaccumulation in the food chain, through various trophic levels, and the interaction of nanoparticles with other agrochemicals and/or pollutants in the environment. To summarize, NPs can affect the responses or the capacity of adaptation of the plant and the belated soil microbiota, being essential to discuss the potential risks to the plant, microbiota, and consumers through the trophic chain (Ma et al. 2010).

17.4.2 Growth of Cultivated Plants and Its Ecotoxicological Sustainability

As stated before, metal-based NPs may be accumulated into the plant host, but nanoparticles may be produced naturally; for example, iron oxide nanoparticles (Fe_2O_4-NPs), which are a magnetic form from iron ore, bioaccumulate in garden cress (*Lepidum sativum*) and peas (*Pisum sativum*) plants, demonstrating that NPs are already present in the natural environment (Abbas et al. 2016). Also, soil microbiota must be considered because of its role in keeping healthy soils and crop productivity. Thus, the use of NPs in crops must be eco-friendly and its toxicity evaluated (Prasad et al. 2017).

Toxicity, perhaps the most important parameter in sustainability, can be evaluated by crop yields, hormone concentration, physiological indexes, cytotoxicity, and genotoxicity, referred to as damage at the cellular and genetic levels, respectively (U.S. Environmental Protection Agency 1988; OECD 2006). Also, NPs' effect may be detected through oxidative stress, alteration in nutrient and water transport, changes in genetic expression, among others. One challenge is to replicate the results obtained in controlled conditions, such as a defined temperature, lack of stressors, controlled soil and plant growth parameters, among others, but the scale-up from laboratory conditions to field crops must consider the interactions between nanomaterials, soil components, and the ecosystem, in short- and long term (Aschberger et al. 2015).

But what happens with the third pillar of sustainable agriculture, namely, social perception? As everything new, consumers have doubts about the use of engineered nanomaterials on crops. Because of this, safety and regulatory measures are a must, as well as acceptance of nanobiotechnology by the agricultural worker and the consumer. Increasing the amounts of NPs entering food plants, such as lettuce and cabbage, or concentrating them in grains and fruits, like wheat and tomatoes, respectively, as a solution to avoid nutrient loss may need an adequate risk assessment and regulation, but will be complicated if organisms, microbiomes, climate, soil properties, and engineered nanomaterials must be taken all into account (Lowry et al. 2019). Despite all the information about research-based on NPs, little has been done about the regulation of the transport and commercialization of nanomaterials as agricultural products, but nanotechnology must be a key strategy to increase crop yields in developing and developed countries.

17.5 Genetically Modified Organisms in Sustainable Agriculture

The Food Agriculture Association (FAO) and the European Commission defined GMOs, and the products derived from, as "*plants or animals produced through techniques, in which genetic material has been altered in a way that could not happen naturally, such as by mating and/or natural recombination.*" All our current crops and livestock are GMOs, in one way or another, because their genetics have been manipulated by the human for more than 10,000 years by artificial selection, aiming to have bigger, more nutritive, softer fruits, among others, and at the same time, to have high production yields, in comparison to their predecessors. Oliver (2014) mentions Triticale (×*Triticosecale*), a crop that would be defined as GMO by FAO, but it is currently accepted as conventional (Oliver 2014). This crop is a hybrid of wheat (*Triticum*) and rye (*Secale*), developed in the laboratory in 1884 to possess higher protein amounts and less gluten, being currently used in bread and pasta. Because the wheat × rye crossing was sterile, fertility was restored to obtain a polyploid plant with two copies of each genome (Ammar et al. 2004).

Another GMO definition, perhaps more suitable, was proposed by the Cartagena Protocol (16) for "*living modified organisms,*" which refers to "*any living organisms possessing foreign genetic material obtained through the use of modern biotechnology*" (Secretariat of the Convention on Biological Diversity 2003).

The impact of GMO has been enormous: To overcome limited crop resources and resist climate change, the total cultivated land dedicated to GMO in 2012 was around 170 million hectares in 28 countries (James 2014), where the commercial cultivation of GM crops in 2016 reached a landmass of two billion hectares over the last two decades, generating benefits of 150.3 billion U.S. Dollars (Brookes and Barfoot 2016).

One of the key points of agricultural sustainability is based on the rapid population growth and human development because the anthropogenic action has greatly affected negatively the environment and loss of biodiversity (Oliver 2014). GM crops have produced increased yields, less land required for agricultural purposes, and crops tolerant to herbicides, among other modifications (Brookes and Barfoot 2014). GMOs may also recover soil quality, reduce water runoff and greenhouse gases, conserve nutrients, and increase water infiltration.

Genetic codification of the plant can be made by the insertion of novel genes from other species to increase their resistance to pests, droughts, and even their nutritional quality, usually named as genetic transformation or insertion. Also, genome or gene editing can be made to modify specific sequences within a gene of the plant to increase or suppress its expression, by the addition or removal of one or two nucleotides, enough to produce aberrant proteins and/or disrupt their function. In general terms, two types of GMO plants can be found: (i) Herbicide-tolerant crops, such as glyphosate and/or with enzymes that degrade the herbicide; (ii) Insect-resistant, in crops that express the insecticidal crystal proteins (CRY) produced by *Bacillus thuringiensis*. In any of the cases, the main objective is to improve crop yields by limiting losses produced by weed competition and phytopathogens, also reducing costs by better management by the farmer, and the reduction of agrochemicals, as herbicides and insecticides (Oliver 2014).

17.5.1 Genetic Transformation Basics

Genetic transformation can be defined as a process that compasses the introduction and effective expression of foreign genes in a host organism, which could be bacteria, yeast, and plants, among others. The expression may result from extrachromosomal genes inserted in plasmid vectors that are directed to the nuclei and persist if the foreign DNA has a way to replicate itself or, if it integrates within somatic chromosomes, it may persist through the lifetime of the organism, but cannot be inherited. To avoid this, foreign genes may be incorporated into the germ-line genome to allow the passing of the gene through further generations (Handler 2008). If the gene comes from a different organism than the host, it will be termed as a transgene, and by using gene regulators, its expression can be modulated to obtain the best outcomes.

Gene transfer to a crop plant for generating a transgenic plant (or GMO) requires two main steps:

- A successful delivery system to insert the gene into the plant cell
- Regeneration of the transgenic plant from the transformed cell(s)

Gene delivery in plants is usually achieved by the direct transfer of "naked" DNA, or by using a bacterial vehicle, namely *Agrobacterium tumefaciens*, which will insert the DNA from the bacterial strain to the plant. Foreign DNA is initially inserted in plasmids, a circular extrachromosomal DNA capable to replicate in a bacterial host (as *A. tumefaciens*) during the transformation process (Oliver 2014).

Starting with the direct transfer of DNA to the plant cells (i.e., transformation) may be achieved by electroporation into protoplasts (plant cell without the cell wall), which consists in the application of an electrical field to the cell, aiming to increase the membrane's permeability, permitting the introduction of chemicals, drugs, or nucleic acids. Also may be used microinjection, based on delivering foreign DNA into a cell using a glass micropipette, chloroplast transformation, silicon nanoparticles, and biolistics/particle bombardment, with the latter being the most common of all the aforementioned techniques (Day and Goldschmidt-Clermont 2011).

Microparticle bombardment, or "*gene gun*," was firstly developed in 1980 using pressurized helium to shoot gold or tungsten microparticles coated with the gene of interest within the plasmid as naked DNA into the plant cells at high velocities (Sanford 1990), integrating itself into the plant genome. By using this technique, cereals, tuber crops, and trees have been transformed into transgenic plants. Its advantage relies on the ability to transfer large DNA fragments and even whole chromosomes (Schmidt et al. 2008).

Because of the high price of gold nanoparticles, *A. tumefaciens* transformation, first described in the late 70s, has been more popular (Figure 17.4). This bacterial strain is known for the ability to infect plants, specifically in the line between roots and stem, altering the genome of the plant by inserting genes that produce cell proliferation, forming ultimately a gall or tumor, where the bacteria live. The desired gene must be inserted into a Ti (tumor-inducing) plasmid, containing a section known as T-DNA (transfer DNA),

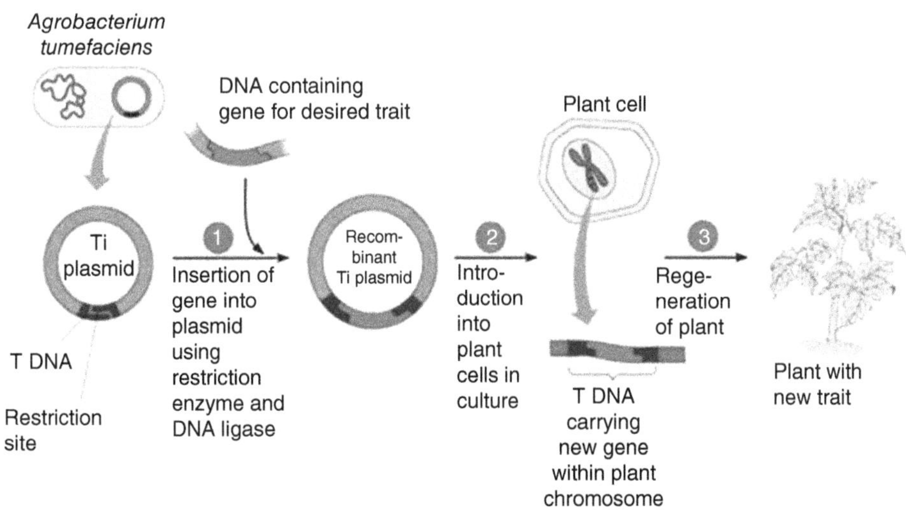

Figure 17.4 A simple pictorial representation of *Agrobacterium*-mediated transformation in sugarcane. *Source:* Sunil Kumar and Immanuel Selvaraj (2019).

which contains genes that encode for enzymes that synthesize auxin and cytokinin, plant hormones that produce cell proliferation and tumor formation.

The Ti plasmid has also the virulence genes that control the capacity of the bacteria to infect the plant tissue and transfer the T-DNA into the nucleus of the plant cell (Barampuram and Zhang 2011). Currently, binary vectors are used, separating the T-DNA region onto a smaller plasmid, which can replicate in *A. tumefaciens* and *Escherichia coli*, different from the Ti plasmid, facilitating the construction of insertion of target genes of interest (Barampuram and Zhang 2011). Afterward, the replacement of T-DNA with genes capable of expression in plants has allowed to develop methods that target the delivery of *A. tumefaciens* to transform ovule *in situ*, avoiding the need for regeneration in the tissue culture. Variations of this methodology have been used to transform wheat, Pak-Choi, corn, and cotton, among others (Chumakov and Moiseeva 2012).

To select for successful transformants, it is very common to insert selectable markers to allow selection between cells that effectively have the inserted gene from the others that did not receive the foreign DNA. The gene markers may be next to the gene of interest or within the plasmid, depending on the host. A plasmid may have one or more markers, and in plants, the most used selectable markers are the ones that give antibiotic resistance and/or herbicide tolerance, and that is one of the key points against GMOs: The potential transfer of the herbicide genes from GMO to non-GMO crops is considered, according to Hare and Chua, as an unacceptable environmental risk, damaging the possibility of sustainable agriculture (Hare and Chua 2002). Because of that, several techniques have been developed to remove the marker gene after transformation, with the most effective being the use of site-specific recombinases enzymes which catalyze DNA-exchange reactions (or recombination) between short target-site sequences unique for the recombinase, located between two enzyme-specific recognition sites, originally used to control gene expression by cassette exchange, excision/insertion, inversion, or translocation (Nern et al. 2011). The most common site-specific recombinases are CRE-LOX and FLP-FRT systems, where the first triplet refers to the enzyme and the second one the recognition sites (Oliver 2014).

By using site-recombinases, the markers can be successfully removed, while the target inserted gene remains in the plant. Oliver (2014) uses a simple diagram to explain the removal strategy (Figure 17.5).

The CRE-LOX system involves three steps: the addition of LOX sites on both sides of the marker in the binary vector gene construct, the expression of the CRE recombinase in the plant after selection, and the multiplication of the marker-free transgenic plants for the development of the GM crop. Figure 17.5 shows that LOX sites must flank the marker gene to allow for CRE to remove it, and then it is degraded and lost, while the transgene (or foreign gene) remains in the GM crop.

Despite the controversies, there are no current proofs that demonstrate that marker genes are a risk to animals and microorganisms from the soil because the DNA that encodes the antibiotic or herbicide gene is a very low proportion of the gram of DNA consumed per day, and goes under degradation as all DNA, so the chance to get any gene incorporated into the host is very unlikely (Rizzi et al. 2012).

GMOs may also be created by genome edition, which allows for mutagenesis of single base pairs, introduction (insertion) and/or deletion of a few base pairs, and DNA fragment substitution (Naso and Petrova 2019). The most known genome-editing techniques are

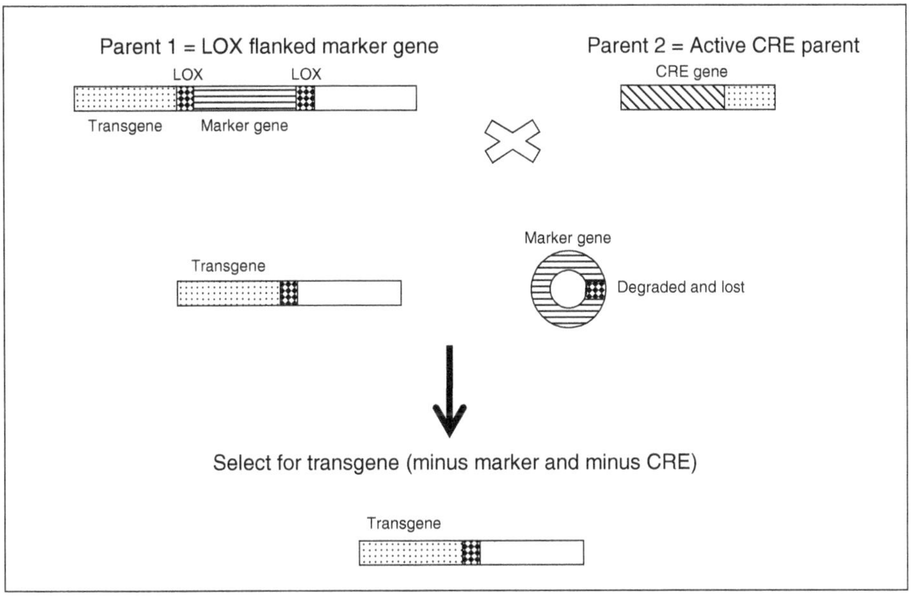

Figure 17.5 CRE-LOX system to remove selection marker genes from genetically modified plants. Checkered boxes: LOX recognition sites for the CRE recombinase, Block: Plant genomic DNA. *Source:* Oliver (2014).

oligonucleotide-directed mutagenesis (ODM; Sauer et al. 2016), transcription activator-like effector nucleases (TALENs; Chen et al. 2015), zinc finger nucleases (ZFNs; Zhang et al. 2020), and clustered regularly interspaced short palindromic repeats and CRISPR-associated proteins (CRISPR-Cas; Doudna and Charpentier 2014).

TALENs and ZFNs, due to the high design complexity and low transfection efficiency (Doudna and Charpentier 2014), have lost popularity in front of the CRISPR-Cas system, the most powerful genome-editing technique that allows high efficiency and precision (Ledford 2015), which was originally discovered as a form of adaptative immunity system in microorganisms as a defense against virus, plasmids, and phage DNA (Albitar et al. 2018).

Charpentier and Doudna, Wang et al. (2019), and Zhang et al. (2020) provide very detailed explanations on how the CRISPR-Cas works (Doudna and Charpentier 2014; Wang et al. 2019; Zhang et al. 2020), but in general terms, the system is comprised of the Cas proteins and the CRISPR system and works in a specified-sequence way recognizing and cleaving foreign nucleotides, when certain requirements are met, such as a high content of guanine and cytosine (G and C nucleotides, respectively) as recognition and edition site (Bernabé-Orts et al. 2019) (Figure 17.6).

To differentiate between GM and non-GM plants, a range of different methods is found, based on the presence/absence of the marker and/or the inserted genes, as well as the detection of any products that are only produced by the GM crop. Some of the techniques used are two-dimensional electrophoresis (Kim et al. 2006), protein capillary electrophoresis (López et al. 2009), high-pressure liquid chromatography (HPLC; López et al. 2009), and the enzyme-linked immunosorbent assay (ELISA; Wentao et al. 2009). The first two are based on the technique named polymerase chain reaction (PCR), which requires further

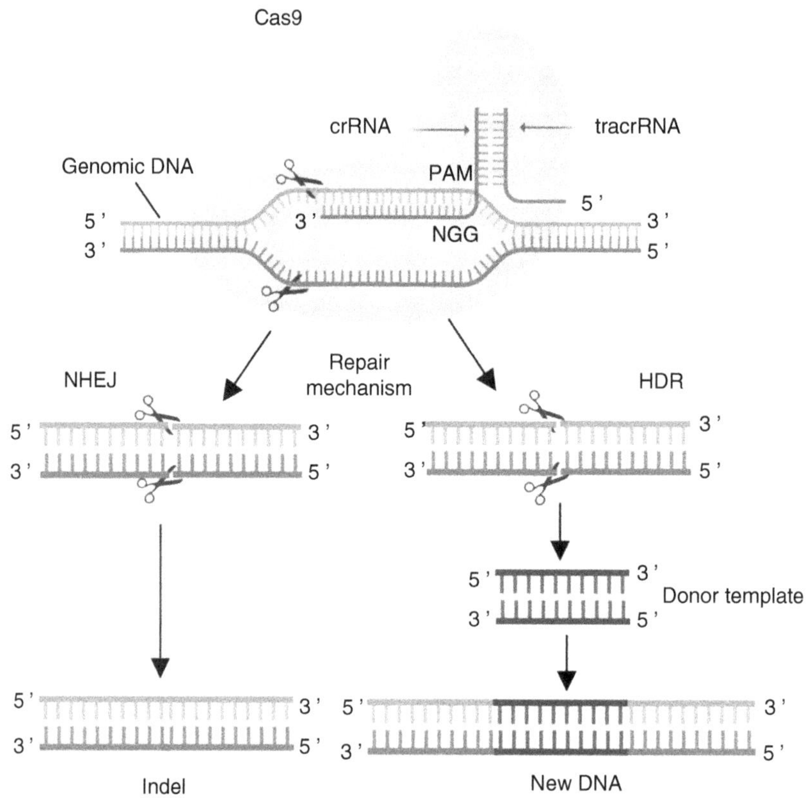

Figure 17.6 Diagrammatic representation of CRISPR-Cas9 system. In the presence of single guide RNA (sgRNA), Cas9 binds to the target site, close to a short DNA sequence motif called PAM, and creates a double-stranded break. As a result, the cell initiates two mechanisms to repair the break: Nonhomologous end-joining (NHEJ) produces insertions, deletions (indel), and loss-of-function mutations, and the other mechanism, homology-directed repair (HDR), corrects a preexisting mutation by introducing a template DNA sequence that helps to repair the break by copying the template sequence. *Source:* Zhang et al. (2020).

electrophoresis or enzymatic digestions, or by real-time PCR (RT-PCR), which does not require post-PCR manipulations, but the equipment is more expensive. In both cases, the techniques present high sensitivity and specificity (Marmiroli et al. 2008).

Many common genetic elements of GMOs can be found by PCR, such as promoters, terminators, or the same transgene (Guo et al. 2012). In 2013, Cottenet et al. developed a multiplex quantitative PCR (qPCR), allowing the detection of 47 target genes (Cottenet et al. 2013). In the same year, Block et al. (2013) developed a database named GMOseek (http://kt.ijs.si/software/GMOtrack/GMOseek.html), an open-access software that reunited information about GMO events and genetic markers, aiming to ease GMO detection from plant origin at different phases (Block et al. 2013), but the last update was on 2017. Nevertheless, the information within the database can be used to search for specific markers, select screening targets, check their occurrence, and assist in the interpretation of the results.

Peng et al. (2016) developed an RT-PCR screening method for GMOs, covering simultaneously 19 genetic elements, with high sensitivity and specificity, offering a simple and cost-efficient strategy to analyze GM plants (Peng et al. 2016). Information on gene markers may be found in several databases, such as the National Center for Biotechnology Information (NCBI) database (https://www.ncbi.nlm.nih.gov) (Geer et al. 2009) and the GenBit database (https://www.genbitgroup.com/en/gmo).

17.6 GMO Uses in Agricultural Sustainability

17.6.1 Golden Rice

The chronic deficiency of vitamin A (or xerophthalmia, its clinical manifestation), essential in the light-absorbing pigment rhodopsin in the human eye, may lead to immunodepression and blindness. The deficiency is associated with the diet, i.e., malnutrition and poverty, so there was required a cheap and popular food source that could have a high concentration of vitamin A (Oyunga et al. 2016). The carotenoid beta-carotene is a precursor of vitamin A, and its synthesis pathway has been fully understood. Due to technology availability and understanding of the plant transformation and carotene synthesis process, rice was selected to receive the full b-carotene biosynthetic pathway (Oliver 2014).

According to Kaput et al. (2015), Golden Rice is the "poster child" to an artificially enhanced crop (Kaput et al. 2015): The first Golden Rice generation introduced the whole beta-carotene pathway (which will be metabolized into vitamin A in the human body) via a pair of vectors using an *A. tumefaciens*-mediated transformation (Ye and Beyer 2000), one encoding for the phytoene synthase (*psy*) with a transit peptide (*tp*) gene from *Narcissus pseudonarcissus*, regulated under the control of glutein (Gt1) promoter, as well as the bacterial phytoene desaturase (*crtl*) from *Erwinia uredovora* and *tp* gene controlled by the cauliflower mosaic virus promoter (CaMV 35S).

The second vector contains the lycopene beta-cyclase (*lcy*) also from *N. pseudonarcissus* controlled by the Gt promoter and the a*phIV* gene (hygromycin resistance gene) controlled also by CaMV 35S. Initially, 1.6 μg of beta-carotene per gram of rice was produced, but by altering the source of the *psy* gene, yield increased almost 10-fold, but because of the debates about GM technologies, it has not been fully commercialized (Kaput et al. 2015). Golden Rice has been one of the first examples of GMOs used as a sustainable solution for malnutrition but also has been used to evaluate food security (Kettenburg et al. 2018).

17.6.2 Pest and Viral Resistance

The first commercialized GM plant was soybean, which contained the 5-enolpyruvylshiki mate-3-phosphate synthase (EPSPS) gene from *A. tumefaciens*, with resistance to the herbicide glyphosate. The EPSPS reaction produces the essential amino acids (aa) phenylalanine and tryptophan, and they are essential because the human does not possess the EPSPS pathway to synthesize aromatic aa. This reason supports the absence of side effects in humans (Kaput et al. 2015).

Rice (*Oryza sativa*) has been very well studied and transformed, conferring resistance to insects and herbicides, as an improvement in grain quality (Huang et al. 2005). Two rice lines that received genes from *Bacillus thuringiensis* for insect resistance received biosafety certificates in 2009 (Liu et al. 2015), permitting the uses of GM crops together with non-GM rice, one of the big concerns being the possibility for gene transfer from GM to non-GM crops.

Another basic tuber crop, potato (*Solanum tuberosum*) is plagued by late blight disease, producing estimated annual losses of seven billion euros (Haverkort et al. 2016) and the current controlling measures require a high number of fungicides. Hence, a late-blight-resistant potato would be considered one of the most sustainable ways to fight the disease (Rietman et al. 2012).

Late blight disease is a potato and tomato (*Solanum lycopersicum)* disease produced by the oomycete *Phytophthora infestans*, and by using biotechnology, the late blight-resistant (LBR) GM potato varieties have been done by the identification and characterization of genes that provide late blight resistance and transfer those genes from wild potato species to the commercial ones (Vleeshouwers et al. 2011). Despite successfully obtaining late-blight resistant plants, the molecular mechanisms involved in the resistance of the potato plant are not fully understood (Mundt 2014), as well as the molecular mechanisms involved in *P. infestans* virulence (Jiang and Tyler 2012).

A report written by European researchers, instead of focusing on the uncertainties of GMO effect on the crops (van Hove and Gillund 2017), preferred to state that GM potatoes contribute to a more sustainable potato production by not requiring any pesticides against the blight disease (Jones et al. 2014). They also even propose to add several resistance genes in one potato variety to fight against *P. infestans,* arguing that the pathogen will not be able to surpass every resistance trait at once. But the durability of the GM potato is unknown, and the prediction models are not prepared to predict what would happen with the GM potato is put into a large-scale production (van Hove and Gillund 2017).

The aforementioned GM potato, surprisingly, is not transgenic, but a cisgenic, i.e., modification of a plant host with genes from varieties of the same or closely related species (Haverkort et al. 2008), being also one of the concerns about what is a GMO. It is believed that the cisgenic approach, based on the search and introduction of resistance genes in the *same* species, may have a better acceptance by the society, instead of the transgenic approach (genetic modification of a host by using genes from an unrelated organism).

The CRISPR-Cas system has been widely used to modify several crops: Tashkandi et al. (2018) engineered tomato and *Nicotiana benthamiana* (plant closely related to tobacco) to induce resistance to the tomato yellow leaf curl virus (TYLCV) (Tashkandi et al. 2018), Yanpeng et al. (2014) produced bread wheat immune to powdery mildew, a fungal disease produced by a fungus of the order *Erysiphales*, by modifying the genes TaMLO-A1 and TamLO-B1 (Yanpeng et al. 2014). Chen et al. (2017) produced cotton resistant to the ascomycota *Verticillium* wilt (Chen et al. 2017). In tomato plants and *Arabidopsis thaliana*, the CRISPR-Cas technique enhanced resistance against phytopathogens, such as *Phytospora capsica*, *Pseudomonas syringae,* and *Xanthomonas spp* (Paula de Toledo Thomazella et al. 2016). Foster et al. (2018) used a plasmid-free CRISPR-Cas9 system, together with ribonucleoproteins (RNPs), to produce resistance against the rice blast fungus (*Magnaporthe oryzae*) (Foster et al. 2018). Other applications of the CRISPR system can be found in Table 17.2.

Table 17.2 Examples of uses of CRISPR in plants.

Plant	Target gene	Target trait
Rice	Stomatal development gene	Abiotic stress tolerance
Rice	Phytoene desaturase; Bentazon-sensitive-lethal	Careteniud biosynthetic pathway; Herbicide resistance
Rice	Chlorophylide-a oxigenase gene	Photosynthesc efficiency
Rice	Acetolactate synthase	Herbicide resistance
Rice	Dense and Erect Panicle 1: Rice oytermost cell-specific gene5	Regulating carbon-nitrogen metabolism; yield; leaf rolling controlling
Arabidopsis and rice	Phytoene desaturase	Carotenoid biosynthetic pathway
Soybean	Fatty acid desaturase	Increase oleic acid levels; Yield improvement
Tobacco	Allene oxide cyclase	Jasmonic acid biosynthesis metabolism
Maize	Maize Gloosy gene2	Epicuticular wax formton—Regulation of fatty acid elongase pathway
Tomato	Salt tolerance gene 1; 2	Salt tolerance; Abiotic stress tolerant
Allotetraploid cotton	Chloroplast developmental gene	Altered chloroplasts
Citrus	Phytoene desaturase	Carotenoid biosynthetic pathway

Source: Information adapted from Bandyopadhyay et al. (2020).

Because of the herbicide-tolerant crops, it allows the uses of chemical less toxic and more environmentally friendly, such as the herbicides glyphosate and glufosinate, both classified as Class III herbicides by EPA, meaning that are mildly toxic and have a low persistence in soil and environment, replacing the most toxic herbicides and/or the ones known to pollute and persist in groundwater (Ronald 2011).

In GM plants, the crop synthesizes a bacterial protein that, when ingested and digested by the insect, produces a toxin that binds to a cell adhesion receptor in the insect gut (Hernández-Rodríguez et al. 2013). Transgenic proteins degrade similarly to native proteins, due to food production methods and *in vivo* digestion enzymes (Hammond et al. 2013)

One common gene insertion to produce insect-resistant GM crops is the gene that encodes for the CRT protein toxin from the bacterial strain *Bacillus thuringensis* (Bt), which attacks selectively caterpillars and beetle pests that feed on the crop plants, ultimately affecting the gut cells of the insect and impeding digestion. Additionally, CRY toxins do not harm vertebrates or plants, as well as being biodegradable, avoiding their accumulation in the environment (Bravo et al. 2007), making these types of GMO crops safe for human and animal consumption (Ronald 2011). The Bt GM crop just delivers the CRY toxin to the insect phytopathogen that eats the plant, making it less likely to cause environmental problems than spraying the toxin directly to the plant. However, concerns about using GM crops are related to the possible development of Bt-resistant insects and gene transfer between GM and non-GM crops; yet, after 20 years of cultivation, there is no proof of negative impacts of gene transfer from Bt GM crop to a wild type (Baulcombe et al. 2014).

In United States, Bt-GM maize reduced the quantities of pesticides used on corn by 80%. This type of reduction has occurred in all countries that used Bt-maize crop, same situation with Bt-cotton and other crops (Brookes and Barfoot 2014). The Bt-toxin is contained solely in the tissues of the GM crop, but despite that, it has been already described Bt-resistance by the western corn rootworm (Gassmann et al. 2011).

The reduction of pesticide use may be directly attributed to the GM technologies to keep the pesticide inside the plant, and that target specifically certain insects. In the sustainability area, agro workers and the ecosystem do not have direct contact with the pesticide (such as the spraying technique), becoming a very positive outcome of the technology (Pimentel and Burgess 2014). A report from the National Academy of Sciences (NAS), in 2016, established that GM crops do not have any negative impact on the environment, ecosystems, biodiversity, and living health, mainly because GMOs have produced the use of fewer herbicides, avoiding their bioaccumulation, but it warns about the presence of novel secondary metabolites produced by genetic modifications (National Academies of Sciences Engineering and Medicine et al. 2016).

Nonetheless, Liu (2015) described that the joint growth of transgenic and conventional rice effectively affects growth and reproduction, which will depend on the environment and competition (Liu et al. 2015). For example, in an ecosystem with the presence of non-specific pesticides to control pests, transgenic rice produced more aborted and less viable seeds than non-GMO plants, but there was no difference between plants if the ecosystem was not affected by insecticide application. Another study also described a difference in yield loss in Bt-transgenic rice, compared to the non-Bt crop under low insect pressure (James 2014), but other GM varieties did not present almost any yield difference (Brookes and Barfoot 2014). An explanation may be that the fitness cost to express the Bt transgene may affect yield with any pressure coming from the absence of insects, as seen in rice (Mannion and Morse 2013) and rice hybrids (Schell and Van Montagu 1977), but this situation does not occur in the *Brassica* genera or sunflower, so in general terms, Bt-transgenic and non-GMO rice will coexist where transgenic crops are permitted (Tabashnik 1994).

17.6.3 Biofortification

According to the World Health Organization (WHO), biofortification is the improvement of the nutritional quality of food crops by using agronomic practices, conventional plant breeding, or modern biotechnology (Bouis et al. 2013).

We saw Golden Rice as the most popular and known of the GMOs, being an excellent example of biofortification, and currently, modern breeding programs are focused also on the improvement of the nutritional and compositional qualities of the crops and the final product. Before the apparition of biotechnology, variations were made by using chemical or radiation mutagenesis, but lack of selectivity and can introduce undesired changes. Now, breeding efforts are based on molecular biology and genomics, allowing for marker-assisted breeding and genotyping by sequencing, but biotech has been a tool to optimize conventional breeding (Oliver 2014).

The combination of conventional breeding techniques with genomic technologies has allowed the identification of plant varieties with naturally higher nutrients, as well as the addition of foreign DNA to enhance nutritional properties. Some examples of biofortified crops are rice with β-carotene, iron or folate, maize as ascorbate (fivefold increase), soybean with four times more oleic acid, canola with four times more omega-3 fatty acid, and wheat with three times more amylase, among others (Oliver 2014).

Gene editing, such as CRISPR-Cas9, has allowed increasing the nutrient content of potatoes by turning off the granule-bound starch synthase (*gbss*) gene (Andersson et al. 2017). Sun et al. (2017) modified the *sbeI* and *sbeIIb* genes in rice, increasing amylose and starch levels, obtaining grains with an increased nutritional value (Sun et al. 2017). Li et al., in 2018, developed a multiplex vector system in tomato to enhance the production of gamma-aminobutyric acid (Li et al. 2018).

A novel approach to enhance crop properties is by the introduction of novel pathways into the plant: Lin et al. (2016) introduced the C4 photosynthesis pathway in rice, estimating that crop yield would be increased by 50%, as well as potentially being applied in soybean, wheat, and potatoes (Lin et al. 2016). Another approach is to give the ability to the plant to perform symbiosis with nitrogen-fixing bacteria, decreasing the need for nitrogen-based fertilizers (Charpentier and Oldroyd 2010).

The bioavailability of micronutrients in biofortified crops will depend on the plant species; for example, betacarotene is well absorbed from biofortified cereals, roots, and tubers, but in grain crops, the presence of phthalates delays the absorption of iron and zinc (La Frano et al. 2014). One problem is if the target nutrient requires more than one single genetic locus, i.e., which gene or genes are involved in the biosynthetic pathway to be able to modify and improve the levels of the nutrient. As a solution, Bohra (2013) reviewed several software that allows researchers to look for analysis of quantitative trait loci (QTL) related to complex phenotypes (Bohra 2013).

17.6.4 Biofuels

In terms of sustainability, biofuels have passed under several generations to improve crop yield and use: The first generation of biofuels required a huge crop area of oilseed rape (OSR—biodiesel) and competed with food production, rising food prices due to fewer crop areas (Rathmann et al. 2010). One proposed solution was to increase new farmland with the promise of decreasing greenhouse gases, in comparison to fossil fuels, which are known to be one of the biggest sources of gas (Searchinger et al. 2008), but if it is not possible, greenhouse emissions may be reduced if the yields are increased using the same land area to have crop products for food, and also for fuels (Moser et al. 2013). As stated previously, GM plants may increase crop production, hence a great solution for biofuels. If the crops are destined solely for biofuels, and not intended for feeding humans nor animals, maybe public opinion will not be as harsh as the usual GMO opinions (Gressel 2008).

17.7 Consequences of the Use of GMO in Sustainable Agriculture

Fitness, according to Primack and Kang, is the relative number of offspring contributed to the next generation by particular individuals or genotypes (Primack and Kang 1989). Tsatsakis (2017) defines it as the survival of a hybrid with a good reproductive ability in a given environment (Tsatsakis et al. 2017). In hybrids, the factors that can affect fitness are pleiotropy, i.e., when one gene influences two or more unrelated phenotypic traits, receiving the name of the pleiotropic gene, natural selection, life cycle, seed dormancy, fecundity, and geography, among others (Watrud et al. 2004).

The literature describes that the potential for cross-fertilization is high in areas where sexually cross-compatibility exists between GM and non-GM crops. GM-wild-type hybrids were reported in canola, corn, oilseed rape, rice, soybean, sunflower, and even *Arabidopsis* (Sánchez et al. 2016; Tsatsakis et al. 2017). If the hybrid passes to the wild, its persistence may be a serious environmental threat: Beckie and Warwick (2010) described that the Oxy235-transgene, which gives tolerance to the oxynil resistance, from GM oilseed rape persisted for years in Canada (Beckie and Warwick 2010).

Schulze et al. observed, in oilseed rape, genes from GM canola (*CP-EPSPS* and *goxy-247* genes) for two consecutive years (Schulze et al. 2014), and in Australia, similar reports suggested that GM canola, resistant to glyphosate, persisted in the natural ecosystem outside cultivated crops (Busi and Powles 2016). Guan et al. (2015) described herbicide resistance genes from GM to wild soybean, without affecting growth (Guan et al. 2015), while Cao et al. (2009) even demonstrated the superior performance of F1 wild-type-Bt GM hybrids, compared to the wild type, with an increased height, number of tillers and spikelets, as well as seed weight (Cao et al. 2009).

Chandler and Dunwel (2008) describe that gene transfer from GMO crop to a wild type will depend on several factors: open or self-pollination as the reproductive strategy of the crop, proximity of GM and non-GM crops, and proper characteristics of the hybrid, such as fertility, fecundity, and fitness, the latter being the most important parameter to establish the presence of the transgene in the wild relative (Chandler and Dunwell 2008). In Oregon, the glyphosate tolerance gene from GM grasses escaped from a field trial and was incorporated into native creeping bentgrass (Reichman et al. 2006), and Schafer et al. (2011) described that a variety of GM Canola led to transgene transfer into no-GMO wild-type canola in North Dakota, but no negative effect was described in these populations (Schafer et al. 2011).

Oliver (2014), in an excellent review, obtains the following conclusion about Bt-GM (Oliver 2014):

- Bt-GM crops have (almost) no effect on the biodiversity of soils. Some papers describe changes but can be explained by other parameters besides the modification
- There are no significant negative/positive effects on nontarget organisms in the field
- An effect may happen into the natural predators and/or parasites of the target pest, but they were only seen under controlled lab or greenhouse studies. In the field, these effects were not observed.
- In areas where Bt-GM crops are predominating, the population of the target pest declines that even benefits farms that grow non-GM crops.

Supporting the last point, Lu et al. (2012) show that the use of Bt-cotton, resulting in a decrease in insecticide use, reduced the population of the cotton bollworm, and increased the numbers of generalist predators, such as ladybugs and spiders (Lu et al. 2012). The increase of these predators produced a decrease in cotton aphid populations, which are not affected by the Bt toxin, which may damage the plant. This strategy may be used for also as a biocontrol of plant pests that affect neighboring non-GM crops.

17.8 General Considerations for the Use of GMOs and Nanotechnology in Sustainable Agriculture

GM crops support sustainable agriculture by not harming animals, humans, and the environment, as established by the EFSA GMO Panel Working Group on Animal Feeding Trials (2008) and in humans, by the Committee on Identifying and Assessing Unintended Effects of Genetically Engineered Foods on Human Health (Ammann 2014; Kaput et al. 2015). Additionally, there are no clear differences in the nutritional content of GM and non-GM crops when plant species and environmental variations are considered (Bøhn et al. 2014).

All the available scientific evidence on the uses of GMOs and nanotechnology must be continuously evaluated and taken into account when any novel research is done, from monitoring experimental studies to GM crop cultivation (Bock et al. 2002). Genetic modifications, such as CRISPR-Cas applications, have several implications related to morals and ethics in agriculture, biomedicine, and industrial biotechnology (Mulvihill et al. 2017). For example, Tsatsakis (2017) exposes that the toxicity tests, at that moment, presented several limitations and were case-specific (Tsatsakis et al. 2017). Also, it is barely impossible to specify which component of the diet may be potentially hazardous, because currently, people are under a mixture of GM foods, rather than just one (Hernández and Tsatsakis 2017).

Fagerstrom raises a simple, but valid question: What would happen if we do not accept GM crops or nanotechnology? Farmers will lose revenues, technologies, and resources lost, a reduction in agricultural productivity and yields, low sustainability because of the high use of pesticides, fertilizers, and other agrochemicals, varieties with reduced vitamins and/or nutrients, and many more that can be solved by using current technologies (Fagerström et al. 2012). The same researcher states that research is highly influenced by the possible environmental impacts of the application of the novel technology to crops; the paradigm wrongly establishes that biotechnology is inherently risky, and because of this, any new biotech application will be considered as a potential risk, requiring to overcome several regulatory hurdles that may diminish the intention to realize research that may be too polemic to be done (Apel 2010).

GM crops and nanotech have diminished the number of pesticides and agrochemicals used in agriculture (Perry et al. 2016), increased crop yields without the expansion of the cultivation areas, and reduced greenhouse gas emissions. Barrows et al. (2014) even state that an additional 13 million hectares of cultivation area would have been required in 2010 if farmers did not use GMOs, which would have come from the expense of rainforests, ultimately increasing CO_2 emissions (Barrows et al. 2014). Yield and production increase and the cost savings associated with the use of GMO crops resulted, in 2015, in economic benefits of more than US$15 billion and an estimated of approximately US$168 billion between 1996 and 2015. Also, the shortening of the production cycles from GMO crops enables farmers to plant a second crop within the same season (Brookes and Barfoot 2017).

Nano and GMO-crops may not be the most optimal solution, but it is a big advance in sustainability terms; anyhow, it has been already demonstrated resistance to biopesticides in, at least, 27 pest species (Bravo et al. 2011), reflecting the importance to keep a follow-up of any previous and current researches (Siegwart et al. 2015).

Kettenburg (2018), using Golden Rice as an example, describes that the use of novel technologies, such as nanotechnology and GMO crops, is strongly related to the integration of the available knowledge, disciplinary affiliation, and personal values, proposing three steps to overcome the problem and to achieve the desired sustainability (Kettenburg et al. 2018):

a) To recognize why a situation or technology is considered a problem, and how it contributes to sustainability
b) To work multi- and transdisciplinary, always prioritizing the well-being and thinking about future outcomes
c) To improve communication among research institutions

Gene transfer and genome editing have demonstrated to be a very strong tool to overcome pests and malnutrition, by engineering biological systems and introduced complete metabolic pathways, but it has also some restraints: As with any novel technology, there are doubts about what happens next. For CRISPR-Cas9, as an example, the recommendation is that seed stocks should not be maintained with active Cas-9 proteins in the genome, to avoid off-target recombination, based on if the primary target becomes successfully modified, the remanent gDNA may identify a new target and recombine (Graham et al. 2020). In plants, this problem may be solved by successive backcrossing, diminishing the mutational load associated with the CRISPR/Cas system (Bandyopadhyay et al. 2020).

Tsatsakis (2017) divides the consequences of using GM plants into two categories (Tsatsakis et al. 2017): The first one falls into the unintended impacts on the ecosystem, such as gene flow, a decrease in genetic diversity and biodiversity, effects on nontarget species, reduction in the efficiency of pesticides and herbicides. The second category is based on health hazards as food (Key et al. 2008). In 2016, the National Academy of Sciences (NAS) published a public letter signed by over 150 Nobel Laureates (https://www.supportprecisionagriculture.org) supporting precision agriculture, highlighting that there are not any alternatives to be able to feed the increasing population (Tsatsakis et al. 2017).

GMOs bring concerns about health, environmental impact, and socioeconomic hazards, resulting in strong opposition to these types of organisms (Gaskell et al. 2010). For example, in India, the local government prohibited the culture of modified eggplant (*Solanum melongena* engineered with *Bacillus thuringensis*), despite granting permission for its commercialization (Nature India 2010).

Europe developed from 1999 to 2004 an extremely strict regulatory framework involving import and cultivation of GM crops (Devos et al. 2006). But current research has shown that GM crop cultivation does not present health or environmental risks, only benefits to the agricultural farmers (Brookes and Barfoot 2016). Other countries, like Norway, have considered also GMO aspects not linked to safety, such as socioeconomic consideration, ethics, and potential contribution to sustainable development (Agapito-Tenfen et al. 2018).

Nanomaterials have the same problem as GMOs: They have demonstrated their efficiency as an antimicrobial agent, as biostimulant, and as a delivery system, giving a great advantage against conventional products, but there is a lack of knowledge about what happens to the nanomaterial in the environment (Hao et al. 2018) and its fate/bioaccumulation along the food chains (Rui et al. 2017). To achieve sustainability, nanomaterials must seek specific properties, such as biocompatibility, biodegradability, and no toxicity, as well as the impact of the nanoparticle by modifying size, shape, and surface functionalization (Fu et al. 2020).

The interaction between plants and NMs as delivery systems is not fully understood, as well as the effect of nanomaterial on plant growth, the effect on nutritional and quality values, being a big hurdle that prevents the use of nano-based fertilizers in agriculture (Bindraban et al. 2015). This chapter has been based on the improvement of crop yields and its defense against pests, but also it must be considered nanomaterials for food packaging as protection and to improve shelf-life, which ultimately will get into the final consumer (Bradley et al. 2011), but as usual, the exact antipathogen mechanism is not fully understood. Also, research groups have described adverse effects of nanoparticles on plants: Boonyanitipong et al., for example, described the phytotoxicity on ZnO_2- and TiO_2-nanoparticles on rice roots by demonstrating a detrimental effect of ZnO_2-NPs on rice roots at the early seedling stage by reducing the number and root length (Boonyanitipong et al. 2011), showing that it is not just the technology, but also to choose the correct type of nanoparticle may be the difference into obtaining positive results (Duhan et al. 2017).

Chai et al. tested nanoparticles effect on agricultural soil by evaluating thermal metabolism, amount and diversity of functional bacteria, and enzymatic activity: Zn- and Cesium-based nanoparticles reduced bacterial numbers, hindered thermogenic metabolism, and inhibited enzymatic activities, while titanium-based NPs reduced also the population of functional bacteria (Chai et al. 2015). He and collaborators (2019) mention that the toxicity of NMs may be altered in long-term exposures, recommending taking precaution, considering the low information about bioavailability and biodistribution of the material.

France reevaluated the safety of TiO_2(E171) nanoparticles as a food additive, and in 2017, also a French group led by the French Agricultural Research Institute reported pre-tumorous damages in rats' colon fed with TiO_2 nanoparticles after 100 days of treatment (Bettini et al. 2017). The REACH agency (from Registration, Evaluation, Authorisation, and Restriction of Chemicals) is the most active regulation agency in the European Union for the use of nanotechnology in the food industry, while this function is done in the United States by the Food and Drug Administration (FDA) (He et al. 2019).

According to Mishra (2017), there is an urgent need for risk assessments and management factors associated with NPs application in agriculture (Mishra et al. 2017). In this regard, regulatory bodies such as FDA, the Organization for Economic Cooperation and Development (OECD), and the International Standard Organization (ISO) exist, with the last two providing guidelines and suggestions to regulatory bodies. The EU and Switzerland have been able to establish nano-specific legislative provisions for agriculture, feed, and food, while others follow nonmandatory frameworks binding with nonlegal guidance (OECD 2013). This lack of standard regulatory frameworks has affected negatively the acceptance and reaching of nanotech-based products into the market. An analysis made by Almeida (2020) shows that between 2003 and 2011, the focus of studies associated with NMs was mainly on laws and regulations, where from 2012 to 2020, studies were made on risk assessments on NMs, consolidating the importance of regulatory studies on the effect of NMs on the environment and ecosystem (Almeida et al. 2020).

Any technology must consider the regulatory environment associated with it, comparing the potential benefits against the potential risks. Good results may improve public acceptance, but there will be always specific groups that will oppose the technology, no matter if the risk is low, or if the benefits are big. The keyword would be "transparency," publishing results and following the regulatory frameworks in every situation but also considering the public opinion by participating in public forums, when required (Bandyopadhyay et al. 2020).

17.9 Conclusions

Agriculture is essential to achieve food and feed quantities to sustain the growing population. Biotechnological tools, such as GMOs and nanotechnology, may help to reach sustainable agriculture that will allow us to improve crop yields, enhance nutrients quantity and quality, promote healthy plant growth, and protect crops from pests. Technologies exist, but the main challenge is to predict which will be the consequences to use them: the lack of knowledge is the big hurdle. Luckily, several agencies and research centers have taken into consideration the environmental impact, and the development of a comprehensive database and alarm system and a standardized regulation and legislation is mandatory to use these biotechnological tools to achieve sustainable agriculture.

As may be concluded, agricultural sustainability is not only a matter of technology regulation but also of industrial development, regional and national policies and regulations, imports and exports regulations, among others, requiring medium- and long-term policies that allow for discussion and for any specific contexts, which may vary from one place to the other (Bogner and Torgersen 2018); however, the future is promising: With the advances in gene editing and nanotechnology, crops and microbes can be modified at a genetic and cellular level, and together with the development of tools and systems to identify and monitor the crops, giant leaps have been done to achieve sustainable agriculture.

References

Abbas, S.S., Haneef, M., Lohani, M. et al. (2016). Nanomaterials used as a plants growth enhancer: an update. *International Journal of Pharma Research & Review* 5 (7): 17–23.

Abdel-Aziz, H.M.M., Hasaneen, M.N.A., and Omer, A.M. (2016). Nano chitosan-NPK fertilizer enhances the growth and productivity of wheat plants grown in sandy soil. *Spanish Journal of Agricultural Research* 14 (1): e0902. https://doi.org/10.5424/sjar/2016141-8205.

Afsharinejad, A., Davy, A., Jennings, B., and Brennan, C. (2016). Performance analysis of plant monitoring nanosensor networks at THz frequencies. *IEEE Internet of Things Journal* 3 (1): 59–69. https://doi.org/10.1109/JIOT.2015.2463685.

Agapito-Tenfen, S.Z., Okoli, A.S., Bernstein, M.J. et al. (2018). Revisiting risk governance of GM plants: the need to consider new and emerging gene-editing techniques. *Frontiers in Plant Science* 9: 1874. https://doi.org/10.3389/fpls.2018.01874.

Albitar, A., Rohani, B., Will, B. et al. (2018). The application of CRISPR/Cas technology to efficiently model complex cancer genomes in stem cells. *Journal of Cellular Biochemistry* 119 (1): 134–140. https://doi.org/10.1002/jcb.26195.

Almeida, L., Felzenszwalb, I., Marques, M., and Cruz, C. (2020). Nanotechnology activities: environmental protection regulatory issues data. *Heliyon* 6 (10): e05303. https://doi.org/10.1016/j.heliyon.2020.e05303.

Ammann, K. (2014). Genomic misconception: a fresh look at the biosafety of transgenic and conventional crops a plea for a process agnostic regulation. *New Biotechnology* 31 (1): 1–17. https://doi.org/10.1016/j.nbt.2013.04.008.

Ammar, K., Mergoum, M., and Rajaram, S. (2004). The history and evolution of triticale. *Triticale Improvement and Production* 1: 1–10.

Andersson, M., Turesson, H., Nicolia, A. et al. (2017). Efficient targeted multiallelic mutagenesis in tetraploid potato (Solanum tuberosum) by transient CRISPR-Cas9 expression in protoplasts. *Plant Cell Reports* https://doi.org/10.1007/s00299-016-2062-3.

Apel, A. (2010). The costly benefits of opposing agricultural biotechnology. *New Biotechnology* 27 (5): 635–640.

Aschberger, K., Gottardo, S., Amenta, V. et al. (2015). Nanomaterials in food – current and future applications and regulatory aspects. *Journal of Physics* 617 (1): 12032.

Atzberger, C. (2013). Advances in remote sensing of agriculture: context description, existing operational monitoring systems and major information needs. *Remote Sensing* 5 (2): 5020949.

Bacchin, P., Marty, A., Duru, P. et al. (2011). Colloidal surface interactions and membrane fouling: investigations at pore scale. *Advances in Colloid and Interface Science* 164 (1–2): 2–11.

Bandyopadhyay, A., Kancharla, N., Javalkote, V.S. et al. (2020). CRISPR-Cas12a (Cpf1): a versatile tool in the plant genome editing tool box for agricultural advancement. *Frontiers in Plant Science* 11: 584151. https://doi.org/10.3389/fpls.2020.584151.

Banerjee, J. and Kole, C. (2016). Plant nanotechnology: an overview on concepts, strategies, and tools. In: *Plant Nanotechnology: Principles and Practices* (ed. C. Kole, D. S. Kumar, and M. V. Khodakovskaya), 1–14. Cham, Switzerland: Springer. https://doi.org/10.1007/978-3-319-42152-0_1.

Barampuram, S. and Zhang, Z.J. (2011). Recent advances in plant transformation. *Methods in Molecular Biology* https://doi.org/10.1007/978-1-61737-957-4_1.

Barrows, G., Sexton, S., and Zilberman, D. (2014). The impact of agricultural biotechnology on supply and land-use. *Environment and Development Economics* https://doi.org/10.1017/S1355770X14000400.

Baulcombe, D., Dunwell, J., Jones, J. et al. (2014). GM science update: a report to the Council for Science and Technology. *Council for Science and Technology*. https://www.gov.uk/government/publications/genetic-modification-gm-technologies.

Bebber, D.P., Ramotowski, M.A.T., and Gurr, S.J. (2013). Crop pests and pathogens move polewards in a warming world. *Nature Climate Change* https://doi.org/10.1038/nclimate1990.

Beckie, H.J. and Warwick, S.I. (2010). Persistence of an oilseed rape transgene in the environment. *Crop Protection* https://doi.org/10.1016/j.cropro.2009.11.013.

Bernabé-Orts, J.M., Casas-Rodrigo, I., Minguet, E.G. et al. (2019). Assessment of Cas12a-mediated gene editing efficiency in plants. *Plant Biotechnology Journal* https://doi.org/10.1111/pbi.13113.

Bettini, S., Boutet-Robinet, E., Cartier, C. et al. (2017). Food-grade TiO_2 impairs intestinal and systemic immune homeostasis, initiates preneoplastic lesions and promotes aberrant crypt development in the rat colon. *Scientific Reports* https://doi.org/10.1038/srep40373.

Bindraban, P.S., Dimkpa, C., Nagarajan, L. et al. (2015). Revisiting fertilisers and fertilisation strategies for improved nutrient uptake by plants. *Biology and Fertility of Soil* 51: 897–911.

Block, A., Debode, F., Grohmann, L. et al. (2013). The GMOseek matrix: a decision support tool for optimizing the detection of genetically modified plants. *BMC Bioinformatics* 14: 256. https://doi.org/10.1186/1471-2105-14-256.

Bock A, Lheureux K, Libeau-Dulos M, Nilsagard H (2002) Scenarios for Co-Existence of Genetically Modified, Conventional and Organic Crops in European Agriculture. Retrieved May

Bogner, A. and Torgersen, H. (2018). Precaution, responsible innovation and beyond - in search of a sustainable agricultural biotechnology policy. *Frontiers in Plant Science* 9: 1884. https://doi.org/10.3389/fpls.2018.01884.

Bøhn, T., Cuhra, M., Traavik, T. et al. (2014). Compositional differences in soybeans on the market: glyphosate accumulates in roundup ready GM soybeans. *Food Chemistry* 153: 207–215. https://doi.org/10.1016/j.foodchem.2013.12.054.

Bohra, A. (2013). Emerging paradigms in genomics-based crop improvement. *Scientific World Journal* 17: 585467.

Boonyanitipong, P., Kositsup, B., Kumar, P. et al. (2011). Toxicity of ZnO and TiO_2 nanoparticles on germinating Rice seed Oryza sativa L. *International Journal of Bioscience, Biochemistry and Bioinformatics* https://doi.org/10.7763/ijbbb.2011.v1.53.

Bouis, H., Low, J., McEwan, M., and Tanumihardjo, S. (2013). *Biofortification: Evidence and Lessons Learned Linking Agriculture and Nutrition*. FAO WHO.

Bradley, E.L., Castle, L., and Chaudhry, Q. (2011). Applications of nanomaterials in food packaging with a consideration of opportunities for developing countries. *Trends in Food Science and Technology* https://doi.org/10.1016/j.tifs.2011.01.002.

Bravo, A., Gill, S.S., and Soberón, M. (2007). Mode of action of bacillus thuringiensis cry and cyt toxins and their potential for insect control. *Toxicon* 49 (4): 423–435.

Bravo, A., Likitvivatanavong, S., Gill, S.S., and Soberón, M. (2011). Bacillus thuringiensis: a story of a successful bioinsecticide. *Insect Biochemistry Molecular Biology* 41 (7): 423–431.

Brookes, G. and Barfoot, P. (2016). Global income and production impacts of using GM crop technology 1996–2014. *GM Crops & Food* 7 (1): 38–77. https://doi.org/10.1080/21645698.2016.1176817.

Brookes, G. and Barfoot, P. (2014). Economic impact of GM crops: the global income and production effects 1996–2012. *GM Crops & Food* https://doi.org/10.4161/gmcr.28098.

Brookes, G. and Barfoot, P. (2017). Farm income and production impacts of using GM crop technology 1996–2015. *GM Crops & Food* 9 (2): 59–89. https://doi.org/10.1080/21645698.2017.1317919.

Busi, R. and Powles, S.B. (2016). Transgenic glyphosate-resistant canola (Brassica napus) can persist outside agricultural fields in Australia. *Agriculture, Ecosystems and Environment* https://doi.org/10.1016/j.agee.2015.12.028.

Cai, D., Wu, Z., Jiang, J. et al. (2015). Controlling nitrogen migration through micro-nano networks. *Scientific Reports* https://doi.org/10.1038/srep03665.

Cao, Q.J., Xia, H., Yang, X., and Lu, B.R. (2009). Performance of hybrids between weedy rice and insect-resistant transgenic rice under field experiments: implication for environmental biosafety assessment. *Journal of Integrative Plant Biology* https://doi.org/10.1111/j.1744-7909.2009.00877.x.

Cassen, R.H. (1987). Our common future: report of the world commission on environment and development. *International Affairs* https://doi.org/10.2307/2621529.
Chai, H., Yao, J., Sun, J. et al. (2015). The effect of metal oxide nanoparticles on functional bacteria and metabolic profiles in agricultural soil. *Bulletin of Environmental Contamination and Toxicology* https://doi.org/10.1007/s00128-015-1485-9.
Chandler, S. and Dunwell, J.M. (2008). Gene flow, risk assessment and the environmental release of transgenic plants. *Critical Reviews in Plant Sciences* 27 (1): 25–49.
Charpentier, M. and Oldroyd, G. (2010). How close are we to nitrogen-fixing cereals? *Current Opinion in Plant Biology* 13 (5): 556–564.
Chen, S., Sanjana, N.E., Zheng, K. et al. (2015). Genome-wide CRISPR screen in a mouse model of tumor growth and metastasis. *Cell* https://doi.org/10.1016/j.cell.2015.02.038.
Chen, X., Lu, X., Shu, N. et al. (2017). Targeted mutagenesis in cotton (Gossypium hirsutum L.) using the CRISPR/Cas9 system. *Scientific Reports* https://doi.org/10.1038/srep44304.
Chumakov, M.I. and Moiseeva, E.M. (2012). Technologies of agrobacterium plant transformation in planta. *Applied Biochemistry and Microbiology* 48: 657–666.
Congreves, K.A. and Van Eerd, L.L. (2015). Nitrogen cycling and management in intensive horticultural systems. *Nutrient Cycling in Agroecosystems* 102: 299–318.
Cottenet, G., Blancpain, C., Sonnard, V., and Chuah, P.F. (2013). Development and validation of a multiplex real-time PCR method to simultaneously detect 47 targets for the identification of genetically modified organisms. *Analytical and Bioanalytical Chemistry* https://doi.org/10.1007/s00216-013-7125-5.
Cox, A., Venkatachalam, P., Sahi, S., and Sharma, N. (2016). Silver and titanium dioxide nanoparticle toxicity in plants: a review of current research. *Plant Physiology and Biochemistry* 107: 147–163.
Das, G., Patra, J.K., Paramithiotis, S., and Shin, H.-S. (2019). The sustainability challenge of food and environmental nanotechnology: current status and imminent perceptions. *International Journal of Environmental Research and Public Health* 16 (23): https://doi.org/10.3390/ijerph16234848.
Day, A. and Goldschmidt-Clermont, M. (2011). The chloroplast transformation toolbox: selectable markers and marker removal. *Plant Biotechnology Journal* 9 (5): 540–553. https://doi.org/10.1111/j.1467-7652.2011.00604.x.
De, A., Bose, R., Kumar, A., and Mozumdar, S. (2014). *Targeted Delivery of Pesticides using Biodegradable Polymeric Nanoparticles*. New Delhi, India: Springer. https://doi.org/10.1007/978-81-322-1689-6.
de Fraiture, C., Molden, D., and Wichelns, D. (2010). Investing in water for food, ecosystems, and livelihoods: an overview of the comprehensive assessment of water management in agriculture. *Agricultural Water Management* https://doi.org/10.1016/j.agwat.2009.08.015.
Devos, Y., Reheul, D., De Waele, D., and Van Speybroeck, L. (2006). The interplay between societal concerns and the regulatory frame on GM crops in the European Union. *Environmental Biosafety Research* https://doi.org/10.1051/ebr:2007002.
di Silvestro, R., Marotti, I., Bosi, S. et al. (2012). Health-promoting phytochemicals of Italian common wheat varieties grown under low-input agricultural management. *Journal of the Science of Food and Agriculture* https://doi.org/10.1002/jsfa.5590.
Dixit, R., Wasiullah, M.D., Pandiyan, K. et al. (2015). Bioremediation of heavy metals from soil and aquatic environment: an overview of principles and criteria of fundamental processes. *Sustainability* 7 (2): 2189–2212.

Doudna, J.A. and Charpentier, E. (2014). The new frontier of genome engineering with CRISPR-Cas9. *Science* 346 (6213): 1258096.

Dubey, A. and Mailapalli, D.R. (2016). Nanofertilisers, nanopesticides, nanosensors of pest and nanotoxicity in agriculture. *Agricultural and Food Sciences, Environmental Science, Materials Science* https://doi.org/10.1007/978-3-319-26777-7_7.

Duhan, J.S., Kumar, R., Kumar, N. et al. (2017). Nanotechnology: the new perspective in precision agriculture. *Biotechnol Reports (Amsterdam, Netherlands)* 15: 11–23. https://doi.org/10.1016/j.btre.2017.03.002.

EFSA GMO Panel Working Group on Animal Feeding Trials. (2008). Safety and nutritional assessment of GM plants and derived food and feed: the role of animal feeding trials. *Food and Chemical Toxicology* 46 (Suppl. 1): S2–S70. https://doi.org/10.1016/j.fct.2008.02.008.

Elmer, W. and White, J.C. (2018). The future of nanotechnology in plant pathology. *Annual Review of Phytopathology* 56: 111–133. https://doi.org/10.1146/annurev-phyto-080417-050108.

Espinoza-Orias, N., Roulin, A., Watzke, H., and Cooper, K. (2014). Connecting the dots: assessing sustainable nutrition at Nestlé. In: *Proceedings of the 9th international conference LCA of food*. LCA Food.

Esquinas-Alcázar, J. (2005). Protecting crop genetic diversity for food security: political, ethical and technical challenges. *Nature Reviews Genetics* 6: 946–953.

Fagerström, T., Dixelius, C., Magnusson, U., and Sundström, J.F. (2012). Stop worrying; start growing. Risk research on GM crops is a dead parrot: it is time to start reaping the benefits of GM. *EMBO Reports* 13 (6): 493–497. https://doi.org/10.1038/embor.2012.59.

Fang, A., Chen, H., Li, H. et al. (2017). Glutathione regulation-based dual-functional upconversion sensing-platform for acetylcholinesterase activity and cadmium ions. *Biosensors & Bioelectronics* https://doi.org/10.1016/j.bios.2016.08.111.

Fincheira, P., Tortella, G., Duran, N. et al. (2020). Current applications of nanotechnology to develop plant growth inducer agents as an innovation strategy. *Critical Reviews in Biotechnology* 40 (1): 15–30. https://doi.org/10.1080/07388551.2019.1681931.

Fischer, E.M. and Knutti, R. (2015). Anthropogenic contribution to global occurrence of heavy-precipitation and high-temperature extremes. *Nature Climate Change* https://doi.org/10.1038/nclimate2617.

Foster, A.J., Martin-Urdiroz, M., Yan, X. et al. (2018). CRISPR-Cas9 ribonucleoprotein-mediated co-editing and counterselection in the rice blast fungus. *Scientific Reports* https://doi.org/10.1038/s41598-018-32702-w.

Frenk, S., Ben-Moshe, T., Dror, I. et al. (2013). Effect of metal oxide nanoparticles on microbial community structure and function in two different soil types. *PLoS One* https://doi.org/10.1371/journal.pone.0084441.

Fu, L., Wang, Z., Dhankher, O.P., and Xing, B. (2020). Nanotechnology as a new sustainable approach for controlling crop diseases and increasing agricultural production. *Journal of Experimental Botany* 71 (2): 507–519. https://doi.org/10.1093/jxb/erz314.

Galbraith, D.W. (2007). Nanobiotechnology: Silica Breaks through in Plants. *Nature Nanotechnology* https://doi.org/10.1038/nnano.2007.118.

Ganeshkumar, R., Sopiha, K.V., Wu, P. et al. (2016). Ferroelectric $KNbO_3$ nanofibers: synthesis, characterization and their application as a humidity nanosensor. *Nanotechnology* https://doi.org/10.1088/0957-4484/27/39/395607.

Gaskell, G., Stares, S., Allansdottir, A. et al. (2010). Europeans and biotechnology in 2010: winds of change? A report to the European Commission's Directorate-General for Research. Publications Office of the European Union. https://doi.org/10.2777/23393.

Gassmann, A.J., Petzold-Maxwell, J.L., Keweshan, R.S., and Dunbar, M.W. (2011). Field-evolved resistance to Bt maize by Western corn rootworm. *PLoS One* https://doi.org/10.1371/journal.pone.0022629.

Geer, L.Y., Marchler-Bauer, A., Geer, R.C. et al. (2009). The NCBI BioSystems database. *Nucleic Acids Research* https://doi.org/10.1093/nar/gkp858.

Ghorbanpour, M. and Fahimirad, S. (2017). Plant nanobionics a novel approach to overcome the environmental challenges. In: *Medicinal Plants and Environmental Challenges*. Springer International Publishing https://doi.org/10.1007/978-3-319-68717-9_14.

Ghormade, V., Deshpande, M.V., and Paknikar, K.M. (2011). Perspectives for nano-biotechnology enabled protection and nutrition of plants. *Biotechnology Advances* https://doi.org/10.1016/j.biotechadv.2011.06.007.

Giannousi, K., Avramidis, I., and Dendrinou-Samara, C. (2013). Synthesis, characterization and evaluation of copper based nanoparticles as agrochemicals against Phytophthora infestans. *RSC Advances* https://doi.org/10.1039/c3ra42118j.

Giraldo, J.P., Landry, M.P., Faltermeier, S.M. et al. (2014). Plant nanobionics approach to augment photosynthesis and biochemical sensing. *Nature Materials* https://doi.org/10.1038/nmat3890.

Godoy-Navajas, J., Aguilar Caballos, M.P., and Gómez-Hens, A. (2011). Heterogeneous immunoassay for soy protein determination using nile blue-doped silica nanoparticles as labels and front-surface long-wavelength fluorimetry. *Analytica Chimica Acta* https://doi.org/10.1016/j.aca.2011.06.005.

Gogos, A., Knauer, K., and Bucheli, T.D. (2012). Nanomaterials in plant protection and fertilization: current state, foreseen applications, and research priorities. *Journal of Agricultural and Food Chemistry* https://doi.org/10.1021/jf302154y.

Graham, N., Patil, G.B., Bubeck, D.M. et al. (2020). Plant genome editing and the relevance of off-target changes1[OPEN]. *Plant Physiology* https://doi.org/10.1104/pp.19.01194.

Gressel, J. (2008). Transgenics are imperative for biofuel crops. *Plant Science* https://doi.org/10.1016/j.plantsci.2007.11.009.

Guan, Z.-J., Zhang, P.-F., Wei, W. et al. (2015). Performance of hybrid progeny formed between genetically modified herbicide-tolerant soybean and its wild ancestor. *AoB Plants* https://doi.org/10.1093/aobpla/plv121.

Guo, J., Chen, L., Liu, X. et al. (2012). A multiplex degenerate PCR analytical approach targeting to eight genes for screening GMOs. *Food Chemistry* https://doi.org/10.1016/j.foodchem.2011.11.096.

Haddad, L., Achadi, E., Bendech, M.A. et al. (2015). The global nutrition report 2014: actions and accountability to accelerate the world's progress on nutrition. *The Journal of Nutrition* https://doi.org/10.3945/jn.114.206078.

Haghighi, M., Afifipour, Z., and Mozafarian, M. (2012). The alleviation effect of silicon on seed germination and seedling growth of tomato under salinity stress. *Journal of Fruit and Ornamental Plant Research* https://doi.org/10.2478/v10032-012-0008-z.

Hammond, B., Kough, J., Herouet-Guicheney, C., and Jez, J.M. (2013). Toxicological evaluation of proteins introduced into food crops. *Critical Review in Toxicology* https://doi.org/10.3109/10408444.2013.842956.

Handler, A.M. (2008). Genetic transformation. In: *Encyclopedia of Entomology* (ed. J.L. Capinera), 1597–1599. Netherlands, Dordrecht: Springer.

Hao, Y., Yuan, W., Ma, C. et al. (2018). Engineered nanomaterials suppress turnip mosaic virus infection in tobacco (Nicotiana benthamiana). *Environmental Science Nano* https://doi.org/10.1039/c8en00014j.

Hare, P.D. and Chua, N.H. (2002). Excision of selectable marker genes from transgenic plants. *Nature Biotechnology* 20 (6): 575–580.

Haverkort, A.J., Boonekamp, P.M., Hutten, R. et al. (2008). Societal costs of late blight in potato and prospects of durable resistance through cisgenic modification. *Potato Research* 51: 47–57.

Haverkort, A.J., Boonekamp, P.M., Hutten, R. et al. (2016). Durable late blight resistance in potato through dynamic varieties obtained by Cisgenesis: scientific and societal advances in the DuRPh project. *Potato Research* https://doi.org/10.1007/s11540-015-9312-6.

He, L., Haynes, C.L., Diez-Gonzalez, F., and Labuza, T.P. (2011). Rapid detection of a foreign protein in milk using IMS-SERS. *Journal of Raman Specroscopy* https://doi.org/10.1002/jrs.2880.

He, X., Deng, H., and Hwang, H.-M. (2019). The current application of nanotechnology in food and agriculture. *Journal of Food and Drug Analysis* 27 (1): 1–21. https://doi.org/10.1016/j.jfda.2018.12.002.

Herforth, A., Frongillo, E.A., Sassi, F. et al. (2014). Toward an integrated approach to nutritional quality, environmental sustainability, and economic viability: research and measurement gaps. *Annals of the New York Academy of Sciences* https://doi.org/10.1111/nyas.12552.

Hernández-Rodríguez, C.S., Hernández-Martínez, P., Van Rie, J. et al. (2013). Shared midgut binding sites for Cry1A.105, Cry1Aa, Cry1Ab, Cry1Ac and Cry1Fa proteins from bacillus thuringiensis in two important corn pests, Ostrinia nubilalis and Spodoptera frugiperda. *PLoS One* https://doi.org/10.1371/journal.pone.0068164.

Hernández, A.F. and Tsatsakis, A.M. (2017). Human exposure to chemical mixtures: challenges for the integration of toxicology with epidemiology data in risk assessment. *Food and Chemical Toxicology* https://doi.org/10.1016/j.fct.2017.03.012.

Huang, J., Hu, R., Rozelle, S., and Pray, C. (2005). Plant science: insect-resistant GM rice in farmers' fields: assessing productivity and health effects in China. *Science* 80: https://doi.org/10.1126/science.1108972.

Hunter, M.C., Smith, R.G., Schipanski, M.E. et al. (2017). Agriculture in 2050: recalibrating targets for sustainable intensification. *BioScience* https://doi.org/10.1093/biosci/bix010.

Iavicoli, I., Leso, V., Beezhold, D.H., and Shvedova, A.A. (2017). Nanotechnology in agriculture: opportunities, toxicological implications, and occupational risks. *Toxicology and Applied Pharmacology* 329: 96–111. https://doi.org/10.1016/j.taap.2017.05.025.

James, C. (2014). *Brief 49 Dlobal status of Commercialized biotech/GM Crops: 2014*. The International Service for the Acquisition of Agri-biotech Applications (ISAAA).

Jampílek, J. and KráL'Ová, K. (2015). Application of nanotechnology in agriculture and food industry, its prospects and risks. *Ecological Chemistry and Engineering* https://doi.org/10.1515/eces-2015-0018.

Jiang, R.H.Y. and Tyler, B.M. (2012). Mechanisms and evolution of virulence in oomycetes. *Annual Review Phytopathology* https://doi.org/10.1146/annurev-phyto-081211-172912.

Jones, J.D.G., Witek, K., Verweij, W. et al. (2014). Elevating crop disease resistance with cloned genes. *Philosophical Transactions of the Royal Society of London B Biological Science* https://doi.org/10.1098/rstb.2013.0087.

Kah, M. (2015). Nanopesticides and nanofertilizers: emerging contaminants or opportunities for risk mitigation? *Frontiers in Chemistry* https://doi.org/10.3389/fchem.2015.00064.

Kaput, J., Kussmann, M., Mendoza, Y. et al. (2015). Enabling nutrient security and sustainability through systems research. *Genes & Nutrition* 10 (3): 462. https://doi.org/10.1007/s12263-015-0462-6.

Kashyap, P.L., Rai, P., Srivastava, A.K., and Kumar, S. (2017). Trichoderma for climate resilient agriculture. *World Journal of Microbiology and Biotechnology* https://doi.org/10.1007/s11274-017-2319-1.

Kaur, P., Thakur, R., Duhan, J.S., and Chaudhury, A. (2018). Management of wilt disease of chickpea in vivo by silver nanoparticles biosynthesized by rhizospheric microflora of chickpea (*Cicer arietinum*). *Journal of Chemical Technology and Biotechnology* https://doi.org/10.1002/jctb.5680.

Kerfahi, D., Tripathi, B.M., Singh, D. et al. (2015). Effects of functionalized and raw multi-walled carbon nanotubes on soil bacterial community composition. *PLoS One* https://doi.org/10.1371/journal.pone.0123042.

Kettenburg, A.J., Hanspach, J., Abson, D.J., and Fischer, J. (2018). From disagreements to dialogue: unpacking the Golden Rice debate. *Sustainability Science* 13 (5): 1469–1482. https://doi.org/10.1007/s11625-018-0577-y.

Key, S., Ma, J.K.C., and Drake, P.M.W. (2008). Genetically modified plants and human health. *Journal of the Royal Society of Medicine* https://doi.org/10.1258/jrsm.2008.070372.

Khan, M.R. and Rizvi, T.F. (2014). Nanotechnology: scope and application in plant disease management. *Plant Pathology Journal* https://doi.org/10.3923/ppj.2014.214.231.

Kim, Y.H., Choi, S.J., Lee, H.A., and Moon, T.W. (2006). Quantitation of CP4 5-enolpyruvylshikimate-3-phosphate synthase in soybean by two-dimensional gel electrophoresis. *Journal Microbiological Biotechnology* 16 (1): 25–31.

La Frano, M.R., de Moura, F.F., Boy, E. et al. (2014). Bioavailability of iron, zinc, and provitamin a carotenoids in biofortified staple crops. *Nutrition Reviews* https://doi.org/10.1111/nure.12108.

Lahiani, M.H., Dervishi, E., Chen, J. et al. (2013). Impact of carbon nanotube exposure to seeds of valuable crops. *ACS Applied Materials & Interfaces* https://doi.org/10.1021/am402052x.

Ledford, H. (2015). CRISPR, the disruptor. *Nature* https://doi.org/10.1038/522020a.

Li, R., Li, R., Li, X. et al. (2018). Multiplexed CRISPR/Cas9-mediated metabolic engineering of γ-aminobutyric acid levels in Solanum lycopersicum. *Plant Biotechnology Journal* https://doi.org/10.1111/pbi.12781.

Lin, H., Karki, S., Coe, R.A. et al. (2016). Targeted knockdown of GDCH in Rice leads to a photorespiratory-deficient phenotype useful as a building Block for C4 Rice. *Plant & Cell Physiology* https://doi.org/10.1093/pcp/pcw033.

Liu, R., Kang, Y., Pei, L. et al. (2016). Use of a new controlled-loss-fertilizer to reduce nitrogen losses during winter wheat cultivation in the Danjiangkou reservoir area of China. *Communications in Soil Science and Plant Analysis* https://doi.org/10.1080/00103624.2016.1166245.

Liu, Y., Ge, F., Liang, Y. et al. (2015). Characterization of competitive interactions in the coexistence of Bt-transgenic and conventional rice. *BMC Biotechnology* 15: 27. https://doi.org/10.1186/s12896-015-0141-0.

López, M.C.G., Garcia-Cañas, V., and Alegre, M.L.M. (2009). Reversed-phase high-performance liquid chromatography-electrospray mass spectrometry profiling of transgenic and non-transgenic maize for cultivar characterization. *Journal of Chromatography A* https://doi.org/10.1016/j.chroma.2009.08.050.

Lowry, G.V., Avellan, A., and Gilbertson, L.M. (2019). Opportunities and challenges for nanotechnology in the Agri-tech revolution. *Nature Nanotechnology* 14 (6): 517–522. https://doi.org/10.1038/s41565-019-0461-7.

Lu, C., Zhang, C., Wen, J. et al. (2002). Research of the effect of nanometer materials on germination and growth enhancement of Glycine max and its mechanism. *Soybean Science* 21 (3): 168–171.

Lu, Y., Wu, K., Jiang, Y. et al. (2012). Widespread adoption of Bt cotton and insecticide decrease promotes biocontrol services. *Nature* https://doi.org/10.1038/nature11153.

Ma, X., Geiser-Lee, J., Deng, Y., and Kolmakov, A. (2010). Interactions between engineered nanoparticles (enps) and plants: phytotoxicity, uptake and accumulation. *Science of The Total Environment* 408 (16): 3053–3061.

Manjaiah, K. M., Mukhopadhyay, R., Paul, R. et al. (2018). Clay minerals and zeolites for environmentally sustainable agriculture. In: *Modified Clay and Zeolite Nanocomposite Materials: Environmental and Pharmaceutical Applications* (ed. B. Sarkar, Y.F. Tsang, and K.M. Manjaiah), 327–329. Amsterdam, Netherlands: Elsevier. https://doi.org/10.1016/B978-0-12-814617-0.00008-6.

Mannion, A.M. and Morse, S. (2013). Gm Crops 1996–2012: a review of agronomic, environmental and socio-economic impacts. *University of Surrey, Centre for Environmental Strategy Working Paper* 4 (13): 1–40.

Marchiol, L. (2018). Nanotechnology in agriculture: new opportunities and perspectives. *New Visions Plant Science* 9 (4): 161.

Marmiroli, N., Maestri, E., Gullì, M. et al. (2008). Methods for detection of GMOs in food and feed. *Analytical and Bioanalytical Chemistry* 392 (3): 369–384.

Mastronardi, E., Tsae, P., Zhang, X. et al. (2015). Strategic role of nanotechnology in fertilizers: potential and limitations. In: *Nanotechnologies in Food and Agriculture*, 25–67. Cham: Springer International Publishing.

McClements, D.J. (2020). Nanotechnology approaches for improving the healthiness and sustainability of the modern food supply. *ACS Omega* 5 (46): 29623–29630. https://doi.org/10.1021/acsomega.0c04050.

McClements, D.J. (2019). *Future Foods: How Modern Science is Transforming the Way We Eat.* Springer.

McLamore, E.S., Diggs, A., Calvo Marzal, P. et al. (2010). Non-invasive quantification of endogenous root auxin transport using an integrated flux microsensor technique. *The Plant Journal* https://doi.org/10.1111/j.1365-313X.2010.04300.x.

Miller, J.B., Zhang, S., Kos, P. et al. (2017). Non-viral CRISPR/Cas gene editing in vitro and in vivo enabled by synthetic nanoparticle co-delivery of Cas9 mRNA and sgRNA. *Angew Chemie - Int Ed.* https://doi.org/10.1002/anie.201610209.

Mishra, S., Keswani, C., Abhilash, P.C. et al. (2017). Integrated approach of Agri-nanotechnology: challenges and future trends. *Frontiers in Plant Science* 8: 471. https://doi.org/10.3389/fpls.2017.00471.

Mishra, S., Singh, B.R., Singh, A. et al. (2014). Biofabricated silver nanoparticles act as a strong fungicide against Bipolaris sorokiniana causing spot blotch disease in wheat. *PLoS One* https://doi.org/10.1371/journal.pone.0097881.

Moser, D., Eckerstorfer, M., Pascher, K. et al. (2013). Potential of genetically modified oilseed rape for biofuels in Austria: land use patterns and coexistence constraints could decrease domestic feedstock production. *Biomass Bioenergy* 50: 35–44. https://doi.org/10.1016/j.biombioe.2012.10.004.

Mout, R., Ray, M., Yesilbag Tonga, G. et al. (2017). Direct cytosolic delivery of CRISPR/Cas9-ribonucleoprotein for efficient gene editing. *ACS Nano* https://doi.org/10.1021/acsnano.6b07600.

Mulvihill, J.J., Capps, B., Joly, Y. et al. (2017). Ethical issues of CRISPR technology and gene editing through the lens of solidarity. *British Medical Bulletin* 122 (1): 17–29.

Mundt, C.C. (2014). Durable resistance: a key to sustainable management of pathogens and pests. *Infection, Genetics and Evolution* 27: 446–455.

Myers, S.S., Zanobetti, A., Kloog, I. et al. (2014). Increasing CO_2 threatens human nutrition. *Nature* https://doi.org/10.1038/nature13179.

Nair, R., Varghese, S.H., Nair, B.G. et al. (2010). Nanoparticulate material delivery to plants. *Plant Science* https://doi.org/10.1016/j.plantsci.2010.04.012.

Naso, G. and Petrova, A. (2019). CRISPR/Cas9 gene editing for genodermatoses: Progress and perspectives. *Emerging Topics in Life Science* https://doi.org/10.1042/ETLS20180148.

National Academies of Sciences Engineering and Medicine, Committee on Genetically Engineered Crops, Board on Agriculture and Natural Resources, National Division on Earth and Life Sciences, Academies of Sciences, Engineering and Medicine. (2016). Genetically engineered crops: experiences and prospects. https://doi.org/10.17226/23395.

National Research Council. (2010). *Toward Sustainable Agricultural Systems in the 21st Century*. National Academies.

Nature India (2010). India bows to negative sentiments on Bt brinjal. *Nature India* https://doi.org/10.1038/nindia.2010.10.

Nern, A., Pfeiffer, B.D., Svoboda, K., and Rubin, G.M. (2011). Multiple new site-specific recombinases for use in manipulating animal genomes. *Proceedings of the National Academy of Sciences of the United States of America* https://doi.org/10.1073/pnas.1111704108.

OECD (2006). OECD test guideline 208: terrestrial plant test – seedling emergence and seedling growth test. In: *Terrestrial Plant Test: Seedling Emergence and Seedling Growth Test*. OECD.

OECD (2013). Regulatory frameworks for nanotechnology in foods and medical products: summary results of a survey activity. In: *OECD Science, Technology and Industry Policy Papers*. OECD.

Oliver, M.J. (2014). Why we need GMO crops in agriculture. *Missouri Medicine* 111 (6): 492–507.

Oyunga, M.A., Grant, F.K.E., Omondi, D.O. et al. (2016). Prevalence and predictors of vitamin a deficiency among infants in western kenya using a cross-sectional analysis. *African Journal of Food, Agriculture, Nutrition and Development* https://doi.org/10.18697/ajfand.73.16190.

Kale, A.P. and Gawade, S.N. (2016). Studies on nanoparticle induced nutrient use eficiency of fertilizer and crop productivity. *Green Chemistry & Technology Letters* https://doi.org/10.18510/gctl.2016.226.

Paula de Toledo Thomazella D, Brail Q, Dahlbeck D, Staskawicz B (2016) CRISPR-Cas9 mediated mutagenesis of a DMR6 ortholog in tomato confers broad-spectrum disease resistance. https://doi.org/10.1101/064824

Peng, C., Wang, P., Xu, X. et al. (2016). Development of a qualitative real-time PCR method to detect 19 targets for identification of genetically modified organisms. *Springerplus* 5 (1): 889. https://doi.org/10.1186/s40064-016-2395-y.

Perry, E.D., Ciliberto, F., Hennessy, D.A., and Moschini, G.C. (2016). Genetically engineered crops and pesticide use in U.S. maize and soybeans. *Science Advances* https://doi.org/10.1126/sciadv.1600850.

Pimentel, D. and Burgess, M. (2014). Environmental and economic costs of the application of pesticides primarily in the United States. *Integrated Pest Management Reviews* https://doi.org/10.1007/978-94-007-7796-5_2.

Pollet, J., Delport, F., Janssen, K.P.F. et al. (2011). Fast and accurate peanut allergen detection with nanobead enhanced optical fiber SPR biosensor. *Talanta* https://doi.org/10.1016/j.talanta.2010.11.032.

Prasad, R., Bhattacharyya, A., and Nguyen, Q.D. (2017). Nanotechnology in sustainable agriculture: recent developments, challenges, and perspectives. *Food Microbiology* https://doi.org/10.3389/fmicb.2017.01014.

Pretty, J. (2008). Agricultural sustainability: concepts, principles and evidence. *Philosophical Transaction Royal Society B Biological Science* 363 (1491): 447–465.

Primack, R.B. and Kang, H. (1989). Measuring fitness and natural selection in wild plant populations. *Annual Review of Ecology and Systematics* 20: https://doi.org/10.1146/annurev.es.20.110189.002055.

Raliya, R., Biswas, P., and Tarafdar, J.C. (2015). TiO_2 nanoparticle biosynthesis and its physiological effect on mung bean (Vigna radiata L.). *Biotechnology Reports* https://doi.org/10.1016/j.btre.2014.10.009.

Raliya, R., Saharan, V., Dimkpa, C., and Biswas, P. (2018). Nanofertilizer for precision and sustainable agriculture: current state and future perspectives. *Journal of Agricultural and Food Chemistry* https://doi.org/10.1021/acs.jafc.7b02178.

Raliya, R. and Tarafdar, J.C. (2013). ZnO nanoparticle biosynthesis and its effect on phosphorous-mobilizing enzyme secretion and gum contents in Clusterbean (Cyamopsis tetragonoloba L.). *Agricultural Research* https://doi.org/10.1007/s40003-012-0049-z.

Rathmann, R., Szklo, A., and Schaeffer, R. (2010). Land use competition for production of food and liquid biofuels: an analysis of the arguments in the current debate. *Renewable Energy* https://doi.org/10.1016/j.renene.2009.02.025.

Reichman, J.R., Watrud, L.S., Lee, E.H. et al. (2006). Establishment of transgenic herbicide-resistant creeping bentgrass (Agrostis stolonifera L.) in nonagronomic habitats. *Molecular Ecology* https://doi.org/10.1111/j.1365-294X.2006.03072.x.

Rietman, H., Bijsterbosch, G., Cano, L.M. et al. (2012). Qualitative and quantitative late blight resistance in the potato cultivar Sarpo Mira is determined by the perception of five distinct RXLR effectors. *Molecular Plant-Microbe Interactions* https://doi.org/10.1094/MPMI-01-12-0010-R.

Rizzi, A., Raddadi, N., Sorlini, C. et al. (2012). The stability and degradation of dietary DNA in the gastrointestinal tract of mammals: implications for horizontal gene transfer and the biosafety of GMOs. *Critical Reviews in Food Science and Nutrition* https://doi.org/10.1080/10408398.2010.499480.

Ronald, P. (2011). Plant genetics, sustainable agriculture and global food security. *Genetics* https://doi.org/10.1534/genetics.111.128553.

Rui, M., Ma, C., Tang, X. et al. (2017). Phytotoxicity of silver nanoparticles to peanut (arachis hypogaea l.): physiological responses and food safety. *ACS Sustainable Chemistry & Engineering* https://doi.org/10.1021/acssuschemeng.7b00736.

Sabir, S., Arshad, M., and Chaudhari, S.K. (2014). Zinc oxide nanoparticles for revolutionizing agriculture: synthesis and applications. *The Scientific World Jouranl* https://doi.org/10.1155/2014/925494.

Sánchez, M.A., Cid, P., Navarrete, H. et al. (2016). Outcrossing potential between 11 important genetically modified crops and the Chilean vascular flora. *Plant Biotechnology Journal* https://doi.org/10.1111/pbi.12408.

Sanford, J.C. (1990). Biolistic plant transformation. *Physiologia Plantarum* https://doi.org/10.1111/j.1399-3054.1990.tb05888.x.

Sauer, N.J., Mozoruk, J., Miller, R.B. et al. (2016). Oligonucleotide-directed mutagenesis for precision gene editing. *Plant Biotechnology Journal* https://doi.org/10.1111/pbi.12496.

Sayes, C.M., Fortner, J.D., Guo, W.V. et al. (2004). The differential cytotoxicity of water-soluble fullerenes. *Nano Letters* 4 (10): 1881–1887. https://doi.org/10.1021/nl0489586.

Schafer, M.G., Ross, A.A., Londo, J.P. et al. (2011). The establishment of genetically engineered canola populations in the U.S. *PLoS One* https://doi.org/10.1371/journal.pone.0025736.

Schell, J. and Van Montagu, M. (1977). The Ti-plasmid of agrobacterium tumefaciens, a natural vector for the introduction of nif genes in plants? *Basic Life Sciences* https://doi.org/10.1007/978-1-4684-0880-5_12.

Schmidt, M.A., LaFayette, P.R., Artelt, B.A., and Parrott, W.A. (2008). A comparison of strategies for transformation with multiple genes via microprojectile-mediated bombardment. *In Vitro Cellular & Development Biology – Plant* https://doi.org/10.1007/s11627-007-9099-5.

Schulze, J., Frauenknecht, T., Brodmann, P., and Bagutti, C. (2014). Unexpected diversity of feral genetically modified oilseed rape (Brassica napus L.) despite a cultivation and import ban in Switzerland. *PLoS One* https://doi.org/10.1371/journal.pone.0114477.

Scott, N.R., Chen, H., and Cui, H. (2018). Nanotechnology applications and implications of agrochemicals toward sustainable agriculture and food systems. *Journal of Agricultural and Food Chemistry* 66: 6451–6456.

Searchinger, T., Heimlich, R., Houghton, R.A. et al. (2008). Use of U.S. croplands for biofuels increases greenhouse gases through emissions from land-use change. *Science* 80: https://doi.org/10.1126/science.1151861.

Secretariat of the Convention on Biological Diversity (2003). Cartagena protocol on biosafety convention on biological diversity. *Biological Conservation* https://doi.org/10.1016/S0006-3207(03)00053-3.

Servin, A., Elmer, W., Mukherjee, A. et al. (2015). A review of the use of engineered nanomaterials to suppress plant disease and enhance crop yield. *Journal of Nanoparticle Research* https://doi.org/10.1007/s11051-015-2907-7.

Shah, V., Collins, D., Walker, V.K., and Shah, S. (2014). The impact of engineered cobalt, iron, nickel and silver nanoparticles on soil bacterial diversity under field conditions. *Environmental Research Letters* https://doi.org/10.1088/1748-9326/9/2/024001.

Shang, Y., Hasan, M.K., Ahammed, G.J. et al. (2019). Applications of nanotechnology in plant growth and crop protection: a review. *Molecules* 24 (14): https://doi.org/10.3390/molecules24142558.

Shojaei, T.R., Salleh, M.A.M., Tabatabaei, M. et al. (2018). Applications of nanotechnology and carbon nanoparticles in agriculture. In: *Synthesis, Technology and Applications of Carbon Nanomaterials*. Elsevier.

Shoults-Wilson, W.A., Reinsch, B.C., Tsyusko, O.V. et al. (2011). Role of particle size and soil type in toxicity of silver nanoparticles to earthworms. *Soil Science Society of America Journal* https://doi.org/10.2136/sssaj2010.0127nps.

Siegwart, M., Graillot, B., Blachere Lopez, C. et al. (2015). Resistance to bio-insecticides or how to enhance their sustainability: a review. *Frontiers in Plant Science* 6: 381. https://doi.org/10.3389/fpls.2015.00381.

Siemiatycki, J., Richardson, L., Straif, K. et al. (2004). Listing occupational carcinogens. *Environmental Health Perspectives* https://doi.org/10.1289/ehp.7047.

Simonin, M., Guyonnet, J.P., Martins, J.M.F. et al. (2015). Influence of soil properties on the toxicity of TiO_2 nanoparticles on carbon mineralization and bacterial abundance. *Journal of Hazardous Materials* https://doi.org/10.1016/j.jhazmat.2014.10.004.

Singh, R., Singh, R., Singh, D. et al. (2010). Effect of weather parameters on Karnal bunt disease in wheat in Karnal region of Haryana. *Journal of Agrometeorology* https://doi.org/10.54386/jam.v12i1.1281.

Solanki, P., Bhargava, A., Chhipa, H. et al. (2015). Nano-fertilizers and their smart delivery system. In: *Nanotechnologies in Food and Agriculture* (ed. M. Rai, C. Ribeiro, L. Mattoso, and N. Duran). Springer International Publishing.

Srivastava, A. and Rao, D.P. (2014). Enhancement of seed germination and plant growth of wheat, maize, peanut and garlic using multiwalled carbon nanotubes. *European Chemical Bulletin* 3 (5): 502–504.

Strayer-Scherer, A., Liao, Y.Y., Young, M. et al. (2018). Advanced copper composites against copper-tolerant xanthomonas perforans and tomato bacterial spot. *Phytopathology* https://doi.org/10.1094/PHYTO-06-17-0221-R.

Sun, Y., Jiao, G., Liu, Z. et al. (2017). Generation of high-amylose rice through CRISPR/Cas9-mediated targeted mutagenesis of starch branching enzymes. *Frontiers in Plant Science* https://doi.org/10.3389/fpls.2017.00298.

Sunil Kumar, B. and Immanuel Selvaraj, C. (2019). Agrobacterium-mediated transformation for insect-resistant plants. In: *Experimental Techniques in Host-Plant Resistance* (ed. A. Kumar Chakravarthy and V. Selvanarayanan), 275–283. Singapore: Springer Singapore.

Tabashnik, B.E. (1994). Evolution of resistance to bacillus thuringiensis. *Annual Review of Entomology* 39: https://doi.org/10.1146/annurev.ento.39.1.47.

Tashkandi, M., Ali, Z., Aljedaani, F. et al. (2018). Engineering resistance against tomato yellow leaf curl virus via the CRISPR/Cas9 system in tomato. *Plant Signaling & Behavior* https://doi.org/10.1080/15592324.2018.1525996.

Tilman, D. and Clark, M. (2014). Global diets link environmental sustainability and human health. *Nature* https://doi.org/10.1038/nature13959.

Torney, F., Trewyn, B.G., Lin, V.S.Y., and Wang, K. (2007). Mesoporous silica nanoparticles deliver DNA and chemicals into plants. *Nature Nanotechnology* https://doi.org/10.1038/nnano.2007.108.

Trenkel, M.E. (1997). *Controlled-Release and Stabilized Fertilizers in Agriculture*. International Fertilizer Industry Association.

Tripathi, D.K., Singh, S., Singh, V.P. et al. (2016). Silicon nanoparticles more efficiently alleviate arsenate toxicity than silicon in maize cultiver and hybrid differing in arsenate tolerance. *Frontiers in Environmental Science* https://doi.org/10.3389/fenvs.2016.00046.

Tripathi, S. and Sarkar, S. (2015). Influence of water soluble carbon dots on the growth of wheat plant. *Applied Nanoscience* https://doi.org/10.1007/s13204-014-0355-9.

Tsatsakis, A.M., Nawaz, M.A., Tutelyan, V.A. et al. (2017). Impact on environment, ecosystem, diversity and health from culturing and using GMOs as feed and food. *Food and Chemical Toxicology* 107 (Pt A): 108–121. https://doi.org/10.1016/j.fct.2017.06.033.

U.S. Environmental Protection Agency. (1988). Integrated Risk Information System (IRIS) Chemical Assessment Summary: Silver; CASRN 7440-22-4. US Environmental Protection Agency.

US Environmental Protection Agency (2007). *What Is a Pesticide*. EPA Gov retrieved Sept.

van Hove, L. and Gillund, F. (2017). Is it only the regulatory status? Broadening the debate on cisgenic plants. *Environmental Sciences Europe* 29 (1): 22. https://doi.org/10.1186/s12302-017-0120-2.

Vannini, C., Domingo, G., Onelli, E. et al. (2014). Phytotoxic and genotoxic effects of silver nanoparticles exposure on germinating wheat seedlings. *Journal of Plant Physiology* https://doi.org/10.1016/j.jplph.2014.05.002.

Velásquez, A.C., Castroverde, C.D.M., and He, S.Y. (2018). Plant–pathogen warfare under changing climate conditions. *Curr Biol* https://doi.org/10.1016/j.cub.2018.03.054.

Venkatachalam, P., Jayaraj, M., Manikandan, R. et al. (2017). Zinc oxide nanoparticles (ZnONPs) alleviate heavy metal-induced toxicity in Leucaena leucocephala seedlings: a physiochemical analysis. *Plant Physiology and Biochemistry* https://doi.org/10.1016/j.plaphy.2016.08.022.

Vijayakumar, P.S., Abhilash, O.U., Khan, B.M., and Prasad, B.L.V. (2010). Nanogold-loaded sharp-edged carbon bullets as plant-gene carriers. *Advanced Functional Materials* https://doi.org/10.1002/adfm.200901883.

Vleeshouwers, V.G.A.A., Raffaele, S., Vossen, J.H. et al. (2011). Understanding and exploiting late blight resistance in the age of effectors. *Annual Review of Phytopathology* https://doi.org/10.1146/annurev-phyto-072910-095326.

Wang, F., Wang, L., Zou, X. et al. (2019). Advances in CRISPR-cas systems for RNA targeting, tracking and editing. *Biotechnology Advances* https://doi.org/10.1016/j.biotechadv.2019.03.016.

Wang, J., Koo, Y., Alexander, A. et al. (2013). Phytostimulation of poplars and Arabidopsis exposed to silver nanoparticles and ag+ at sublethal concentrations. *Environmental Science & Technology* https://doi.org/10.1021/es4004334.

Wang, P., Lombi, E., Zhao, F.J., and Kopittke, P.M. (2016). Nanotechnology: a new opportunity in plant sciences. *Trends in Plant Science.* https://doi.org/10.1016/j.tplants.2016.04.005.

Wang, S., Wang, F., and Gao, S. (2015). Foliar application with nano-silicon alleviates cd toxicity in rice seedlings. *Environmental Science and Pollution Research* https://doi.org/10.1007/s11356-014-3525-0.

Wang, Z., Wei, F., Liu, S.Y. et al. (2010). Electrocatalytic oxidation of phytohormone salicylic acid at copper nanoparticles-modified gold electrode and its detection in oilseed rape infected with fungal pathogen sclerotinia sclerotiorum. *Talanta* https://doi.org/10.1016/j.talanta.2009.09.023.

Watrud, L.S., Lee, E.H., Fairbrother, A. et al. (2004). Evidence for landscape-level, pollen-mediated gene flow from genetically modified creeping bentgrass with CP4 EPSPS as a marker. *Proceedings of the National Academy of Sciences of the United States of America* https://doi.org/10.1073/pnas.0405154101.

Wentao, X., Kunlun, H., Zhihong, L. et al. (2009). Application of stepwise ammonium sulfate precipitation as cleanup tool for an enzyme-linked immunosorbent assay of glyphosate oxidoreductase in genetically modified rape of GT73. *Journal of Agricultural and Food Chemistry* 58: 11.

Willett, W., Rockström, J., Loken, B. et al. (2019). Food in the Anthropocene: the EAT–lancet commission on healthy diets from sustainable food systems. *Lancet* https://doi.org/10.1016/S0140-6736(18)31788-4.

Wong, M.H., Giraldo, J.P., Kwak, S.Y. et al. (2017). Nitroaromatic detection and infrared communication from wild-type plants using plant nanobionics. *Nature Materials* https://doi.org/10.1038/nmat4771.

Wu, L. and Liu, M. (2008). Preparation and properties of chitosan-coated NPK compound fertilizer with controlled-release and water-retention. *Carbohydrate Polymers* https://doi.org/10.1016/j.carbpol.2007.08.020.

Yanpeng, W., Xi, C., Qiwei, S. et al. (2014). Simultaneous editing of three homoeoalleles in hexaploid bread wheat confers heritable resistance to powdery mildew. *Nature Biotechnology*.

Ye, X. and Beyer, P. (2000). Engineering the provitamin a (β-carotene) biosynthetic pathway into (carotenoid-free) rice endosperm. *Science* 80: https://doi.org/10.1126/science.287.5451.303.

Zhang, D., Hussain, A., Manghwar, H. et al. (2020). Genome editing with the CRISPR-Cas system: an art, ethics and global regulatory perspective. *Plant Biotechnology Journal* 18 (8): 1651–1669. https://doi.org/10.1111/pbi.13383.

Zhang, M., Gao, B., Chen, J. et al. (2014). Slow-release fertilizer encapsulated by graphene oxide films. *Chemical Engineering Journal* https://doi.org/10.1016/j.cej.2014.06.023.

Zhang, X., Zhang, J., and Zhu, K.Y. (2010). Chitosan/double-stranded RNA nanoparticle-mediated RNA interference to silence chitin synthase genes through larval feeding in the African malaria mosquito (Anopheles gambiae). *Insect Molecular Biology* https://doi.org/10.1111/j.1365-2583.2010.01029.x.

18

Global Nanotechnology Research for Agriculture

Kounaina Khan[1,2], Farhan Zameer[1,2], Pratheek Jain[1], Subrahmanya Padyana[1], and Ajar Nath Yadav[3]

[1] *PathoGutOmics Laboratory, Alva's Traditional Medicinal Archive (ATMA), Department of Ayurveda Pharmacology, Alva's Ayurveda Medical College, Vidyagiri, Moodubidire, Dakshina Kannada, Karnataka, India*
[2] *TheOmic School Pvt. Ltd., Vijaynagar, Mysore, Karnataka, India*
[3] *Department of Genetics, Plant Breeding and Biotechnology, Dr. Khem Singh Gill Akal College of Agriculture, Eternal University, Baru Sahib, Sirmour, Himachal Pradesh, India*

18.1 Introduction

Agriculture, the cornerstone of human survival, is constantly challenged by increasing population demands, climate change, soil degradation, and water scarcity (Acharya et al. 2023a). These factors necessitate the development of innovative solutions to enhance productivity while maintaining sustainability. One promising frontier is nanotechnology, an interdisciplinary field that operates on the nanoscale, typically between 1 and 100 nm. Nanotechnology holds immense potential to revolutionize agricultural practices, offering breakthroughs in precision farming, resource efficiency, pest control, and crop enhancement (An et al. 2022). Nanotechnology is still in its relative infancy in the agricultural sector, but its potential is immense. As global research continues, the focus will likely shift toward developing eco-friendly nanomaterials that degrade harmlessly in the environment and finding ways to reduce the production costs of nanomaterials to make them more accessible to smallholder farmers (Acharya et al. 2023b). Furthermore, advances in artificial intelligence (AI) and machine learning could be integrated with nanosensor data to create fully automated, responsive farming systems that optimize crop growth and resource use (Balusamy et al. 2023).

Nanomaterials, the key players in nanotechnology, exhibit unique chemical and physical properties that differentiate them from their bulk counterparts. Their high surface area-to-volume ratio, enhanced reactivity, and tunable optical properties make them highly effective in agricultural applications. Broadly, nanomaterials used in agriculture fall into several categories: carbon-based nanomaterials (such as carbon nanotubes and graphene), metal and metal oxide nanoparticles (like silver, gold, and zinc oxide), and polymeric nanoparticles. Smart delivery systems are one of the most exciting applications of nanotechnology in agriculture (Basit et al. 2022).

Traditional fertilizers and pesticides often have low efficiency, with significant quantities lost to volatilization, leaching, or runoff, contributing to environmental pollution (Manjunatha et al. 2013). Nanomaterials can address this by enabling controlled release and targeted delivery of agrochemicals. For example, nano-fertilizers can slowly release nutrients, ensuring efficient absorption by crops while minimizing wastage. Nanopesticides, on the other hand, can be designed to release their active ingredients only when triggered by environmental conditions, reducing the need for frequent applications and lowering chemical residues in the soil and water. Nanomaterials are also showing promise in soil health and water management. Nanoclays, for instance, can improve soil structure and increase its water retention capacity, which is particularly beneficial in arid regions (Biswas et al. 2019). Metal oxide nanoparticles, such as iron oxide, are effective in the remediation of contaminated soils, helping to bind and neutralize toxic substances, such as heavy metals. Similarly, nanotechnology-based filtration systems can purify irrigation water, removing contaminants like pathogens and agricultural runoff before it is used for crop production.

Precision agriculture is an emerging approach that uses advanced technology to monitor and manage crop inputs more efficiently. Nanotechnology plays a pivotal role in this domain by enabling the development of nanosensors, which provide real-time data on various environmental parameters, including soil moisture, pH, nutrient levels, and the presence of pathogens (Shafi et al. 2019). Nanosensors, due to their small size and sensitivity, can be embedded directly in the soil or attached to plant surfaces to continuously monitor crop health. For instance, carbon-based nanosensors can detect minute changes in soil chemistry and alert farmers to nutrient deficiencies or imbalances. This information allows for site-specific farming—a practice that optimizes the use of fertilizers and water only where and when needed, reducing waste and improving crop yields. In addition, nanobiosensors are being developed to detect early signs of plant stress or disease. By identifying pathogens before symptoms appear, farmers can take preventive measures, minimizing crop losses and reducing the need for chemical treatments. Nanotechnology is also making strides in seed science (Zhu et al. 2022).

Nanoparticle coatings applied to seeds can enhance germination rates, protect against pests, and promote early seedling growth. For example, zinc oxide nanoparticles have been shown to improve seedling vigor by enhancing the uptake of essential nutrients, such as zinc, during the early stages of development. Additionally, nanomaterial-based seed treatments can provide protection against fungal infections, improving crop resilience without the need for harmful fungicides (Patil et al. 2021a; Patil et al. 2021b). Nanotechnology can also play a role in genetic engineering and plant breeding. Nanoparticles can be used as carriers for deoxyribonucleic acid (DNA) and ribonucleic acid (RNA), facilitating the delivery of genetic material into plant cells with greater efficiency than traditional methods. This opens up new avenues for developing crops with enhanced traits, such as improved drought resistance, salinity tolerance, and pest resistance (Yadav 2021).

Despite the potential benefits, the application of nanotechnology in agriculture raises important questions about safety and environmental impact. Nanoparticles, due to their small size, may behave differently in biological systems compared to bulk materials. Their potential to accumulate in the food chain and their long-term effects on soil and water ecosystems need thorough investigation (Wang et al. 2024). Studies have shown that certain nanoparticles, such as silver nanoparticles, exhibit antimicrobial properties, which can be beneficial in controlling plant pathogens (Zameer et al. 2016; Satapathy et al. 2020; Avinash et al. 2023). However, these same properties raise concerns about their effects on beneficial microorganisms in the soil, which play a crucial role in nutrient cycling and soil health (Zameer et al. 2010a). Additionally, the toxicity of nanomaterials to humans and animals is a growing field of study. It is critical to establish safe concentration limits and assess potential risks before wide-scale implementation of nanotechnology in agriculture (Zameer et al. 2016; Zhang and Goss 2022).

18.1.1 Brief Historical Background of Nanotechnology in Agriculture

The history of nanotechnology in agriculture is relatively recent, emerging in the late 20th century, but its roots are tied to broader advancements in nanoscience. The term "nanotechnology" was first coined by physicist Richard Feynman in 1959, although practical developments did not occur until the 1980s and 1990s with the invention of the scanning tunneling microscope and atomic force microscopy, enabling scientists to manipulate materials at the nanoscale (Oberemok et al. 2018). Early research on nanotechnology focused on fields like materials science and medicine, but by the late 1990s, its potential applications in agriculture began to be recognized. Initial studies explored the use of nanoparticles for more efficient delivery of fertilizers and pesticides, aimed at addressing agricultural challenges such as nutrient leaching and pest resistance. This period saw the development of nanoencapsulation technologies, which enabled slow and controlled release of agrochemicals, reducing environmental pollution and improving crop yields (Carbone et al. 2022; Sahu et al. 2024).

The early 2000s marked a turning point, with significant global investments in nanotechnology R&D, leading to further agricultural applications. Governments and research institutions, particularly in the United States, Europe, and Asia, began funding projects to explore nanomaterials for soil improvement, water purification, and precision farming. For example, the use of nanosensors in agriculture was explored to monitor soil conditions and crop health in real-time, improving resource efficiency (Cummings et al. 2021). Since then, the integration of nanotechnology in agriculture has expanded to include genetic modification tools, seed treatments, and smart farming systems. While still in its developmental phase, the field continues to grow, offering promising solutions for sustainable agriculture and global food security in the face of environmental challenges and a growing global population.

18.1.2 Importance of Nanotechnology for Global Agriculture

Nanotechnology holds immense importance for global agriculture as it offers innovative solutions to some of the most pressing challenges in food production. With the global population expected to reach nearly 10 billion by 2050, the demand for food is rapidly increasing. Simultaneously, agriculture faces serious issues like declining arable land, water scarcity, climate change, and the degradation of ecosystems (Vaidya et al. 2024). In this context, nanotechnology emerges as a key player in ensuring sustainable, efficient, and resilient agricultural practices. One of the significant contributions of nanotechnology is its ability to enhance resource efficiency. Nanomaterials, such as nano-fertilizers and nano-pesticides, allow for targeted delivery of nutrients and crop protection agents, significantly reducing waste (Tang et al. 2023). Traditional fertilizers often suffer from nutrient loss through leaching and volatilization, but nanoparticles can be engineered for slow, controlled release, ensuring plants absorb a higher proportion of the applied nutrients. This not only improves crop yields but also minimizes environmental pollution from excess agrochemical use.

Nanotechnology also plays a pivotal role in precision agriculture, enabling farmers to manage their resources more effectively. Nanosensors can provide real-time data on soil conditions, water levels, and crop health, allowing for site-specific interventions. These sensors help optimize water usage, reducing waste in irrigation, and ensure the right amount of fertilizers and pesticides are applied at the right time, improving overall productivity while lowering costs. Additionally, nanotechnology can improve crop resilience. Nanoparticle-based treatments, including seed coatings and nanoformulated pesticides, can protect crops from pests, diseases, and harsh environmental conditions, such as drought or soil salinity (Sushil et al. 2021). This enhances food security, particularly in regions prone to extreme weather or resource scarcity. Nanotechnology's ability to improve efficiency, precision, and resilience in agriculture makes it essential for addressing the global food security challenge, ensuring both productivity and sustainability.

Global agriculture faces a host of interconnected challenges, including food security, climate change, and land degradation (Meyfroidt et al. 2022). Addressing these issues is critical for ensuring a sustainable and equitable future. (i) Food security is one of the most pressing challenges in agriculture. With population growth, the demand for food continues to rise, but agricultural productivity struggles to keep pace. Factors such as limited arable land, declining crop yields, and inadequate infrastructure contribute to food scarcity in many regions. Additionally, inefficient farming practices and post-harvest losses further exacerbate the issue. To meet the growing demand for food, agriculture must produce more from fewer resources, all while maintaining sustainability. This requires innovations that increase yields, improve resource efficiency, and reduce waste. (ii) Climate change compounds the difficulties faced by agriculture, affecting crop productivity and the stability of food systems worldwide. Rising temperatures, changing precipitation patterns, and increased frequency of extreme weather events (droughts, floods, and storms) pose significant threats to farming. In regions already vulnerable to food insecurity, these environmental stresses can devastate agricultural production.

Farmers need adaptive practices and technologies, such as drought-resistant crops, improved irrigation systems, and more efficient water management, to cope with these changes and maintain a consistent food supply. (iii) Land degradation, including soil erosion, desertification, and the depletion of soil fertility, threatens global agricultural productivity. Overuse of chemical inputs, deforestation, and unsustainable farming practices degrade the land, reducing its capacity to support crops. This problem is particularly acute in developing countries where farmers rely heavily on the land for their livelihoods. Sustainable land management practices, soil rehabilitation, and conservation efforts are vital to reversing degradation and restoring the productivity of agricultural ecosystems. To overcome these challenges, innovative agricultural technologies, including nanotechnology, precision farming, and sustainable land-use practices, are essential for creating a resilient global agricultural system.

18.1.3 Global Research in Nano-Agritech

The global research landscape for nanotechnology in agriculture is rapidly evolving, as various countries and institutions recognize the potential of nanotechnology to address critical agricultural challenges such as food security, environmental sustainability, and resource efficiency. This integration is gaining momentum due to its ability to offer innovative solutions, including smart delivery systems for agrochemicals, precision farming technologies, and soil health improvement (Ijaz et al. 2023). Several regions are leading the charge in nanotechnology research for agriculture. The United States is a major player in agricultural nanotechnology, with initiatives funded by the U.S. Department of Agriculture (USDA), National Science Foundation (NSF), and various universities. Research focuses on developing nano-based fertilizers, pesticides, and nanosensors to improve efficiency and reduce environmental impact.

In Europe, the European Union (EU) has funded numerous projects under its Horizon 2020 and Horizon Europe programs. The focus here is on sustainable farming practices, nanotechnology-enabled smart farming systems, and environmental protection. In Asia-Pacific countries like China, India, and Japan, are heavily investing in nanotechnology for agriculture (Mubeen et al. 2023). China is focusing on nanoformulations for pesticides and fertilizers, while India's research is geared toward addressing challenges like drought, soil degradation, and food security through nanotechnology-based solutions. The key areas of research concentrate on nano-fertilizers and nano-pesticides, nanosensors for precision agriculture, soil, and water remediation.

18.1.4 International Collaborations and Funding Initiatives

International collaboration plays a vital role in the development of agricultural nanotechnology. Research consortia, such as the Food and Agriculture Organization (FAO) and the Global Forum on Agricultural Research (GFAR), promote cross-border collaborations and knowledge exchange. These organizations support nanotechnology research focused on achieving global food security and sustainable agriculture. In addition, there are increasing public-private partnerships where government institutions collaborate with corporations and startups to commercialize nanotechnology innovations (Brooks 2011).

Many global conferences and symposiums, such as the International Conference on Nanoagriculture, highlight the latest research trends and breakthroughs in the field (Datta et al. 2023). Further, the Government of India (GoI) has launched several initiatives through the Indian Council for Agriculture Research (ICAR) to promote nano-agritechnology. Despite the promising developments, there are significant challenges in scaling nanotechnology's integration into agriculture. Issues like high production costs, regulatory hurdles, and limited infrastructure in developing regions slow widespread adoption. Furthermore, concerns about the potential long-term environmental and health impacts of nanomaterials call for more comprehensive risk assessments and regulatory frameworks. Looking ahead, the global research landscape is expected to witness more interdisciplinary collaborations, integrating AI, machine learning, and big data with nanotechnology for smart farming solutions. The focus is likely to shift toward creating eco-friendly, biodegradable nanomaterials that offer high efficiency without harming ecosystems. As global efforts to combat climate change and ensure food security intensify, the integration of nanotechnology into agriculture will likely play a critical role in transforming the future of farming.

18.2 Nanomaterials in Agriculture

18.2.1 Types of Nanomaterials Used in Agriculture

Nanomaterials offer unique properties, such as high surface area, reactivity, and tunable characteristics, which make them highly effective for agricultural applications. Various types of nanomaterials are used in agriculture to improve efficiency and sustainability. (i) Carbon-based nanomaterials (carbon nanotubes and graphene): These materials possess remarkable mechanical strength, electrical conductivity, and large surface areas (Mathew et al. 2020). Carbon nanotubes (CNTs) have been explored for their ability to improve seed germination, plant growth, and water uptake (Díez-Pascual 2021). Graphene, a two-dimensional carbon structure, is being studied for its potential in enhancing photosynthesis, increasing plant resistance to stress, and delivering agrochemicals efficiently. (ii) Metal/metal oxide nanoparticles (silver, gold, and zinc oxide): Metal nanoparticles, such as silver and gold, are known for their antimicrobial properties, making them effective in combating plant pathogens (Prasad et al. 2019; Arora et al. 2022). Zinc oxide nanoparticles play a crucial role in promoting plant growth by enhancing nutrient uptake, while also acting as fertilizers and pest-control agents. (iii) Polymeric nanoparticles and nanoclays: Polymeric nanoparticles are often used to encapsulate agrochemicals for controlled release. Nanoclays, which consist of layered silicates, can be used to improve soil structure, water retention, and nutrient delivery, especially in arid regions (Pavlicevic et al. 2023; Dinesh et al. 2024).

18.2.2 Applications of Nanomaterials in Agri-technology

Nanomaterials find multiple applications in agriculture, contributing to increased crop productivity, resource efficiency, and environmental protection. Smart delivery systems for agrochemicals (fertilizers and pesticides): One of the primary applications of nanotechnology in agriculture is the development of smart delivery systems. Nano-fertilizers and nanopesticides can be engineered for controlled, slow release and targeted delivery, ensuring that plants receive nutrients and protection when and where needed. This minimizes the leaching of chemicals into the environment and enhances plant absorption, reducing the frequency of applications. Nanomaterials for soil enhancement and water purification: Nanoclays and other nanomaterials improve soil health by increasing its water-holding capacity and nutrient retention. Additionally, nanomaterials like titanium dioxide and iron oxide nanoparticles are used for soil and water purification. These materials can remove contaminants, such as heavy metals and organic pollutants, improving water quality for irrigation and reducing the environmental impact of farming (Singh et al. 2016). Nanosensors for environmental monitoring and crop health: Nanosensors are highly sensitive devices that monitor environmental conditions like soil pH, moisture, nutrient levels, and temperature. These sensors can detect early signs of plant stress or disease, enabling farmers to take timely corrective actions. The use of nanosensors improves precision agriculture, optimizing resource use and reducing input waste (Gamage et al. 2022; Akki et al. 2024).

18.2.3 Mechanisms of Action

Nanomaterials interact with plant and soil systems through various mechanisms, which influence their effectiveness in agricultural applications. Interaction mechanisms between nanomaterials and plant/soil systems: Nanomaterials can enter plant systems through roots, leaves, or seeds. Once inside the plant, they may influence cellular functions, including photosynthesis, nutrient uptake, and stress resistance. For instance, carbon nanotubes can penetrate plant cell walls, allowing them to act as carriers for agrochemicals or genetic material. In the soil, nanoparticles can improve nutrient availability, enhance water retention, or bind to contaminants, aiding in soil remediation. Controlled release and targeted delivery mechanisms: Nanomaterials are often designed to release agrochemicals in response to specific triggers, such as changes in temperature, pH, or moisture levels. This controlled release ensures that fertilizers or pesticides are delivered at the right time and location, minimizing losses and environmental damage. For example, polymeric nanoparticles can encapsulate nutrients or pesticides, releasing them gradually over time as crops need them, leading to higher efficiency and reduced chemical runoff (Beulah et al. 2015; Sahu et al. 2023). These diverse types, applications, and mechanisms of action underline the versatility and potential of nanomaterials in transforming modern agriculture, making it more efficient, sustainable, and responsive to global challenges (He et al. 2019; More et al. 2021).

18.3 Nanotechnology for Precision Agriculture

18.3.1 Concept of Precision Agriculture

Precision agriculture is a modern farming approach that optimizes agricultural inputs—such as water, fertilizers, and pesticides—based on real-time data. By tailoring practices to the specific needs of crops and local environmental conditions, precision farming enhances resource efficiency, increases crop yields, and reduces environmental impact. Key technologies like remote sensing, Global Positioning System (GPS), and AI-based systems play pivotal roles by providing farmers with accurate data to make informed decisions (Santás-Miguel et al. 2023). For instance, satellite imagery and sensors can assess soil conditions and plant health across vast areas, allowing farmers to apply inputs only where needed, avoiding overuse and waste (Figure 18.1).

18.3.2 Nanotechnology-Enabled Precision Tools

Nanotechnology significantly enhances precision agriculture by providing tools like nanosensors that monitor soil and crop conditions with exceptional accuracy. These

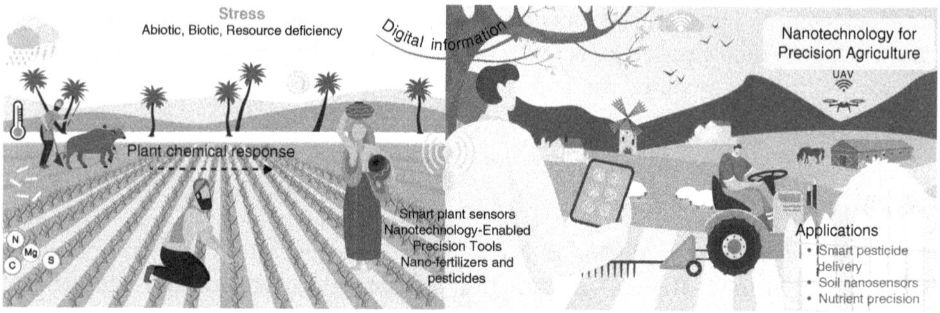

Figure 18.1 Transformation from traditional farming to precision agriculture.

sensors detect variables such as soil moisture, pH, and nutrient levels in real time, enabling farmers to make timely adjustments. Additionally, nano-biosensors can identify early signs of plant diseases or pest infestations by detecting specific pathogens or chemical changes in plants. This early detection allows for targeted interventions, reducing crop losses and minimizing pesticide use (Li et al. 2024).

18.3.3 Drones and Smart Devices for Agriculture

The integration of nanotechnology with drones and Internet of Things (IoT) devices further advances precision agriculture. Drones equipped with nanosensors can monitor crop health from above, while smart devices connected via IoT can relay data for site-specific management. These technologies enable the precise application of water, fertilizers, and pesticides, reducing waste and increasing productivity. This synergy of nanotechnology, drones, and smart devices helps farmers manage crops more effectively, ensuring sustainability and efficiency in agriculture (Guevara et al. 2024).

18.4 Sustainable Agriculture: Nanotechnology's Role in Reducing Environmental Impact

Nanotechnology offers transformative solutions in agriculture, enabling sustainable practices that minimize environmental damage while maintaining or enhancing crop productivity. Its role in reducing the negative impact of conventional agricultural methods is multifaceted, addressing agrochemical use, soil and water remediation, and energy efficiency (Figure 18.2).

18.4.1 Reduction in Agrochemical Usage

Nanoformulated fertilizers and pesticides significantly reduce the environmental contamination associated with traditional agrochemicals. These nano-based formulations improve the bioavailability of nutrients and active ingredients, ensuring more efficient uptake by plants. For instance, nano-fertilizers enable controlled, slow-release mechanisms that deliver nutrients directly to the plant root zone, minimizing leaching into groundwater and reducing the need for excessive fertilizer applications. Similarly, nano-pesticides enhance

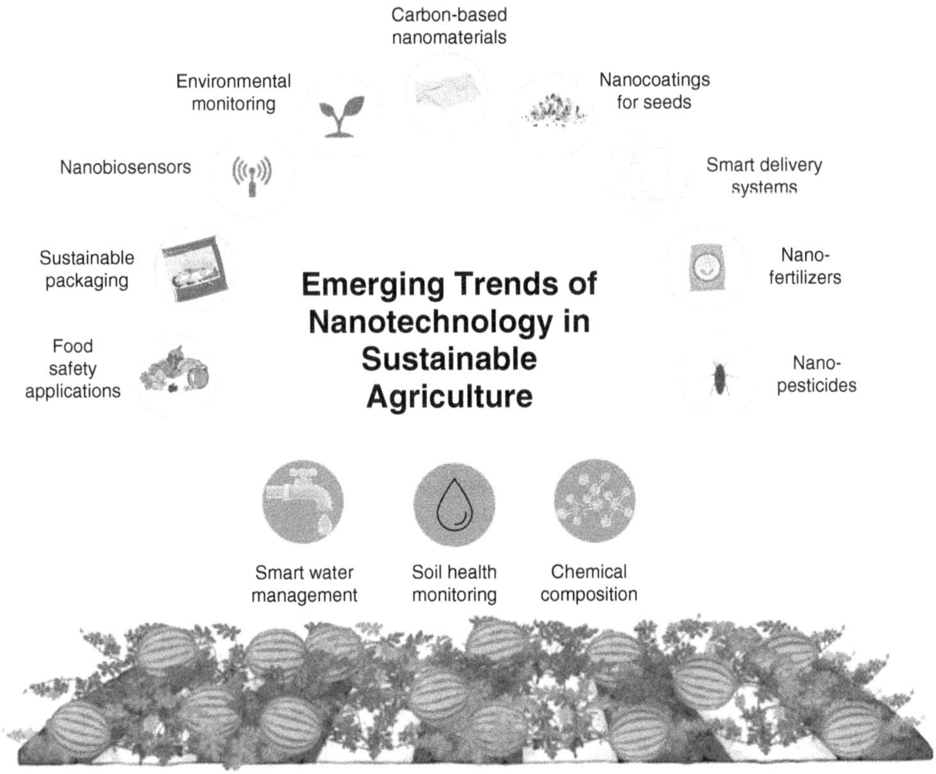

Figure 18.2 Impact of nanotechnology for sustainable agriculture.

the targeted delivery of active compounds to pests, thereby lowering the overall chemical load released into ecosystems (Li et al. 2024; Guevara et al. 2024). This reduction in agrochemical usage decreases harmful runoff into water bodies, diminishing pollution and preserving biodiversity.

18.4.2 Soil and Water Remediation

Nanomaterials also play a vital role in remediating contaminated soil and water systems. Engineered nanomaterials, such as iron oxide nanoparticles and nanoclays, have shown great promise in removing pollutants, including heavy metals (e.g., cadmium, lead) and organic toxins, from soils and irrigation water. These nanomaterials can bind to contaminants or catalyze their breakdown, reducing toxicity and restoring soil health. Moreover, nanomaterials enhance the filtration of harmful agricultural pollutants, like pesticides and herbicides, from water sources, preventing their accumulation in ecosystems and contributing to cleaner water cycles.

18.4.3 Carbon Footprint and Energy Efficiency

Nanotechnology also contributes to reducing agriculture's carbon footprint and improving energy efficiency. Precision farming enabled by nanosensors and smart delivery systems reduces the overuse of water, fertilizers, and pesticides, cutting down on the energy required for producing and applying these inputs. Additionally, the enhanced efficiency of nanomaterials in nutrient delivery and crop protection reduces the frequency of applications, lowering fuel consumption for agricultural machinery. These result in a significant decrease in greenhouse gas emissions and helps farmers adopt more sustainable and energy-efficient farming practices. Together, these advancements position nanotechnology as a key enabler of environmentally responsible agriculture, promoting sustainability without compromising food security (Sun et al. 2022).

18.5 Nanotechnology in Seed Science and Crop Enhancement

Nanotechnology is revolutionizing seed science and crop enhancement by improving seed quality, genetic manipulation, and crop growth, thereby boosting agricultural productivity.

18.5.1 Seed Coating and Germination

Nanoparticles are increasingly being used to enhance seed germination and seedling vigor. Nanoparticle-based seed coatings can create a protective layer around the seed, which promotes better water absorption and faster germination. This layer also protects the seed from pathogens, improving overall seedling health. Nanomaterials such as silicon dioxide nanoparticles have been shown to stimulate germination and root elongation, leading to more robust seedlings (do Espirito Santo Pereira et al. 2021). Additionally, nanoparticle-based seed priming can enhance the seed's resistance to abiotic stressors such as drought or salinity, leading to more resilient crops. This method prepares seeds for rapid, even growth and improves the chances of successful crop establishment under suboptimal conditions.

18.5.2 Genetic Improvement and Plant Breeding

Nanotechnology plays an emerging role in genetic improvement and plant breeding. Nanoparticles can act as carriers for DNA and RNA molecules, aiding in the delivery of genetic material for transgenic crop development and gene editing technologies like CRISPR-Cas9 (Tyumentseva et al. 2023). This method enhances the efficiency of genetic engineering, enabling the development of crops with improved traits such as disease resistance and stress tolerance (e.g., to drought or salinity). Nanomaterials, such as carbon nanotubes, have also been explored for delivering gene-editing tools directly into plant cells, bypassing traditional breeding methods and accelerating the process of creating enhanced crop varieties.

18.5.3 Crop Growth and Yield Enhancement

Nanotechnology improves crop growth and yield through enhanced nutrient delivery. Nanoparticles can improve the availability of essential nutrients, such as nitrogen and phosphorus, in the soil, enabling plants to absorb them more efficiently. Studies have shown that nanofertilizers increase root growth, biomass, and overall yield. Field trials have demonstrated significant yield increases in crops like wheat and rice with the application of nanotechnology-based treatments, highlighting its potential to boost food production sustainably (Zhang et al. 2021a).

18.6 Safety, Toxicology, and Environmental Impact of Nanomaterials

While nanotechnology holds great potential in agriculture, its safety, toxicology, and environmental impact require thorough examination to ensure responsible use. Understanding the potential risks posed by nanomaterials to plants, animals, and humans is critical for their safe application.

18.6.1 Toxicity of Nanomaterials to Plants, Animals, and Humans

Nanomaterials exhibit unique physical and chemical properties that may result in toxicity to biological systems. Research has shown that certain nanoparticles, such as silver and zinc oxide, can be toxic to plants, affecting root development, photosynthesis, and overall plant health. The interaction of nanoparticles with plant cells, including their ability to penetrate the plant cell wall, may disrupt cellular processes at higher concentrations. Similarly, the toxicity of nanoparticles in animals and humans depends on factors such as size, shape, chemical composition, and concentration. Inhalation or ingestion of nanoparticles by humans or animals can lead to oxidative stress, inflammation, and cellular damage, raising concerns about long-term exposure (Zielińska et al. 2020). However, establishing safe concentration limits is key to mitigating these risks. Studies suggest that, at low concentrations, nanoparticles can be beneficial for plant and crop growth without causing harm. Hence, determining thresholds for safe use is essential for preventing adverse effects.

18.6.2 Nanoparticle Accumulation in the Food Chain

A significant concern is the potential for bioaccumulation of nanoparticles in crops and livestock, which could transfer up the food chain and pose risks to human health. Nanoparticles used in fertilizers, pesticides, and soil amendments may be taken up by plant roots and accumulate in edible plant tissues. Similarly, livestock consuming contaminated feed or water could accumulate nanoparticles in their tissues. While many studies indicate that nanoparticles can be metabolized or excreted, others raise concerns about long-term bioaccumulation and chronic exposure risks to humans consuming food products containing nanoparticle residues. Nanoparticle residues in food products have potential health risks, including gastrointestinal disorders, immune responses, and DNA damage. Thus, there is a pressing need for comprehensive toxicological assessments to determine acceptable levels of nanoparticle residues in food (Li et al. 2022).

18.6.3 Regulatory Framework for Nanomaterials

The regulatory landscape for the use of nanomaterials in agriculture is still evolving. Currently, global and regional regulatory bodies, such as the U.S. Environmental Protection Agency (EPA) and the European Food Safety Authority (EFSA), are working to establish safety guidelines for the use of nanotechnology in farming. These guidelines focus on conducting risk assessments, setting safe exposure limits, and ensuring that nanoparticle applications in agriculture do not pose undue risks to humans or the environment. However, policy development remains in its early stages, and many countries lack specific regulations governing the use of nanomaterials in agriculture. As research advances, a more comprehensive and standardized regulatory framework is needed to address the risks associated with the deployment of nanomaterials in agricultural systems (Shandilya et al. 2023).

18.7 Global Trends in Nanotechnology Research for Agriculture

Nanotechnology research in agriculture is gaining global momentum, with key regions making significant contributions. Governments and private sectors are investing heavily in developing nanotechnologies aimed at solving critical challenges such as food security, resource efficiency, and environmental sustainability (Figure 18.3).

18.7.1 Regional Research and Development Focus

North America, Europe, and the Asia-Pacific are leading regions in agricultural nanotechnology research. In North America, the United States has emerged as a pioneer, with numerous initiatives backed by the USDA, the NSF, and private companies. Research focuses on developing smart delivery systems for fertilizers and pesticides, as well as precision farming tools like nanosensors. In Europe, the focus is more on sustainable farming practices, driven by the

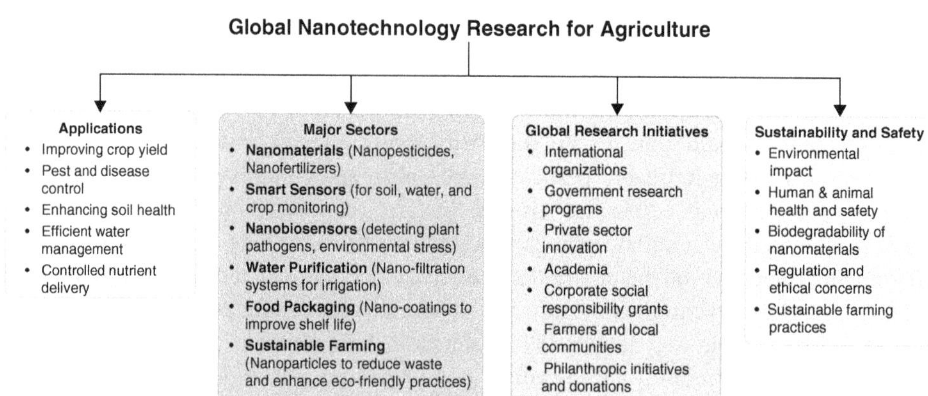

Figure 18.3 Verticals in global nanotechnology research and its applications in agriculture.

European Union's Horizon Europe program. The region has been investing in nanomaterials for reducing chemical inputs and environmental pollution, aligning with the EU's "Green Deal" policy for sustainable agriculture. Asia-Pacific—particularly China, India, and Japan—is another hotspot for agricultural nanotechnology. China is heavily invested in developing nano-fertilizers and nano-pesticides aimed at improving resource efficiency and combating food insecurity. India's focus is on nanotechnology solutions for challenges such as drought, salinity, and land degradation (Conrad et al. 2018).

18.7.2 Funding and Collaborations

Global funding for agricultural nanotechnology is on the rise, with several major initiatives driving research forward. The Horizon Europe and Horizon 2020 programs in Europe have allocated substantial funds toward nanotechnology research in agriculture. In the United States, the Agriculture and Food Research Initiative (AFRI) funds projects exploring nanomaterial applications in farming. International collaborations are also increasing. Research consortiums like the Global Forum on Agricultural Research (GFAR) and the Food and Agriculture Organization (FAO) facilitate cross-border partnerships and knowledge transfer (Savelli et al. 2019). Many countries are partnering with global tech firms and startups to accelerate the commercialization of agricultural nanotechnology innovations. These collaborations often focus on technology transfer, where developed nations share advancements with emerging economies to address agricultural challenges on a global scale.

18.7.3 Trends in Patent and Publication Data

Recent years have seen a surge in both patents and scientific publications related to nanotechnology in agriculture. According to patent databases, innovations in nano-fertilizers, nano-pesticides, and nanobiosensors are leading areas of intellectual property filings. The United States and China are leading in patent applications, reflecting their robust R&D ecosystems. In terms of scientific publications, studies have increasingly focused on field trials, demonstrating the efficacy of nanotechnology in improving crop yields and reducing chemical usage. Publications have also explored the environmental impact of nanomaterials, contributing to growing interest in sustainable agriculture (Abd-Elsalam 2024). Breakthroughs include the development of biodegradable nanomaterials and carbon-based nanomaterials like graphene for enhancing plant growth and disease resistance, signaling a shift toward eco-friendly nanotech solutions. Overall, the global research landscape is dynamic, with continued growth in funding, collaborations, and innovations that promise to transform modern agriculture.

18.8 Case Studies: Success Stories and Field Trials

The integration of nanotechnology in agriculture has already led to numerous success stories and promising field trials worldwide. These examples highlight the potential for nanotechnology to improve agricultural productivity and sustainability.

18.8.1 Successful Applications of Nanotechnology in Agriculture

In India, nanotechnology has been applied to develop nano-fertilizers that significantly increase crop yields while reducing the need for traditional chemical fertilizers. A notable example is the development of nano-urea by the Indian Farmers Fertilizer Cooperative (IFFCO), which has shown a reduction in nitrogen loss, leading to higher nitrogen use efficiency and improved crop productivity. In large-scale trials, nano-urea has enhanced yields for crops like wheat and rice by up to 15–20% (Kumar et al. 2022). In Brazil, the use of nano-pesticides has proven effective in controlling insect pests and diseases in crops like soybeans. Nano-formulated insecticides have demonstrated better target specificity, minimizing damage to non-target species and reducing overall pesticide use. These advancements not only improve crop yields but also minimize environmental contamination, a significant benefit for a country with a large agricultural export market (dos Santos et al. 2023). In Africa, pilot projects involving nanosensors for soil nutrient monitoring have helped smallholder farmers optimize fertilizer usage. By detecting nutrient deficiencies in real time, these sensors enable precise application, resulting in improved crop yields and reduced costs (Bonilla-Cedrez et al. 2021).

18.8.2 Field Trials and Pilot Projects

Several large-scale field trials have been conducted to evaluate the effectiveness of nanotechnology-enabled agricultural inputs. In the United States, field trials using nano-fertilizers have demonstrated enhanced nutrient delivery and crop growth. These trials have shown a reduction in the amount of fertilizers needed by up to 30%, translating into cost savings for farmers and less environmental pollution (Rahman et al. 2024). Similarly, pilot projects in China have explored the use of nano-pesticides in large-scale rice and wheat farming. These projects have resulted in improved pest control with fewer applications, reducing labor costs and the environmental impact of excessive pesticide use (Verma et al. 2024). The use of nanosensors for precision agriculture has also been tested in several countries, providing valuable data on soil health and crop stress that helps optimize farming practices.

18.8.3 Lessons Learned from Real-World Implementations

While the potential of nanotechnology in agriculture is clear, real-world implementations have revealed both challenges and successes. One of the main challenges is scalability—while nanotechnology has shown great promise in controlled environments, scaling these innovations for widespread use remains difficult due to cost and infrastructure limitations, particularly in developing countries. Regulatory hurdles and safety concerns surrounding nanomaterials are also key challenges. Many countries lack a robust regulatory framework for nanotechnology in agriculture, creating uncertainties around long-term environmental and health impacts. However, successful case studies highlight the importance of collaboration between governments, research institutions, and private sectors to address these challenges. The positive outcomes of field trials demonstrate that, with continued research and investment, nanotechnology has the potential to revolutionize agriculture on a global scale, improving both economic and environmental outcomes.

18.9 Challenges and Limitations of Nanotechnology in Agriculture

Despite the immense potential of nanotechnology to revolutionize agriculture, several challenges and limitations must be addressed to ensure its successful integration into global farming systems. These challenges encompass technical, economic, and ethical aspects.

18.9.1 Technical and Scientific Challenges

One of the primary technical challenges in agricultural nanotechnology is scaling up solutions from the lab to the field. While many nanotechnologies have demonstrated efficacy in controlled laboratory settings, translating these innovations into large-scale farming practices can be difficult. Factors such as environmental variability, crop diversity, and regional soil differences can alter the behavior and effectiveness of nanomaterials, making field applications more complex than anticipated. There are also significant knowledge gaps regarding the long-term effects of nanomaterials on ecosystems and biodiversity. While many studies focus on the short-term benefits, little is known about the accumulation of nanoparticles in soil and water and their potential impact on microorganisms, plants, animals, and overall ecosystem health (Zameer and Gopal 2010). This uncertainty poses risks for ecosystem stability and highlights the need for comprehensive research into the long-term environmental fate of nanomaterials.

18.9.2 Economic Barriers

The high production costs associated with manufacturing nanomaterials present another barrier, particularly for small-scale and resource-limited farmers. The processes required to produce nanoparticles, nanosensors, and other nanotechnology-enabled tools are often expensive, raising concerns about their affordability and accessibility for low-income farmers, especially in developing countries. Without significant price reductions or financial support, smallholders may be excluded from the benefits of nanotechnology, exacerbating existing inequalities in agricultural productivity. Intellectual property (IP) challenges further complicate the landscape. The patenting of nanotechnology innovations can be expensive, and stringent IP regulations may restrict access to cutting-edge technologies. This could hinder collaborative research efforts and limit the dissemination of nanotechnologies, particularly to regions where agricultural innovation is most needed (Jain et al. 2018).

18.9.3 Ethical and Social Considerations

Public perception of nanotechnology in agriculture remains mixed, with concerns about the safety of food products produced using nanomaterials. Despite scientific advancements, there is resistance to the use of nanotechnology in food production, often driven by a lack of awareness or understanding of its benefits and safety. Public resistance could slow the adoption of nanotechnologies, particularly in regions where consumer trust in agricultural innovation is already fragile. Additionally, ethical concerns arise regarding the potential environmental impacts of widespread nanoparticle use. If not carefully managed, nanomaterials could contribute to pollution or biodiversity loss. Moreover, there are questions about social equity—particularly whether smallholders and marginalized communities will have equal access to these technologies, or if the benefits of nanotechnology will remain concentrated among wealthier farmers and agribusinesses (Grobler et al. 2021). Addressing these challenges will require a multidisciplinary approach involving continued research, stakeholder engagement, and equitable policy frameworks to ensure that the benefits of nanotechnology in agriculture are widely distributed and responsibly managed.

18.10 Future Outlook: Advancing Nanotechnology for Global Agriculture

Nanotechnology is poised to play a transformative role in agriculture, addressing critical challenges such as food security, resource efficiency, and environmental sustainability. Looking ahead, the integration of emerging technologies, evolving policy frameworks, and interdisciplinary research will shape the future of agricultural nanotechnology and its potential impact on global farming systems.

18.10.1 Emerging Technologies in Agricultural Nanotechnology

The future of agricultural nanotechnology is expected to witness advancements in nanoscale materials and systems, including innovations such as bio-nanotechnology and quantum dots. Bio-nanotechnology, which merges biological molecules with nanomaterials, holds promise for developing highly selective and efficient delivery systems for fertilizers, pesticides, and genetic materials (Shao et al. 2022). For instance, bio-nanocarriers could enable more precise delivery of agrochemicals, reducing waste and minimizing environmental contamination. Additionally, bio-nanotechnology has the potential to enhance crop resilience by promoting plant growth and improving stress tolerance through interactions with plant cells at the molecular level. Quantum dots, another emerging nanomaterial, offer unique properties that can be leveraged for applications such as precision sensing in agriculture.

Quantum dots have high sensitivity and can be used to develop ultra-sensitive nanosensors for detecting specific plant diseases, nutrient deficiencies, or environmental conditions. These sensors will enable farmers to make real-time decisions, optimize resource use, and reduce inputs, all while increasing productivity and sustainability. In the coming years, nanotechnology is expected to integrate with AI, machine learning (ML), and big data. AI and ML can process vast amounts of data collected by nanosensors to provide predictive insights and recommendations for crop management, disease prevention, and soil health (Zhang et al. 2021b). For instance, AI algorithms could analyze soil sensor data to determine optimal fertilizer application times, maximizing crop yields. The combination of nanotechnology and digital technologies will lead to more automated, data-driven farming practices, enhancing efficiency and precision.

18.10.2 Policy and Regulatory Developments

As nanotechnology advances, policy and regulatory frameworks must evolve to keep pace with the innovations. Currently, the regulatory landscape surrounding nanomaterials in agriculture is fragmented, with different countries having varying guidelines on the use of nanoparticles in food and farming systems. There is a pressing need for harmonized international standards to ensure the safe and effective application of nanotechnology globally. Risk assessments and safety evaluations for nanomaterials must be standardized, with clear guidelines for manufacturers, farmers, and consumers. This includes determining acceptable thresholds for nanoparticle residues in food, soil, and water, as well as understanding their long-term environmental and health impacts. Global cooperation will be essential in creating a robust regulatory framework that addresses these concerns while fostering innovation. International organizations such as the FAO, the World Health Organization (WHO), and the UN will play a critical role in facilitating discussions and collaborations between governments, industry, and academia (Nuttall et al. 2014). By establishing unified safety standards and promoting responsible development, policymakers can ensure that nanotechnology enhances agricultural productivity without compromising environmental or human health (Sulthana et al. 2018).

18.10.3 Vision for the Next Decade

The next decade will likely see significant advancements in the field of agricultural nanotechnology, with the potential for breakthroughs in several key areas. First, the scalability of nanomaterials and nanotechnologies will improve, making them more accessible to farmers worldwide, including smallholders in developing countries. As production techniques become more efficient, the cost of nanotechnology-enabled tools such as nanofertilizers, nano-pesticides, and nanosensors is expected to decrease, making them more affordable. Furthermore, interdisciplinary research will be crucial in driving innovations. Collaboration between agricultural scientists, nanotechnologists, environmental scientists, and data scientists will lead to the development of holistic solutions that address complex agricultural challenges. For example, integrating nanotechnology with sustainable farming practices, such as regenerative agriculture and agroecology, can help mitigate the environmental impact of intensive farming while increasing productivity. Nanomaterials can also be incorporated into biodegradable agricultural inputs, reducing the carbon footprint and contributing to circular farming economies.

In the realm of global food security, nanotechnology has the potential to significantly enhance crop yields, particularly in regions prone to food shortages and climate-induced agricultural stress. Innovations such as nanoparticle-based seed treatments and nano-enabled plant breeding techniques can boost crop resilience to drought, salinity, and pests, helping farmers in vulnerable regions adapt to changing climatic conditions. The use of nanosensors for real-time monitoring will further support sustainable water and nutrient management, contributing to more resilient food production systems. Looking ahead, agricultural nanotechnology will be instrumental in achieving the United Nations Sustainable Development Goals (SDGs), particularly those related to zero hunger (SDG 2) and climate action (SDG 13). By enhancing resource efficiency, reducing environmental impacts, and increasing food production, nanotechnology can help ensure sustainable food systems that meet the needs of a growing global population (Foong et al. 2022).

18.11 Conclusions

While challenges remain in scaling up nanotechnology and ensuring its safe and equitable use, the future outlook is promising. With continued investment in research, technological innovation, and the development of robust regulatory frameworks, nanotechnology is set to become a cornerstone of sustainable agriculture, contributing to global food security and environmental stewardship.

Acknowledgments

FZ expresses gratitude toward Vision Group of Science and Technology (VGST); Karnataka Science and Technology Promotion Society (KSTePS); and Department of Science and Technology, Government of Karnataka (GoK), Bangalore, for the K-FIST L1 grant with GRD No. 1051/2023-2025. All authors thank Dr. M. Mohan Alva, Chairman; Mr. Vivek Alva; Dr. Vinay Alva, Managing trustee, Alva's Education Foundation (AEF); and Dr. Sajith M., the Principal, Alva's Ayurveda Medical College, for their constant support and motivation. All authors credit BioRender.com for the design of the figures in this chapter.

References

Abd-Elsalam, K.A. (2024). Special issue: agricultural nanotechnology. *Plants* 13 (4): 489. https://doi.org/10.3390/plants13040489.

Acharya, M., Shobha, R., Krishna, C.R., and Zameer, F. (2023a). Chapter 1: Origin of cultivated plants. In: *Agrobotany Text Book in Accordance to NEP Scheme, Govt. of India*, vol. 1, 1–44. Karnataka, India: United Agencies.

Acharya, M., Shobha, R., Krishna, R.C., and Zameer, F. (2023b). Chapter 8: Propagation of specific plant types. In: *Basics of Gardening Text Book in Accordance to NEP Scheme, Govt. of India*, vol. 1, 378–435. Karnataka, India: Chethan Books.

Akki, A.J., Jain, P., Kulkarni, R. et al. (2024). Microbial biotechnology alchemy: transforming bacterial cellulose into sensing disease - A review. *Sensors International* 5: 100277. https://doi.org/10.1016/j.sintl.2023.100277.

An, C., Sun, C., Li, N. et al. (2022). Nanomaterials and nanotechnology for the delivery of agrochemicals: strategies towards sustainable agriculture. *Journal of Nanobiotechnology* 20 (1): 11. https://doi.org/10.1186/s12951-021-01214-7.

Arora, S., Murmu, G., Mukherjee, K. et al. (2022). A comprehensive overview of nanotechnology in sustainable agriculture. *Journal of Biotechnology* 355: 21–41. https://doi.org/10.1016/j.jbiotec.2022.06.007.

Avinash, M.G., Aishwarya, S., Zameer, F., and Gopal, S. (2023). *Pseudomonas aeruginosa* biofilm and their molecular escape strategies. *Journal of Applied Biology & Biotechnology* 11 (3): 28–37. https://doi.org/10.7324/jab.

Balusamy, S.R., Joshi, A.S., Perumalsamy, H. et al. (2023). Advancing sustainable agriculture: a critical review of smart and eco-friendly nanomaterial applications. *Journal of Nanobiotechnology* 21 (1): 372. https://doi.org/10.1186/s12951-023-02135-3.

Basit, F., Asghar, S., Ahmed, T. et al. (2022). Facile synthesis of nanomaterials as nanofertilizers: a novel way for sustainable crop production. *Environmental Science and Pollution Research International* 29 (34): 51281–51297. https://doi.org/10.1007/s11356-022-20950-3.

Beulah, K.C., Aishwarya, T., Waseem, K. et al. (2015). Phyto-antiquorumones: an herbal approach for blocking bacterial trafficking and pathogenesis. *International Journal of Pharmacy and Pharmaceutical Sciences* 7 (1): 29–34. https://journals.innovareacademics.in/index.php/ijpps/article/view/8543.

Biswas, B., Warr, L.N., Hilder, E.F. et al. (2019). Biocompatible functionalisation of nanoclays for improved environmental remediation. *Chemical Society Reviews* 48 (14): 3740–3770. https://doi.org/10.1039/c8cs01019f.

Bonilla-Cedrez, C., Chamberlin, J., and Hijmans, R.J. (2021). Fertilizer and grain prices constrain food production in sub-Saharan Africa. *Nature Food* 2 (10): 766–772. https://doi.org/10.1038/s43016-021-00370-1.

Brooks, S. (2011). Is international agricultural research a global public good? The case of rice biofortification. *The Journal of Peasant Studies* 38 (1): 67–80. https://doi.org/10.1080/03066150.2010.538581.

Carbone, C., Caddeo, C., and Musumeci, T. (2022). Nanoencapsulation strategies for active compounds delivery. *Nanomaterials* 12 (8): 1319. https://doi.org/10.3390/nano12081319.

Conrad, Z., Niles, M.T., Neher, D.A. et al. (2018). Relationship between food waste, diet quality, and environmental sustainability. *PLoS One* 13 (4): e0195405. https://doi.org/10.1371/journal.pone.0195405.

Cummings, C.L., Kuzma, J., Kokotovich, A. et al. (2021). Barriers to responsible innovation of nanotechnology applications in food and agriculture: a study of US experts and developers. *NanoImpact* 23: 100326. https://doi.org/10.1016/j.impact.2021.100326.

Datta, S., Hamim, I., Jaiswal, D.K., and Sungthong, R. (2023). Sustainable agriculture. *BMC Plant Biology* 23 (1): 588. https://doi.org/10.1186/s12870-023-04626-9.

Díez-Pascual, A.M. (2021). Carbon-based nanomaterials. *International Journal of Molecular Sciences* 22 (14): 7726. https://doi.org/10.3390/ijms22147726.

Dinesh, B., Poyya, J., Zameer, F. et al. (2024). Effect of polyvinylpyrrolidone on antioxidant and antibacterial activity of silver metal nanoparticles: a comparative analysis. *Proceedings of the National Academy of Sciences, India Section A: Physical Sciences* 94: 359–368. https://doi.org/10.1007/s40010-024-00882-y.

do Espirito Santo Pereira, A., Caixeta Oliveira, H., Fernandes Fraceto, L., and Santaella, C. (2021). Nanotechnology potential in seed priming for sustainable agriculture. *Nanomaterials (Basel, Switzerland)* 11 (2): 267. https://doi.org/10.3390/nano11020267.

dos Santos, M.S., Antunes Filho, S., and Backx, B.P. (2023). Bionanotechnology in agriculture: a one health approach. *Life (Basel, Switzerland)* 13 (2): 509. https://doi.org/10.3390/life13020509.

Foong, S.Y., Chan, Y.H., Loy, A.C.M. et al. (2022). The nexus between biofuels and pesticides in agroforestry: pathways toward United Nations sustainable development goals. *Environmental Research* 214 (pt 1): 113751. https://doi.org/10.1016/j.envres.2022.113751.

Gamage, A., Thiviya, P., Mani, S. et al. (2022). Environmental properties and applications of biodegradable starch-based nanocomposites. *Polymers* 14 (21): 4578. https://doi.org/10.3390/polym14214578.

Grobler, L., Suleman, E., Raj, T.G., and D. B. (2021). Patents and technology transfer in CRISPR technology. *Progress in Molecular Biology and Translational Science* 180: 153–182. https://doi.org/10.1016/bs.pmbts.2021.01.009.

Guevara, I., Ryan, S., Singh, A. et al. (2024). Edge IoT prototyping using model-driven representations: a use case for smart agriculture. *Sensors (Basel)* 24 (2): 495. https://doi.org/10.3390/s24020495.

He, X., Deng, H., and Hwang, H.M. (2019). The current application of nanotechnology in food and agriculture. *Journal of Food and Drug Analysis* 27 (1): 1–21. https://doi.org/10.1016/j.jfda.2018.12.002.

Ijaz, M., Khan, F., Ahmed, T. et al. (2023). Nanobiotechnology to advance stress resilience in plants: current opportunities and challenges. *Materials Today Bio* 22: 100759. https://doi.org/10.1016/j.mtbio.2023.100759.

Jain, A., Ranjan, S., Dasgupta, N., and Ramalingam, C. (2018). Nanomaterials in food and agriculture: An overview on their safety concerns and regulatory issues. *Critical Reviews in Food Science and Nutrition* 58 (2): 297–317. https://doi.org/10.1080/10408398.2016.1160363.

Kumar, A., Singh, K., Verma, P. et al. (2022). Effect of nitrogen and zinc nanofertilizer with the organic farming practices on cereal and oil seed crops. *Scientific Reports* 12 (1): 6938. https://doi.org/10.1038/s41598-022-10843-3.

Li, B., Sun, C., Li, J., and Gao, C. (2024). Targeted genome-modification tools and their advanced applications in crop breeding. *Nature Reviews Genetics* 25 (9): 603–622. https://doi.org/10.1038/s41576-024-00720-2.

Li, M., Zhang, Y., Feng, S. et al. (2022). Bioaccumulation and biomagnification effects of nano-TiO_2 in the aquatic food chain. *Ecotoxicology (London, England)* 31 (6): 1023–1034. https://doi.org/10.1007/s10646-022-02572-0.

Manjunatha, H.P., Singh, H., Chauhan, J.B. et al. (2013). Induction of resistance against sorghum downy mildew by seed treatment with *Duranta repens* extracts. *IOSR Journal of Agriculture and Veterinary Science* 3 (6): 37–44.

Mathew, T., Aishwarya Sree, R., Aishwarya, S. et al. (2020). Graphene-based functional nanomaterials for biomedical and bioanalysis applications. *FlatChem* 23: 100184. https://doi.org/10.1016/j.flatc.2020.100184.

Meyfroidt, P., de Bremond, A., Ryan, C.M. et al. (2022). Ten facts about land systems for sustainability. *Proceedings of the National Academy of Sciences of the United States of America* 119 (7): e2109217118. https://doi.org/10.1073/pnas.2109217118.

More, V.S., Ebinesar, A., Prakruthi, A. et al. (2021). Isolation and purification of microbial exopolysaccharides and their industrial application. In: *Microbial Polymers* (ed. A. Vaishnav and D.K. Choudhary). Singapore: Springer https://doi.org/10.1007/978-981-16-0045-6_3.

Mubeen, I., Fawzi Bani Mfarrej, M., Razaq, Z. et al. (2023). Nanopesticides in comparison with agrochemicals: outlook and future prospects for sustainable agriculture. *Plant Physiology and Biochemistry* 198: 107670. https://doi.org/10.1016/j.plaphy.2023.107670.

Nuttall, I., Miyagishima, K., Roth, C., and de La Rocque, S. (2014). The United Nations and one health: the international health regulations (2005) and global health security. *Revue Scientifique et Technique (International Office of Epizootics)* 33 (2): 659–668. https://doi.org/10.20506/rst.33.2.2303.

Oberemok, V.V., Laikova, K.V., Repetskaya, A.I. et al. (2018). A half-century history of applications of antisense oligonucleotides in medicine, agriculture and forestry: we should continue the journey. *Molecules (Basel, Switzerland)* 23 (6): 1302. https://doi.org/10.3390/molecules23061302.

Patil, A.G., Kounaina, K., Aishwarya, S. et al. (2021a). Myco-nanotechnology for sustainable agriculture: challenges and opportunities. In: *Recent Trends in Mycological Research (Fungal Biology)* (ed. A.N. Yadav). Cham: Springer https://doi.org/10.1007/978-3-030-60659-6_20.

Patil, A.G., Khan, K., Aishwarya, S. et al. (2021b). Fungal amylases and their industrial applications. In: *Industrially Important Fungi for Sustainable Development. Fungal Biology* (ed. A.M. Abdel-Azeem, A.N. Yadav, N. Yadav, and M. Sharma). Cham: Springer https://doi.org/10.1007/978-3-030-85603-8_11.

Pavlicevic, M., Elmer, W., Zuverza-Mena, N. et al. (2023). Nanoparticles and biochar with adsorbed plant growth-promoting rhizobacteria alleviate Fusarium wilt damage on tomato and watermelon. *Plant Physiology and Biochemistry* 203: 108052. https://doi.org/10.1016/j.plaphy.2023.108052.

Prasad, A., Baker, S., Nagendra Prasad, M.N. et al. (2019). Phytogenic synthesis of silver nanobactericides for anti-biofilm activity against human pathogen *H. pylori*. *SN Applied Sciences* 1: 341. https://doi.org/10.1007/s42452-019-0362-2.

Rahman, N.M.F., Malik, W.A., Baten, M.A. et al. (2024). Sustaining rice productivity through weather-resilient agricultural practices. *Journal of the Science of Food and Agriculture* 104 (4): 2303–2313. https://doi.org/10.1002/jsfa.13119.

Sahu, B., Ramesh, A., and Zameer, F. (2024). Impact of nanomaterials on leather: a nano-Saga from processing to application. *Clean Technologies and Environmental Policy* 27. https://doi.org/10.1007/s10098-024-02912-0: 5855–5882.

Sahu, B., Sharma, D.D., Jayakumar, G.C. et al. (2023). A review on an imperative by-product: glycosaminoglycans – a holistic approach. *Carbohydrate Polymer Technologies and Applications* 5: 100275.

Santás-Miguel, V., Arias-Estévez, M., Rodríguez-Seijo, A., and Arenas-Lago, D. (2023). Use of metal nanoparticles in agriculture. A review on the effects on plant germination. *Environmental Pollution (Barking, Essex: 1987)* 334: 122222. https://doi.org/10.1016/j.envpol.2023.122222.

Satapathy, P., Aishwarya, S., Shetty, M.R. et al. (2020). Phyto-nano-antimicrobials: Synthesis, characterization, discovery, and advances. In: *Frontiers in Anti-Infective Drug Discovery*, vol. 8, 196). Bentham Science Publishers https://doi.org/10.2174/9789811412387119080l.

Savelli, C.J., Bradshaw, A., Ben Embarek, P., and Mateus, C. (2019). The FAO/WHO international food safety authorities network in review, 2004-2018: learning from the past and looking to the future. *Foodborne Pathogens and Disease* 16 (7): 480–488. https://doi.org/10.1089/fpd.2018.2582.

Shafi, U., Mumtaz, R., García-Nieto, J. et al. (2019). Precision agriculture techniques and practices: from considerations to applications. *Sensors (Basel, Switzerland)* 19 (17): 3796. https://doi.org/10.3390/s19173796.

Shandilya, N., Barreau, M.S., Suarez-Merino, B. et al. (2023). TRAAC framework to improve regulatory acceptance and wider usability of tools and methods for safe innovation and sustainability of manufactured nanomaterials. *NanoImpact* 30: 100461. https://doi.org/10.1016/j.impact.2023.100461.

Shao, C., Zhao, H., and Wang, P. (2022). Recent development in functional nanomaterials for sustainable and smart agricultural chemical technologies. *Nano Convergence* 9 (1): 11. https://doi.org/10.1186/s40580-022-00302-0.

Singh, H., Zameer, F., Khanum, S.A. et al. (2016). Durantol - a phytosterol antifungal contributor from *Duranta repens* Linn. For organic Management of Sorghum Downy Mildew. *European Journal of Plant Pathology* 146: 671–682. https://doi.org/10.1007/s10658-016-0951-5.

Sulthana, R., Taqui, S.N., Zameer, F. et al. (2018). Adsorption of ethidium bromide from aqueous solution onto nutraceutical industrial fennel seed spent: kinetics and thermodynamics modeling studies. *International Journal of Phytoremediation* 20 (11): 1075–1086. https://doi.org/10.1080/15226514.2017.1365331.

Sun, J., Wang, Z., Du, Y. et al. (2022). Optimized tillage improves yield and energy efficiency while reducing carbon footprint in winter wheat-summer maize rotation systems. *The Science of the Total Environment* 820: 153278. https://doi.org/10.1016/j.scitotenv.2022.153278.

Sushil, A., Kamla, M., Nisha, K. et al. (2021). Nano-enabled pesticides in agriculture: budding opportunities and challenges. *Journal of Nanoscience and Nanotechnology* 21 (6): 3337–3350. https://doi.org/10.1166/jnn.2021.19018.

Tang, Y., Zhao, W., Zhu, G. et al. (2023). Nano-pesticides and fertilizers: solutions for global food security. *Nanomaterials* 14 (1): 90. https://doi.org/10.3390/nano14010090.

Tyumentseva, M., Tyumentsev, A., and Akimkin, V. (2023). CRISPR/Cas9 landscape: current state and future perspectives. *International Journal of Molecular Sciences* 24 (22): 16077. https://doi.org/10.3390/ijms242216077.

Vaidya, S., Deng, C., Wang, Y. et al. (2024). Nanotechnology in agriculture: a solution to global food insecurity in a changing climate? *NanoImpact* 34: 100502. https://doi.org/10.1016/j.impact.2024.100502.

Verma, K.K., Song, X.P., Kumari, A. et al. (2024). Climate change adaptation: challenges for agricultural sustainability. *Plant, Cell & Environment* 48: 2522–2533. https://doi.org/10.1111/pce.15078.

Wang, C., Jiang, Y., He, K., and Wāng, Y. (2024). Eco-friendly synthesis of silver nanoparticles against mosquitoes: pesticidal impact and indispensable biosafety assessment. *The Science of the Total Environment* 953: 176006. https://doi.org/10.1016/j.scitotenv.2024.176006.

Yadav, A.N. (2021). *Recent Trends in Mycological Research*. Springer International Publishing https://doi.org/10.1007/978-3-030-68260-6.

Zameer, F. and Gopal, S. (2010). Evaluation of antibiotic susceptibility in mixed culture biofilms. *International Journal of Biotechnology and Biochemistry* 6 (1): 93–99. https://www.researchgate.net/publication/242100772.

Zameer, F., Kreft, J., and Gopal, S. (2010a). Interaction of the dual species biofilms of Listeria monocytogenes and Staphylococcus epidermidis. *Journal of Food Safety* 30 (4): 954–968. https://doi.org/10.1111/j.1745-4565.2010.00254.x.

Zameer, F., Meghashri, S., Gopal, S., and Rao, B.R. (2010b). Chemical and microbial dynamics during composting of herbal pharmaceutical industrial waste. *Journal of Chemistry* 7: 645978. https://doi.org/10.1155/2010/645978.

Zameer, F., Rukmangada, M.S., Chauhan, J.B. et al. (2016). Evaluation of adhesive and anti-adhesive properties of *Pseudomonas aeruginosa* biofilms and their inhibition by herbal plants. *Iranian Journal of Microbiology* 8 (2): 108–119. https://pubmed.ncbi.nlm.nih.gov/27307976/.

Zhang, D., Zhang, Z., Unver, T., and Zhang, B. (2021a). CRISPR/Cas: a powerful tool for gene function study and crop improvement. *Journal of Advanced Research* 29: 207–221. https://doi.org/10.1016/j.jare.2020.10.003.

Zhang, P., Guo, Z., Ullah, S. et al. (2021b). Nanotechnology and artificial intelligence to enable sustainable and precision agriculture. *Nature Plants* 7 (7): 864–876. https://doi.org/10.1038/s41477-021-00946-6.

Zhang, Y. and Goss, G.G. (2022). Nanotechnology in agriculture: comparison of the toxicity between conventional and nano-based agrochemicals on non-target aquatic species. *Journal of Hazardous Materials* 439: 129559. https://doi.org/10.1016/j.jhazmat.2022.129559.

Zhu, L., Chen, L., Gu, J. et al. (2022). Carbon-based nanomaterials for sustainable agriculture: their application as light converters, nanosensors, and delivery tools. *Plants* 11 (4): 511. https://doi.org/10.3390/plants11040511.

Zielińska, A., Carreiró, F., Oliveira, A.M. et al. (2020). Polymeric nanoparticles: production, characterization, toxicology and ecotoxicology. *Molecules* 25 (16): 3731. https://doi.org/10.3390/molecules25163731.

19

Regulatory Aspects of Nanotechnology in the Agriculture Sector

Ayomide Emmanuel Fadiji[1,2], *Adegboyega Adeniji*[3], *and Adedayo Ayodeji Lanrewaju*[4]

[1] *Plant Biosecurity Unit, Department of Agriculture and Fisheries, Darwin, Northern Territory, Australia*
[2] *Food Security and Safety Focus Area, Faculty of Natural and Agricultural Sciences, North-West University, Private Bag, Mmabatho, South Africa*
[3] *State Key Laboratory for Biology of Plant Diseases and Insect Pests, Institute of Plant Protection, Chinese Academy of Agricultural Sciences, Beijing, China*
[4] *Biotechnology and Food Science, Faculty of Applied Science, Durban University of Technology, Durban, South Africa*

19.1 Introduction

Nanotechnology has emerged as a disruptive force that could reshape the agricultural industry. Scientists are discovering novel methods to boost agricultural yield, soil health, and minimize environmental impact by altering materials at the nanoscale (Parisi et al. 2015). Nano-fertilizers, nano-pesticides, and nanosensors pave the way for precision agriculture, giving farmers unparalleled control over their inputs and outputs (Khot et al. 2012). As the world's population grows and climate change constitutes new problems for food security, using nanotechnology in agriculture represents both an opportunity and a requirement.

However, the fast growth and application of nanotechnology in agriculture presents substantial regulatory concerns. The distinct features of nanomaterials, such as enhanced reactivity, bioavailability, and possible toxicity, need a rethinking of existing regulatory frameworks (Scott and Chen 2012). Current rules may fail to appropriately address the unique dangers associated with nanoparticles, resulting in uncertainty in safety evaluations, labeling, and environmental impact analysis (Schoonjans et al. 2023). The absence of standardized methods for testing and monitoring nanomaterials complicates regulatory supervision (Allan et al. 2021).

Therefore, this review examines the current regulatory landscape governing the use of nanotechnology in the agriculture sector. It highlights the gaps and inconsistencies in international and national regulations, discusses the implications of these shortcomings, and proposes strategies for developing comprehensive regulatory frameworks. By addressing the ethical, environmental, and health-related concerns associated with agricultural nanotechnology, policymakers can facilitate safe innovation while protecting consumers and ecosystems. A harmonized and proactive regulatory approach is essential to fully realize the benefits of nanotechnology in agriculture and to foster public trust in these emerging technologies.

19.2 Overview of Nanotechnology in Agriculture

Nanotechnology is quickly developing as a revolutionary tool in agriculture, with potential applications spanning crop productivity, protection, and sustainability. Working at the nanoscale (1–100 nm), researchers may develop innovative materials and technologies with exceptional accuracy and efficiency, tackling many of the issues associated with traditional agriculture operations and thus being applied as nanofertilizers, nanopesticides, nanominerals, nanosensors, and nanocatalysts, among others (Figure 19.1). Nanofertilizers provide more efficient and regulated delivery of nutrients than traditional fertilizers. This method minimizes nutrient losses from leaching and volatilization while ensuring that plants receive nutrients as needed. Nanoparticles can encapsulate fertilizers, releasing them slowly in response to environmental factors such as soil pH or moisture (Ditta et al. 2015).

Studies have shown that nanofertilizer can enhance crop yield and reduce environmental pollution, improving productivity and sustainability (Adisa et al. 2019). Furthermore, nanopesticides provide a targeted, effective approach to managing pests, reducing the need for large quantities of chemicals. Nanoparticles can carry pesticides directly to the site of infection or infestation, enhancing their efficacy and reducing off-target effects (Kah et al. 2018). Nanoformulations can also improve the stability of active ingredients, protecting them from degradation due to environmental factors such as UV light and ensuring their slow release, which minimizes the risk of pest resistance development.

Also, nanotechnology enables the development of smart delivery systems for nutrients and agrochemicals. These devices may respond to environmental cues like moisture, temperature, or pH to release their contents at the right moment, improving the accuracy of agricultural operations. Nanoparticles, for example, can be designed to release nutrients when soil moisture levels are low, or the plant is actively developing, reducing waste and increasing efficiency (Shang et al. 2019). Nanoparticles have also been shown to boost plant development and resistance to abiotic challenges such as drought, salt, and severe temperatures. Silicon and zinc oxide nanoparticles, for example, have been demonstrated to boost

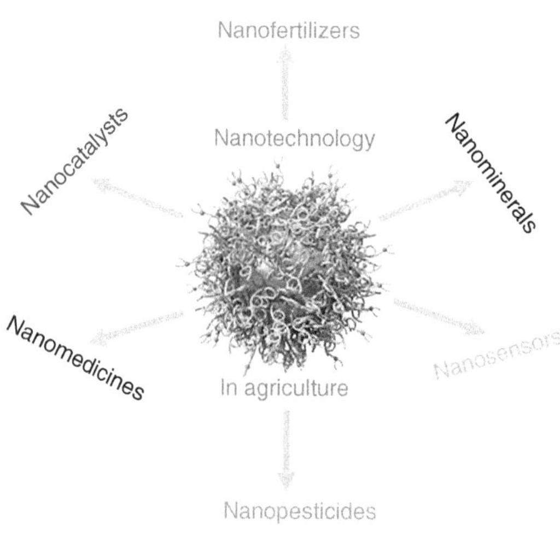

Figure 19.1 Different expressions of nanotechnology in agriculture.

water absorption, photosynthesis, and antioxidant defence systems in plants, making them more tolerant to stress (Jalil and Ansari 2019; Wang et al. 2023). Furthermore, silver and iron oxide nanoparticles have enhanced plant immune responses to pathogens (Rastogi et al. 2017). Moreover, nanotechnology may be used to promote soil health and assist in rehabilitating damaged settings. Nanoparticles can detoxify pollutants like heavy metals by binding and removing them from the soil. This use is especially useful in places impacted by industrial pollution or excessive pesticide usage, where soil deterioration is a major problem (Roy et al. 2021).

Nanoparticles have also been shown to enhance seed germination and vigor (Table 19.1). Nano-coatings on seedlings can promote germination, defend against infections, and increase water retention (Jampílek et al. 2017). These coatings can also give nutrients or growth stimulants directly to the germinating seed, improving early-stage development, especially in harsh environments. One of the most cutting-edge uses of nanotechnology is the creation of nanobiosensors for precision agriculture. These sensors can detect even minute environmental changes, such as nutrient levels, soil moisture, pests, and diseases. This real-time data enables farmers to make educated crop management decisions while optimizing inputs such as water, fertilizer, and pesticides (Antonacci et al. 2018). Nanobiosensors can also be used with drones and other automated systems for large-scale monitoring and data collection. Nanotechnology is also being researched to provide more effective crop protection solutions. This comprises nanoemulsions, nanoparticles, and nanotubes, which may deliver bioactive substances like insecticides and herbicides with great accuracy. Nanoparticles' tiny size allows them to enter plant tissues more efficiently,

Table 19.1 Effect of nanoparticles on seed germination.

Nanoparticle	Plant	Germination enhancement (%)	References
Ferric oxide	*Triticum aestivum*	41.60	Sundaria et al. (2019)
Silver	*Trigonella foenum-graecum*	5.30	Hojjat and Kamyab (2017)
Silver	*Capsicum annuum*	90–100	Mawale and Giridhar (2024)
Nano phosphorus	*Vigna radiata*, *Vigna mungo* and *Vigna unguiculata*	20.83, 38.1 and 20.83	Priya et al. (2015)
Silicon dioxide	*Triticum aestivum*	16.78	Akhtar and Ilyas (2022)
Titanium dioxide	*Raphanus sativus*	20.00	Manesh et al. (2018)
Titanium dioxide	*Triticum aestivum*	16.30	Zahra et al. (2019)
Titanium dioxide	*Prunus mahaleb*	65.00	Goodarzi et al. (2017)
Chitosan and zinc oxide	*Oryza sativa*	20.00	Vemula (2022)
Zinc oxide	*Vigna unguiculata*	3.18	Srinivasan et al. (2017)
Zinc oxide	*Brassica napus*	7.23	Alhammad et al. (2023)
Zinc oxide	*Triticum aestivum*	13.80	Bayat et al. (2022)

delivering protective chemicals where they are most required. Nanopesticides have been demonstrated in studies to require less active ingredients to produce the same degree of protection as traditional treatments, resulting in less environmental pollution (Kah et al. 2018).

Nanotechnology has the potential to improve crop post-harvest preservation significantly. Nanoparticles and nanocomposites can be utilized in packaging to extend the shelf life of agricultural goods by preventing microbial development and retarding deterioration. For example, silver and zinc oxide nanoparticles are recognized for their antibacterial characteristics and are being explored in active packaging solutions to minimize fruit and vegetable degradation during storage (Sadeghi et al. 2021; Mirsharifi et al. 2023). Water shortage is a major concern in agriculture, and nanotechnology is used to increase water efficiency. Nanomaterials like carbon nanotubes and hydrogels can be utilized in irrigation systems to keep water in the soil longer, eliminating the need for regular watering.

Additionally, nanotechnology-based filtering systems can cleanse wastewater, allowing it to be reused in agriculture (Kuhn et al. 2022).

These technologies can potentially alleviate water scarcity, especially in dry locations. Nanoparticles can be employed to improve soil health and water retention. Nanoclays and other nanoscale soil additives can enhance soil structure, boost aeration, and encourage beneficial microbial activity. Some nanoparticles can also reduce soil salinity and boost nutrient bioavailability, addressing soil degradation concerns that impact agricultural yield (Hasan et al. 2020). Nanotechnology can also help with genetic alteration by making CRISPR/Cas9 systems more easily delivered into plant cells. This precise genome-editing method may be improved by employing nanocarriers to transfer genetic material to specific plant tissues, making gene editing more accessible to a wider range of crops (Bharathan and Turarbekova 2024).

Nanotechnology is becoming increasingly essential in animal husbandry since it improves health management, feeding, and production efficiency. Nanotechnology is being utilized to produce nanomedicines for detecting, treating, and preventing cattle illnesses. Nanoparticles can transport and distribute medications in a more focused way, resulting in controlled release and improved bioavailability. This decreases the dose necessary while also minimizing negative effects. Nanoparticles can also be employed in vaccines to boost immune responses, resulting in more effective and long-lasting protection against diseases (Prasad et al. 2023). Nanobiosensors are increasingly being used to identify illnesses in animals at an early stage, which is critical for stopping infections from spreading. These nanoparticle-based biosensors detect biological indicators in blood, milk, or saliva infections. Rapid, on-site diagnostics enable prompt treatments, increasing disease management in herds and reducing economic losses (Péter et al. 2022).

Nanotechnology improves the bioavailability of nutrients in animal feed. Vitamins, minerals, and other supplements can be nanoencapsulated to increase absorption in the gastrointestinal system, ensuring that animals get the most out of the nutrients they ingest. For example, nanominerals like zinc, selenium, and iron can be supplied to animals more effectively, boosting growth, reproductive health, and overall output (Hassan et al. 2020). Nanoparticles are utilized to improve animal growth rates by increasing feed efficiency. Nanomaterials can also be used as growth promoters to increase weight gain and shorten the time to market for meat. This might help meet the increased demand for meat products while using fewer resources (Reddy et al. 2020).

Nanotechnology can improve animal reproductive health by increasing sperm quality and fertility. Nanoparticles can be employed in cryopreservation to keep sperm safe throughout freezing and thawing operations. The inclusion of nano-additives in semen extenders can improve sperm motility and lifespan, increasing fertility rates in artificial insemination programs (Hashem and Gonzalez-Bulnes 2021). Nanopesticides provide an innovative approach to managing pests and parasites in animal husbandry. Nanoscale formulations of pesticides offer greater efficacy and longer-lasting effects than conventional pesticides. This is particularly useful in controlling ectoparasites like ticks, lice, and mites, which can harm livestock health and reduce productivity. Furthermore, nanoacaricides and nanoinsecticides help minimize chemical residues in the environment, offering a more sustainable solution (Zaheer et al. 2024).

Nanotechnology is also being investigated to remediate animal waste. Nanoparticles can help break down contaminants in manure, lowering greenhouse gas emissions and increasing the efficiency of waste-to-energy systems. For example, nanocatalysts can speed up the decomposition of organic waste, resulting in more efficient biogas production. This promotes sustainable farming methods by tackling waste management and energy production issues (Sharma et al. 2023). Nanotechnology can help monitor food safety in animal products, including meat, milk, and eggs. Nanosensors may detect hazardous infections such as *Salmonella* and *E. coli* in animal-derived goods, preventing these toxins from reaching consumers. Furthermore, nano-packaging materials with antimicrobial capabilities are being developed to increase the shelf life of animal products, reduce food waste, and improve food safety (Zhou et al. 2023).

19.3 Regulatory Frameworks for Nanotechnology

19.3.1 Global Regulatory Landscape

Nanotechnology's rapid evolution in agriculture has led to groundbreaking applications, such as enhanced nutrient delivery, precision farming, and advanced plant protection mechanisms (Indira et al. 2021; Singh et al. 2023). However, the application of these technologies has highlighted significant regulatory challenges. Indira et al. (2021) emphasized that nanocarriers, nanocapsules, and nanocomposites offer controlled release mechanisms that reduce environmental contaminants, making them promising tools for sustainable agriculture. Similarly, Singh et al. (2023) find that nanoparticles like zinc, titanium dioxide, and silicon dioxide boost photosynthesis and improve seed germination and plant growth. While these advances demonstrate the potential of nanomaterials to revolutionize agricultural practices, they also raise complex questions regarding their safety and long-term effects on human health and the environment. Allan et al. (2021) provide a global perspective on the regulatory challenges arising from these innovations. The rapid pace at which new nanomaterials are developed far outpaces existing regulatory frameworks. While progress has been made in analytical methods for labeling and safety assessment, significant gaps remain in harmonizing definitions and risk assessment methodologies across international boundaries. This issue is exacerbated by the complexity and diversity of nanomaterials, which vary widely in structure, behavior, and potential risks.

As Allan et al. (2021) opined, these differences complicate the creation of unified global standards, which could facilitate international collaboration on nanomaterial safety. For example, Amenta et al. (2015) analyze the regulatory landscape of nanotechnology in agriculture, feed, and food sectors across EU and non-EU countries. The study highlights that nano-enabled agrochemicals, nutrients, and intelligent food packaging offer enhanced efficacy and controlled release mechanisms but pose potential health and environmental risks due to their unique properties. Mechanistically, the study emphasizes that EU legislation explicitly incorporates nano-specific provisions, unlike most non-EU regions, where regulations rely on general chemical safety frameworks. They stressed the need for harmonized international regulations, improved risk assessment methodologies, and transparency measures such as labeling and product registration to ensure safe nanotechnology applications. Despite these challenges, regulatory bodies like the U.S. Food and Drug Administration (FDA), the European Food Safety Authority (EFSA), and Environmental Protection Agency (EPA) have attempted to address nanotechnology's regulatory needs. However, the current regulatory mechanisms are often inadequate in addressing nanoparticle risks, particularly regarding bioaccumulation and trophic transfer (Mishra et al. 2019). These risks remain poorly understood due to a lack of long-term exposure data, making it difficult for regulatory bodies to establish robust and comprehensive safety guidelines.

19.3.2 Regulatory Bodies and their Roles

Regulatory agencies play a crucial role in ensuring the safe and effective use of nanotechnology in agriculture. In particular, organizations such as the FDA, the EFSA, and the EPA are responsible for evaluating the safety of nanomaterials used in food, agriculture, and environmental applications (Singh and Kumar 2023). In the United States, the FDA oversees the safety of food additives, food contact materials, and feed additives under the Federal Food, Drug, and Cosmetic Act (FFDCA). Food and color additives are subject to mandatory pre-market approval, whereas certain food substances, such as those classified as Generally Recognized as Safe (GRAS), do not require prior authorization. The FFDCA does not include specific provisions for nanotechnology-based products, and the FDA has not yet established a formal regulatory definition for nanomaterials. Instead, the agency has adopted a broad, science-based approach to evaluating nanotechnology applications. Recognizing both the potential benefits and uncertainties associated with emerging technologies, the FDA has chosen not to introduce new, dedicated regulations for nanomaterials, relying instead on existing regulatory frameworks. This strategy incorporates horizon scanning and internal reviews of product submissions to monitor developments in nanotechnology (Allan et al. 2021).

The EFSA, as an independent European agency, assesses risks related to food and feed safety, including nutrition, health claims, biological hazards, contaminants, and plant protection products. It provides scientific guidance to policymakers and supports the regulation and implementation of policies for human, animal, and plant health. EFSA also evaluates the environmental risks associated with food production, considering potential impacts on biodiversity. To address the complexities of nanotechnology applications in food and feed, EFSA has established a specialized scientific network to facilitate knowledge exchange among EU member states and prioritize risk assessment activities (Ververis et al. 2020). Additionally, EFSA maintains a dedicated working group on nanotechnology in food and feed and has published guidance on assessing nanotechnology-related risks. The regulatory landscape for nanotechnology applications in the EU and non-EU countries has been extensively reviewed by Amenta et al. (2015), emphasizing the increasing importance of the "safe-by-design" approach. This concept aims to develop safe nanomaterials throughout their life cycle, from production to disposal or recycling and was first formulated for nanomaterials under the EU flagship project NANoREG (Gottardo et al. 2021).

Several non-EU European countries have also been active in nanotechnology regulation. Switzerland, for example, applies its existing chemical regulations to nanomaterials. For pesticide registration, companies must disclose detailed information on the composition, particle size, surface area, aggregation state, coatings, and functionalization of nanomaterials, as stipulated by a 2010 Swiss Ordinance (Bundesrat 2010). As of 2013, Switzerland's Federal Office of Public Health (FOPH) had not received applications for food additives containing nanomaterials, nor had nanotechnology-based pesticides been submitted for approval (Peters et al. 2014; Amenta et al. 2015). In Asia, several nations are actively involved in producing and regulating nanomaterials. Countries like Japan and South Korea have developed standards and certification systems for nano-enabled products and participate in the OECD Working Party on Manufactured Nanomaterials (WPMN).

Japan's regulatory science framework, overseen by institutions such as the Ministry of Health, Labour and Welfare (MHLW), the Pharmaceutical and Medical Devices Agency (PMDA), the National Institute of Health Sciences (NIHS), and the Agency for Medical Research and Development (AMED), is designed to ensure that research findings are rapidly translated into practical applications. The PMDA established a Regulatory Science Centre in 2018 to support the development of advanced therapies and technologies, including nanotechnology, through horizon scanning (Rodríguez-Gómez et al. 2024). This proactive approach allows regulators to stay informed about technological advancements and leverage clinical data and electronic health records for safety assessments (Allan et al. 2021).

Despite these regulatory efforts, significant challenges remain in adapting existing frameworks to the unique properties of nanomaterials. As Mahra et al. (2024) noted many emerging nanotechnology applications in agriculture, such as nano-encapsulation for controlled nutrient release and seed nanopriming for enhanced resilience, do not fit neatly into conventional agrochemical regulatory categories. This underscores the need for more targeted evaluation frameworks that can effectively assess the safety, efficacy, and long-term environmental impact of nanotechnology-driven agricultural innovations.

19.3.3 Regulatory Challenges

One of the foremost concerns is the inconsistency in nanotechnology regulation across different jurisdictions (Mitter and Hussey 2019; Allan et al. 2021). While the European Union has incorporated nano-specific provisions within its legislative frameworks, most non-EU regions still rely on generalized chemical safety regulations that lack specificity regarding nanoparticle risk assessment. This lack of harmonized global policies creates regulatory uncertainty, potentially hindering the commercialization of nano-enabled agricultural products and complicating international trade (Amenta et al. 2015; Rodriguez et al. 2019). Differences in nanoparticle classification, safety assessment methodologies, and permissible exposure limits further exacerbate the challenge, making it difficult to establish universal guidelines for nano-based agricultural applications.

Another significant regulatory hurdle lies in the risk assessment of engineered nanoparticles (ENPs) used in agrochemicals. Unlike conventional fertilizers and pesticides, nanoformulated products exhibit unique physicochemical properties such as increased surface reactivity, bioavailability, and persistence, which may lead to unintended environmental and ecological consequences (Iavicoli et al. 2017; Bathi et al. 2022). Research highlights that nanoparticles can bioaccumulate in soil ecosystems, undergo trophic transfer in food chains, and induce oxidative stress in nontarget organisms due to reactive oxygen species (ROS) generation (Servin and White 2016; Mahra et al. 2024). However, current risk assessment models often fail to capture these complexities, as they primarily rely on short-term, high-dose toxicity studies that may not accurately reflect real-world agricultural exposures (Mishra et al. 2019). The long-term effects of chronic low-dose nanoparticle exposure on soil microbial communities, plant metabolism, and human health remain largely unexplored, necessitating more sophisticated assessment tools and environmental monitoring strategies.

Furthermore, the ethical dimensions of nanotechnology in agriculture remain a contentious issue, particularly regarding public perception, transparency, and labeling requirements. Mukherjee et al. (2019b) and Coles and Frewer (2013) emphasize that low public awareness and knowledge gaps contribute to widespread skepticism toward nano-enabled agricultural products. Risk perception theories suggest that media narratives, cultural cognition, and regulatory transparency often shape public attitudes toward nanotechnology. Without clear labeling and public engagement initiatives, consumer apprehension may impede the adoption of nano-based fertilizers and pesticides, despite their potential benefits in improving crop yields and reducing chemical runoff. Ethical considerations also extend to intellectual property rights issues and access to nanotechnology, as proprietary nanoformulations may disproportionately benefit large agribusinesses while limiting access for smallholder farmers in developing regions (Chowdhury et al. 2017).

To address these challenges, a multi-pronged regulatory strategy is needed. First, international harmonization of nanotechnology regulations should be prioritized to establish consistent safety standards and risk assessment protocols. Efforts to create globally recognized definitions of nanomaterials, such as those proposed by the Organisation for Economic Cooperation and Development (OECD) and the EFSA, must be expanded to encompass agricultural applications (Amenta et al. 2015; Kumari et al. 2023). Second, risk governance should incorporate life-cycle analysis and real-world exposure studies to assess nanoparticles' fate, persistence, and bioaccumulation potential in agricultural environments (Servin and White 2016). Third, transparency measures—including nano-specific labeling, public disclosure of safety data, and participatory decision-making processes—are essential to build public trust and facilitate informed consumer choices (Mukherjee et al. 2019b; Meena and Chandel 2024). Finally, investment in green nanotechnology approaches, such as biodegradable nanocarriers and eco-friendly synthesis methods, can help mitigate the environmental risks associated with ENPs while maintaining the agricultural benefits of nanotechnology (Abobatta 2018; Manjunatha et al. 2019). While nanotechnology offers groundbreaking opportunities for agricultural sustainability, its safe and ethical deployment hinges on robust regulatory oversight, interdisciplinary risk assessment, and proactive public engagement. Without these safeguards, the rapid commercialization of nano-enabled agricultural products risks outpacing our understanding of their long-term ecological and societal impacts. Thus, a forward-thinking regulatory approach that balances innovation with precaution is essential to unlock the full potential of nanotechnology in agriculture while ensuring environmental and consumer safety.

19.4 Specific Regulations and Standards for Nanotechnology in Agriculture

19.4.1 Regulation of Nano-fertilizers and Nano-pesticides

The regulation of nano-fertilizers and nano-pesticides is a critical area requiring well-defined regulatory frameworks to ensure their safe and effective integration into agricultural systems. Studies by Mahra et al. (2024) and Mishra et al. (2019) highlight the significant benefits of these nano-enabled products, including improved nutrient uptake, reduced chemical wastage, and targeted delivery, which collectively enhance agricultural efficiency. Meena and Chandel (2024) further emphasize that nano-fertilizers, nano-pesticides, and nanosensors contribute to precision agriculture by minimizing environmental contamination and optimizing resource utilization. Despite these advantages, concerns over nanoparticle toxicity, bioaccumulation, and long-term environmental persistence remain, necessitating stringent regulatory oversight. The potential ecological disruptions associated with these materials underscore the need for comprehensive risk assessment protocols, robust labeling requirements, and public awareness initiatives to promote responsible use. Current regulatory processes, which predominantly rely on conventional chemical-based criteria, may not be sufficient to assess the unique properties of nanomaterials. For instance, nano-pesticide regulatory challenges demand modifications to conventional toxicity assessment protocols (Kah 2015).

Unlike traditional pesticides, nano-formulations require specialized reference materials and controls—such as bulk counterparts or metal salts—to facilitate accurate risk comparisons. Ensuring a homogenous distribution of nano-pesticides in test systems is essential, which may necessitate novel dosing strategies, increased mixing, or periodic water changes to maintain stable exposure concentrations in aquatic environments. Moreover, the distinct physicochemical properties of nano-pesticides influence their uptake and degradation kinetics, requiring adjustments in study durations to account for potential long-term effects. Advanced characterization techniques are necessary to monitor particle behavior, transformation, and bioavailability throughout toxicity studies, often demanding more frequent assessments than those typically conducted for conventional pesticides.

Additional regulatory challenges stem from the need to establish appropriate dose metrics and address issues related to nanoparticle aggregation or agglomeration, both of which can affect the effective concentration of nanoparticles in test environments

(Kookana et al. 2014; Fadiji et al. 2022). Given these complexities, regulatory bodies must develop and implement testing protocols tailored to the unique properties and behaviors of nano-enabled agricultural inputs. This includes refining exposure assessments, updating risk evaluation frameworks, and incorporating long-term ecological monitoring strategies. Establishing a standardized regulatory approach will be essential in ensuring that nano-fertilizers and nano-pesticides deliver their intended agricultural benefits while minimizing potential risks to ecosystems and human health.

19.4.2 Standards for Nanomaterials in Soil and Feed

In addition to product-specific regulation, there is a need for clear standards governing the use of nanomaterials in soil and animal feed. As described by Srilatha (2011) and Mahra et al. (2024), introducing nanoparticles into soils can have unintended consequences, including changes to soil microbial communities and nutrient cycling. Regulatory bodies should establish standards for the acceptable concentration of nanoparticles in soils and monitor their potential impact on ecosystem health. Similarly, the use of nanoparticles in animal feed must be carefully regulated to prevent bioaccumulation and ensure food safety. According to Meena and Chandel (2024), these standards should include labeling and traceability requirements, ensuring that consumers and farmers alike are fully informed about the presence and potential risks of nanomaterials in the food and agricultural supply chain. In conclusion, while there are examples of successful regulatory approval for some nano-enabled products, such as those used in precision farming and seed enhancement (Singh and Kumar 2023; Mahra et al. 2024), there are also notable instances where regulatory processes have been delayed or complicated by insufficient data on the long-term impacts of nanomaterials. As the body of evidence grows, it will be crucial for regulatory bodies to adapt their frameworks to ensure that these innovative products can be safely and effectively integrated into agricultural systems.

19.5 Ethical and Societal Considerations

19.5.1 Potential Risks and Impacts of Nanomaterials on Human Health and Environment

The advanced utilization of nanotechnology in agricultural practices and food products raises significant concerns among a substantial segment of society due to the various adverse effects of different nanoparticles (Baruah and Dutta 2009). While the characteristics and safeguarding of bulk substances are typically clear, nanoscale materials consistently exhibit diverse properties in contrast to their macroscale counterparts (De Azeredo 2009). The risks linked to nanotechnology primarily stem from the diminutive scale of nanoparticles, which possess extensive surface areas that facilitate their dispersion. This characteristic enables them to infiltrate cells and access remote regions within the body, potentially leading to toxicity (Dasgupta et al. 2015). Owing to their comparable dimensions to DNA, nanomaterials have the potential to interact with biological specimens (Baruah and Dutta 2009).

The degradation of nanocomposites may occur due to environmental conditions, leading to the release of embedded nanomaterials from the polymeric matrix into the surrounding environment (Moustafa et al. 2019). The integrity of certain food packaging materials, such as low-density polyethylene, is altered following exposure to environmental factors, including ultraviolet radiation or ozone in humid environments. Han et al. (2018). indicate that polyethylene samples characterized by low density experienced oxidation when exposed to UV light or ozone, significantly altering their structural, physical, and thermal properties. The application of pesticides and nano fertilizers is prevalent in the agricultural domain, leading to their dispersion into water, soil, or the atmosphere, which poses significant health risks for farmers (Roohinejad and Greiner 2017). The presence of nanoparticles in the soil is anticipated to hinder plant development and lead to their accumulation in consumable plant tissues (Rajput et al. 2020).

The introduction of nanoparticles into the environment initiates a sequence of transformations, and their interactions within the human body present considerable concerns. Nanoparticles can induce oxidative damage and elicit responses, while discarded nanoparticles may harbor the potential for toxicity (Lewicka et al. 2013; Han et al. 2018; Narei et al. 2018). Nanotoxicity is primarily mediated through the significant generation of free radicals, leading to oxidative stress, which impairs the cells' ability to maintain normal biological redox-regulated functions (Pathakoti et al. 2017). The deterioration of nano-clay found in low-density polyethylene aggregates can potentially induce carcinoma in alveolar basal epithelial cells in humans (Kumar et al. 2018).

Many studies have been undertaken to explore the antibacterial mechanisms of silver. However, as heavy metal, it can potentially cause toxicity by denaturing proteins and enzymes within the body when found in elevated concentrations; the risk of its migration can be assessed (Li et al. 2017). Research regarding Ag and TiO_2 nanoparticles, along with carbon nanotubes, has demonstrated that these substances possess the ability to infiltrate blood circulation, and their insoluble characteristics may lead to accumulation within various organs (Ramachandraiah et al. 2015, Sharma et al. 2017). Incorporating TiO_2 as a food additive or in alternative forms within the human body can induce inflammation through oxidative stress alongside genotoxic effects that may lead to chromosomal instability (Baranowska-Wójcik et al. 2020).

Nanoparticles can enter the body through ingestion, inhalation, or skin contact. The extensive use of nanomaterials in food packaging raises significant concerns regarding their potential release from contaminated food or the surrounding environment (Han et al. 2018). Nevertheless, there is a paucity of data concerning the migration of nanoparticles from packaging materials into food, as well as their ultimate toxicological effects (De Azeredo 2009; Pathakoti et al. 2017). The selection of green nanofillers in nanocomposite research is important for preserving environmental integrity, human well-being, and animal welfare. Furthermore, factors such as concentration, molecular weight, particle size, diffusivity, and the stability of specific substances within polymer blends, alongside temperature, pH levels, polymeric structure, viscosity, contact duration, mechanical pressure, and food composition, are critical (Sharma et al. 2017). Precisely evaluating nanoparticles released into the environment, their accumulation in flora, and the subsequent health implications is equally vital (Rajput et al. 2020).

It is essential to establish robust guidelines, policies, and regulatory frameworks to ensure the safe application of nanoparticles within the food industry. For example, the FDA plays a crucial role in regulating nanofoods and food packaging in the United States, while the European Union oversees the control of nanotechnology-based food ingredients in Europe (Nile et al. 2020). Nonetheless, most nations producing nanomaterials lack adequate regulations specifically tailored to nanotechnology. Consequently, comprehensive governmental guidelines, legislation, and stringent toxicological screening methodologies are imperative for the lawful application of nanotechnology (Nile et al. 2020).

19.5.2 Public Perception of Nanotechnology

Every technological advancement possesses both beneficial and detrimental aspects. Comparably, the advantages of nanotechnology in agriculture and food security are accompanied by certain harmful effects. To this day, understanding remains limited regarding the impact of nanomaterial exposure on health and the environment. However, given its recent emergence, extensive research is underway to amass a substantial volume of data. As nanotechnology evolves, individuals will increasingly acknowledge its significance and potential adverse effects. This will probably influence the populace, as evidenced by the widespread public demonstrations regarding genetically modified crops in the past. The collective readiness to permit research and experimentation before uncertain trials may result in potential risks to the environment and/or advantages for humanity (Mukherjee et al. 2019a).

Generally, individuals responding to the repercussions may lack awareness of the underlying facts. They tend to be swayed more by the opinions of their neighbors and friends than by empirical evidence. Surveys of public opinion were carried out to gain insights into community perceptions regarding nanotechnology and its associated risks, as evidenced by studies conducted in various regions, including Japan, Sweden, and the US, among others (Fujita et al. 2006; Siegrist et al. 2008; Kahan et al. 2009; Scheufele et al. 2009). The study of perception holds significant importance in any emerging field of research, as it is posited that risk perception arises from social and psychological influences rather than merely a lack of knowledge (Stanovich and West 2008). Another reason is that human behavior stems from what individuals believe or perceive to be true; perceptions and biases are not readily subject to alteration in light of new knowledge (Irwin and Wynne 1996). Kahan et al. (2009) indicated that public attitudes are susceptible to the influences of psychological dynamics linked to cultural cognition.

The mere distribution of knowledge is insufficient; engaging with socio-psychological factors and providing education to transform individuals' attitudes, behaviors, and belief systems regarding new technology is essential. Establishing a framework for fostering awareness of nanotechnology through education is essential. It is imperative that seasoned researchers, educators, and scholars disseminate precise information regarding nanotechnologies and their various applications to the general populace (Mukherjee et al. 2019a). Historical evidence indicates that endeavors to manipulate risk perceptions through educational initiatives can yield negative outcomes when the advantages are excessively emphasized while the associated risks remain obscured (Satterfield et al. 2009). Society is composed of individuals with varied socio-psychological backgrounds. The current body of research indicates that predisposing characteristics, including race, cultural values, gender, and religion, shape perceptions regarding the risks and benefits linked to new technologies. These factors exert a direct influence through emotional responses and an indirect influence via the interplay of emotions and individual tendencies (Kahan et al. 2009).

Providing precise information regarding the potential and risks associated with nanotechnology is essential. This should encompass open-access research reports and facilitate discussions and debates surrounding its advantages and disadvantages. Furthermore, incorporating nanotechnology as a subject in both school and graduate and postgraduate educational curricula is imperative. This will render information accessible to society, fostering greater familiarity with the technology and ultimately leading to favorable assessments of its applications. The theory posits that, in general, moderate attitudes toward nanotechnology present an optimal opportunity for a deeper comprehension of its benefits rather than solely focusing on the associated risks (Satterfield et al. 2009). Altering individuals' perspectives is a formidable endeavor. The impact is contingent upon the nature of the media content, its frequency, the extent of its influence, and, most critically, the degree of public exposure to it.

The current issue pertains to the public's acceptance of such products. Kim et al. (2014) examined the awareness of nanotechnology among consumers and experts in South Korea, revealing that consumer groups exhibited a more favorable disposition towards nanotechnology despite possessing limited knowledge. Researchers identified a pressing need for consumer education regarding nanotechnology and nanomaterials, which arise from the premise that increased awareness of nanotechnology would likely foster a more positive attitude towards nano products.

Studies have shown that knowledge exerts a moderately constrained influence on attitudes toward new technologies (Sturgis and Allum 2004; Bauer et al. 2007). Awareness has been reported to be associated with more favorable views on nanotechnology after considering statistical controls and additional illustrative factors (Vandermoere et al. 2010; Vandermoere et al. 2011). Nonetheless, the investigations utilizing more comprehensive test-based assessments of knowledge, rather than self-reported awareness, demonstrated that proficiency in fundamental science, rather than education specific to nanotechnology, serves as the more crucial indicator of favorable attitudes towards nanotechnology (Lee et al. 2005; Scheufele and Lewenstein 2005; Scheufele et al. 2007; Ho et al. 2010).

19.5.3 Communication and Transparency Issues

The issue of broader transparency in policymaking, as viewed by various policymakers, is crucial for restoring public trust (Wardman and Lofstedt 2009). Transparency is fundamentally linked to openness and is vital for effective risk communication. Methods to enhance transparency involve ensuring that all policy decisions and the relationships between policymakers and key stakeholders are available to the public. Acknowledging and articulating all uncertainty is essential to enhance transparency (Charlebois and Summan 2015). Effective risk communication must occur in a transparent setting that encourages questions and dialogue with presenters who understand the risks and can convey them clearly (Sukted et al. 2017). Transparency plays a vital role in upholding the reputation of an organization and fosters trust development. It ensures that crucial documents are readily available; thus, stakeholders can access scientific outputs, recognize all genuine risks, and address concerns regarding perceived risks (Charlebois and Summan 2015).

Timely communication is essential for safeguarding public health, fostering the development and maintenance of trust, and averting the proliferation of rumours and misinformation (Baba and Esfandiari 2023). Responsiveness pertains to the degree to which individuals responsible for food safety address the requirements of risk communication and meet the expectations of the intended audience in their interactions. For instance, individuals may harbor scepticism towards risk communications if these fail to address their perceptions and concerns, instead offering only technical data related to risk assessments. Therefore, to facilitate effective risk communication, it is essential to recognize and understand the information needs and communication expectations of the target audience and to address these issues within communication activities (Organization 2018). A prompt response is crucial as it reflects the organization's agility in addressing concerns, thereby enhancing its credibility and dependability. The erosion of trust in regulatory agencies may occur when there is a lag in the dissemination of information. Governments have historically faced scrutiny for their sluggishness in response and communication and a notable lack of coordination among different agencies (Charlebois and Summan 2015).

19.5.4 Balancing Innovation with Safety

The emergence of novel technology frequently presents fresh challenges for lawmakers, particularly when its advertised advantages provoke apprehensions regarding human health and ecological safety. In these instances, the formulation of new legislation or the modification of current statutes may become imperative. Regulators must remain cognizant of forthcoming materials, technologies, and innovations from the initial phases of the innovation process to ascertain whether existing legislation adequately encompasses all pertinent human and environmental safety considerations. For instance, a recent survey indicates that the industry has shown reluctance to fully commit to the development of nanotechnology due to the ambiguous regulatory environment and concerns that it may lead to provisions that lack scientific rigor (Sadowski and Guston 2016). The absence of suitable risk assessment and regulation to address safety issues is a significant obstacle to adopting innovative nanomaterials in agriculture, a field that is just beginning to emerge (Lowry et al. 2019). To address this challenge, it is essential to identify, during the design phase, the nano-specific characteristics that yield beneficial outcomes and could result in negative consequences. Furthermore, evaluating the potential impacts of these novel solutions throughout the life cycle compared to traditional alternatives could promote their sustainable use (Lowry et al. 2019).

In their evaluation of the necessity for adaptation or transformation of existing risk frameworks about chemicals and nanomaterials concerning active nanostructures, Kuzma and Roberts (2016) assert that these frameworks must incorporate life cycle thinking, systems thinking, and probabilistic analysis (engineering concepts, system dynamics models, fault trees, and event trees) and transition from a deterministic to a probabilistic interpretation of risk. They emphasize the need to expand the concept of risk to encompass the social and ethical ramifications that the emerging risks associated with active nanostructures may entail. Indeed, the pursuit of safety is intrinsically linked to sustainability, and there exists a burgeoning political impetus for a systemic and transformative shift towards a circular economy within the EU and potentially beyond.

The European Chemicals Agency (ECHA) has thoroughly evaluated the relevance of the Registration, Evaluation, Authorisation and Restriction of Chemicals (REACH) Regulation about next-generation nanomaterials, specifically those that are stimuli-responsive. Their findings indicate that the identification and characterization criteria for nanoforms outlined in Annex VI of the REACH Regulation fail to address the dynamic characteristics inherent in these innovative applications adequately. Consequently, there is a pressing need for additional guidance on how to incorporate the intended functionality and the alterations that arise in response to external stimuli within a registration dossier, a necessity that extends beyond nanomaterials to encompass other stimuli-responsive materials as well (Camboni et al. 2019). In pursuit of this objective, developing novel analytical methodologies may be imperative for precisely characterizing nanostructures and nanostructured materials. The ECHA has expressed apprehension regarding the potential inadequacy of the proposed definition of nanomaterial of the European Commission, which is employed to identify nanoforms within the framework of REACH. This definition relies solely on the external dimensions of constituent particles as the criterion for determining a nanomaterial. Notably, only solid particles are encompassed by this definition, as the inherently dynamic characteristics of fluid particles preclude the accurate determination of their external dimensions (Rauscher et al. 2019).

19.6 Future Directions and Recommendations

The application of nanotechnology in agriculture encompasses products such as nano-fertilizers, nano-pesticides, and nanosensors, all of which contribute to increased productivity and sustainability. Furthermore, innovative nanomaterials are at the forefront of nanotechnology advancements. Various external stimuli, such as pH, temperature, and light, can activate these multifunctional entities to modify their functionalities (Gottardo et al. 2021). A safe and sustainable-by-design methodology tackles the health and environmental issues associated with the rise of these smart nanomaterials (Mech et al. 2022). This strategy mitigates potential adverse effects during the design phase and guarantees that nanomaterials aligned with circular and green economy principles are prepared for market introduction.

Nonetheless, significant regulatory concerns regarding the safety, environmental implications, and enduring consequences of these nanomaterials exist. To effectively tackle these concerns, it is imperative to establish comprehensive regulatory frameworks and standardized testing protocols to guarantee the safe implementation and societal acceptance of nanotechnology products. The readiness of regulatory frameworks is essential for fully realizing the capabilities of nanotechnology. Through foresight and a proactive stance toward safety challenges, regulatory entities can facilitate the conscientious advancement and implementation of nanotechnology (Rana et al. 2024). While the directives mentioned above offer a degree of regulation, the evolving field of nanotechnology necessitates more comprehensive and conclusive legislation for regional and global industries.

Nanotechnology presents many advantages in agriculture, such as improved efficiency and precision. However, it is not without drawbacks, which encompass substantial development costs, possible environmental and health hazards, and regulatory hurdles. The potential of nanotechnology could be limited by the necessity for comprehensive research to grasp long-term implications and the demand for appropriate infrastructure and specialized knowledge to execute nano-based solutions proficiently. Moreover, integrating nanotechnology into agricultural practices may encounter opposition from apprehensions regarding its effects on biodiversity and the equilibrium of ecosystems. As advancements in research and development progress in this domain, integrating nanotechnology within agriculture is poised to significantly contribute to establishing sustainable and ecologically sound agricultural practices.

The implications of nanotechnology encompass a range of economic, legal, social, and risk factors that warrant thorough examination. From a financial standpoint, nanotechnology possesses the potential to transform a multitude of sectors, including healthcare, energy, and agriculture. It has the potential to enhance efficiency, reduce production expenses, and introduce novel materials characterized by distinctive properties. Nevertheless, the substantial costs associated with research and development must be considered, in addition to the potential risks of liability (Yadav et al. 2023).

From a legal standpoint, nanotechnology presents critical inquiries regarding property rights, liability for products, and the framework of regulatory supervision. The emergence of nanotechnology indicates that conventional regulatory frameworks might be inadequate to tackle the distinct risks linked to these materials. Consequently, governments and regulatory bodies must collaborate in establishing suitable legal frameworks that will guarantee the safe and responsible advancement and application of nanotechnology (Bowman and Hodge 2008). The advantages of nanotechnology for society should be carefully considered alongside its possible disadvantages, which encompass implications for human health, environmental effects, and ethical issues. Also, the possibility of unintended repercussions, including releasing nanoparticles into the ecosystem or unexpected health hazards, must be considered. Furthermore, it is imperative to consider inquiries regarding fairness, accessibility, and the allocation of advantages and disadvantages linked to nanotechnology. Additionally, it is essential to thoroughly evaluate the risks linked to nanotechnology, which encompasses the assessment of potential health hazards, environmental consequences, and societal ramifications, as well as formulating suitable risk management strategies. Continuous inquiry and advancement, coupled with open dialogue and cooperation among all parties involved, will be essential in mitigating these risks and guaranteeing the prudent and ethical progression and application of nanotechnology.

Incorporating nanotechnology within agricultural practices fosters sustainability by improving resource utilization efficiency, minimizing environmental pollution, and safeguarding natural ecosystems. For example, implementing nanosensors in precision agriculture facilitates immediate observation of crop vitality, soil parameters, and ecological influences. This knowledge empowers farmers to engage in informed decision-making, enhancing resource efficiency and reducing ecological consequences. Furthermore, the advancement of biodegradable and eco-friendly nanomaterials guarantees that the advantages of nanotechnology are not achieved at the cost of environmental well-being. It possesses considerable promise to revolutionize agriculture by alleviating the ecological consequences of conventional farming methods. Through optimizing resource utilization and minimizing chemical inputs, nanotechnology can safeguard soil integrity, avert water contamination, maintain biodiversity, and diminish greenhouse gas emissions.

Scientific investigations must address the deficiency of information regarding toxicological and environmental effects by intensifying research into the hazardous characteristics and biological reactivity of nanomaterials, as well as the critical physicochemical properties. This should be achieved by utilizing experimental conditions that more accurately reflect the exposure scenarios encountered in agricultural settings. Research on human health and ecotoxicology must collaboratively explore the intricate relationships among agro-ecosystems, nanoscale materials, exposure levels, and human populations (Gupta et al. 2024). This may encompass elucidating the impacts that the environmental bio-physico and chemical attributes exert on the biological reactivity of nanomaterials when considered from a lifecycle standpoint. The dissemination of this information is imperative among the scientific community, industry, and governmental regulatory agencies to accurately delineate the hazardous profile of nanomaterials across diverse exposure scenarios. This information may subsequently be utilized to formulate appropriate risk assessment protocols and precautionary management strategies, encompassing the determination of sufficient environmental exposure thresholds and biological limit values as benchmarks to aid in mitigating potential health hazards.

19.7 Conclusion

The integration of nanotechnology into agriculture presents transformative opportunities to enhance productivity, resource efficiency, and sustainability. However, realizing these benefits necessitates the establishment of comprehensive regulatory frameworks that address potential environmental, health, and safety concerns. A critical analysis of existing policies reveals significant gaps and inconsistencies, underscoring the need for harmonized international standards and robust governance structures. By proactively developing and implementing such regulations, stakeholders can ensure that adopting nanotechnology in agriculture aligns with public health objectives and environmental sustainability, fostering responsible innovation that meets the global demand for food security.

References

Abobatta, W.F. (2018). Nanotechnology application in agriculture. *Acta Scientific Agriculture* 2: 99–102.

Adisa, I. O., Pullagurala, V. L. R., Peralta-Videa, J. R., et al. (2019). Recent advances in nano-enabled fertilizers and pesticides: a critical review of mechanisms of action. *Environmental Science: Nano* 6(7): 2002–2030.

Akhtar, N. and Ilyas, N. (2022). Role of nanosilicab to boost the activities of metabolites in Triticum aestivum facing drought stress. *Plant and Soil* 477: 99–115.

Alhammad, B.A., Ahmad, A., Seleiman, M.F., and Tola, E. (2023). Seed priming with nanoparticles and 24-epibrassinolide improved seed germination and enzymatic performance of *Zea mays* L. in salt-stressed soil. *Plants* 12: 690.

Allan, J., Belz, S., Hoeveler, A. et al. (2021). Regulatory landscape of nanotechnology and nanoplastics from a global perspective. *Regulatory Toxicology and Pharmacology* 122: 104885.

Amenta, V., Aschberger, K., Arena, M. et al. (2015). Regulatory aspects of nanotechnology in the agri/feed/food sector in EU and non-EU countries. *Regulatory Toxicology and Pharmacology* 73: 463–476.

Antonacci, A., Arduini, F., Moscone, D. et al. (2018). Nanostructured (bio) sensors for smart agriculture. *Trends in Analytical Chemistry* 98: 95–103.

Baba, F.V. and Esfandiari, Z. (2023). Theoretical and practical aspects of risk communication in food safety: a review study. *Heliyon* 9: e18141.

Baranowska-Wójcik, E., Szwajgier, D., Oleszczuk, P., and Winiarska-Mieczan, A. (2020). Effects of titanium dioxide nanoparticles exposure on human health—a review. *Biological Trace Element Research* 193: 118–129.

Baruah, S. and Dutta, J. (2009). Nanotechnology applications in pollution sensing and degradation in agriculture: a review. *Environmental Chemistry Letters* 7: 191–204.

Bathi, J.R., Wright, L., and Khan, E. (2022). Critical review of engineered nanoparticles: environmental concentrations and toxicity. *Current Pollution Reports* 8: 498–518.

Bauer, M.W., Allum, N., and Miller, S. (2007). What can we learn from 25 years of PUS survey research? Liberating and expanding the agenda. *Public Understanding of Science* 16: 79–95.

Bayat, M., Zargar, M., Murtazova, K.M.-S. et al. (2022). Ameliorating seed germination and seedling growth of nano-primed wheat and flax seeds using seven biogenic metal-based nanoparticles. *Agronomy* 12: 811.

Bharathan, N. and Turarbekova, Z. (2024). Application of CRISPR technology and nanomaterials to advance food supply. In: *Applications of Nanotechnology in Microbiology* (ed. R.S. Chaughule and A.S. Lokur), 157–191. Cham: Springer. https://doi.org/10.1007/978-3-031-49933-3_7.

Bowman, D.M. and Hodge, G.A. (2008). 'Governing'nanotechnology without government? *Science and Public Policy* 35: 475–487.

Bundesrat, S. (2010). Millenniums-entwicklungsziele–Zwischenbericht der Schweiz 2010. Juni.

Camboni, M., Hanlon, J., Pérez-García, R., and Floyd, P. (2019). A state of play study of the market for so called 'next generation' nanomaterials.

Charlebois, S. and Summan, A. (2015). A risk communication model for food regulatory agencies in modern society. *Trends in Food Science & Technology* 45: 153–165.

Chowdhury, P., Gogoi, M., Borchetia, S., and Bandyopadhyay, T. (2017). Nanotechnology applications and intellectual property rights in agriculture. *Environmental Chemistry Letters* 15: 413–419.

Coles, D. and Frewer, L.J. (2013). Nanotechnology applied to European food production—a review of ethical and regulatory issues. *Trends in Food Science & Technology* 34: 32–43.

Dasgupta, N., Ranjan, S., Mundekkad, D. et al. (2015). Nanotechnology in agro-food: from field to plate. *Food Research International* 69: 381–400.

De Azeredo, H.M. (2009). Nanocomposites for food packaging applications. *Food Research International* 42: 1240–1253.

Ditta, A., Arshad, M., and Ibrahim, M. (2015). Nanoparticles in sustainable agricultural crop production: applications and perspectives. In: *Nanotechnology and Plant Sciences* (ed. M. Siddiqui, M. Al-Whaibi, and F. Mohammad), 55–75. Cham: Springer. https://doi.org/10.1007/978-3-319-14502-0_4.

Fadiji, A.E., Mortimer, P.E., Xu, J. et al. (2022). Biosynthesis of nanoparticles using endophytes: a novel approach for enhancing plant growth and sustainable agriculture. *Sustainability* 14: 10839.

Fujita, Y., Yokoyama, H., and Abe, S. (2006). Perception of nanotechnology among the general public in Japan—of the NRI nanotechnology and society survey project. *Asia Pacific Nanotech Weekly* 4.

Goodarzi, G.R., Noor, V.P., and Ahmadloo, F. (2017). Effects of nanoparticle treatments on propagation of Prunus mahaleb L. by seed.

Gottardo, S., Mech, A., Drbohlavová, J. et al. (2021). Towards safe and sustainable innovation in nanotechnology: state-of-play for smart nanomaterials. *NanoImpact* 21: 100297.

Gupta, S., Kumar, D., Aziz, A. et al. (2024). Nanoecology: exploring engineered nanoparticles' impact on soil ecosystem health and biodiversity. *Egyptian Journal of Soil Science* 64: 1637–1655.

Han, C., Zhao, A., Varughese, E., and Sahle-Demessie, E. (2018). Evaluating weathering of food packaging polyethylene-nano-clay composites: release of nanoparticles and their impacts. *NanoImpact* 9: 61–71.

Hasan, M.K., Shopan, J., and Ahammed, G.J. (2020). Nanomaterials and soil health for agricultural crop production: current status and future prospects. *Nanomaterials for Agriculture Forestry Applications* 2020: 289–312.

Hashem, N.M. and Gonzalez-Bulnes, A. (2021). Nanotechnology and reproductive management of farm animals: challenges and advances. *Animals* 11: 1932.

Hassan, S., Hassan, F.-U., and Rehman, M.S.-U. (2020). Nano-particles of trace minerals in poultry nutrition: potential applications and future prospects. *Biological Trace Element Research* 195: 591–612.

Ho, S.S., Scheufele, D.A., and Corley, E.A. (2010). Making sense of policy choices: understanding the roles of value predispositions, mass media, and cognitive processing in public attitudes toward nanotechnology. *Journal of Nanoparticle Research* 12: 2703–2715.

Hojjat, S.S. and Kamyab, M. (2017). The effect of silver nanoparticle on Fenugreek seed germination under salinity levels. *Russian Agricultural Sciences* 43: 61–65.

Iavicoli, I., Leso, V., Beezhold, D.H., and Shvedova, A.A. (2017). Nanotechnology in agriculture: opportunities, toxicological implications, and occupational risks. *Toxicology and Applied Pharmacology* 329: 96–111.

Indira, M., Krupanidhi, S., Venkateswarulu, T. C., Pallavi, G., Peele, K. A. (2021). Current Aspects of Nanotechnology: Applications in Agriculture. In: *Biobased Nanotechnology for Green Applications. Nanotechnology in the Life Sciences* (ed. H. Sarma, S.J. Joshi, R. Prasad, and J. Jampilek), 73–99. Springer, Cham. https://doi.org/10.1007/978-3-030-61985-5_3

Irwin, A. and Wynne, B. (1996). *Misunderstanding Science?: The Public Reconstruction of Science and Technology*. Cambridge University Press.

Jalil, S.U. and Ansari, M.I. (2019). Nanoparticles and abiotic stress tolerance in plants: synthesis, action, and signaling mechanisms. *Plant Signaling Molecules* 2019: 549–561.

Jampílek, J., Kráľová, K., Prasad, R. et al. (2017). Nanomaterials for delivery of nutrients and growth-promoting compounds to plants. In: *Nanotechnology* (ed. R. Prasad, M. Kumar, and V. Kumar), 177–226. Singapore: Springer. https://doi.org/10.1007/978-981-10-4573-8_9.

Kah, M. (2015). Nanopesticides and nanofertilizers: emerging contaminants or opportunities for risk mitigation? *Frontiers in Chemistry* 3: 64.

Kah, M., Kookana, R.S., Gogos, A., and Bucheli, T.D. (2018). A critical evaluation of nanopesticides and nanofertilizers against their conventional analogues. *Nature Nanotechnology* 13: 677–684.

Kahan, D.M., Braman, D., Slovic, P. et al. (2009). Cultural cognition of the risks and benefits of nanotechnology. *Nature Nanotechnology* 4: 87–90.

Khot, L.R., Sankaran, S., Maja, J.M. et al. (2012). Applications of nanomaterials in agricultural production and crop protection: a review. *Crop Protection* 35: 64–70.

Kim, Y.-R., Lee, E.J., Park, S.H. et al. (2014). Comparative analysis of nanotechnology awareness in consumers and experts in South Korea. *International Journal of Nanomedicine* 9: 21–27.

Kookana, R.S., Boxall, A.B., Reeves, P.T. et al. (2014). Nanopesticides: guiding principles for regulatory evaluation of environmental risks. *Journal of Agricultural and Food Chemistry* 62: 4227–4240.

Kuhn, R., Bryant, I.M., Jensch, R., and Böllmann, J. (2022). Applications of environmental nanotechnologies in remediation, wastewater treatment, drinking water treatment, and agriculture. *Applied Nanoscience* 3: 54–90.

Kumar, S., Shukla, A., Baul, P.P. et al. (2018). Biodegradable hybrid nanocomposites of chitosan/gelatin and silver nanoparticles for active food packaging applications. *Food Packaging and Shelf Life* 16: 178–184.

Kumari, R., Suman, K., Karmakar, S. et al. (2023). Regulation and safety measures for nanotechnology-based agri-products. *Frontiers in Genome Editing* 5: 1200987.

Kuzma, J. and Roberts, J.P. (2016). Is adaptation or transformation needed? Active nanomaterials and risk analysis. *Journal of Nanoparticle Research* 18: 1–18.

Lee, C.-J., Scheufele, D.A., and Lewenstein, B.V. (2005). Public attitudes toward emerging technologies: examining the interactive effects of cognitions and affect on public attitudes toward nanotechnology. *Science Communication* 27: 240–267.

Lewicka, Z.A., William, W.Y., Oliva, B.L. et al. (2013). Photochemical behavior of nanoscale TiO2 and ZnO sunscreen ingredients. *Journal of Photochemistry and Photobiology A: Chemistry* 263: 24–33.

Li, L., Zhao, C., Zhang, Y. et al. (2017). Effect of stable antimicrobial nano-silver packaging on inhibiting mildew and in storage of rice. *Food Chemistry* 215: 477–482.

Lowry, G.V., Avellan, A., and Gilbertson, L.M. (2019). Opportunities and challenges for nanotechnology in the agri-tech revolution. *Nature Nanotechnology* 14: 517–522.

Mahra, S., Tripathi, S., Tiwari, K. et al. (2024). Harnessing nanotechnology for sustainable agriculture: from seed priming to encapsulation. *Plant Nano Biology* 9: 100124.

Manesh, R.R., Grassi, G., Bergami, E. et al. (2018). Co-exposure to titanium dioxide nanoparticles does not affect cadmium toxicity in radish seeds (*Raphanus sativus*). *Ecotoxicology and Environmental Safety* 148: 359–366.

Manjunatha, R., Naik, D., and Usharani, K. (2019). Nanotechnology application in agriculture: a review. *Journal of Pharmacognosy and Phytochemistry* 8: 1073–1083.

Mawale, K.S. and Giridhar, P. (2024). Green approaches for the synthesis of silver nanoparticle and its augmentation in Seed. germination, growth, and antioxidant level in Capsicum annuum L. *Plant Nano Biology* 10: 100107.

Mech, A., Gottardo, S., Amenta, V. et al. (2022). Safe-and sustainable-by-design: the case of Smart Nanomaterials. A perspective based on a European workshop. *Regulatory Toxicology and Pharmacology* 128: 105093.

Meena, L.K. and Chandel, A. (2024). Ethical and regulatory perspective on nanotechnology in agriculture. *Just Agriculture* 5: 94–104.

Mirsharifi, S.M., Sami, M., Jazaeri, M., and Rezaei, A. (2023). Production, characterization, and antimicrobial activity of almond gum/polyvinyl alcohol/chitosan composite films containing thyme essential oil nanoemulsion for extending the shelf-life of chicken breast fillets. *International Journal of Biological Macromolecules* 227: 405–415.

Mishra, M., Dashora, K., Srivastava, A. et al. (2019). Prospects, challenges and need for regulation of nanotechnology with special reference to India. *Ecotoxicology and Environmental Safety* 171: 677–682.

Mitter, N. and Hussey, K. (2019). Moving policy and regulation forward for nanotechnology applications in agriculture. *Nature Nanotechnology* 14: 508–510.

Moustafa, H., Youssef, A.M., Darwish, N.A., and Abou-Kandil, A.I. (2019). Eco-friendly polymer composites for green packaging: future vision and challenges. *Composites Part B: Engineering* 172: 16–25.

Mukherjee, A., Maity, A.A., Pramanik, P., et al.(2019). Public perception about use of nanotechnology in agriculture. In: *Advances in phytonanotechnology* (ed. M. Ghorbanpour and S.H. Wani), 405–418. Academic Press. https://doi.org/10.1016/B978-0-12-815322-2.00019-5.

Mukherjee, A., Maity, A., Pramanik, P. et al. (2019b). Public perception about use of nanotechnology in agriculture. In: *Advances in Phytonanotechnology From Synthesis to Application* (ed. M. Ghorbanpour and S.H. Wani), 405–418. https://doi.org/10.1016/C2017-0-02650-2.

Narei, H., Ghasempour, R., and Akhavan, O. (2018). Toxicity and safety issues of carbon nanotubes. In: Rafiee, R. (eds) *Carbon Nanotube-reinforced Polymers*. Elsevier, pp. 145–171.

Nile, S.H., Baskar, V., Selvaraj, D. et al. (2020). Nanotechnologies in food science: applications, recent trends, and future perspectives. *Nano-Micro Letters* 12: 1–34.

Organization, W.H (2018). *Risk Communication Applied to Food Safety: Handbook*. Food & Agriculture Org.

Parisi, C., Vigani, M., and Rodríguez-Cerezo, E. (2015). Agricultural nanotechnologies: what are the current possibilities? *Nano Today* 10: 124–127.

Pathakoti, K., Manubolu, M., and Hwang, H.-M. (2017). Nanostructures: current uses and future applications in food science. *Journal of Food and Drug Analysis* 25: 245–253.

Péter, B., Farkas, E., Kurunczi, S. et al. (2022). Review of label-free monitoring of bacteria: from challenging practical applications to basic research perspectives. *Biosensors* 12: 188.

Peters, R., Brandhoff, P., Weigel, S. et al. (2014). Inventory of nanotechnology applications in the agricultural, feed and food sector. *EFSA Supporting Publications* 11: 621E.

Prasad, R.D., Prasad, N.R., Prasad, R.B. et al. (2023). A critical review on bio-mimetic synthesis of transition metal nanoparticles: their biomedical applications in veterinary medicine. *ES Food Agroforestry* 15: 1027.

Priya, B., Srinivasarao, M., and Mukherjee, S. (2015). Screening of phosphorus nanoparticle concentration based on their effects at germination & seedling level in mung, urd and cowpea. *Vegetos- An International Journal of Plant Research* 28: 169.

Rajput, V., Minkina, T., Mazarji, M. et al. (2020). Accumulation of nanoparticles in the soil-plant systems and their effects on human health. *Annals of Agricultural Sciences* 65: 137–143.

Ramachandraiah, K., Han, S.G., and Chin, K.B. (2015). Nanotechnology in meat processing and packaging: potential applications—a review. *Asian-Australasian Journal of Animal Sciences* 28: 290–302.

Rana, L., Kumar, M., Rajput, J. et al. (2024). Nexus between nanotechnology and agricultural production systems: challenges and future prospects. *Discover Applied Sciences* 6: 1–19.

Rastogi, A., Zivcak, M., Sytar, O. et al. (2017). Impact of metal and metal oxide nanoparticles on plant: a critical review. *Frontiers in Chemistry* 5: 78.

Rauscher, H., Roebben, G., Mech, A. et al. (2019). *An Overview of Concepts and Terms Used in the European Commission's Definition of Nanomaterial*, 1–44. *Publications Office of the European Union.*

Reddy, P.R.K., Yasaswini, D., Reddy, P.P.R. et al. (2020). Applications, challenges, and strategies in the use of nanoparticles as feed additives in equine nutrition. *Veterinary World* 13: 1685.

Rodriguez, P.A., Rothballer, M., Chowdhury, S.P. et al. (2019). Systems biology of plant-microbiome interactions. *Molecular Plant* 12: 804–821.

Rodríguez-Gómez, F.D., Monferrer, D., Penon, O., and Rivera-Gil, P. (2024). Implementing Horizon Scanning as a tool for the strategic development of regulatory guidelines for nanotechnology-enabled health products. *Frontiers in Medicine* 10: 1308047.

Roohinejad, S. and Greiner, R. (2017). Nanoscience: relevance for agriculture and the food sector. In: *Nanotechnology in Agriculture and Food Science*, 347–362. Wiley.

Roy, A., Sharma, A., Yadav, S. et al. (2021). Nanomaterials for remediation of environmental pollutants. *Bioinorganic Chemistry and Applications* 2021: 1764647.

Sadeghi, K., Shahedi, M., Najafi, M. et al. (2021). Types of nanomaterials in food packaging: a review. *International Journal Of Nanoparticles* 13: 63–95.

Sadowski, J. and Guston, D.H. (2016). 'You caught me off guard': probing the futures of complex engineered nanomaterials. *Journal of Nanoparticle Research* 18: 1–13.

Satterfield, T., Kandlikar, M., Beaudrie, C.E. et al. (2009). Anticipating the perceived risk of nanotechnologies. *Nature Nanotechnology* 4: 752–758.

Scheufele, D.A. and Lewenstein, B.V. (2005). The public and nanotechnology: how citizens make sense of emerging technologies. *Journal of Nanoparticle Research* 7: 659–667.

Scheufele, D.A., Corley, E.A., Dunwoody, S. et al. (2007). Scientists worry about some risks more than the public. *Nature Nanotechnology* 2: 732–734.

Scheufele, D.A., Corley, E.A., Shih, T.-J. et al. (2009). Religious beliefs and public attitudes toward nanotechnology in Europe and the United States. *Nature Nanotechnology* 4: 91–94.

Schoonjans, R., Castenmiller, J., Chaudhry, Q. et al. (2023). Regulatory safety assessment of nanoparticles for the food chain in Europe. *Trends in Food Science and Technology in Agriculture* 134: 98–111.

Scott, N. and Chen, H. (2012). Nanoscale science and engineering for agriculture and food systems. *Industrial Biotechnology* 8: 340–343.

Servin, A.D. and White, J.C. (2016). Nanotechnology in agriculture: next steps for understanding engineered nanoparticle exposure and risk. *NanoImpact* 1: 9–12.

Shang, Y., Hasan, M.K., Ahammed, G.J. et al. (2019). Applications of nanotechnology in plant growth and crop protection: a review. *Molecules* 24: 2558.

Sharma, C., Dhiman, R., Rokana, N., and Panwar, H. (2017). Nanotechnology: an untapped resource for food packaging. *Frontiers in Microbiology* 8: 1735.

Sharma, A., Kashyap, B.K., Patel, O.P., and Pareek, A. (2023). Nanotechnology: opportunity and challenges in waste management. In: *Current Research Trends and Applications in Waste Management* (ed. B.K. Kashyap and M.K. Solanki), 341–369. Singapore: Springer. https://doi.org/10.1007/978-981-99-3106-4_13.

Siegrist, M., Stampfli, N., Kastenholz, H., and Keller, C. (2008). Perceived risks and perceived benefits of different nanotechnology foods and nanotechnology food packaging. *Appetite* 51: 283–290.

Singh, R. and Kumar, S. (2023). Regulatory and safety concerns regarding the use of active nanomaterials in food industry. In: *Nanotechnology Advancement in Agro-food Industry*. Springer.

Singh, A., Rajput, V.D., Varshney, A. et al. (2023). Small tech, big impact: agri-nanotechnology journey to optimize crop protection and production for sustainable agriculture. *Plant Stress* 10: 100253.

Srilatha, B. (2011). Nanotechnology in agriculture. *Jouurnal of Nanomedicine and Nanotechnology* 2: 1–5.

Srinivasan, R., Maity, A., Singh, K. et al. (2017). Influence of copper oxide and zinc oxide nano-particles on growth of fodder cowpea and soil microbiological properties. *Range Management and Agroforestry* 38: 208–214.

Stanovich, K.E. and West, R.F. (2008). On the relative independence of thinking biases and cognitive ability. *Journal of Personality and Social Psychology* 94: 672–695.

Sturgis, P. and Allum, N. (2004). Science in society: re-evaluating the deficit model of public attitudes. *Public Understanding of Science* 13: 55–74.

Sukted, N., Tuitemwong, P., Erickson, L.E. et al. (2017). Reducing risk of campylobacteriosis from poultry: a mini review. *Infection* 15: 16.

Sundaria, N., Singh, M., Upreti, P. et al. (2019). Seed priming with iron oxide nanoparticles triggers iron acquisition and biofortification in wheat (*Triticum aestivum* L.) grains. *Journal of Plant Growth Regulation* 38: 122–131.

Vandermoere, F., Blanchemanche, S., Bieberstein, A. et al. (2010). The morality of attitudes toward nanotechnology: about God, techno-scientific progress, and interfering with nature. *Journal of Nanoparticle Research* 12: 373–381.

Vandermoere, F., Blanchemanche, S., Bieberstein, A. et al. (2011). The public understanding of nanotechnology in the food domain: the hidden role of views on science, technology, and nature. *Public Understanding of Science* 20: 195–206.

Vemula, A. (2022). Chitosan bionanocomposite: a potential approach for sustainable agriculture. *Medical, Agriculture and Environment Science* 2: 41–46.

Ververis, E., Ackerl, R., Azzollini, D. et al. (2020). Novel foods in the European Union: scientific requirements and challenges of the risk assessment process by the European Food Safety Authority. *Food Research International* 137: 109515.

Wang, Z., Wang, S., Ma, T. et al. (2023). Synthesis of zinc oxide nanoparticles and their applications in enhancing plant stress resistance: a review. *Agronomy* 13: 3060.

Wardman, J. and Lofstedt, R. (2009). European food safety authority-risk communication annual review.

Yadav, A., Yadav, K., Ahmad, R., and Abd-Elsalam, K.A. (2023). Emerging frontiers in nanotechnology for precision agriculture: advancements, hurdles and prospects. *Agrochemicals* 2: 220–256.

Zaheer, T., Abbas, R.Z., Rehman, T.U. et al. (2024). Novel insights regarding the safety and efficacy of pyrethroid-coated nanoparticles against *Hyalomma* ticks. *Toxicology Mechanisms and Methods* 34: 148–163.

Zahra, Z., Ali, M.A., Parveen, A. et al. (2019). Exposure–response of wheat cultivars to TiO2 nanoparticles in contrasted soils. *Soil and Sediment Contamination: an International Journal* 28: 184–199.

Zhou, Y.-H., Mujumdar, A.S., Vidyarthi, S.K. et al. (2023). Nanotechnology for food safety and security: a comprehensive review. *Food Reviews International* 39: 3858–3878.

20

Functional Annotation of Microbes and Nanotechnology for Agricultural Productivity

Manali Singh[1], Shivani Bhutani[2], and Sagar Chhabra[3]

[1] *Department of Life Sciences, Parul Institute of Applied Sciences, Parul University, Post Waghodia, Vadodara, Gujarat, India*
[2] *Department of Biotechnology, Invertis Village, Bareilly-Lucknow National Highway, NH-24, Bareilly, Uttar Pradesh*
[3] *Founder & Independent Agronomy Advisor, AgCrest Bio Solutions, Saskatoon, Saskatchewan, Canada Professional Agrologist (PAg), Saskatchewan Institute of Agrologists (SIA) Director (Volunteer), SIA Saskatoon Branch, Saskatchewan Institute of Agrologists*

20.1 Introduction

Plants and crops are the source of food and nutrition for all life forms. They are exploited at a large scale for the fulfillment of commercial demands in pharmaceuticals, as agricultural crops, secondary metabolites, timber, fuel, etc (Singh et al. 2017; Bhasin et al. 2019; Mishra et al. 2019). Agricultural practices basically rely on the input of chemical-based fertilizers and pesticides that are costly, labor-intensive, and accompanied by deteriorating soil and human health (Chhabra et al. 2010; Chhabra et al. 2013). Chemicals used in agricultural practices are the major pollutants of the environment (Singh et al. 2019; Rai et al. 2020). The increased global temperatures due to pollution can lead to shift of the available lands for agriculture (Asawa et al. 2019). Indiscriminate uprooting of plants has also led to the extinction of elite plant varieties that were commercially very important (Singh et al. 2017; Singh et al. 2018; Gupta and Pandey 2019). Through proper agricultural practices, these stresses can be reduced. Fertilizers in appropriate amounts, if given to plants and crops have been shown to enhance the plant productivity (Tripathi et al. 2020). Plant growth promoting micro-organisms (PGPMs) have been known to enhance the quality of crops by using minerals in lesser complex forms from the soil (Singh et al. 2021). Many molecular mechanisms like enzymatic digestion, glycosylation, DNA methylations, etc have been known to enhance the host-pathogen interactions (Mueed et al. 2021). Foliar spray of elicitors enhanced the yield of secondary metabolites of Ashwagandha (Singh et al. 2020). Use of bioformulations as a modern sustainable approach for agriculture has also enhanced the crop productivity.

Inoculation of plant growth-promoting microorganisms or consortia increases crop yield in several ways along with maintaining soil health. Thus, the application of microbial inoculants to agricultural fields has come out as a hopeful implement for improving crop productivity due to its cost-effectiveness and sustainable nature (Gupta and Pandey 2019). Microbes play a major role in revitalizing agriculture. PGPMs, an important component of the soil microbiome, increase plant growth and productivity by providing multifarious benefits conferring abiotic and biotic stress tolerance through various direct or indirect mechanisms.

The direct mechanism implies the production of microbial compounds that either increase the concentrations or the bioavailability of compounds that ameliorate soil fertility by supplying growth regulators, mobilizing soil minerals such as potassium or phosphorous (Naik et al. 2019). For example of the important direct mechanisms of biological nitrogen fixation that convert unusable forms of nitrogen (e.g. atmospheric N_2) to a usable form of nitrogen (e.g. ammonia), for plant growth are well known. Similarly, microbial production of phytohormones like auxin, cytokinin, ethylene, etc. promotes root growth (Kaur et al. 2016). Microbial ACC deaminase activity reduces the harmful effect of ethylene by degrading 1-aminocyclopropane-1-carboxylate (ACC), a precursor of ethylene, thereby promoting plant growth (Glick 2012). Other mechanisms include phosphate solubilization that converts insoluble phosphate into soluble forms, i.e., monobasic and dibasic ions, making it available to plants and siderophore production for sequestering iron (Chhabra and Dowling 2017; Chhabra 2019).

Some microbes indirectly assist plant growth by acting as biocontrol agents that either kill or suppress the growth of phytopathogens (Naik et al. 2019). They do so by producing various metabolites like lytic enzymes (chitinases and glucanases), volatile compounds, siderophores, antibiotics, and toxins like phenazines, amphysines, diacetylfloroglucinol, pyrrolnitrin, HCN, colistin, circulin, oomycins, tropolone, polymyxin, tensin, and cyclic lipopeptides. Other indirect mechanisms involve increasing innate immunity in plants against phytopathogens by inducing plant-induced systemic resistance (Conrath et al. 2015), rhizoremediation (Mani et al. 2016), and interfering with quorum sensing in pathogenic microbes. In recent years, nanotechnology emerged as a new field because of its extensive applications in various areas like medicine, medical drugs, catalysis, energy, and materials. Scientists have implemented considerable efforts toward the production of nanoparticles by various methods, including chemical, physical, and biological methods (Ghidan et al. 2017). However, these methods also have some drawbacks regarding the difficulty of upgrading the process, segregation, and refinement of nanoparticles from the microemulsion (oil, aqueous phase, surfactant, and co-surfactant) and the huge amount of surfactants required (Pilarska et al. 2013) Eco-friendly techniques for manufacturing nanoparticles by plant extracts or microbes are valuable due to their simplicity, suitability, and necessity for smaller reaction times. Nanomaterials synthesized by eco-friendly and green methods proved very beneficial for enhancing the plant growth regulators, fertilization process, and pesticides (Ghidan et al. 2017; Chhabra and Prasad 2021).

Equally, it is helpful in controlling pollution and because of the large surface area offered by the tiny nanoparticles, it makes them valuable to tackle problems not fulfilled by physical, chemical pesticides, and biological control methods. Nanotechnology can enhance the production of agriculture, and its uses comprise: (i) nanoformulations used in agrochemicals in fertilizers and pesticides for crop enhancement; (ii) nanosensors used for the security of crops by the recognition of diseases and parts of agrochemicals; (iii) genetic engineering of plants that can be done by nanodevices; (iv) helpful in animal breeding, poultry production, animal health; and (v) postharvest management. Nanotechnology approaches having nanoparticle-mediated DNA or gene transfer in plants for the enhancement of agrifood and storage, insect-resistant varieties, and enhanced product shelf life. Nanotechnology also enhanced the progress of fuel production from biomass. Nanomaterials are familiar for having precise physical, chemical, and mechanical properties. In this chapter, we review the important aspects and functions with a focus on microbes in general and in producing nanomaterials and the nanotechnologies for agriculture. The microbial functions, functional aspects and applications of microorganisms and microbial nanomaterials for improving agricultural yield, disease resistance of plants and recent developments and the use of genomic and metagenomic tools and technologies, approaches in studying microorganisms and microbial diversity and the potential for crop improvement are covered and reviewed here.

20.2 Microbes and their Classification Based on Location

Soil is a highly complex, dynamic, and richest habitat for a diverse group of microorganisms on Earth that includes bacteria, archaea, fungi, algae, and protozoa. They significantly add to nutrient cycling, making them available to plants. They continually interact with plants some being commensalistic, some being mutualistic, while others being pathogenic or parasitic. Soil microbiome forms an indispensable component of the rhizosphere, where exudates secreted by plant roots create a selective environment for the establishment of PGPRs in its vicinity (Lucas et al. 2014). These exudates are organic compounds like sugars, phenolics, amino acids, proteins, and signal peptides that serve as a nutritional requirement for the soil microbiota. Plant-microbe interaction is a two-way conversation, where firstly plant roots release signal compounds that are recognized by microbes, triggering the discharge of microbial compounds, which is recognized by plant leading to the establishment of plant-microbe relationship (Figure 20.1). Based on location, these microbes or microbiome in any plant association can be of two types:

20.2.1 Epiphytes

Microbes that inhabit the surface of the plant are known as epiphytes. They may inhabitat the rhizosphere, rhizoplane, or phylloplane of the host plant. The phyllosphere includes all parts of a plant that are aboveground. The rhizosphere belongs to the region around plant roots into which plant exudates are released, while the rhizoplane represents the plant root surface. A range of microbes growing in these regions is recognized to encourage plant growth and are called extracellular plant growth-promoting microorganism (ePGPM). Members of genera *Bacillus, Microcoleus, Caldilinea, Clostridium, Chloroflexi, Actinobacteria*, etc. are some prominent rhizospheric bacteria (Mukhtar et al. 2016). Mahyarudin and Lestari (2015) reported the presence of species of *Actinoplanes, Actinokineospora, Streptomyces, Geodermatophilus*, and *Kocuria* from the rhizosphere of rice.

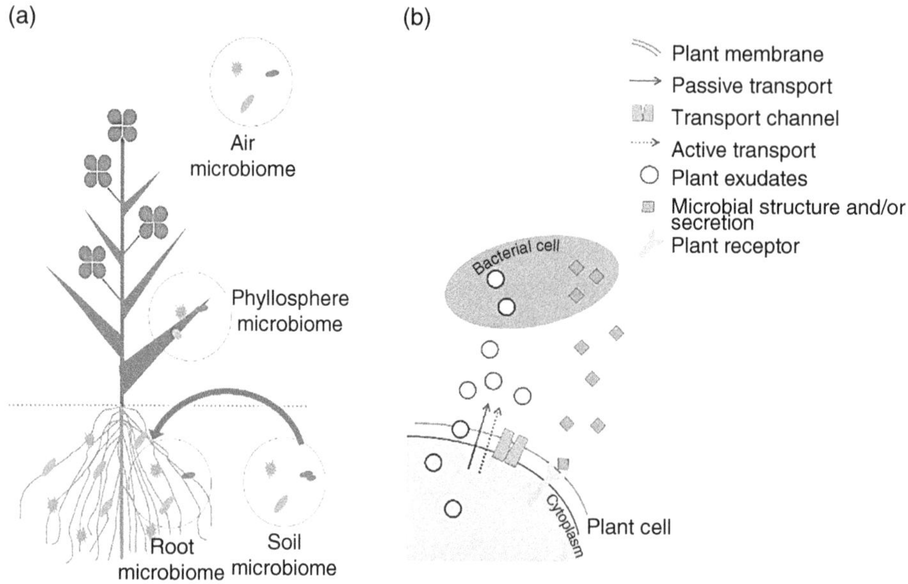

Figure 20.1 Plant-microbe interaction. (a) Basically, the plant microbiome originates from soil, i.e., soil microbiome. In the soil microbiome, the microorganisms colonized adjacent to the root in the rhizosphere due to secretions/exudates, which is the root microbiome. In addition, the microorganisms potentially migrate from the soil and air to the leaves and are colonized to form a phyllosphere microbiome. (b) The secretion (root exudates) in the rhizosphere attracts specific microorganisms to trigger a specific response. The plant secretions could be either active (through transport channels) or passive (through diffusion). The microbial signaling and recognition are considered by the plant through receptors (for microbial structure and/or secretion) present in the plant cell membrane.

20.2.2 Endophytes

Microbes inhabiting the internal tissues of plants asymptomatically are regarded as endophytes. They co-exist with plants by penetrating host tissue by synthesizing an array of hydrolytic enzymes, viz., cellulase, pectinase, proteinase, and xylanase. Those endophytic microbes known to promote plant growth are regarded as intracellular plant growth-promoting microorganism (iPGPM). They foster plant growth by symbiotically fixing atmospheric nitrogen, solubilizing macronutrients, synthesizing phytohormones, siderophores, HCN, and acting as a biocontrol agent against phytopathogens. Microbes belonging to the phyla Proteobacteria, Firmicutes, Actinobacteria, and Ascomycota have been studied as the most dominant endophytes in leguminous and nonleguminous plants (Rana et al. 2020). Species belonging to *Rhizobium* (Yan et al. 2017), *Bacillus* (Yang et al. 2020), *Pedobacter* (Zhang et al. 2018), *Chryseobacterium phosphatilyticum* (Jeong et al. 2019), *Burkholderia, Bacillus* (Chung et al. 2015), *Paenibacillus* (Gao et al. 2015), *Amnibacterium* (Kim et al. 2019), and *Streptomyces* (Kaewkla and Franco 2017) are the most dominant endophytic genera isolated from rice, wheat, maize, foxtail, and cucumber. The following section highlights role of microbes in revitalizing agriculture:

20.2.2.1 As Biofertilizers

Any substance that contains live microorganisms that provides various macro- or micro-elements (N, P, K) essential for the growth of plants is called a biofertilizer (Prashar and Shah 2016). The use of microbes by direct mechanism for enhancing plant growth through biological nitrogen fixation, phosphate solubilization, etc. Nitrogen fixation is a sustainable and economical substitute for chemical nitrogen fertilizers. The approaches of biofertilizers have gradually gained impetus in revitalizing agriculture by maintaining soil health and reducing pollutants in the environment. The list of microbial biofertilizers used for diverse crops is listed in Table 20.1.

Table 20.1 Beneficial roles of microbes in revitalizing agriculture.

Effect/response/ improvement	Plant	Symbiont/microbe	Reference
Amelioration of salinity stress	Barley	*Azotobacterchroococcum, Bacillus amyloliquefaciens, Bacillus circulans, Bacillus megaterium, Paenibacilluspolymyxa, Pseudomonas fluorescens*	Kasim et al. (2016)
	Mung bean	*Rhizobium phaseoli, Pseudomonas syringae, Pseudomonasfluorescens*	Ahmad et al. (2011)
	Wheat	*Brevibacterium halotolerans*	Ansari and Ahmad (2018)
	Faba bean	*Pseudomonas anguilliseptica*	Mohammed (2018)
	Barley, Oats	*Pseudomonas sp.*	Chang et al. 2014
	Chickpea	*Halomonasvariabilis, Planococcusrifietoensis*	Qurashi and Sabri (2012)
Amelioration of drought stress	Rice	*Trichoderma harzianum*	Shukla (2012)
Growth in saline and drought condition	Soybean	*Psedomonas putida*	Kang et al. (2014)
Induced systemic resistance	Panax	*Bacillus amyloliquefaciens*	Lee (2015)
	Pepper	*Brevibacterium iodinum*	Son (2014)
	Rice	*Bacillus* sp.	Udayashankar et al. (2011)
	Cucumber, Arabidopsis	*Trichoderma harzianum, Pseudomonas* sp.	Alizadeh et al. (2013)
Higher expression of stress-related genes, HSPs and CaPR-10	Pepper	*Bacillus licheniformis*	Lim and Kim (2013)

(Continued)

Table 20.1 (Continued)

Effect/response/ improvement	Plant	Symbiont/microbe	Reference
Growth in drought condition	Wheat	*Bacillus amyloliquefaciens, Azospirillum bzasilense*	Kasim et al. (2013)
PGP, antifungal activity, enhanced enzymatic activity	Chickpea	*Trichoderma viride*	Das et al. (2017)
		Mesorhizobium ciceri	
Enhanced plant biomass, grain yield, leghaemoglobin content of nodules	Chickpea	*Anabaena laxa*	Bidyarani et al. (2016)
Increased nodulation, nutrient uptake and yield	Chickpea	*Mesorhizobium* sp.	Verma et al. (2013)
Plant growth promotion and biocontrol	Cotton	*Trichoderma viride, Anabaena torulosa*	Triveni et al. (2015)
Disease suppression, improved mineral content	Rice	*Alcaligenes faecalis*	Kakar et al. (2018)
		Bacillus amyloliquefaciens	
Growth and decreased uptake and accumulation of arsenic in plants	Rice	*Kocuriaflava, Bacillus vietnamensis*	Mallick et al. (2018)
Phytoremediation of heavy metals	*Brassica juncea*	*Rhizobium leguminosarum, Pseudomonas brassicacearum*	Adediran (2015)
	Alnus firma	*Baclllusthuringiensis*	Babu et al. (2013)
	Gladiolus grandiflorus	*Pseudomonas putida, Thiobacillusthiooxidans*	Mani et al. (2016)
	Solanumnigrum	*Pseudomonas* sp.	Chen et al. (2014)
	Sedum	*Bacillus pumilus*	Ma et al. (2015)
Growth, yield and nutrient content	Banana	*Pseudomonasfluorescens*	Kavino et al. (2010)
Stress amelioration, dry weight, fresh weight, shoot length, root length, available nitrogen, yield index	Mung bean Soybean	*Anabaena, Azotobacter chroococcum, Bradyrhizobium, Anabaena laxa, Calothrix, Trichoderma viride*	Prasanna et al. (2014)
	Maize	*Azospirillum lipoferum*	Bano et al. (2013)
Growth quality, adaptation of the seedlings	Oil palm	*Bacillus salmalaya*	Azri et al. (2018)
Phosphate solubilization	Oil palm	*Burkholderia gladioli, Penicilliumaculeatum*	Istina et al. (2015)
Potassium solubilization	Panax	*Bacillus macilaginosus, Bacillus edaphicus, Bacillus circulans*	Meena et al. (2014)

Table 20.1 (Continued)

Effect/response/ improvement	Plant	Symbiont/microbe	Reference
Stress amelioration, growth, higher K+/Na+ ratio and higher inorganic contents	Groundnut	*Brachybacterium saurashtrense, Brevibacteriumcasei, Haererohalobacter*	Shukla et al. (2012)
	Maize	*Pseudomonas and Enterobacter*	Nadeem (2009)
Stress amelioration, reduced ROS, lipid peroxidation and superoxide dismutase activity	Rice	*Pseudomonas pseudoalcaligenes, Bacillus pumilus*	Jha and Subramanian (2014)
Decreased volatile emissions and enhanced photosynthesis	Wheat	*Bacillus thuringiensis*	Timmusk et al. (2014)
Growth, biomass and drought tolerance index	Wheat	*Rhizobium leguminosarum, Mesorhizobium ciceri, Rhizobium phaseoli*	Hussain et al. (2014)
Efficient uptake of insoluble phosphate from soil	*Coffeaarabica* L.	*Erwinia sp.*	Muleta et al. (2013)
		Pseudomonas chlororaphis	
Drought stress amelioration, increased potassium and proline content, decreased ascorbate and glutathione reductase	*Lavandula dentate*	*Bacillus thuringiensis*	Armada et al. (2014)
Inhibition of crown root rot	Groundnut	*Paenibacillus polymyxa*	Haggag and Timmusk (2008)
Induce resistance against black rot	Cabbage	*Paenibacillus* sp.	Ghazalibigla et al. (2016)
Biocontrol of *Rhizoctonia solani* and *Collectotrichum gloeosporioides*	Chilli	*Pseudomonas aeruginosa*	Sasirekha and Srividya (2016)
Production of pathogen-related (PR) protein	Cucumber	*Bacillus subtilis, Pseudomonas fluorescens*	El-Borollosy and Oraby (2012)
Growth and biocontrol	—	*Bacillus* sp., *Pseudomonas* sp.	Praveen Kumar et al. (2014)

20.2.2.2 As Phytostimulators

Any exogenous substance, synthesized outside the plant, that boosts plant growth and development is considered a phtyostimulator. Microbes act as phytostimulators through the creation of phytohormones that regulate various plant physiological processes like, flowering, fruit maturation, senescence, etc. For example. indole acetic acid and cytokinin regulate the extension of lateral roots and root hairs, thereby promoting efficient water and mineral absorption, signifying their importance in increased drought tolerance. *Bacillus subtilis*, a cytokinin producer of imparted tolerance to drought in oriental Thuja seedlings (Liu et al. 2013). Gibberellic acid regulates shoot elongation, seed germination, floral induction, and fruit set. However, microbial gibberellins have rare reports, with *Bacillus licheniformis* and *Bacillus pumilus* being documented to produce the same (Jha and Saraf 2015).

20.2.2.3 As Biopesticides

Some microorganisms have antagonistic effects on phytopathogenic microbes, thereby acting as biopesticides and providing resistance against biotic stress. They do so by indirect mechanisms as mentioned in the previous section like production of antibiotics, siderophores, hydrolytic enzymes, HCN, and imparting acquired and induced systemic resistance (Prashar and Shah 2016).

20.2.2.4 As Bioremediator

Soil contaminated with heavy metals adversely affects plant growth. Microorganisms able to metabolize heavy metals decrease the toxicity of heavy metals, resulting in enhanced growth and development. Ma et al. (2011) reviewed the role of rhizo- and endophytic bacteria in promoting phytoremediation of metalliferous soils, where *Burkholderia* sp. J62, *Pseudomonas aeruginosa* MKRh3, *Pseudomonas monteillii*, *Bacillus subtilis*, *Brevibacterium halotolerans*, *B. pumilus*, *Pseudomonas pseudoalcaligenes*, etc. were several of the prominent bacteria species reported.

20.2.2.5 As Abiotic Stress Tolerance

Plants undergo various kinds of abiotic stresses that not only hamper crop productivity but also pose a serious threat to food security. Different abiotic stresses, viz., drought stress due to poor or erratic rainfall, heat stress because of climate change resulting in a consistent increase in local and global temperatures, salinity, heavy metal stress, etc. limit the growth and consequently the productivity of major food crops (Hussain et al. 2019). They do so by various direct and indirect mechanisms like the accumulation of osmolytes, antioxidants, indole acetic acid, etc. (Figure 20.2).

20.3 Nanotechnology: Opportunities and Challenges for Agriculture

Nanotechnology was first introduced by Norio Taniguchi in 1974. The term “nano” is a Greek word meaning “dwarf” and it means 10^{-9} or one-billionth part of a meter (Thakkar et al. 2010). Because of their small size, nanoparticles have some distinctive characteristic like strength and reactivity. Nanoparticles (NPs) are useful for agriculture in many ways such as in upholding the soil’s fertility; in improving disease resistance in plants or crops and in plant growth promotion activities. Some of the examples of plant growth promotion activity and important nanomaterials are detailed in Tables 20.2 and 20.3.

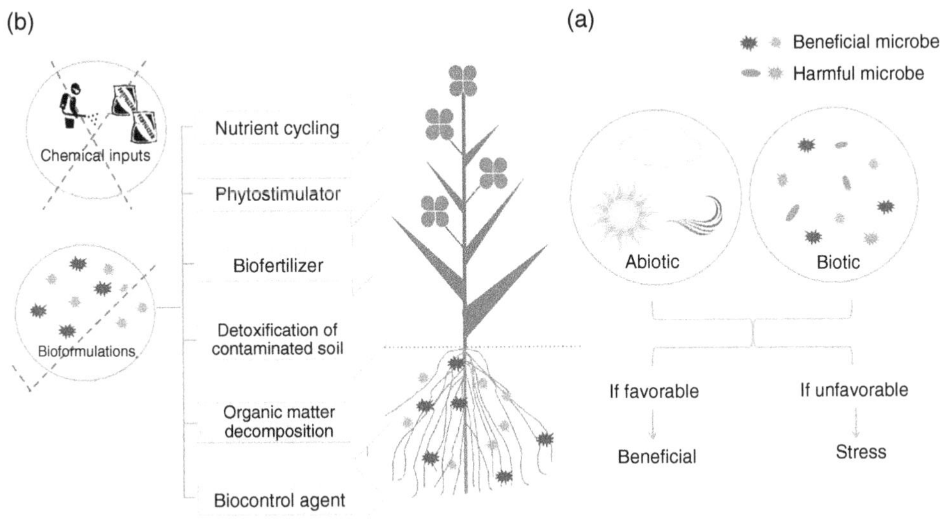

Figure 20.2 In an ecosystem, biotic and abiotic factors that influence plant growth and development. (a) Foliar application of different harmful chemicals must be avoided as they limit the productivity of plant. (b) Exposure of plant to different types of Biotic and Abiotic factors also affect plant productivity.

Table 20.2 Effect of nanomaterials on crop physiology and plant protection.

Nanomaterials	Crop Species	Mode of application	Responses	Reference
ZnO	*Triticum aestivum.*	Mixed with growth substrate	Improved grain yield and biomass buildup.	Du et al. (2019)
	Cyamopsis Tetragonoloba,	Foliar spray	Increased nutrient content	Tirani et al. (2019)
	Nicotiana tabacum	Hydroponics	Increased growth physiology, increased.	Iqbal et al. (2019)
	Triticum aestivum	Mixed with pot soils	metabolites, enzymatic activities	Pallavi et al. (2016)
AgNPs	*Vigna sinensis*	Foliar application	Improved growth and tolerance to heat stress.	Vanti et al. (2019)
	Vigna unguiculata	Foliar application	Enhanced growth and biomass by exciting root nodulation and soil bacterial diversity.	Malandrakis et al. (2019)
	Prunus domestica fruits	Fruit spray	*Xanthomonas axonopodis* pv. *malvacearum* and *Xanthomonas campestris* pv. *campestris* in vitro.	Giannousi et al. (2013)
ZnO, CuO and Ag NPs	*Solanum lycopersicum*	Foliar application	Concealed grey mold symptoms caused *by B. cinerea* and soil-borne diseases.	Imada et al. (2016)

(*Continued*)

Table 20.2 (Continued)

Nanomaterials	Crop Species	Mode of application	Responses	Reference
CuO	*Solanum lycopersicum*	Drenching	Effectively controlled late blight disease caused by *Phytophthora infestans.*	Wang et al. (2016)
MgO	*Oryza sativa*	Foliar application	Controlled bacterial wilt disease by repressing pathogen *Ralstonia solanacearum.*	Yang et al. (2007)
SiO_2 NPs	*Spinacia oleracea*	Foliar application	Reduced heavy metal toxicity and enhanced growth by declining bioconcentration and translocation in plants.	Das et al. (2016)
TiO_2	*Oryza sativa*	Foliar application	Increased biomass accumulation, chlorophyll, nitrogen, and protein content.	Wang et al. (2016)

Source: Adapted from Shang et al. (2019).

20.3.1 Nanomaterials Classification

Nanomaterials can be categorized according to their dimensions in the macroscale like in 0D having those NPs whose dimensions are all on the nanoscale; 1D are the nanowires and nanofibers that only have one dimension in the macroscale; 2D covers thin films and nanosheets, and finally, 3D constitutes the materials in bulk (Singh 2016). Another categorized basis of nanomaterials classification is based on chemical structure: the examples or group include carbon, ceramic (metal oxides), metal, and polymeric compounds (Khan et al. 2019).

20.3.1.1 Carbon-based NPs

The carbon-based nanomaterials include structures such as graphene, fullerenes, and carbon nanotubes (CNTs) (Ealias and Saravanakumar 2017). These CNTs are used as delivery system of various agrochemicals or other substances to hosts, thus leading to decrease in the quantity of chemicals released into the environment (Hajirostamlo et al. 2015). Usually, the use of CNT as nano-sponges having iron and sulfur enhances effectiveness in soaking up water contaminants like fertilizers, oil, pesticides, and pharmaceuticals.

20.3.1.2 Ceramic-based NPs

Ceramic are inorganic solids made of metal-oxide compounds such as ZnO, FeO_2, and TiO_2 (Thomas et al. 2015). These nanomaterials have been associated with increasing plant growth promotion activity.

20.3.1.3 Metal Oxide NPs

The metal oxide compounds offer photocatalytic characteristics, such as in the form of TiO_2, ZnO, WO_x, SnO_2, Fe_2O_3, CuO, ZrO_2, and MoO_3 (Sharma et al. 2018). The oxide NPs stimulate the antioxidant enzymes such as peroxidase (POX), superoxide dismutase (SOD),

Table 20.3 Microbial synthesized nanoparticles against plant pathogen.

Microbes	Source of isolation	Nanoparticles	Application	Plant disease control	Host	Reference
Bacteria						
Pseudomonas rhodesiae	Rhizospheric soil of cotton	Ag	Antibacterial agent	Dickeya dadanti	Sweet potato	Hossain et al. (2019)
Aeromonas hydrophila	Unknown	ZnO	-	Aspergillus flavus	Maize	Jayaseelan et al. (2012)
Streptomyces spp.	Oxalis corniculata leaves	CuO	Antifungal agent	Alternaria alternata, Fusarium oxysporum, Pythium ultimum, and Aspergillus niger	Multiple crops	Hassan et al. (2019)
Streptomyces capillispiralis	Convolvulusarvensis leaves	Cu	Antifungal agent	Alternaria spp., Aspergillus niger, Pythium spp., and Fusarium spp.	Multiple crops	
Streptomyces griseus	Rhizospheric soil of tea	Cu	Antifungal agent	Poria hypolateritia	Tea	Ponmurugan et al. (2016)
Bacillus licheniformis	Soil	Ag	Antiviral agent	Bean yellow mosaic virus	Faba bean	Elbeshehy et al. (2015)
Bacillus cereus	Wastewater contaminated soil	Ag	Antibacterial agent	Xanthomonas oryzae pv. oryzae	Rice	Ahmed et al. (2020)

Source: Adapted from Ali et al. (2020).

and catalase (CAT), as well as initiating the production of bioactive compounds (Ghorbanpour 2015). The application of titanium dioxide, for example, helped in stimulating an elicitation in *Salvia officinalis* of secondary metabolism and resulted in the highest phenolic and flavonoid contents at 200 and 100 mg/L, respectively (Ghorbanpour 2015).

20.3.1.4 Metallic NPs

These NPs include metals such as Zn, Au, Ag, Pt, and Ni and alloys (Dolez 2015). Metallic NPs have various roles—for example, as antibacterial, medical diagnostic, electrodes, and optics.

The wide acceptance of silver nanoparticles (AgNPs) under this category is well known and is attributed to their antibacterial activity (Radhakrishnan 2017). AgNPs have an important role in the agricultural sector in plant productivity. For example, Jasim (2017) recognized fenugreek seedlings treated with biosynthesized AgNPs and also found improved root length, shoot length, the number of leaves, and overall weight in plant. The use of Titania in the form of titanium dioxide (TiO_2) tested has beneficial effects on plants, such as by modulating Reactive Oxygen Species (ROS) signaling and nitrogen metabolism in plant.

At low concentrations of nano-TiO_2 applied to leaves or roots can enhance chlorophyll content, and thereby photosynthesis has also been recognized (Gao 2008), as well as the stimulation of various enzyme activity in crops (Lyu et al. 2017).

20.4 Bacteria in Biogenesis of Nanoparticles

Due to cost-effectiveness many of the bacteria have been recognized to be useful in nanoparticle synthesis, such as *Bacillus*, *Pseudomonas*, *Shewanella* and *Cornybacterium*, etc. These microorganisms have been used as sources of NP creation and can help improve agricultural production or productivity (Timmusk et al. 2018). There are many applications or functions of bacterially synthesized nanoparticles in agriculture; some of them are described here.

20.4.1 Bacteria as Nanobiopesticide/Biocontrol Agent

The bacterially synthesized nanoparticles have been useful as biocontrol agent for many plant diseases, such as *Alternaria* species, *Aspergillus niger*, *Pythium* species, and *Fusarium* species, etc. Several nanomaterials, such as Ag, Zn, ZnO, and CuO synthesized by bacteria with antifungal and antibacterial properties and reported or tested in host plants, such as in maize, sweet potato, faba bean, tea, rice, etc (Ali et al. 2014). Many bacteria, such as *Pseudomonas*, *Aeromonas hydrophila*, *Streptomyces* sp., and *Bacillus* sp., with nanomaterial biosynthesis and in antagonistic or biocontrol function against several pathogens have been described in studies.

20.4.2 Bacterial Nanoparticles as Nanobiofertilizers

PGPR (*B. subtilis, Paenibacillus elgii, Pseudomonas putida, and Pseudomonas fluorescens*) treated with silver, gold, and aluminum NPs has been described as sustaining plant growth yield and enhancing pathogen resistance. These NP-based formations enhance the union of valuable bacteria in the root rhizosphere. Nanoparticle ZnO resulted in enhancements in the rate of seed germination in diverse monocots and dicots (Mishra and Kumar 2009). De Gregorio et al. (2017) showed that immobilization of *Burkholderia caribensis* and *Pantoea agglomerans* in nanofibers did not modify the valuable properties of either rhizobacteria. The dry weight of soya bean roots and seed germination was studied to improve on immobilization with PGPR. Nanofibers and other nanomaterials can be a hopeful ecofriendly cause for inoculum development in which PGPR delivers its valuable plant growth-promoting character, and nanofibers defend bacteria and seeds from abiotic stresses. Slow-release nanofertilizers, which release nutrients and plant growth-promoting rhizobacteria were formulated by Mala (2017); blending of nanofertilizer with neem cake, a byproduct of neem (*Azadirachta indica*), and PGPR was carried out. Seed vigor index and stimulation of various enzymes get accelerated due to this slow-release fertilizer system (Mala 2017).

20.4.3 Bacterial Nanoparticles as Biosensors

Nanobiosensors are particular sensors formulated by a gathering of diverse transducers (biological, physical, chemical, electrochemical, etc.). They are classified on the foundation of their transduction principle, such as optical, piezoelectric, or electrochemical. Biosensors have also been characterized established on their most responsive carrier elements, and could be aptasensors, genosensors, enzymatic biosensors, and immunosensors (Oliveira et al. 2018). Biologically/Bacterially formulated biosensors are now manufactured by using various NPs like Ag, Au, Zn, Cu, etc. in microbes like bacteria, virus, and fungus. These NPs-based devices (nanoformulated particles, nanoencapsulated beds, and nanowires) are particular in the categorization of diseases in plants and play a vital role connected to pesticide and insecticide buildup in agricultural system. Gui et al. (2017) showed that *Pseudomonas putida* (BMM-PL) could function as entire cell-based biosensors. They accomplished the function of these bacteria in finding phenanthrene in contaminated soil.

20.4.4 Plant Absorption and Uptake of Nanoparticles

The nature of nanoparticles decides the ability of uptake of nanoparticles by plant, but it also depends on other factors also like the plant physiology and the interaction of the nanomaterials with the environment. The size of NPs is one of the major limitations for penetration into plant tissues, and various studies concluded about the utmost dimensions that plants permit for nanoparticles to move and collect inside the cells, usually with 40–50 nm as a size exclusion limit (Taylor et al. 2014). The methods of application are also a very important factor in establishing how successfully a plant will internalize the nanomaterials: roots are particular in the assimilation of nutrients and water. Once the nanoparticles get penetrated into the plant cell, there are two ways for them to progress through tissues: the apoplast and the symplast. Apoplastic transport occurs outside the plasma membrane through the extracellular spaces, cell walls of adjacent cells, and xylem vessels (Sattelmacher 2001), while symplastic transport involves movement of water and substances between the cytoplasm of adjacent cells by the particular structures called plasmodesmata and sieve plates (Roberts and Oparka 2003). The apoplastic pathway is significant for radial movement within plant tissues, and permits nanomaterials to attain the root central cylinder and the vascular tissues for forward movement to the aerial part (Sun et al. 2014). When allowing foliar applications, nanomaterials should cross the hurdle of the cuticle present, following the lipophilic or the hydrophilic pathway (Schönherr 2002). In lipophilic diffusion occurs through cuticular waxes, while the hydrophilic pathway is enabled by polar aqueous pores accessible in the cuticle or stomata. Since the diameter of cuticular pores has been predictably approximately 2 nm (Eichert and Goldbach 2008), with a size exclusion limit above 10 nm, the stomatal pathway seems to be a more efficient route for nanoparticle penetration (Eichert et al. 2008). If nanoparticles get transported solely through the xylem and not the phloem, they will move only from root to shoot and leaves, and not downwards, because of that the NPs should be applied to the roots to get a good allocation in the plant. Remarkably, organs like flowers, fruitss and seeds have an ability to import fluids from the phloem (sink activity) and are likely to gather NMs. Besides plant toxicity, NP accumulation in specialized organs raises another important issue related to their safe use in human and animal consumption (Pérez-de-Luque 2017).

20.5 Functional Annotation and the Importance of Studying Genome Metagenomes and Microbial Structure and Nanomaterial-producing Microbiomes

Microorganisms show wide arrays of beneficial attributes for enhanced and sustainable agricultural productivity. Bacteria dwelling in the rhizosphere are termed as "Rhizospheric bacteria" (Dong et al. 2019) and provide benefit to plants through exhibiting numerous plant growth-promoting traits, including mineral phosphate solubilization (Vyas and Gulati 2009), organic phosphorus mineralization (Singh and Satyanarayana 2011), nitrogen fixation (Kuan et al. 2016), production of iron-chelating molecule (siderophore) (Sayyed et al. 2013), phytohormones especially indole acetic acid (IAA) (Chandra et al. 2018), exopolysachharides (Naseem et al. 2018). A number of microorganisms with nanomaterial-producing potential traits and plant-promoting potential and as such with antagonistic activities have being characterized and utilized in agriculture (as mentioned in the last section).Plant roots are special habitat for huge range of bacteria, and the majority of them assist plants through increasing plant nutrition (both macro- and micronutrients), and amelioration of stress tolerance (Vacheron et al. 2013). Along with bearing multiple plant probiotic traits, the agriculturally important microorganisms increase plant growth and yield with less reliance on chemical fertilizers. Therefore, using microorganisms as plant probiotics and plant growth promotion can be an environmentally friendly and less costlyr approach (de Souza Vandenberghe et al. 2017).

The functional traits associated with plant growth are due to the expression of certain important genes in plant growth-promoting bacteria, such as nifH (nitrogen fixation), pqq (pyrroloquinoline quinone synthesis), ipdC/ppdC (associated with indole acetic acid production), etc. have been recognized (Bruto et al. 2014). The important microbial genes related with their plant growth-promoting importance are illustrated in Table 20.4.

Table 20.4 Functional annotation of the genes/enzymes in bacteria responsible for plant growth promotion.

S. no.	Annotated mechanisms of plant growth promotion	Functional gene/enzyme in bacteria responsible for plant growth promotion	Bacterial isolates	Reference
1	Nitrogen Fixation	*nifH* gene (nitrogenase-encoding genes)	*Azospirillum brasilense*	Santos et al. (2017)
			Bacillus megaterium, Bacillus mycoides	Singh et al. (2020)
			Burkholderia	Suárez-Moreno et al. (2012),
			Pseudomonas spp.	Li et al. (2017)
2	Mineral phosphate solubilization	*pqqBCDEFG* (pyrroloquinoline quinone-encoding genes)	*Pseudomonas fluorescens F113*	Miller et al. (2010)
			Pseudomonas fluorescens B16	Choi et al. (2008)
			Erwinia herbicola	Liu et al. (1992)
			Enterobacter intermedium	Kim et al. (2003)
3	Organic phosphorus mineralization	Phytases	*Pseudomonas* sp, *Pantoea* sp, *Enterobacter*	Maougal et al. (2014)
4	Indole acetic acid (IAA) production	*ipdC*/*ppdC* (indole-3-pyruvate decarboxylase/ phenylpyruvate decarboxylase gene)	*Azospirillum brasilense*	Spaepen et al. (2007)
			Enterobacter cloacae UW5	Ryu and Patten (2008)
			Pantoea agglomerans	Luziatelli et al. (2020)
5	Siderophore production	pvd, pyoverdine homologous genes, fpvA, mbtH, acrA,B, fhu	Members of *Enterobacteriaceae* and *Pseudomonadaceae* family	Gupta et al. (2014)
6	Induction of abiotic stress tolerance in plant	*acdS* (1-aminocyclopropane-1-carboxylate (ACC) deaminase gene)	Fluorescent *Pseudomonas* sp	Ali et al. (2014)
			Pseudomonas sp. UW4	Orozco-Mosqueda et al. (2019)
			Bacillus subtilis Rhizo SF 48	Gowtham et al. (2020)

(Continued)

Table 20.4 (Continued)

S. no.	Annotated mechanisms of plant growth promotion	Functional gene/enzyme in bacteria responsible for plant growth promotion	Bacterial isolates	Reference
7	Synthesis of antimicrobial compounds	*hcnABC* (hydrogen cyanide) and *phlACBD* (2,4-diacetylphloroglucinol)	*P. fluorescens* F113, *P. protegens* CHA0	Haas et al. (2000)
8	Nitrous oxide production	NirK (copper nitrite reductase gene)	*Azospirillum brasilense* Sp245	Creus et al. (2005)
9	Induced systemic resistance	acetoine genes *budAB* and 2,3-butanediol gene *budC*	*Enterobacter* sp. 638	Taghavi et al. (2010)
10	*Production of* γ-aminobutyric acid (GABA)	gabD, T	*Pseudomonas fluorescens*	Loper et al. (2012)
11	Trehalose metabolism	Trehalose synthase gene homolog	*Bradyrhizobium japonicum*	Sugawara et al. (2010)
12	Detoxification of arsenic	*ars*C gene	*Bacillus flexus* and *Acinetobacter junii*	Marwa et al. (2019)

A few microbial nanomaterials and pathways in plant growth promotion have also been described (Ali et al. 2019). Most of the microorganisms live in rhizospheric regions and soil, but are unable to grow in a lab. Therefore, the culturally independent tactic such as metagenomics is employed to decode the microbiome of the rhizosphere. Moreover, the metabolomic tactics in order to annotate the functional genes that are accountable for a number of plant growth-promoting traits are known (Bruto et al. 2014). The designing and isolation of efficient microorganisms, bioinoculants, or microbial consortium utilize both genome- and metagenome-based approaches, which can be highly useful to study bacterial diversity and functional genes and pathways or for utilizing microorganisms or microbiomes in agriculture.

A number of pipelines and functional annotation tools are described for microbial genome study, and functional trait recognition has been described in literature (Dudhagara et al. 2015; Karp et al. 2019). Some of the molecular tools and culturable and nonculturable approaches that can be utilized to recognize the functional traits of microorganisms isolated in laboratory or study microbial interactions via genomics and metagenomic approaches, which are useful for agicultural productivity and plant growth promotion in agroecosystems and utilizing plant growth-promoting microorganisms with multiple traits in agriculture (Figure 20.3).

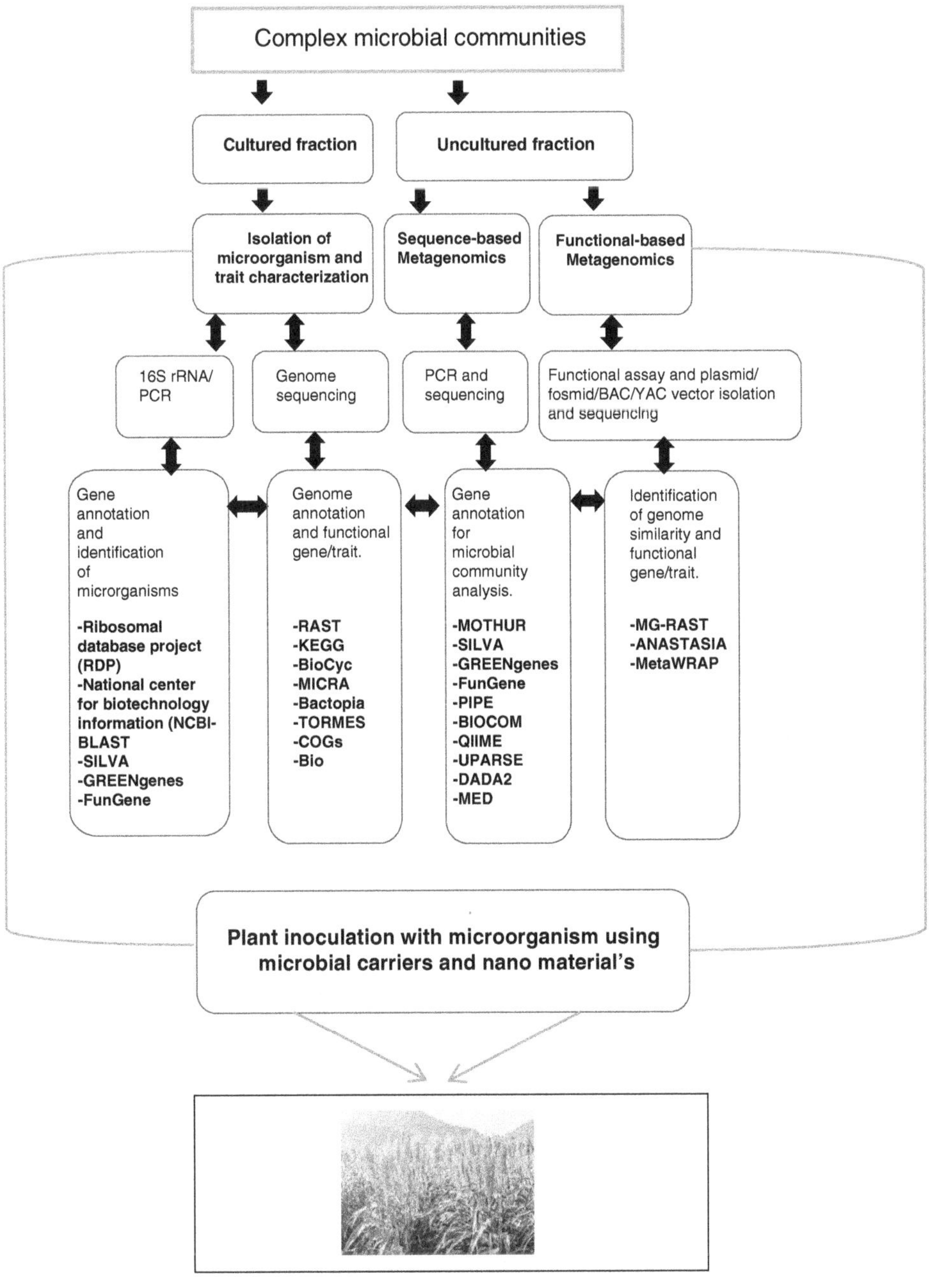

Figure 20.3 Workflow on studying plant growth promoting microbial diversity and its trait recognition using genomic and metagenomic tools and its utilization in agriculture. *Source:* Singha songsak/Adobe Stock Photos.

20.6 Conclusions

A number of microbial traits are useful and utilizing microorganisms and microbiomes in agriculture relies on effective strategies for the isolation and bioformulation of microbial consortia. The utilization of genomics and metagenomic approaches can be useful in this regard to understand plant and soil microbiomes and their interactions in order to utilize the microbiomes in agriculture for increasing productivity of agroecosystems.

Acknowledgment

The authors are grateful to the research group members, RDC and Department of Life Sciences, Parul Institute of Applied Sciences, Parul University, Post-Waghodia, Vadodara, Gujarat, PIN 391760, India, for providing infrastructural facilities and constant encouragement.

References

Adediran, G.A. (2015). Mechanisms behind bacteria induced plant growth promotion and Zn accumulation in Brassica juncea. *Journal of Hazardous Materials* 283: 490–499.

Ahmad, M., Zahir, Z.A., Asghar, H.N., and Asghar, M. (2011). Inducing salt tolerance in mung bean through coinoculation with rhizobia and plant-growth-promoting rhizobacteria containing 1-aminocyclopropane-1-carboxylate deaminase. *Canadian Journal of Microbiology* 57 (7): 578–589.

Ahmed, T., Shahid, M., Noman, M. et al. (2020). Silver nanoparticles synthesized by using bacillus cereus SZT1 ameliorated the damage of bacterial leaf blight pathogen in rice. *Pathogens* 9 (3): 160.

Ali, S.Z., Sandhya, V., and Venkateswar Rao, L. (2014). Isolation and characterization of drought-tolerant ACC deaminase and exopolysaccharide-producing fluorescent Pseudomonas sp. *Annals of Microbiology 64* (2): 493–502. https://doi.org/10.1007/s13213-013-0680-3.

Ali, J., Ali, N., Wang, L. et al. (2019). Revisiting the mechanistic pathways for bacterial mediated synthesis of noble metal nanoparticles. *Journal of Microbiological Methods* 159: 18–25. https://doi.org/10.1016/j.mimet.2019.02.010.

Ali, M.A., Ahmed, T., Wu, W., Hossain, A., Hafeez, R., Masum, M.M.I., Wang, Y., An, Q., Sun, G., and Li, B. (2020). Advancements in plant and microbe-based synthesis of metallic nanoparticles and their antimicrobial activity against plant pathogens. *Nanomaterials* 10 (6): 1146. https://doi.org/10.3390/nano10061146.

Alizadeh, H., Behboudi, K., Ahmadzadeh, M. et al. (2013). Induced systemic resistance in cucumber and *Arabidopsis thaliana* by the combination of *Trichodermaharzianum* Tr6 and *Pseudomonas* sp. Ps14. *Biological Control* 65 (1): 14–23.

Ansari, F.A. and Ahmad, I. (2018). Plant growth promoting attributes and alleviation of salinity stress to wheat by biofilm forming *Brevibacterium*sp. FAB3 isolated from rhizospheric soil. *Saudi Journal of Biological Sciences* https://doi.org/10.1016/j.sj.s.2018.08.003.

Armada, E., Roldán, A., and Azcon, R. (2014). Differential activity of autochthonous bacteria in controlling drought stress in native *Lavandula* and *Salvia* plants species under drought conditions in natural arid soil. *Microbial Ecology* 67 (2): 410–420.

Asawa, P., Varshney, N., Pandey, V. et al. (2019). Efforts to reduce greenhouse gases, air pollution and climatechange: global and regional perspectives. *International Journal of Agriculture, Environment, and Bioresearch* 4 (3): 129–144.

Azri, M.H., Ismail, S., and Abdullah, R. (2018). An endophytic*Bacillus* strain promotes growth of oil palm seedling by fine root biofilm formation. *Rhizosphere* 5: 1–7.

Babu, A.G., Kim, J.D., and Oh, B.T. (2013). Enhancement of heavy metal phytoremediation by Alnus firma with endophytic Bacillus thuringiensis GDB-1. *Journal of Hazardous Materials* 250: 477–483.

Bano, Q.U.D.S.I.A., Ilyas, N., Bano, A. et al. (2013). Effect of Azospirillum inoculation on maize (Zea mays L.) under drought stress. *Pakistan Journal of Botany* 45 (S1): 13–20.

Bhasin, S., Singh, M., and Singh, D. (2019). Review on bioactive metabolites of *Withania somnifera*. (L.) Dunal and its pharmacological significance. *Research & Reviews: Journal of Pharmacognosy and Phytochemistry* 8 (3): 3906–3909.

Bidyarani, N., Prasanna, R., Babu, S. et al. (2016). Enhancement of plant growth and yields in chickpea (*Cicerarietinum* L.) through novel cyanobacterial and biofilmed inoculants. *Microbiological Research* 188: 97–105.

Bruto, M., Prigent-Combaret, C., Muller, D., and Moënne-Loccoz, Y. (2014). Analysis of genes contributing to plant-beneficial functions in plant growth-promoting rhizobacteria and related proteobacteria. *Scientific Reports* 4: 6261. https://doi.org/10.1038/srep06261.

Chandra, S., Askari, K., and Kumari, M. (2018). Optimization of indole acetic acid production by isolated bacteria from *Stevia rebaudiana* rhizosphere and its effects on plant growth. *Journal, Genetic Engineering & Biotechnology 16* (2): 581–586. https://doi.org/10.1016/j.jgeb.2018.09.001.

Chang, P., Gerhardt, K.E., Huang, X.D. et al. (2014). Plant growth-promoting bacteria facilitate the growth of barley and oats in salt-impacted soil: implications for phytoremediation of saline soils. *International Journal of Phytoremediation* 16: 1133–1147.

Chen, L., Luo, S., Li, X., Wan, Y., Chen, J., & Liu, C. (2014).Interaction of Cd-Hyperaccumulator *Solanum nigrum* L. and functional endophyte Pseudomonas sp. Lk9 on soil heavy metals uptake. *Soil Biology and Biochemistry* 68: 300–308.

Chhabra, S. (2019). Phosphorus management in agroecosytems and role and relevance of microbes in environmental sustainability, pp 53–66, In: *Sustainable Green Technologies for Environmental Management* (ed. S. Shah, V. Venkatramanan, and R. Prasad). Singapore: Springer. https://doi.org/10.1007/978-981-13-2772-8_3

Chhabra, S., and Dowling, D. N. (2017). Endophyte-promoted nutrient acquisition: phosphorus and iron. In: *Functional Importance of the Plant Microbiome: Implications for Agriculture, Forestry and Bioenergy*, pp. 21-42. Springer. https://doi.org/10.1007/978-3-319-65897-1_3

Chhabra, S. and Prasad, R. (2021). Sustainable green technologies for environmental management- a book review. *Journal of Cleaner Production* 284: 124754. https://doi.org/10.1016/j.jclepro.2020.124754.

Chhabra, S., Hogan, J., Ryan, D., et al. (2010). Analysis of microbial communities under long-term conventional and reduced-input management of tillage soil. In: *Microorganisms in Industry and Environment, from Scientific and Industrial Research to Consumer Products*, 8–11. World Scientific Pub Co. https://doi.org/10.1142/9789814322119_0002.

Chhabra, S., Brazil, D., Morrissey, J. et al. (2013). Fertilization management affects the alkaline phosphatase bacterial community in barley rhizosphere soil. *Biology and Fertility of Soils* 49: 31–39. https://doi.org/10.1007/s00374-012-0693-2.

Choi, O., Kim, J., Kim, J.G. et al. (2008). Pyrroloquinoline quinone is a plant growth promotion factor produced by Pseudomonas fluorescens B16. *Plant Physiology* 146 (2): 657–668. https://doi.org/10.1104/pp.107.112748.

Chung, E.J., Hossain, M.T., Khan, A. et al. (2015). *Bacillus oryzicola* sp. nov., an endophytic bacterium isolated from the roots of rice with antimicrobial, plant growth promoting, and systemic resistance inducing activities in rice. *Plant Pathology Journal* 31: 152.

Conrath, U., Beckers, G.J.M., Langenbach, C.J.G., and Jaskiewicz, M.R. (2015). Priming for enhanced defense. *Annual Review of Phytopathology* 53: 97–119. https://doi.org/10.1146/annurev-phyto-080614-120132.

Creus, C.M., Graziano, M., Casanovas, E.M. et al. (2005). Nitric oxide is involved in the Azospirillum brasilense-induced lateral root formation in tomato. *Planta 221* (2): 297–303. https://doi.org/10.1007/s00425-005-1523-7.

Das, C.K., Srivastava, G., Dubey, A. et al. (2016). Nano-iron pyrite seed dressing: a sustainable intervention to reduce fertilizer consumption in vegetable (beetroot, carrot), spice (fenugreek), fodder (alfalfa), and oilseed (mustard, sesamum) crops. *Nanotechnology For Environmental Engineering* 1: 2. https://doi.org/10.1007/s41204-016-0002-7.

Das, K., Rajawat, M.V.S., Saxena, A.K., and Prasanna, R. (2017). Development of *Mesorhizobiumciceri*-Based biofilms and analyses of their antifungal and plant growth promoting activity in chickpea challenged by *Fusarium* Wilt. *Indian Journal of Microbiology* 57: 48–59.

De Gregorio, P.R., Michavila, G., Muller, L.R. et al. (2017). Beneficial rhizobacteria immobilized in nanofibers for potential application as soybean seed bioinoculants. *PLoS One* 12 (5): e0176930.

Dolez, P.I. (2015). Nanomaterials definitions, classifications, and applications. In: *Nanoengineering* (ed. P.I. Dolez), 3–40. Amsterdam, The Netherlands: *Elsevier*.

Dong, C.J., Wang, L.L., Li, Q., and Shang, Q.M. (2019). Bacterial communities in the rhizosphere, phyllosphere and endosphere of tomato plants. *PLoS One 14* (11): e0223847. https://doi.org/10.1371/journal.pone.0223847.

Du, W., Yang, J., Peng, Q. et al. (2019). Comparison study of zinc nanoparticles and zinc sulphate on wheat growth: from toxicity and zinc biofortification. *Chemosphere* 227: 109–116.

Dudhagara, P., Bhavsar, S., Bhagat, C. et al. (2015). Web resources for metagenomics studies. *Genomics, Proteomics & Bioinformatics* 13 (5): 296–303. https://doi.org/10.1016/j.gpb.2015.10.003.

Ealias, A.M. and Saravanakumar, M.P. (2017). A review on the classification, characterization, synthesis of nanoparticles and their application. *IOP Conference Series: Materials Science and Engineering* 263: 1–15.

Eichert, T. and Goldbach, H.E. (2008). Equivalent pore radii of hydrophilic foliar uptake routes in stomatous and astomatous leaf surfaces–further evidence for a stomatal pathway. *Physiologia Plantarum* 132: 491–502. https://doi.org/10.1111/j.1399-3054.2007.01023.x.

Eichert, T., Kurtz, A., Steiner, U., and Goldbach, H.E. (2008). Size exclusion limits and lateral heterogeneity of the stomatal foliar uptake pathway for aqueous solutes and water-suspended nanoparticles. *Physiologia Plantarum* 134: 151–160. https://doi.org/10.1111/j.1399-3054.2008.01135.x.

Elbeshehy, E.K.F., Elazzazy, A.M., and Aggelis, G. (2015). Silver nanoparticles synthesis mediated by new isolates of Bacillus spp.; nanoparticle characterization and their activity against bean yellow mosaic virus and human pathogens. *Frontiers in Microbiology* 6: 453.

El-Borollosy, A.M. and Oraby, M.M. (2012). Induced systemic resistance against Cucumber mosaic cucumovirus and promotion of cucumber growth by some plant growth-promoting rhizobacteria. *Annals of Agricultural Science* 57: 91–97.

Gao, F. (2008). Was improvement of spinach growth by nano-TiO(2) treatment related to the changes of Rubisco activase? *Biometals* 21: 211–217. https://doi.org/10.1007/s10534-007-9110-y.

Gao, J.L., Lv, F.Y., Wang, X.M. et al. (2015). *Paenibacilluswenxiniae* sp. nov., a nifH gene-harbouringendophytic bacterium isolated from maize. *Antonie Van Leeuwenhoek* 108 (5): 1015–1022.

Ghazalibigla, H., Hampton, J.G., de Jong, E.V.Z., and Holyoake, A. (2016). Is induced systemic resistance the mechanism for control of black rot in *Brassica oleracea* by a *Paenibacillus* sp.? *Biological Control* 92: 195–201.

Ghidan, A.Y., Al Antary, T.M., and Awwad, A.M. (2017). Aphidicidal potential of green synthesized magnesium hydroxide nanoparticles using Olea europaea leaves extract. *Asian Research Publishing Network Journal of Agricultural and Biological Science* 12 (10): 293–301.

Ghorbanpour, M. (2015). Major essential oil constituents, total phenolics and flavonoids content and antioxidant activity of Salvia o_cinalis plant in response to nano-titanium dioxide. *Indian Journal of Plant Physiology* 20: 249–256.

Giannousi, K., Avramidis, I., and Dendrinou-Samara, C. (2013). Synthesis, characterization and evaluation of copper based nanoparticles as agrochemicals against *Phytophthora infestans*. *Royal Society of Chemistry, and Advances* 3: 21743–21752.

Glick, B.R. (2012). Plant growth-promoting bacteria: mechanisms and applications. *Scientifica* 2012: 1–15.

Gowtham, H.G., Brijesh, S., Mahadevamurthy, M. et al. (2020). Induction of drought tolerance in tomato upon the application of ACC deaminase producing plant growth promoting rhizobacterium *Bacillus subtilis* Rhizo SF 48. *Microbiological Research* 234: 126422. https://doi.org/10.1016/j.micres.2020.126422.

Gui, Q., Lawson, T., Shan, S. et al. (2017). The application of whole cell-based biosensors for use in environmental analysis and in medical diagnostics. *Sensors* 17: 1623. https://doi.org/10.3390/s17071623.

Gupta, S. and Pandey, S. (2019). ACC Deaminase producing bacteria with multifarious plant growth promoting traits alleviates salinity stress in French Bean (*Phaseolus vulgaris*) plants. *Frontiers in Microbiology* 10: 1506. https://doi.org/10.3389/fmicb.2019.01506.

Gupta, A., Gopal, M., Thomas, G.V. et al. (2014). Whole genome sequencing and analysis of plant growth promoting bacteria isolated from the rhizosphere of plantation crops coconut, cocoa and arecanut. *PLoS One* 9 (8): e104259. https://doi.org/10.1371/journal.pone.0104259.

Haas, D., Blumer, C., and Keel, C. (2000). Biocontrol ability of fluorescent pseudomonads genetically dissected: importance of positive feedback regulation. *Current Opinion in Biotechnology* 11: 290–297.

Haggag, W. and Timmusk, S. (2008). Colonization of peanut roots by biofilm-forming *Paenibacilluspolymyxa* initiates biocontrol against crown rot disease. *Journal of Applied Microbiology* 104: 961–969.

Hajirostamlo, B., Mirsaeedghazi, N., Arefnia, M. et al. (2015). The role of research and development in agriculture and its dependent concepts in agriculture. *Asian Journal of Applied Science and Engineering* 4: 79–81. https://doi.org/10.1016/j.cocis.2008.01.005.

Hassan, S.E., Fouda, A., Radwan, A.A. et al. (2019). Endophytic actinomycetes Streptomyces spp mediated biosynthesis of copper oxide nanoparticles as a promising tool for biotechnological applications. *Journal of Biological Inorganic Chemistry* 24: 377–393.

Hossain, A., Hong, X., Ibrahim, E. et al. (2019). Green synthesis of silver nanoparticles with culture supernatant of a bacterium pseudomonas rhodesiae and their antibacterial activity against soft rot pathogen dickeya dadantii. *Molecules* 24: 2303.

Hussain, M.B., Zahir, Z.A., Asghar, H.N., and Asgher, M. (2014). Can catalase and exopolysaccharides producing rhizobia ameliorate drought stress in wheat? *International Journal of Agriculture and Biology* 16: 3–13.

Hussain, H.A., Men, S., Hussain, S. et al. (2019). Interactive effects of drought and heat stresses on morpho-physiological attributes, yield, nutrient uptake and oxidative status in maize hybrids. *Scientific Reports* 9 (1): 1–12.

Imada, K., Sakai, S., Kajihara, H. et al. (2016). Magnesium oxide nanoparticles induce systemic resistance in tomato against bacterial wilt disease. *Plant Pathology* 65: 551–560.

Iqbal, M., Raja, N.I., Hussain, M. et al. (2019). Effect of silver nanoparticles on growth of wheat under heat stress. *Indian Journal of Science and Technology* 43: 387–395.

Istina, I.N., Widiastuti, H., Joy, B., and Antralina, M. (2015). Phosphate-solubilizing microbe from Saprists peat soil and their potency to enhance oil palm growth and p uptake. *Procedia Food Science* 3: 426–435.

Jasim, B. (2017). Plant growth and diosgenin enhancement effect of silver nanoparticles in Fenugreek (Trigonella foenum-graecum L.). *Saudi Pharmaceutical Journal* 25 (3): 443–447. https://doi.org/10.1016/j.jsps.2016.09.012.

Jayaseelan, C., Rahuman, A.A., Kirthi, A.V. et al. (2012). Novel microbial route to synthesize ZnO nanoparticles using Aeromonas hydrophila and their activity against pathogenic bacteria and fungi. *Spectrochimica Acta. Part A, Molecular and Biomolecular Spectroscopy* 90: 78–84.

Jeong, J.J., Sang, M.K., Lee, D.W. et al. (2019). *Chryseobacterium phosphatilyticum* sp. nov., a phosphate solubilizing endophyte isolated from cucumber (*Cucumissativus* L.) root. *International Journal of Systematic and Evolutionary Microbiology* 69: 610–615.

Jha, C.K. and Saraf, M. (2015). Plant growth promoting Rhizobacteria (PGPR): a review. *Journal of Agricultural Research and Development* 5 (2): 0108–0119.

Jha, Y. and Subramanian, R.B. (2014). PGPR regulate caspase-like activity, programmed cell death, and antioxidant enzyme activity in paddy under salinity. *Physiology and Molecular Biology of Plants* 20: 201–207.

Kaewkla, O. and Franco, C.M.M. (2017). *Streptomyces roietensis* sp. nov., an endophyticactinobacterium isolated from the surface-sterilized stem of jasmine rice, *Oryza sativa* KDML 105. *International Journal of Systematic and Evolutionary Microbiology* 67: 4868–4872.

Kakar, K., Nawaz, Z., Cui, Z. et al. (2018). Rhizosphere-associated *Alcaligenes* and *Bacillus* strains that induce resistance against blast and sheath blight diseases, enhance plant growth and improve mineral content in rice. *Journal of Applied Microbiology* 124: 779–796.

Kang, S.M., Radhakrishnan, R., Khan, A.L. et al. (2014). Gibberellin secreting rhizobacterium, *Pseudomonas putida* H-2-3 modulates the hormonal and stress physiology of soybean to improve the plant growth under saline and drought conditions. *Plant Physiology and Biochemistry* 84: 115–124.

Karp, P.D., Ivanova, N., Krummenacker, M. et al. (2019). A comparison of microbial genome web portals. *Frontiers in Microbiology* 10: 208. https://doi.org/10.3389/fmicb.2019.00208.

Kasim, W.A., Osman, M.E., Omar, M.N. et al. (2013). Control of drought stress in wheat using plant-growth-promoting bacteria. *Journal of Plant Growth Regulation 32* (1): 122–130.

Kasim, W.A., Reda, M., Gaafar Rania, M. et al. (2016). Effect of biofilm forming plant growth promoting rhizobacteria on salinity tolerance in barley. *Annals of Agricultural Sciences* 61 (2): 217–227.

Kaur, H., Kaur, J., and Gera, R. (2016). Plant growth promoting rhizobacteria: a boon to agriculture. *International Journal of Cell Science and Biotechnology* 5: 17–22. https://doi.org/10.1007/978-981-13-6040-4_6.

Kavino, M., Harish, S., Kumar, N. et al. (2010). Effect of chitinolytic PGPR on growth, yield and physiological attributes of banana (*Musa* spp.) under field conditions. *Applied Soil Ecology* 45: 71–77.

Khan, I., Saeed, K., and Khan, I. (2019). Nanoparticles: properties, applications and toxicities. *Arabian Journal of Chemistry* 12: 908–931.

Kim, C.H., Han, S.H., Kim, K.Y. et al. (2003). Cloning and expression of pyrroloquinoline quinone (PQQ) genes from a phosphate-solubilizing bacterium Enterobacter intermedium. *Current Microbiology 47* (6): 457–461. https://doi.org/10.1007/s00284-003-4068-7.

Kim, I., Chhetri, G., Kim, J., and Seo, T. (2019). *Amnibacteriumsetariae* sp. nov., an endophyticactinobacterium isolated from dried foxtail. *Antonie Van Leeuwenhoek* 112: 1731–1738.

Kuan, K.B., Othman, R., Abdul Rahim, K., and Shamsuddin, Z.H. (2016). Plant growth-promoting rhizobacteria inoculation to enhance vegetative growth, nitrogen fixation and nitrogen remobilisation of maize under greenhouse conditions. *PLoS One 11* (3): e0152478. https://doi.org/10.1371/journal.pone.0152478.

Lee, B.D. (2015). Induction of systemic resistance in Panax ginseng against *Phytophthoracactorum* by native *Bacillus amyloliquefaciens* HK34. *Journal of Ginseng Research* 39: 213–220.

Li, H.B., Singh, R.K., Singh, P. et al. (2017). Genetic diversity of nitrogen-fixing and plant growth promoting *pseudomonas* species isolated from sugarcane rhizosphere. *Frontiers in Microbiology* 8: 1268. https://doi.org/10.3389/fmicb.2017.01268.

Lim, J.H. and Kim, S.D. (2013). Induction of drought stress resistance by multi-functional PGPR *Bacillus licheniformis* K11 in pepper. *Plant Pathology Journal* 29 (2): 201–208.

Liu, S.T., Lee, L.Y., Tai, C.Y. et al. (1992). Cloning of an Erwinia herbicola gene necessary for gluconic acid production and enhanced mineral phosphate solubilization in Escherichia coli HB101: nucleotide sequence and probable involvement in biosynthesis of the coenzyme pyrroloquinoline quinone. *Journal of Bacteriology 174* (18): 5814–5819. https://doi.org/10.1128/jb.174.18.5814-5819.1992.

Liu, Q., Zhang, X., Zhao, Y. et al. (2013). Fullerene-induced increase of glycosyl residue on living plant cell wall. *Environmental Science & Technology* 47: 7490–7498.

Loper, J.E., Hassan, K.A., Mavrodi, D.V. et al. (2012). Comparative genomics of plant-associated Pseudomonas spp.: insights into diversity and inheritance of traits involved in multitrophic interactions. *PLoS Genetics 8* (7): e1002784. https://doi.org/10.1371/journal.pgen.1002784.

Lucas, J.A., García-Cristobal, J., Bonilla, A. et al. (2014). Beneficial rhizobacteria from rice rhizosphere confers high protection against biotic and abiotic stress inducing systemic resistance in rice seedlings. *Plant Physiology and Biochemistry* 82: 44–53.

Luziatelli, F., Ficca, A.G., Bonini, P. et al. (2020). A genetic and metabolomic perspective on the production of indole-3-acetic acid by *pantoea agglomerans* and use of thei metabolites as biostimulants in plant nurseries. *Frontiers in Microbiology* 11: 1475. https://doi.org/10.3389/fmicb.2020.01475.

Lyu, S., Wei, X., Chen, J. et al. (2017). Titanium as a beneficial element for crop production. *Frontiers in Plant Science* 8: 1–19. https://doi.org/10.3389/fpls.2017.00597.

Ma, Y., Prasad, M.N., Rajkumar, M., and Freitas, H. (2011). Plant growth promoting rhizobacteria and endophytes accelerate phytoremediation of metalliferous soils. *Biotechnology Advances* 29 (2): 248–258.

Ma, Y., Oliveira, R.S., Nai, F. et al. (2015). The hyperaccumulator Sedum plumbizincicola harbors metal-resistant endophytic bacteria that improve its phytoextraction capacity in multi-metal contaminated soil. *Journal of Environmental Management* 156: 62–69.

Mahyarudin, R.I. and Lestari, Y. (2015). Metagenomic of actinomycetes based on 16S rRNA and nifH genes in soil and roots of four Indonesian rice cultivars using PCR-DGGE. *HAYATI Journal of Biosciences 22*: 113–121.

Mala, R. (2017). Evaluation of Nano structured slow release fertilizer on the soil fertility, yield and nutritional profile of Vigna radiata. *Recent Patents on Nanotechnology* 11 (1): 50–62. https://doi.org/10.2174/1872210510666160727093554.

Malandrakis, A.A., Kavroulakis, N., and Chrysikopoulos, C.V. (2019). Use of copper, silver and zinc nanoparticles against foliar and soil-borne plant pathogens. *Science of the Total Environment* 670: 292–299.

Mallick, I., Bhattacharyya, C., Mukherji, S. et al. (2018). Effective rhizoinoculation and biofilm formation by arsenic immobilizing halophilic plant growth promoting bacteria (PGPB) isolated from mangrove rhizosphere: a step towards arsenic rhizoremediation. *Science of the Total Environment* 610: 1239–1250.

Mani, D., Kumar, C., and Patel, N.K. (2016). Integrated micro-biochemical approach for phytoremediation of cadmium and lead contaminated soils using Gladiolus grandiflorus L. cut flower. *Ecotoxicology and Environmental Safety* 124: 435–446.

Maougal, R.T., Brauman, A., Plassard, C. et al. (2014). Bacterial capacities to mineralize phytate increase in the rhizosphere of nodulated common bean (Phaseolus vulgaris) under P deficiency. *European Journal of Soil Biology 62*: 8–14.

Marwa, N., Singh, N., Srivastava, S. et al. (2019). Characterizing the hypertolerance potential of two indigenous bacterial strains (Bacillus flexus and Acinetobacter junii) and their efficacy in arsenic bioremediation. *Journal of Applied Microbiology 126* (4): 1117–1127. https://doi.org/10.1111/jam.14179.

Meena, V.S., Maurya, B.R., and Verma, J.P. (2014). Does a rhizospheric microorganism enhance K^+ availability in agricultural soils? *Microbiological Research* 169 (5–6): 337–347.

Miller, S.H., Browne, P., Prigent-Combaret, C. et al. (2010). Biochemical and genomic comparison of inorganic phosphate solubilization in Pseudomonas species. *Environmental Microbiology Reports 2* (3): 403–411. https://doi.org/10.1111/j.1758-2229.2009.00105.x.

Mishra, V.K. and Kumar, A. (2009). Impact of metal nanoparticles on the plant growth promoting Rhizobacteria. *Digest Journal of Nanomaterials and Biostructures* 4: 587–592.

Mishra, A.K., Mishra, P., Singh, A.V.V., and Singh, M. (2019). Production of ENA (extra neutral alcohol) from molasses of sugarcane. *International Journal of Advanced Science and Research* 4 (3): 56–60.

Mohammed, A.F. (2018). Effectiveness of exopolysaccharides and biofilm forming plant growthpromoting rhizobacteria on salinity tolerance of faba bean (Vicia faba L*.). African Journal of Microbiology Research* 12: 399–404.

Mueed, Z., Singh, M., Mishra, A., and Poddar, N.K. (2021). Glycome in immunological processes: current scenario and future prospects glycome in immunological processes.

In: *The Glycome: Understanding the Diversity and Complexity of Glycobiology Glycome: The sweet enigma Apple Academic Press*. USA.

Mukhtar, S., Mirza, M.S., Awan, H.A. et al. (2016). Microbial diversity and metagenomic analysis of the rhizosphere of para grass (*Urochloamutica*) growing under saline conditions. *Pakistan Journal of Botany 48*: 779–791.

Muleta, D., Assefa, F., Börjesson, E., and Granhall, U. (2013). Phosphate-solubilisingrhizobacteria associated with Coffeaarabica L. in natural coffee forests of southwestern Ethiopia. *Journal of the Saudi Society of Agricultural Sciences* 12 (1): 73–84.

Nadeem, S.M. (2009). Rhizobacteria containing ACC-deaminase confers salt tolerance in maize grown on salt-affected fields. *Canadian Journal of Microbiology* 55: 1302–1309.

Naik, K., Mishra, S., Srichandan, H. et al. (2019). Plant growth promoting microbes: potential link to sustainable agriculture and environment. *Biocatalysis and Agricultural Biotechnology* 21: 101326. https://doi.org/10.1016/j.bcab.2019.101326.

Naseem, H., Ahsan, M., Shahid, M.A., and Khan, N. (2018). Exopolysaccharides producing rhizobacteria and their role in plant growth and drought tolerance. *Journal of Basic Microbiology 58* (12): 1009–1022. https://doi.org/10.1002/jobm.201800309.

Oliveira, J.L.D., Campos, E.V.R., Pereira, A.E.S. et al. (2018). Zein nanoparticles as eco-friendly carrier systems for botanical repellents aiming sustainable agriculture. *Journal of Agricultural and Food Chemistry* 66: 1330–1340. https://doi.org/10.1021/acs.jafc.7b05552.

Orozco-Mosqueda, M., Duan, J., DiBernardo, M. et al. (2019). The production of ACC deaminase and trehalose by the plant growth promoting bacterium *Pseudomonas* sp. UW4 synergistically protect tomato plants against salt stress. *Frontiers in Microbiology* 10: 1392. https://doi.org/10.3389/fmicb.2019.01392.

Pallavi, M.C.M., Srivastava, R., Arora, S., and Sharma, A.K. (2016). Impact assessment of silver nanoparticles on plant growth and soil bacterial diversity. *3 Biotech* 6: 254. https://doi.org/10.1007/s13205-016-0567-7.

Pérez-de-Luque, A. (2017). Interaction of nanomaterials with plants: what do we need for real applications in agriculture? *Frontiers in Environmental Science* 5: 12. https://doi.org/10.3389/fenvs.2017.00012.

Pilarska, A., Wysokowski, M., Markiewicz, E., and Jesionowski, T. (2013). Synthesis of magnesium hydroxide and its calcinates by a precipitation method with the use of magnesium sulfate and poly (ethylene glycols). *Powder Technology* 235: 148–157.

Ponmurugan, P., Manjukarunambika, K., Elango, V., and Gnanamangai, B.M. (2016). Antifungal activity of biosynthesised copper nanoparticles evaluated against red root-rot disease in tea plants. *Journal of Experimental Nanoscience* 11: 1019–1031.

Prasanna, R., Triveni, S., Bidyarani, N. et al. (2014). Evaluating the efficacy of cyanobacterial formulations and biofilmed inoculants for leguminous crops. *Archives of Agronomy and Soil Science* 60: 349–366.

Prashar, P. and Shah, S. (2016). Impact of fertilizers and pesticides on soil microflora in agriculture. In: *Sustainable Agriculture Reviews*, vol. 19 (ed. E. Lichtfouse), 331–361. Springer International Publishing Switzerland. https://doi.org/10.1007/978-3-319-26777-7_8.

Praveen Kumar, G., Mir Hassan Ahmed, S.K., Desai, S. et al. (2014). In vitro screening for abiotic stress tolerance in potent biocontrol and plant growth promoting strains of Pseudomonas and Bacillus spp. *International Journal of Bacteriology* 195946: 6. https://doi.org/10.1155/2014/195946.

Qurashi, A.W. and Sabri, A.N. (2012). Bacterial exopolysaccharide and biofilm formation stimulate chickpea growth and soil aggregation under salt stress. *Brazilian Journal of Microbiology* 43: 1183–1191.

Radhakrishnan, E.K. (2017). Microbially and phytofabricated AgNPs with different mode of bactericidal action were identified to have comparable potential for surface fabrication of central venous catheters to combat Staphylococcus aureus biofilm. *Journal of Photochemistry and Photobiology B: Biology* https://doi.org/10.1016/j.jphotobiol.2017.04.036.

Rai, P.K., Singh, M., Kumar, A. et al. (2020). *Role and Potential Applications of Plant Growth Promoting Rhizobacteria for Sustainable Agriculture New and Future Developments in Microbial Biotechnology and Bioengineering*, 49–60. Elsevier.

Rana, K.L., Kour, D., Kaur, T. et al. (2020). Endophytic microbes: biodiversity, plant growth-promoting mechanisms and potential applications for agricultural sustainability. *Antonie Van Leeuwenhoek* 113: 1075–1107. https://doi.org/10.1007/s10482-020-01429-y.

Roberts, A.G. and Oparka, K.J. (2003). Plasmodesmata and the control of symplastic transport. *Plant, Cell & Environment* 26: 103–124. https://doi.org/10.1046/j.1365-3040.2003.00950.x.

Ryu, R.J. and Patten, C.L. (2008). Aromatic amino acid-dependent expression of indole-3-pyruvate decarboxylase is regulated by TyrR in Enterobacter cloacae UW5. *Journal of Bacteriology* 190 (21): 7200–7208. https://doi.org/10.1128/JB.00804-08.

Santos, K.F.D.N., Moure, V.R., Hauer, V. et al. (2017). Wheat colonization by an Azospirillum brasilense ammonium-excreting strain reveals upregulation of nitrogenase and superior plant growth promotion. *Plant and Soil* 415 (1–2): 245–255.

Sasirekha, B. and Srividya, S. (2016). Siderophore production by Pseudomonas aeruginosa FP6, a biocontrol strain for Rhizoctoniasolani and Colletotrichumgloeosporioides causing diseases in chilli. *Agriculture and Natural Resources* 50 (4): 250–256.

Sattelmacher, B. (2001). The apoplast and its significance for plant mineral nutrition. *The New Phytologist* 149: 167–192. https://doi.org/10.1046/j.1469-8137.2001.00034.x.

Sayyed, R.Z., Chincholkar, S.B., Reddy, M.S. et al. (2013). Siderophore producing PGPR for crop nutrition and phytopathogen suppression. In: *Bacteria in Agrobiology: Disease Management* (ed. D. Maheshwari), 449–471. Berlin, Heidelberg: Springer. https://doi.org/10.1007/978-3-642-33639-3_17.

Schönherr, J. (2002). A mechanistic analysis of penetration of glyphosate salts macross astomatous cuticular membranes. *Pest Management Science* 58: 343–351. https://doi.org/10.1002/ps.462.

Shang, Y., Hasan, M.K., Ahammed, G.J., Li, M., Yin, H., and Zhou, J. (2019). Applications of nanotechnology in plant growth and crop protection: a review. *Molecules* 24 (14): 2558. https://doi.org/10.3390/molecules24142558.

Sharma, A., Ahmad, J., and Flora, S.J.S. (2018). Application of advanced oxidation processes and toxicity assessment of transformation products. *Environmental Research* 167: 223–233.

Shukla, N. (2012). Biochemical and physiological responses of rice (Oryza sativa L.) as influenced by Trichodermaharzianum under drought stress. *Plant Physiology and Biochemistry* 54: 78–88.

Shukla, P.S., Agarwal, P.K., and Jha, B. (2012). Improved salinity tolerance of Arachishypogaea (L.) by the interaction of halotolerant plant-growth-promoting rhizobacteria. *Journal of Plant Growth Regulation* 31: 195–206.

Singh, A.K. (2016). *Introduction to Nanoparticles and Nanotoxicology. Engineered Nanoparticles*, 1–18. Cambridge, MA: Academic Press.

Singh, B. and Satyanarayana, T. (2011). Microbial phytases in phosphorus acquisition and plant growth promotion. *Physiology and Molecular Biology of Plants: An International Journal of Functional Plant Biology* 17 (2): 93–103. https://doi.org/10.1007/s12298-011-0062-x.

Singh, M., Shah, P., Punetha, H. et al. (2017). Isolation and quantification of a potent anti cancerous compound, Withaferin A from the aerial parts of *Withania Somnifera* (Ashwagandha). *Advances in Plant Sciences* 30 (11): 231 235.

Singh, M., Shah, P., Punetha, H., and Agrawal, S. (2018). Varietal comparison of withanolide contents in different tissues of *Withania somnifera* (L.) Dunal (Ashwagandha). *International Journal of Life Sciences* 4: 2455–1716.

Singh, N., Iqbal, Z., Ansari, T.A. et al. (2019). The portent plant with a purpose: Aloe vera. *Journal of Pharmacognosy and Phytochemistry.* 8 (3): 4124–4130.

Singh, M., Poddar, N.K., Singh, D., and Agrawal, S. (2020). Foliar application of elicitors enhanced the yield of withanolide contents in *Withania somnifera* (L.) Dunal (variety, Poshita). *3 Biotech* 10: 157.

Singh, M., Singh, D., Rai, P.K. et al. (2021). Fungi in remediation of hazardous wastes: current status and future outlook springer nature Switzerland AG. In: *Recent Trends in Mycological Research, Fungal Biology.*

Son, J.S. (2014). Screening of plant growth-promoting rhizobacteria as an elicitor of systemic resistance against gray leaf spot disease in pepper. *Applied Soil Ecology* 73: 1–8.

de Souza Vandenberghe, L.P., Garcia, L., Rodrigues, C. et al. (2017). Potential applications of plant probiotic microorganisms in agriculture and forestry. *AIMS Microbiology 3* (3): 629–648. https://doi.org/10.3934/microbiol.2017.3.629.

Spaepen, S., Versées, W., Gocke, D. et al. (2007). Characterization of phenylpyruvate decarboxylase, involved in auxin production of Azospirillum brasilense. *Journal of Bacteriology* 189 (21): 7626–7633. https://doi.org/10.1128/JB.00830-07.

Suárez-Moreno, Z.R., Caballero-Mellado, J., Coutinho, B.G. et al. (2012). Common features of environmental and potentially beneficial plant-associated Burkholderia. *Microbial Ecology 63* (2): 249–266. https://doi.org/10.1007/s00248-011-9929-1.

Sugawara, M., Cytryn, E.J., and Sadowsky, M.J. (2010). Functional role of Bradyrhizobium japonicum trehalose biosynthesis and metabolism genes during physiological stress and nodulation. *Applied and Environmental Microbiology 76* (4): 1071–1081. https://doi.org/10.1128/AEM.02483-09.

Sun, D., Hussain, H.I., Yi, Z. et al. (2014). Uptake and cellular distribution, in four plant species, of fluorescently labeled mesoporous silica nanoparticles. *Plant Cell Reports* 33: 1389–1402. https://doi.org/10.1007/s00299-014-1624-5.

Taghavi, S., van der Lelie, D., Hoffman, A. et al. (2010). Genome sequence of the plant growth promoting endophytic bacterium Enterobacter sp. 638. *PLoS Genetics 6* (5): e1000943. https://doi.org/10.1371/journal.pgen.1000943.

Taylor, A.F., Rylott, E.L., Anderson, C.W., and Bruce, N.C. (2014). Investigating the toxicity, uptake, nanoparticle formation and genetic response of plants to gold. *PLoS One* 9: e93793. https://doi.org/10.1371/journal.pone.0093793.

Thakkar, M.N., Mhatre, S., and Parikh, R.Y. (2010). Biological synthesis of metallic nanoparticles. *Nanomedicine: Nanotechnology, Biology and Medicine* 6: 257–262.

Thomas, S., Harshita, B.S.P., Mishra, P., and Talegaonkar, S. (2015). Ceramic nanoparticles: fabrication methods and applications in drug delivery. *Current Pharmaceutical Design* 21: 6165–6188.

Timmusk, S., Islam, A., Abd El, D. et al. (2014). Drought-tolerance of wheat improved by rhizosphere bacteria from harsh environments: enhanced biomass production and reduced emissions of stress volatiles. *PLoS One 9* (5): e96086.

Timmusk, S., Seisenbaeva, G., and Behers, L. (2018). Titania (TiO2) nanoparticles enhance the performance of growth-promoting rhizobacteria. *Scientific Reports* 8: 1–13. https://doi.org/10.1038/s41598-017-18939-x.

Tirani, M.M., Haghjou, M.M., and Ismaili, A. (2019). Hydroponic grown tobacco plants respond to zinc oxide nanoparticles and bulk exposures by morphological, physiological and anatomical adjustments. *Functional Plant Biology* 46: 360–375.

Tripathi, V., Srivastava, A., and Singh, M. (2020). Fertilizer prescrition based on soil test crop re sponse approach in marigold-urd cropping system grown in soils of Uttarakhand. *Progressive Agriculture* 20 (1 and 2): 80–92.

Triveni, S., Prasanna, R., Kumar, A. et al. (2015). Evaluating the promise of Trichoderma and Anabaena based biofilms as multifunctional agents in Macrophomina phaseolina-infected cotton crop. *Biocontrol Science and Technology* 25: 656–670.

Udayashankar, A.C., Nayaka, S.C., Reddy, M.S., and Srinivas, C. (2011). Plant growth-promoting rhizobacteria mediate induced systemic resistance in rice against bacterial leaf blight caused by *Xanthomonasoryzae*pv. oryzae. *Biological Control* 59: 114–122.

Vacheron, J., Desbrosses, G., Bouffaud, M.L. et al. (2013). Plant growth-promoting rhizobacteria and root system functioning. *Frontiers in Plant Science* 4: 356. https://doi.org/10.3389/fpls.2013.00356.

Vanti, G.L., Nargund, V.B., Basavesha K.N., Vanarchi, R., Kurjogi, M., Mulla, S.I., Tubaki, S., Patil, R.R. (2019) Synthesis of *Gossypium hirsutum*-derived silver nanoparticles and their antibacterial efficacy against plant pathogens. *Applied Organometallic Chemistry33*, e4630, doi:https://doi.org/10.1002/aoc.4630.

Verma, J.P., Yadav, J., Tiwari, K.N., and Kumar, A. (2013). Effect of indigenous *Mesorhizobium* spp. and plant growth promoting rhizobacteria on yields and nutrients uptake of chickpea (*Cicerarietinum* L.) under sustainable agriculture. *Ecological Engineering* 51: 282–286.

Vyas, P. and Gulati, A. (2009). Organic acid production in vitro and plant growth promotion in maize under controlled environment by phosphate-solubilizing fluorescent Pseudomonas. *BMC Microbiology* 9: 174. https://doi.org/10.1186/1471-2180-9-174.

Wang, S.H., Wang, F.Y., Gao, S.C., and Wang, X.G. (2016). Heavy metal accumulation in different rice cultivars as influenced by foliar application of nano-silicon. *Water, Air, and Soil Pollution* 227: 228. https://doi.org/10.1007/s11270-016-2928-6.

Yan, J., Yan, H., Liu, L.X. et al. (2017). *Rhizobium hidalgonense* sp. nov., a nodule endophytic bacterium of *Phaseolus vulgaris* in acid soil. *Archives of Microbiology* 199: 97–104.

Yang, F., Liu, C., Gao, F. et al. (2007). The improvement of spinach growth by nano-anatase TiO2 treatment is related to nitrogen photoreduction. *Biological Trace Element Research* 119: 77–88.

Yang, F., Zhang, R., Wu, X. et al. (2020). An endophytic strain of the genus *Bacillus* isolated from the seeds of maize (*Zea mays* L.) has antagonistic activity against maize pathogenic strains. *Microbial Pathogenesis* 142: 104074.

Zhang, L.L., Gan, L.Z., Xu, Z.B. et al. (2018). *Pedobacterchitinilyticus* sp. nov., a chitin-degrading bacterium isolated from wheat leaf tissue. *International Journal of Systematic and Evolutionary Microbiology* 68: 3713–3719.

21

Nanotechnology and Microbiology for Agricultural Sustainability: Current Research and Future Challenges

Rajeshwari Negi[1], Tanvir Kaur[1], Samiksha Jhamta[2], Neelam Yadav[3], Ashok Yadav[4], Paridhi Puri[5], Neelam Thakur[2], and Ajar Nath Yadav[6]

[1] *Department of Biosciences, Graphic Era Deemed to Be University, Dehradun, Uttarakhand, India*
[2] *Department of Zoology, Akal College of Basic Sciences, Eternal University, Sirmour, Himachal Pradesh, India*
[3] *Centre of Research Impact and Outcome, Chitkara University Institute of Engineering and Technology, Chitkara University, Rajpura, Punjab, India*
[4] *Department of Botany, Banaras Hindu University, Varanasi, Uttar Pradesh, India*
[5] *University Centre for Research and Development, Chandigarh University, Mohali, Punjab, India*
[6] *Department of Genetics, Plant Breeding and Biotechnology, Dr. Khem Singh Gill Akal College of Agriculture, Eternal University, Sirmour, Himachal Pradesh, India*

21.1 Introduction

The main pillar of the developing economy is agriculture, which also produces food for a better quality of life. The agricultural industry is currently dealing with an abundance of challenges, such as the unpredictability of climate change, soil contamination from various hazardous environmental pollutants like pesticides and fertilizers, and, most importantly, rising food demands due to a growing world population (Negi et al. 2024c). According to a new United Nations assessment, the world's population is expected to reach 8.5 billion people by 2030 and roughly 9 billion by 2050 (Siegel and Siegel 2021). Therefore, it is imperative that food production be increased by more than 50% in order to fulfill the demands of the continuously expanding population. Conversely, the population subsistence depends on natural resources, including forests, oceans, and ecological biodiversity, all of which are being alarmingly depleted by industrialization (Raimi et al. 2022; Negi et al. 2024b).

Consequently, it promotes the necessity of greater crop output and improved food security. The delicate equilibrium between the environment and food production depends on ecological diversification because it increases agricultural resistance to environmental stressors that could cause failure of crops (Aguilera et al. 2020). More precisely, the forestry, aquaculture, livestock, and crop systems that support human life make up biodiversity for food and agriculture (Vanbergen et al. 2020; Negi et al. 2024a). The world's population relies heavily on the many animal and plant species found in forests and oceans for fuel, fiber, food, and fodder. Food security is therefore tremendously impacted by any problems that could lead to ecological imbalance (Ali et al. 2023). To address these pertinent concerns, innovative methods and approaches are always being developed. In light of this, the development of goods based on nanomaterials (NM) is one such move that has been taken to transform contemporary agricultural methods (Yadav et al. 2023). These materials exhibit significant reactivity because of their huge surface area to volume ratio and unique physicochemical characteristics, which clearly offer the benefit of necessary change in response to rising demand. Historically, nanotechnology has been used in various fields, including pharmaceutical science, diagnosis, chemistry, and therapeutic (Sahu et al. 2021).

Nanotechnology and biotechnology are two innovations that are increasingly spreading throughout practically every industry that involves biology, engineering, cosmetics, cleanup, biomedicine, agriculture, food, etc. At the vanguard of nanotechnology, nanoparticles (NPs) are incorporated into a variety of consumer goods because of their exceptional potential in comparison to their enormous counterparts. Additionally, the distinctive and promising applications of these new-age materials have drawn unambiguous awareness in the agricultural and food sector (Amutha et al. 2024; Khan et al. 2024).

By employing these cutting-edge materials, modern agriculture is evolving into precision agriculture, which aims to increase output from the resources at hand. Many NM types, including nanotubes, nanoclays, and nanowires, have unique surface chemical, optical, and electrical characteristics that provide them with increased sensitivity, better identification limits, and quicker reaction times (Singh et al. 2024). Despite being necessary to boost crop yields, fertilizers also reduce soil fertility by disrupting the soil mineral balance. Typically, fertilizers, insecticides, and antibiotics are sprayed; these substances can effectively run off (Hesham et al. 2021).

Furthermore, it is necessary to regulate the extremely high cost of producing these fertilizers and insecticides. In agriculture, the use of NMs seeks to minimize production costs to maximize output, decrease nutrient losses to boost yields, and limit product amounts for safeguarding plants (Maity et al. 2022). To reap the full potential from the practical applications of microbial nanotechnology, some work remains, including improved process design, risk assessment of nano-biopesticides and nano-biofertilizers, and legislation for the commercialization of nano-agroproducts. To illustrate the entire spectrum of modern farming practices worldwide, one challenging issue that remains unresolved is how to enhance farming systems to cope with erratic climate fluctuations, including the development of new crop varieties that are resistant to heat, drought, and other environmental stresses. It firmly believes that achieving the objective of supporting global agriculture requires a comprehensive understanding of nanotechnology (Usman et al. 2020).

21.2 Nanoparticles

NPs can be defined as any material with a size between 1 and 100 nm. The term "natural nanoparticles" refers to the several types of NPs that have been found in nature, each having its own size and function (Sekhon 2014). These consist of cytochrome c (3.1 nm), milk lipoproteins (10 nm), and hemoglobin (5.5 nm). These days, regulated laboratory settings are used to create NPs, which are referred to as produced or engineered nanoparticles (ENPs). This category includes carbon nanotubes (CNTs), which have numerous uses in agriculture and other industries. Utilizing such ENPs to restore a functional system with desired properties at the molecular and atomic level is known as nanotechnology.

A multidisciplinary idea, nanotechnology has roots in the fields of medical, textiles, agriculture, electronics, pharmaceuticals, and food (Sangeetha et al. 2017). According to Layman, it is the study of designing, producing, arranging, and modifying materials and products at the nanoscale. It delivers various elements, including drugs, fungicides, herbicides, pesticides, and even a fragment of genetic material, to biological systems using NPs as magic bullets (Ahmed et al. 2013).

21.2.1 Engineered Nanoparticles

ENPs are also referred to as manufactured, industrial, or manufactured NPs. The industrial synthesis of NPs has been growing recently because of their distinctive physiological and chemical characteristics. Their enormous surface area to weight ratio, variety of sizes and shapes, ability to enter cells, and superior adsorption are what set them apart (Phogat et al. 2016). The distinctive characteristics of NPs are caused by their high surface area-to-volume ratio. The behavior and fate of NPs are directly influenced by their complexity (Christian et al. 2008). The nanoscale modifies the atomic density and the periodic boundaries around a crystalline particle. When compared to its bulk substance, the NPs' physicochemical properties are therefore different (Guo et al. 2013). ENPs have a variety of unique qualities that can be used in agricultural and related fields. These include catalysis, high reactivity, antibacterial, optical, mechanical, electrical, magnetic, and sterical qualities (Christian et al. 2008; Pandey and Jain 2020). According to their composition, ENPs are divided into five groups: metal-based nanoparticles, dendrimers, quantum dots (QDs), carbon-based nanoparticles, and nanocomposites. These groups have a wide range of uses in many industries.

21.2.2 Methods of Synthesis of Nanoparticles

One of the most important processes in the realm of nanotechnology is the synthesis of NPs. It establishes the NPs' size, shape, geometry, surface area, homogeneity, dispersion, and chemical makeup, all of which might be extremely important for the application (Nayak et al. 2023). Physical, chemical, and biological approaches are the broad categories into which the synthesis can be divided. Physical synthesis typically employs a top-down strategy that breaks down larger structures into smaller, nanoscale particles, whereas chemical and biological synthesis employs a bottom-up strategy that builds materials layer by layer from their smallest units, such as atoms, molecules, or clusters (Kumari et al. 2023).

Employing mechanical processes including thermal energy, radiation, and mechanical pressure, physical synthesis of NP creates nanoscale materials from bulk sources without the use of chemicals (Al-Harbi and Abd-Elrahman 2024). This method creates uniform and environmentally benign NPs. The main chemical synthesis techniques employed in NPs include co-precipitation, inert gas condensation or molecular condensation, spinning, pyrolysis, solvothermal, hydrothermal, chemical vapor deposition (CVD), sol-gel, and reverse micelle techniques. The utilization of enzymes and secondary metabolites from bacteria, viruses, and fungi to create and work with well-organized NPs is known as microbial-mediated synthesis, and it is thought to be important (Nair et al. 2022). The creation of NPs has also benefited from the plant-mediated techniques. Proteins, polysaccharides, amino acids, and phytochemicals like flavonoids, alkaloids, tannins, and polyphenols are among the components of plant extracts that have demonstrated the ability to stabilize and reduce NPs. Here, a plant extract is utilized, which has the benefit of lowering metallic ions and making it easier for metallic NPs to form and stabilize (Naikoo et al. 2021; Karunakaran et al. 2023). The green synthesis approach provides a number of benefits over chemical and physical approaches, including ease of development, simplicity, efficiency, cost-effectiveness, and low waste formation during the manufacturing of nanoparticles (Karunakaran et al. 2023) (Figure 21.1).

21.3 Nanoparticles Interaction with Plants

Scientific and technological developments contribute to the agricultural industry by providing fresh perspectives and answers to challenging issues. Since the development of nanotechnology, more effective and contaminant-free nanoformulations for sustainable agriculture are consistently being devised (Fraceto et al. 2016). These cutting-edge substances have the capacity to change plant physiology immediately upon entering the intricate plant soil system, which may be readily utilized to comprehend the consequences. Also necessary for the controlled delivery of active chemicals is a vital understanding of the positive or negative interactions that these NMs have with plants. They might therefore offer fresh approaches to creating better products based on NMs. It is also thought that NM concentrations in the natural environment are far below what is deemed dangerous. However, thorough safety inspections are still needed to close these loopholes (Batley et al. 2013).

21.3.1 Size-dependent Uptake of Nanoparticles

The size of the NPs must be taken into account as a crucial factor while researching their uptake in plants because the different barriers that are trapped within them fluctuate in size from micrometres (mm) to nanometres (nm). For example, cells that form a cuticle membrane make up the foliar epidermis. In order to facilitate gaseous exchange, the epidermis is made up of stomata with two guard cells that form a pore that is roughly 3–12 μm broad and 10–30 μm long. Consequently, these stomatal apertures allow NPs to pass through the plants. Furthermore, the trichrome of the stomata and the cuticle layer on the epidermis have quite different permeability characteristics. However, with a size exclusion limit in the nm range, the cuticle layer appears more prominently on the leaf epidermis (Wang et al. 2016). It was reported that NPs between 4 and 100 nm in size may penetrate the cuticle by rupturing the waxy layer, and that fluorescently tagged NPs larger than 50 nm could gather in the epidermis beneath the cuticle in areas without stomata (Larue et al. 2012a). TEM experiments revealed the probable penetration of tiny NPs, such as 20 nm Fe_3O_4 NPs, in the *Nicotiana benthamiana* plant (Cai et al. 2020).

The primary size exclusion restriction was mostly enforced by these plant cell walls, which are typically between 3.5 and 20 nm in size, and more frequently around 5 nm. In one investigation, SWCNT was able to create a structure resembling endocytosis within the

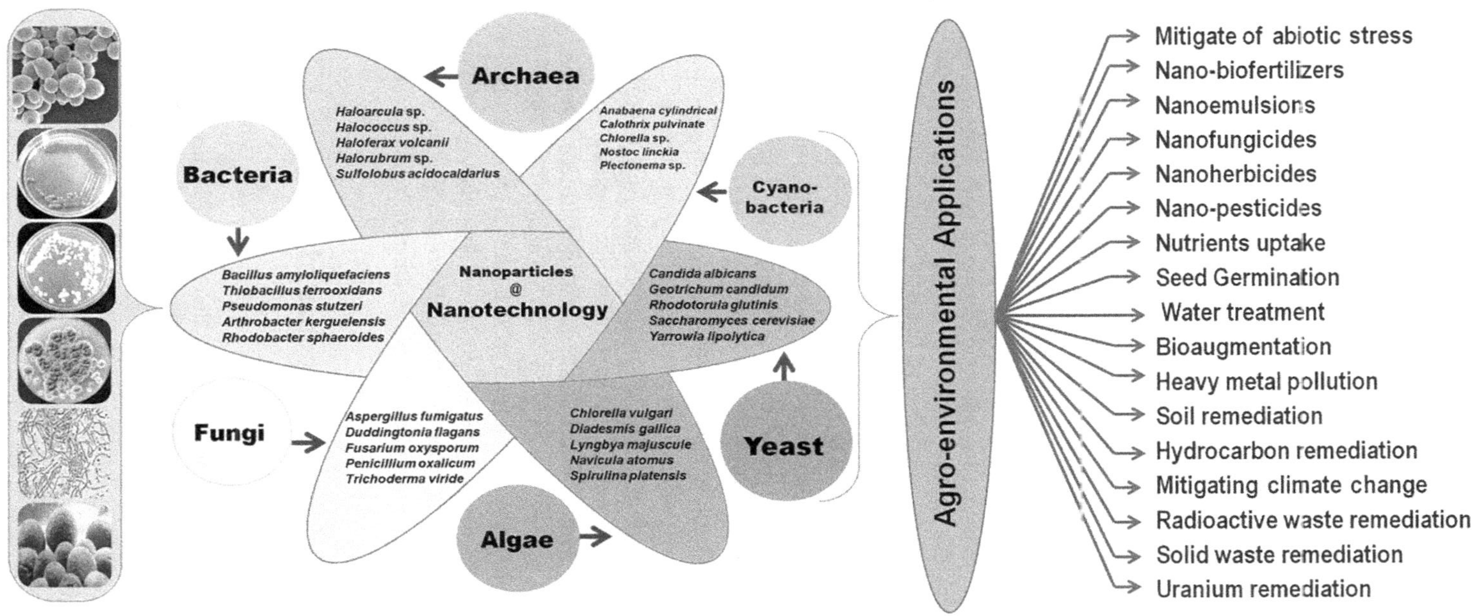

Figure 21.1 Microbial nanoparticles and their biotechnological applications for agro-environmental sustainability. *Source:* Yadav (2021).

complete *Arabidopsis thaliana* plant cell wall (Shen et al. 2010). It is proposed that NPs can pass through the cell wall if they fall within a certain size range. QDs, for instance, are easily exploited to enter plants since they are within the size range of plant cell wall pores (sub-10 nm) (Wu et al. 2017). Li et al. (2020) found that when sprayed foliarly, AgNPs with sizes ranging from 24.8 to 38.6 nm were present in lettuce leaves. Additionally, they illustrated how $AgNO_3$ applied with AgNPs might undergo in-plant transformation, which further clarifies this biological transformation phenomenon.

21.3.2 Surface Charge-dependent Uptake of Nanoparticles

In essence, the high degree of freedom of NPs depends on their uptake, absorption, and trafficking within plants. When NMs interact physic-chemically with plants, their energy and surface charge alter certain membrane proteins, carriers, and receptors on the surface (Juárez-Maldonado et al. 2019). NPs with varying surface charges have distinct surface characteristics and aggregation capacities than their pure counterparts (Hotze et al. 2010). An uneven distribution of fixed negative charges and both hydrophobic and hydrophilic components are found in the biological membrane, or cell wall, found inside leaves (the surface potential of cellulose fibers and lignin is −15 and 45 mV, respectively) (Zeng et al. 2017). Thus, the negatively charged cell wall promotes the tissue's absorption of positively charged NPs, which could be one cause. It is possible that the ion exchange surface of negatively charged plant cell walls facilitates the entry of cationic rather than anionic NPs (Meychik et al. 2005). On the other hand, negatively charged NPs significantly improve transport efficiency. Much greater adsorption of positively charged NPs at root surfaces is demonstrated by the surface charge-dependent absorption and transport of AuNPs. Nonetheless, it was shown that negatively charged NPs had a greater rate of internalization and translocation (Zhu et al. 2012). In this case, positively charged CeO NPs strongly adsorbed onto negatively charged root surfaces, while negatively charged CeO_2 NPs demonstrated poor root accumulation but increased shoot internalization, primarily due to their ability to overcome electrostatic repulsion (Liu et al. 2019). Additionally, the intake of AuNPs in roots and their movement from roots to shoots are dependent on surface charge. Negatively charged NPs are believed to aid in both apoplastic and symplastic trafficking into the vascular system from roots to shoots (Milewska-Hendel et al. 2019). As a result, surface charge alteration greatly alters the NPs transport performance throughout plants.

21.4 The Physiochemical Response of Plants Toward Nanoparticle

NMs internal absorption in plants is mostly determined by their size, chemical makeup, functional groups, and coating type. Plant physiology is impacted by the interaction and uptake of NMs, which results in molecular alterations (Jin et al. 2017). A distinct chemical mechanism underlies the multiwalled carbon nanotubes (MWCNTs) encouraging impact on tobacco cell proliferation (55–64%). Most of the active genes implicated are associated with tobacco aquaporin (NtPIP1), cell wall expansion (NtLRX1), and cell division (CycB) (Khodakovskaya et al. 2012). The key to greater development and vigor is NMs capacity to pierce seeds strong outer layer and permit water importation. Furthermore, a promising tactic is the seed priming approach, which further supports the potential of high-yield value crops by using nanotechnology prior to sowing (Acharya et al. 2020).

The growth of a few major crops, including maize, soybeans, and barley, is impacted by the application of MWCNTs. TEM imaging and Raman spectroscopy verified that MWCNTs congregate inside the endosperm of exposed seeds (Lahiani et al. 2013). In order to stimulate plant growth, NPs were carried to various plant sections and interacted with cellular machinery. By interacting with chloroplasts, mesoporous silica (MSN) NPs can boost photosynthesis, leading to higher total protein, chlorophyll content, and improved seed germination. The fact that none of the plants experienced stress at the maximum concentration (2,000 mg/L) of MSNs suggests that using them as a smart delivery mechanism is safer (Sun et al. 2016). Silica is crucial for nutrition since a lack of it weakens plants and makes them more vulnerable to biotic and abiotic stress. Applying silica to plants under stress also helps them experience less drought stress and produce more biomass (Tahir and Kanwal 2007). In comparison to the bulk counterpart and control, maize plants exhibit a greater response to the application of silica NPs (SiO_2 NPs) made from rice husk and meticulously characterized. The maize plants' roots grew longer, and their silica intake was improved, which helps them resist drought. Additionally, root sectioning shows that SiO_2 NPs are directly absorbed and accumulated in maize epidermal cells (Suriyaprabha et al. 2012).

As demonstrated in canola (*Brassica napus*), another possible NP, TiO_2, similarly stimulates plant growth, leading to increased seed vigor and germination. When compared to the control, the greatest effect was observed at concentrations between 1,200 and 1,500 mg/L, with increased plumule growth and big radicle (Mahmoodzadeh et al. 2013). Depending on its size and additive concentration, each NP can have either a positive or negative effect. Compared to larger particles, the smallest TiO_2 NPs accumulate more in plants (Larue et al. 2012a). Additionally, these NPs can influence miRNA levels to start plant growth-promoting pathways (Boykov et al. 2019). By increasing the amount of chloroplasts and grana stacking, FeNPs were discovered to have a favorable effect on the *Capsicum annum* plant at lower doses. On the other hand, the plant suffered damage when these NPs were sprayed in greater quantities (Yuan et al. 2018). It has been noted that other metal oxide nanoparticles, such as ZnO and Cu, have an impact on black gram (*Vigna mungo*) seed germination (Raja et al. 2019). The growth-promoting and seed-germination abilities of less researched platinum nanoparticles protected by polyvinylpyrrolidone (PVP) in *Pisum sativum* (Rahman et al. 2020). Another study documented that, by increasing the amounts of indole-3-acetic acid (IAA) in wheat even at a low concentration, chitosan, another essential and biocompatible NP, had an impact on seed germination and growth (Li et al. 2019).

21.5 Application of Microbial Nanotechnology in Agriculture

The combined application of two different fields of science, i.e. nanotechnology and microbiology is gradually expanding their roots in the agricultural sector. Microbial biofabricated NPs or combination of NPs with microbial isolates is useful for enhancing the growth of the plants by providing the desired nutrients, protection from diseases caused

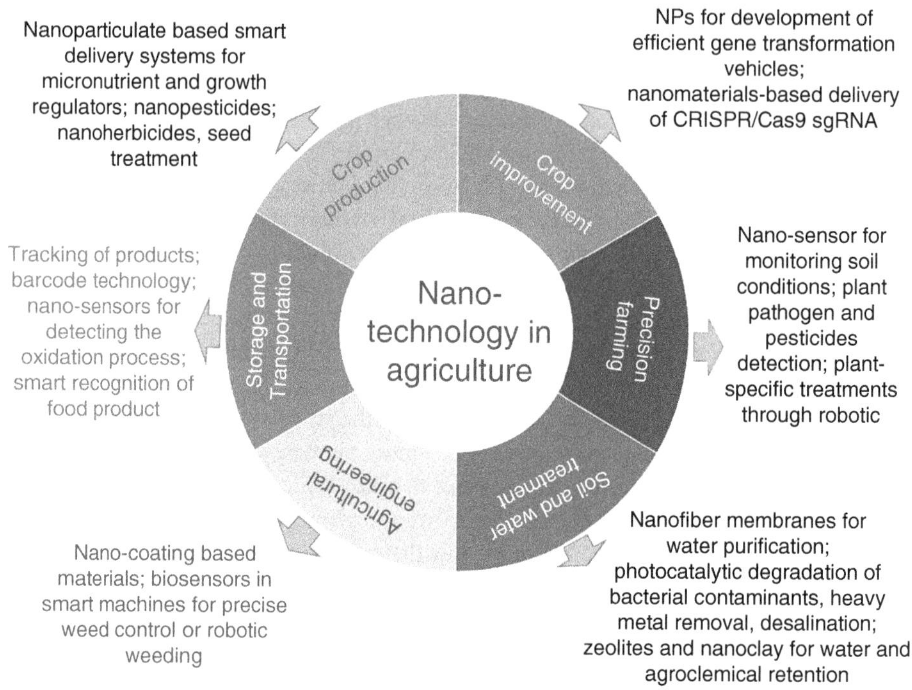

Figure 21.2 Applications of nanotechnology in agricultural sector. *Source:* Acharya and Pal (2020).

by pests and pathogens, alleviation of abiotic and early detection of the disease for timely management (Figure 21.2).

21.5.1 Nano-biofertilizers

Chemical fertilizer replacement is the urge of the correct era due to the hazardous effect on the environment, humans, and other microdiversity of soil. Fertilizers are the need of the agricultural sector, which is indispensable and the search of substituent is needed. Biofertilizers from microbes have taken their place, and massive amount of research has been conducted worldwide, as they provide the required nutrients, supplements, and growth hormones without harnessing the environment's quality. In advances to biofertilizer, the use of NPs along with biofertilizers is gaining much interest worldwide because of their better applications and results in enhancing the growth of the crops. Different studies have been published so far that indicated the better results of the nano-biofertilizer use (Kumari and Singh 2020). The combination of plant growth-promoting bacteria (*Azotobacter* and *Pseudomonas*) and nanofertilizer (Fe, Mn, and Zn) was inoculated in the spring wheat and found to increase seed number, spike number, spike length, peduncle length and decrease the number of emergence spikes and physiological maturity (Mardalipour et al. 2014).

In a report, application of calcium phosphate nanoparticles and fungal strains, i.e. *Piriformospora indica* and *Glomus mosseae* in maize resulted in the augmented performance index and chlorophyll content of the plants (Rane et al. 2015). In a similar report, silver NPs were reported for enhancing the leaf number, root length, and wet weight of fenugreek upon inoculation (Jasim et al. 2017). In a study, application of gold NPs (synthesized from *Bacillus subtilis*) with *Pseudomonas monteilii* having indole-3-acetic acid production capability in *Vigna unguiculata* seedlings was reported to increase the growth of the treated plants (Panichikkal et al. 2019). Application of zinc oxide nanoparticles from *Bacillus subtilis* in maize were found to increase the protein content and leaf area of the treated plants (Sabir et al. 2020). In an investigation, AgNPs biofabricated from *Streptomyces griseus* application in chickpea was found to enhance the growth, nutrient, and yield of the grain (Sambangi and Gopalakrishnan 2023). A study reported that the application of ZnONPs, CuONPs along with *Bacillus* sp. was enhancing overall growth, proline level, nitrogen content, and chlorophyll content of black wheat treated plants (Karunakaran et al. 2024).

21.5.2 Nanomaterials for Soil Remediation

Soil of the earth has been gold for mankind, due to which it has been overexploited and left it with diverse pollutants or contaminants. It has been contaminated with pollutants like pesticides, xenobiotics, heavy metals, polyaromatic hydrocarbons, and crude oil, which have depleted the quality of the soil. The presence of the contaminants in soil resulted in the depleted agronomical production, which is already low as compared to the human's ever-growing pollution. Bioremediation was known for the removal of pollutants, which sustains the environment by using the microbial cell and plant extracts. The microbial-based bioremediation was found as an effective technology, and now combination nanomaterials with microbes have proven to be more effective (Singh et al. 2023). In an investigation, *Pseudomonas fluorescence* and *Bacillus cereus* with Ag-nano particles were reported for the remediation of heavy metals (lead, cadmium, and nickel) in soil, and their inoculation enhances the maize plants' growth (Khan and Bano 2016). In a report, biosynthesized selenium NPs from *Rahnella aquatilis* were found to have bioremediation capability of selenium pollution in soil (Zhu et al. 2018).

Nanoparticles of zinc oxide and iron oxide were reported for alleviating the heat stress in the soil, and the application resulted in a higher microbial count for bacteria belonging to phyla Actinobacteria, *Pseudomonas, Azotobacter* and fungi (Kumar et al. 2019). Bioremediation of crude oil was achieved by the AgNPs biofabricated from *Aspergillus flavus* and *Trichoderma harzianum* (Al-Zaban et al. 2020). In an investigation, diesel oil in soil was bioremediated by the use of myco-synthesized AgNPs (Uba and Obiefuna 2023). ZnONPs and *Pennisetum giganteum* combination was found as a potential mitigator for phytoremediation of Cd contamination in soil and their application in soil resulted in the microbial community (*Chloroflexi, Cellvibro, Stenotrophomonas,* and *Thiobacillus*) associated with rhizocompartment of the plant (Li et al. 2024).

21.5.3 Nanotechnology for Disease and Pest Management

Pests and diseases were the major reason for the decreased production of the agronomical yield, and to control them, diversified chemical inputs have been made, which resulted in the contamination of soil and harnessed the environment's sustainability. Biopesticides are well researched and known worldwide for their application and benefits. Now the advanced and better technology for the control of the disease and pest has been known, i.e. use of the NPs. The use of biofabricated NPs or the combination of the microbial strains has been reported for suppressing the growth of the pathogen, and many studies have been published worldwide (Ali et al. 2020). Biofabricated silver nanoparticles (NPs) from the bacterium *Serratia* sp. were found to have biocontrol activity against the spot blotch disease pathogen (*Bipolaris sorokiniana*), and their application enhances wheat growth (Mishra et al. 2014).

In a study, titania nanoparticles, *Bacillus amyloliquefaciens* and oilseed rape plant extract were reported to have antagonistic activity against fungal phytopathogen *Alternaria brassicae*. Further inoculation of the combination was reported for enhancing the plant growth of oilseed rape plants (Palmqvist et al. 2015). In another investigation, AgNPs from Stenotrophomonas sp. were reported for having biocontrol activity against *Sclerotium rolfsii,* which causes collar rot disease in chickpea. Further, the application of AgNPs was reported for enhancing the treated chickpea plants (Mishra et al. 2017). In a report, a combination of AgNPs, salicylic acid, and *Bacillus cereus* was reported to have antagonistic activity against yellow rust disease, and its application in wheat resulted in enhanced yield, chlorophyll, and carotenoid content (Bano and Ummatul 2020). In a similar report, a ZnNPs and *Trichoderma harzianum* combination was found to enhance the tomato plants' growth by suppressing the growth of the gray mold disease pathogen, i.e. *Botrytis cinerea* (Imran et al. 2023). Combined application of biosynthesized Se and TiO_2 NPs with *Enterobacter cloacae* was found to suppress the growth of phytopathogen *Fusarium culmorum* and their application in wheat resulted in improved wheat growth (Kashisaz et al. 2024).

21.5.4 Nanobiosensor for Detection of Plant Diseases and Biological Components

The timely detection of the plant disease can play a very significant role in the production of more yield of the crops, as sometime >50% of the yield is lost pre- and post-harvest and loss drastically affects the humans economically. The timely detection of the disease may facilitate the timely management practices that can save the yield loss. Different methods have been used for the diagnosis of crop diseases, which include symptom inspection, DNA-based detection, nucleic acid blot assays, next-generation sequencing, and serological assay, but these techniques are less reliable at asymptomatic stages and moreover they are time-consuming, require costly equipment and may generate false results (Kaur et al. 2020). As a substitute for these techniques, nanotools are gaining much attention, as they miniaturize the process in which nanobiosensors are used for the detection of the antibody, volatile compounds and DNA as biosensing receptors. The use of nanobiosensor technology is nondestructive, economically better and easy to use for the detection of the plant pathogens (Kashyap et al. 2019). In an investigation, nano-based biosensors, which were developed from electrochemical approach, were used to detect the leaf scald disease causal agent, i.e., *Xanthomonas albilineans,* which infects the sugarcane. The biosensor detects the disease in three steps, including DNA isolation of the plant, purification of target sequence, and EC detection of the target DNA (Chakraborty et al. 2024).

21.5.5 Seed Nano-priming

Abiotic and biotic stresses are always being the big problem for the humans, as they significantly lower the productivity of the agricultural productivity. Already mankind is under the pressure of producing more yield to satisfy the need of the growing population, and in sometime new approaches will be discovered to cope up with this kind of problem. To protect the plants from the different stress, before the sowing of the seed, they have been coated with salt solution, water, osmotic agents, plant hormone solution and microbial solution (De La Torre-Roche et al. 2020). The coating technology is known as seed priming, which have proven benefits to the crop, as it improves the seed germination rate, growth, stress tolerance towards the biotic (pest and pathogens) and abiotic (drought, salinity, heavy metals, and temperature) factors. With advances in the coating solution, NPs are found to be more beneficial, and this technique is known as nano-priming. NPs with a size less than 100 nm have shown greater tolerance towards the stress and increase the seed germination rate to bigger extent (Nile et al. 2022; Imtiaz et al. 2023). In an investigation, ZnO NPs were coated on the rice seeds, which resulted in the mitigation of the drought stress and improved the tiller number, straw, and paddy yield (Waqas Mazhar et al. 2022). Biosynthesized selenium NPs were seed primed on the rice seeds in another study and results found to be increase the germination rate, seedling growth, and pathogen resistance (Setty et al. 2023). Similarly, silica NPs coating on *Stevia rebaudiana* Bertoni seeds was improving the germination rate, catalase and peroxidase activity, and sucrose in roots and shoots of the treated plants (Hasanaklou et al. 2023).

21.6 Microbial Nanotechnology and Agricultural Sustainable Development

Microbial nanotechnology involves the use of microbes to synthesize nanomaterial or apply nanotechnology to study and manipulate microbial processes. Research in microbial nanotechnology, often referred to as microbial nanobiotechnology, is expanding quickly and has the potential to transform various industries, like bioremediation, energy generation, medicine, and agriculture (Mandeep and Shukla 2020). This cutting-edge technique creates novel materials with unique qualities and functions by combining nanotechnology and microbes. Microbial nanotechnology promotes ecofriendly and sustainable farming by fostering environmental stewardship, optimizing resource use, and reducing reliance on chemical inputs, contributing to a more sustainable agroecosystem (Abd-Elsalam 2024). It plays a vital role in enhancing productivity by regulating nutrient availability and observing pesticides, water quality and supporting sustainable agriculture development (Gruère 2012; Mukhopadhyay 2014; Prasad et al. 2014). Nanotechnology significantly contributes to agriculture and food security by enhancing food processing, development and plant protection the precise delivery of agrochemicals to improve disease resistance and boost plant growth (Kashyap et al. 2018). The NPs, beyond their size have an effect on toxicity, which comprises structure, shape, chemical composition, size, behavior, surface charge, clumping, and disaggregation (Ion et al. 2010; Arora et al. 2022). Different sizes or shapes of NPs with the same chemical makeup can show various degrees of toxicity. Research on nanotechnology's effects on agriculture is increasingly becoming a fundamental component of sustainable development. Relevant uses of nanotubes, fullerenes, biosensors, controlled delivery systems, and nanofiltration in the agri-food industries have been researched (Sabir et al. 2014). This technique has demonstrated potential in maintaining soil fertility, controlling agricultural resources, and delivering medications to plants. Its use in food processing, packaging systems, risk assessment, and the usage of biomass and agricultural waste is also constantly being evaluated (Floros et al. 2010; Gupta et al. 2023). It is also used in agriculture by developing nanosensors, nanofungicides, nanoherbicides, and nanofertilizers. Copper nanoparticles, zinc, silica, titanium dioxide, and polymeric nanoparticles like dendrimers, which function as nanocarriers, can all be used to create nanofertilizers, which improve crop development by facilitating the absorption of vital micronutrients (Kah et al. 2018; López-Valdez et al. 2018; VandeVoort and Arai 2018).

Encapsulating pesticides for controlled release is the main use of nanopesticides, which offer defense against abiotic stressors. This prolongs the life of the active chemical ingredient, eases the coast, and improves the pesticide's stability and selectivity (Pascoli et al. 2018). NM increase the effectiveness of microorganisms while catalyzing the breakdown of waste and hazardous materials. Utilizing living organisms, bioremediation breaks down or eliminates poisons and other dangerous materials from agricultural soil and water. Commonly used words include mycoremediation, phytoremediation, and several helpful microorganisms. Bioremediation promotes sustainable remediation solutions by effectively and environmentally eliminating heavy metals from water and soil. By removing toxic substances from agricultural soil, the nano-nano interaction makes it sustainable (Dixit et al. 2015; Prasad et al. 2017).

21.7 Toxicological Impact of Nanoparticle Risk and Health Hazards in Agriculture Application

NPs are used in agriculture for applications such as nanofertilzers, nanopesticides, and nanosensors. While they offer significant benefits, their toxicological impact raises concern regarding the agricultural environment, ecosystem, and human health applications. As nanotechnologies proliferated, the amount of produced NPs in the environment increased significantly (Du et al. 2017; Tripathi et al. 2017). The toxicity of metal can be determined by various factors such as solubility, specificity of biological sites and so forth. Metal NPs exhibit some activities like antibacterial, antifungal, and anticandidal (Patra and Baek 2017). The efficiency of nanotoxic properties of NPs depends on the targeted cell wall structure, and its cytotoxicity depends on the membrane surface charge. The nanotoxicity attribute has an electrostatic interface between NPs with membrane, and their growth in cytoplasm (Rana and Kalaichelvan 2013; Aziz et al. 2015; Aziz et al. 2016). The rapid growth of nanotechnology raises safety concerns, as many NPs are toxic to microbes, potentially leading to environmental pollution and degradation in nanopesticide and nanofugicide formulations. The accumulation of NPs in soil and access into the food chain must be thoroughly studied to develop protection for whispered use of nanomaterial-based methods. Several studies have reported dose-dependent toxicity of NPs in crop plants (Liu and Lal 2015).

Overexposure to photochemically active NPs such as silver, zinc oxide, silicon dioxide, fullerenes, and titanium dioxide can lead to hindered photosynthesis, interfere with metabolic processes, oxidative stress and hinder development. In the presence of oxygen, the excited electrons produced around them create superoxide radicals by direct electron transfer (Hoffmann et al. 2007; Vannini et al. 2014). While some authors claim NPs, silicon NPs, and fullerenes may provide protection against oxidative stress, *in vitro* investigations on the toxicity of NPs have verified the production of ROS, for instance, by titanium dioxide and fullerenes (Tripathi et al. 2017; Venkatachalam et al. 2017). Alumina (Al_2O_3) is one of the other NPs that acts as an environmental contaminant and inhibits plant development. In a similar vein, it has been proven that the pure aluminum NPs prevent the establishment of roots in several agricultural plants, such as soybean, corn, cucumber, cabbage, and carrot. As the concentration of Al_2O_3 NP exposure rose, tobacco seedlings also showed a slow and notable decline in root length, biomass, and leaf count (Burklew et al. 2012). In numerous plants, including rapeseed, radish, and rye grass, zinc (Zn) and zinc oxide (ZnO) nanoparticles (NPs) have demonstrated an inhibitory impact on root elongation and seed germination (Lin and Xing 2007).

While silver NPs have a cytotoxic impact on onions and zucchini, copper NPs have a cytotoxic effect on mung beans. Numerous studies have documented the harmful effects of high concentrations of multi-walled carbon nanotubes on a variety of plants, including rice and Arabidopsis (Nair et al. 2010; Ghosh et al. 2015; Wani et al. 2024). Nano-ZnO accumulates in the edible parts of soybean plants through root uptake, impacting food quality, while nano-CeO_2 significantly reduces nitrogen fixing and crop yield (Priester et al. 2012; Deng 2016). The size superiority of NM can pose potentially hazard to both plants and animals due to their unique properties, which may lead to unforeseen biological interaction and risks. Due to their exceptionally insignificant size, NPs can block plant stomata, which can significantly impact plant morphology and physiology by disrupting gas exchange, water regulation, and nutrient uptake. Since NM can easily overcome biological barriers, their efficacy, retention, and degradation are high (Vishwakarma et al. 2010; Waheed et al. 2022).

To understand their behavior and potential risks in living organisms, ensuring safe application in various agricultural fields. Reproductive parts such as pollen are relentlessly impacted by NPs interface, thus disrupting the movement and uptake of essential substances like water, sugar, minerals, and their intermediates, which are important for the proper growth and reproduction (Scarpin 2024). Similar effects can occur in animals as well, where NM cause vasculature and blood vessels. Their small size allowed them to easily penetrate tissues, potentially leading to more severe hazards (Hoseinzadeh et al. 2022). The basic nature of these NM can exacerbate their impact by incorporating into genomic content potentially causing genetic toxicity. This may lead to numerous health issues, including respiratory hematological, and inflammatory conditions further amplifying the risk to both plants and animals. Due to this, the properties of NM in biomolecules, chemicals, and agriculture should be protected. It is recommended to follow existing monitoring contexts that confirm the nontoxic application of this swiftly developing technology (Watson et al. 2011; Isibor et al. 2024). Such framework can support sustainable agricultural practices while ensuring proper validation through time resolved trials to assess potential risks and benefits comprehensively.

21.8 Conclusions and Future Perspectives

In outcome, nanotechnology and microbiology are rapidly developing fields that will find extensive use in sustainable agriculture. Large-scale manufacturing and rising demand for nanoproducts have led to a revolution in nanotechnology. Nanotechnology has greatly benefited agriculture with the development of biosensors, nanoscale transporters, and artificial xylem vessels. These are some of the most significant advances in nanotechnology and microbiology that have brought about the modernization of the agricultural industry. However, there is an urgent need to uncover the many unexplored facets of nanotechnology. Understanding the molecular-level interactions between polymers and nutraceuticals, as well as how they affect food sector goods as delivery systems, are further areas that require microbial-based nanotechnology research. Further research on nanopesticides is necessary. Little investigation has been done on how azadirachtin, thymol, and curcumin work in relation to NPs. According to the European Commission, nanotechnology is one of the six "Key enabling technologies" that can support long-term growth and competitiveness across a range of industrial application sectors. It is crucial to incorporate the opinions of scientific, industry, and regulatory experts after providing a brief overview of the work done on NPs in relation to agro nanotechnology throughout the past ten years. The toxicity level of many NPs is still unknown, despite the abundance of information regarding individual NPs. As a result, the use of these materials is restricted since risk assessments and their impacts on human health are not well understood. The utilization of this technology requires the creation of an extensive database and warning system, as well as international collaboration for legislation and regulation.

References

Abd-Elsalam, K.A. (2024). Microbial nanotechnology. *MDPI* 12: 352.

Acharya, A and PK Pal, 2020. Agriculture nanotechnology: translating research outcome to field applications by influencing environmental sustainability. *NanoImpact*https://doi.org/10.1016/j.impact.2020.100232, 19, 100232

Acharya, P., Jayaprakasha, G.K., Crosby, K.M. et al. (2020). Nanoparticle-mediated seed priming improves germination, growth, yield, and quality of watermelons (*Citrullus lanatus*) at multi-locations in Texas. *Scientific Reports* 10: 5037.

Aguilera, E., Díaz-Gaona, C., García-Laureano, R. et al. (2020). Agroecology for adaptation to climate change and resource depletion in the Mediterranean Region. A review. *Agricultural Systems* 181: 102809.

Ahmed, F., Arshi, N., Kumar, S., et al. (2013). Nanobiotechnology: scope and potential for crop improvement. In: *Crop Improvement Under Adverse Conditions* (ed. N. Tuteja, and S.S. Gill), 245-269. New York, NY: Springer New York.

Al-Harbi, N. and Abd-Elrahman, N.K. (2024). Physical methods for preparation of nanomaterials, their characterization and applications: a review. *Journal of Umm Al-Qura University for Applied Sciences* https://doi.org/10.1007/s43994-024-00165-7 11: 356–377.

Ali, M.A., Ahmed, T., Wu, W. et al. (2020). Advancements in plant and microbe-based synthesis of metallic nanoparticles and their antimicrobial activity against plant pathogens. *Nanomaterials* https://doi.org/10.3390/nano10061146 10.

Ali, M.A., Abdellah, I.M., and Eletmany, M.R. (2023). Towards sustainable management of insect pests: protecting food security through ecological intensification. *International Journal of Chemical and Biochemical Sciences* 24: 386.

Al-Zaban, M.I., Mahmoud, M.A., AlHarbi, M.A., and Bahatheq, A.M. (2020). Bioremediation of crude oil by rhizosphere fungal isolates in the presence of silver nanoparticles. *International Journal of Environmental Research and Public Health* https://doi.org/10.3390/ijerph17186564 17.

Amutha, C., Gopan, A., Pushbalatatha, I., et al. (2024). Nanotechnology and governance: regulatory framework for responsible innovation. In: *Nanotechnology in Societal Development. Advanced Technologies and Societal Change* (ed. S.C. George and B. Tawiah). Springer, Singapore. https://doi.org/10.1007/978-981-97-6184-5_14.

Arora, S., Murmu, G., Mukherjee, K. et al. (2022). A comprehensive overview of nanotechnology in sustainable agriculture. *Journal of Biotechnology* 355: 21–41.

Aziz, N., Faraz, M., Pandey, R. et al. (2015). Facile algae-derived route to biogenic silver nanoparticles: synthesis, antibacterial, and photocatalytic properties. *Langmuir* 31: 11605–11612.

Aziz, N., Pandey, R., Barman, I., and Prasad, R. (2016). Leveraging the attributes of *Mucor hiemalis*-derived silver nanoparticles for a synergistic broad-spectrum antimicrobial platform. *Frontiers in Microbiology* 7: 1984.

Bano, A. and Ummatul, H. (2020). Interactive effects of Ag-nanoparticles, salicylic acid, and plant growth promoting rhizobacteria on the physiology of wheat infected with yellow rust. *Journal of Plant Pathology* https://doi.org/10.1007/s42161-020-00626-y 102: 1215–1225.

Batley, G.E., Kirby, J.K., and McLaughlin, M.J. (2013). Fate and risks of nanomaterials in aquatic and terrestrial environments. *Accounts of Chemical Research* 46: 46854.

Boykov, I.N., Shuford, E., and Zhang, B. (2019). Nanoparticle titanium dioxide affects the growth and microRNA expression of switchgrass (Panicum virgatum). *Genomics* 111: 450–456.

Burklew, C.E., Ashlock, J., Winfrey, W.B., and Zhang, B. (2012). Effects of aluminum oxide nanoparticles on the growth, development, and microRNA expression of tobacco (Nicotiana tabacum). *PLoS One* 7: e34783.

Cai, L., Cai, L., Jia, H. et al. (2020). Foliar exposure of Fe3O4 nanoparticles on Nicotiana benthamiana: evidence for nanoparticles uptake, plant growth promoter and defense response elicitor against plant virus. *Journal of Hazardous Materials* 393: 122415.

Chakraborty, M., Bhuiyan, S.A., Strachan, S. et al. (2024). A sensitive, specific and fast electrochemical-based nanobiosensor diagnostic for *Xanthomonas albilineans*, the cause of sugarcane leaf scald disease. *Advanced Sensor Research* https://doi.org/10.1002/adsr.202400103 4.

Christian, P., Von der Kammer, F., Baalousha, M., and Hofmann, T. (2008). Nanoparticles: structure, properties, preparation and behaviour in environmental media. *Ecotoxicology* 17: 326–343.

De La Torre-Roche, R., Cantu, J., Tamez, C. et al. (2020). Seed biofortification by engineered nanomaterials: a pathway to alleviate malnutrition? *Journal of Agricultural and Food Chemistry* https://doi.org/10.1021/acs.jafc.0c04881 68: 12189–12202.

Deng, Y. (2016). Uptake and accumulation of engineered nanomaterials by agricultural crops and associated risks in the environment and food safety.

Dixit, R., Wasiullah, X., Malaviya, D. et al. (2015). Bioremediation of heavy metals from soil and aquatic environment: an overview of principles and criteria of fundamental processes. *Sustainability* 7: 2189–2212.

Du, W., Tan, W., Peralta-Videa, J.R. et al. (2017). Interaction of metal oxide nanoparticles with higher terrestrial plants: physiological and biochemical aspects. *Plant Physiology and Biochemistry* 110: 210–225.

Floros, J.D., Newsome, R., Fisher, W. et al. (2010). Feeding the world today and tomorrow: the importance of food science and technology: an IFT scientific review. *Comprehensive Reviews in Food Science and Food Safety* 9: 572–599.

Fraceto, L., Grillo, R., de Medeiros, G. et al. (2016). Nanotechnology in agriculture: which innovation potential does it have? *Frontiers in Environmental Science* 4: 20.

Ghosh, M., Bhadra, S., Adegoke, A. et al. (2015). MWCNT uptake in Allium cepa root cells induces cytotoxic and genotoxic responses and results in DNA hyper-methylation. *Mutation Research, Fundamental and Molecular Mechanisms of Mutagenesis* 774: 49–58.

Gruère, G.P. (2012). Implications of nanotechnology growth in food and agriculture in OECD countries. *Food Policy* 37: 191–198.

Guo, D., Xie, G., and Luo, J. (2013). Mechanical properties of nanoparticles: basics and applications. *Journal of Physics D: Applied Physics* 47: 013001.

Gupta, A., Rayeen, F., Mishra, R. et al. (2023). Nanotechnology applications in sustainable agriculture: an emerging eco-friendly approach. *Plant Nano Biology* https://doi.org/10.1016/j.plana.2023.100033 4.

Hasanaklou, N.T., Mohagheghi, V., Hasanaklou, H.T. et al. (2023). Seed nano-priming using silica nanoparticles: effects in seed germination and physiological properties of Stevia Rebaudiana Bertoni. *Chemical and Biological Technologies in Agriculture* https://doi.org/10.1186/s40538-023-00445-0 10.

Hesham, A.E.-L., Kaur, T., Devi, R. et al. (2021). Current trends in microbial biotechnology for agricultural sustainability: conclusion and future challenges. In: Yadav, A.N., Singh, J., Singh, C., Yadav, N. (eds) *Current Trends in Microbial Biotechnology for Sustainable Agriculture*. Springer, Singapore. pp. 555–572. https://doi.org/10.1007/978-981-15-6949-4_22.

Hoffmann, M., Holtze, E., and Wiesner, M.R. (2007). *Reactive Oxygen Species Generation on Nanoparticulate Material*. New York, NY: Mcgraw-Hill Professional.

Hoseinzadeh, A., Ghoddusi Johari, H., Anbardar, M.H. et al. (2022). Effective treatment of intractable diseases using nanoparticles to interfere with vascular supply and angiogenic process. *European Journal of Medical Research* 27: 232.

Hotze, E.M., Phenrat, T., and Lowry, G.V. (2010). Nanoparticle aggregation: challenges to understanding transport and reactivity in the environment. *Journal of Environmental Quality* 39: 1909–1924.

Imran, M., Abo-Elyousr, K.A.M., El-Sharnouby, M.E. et al. (2023). Biocontrol potential of trichoderma harzianum and zinc nanoparticles to mitigate gray mold disease of tomato. *Gesunde Pflanzen* https://doi.org/10.1007/s10343-022-00686-3 75: 151–163.

Imtiaz, H., Shiraz, M., Mir, A.R. et al. (2023). Nano-priming techniques for plant physio-biochemistry and stress tolerance. *Journal of Plant Growth Regulation* https://doi.org/10.1007/s00344-023-10981-6 42: 6870–6890.

Ion, A.C., Ion, I., Culetu, A., and Gherase, D. (2010). *Carbon-based Nanomaterials. Environmental Applications*. Romania.

Isibor, P.O. II, Kayode Edwards, and Taiwo, O.S. (2024). Emerging technology and future directions in environmental nanotoxicology. In: Isibor, P.O., Devi, G., Enuneku, A.A. (eds) *Environmental Nanotoxicology: Combatting the Minute Contaminants*, Springer, Cham. pp. 325–346. https://doi.org/10.1007/978-3-031-54154-4_16.

Jasim, B., Thomas, R., Mathew, J., and Radhakrishnan, E.K. (2017). Plant growth and diosgenin enhancement effect of silver nanoparticles in Fenugreek (*Trigonella foenum-graecum* L.). *Saudi Pharmaceutical Journal* https://doi.org/10.1016/j.jsps.2016.09.012 25: 443–447.

Jin, Y., Fan, X., Li, X. et al. (2017). Distinct physiological and molecular responses in Arabidopsis thaliana exposed to aluminum oxide nanoparticles and ionic aluminum. *Environmental Pollution* 228: 517–527.

Juárez-Maldonado, A., Ortega-Ortíz, H., Morales-Díaz, A.B. et al. (2019). Nanoparticles and nanomaterials as plant biostimulants. *International Journal of Molecular Sciences* 20.

Kah, M., Kookana, R.S., Gogos, A., and Bucheli, T.D. (2018). A critical evaluation of nanopesticides and nanofertilizers against their conventional analogues. *Nature Nanotechnology* 13: 677–684.

Karunakaran, G., Sudha, K.G., Ali, S., and Cho, E.-B. (2023). Biosynthesis of nanoparticles from various biological sources and its biomedical applications. *Molecules* 28: 4527.

Karunakaran, A., Fathima, Y., Singh, P. et al. (2024). Next-generation biofertilizers: nanoparticle-coated plant growth-promoting bacteria biofertilizers for enhancing nutrient uptake and wheat growth. *Agriculture* https://doi.org/10.3390/agriculture14040517 14.

Kashisaz, M., Enayatizamir, N., Fu, P., and Eslahi, M. (2024). Co-application of beneficial microorganisms and nanoparticles to improve wheat growth in infected *Fusarium culmorum* soil. *Applied Soil Ecology* https://doi.org/10.1016/j.apsoil.2024.105622 203.

Kashyap, P.L., Rai, P., Kumar, R. et al. (2018). Microbial nanotechnology for climate resilient agriculture. In: *Microbes for Climate Resilient Agriculture*.

Kashyap, P.L., Kumar, S., Jasrotia, P. et al. (2019). Nanosensors for plant disease diagnosis: current understanding and future perspectives. In: *Nanoscience for Sustainable Agriculture* (ed. R.N. Pudake, N. Chauhan, and C. Kole), 189–205. Cham: Springer International Publishing.

Kaur, S.I., Kashyap, P.L., Kang, S.S., and Sharma, A. (2020). Detection and diagnosis of seed-borne viruses and virus-like pathogens. In: *Seed-Borne Diseases of Agricultural Crops: Detection, Diagnosis & Management* (ed. R. Kumar and A. Gupta), 169–199. Singapore: Springer Singapore.

Khan, N. and Bano, A. (2016). Role of plant growth promoting rhizobacteria and Ag-nano particle in the bioremediation of heavy metals and maize growth under municipal wastewater irrigation. *International Journal of Phytoremediation* https://doi.org/10.1080/15226514.2015.1064352 18: 211–221.

Khan, S.S., Kour, D., Kaur, T. et al. (2024). Microbial nanotechnology for precision nanobiosynthesis: innovations, current opportunities and future perspectives for industrial sustainability. *Current Microbiology* 81: 251.

Khodakovskaya, M.V., De Silva, K., Biris, A.S. et al. (2012). Carbon nanotubes induce growth enhancement of tobacco cells. *ACS Nano* 6: 2128–2135.

Kumar, A., Rakshit, R., Bhowmik, A. et al. (2019). Nanoparticle-induced changes in resistance and resilience of sensitive microbial indicators towards heat stress in Soil. *Journal* https://doi.org/10.3390/su11030862 11.

Kumari, R. and Singh, D.P. (2020). Nano-biofertilizer: an emerging eco-friendly approach for sustainable agriculture. *Proceedings of the National Academy of Sciences, India Section B: Biological Sciences* https://doi.org/10.1007/s40011-019-01133-6 90: 733–741.

Kumari, S., Raturi, S., Kulshrestha, S. et al. (2023). A comprehensive review on various techniques used for synthesizing nanoparticles. *Journal of Materials Research and Technology* https://doi.org/10.1016/j.jmrt.2023.09.291 27: 1739–1763.

Lahiani, M.H., Dervishi, E., Chen, J. et al. (2013). Impact of carbon nanotube exposure to seeds of valuable crops. *ACS Applied Materials & Interfaces* 5: 7965–7973.

Larue, C., Laurette, J., Herlin-Boime, N. et al. (2012a). Accumulation, translocation and impact of TiO2 nanoparticles in wheat (Triticum aestivum spp.): influence of diameter and crystal phase. *The Science of the Total Environment* 431: 197–208.

Larue, C., Veronesi, G., Flank, A.-M. et al. (2012b). Comparative uptake and impact of TiO2 nanoparticles in wheat and rapeseed. *Journal of Toxicology and Environmental Health, Part A* 75: 722–734.

Li, R., He, J., Xie, H. et al. (2019). Effects of chitosan nanoparticles on seed germination and seedling growth of wheat (Triticum aestivum L.). *International Journal of Biological Macromolecules* 126: 91–100.

Li, W.-Q., Qing, T., Li, C.-C. et al. (2020). Integration of subcellular partitioning and chemical forms to understand silver nanoparticles toxicity to lettuce (Lactuca sativa L.) under different exposure pathways. *Chemosphere* 258: 127349.

Li, H, A Rehman, S ur Rahman, et al., 2024. Biosynthesized zinc oxide nanoparticles modulate the phytoremediation potential of Pennisetum giganteum and its rhizocompartments associated microbial community structure. *Journal of Cleaner Production*https://doi.org/10.1016/j.jclepro.2023.140346, 434, 140346

Lin, D. and Xing, B. (2007). Phytotoxicity of nanoparticles: inhibition of seed germination and root growth. Environmental pollution.

Liu, R. and Lal, R. (2015). Potentials of engineered nanoparticles as fertilizers for increasing agronomic productions. *The Science of the Total Environment* 514: 131–139.

Liu, M., Feng, S., Ma, Y. et al. (2019). Influence of surface charge on the phytotoxicity, transformation, and translocation of CeO2 nanoparticles in cucumber plants. *ACS Applied Materials & Interfaces* 11: 16905.

Lopez-Valdez, F., Miranda-Arambula, M., Rios-Cortes, A.M., et al. (2018). Nanofertilizers and their controlled delivery of nutrients. In: *Agricultural Nanobiotechnology* (ed. F. Lopez-Valdez, and F. Fernandez-Luqueno). Springer, Cham. https://doi.org/10.1007/978-3-319-96719-6_3.

Mahmoodzadeh, H., Nabavi, M., and Kashefi, H. (2013). Effect of nanoscale titanium dioxide particles on the germination and growth of canola (*Brassica napus*).

Maity, D., Gupta, U., and Saha, S. (2022). Biosynthesized metal oxide nanoparticles for sustainable agriculture: next-generation nanotechnology for crop production, protection and management. *Nanoscale* 14: 13950–13989.

Mandeep and P Shukla, 2020. Microbial nanotechnology for bioremediation of industrial wastewater. Frontiers in Microbiology, 11: 590631

Mardalipour, M., Zahedi, H., and Sharghi, Y. (2014). Evaluation of nano biofertilizer efficiency on agronomic traits of spring wheat at different sowing date. *Biological Forum*, Research Trend 6: 349.

Mcychik, N., Nikolacva, J., and Ycrmakov, I. (2005). Ion exchange properties of the root cell walls isolated from the halophyte plants (*Suaeda altissima* L.) grown under conditions of different salinity. *Plant and Soil* 277: 163–174.

Milewska-Hendel, A., Zubko, M., Stróż, D., and Kurczyńska, E.U. (2019). Effect of nanoparticles surface charge on the Arabidopsis thaliana (L.) roots development and their movement into the root cells and protoplasts. *International Journal of Molecular Sciences* 20: 1650.

Mishra, S., Singh, B.R., Singh, A. et al. (2014). Biofabricated silver nanoparticles act as a strong fungicide against bipolaris sorokiniana causing spot blotch disease in wheat. *PLoS One* https://doi.org/10.1371/journal.pone.0097881 9: e97881.

Mishra, S., Singh, B.R., Naqvi, A.H., and Singh, H.B. (2017). Potential of biosynthesized silver nanoparticles using Stenotrophomonas sp. BHU-S7 (MTCC 5978) for management of soil-borne and foliar phytopathogens. *Scientific Reports* https://doi.org/10.1038/srep45154 7.

Mukhopadhyay, S.S. (2014). Nanotechnology in agriculture: prospects and constraints. *Nanotechnology, Science and Applications* 7: 63–71.

Naikoo, G.A., Mustaqeem, M., Hassan, I.U. et al. (2021). Bioinspired and green synthesis of nanoparticles from plant extracts with antiviral and antimicrobial properties: a critical review. *Journal of Saudi Chemical Society* 25: 101304.

Nair, R., Varghese, S.H., Nair, B.G. et al. (2010). Nanoparticulate material delivery to plants. *Plant Science* 179: 154–163.

Nair, G.M., Sajini, T., and Mathew, B. (2022). Advanced green approaches for metal and metal oxide nanoparticles synthesis and their environmental applications. *Talanta Open* 5: 100080.

Nayak, V., Singh, K.R., Paliwal, R. et al. (2023). Chapter 15—introduction to nanotechnological utility in the pharmaceutical industry. In: *Nanotechnology for Drug Delivery and Pharmaceuticals* (ed. R. Pratap Singh, K.R. Singh, and J. Singh) et al., 337–355. Academic Press.

Negi, R., Sharma, B., Jan, T. et al. (2024a). Microbial consortia: promising tool as plant bioinoculants for agricultural sustainability. *Current Microbiology* 81: 222.

Negi, R., Sharma, B., Kumar, S. et al. (2024b). Plant endophytes: unveiling hidden applications toward agro-environment sustainability. *Folia Microbiologica* 69: 181–206.

Negi, R., Sharma, B., Parastesh, F. et al. (2024c). Microbial consortia mediated regulation of plant defense: a promising tool for sustaining crops protection. *Physiological and Molecular Plant Pathology* 134: 102393.

Nile, S.H., Thiruvengadam, M., Wang, Y. et al. (2022). Nano-priming as emerging seed priming technology for sustainable agriculture—recent developments and future perspectives. *Journal of Nanobiotechnology* https://doi.org/10.1186/s12951-022-01423-8 20: 254.

Palmqvist, N.G.M., Bejai, S., Meijer, J. et al. (2015). Nano titania aided clustering and adhesion of beneficial bacteria to plant roots to enhance crop growth and stress management. *Scientific Reports* https://doi.org/10.1038/srep10146 5.

Pandey, G. and Jain, P. (2020). Assessing the nanotechnology on the grounds of costs, benefits, and risks. *Beni-Suef University Journal of Basic and Applied Sciences* 9: 63.

Panichikkal, J., Thomas, R., John, J.C., and Radhakrishnan, E.K. (2019). Biogenic gold nanoparticle supplementation to plant beneficial pseudomonas monteilii was found to enhance its plant probiotic effect. *Current Microbiology* https://doi.org/10.1007/s00284-019-01649-0 76: 503–509.

Pascoli, M., Lopes-Oliveira, P.J., Fraceto, L.F. et al. (2018). State of the art of polymeric nanoparticles as carrier systems with agricultural applications: a minireview. *Energy, Ecology and Environment* 3: 137–148.

Patra, J.K. and Baek, K.-H. (2017). Antibacterial activity and synergistic antibacterial potential of biosynthesized silver nanoparticles against foodborne pathogenic bacteria along with its anticandidal and antioxidant effects. *Frontiers in Microbiology* 8: 167.

Phogat, N., Khan, S.A., Shankar, S. et al. (2016). Fate of inorganic nanoparticles in agriculture. *Advanced Materials Letters* 7: 3–12.

Prasad, R., Kumar, V., and Prasad, K.S. (2014). Nanotechnology in sustainable agriculture: present concerns and future aspects. *African Journal of Biotechnology* 13: 705.

Prasad, R., Bhattacharyya, A., and Nguyen, Q.D. (2017). Nanotechnology in sustainable agriculture: recent developments, challenges, and perspectives. *Frontiers in Microbiology* 8: 1014.

Priester, J.H., Ge, Y., Mielke, R.E. et al. (2012). Soybean susceptibility to manufactured nanomaterials with evidence for food quality and soil fertility interruption. *National Academy of Sciences of the United States of America* 109: E2451–E2456.

Rahman, M., Chakraborty, A., Mazumdar, S. et al. (2020). Effects of poly (vinylpyrrolidone) protected platinum nanoparticles on seed germination and growth performance of Pisum sativum. *Nano-Structures & Nano-Objects* 21: 100408.

Raimi, M.O., Saliu, A.O., Babatunde, A. et al. (2022). The challenges and conservation strategies of biodiversity: the role of government and non-governmental organization for action and results on the ground. In: *Biodiversity in Africa: Potentials, Threats and Conservation*, 473–504. Springer.

Raja, K., Sowmya, R., Sudhagar, R. et al. (2019). Biogenic ZnO and Cu nanoparticles to improve seed germination quality in blackgram (*Vigna mungo*). *Materials Letters* 235: 164–167.

Rana, S. and Kalaichelvan, P. (2013). Ecotoxicity of nanoparticles. *International Scholarly Research Notices* 2013: 574648.

Rane, M., Bawskar, M., Rathod, D. et al. (2015). Influence of calcium phosphate nanoparticles, Piriformospora indica and Glomus mosseae on growth of Zea mays. *Advances in Natural Sciences: Nanoscience and Nanotechnology* https://doi.org/10.1088/2043-6262/6/4/045014 6.

Sabir, S., Arshad, M., and Chaudhari, S.K. (2014). Zinc oxide nanoparticles for revolutionizing agriculture: synthesis and applications. *The Scientific World Journal* 2014: 1–8.

Sabir, S., Zahoor, M.A., Waseem, M. et al. (2020). Biosynthesis of ZnO nanoparticles using bacillus subtilis: characterization and nutritive significance for promoting plant growth in Zea mays L. *Dose-Response* https://doi.org/10.1177/1559325820958911 18.

Sahu, T., Ratre, Y.K., Chauhan, S. et al. (2021). Nanotechnology based drug delivery system: current strategies and emerging therapeutic potential for medical science. *Journal of Drug Delivery Science and Technology* 63: 102487.

Sambangi, P and S Gopalakrishnan, 2023. Streptomyces-mediated synthesis of silver nanoparticles for enhanced growth, yield, and grain nutrients in chickpea. *Biocatalysis and Agricultural Biotechnology*https://doi.org/10.1016/j.bcab.2022.102567

Sangeetha, J., Thangadurai, D., Hospet, R. et al. (2017). Agricultural nanotechnology: concepts, benefits, and risks. In: *Nanotechnology: An Agricultural Paradigm.*

Scarpin, D. (2024). Innovative technologies enabling plants to increase resilience as a pivotal feature in the conservation of agro-ecosystems.

Sekhon, B.S. (2014). Nanotechnology in agri-food production: an overview. *Nanotechnology, Science and Applications* 7: 31–53.

Setty, J., Samant, S.B., Yadav, M.K. et al. (2023). Beneficial effects of bio-fabricated selenium nanoparticles as seed nanopriming agent on seed germination in rice (Oryza sativa L.). *Scientific Reports* https://doi.org/10.1038/s41598-023-49621-0 13: 22349.

Shen, C.X., Zhang, Q.F., Li, J. et al. (2010). Induction of programmed cell death in Arabidopsis and rice by single-wall carbon nanotubes. *American Journal of Botany* 97: 1602–1609.

Siegel, F.R. and Siegel, F.R. (2021). Economic realities in 2020 populations: what do they portend for 2050? 2100? In: *The Earth's Human Carrying Capacity: Limitations Assessed, Solutions Proposed.*

Singh, A., Tyagi, P., Ranjan, R. et al. (2023). Bioremediation of hazardous wastes using green synthesis of nanoparticles. *Processes* https://doi.org/10.3390/pr11010141 11.

Singh, N., Kumar, B., Usman, U.L., and Susan, M.A.B.H. (2024). Nano revolution: exploring the frontiers of nanomaterials in science, technology, and society. *Nano-Structures & Nano-Objects* 39: 101299.

Sun, D., Hussain, H.I., Yi, Z. et al. (2016). Mesoporous silica nanoparticles enhance seedling growth and photosynthesis in wheat and lupin. *Chemosphere* 152: 81–91.

Suriyaprabha, R., Karunakaran, G., Yuvakkumar, R. et al. (2012). Silica nanoparticles for increased silica availability in maize (*Zea mays*. L) seeds under hydroponic conditions. *Current Nanoscience* 8: 902–908.

Tahir, M.A. and Kanwal, S. (2007). Effect of silicon application on wheat (*Triticum aestivum* L.) growth under water deficiency stress.

Tripathi, D.K., Tripathi, A., Shweta et al. (2017). Uptake, accumulation and toxicity of silver nanoparticle in autotrophic plants, and heterotrophic microbes: a concentric review. *Frontiers in Microbiology* 8: 7.

Uba, B.O. and Obiefuna, G.O. (2023). Aerobically enhanced nano-bioremediation of diesel oil contaminated soil and water using myco-synthesized silver nanoparticle as biostimulating agent. *Science World Journal* 18: 75.

Usman, M., Farooq, M., Wakeel, A. et al. (2020). Nanotechnology in agriculture: current status, challenges and future opportunities. *The Science of the Total Environment* 721: 137778.

Vanbergen, A.J., Aizen, M.A., Cordeau, S. et al. (2020). Transformation of agricultural landscapes in the Anthropocene: nature's contributions to people, agriculture and food security. *Advances in Ecological Research* 63: 193–253.

VandeVoort, A.R. and Arai, Y. (2018). Macroscopic observation of soil nitrification kinetics impacted by copper nanoparticles: implications for micronutrient nanofertilizer. *Nanomaterials (Basel)* 8: 927.

Vannini, C., Domingo, G., Onelli, E. et al. (2014). Phytotoxic and genotoxic effects of silver nanoparticles exposure on germinating wheat seedlings. *Journal of Plant Physiology* 171: 1142–1148.

Venkatachalam, P., Jayaraj, M., Manikandan, R. et al. (2017). Zinc oxide nanoparticles (ZnONPs) alleviate heavy metal-induced toxicity in Leucaena leucocephala seedlings: a physiochemical analysis. *Plant Physiology and Biochemistry* 110: 59–69.

Vishwakarma, V., Samal, S.S., and Manoharan, N. (2010). Safety and risk associated with nanoparticles-a review. *Journal of Minerals and Materials Characterization and Engineering* 09: 455–459.

Waheed, S., Li, Z., Zhang, F. et al. (2022). Engineering nano-drug biointerface to overcome biological barriers toward precision drug delivery. *Journal of Nanobiotechnology* 20: 395.

Wang, P., Lombi, E., Zhao, F.-J., and Kopittke, P.M. (2016). Nanotechnology: a new opportunity in plant sciences. *Trends in Plant Science* 21: 699–712.

Wani, A.K., Khan, Z., Sena, S. et al. (2024). Carbon nanotubes in plant dynamics: unravelling multifaceted roles and phytotoxic implications. *Plant Physiology and Biochemistry* 210: 108628.

Waqas Mazhar, M., Ishtiaq, M., Hussain, I. et al. (2022). Seed nano-priming with zinc oxide nanoparticles in rice mitigates drought and enhances agronomic profile. *PLoS One* https://doi.org/10.1371/journal.pone.0264967 17: e0264967.

Watson, S.B., Gergely, A., and Janus, E.R. (2011). Where is agronanotechnolgoy heading in the United States and European Union. *Natural Resources and Environment* 26: 8.

Wu, H., Santana, I., Dansie, J., and Giraldo, J.P. (2017). In vivo delivery of nanoparticles into plant leaves. *Current Protocols in Chemical Biology* 9: 269–284.

Yadav, A.N. (2021). Nanotechnology for agro-environmental sustainability. *Journal of Applied Biology and Biotechnology* 9: 11.

Yadav, N., Garg, V.K., Chhillar, A.K., and Rana, J.S. (2023). Recent advances in nanotechnology for the improvement of conventional agricultural systems: a review. *Plant Nano Biology* 4: 100032.

Yuan, J., Chen, Y., Li, H. et al. (2018). New insights into the cellular responses to iron nanoparticles in *Capsicum annuum*. *Scientific Reports* 8: 1.

Zeng, Y., Himmel, M.E., and Ding, S.-Y. (2017). Visualizing chemical functionality in plant cell walls. *Biotechnology for Biofuels* 10: 263.

Zhu, Z.-J., Wang, H., Yan, B. et al. (2012). Effect of surface charge on the uptake and distribution of gold nanoparticles in four plant species. *Environmental Science & Technology* 46: 12391–12398.

Zhu, Y., Ren, B., Li, H. et al. (2018). Biosynthesis of selenium nanoparticles and effects of selenite, selenate, and selenomethionine on cell growth and morphology in Rahnella aquatilis HX2. *Applied Microbiology and Biotechnology* https://doi.org/10.1007/s00253-018-9060-z 102: 6191–6205.

Index

e

f

g

h

www.ingramcontent.com/pod-product-compliance
Lightning Source LLC
LaVergne TN
LVHW062203200926
848484LV00006B/1304